OPTICAL LINE SYSTEMS

OPTICAL LINE SYSTEMS

Transmission Aspects

D J H Maclean

Formerly at

Standard Telecommunication Laboratories
Harlow, UK

JOHN WILEY & SONS
Chichester • New York • Brisbane • Toronto • Singapore

Baffins Lane, Chichester,
West Sussex PO19 IUD, England

National (01243) 779777
International (+44) 1243 779777

Other Wiley Editorial Offices

John Wiley & Sons, Inc., 605 Third Avenue,
New York, NY 10158-0012, USA

Jacaranda Wiley Ltd, 33 Park Road, Milton,
Queensland 4064, Australia

John Wiley & Sons (Canada) Ltd, 22 Worcester Road,
Rexdale, Ontario M9W 1L1, Canada

John Wiley & Sons (Asia) Pte Ltd, 2 Clementi Loop #02-01,
Jin Xing Distripark, Singapore 0512

Library of Congress Cataloging-in-Publication Data

Maclean, D.J.H.
Optical line systems/D.J.H. Maclean.
p. cm.
Includes bibliographical references and index.
ISBN 0 471 95083 1
1. Optical communications. I. Title.
TK5103.59.M35 1996
621.382′75 — dc20 95-4908
CIP

British Library Cataloguing in Publication Data

A catalogue record for this book is available from the British Library

ISBN 0 471 95083 1

Typeset in 10/12pt Palatino by Laser Words, Madras, India
Printed and bound in Great Britain by Bookcraft (Bath) Ltd.

To my wife, children and granddaughter,
Sandra, Bruce, Fiona and Sapphire,
and to my parents,
Robert and Elsie Maclean.

A Dream Come True

Since the establishment of telegraph cable routes spanning the globe in the 1870s, generations of transmission engineers have dreamed of the time when the enormous potential of light would be exploited in telecommunications to provide

- ♦ near infinite bandwidth
- ♦ near zero path loss
- ♦ near zero path distortion
- ♦ near zero material usage
- ♦ near zero power consumption
- ♦ near zero cost

Realisation of the dream began in the 1970s. By 1995, installed single-mode fibres had provided a bearer of effectively unlimited bandwidth, and erbium-doped fibres pumped by semiconductor laser diodes had realised optical amplifiers which were ideal for most practical purposes.

CONTENTS

Chapter 12 Single-Mode Cables, Splices and Connectors 409

Chapter 13 General System Considerations 443

PREFACE

Since the invention of the electric telegraph, generations of engineers have concerned themselves with the modelling of systems and networks. Their goal has been, and continues to be, the gaining of fundamental insights and understanding leading to the optimum exploitation of available technology. For over 130 years this has brought about startling advances in the development of transmission systems, switching and networks. We are now within sight of realising a global infrastructure that represents the nervous system of the planet, with telecommunications governing and underpinning all of mankind's activity. It is therefore vital that we continue to expand our understanding of all facets of this global infrastructure, from the constituent parts to market demands.

(From Foreword by Peter Cochrane and David Heatley to 'Advanced applications and technologies', *BT Technol. J.*, **12**, no. 2, Apr. 1994)

The objective of the present book is to provide information on the properties and systems-related characteristics of major transmission hardware components, and on the design of optical fibre digital line systems (ofdls) — and do so in a form directly usable by engineers in the global telecommunications industry.

This industry is in the throes of unprecedented changes which will affect those in it and the wider publics they serve well into the next century. Introductory chapters discuss important issues in the change from copper to glass and the impact of optical fibre technologies on existing networks. The significance of the introduction of the Synchronous Digital Hierarchy (SDH) in the UK context is outlined. The interworking of SDH and existing plesiochronous digital hierarchies (PDH) systems will pose challenges for some time to come, and the migration from PDH to SDH will need to be matched to the long term objectives of the operator. Timing and synchronisation problems will need to be solved and the reliability of mixed SDH/PDH working assured. The introduction of SDH systems into networks will raise many issues and will require the commissioning, testing and evaluation of SDH-specific test equipment. It is too early to include these and other related topics in this book. Although multimode fibres, jointing and systems are used extensively in the local (access) area, chapters on these topics have been omitted for brevity.

Chapter one includes material on the evolving legal, regulatory, competitive and economic facts which are shaping the industry, using the situations in several leading countries for illustration, and considering the changing roles of international organisations. Telecommunications itself is a rapidly growing industry, for example, it has been predicted that by the year 2000 it could be Europe's largest equity sector.

The author's approach in the more technical chapters is to present a coherent selection of information on single-mode fibres and cables, laser diodes and transmitters and an optical receivers, as needed for hardware system design and performance evaluation. The chosen examples of components, subsystems and system performance are derived from results obtained on installed systems, field and laboratory trials and demonstrations, as reported from leading organisations worldwide in hundreds of peer-refereed technical papers published in recent years. The main part of the book thus includes a very practical collection of data from component to system level on many aspects of transmission over optical fibre digital line systems as used in national landline or underwater system and international submerged systems at bit rates up to 5 Gbit/s and beyond. A full length final chapter considers some aspects of measurements.

Within the telecommunications and Cable TV industries, engineers, physicists, material scientists, mathematicians and technologists in other relevant disciplines should find the book a useful source of information in their work. Such staff are employed in all levels from new graduate to senior positions, and are likely to find themselves under increasing pressure at work. Because they will have fewer opportunities to search the literature thoroughly or acquire a broader knowledge of their field, they should welcome this book. In both traditional public and private sectors such staff are found throughout the whole gamut of activities from basic research, to research and development, to manufacture, installation and operation. This work should also prove valuable to large organisations which maintain technical libraries and which provide training at appropriate levels for technical staff.

In addition, international agencies (e.g. ITU, IEC, ISO, ETSI) and banking (e.g. World Bank, European Bank for Reconstruction and Development), governments, regulatory agencies, armed forces, large multinational firms and consultants all employ staff with relevant expertise in telecommunications. As competition and liberalisation spreads, licences may be granted to many new entrants provided, inter alia, they have the necessary experience of the market and the ability to manage a network with a high level of security and integrity. The need for more commercially-aware technical staff and of easily accessible relevant information may therefore grow significantly.

Finally, at universities and specialised institutes with research and/or teaching in the field of telecommunication transmission at graduate level (and perhaps final-year undergraduate level), selected material from the book may enhance existing courses or provide a basis for new modules.

Brief overview of main contents

A total of four chapters are devoted to various aspects of single-mode fibres including theory, fabrication, and optimum designs for the four main types of refractive index profiles in use. A detailed treatment of microbending and macrobending losses in bare, coated and in cabled fibres is given and reference made to fibre and cable installation using viscous air flow. Statistical aspects of the effect of fibre dispersion and of production tolerances on system performance are treated. Types of cables and their reliability, single and multi-fibre splicing and connectors are all considered in appropriate detail.

Three chapters are devoted to some system aspects of laser diodes, including performance specification and characteristics of a 1300 nm type of bulk device widely deployed in landline and submerged systems from the mid-1980s. Examples are given of true time-resolved spectra of modulated lasers, their statistical nature and their system implications. Effects of temperature changes on laser characteristics are considered. Detailed treatment is given of the effects of external feedback which can severely degrade performance. Noise properties of lasers and in particular the system implications of mode competition (or partition) noise are discussed. Features of quantum well devices of particular interest to system designers, and the advantages of mqw lasers as transmitters and as pump diodes for fibre amplifiers are considered.

Three longer chapters deal with many aspects of optical receivers for imdd systems and include examples from the TAT-8 and PTAT-1 transatlantic systems. Detailed performance characterisation and other systems-related aspects of pinfet optical receivers are presented. Conventional and optically preamplified receivers for imdd systems operating at rates up to 10 Gbit/s are included. Both pin and avalanche photodiode time and frequency domain responses, noise and reliability properties of interest are discussed. The third chapter presents an improved design approach for Gbit/s receivers which exploits the advances in characterisation, modelling, computer-aided design and simulation for small signal (and large-signal) and noise analyses achieved since Personick's 1973 approach. Consideration of freedom from self-oscillation and the effects of flicker noise on receiver performance are covered.

Three chapters are devoted to systems considerations in general and then design aspects illustrated by a 565 Mbit/s system and the application of optical amplifiers. The third system covers submerged systems, again using an actual system to illustrate many of the important issues. Both worst-case and statistical designs of long-haul systems are covered. The twin topics of power budgets and margins are considered for optically amplified systems as well as for regenerative systems.

The final chapter on aspects of measurement opens with general philosophy and equipment test methods for factory, installation and for maintainence. Specific types of measurement were mentioned as they occured in earlier chapters, but more detailed descriptions of measurements on fibres, cables and joints are given as well as on sources and on detectors. The chapter ends with jitter measurement techniques and requirements.

The reader may be surprised by the absence of coherant systems and of soliton transmission systems among the chapter contents, despite the fact that both have attracted a great deal of research and development interest. However, growth in the capacity of the fibre pairs used for intensity-modulated direct-detection installed systems from 140 Mbit/s to 5 Gbit/s has outstripped any foreseeable traffic demand, due to advances in component technology, especially the commercial availability of erbium-doped fibre amplifiers operating in the 1550 nm band and of early praseodymium-doped fluoride fibre amplifiers for use in the 1300 nm band. IMDD system therefore represent the most cost effective means of meeting current and predicted traffic forecasts in the mid-1990s. By the end of 1995 the Atlantic and Pacific Oceans will have been spanned by SDH-compatible optical line systems operating at 5 Gbit/s in the 1550 nm band and embodying erbium-doped fibre amplifier repeaters. As with earlier submerged optical systems they are likely to provide service of excellent quality at reduced cost to the customer.

Optical amplifiers, whether fibre or semiconductor, are mentioned in places but not treated in any detail, partly for lack of space and partly because they are covered in depth in several books published in 1993 and 1994. However, the author has written a companion to this present book, using his same approach, which is devoted to the subject of optical fibre amplifiers for use in 1550 and 1300 nm line systems. Applications in coherent, multichannel and soliton transmission systems are all included as well as applications in imdd systems.

The world of telecommunications is becoming increasingly complex and inherently nonlinear, with the interaction of technologies, systems, networks and customers proving extremely difficult to model.' However, '... transmission systems hardware has already undergone a meteoric rise in complexity, followed quite naturally by incredible simplification,...

(From Foreword by Peter Cochrane and David Heatley to 'Advanced applications and technologies', BT Technol. J., **12**, no. 2, Apr. 1994)

ACKNOWLEDGEMENTS

Reviewing the list of sources I have selected as the base of this book, I am very conscious of the debt that I owe to the authors and the organisations which employ them. Both have contributed greatly to publicising advances in the world telecommunications industry, which have and will continue to benefit customers in quality and variety of services. I am also very grateful to many individual copyright holders for their permission to reproduce material from their publications, and their particular form of acknowledgement has been inserted in the appropriate figures and tables.

Many of the contributions are by authors from British Telecom, which I chose for their authoritative treatment and inclusion of practical considerations. From my early days as an electrical engineer in industry working in research and development, I have benefitted from the expertise of British Telecom staff, first at Dollis Hill, then at Martlesham and at other sites elsewhere. In particular, during the early years of what became the present book, I owed much to the help, encouragement and vision of Peter Cochrane, sustained over many years, I must also thank other BT staff who also helped by reviewing and commenting on early versions of chapters or in face-to-face discussions on topics where they were the acknowledged experts. These included Lutz Blank, Mike Brain, Peter Chidgey, Ken Fitchew, Ian Fletcher, Ian Garrett, Ian Hawker, Dave Heatley, Ian Henning, Ray Hooper, Chris Lilly, Andrew Lord, Alan Mitchell, Stephen Mallinson, John Midwinter, Mike O'Mahoney, David Smith, Dave Spirit, Bill Stallard, John Wright; all contributed indirectly in bringing this book to publication.

Staff at Standard Telecommunication Laboratories, Harlow, have made many outstanding R & D contributions to optical fibre systems and components and I am happy to acknowledge my debt to former colleagues and others there. These included Kevin Byron, Chris Cock, Tony Davis, Geof Farrington, Takis Hadifotiou, Tony Hall, Gordon Henshall, Ian Hirst, Ted Irons, John Irven, Alan Jeal, Tony Jessop, Paul Kirkby, John Lees, Bill Powell, Peter Radley, Peter Sothcott, George Swanson and Mike Wright. Lesley Hepden of STC Submarine Systems has also been very helpful.

I also wish to thank Peter Ball and Alan Kent, both former colleagues with whom I worked on the Standard Optical Systems studies.

Finally I should like to thank Ann-Marie Halligan, Publishing Editor and her staff, and Robert Hambrook, Senior Production Editor and his staff, for their advice, help and support in taking my manuscript through to the finished book. Any errors which remain are entirely my own.

My constant companion throughout the years during which this book has occupied much of my leisure time is my wife Sandra, whose support, patience, acceptance of outings deferred, and understanding have been invaluable. Both of us are very happy to see this project come to a successful end.

Douglas Maclean,
Glasgow, January, 1996.

1

FROM COPPER TO GLASS

The linking up of city to city, state to state, and nation to nation has greater possibilities than we know of yet.

Theodore H. Vail,
First General Manager, Bell Telephone Company, 1877

1.1 INTRODUCTION

Engineering differs from other professions in that it uses technical knowledge based on science and mathematics to meet human needs. One such is the desire to talk with others who may be in a far-off town, country, continent; and to do so immediately by means of individually recognisable voices. That the international telephone network is the biggest machine ever designed, constructed and operated continuously bears witness to the potency of this desire. By the end of the 1980s there were an estimated 650 million telephones worldwide for a population of about 5500 million, and the number of telephones per hundred of the population of each country correlated very positively with the corresponding income. The potential for growth in the global network is still immense, and it is up to governments, regulators and the appropriate international bodies to secure the global connectivity necessary for 21st century communications.

This chapter highlights aspects of the realities which have shaped, are shaping and will shape telecommunications in the 21st century. It is difficult for many communication engineers, faced with the pressing needs of yesterday or today, to lift their heads and see the wider realities. Yesterday's skills may not be relevant tomorrow, as many thousands in the industry found when their jobs disappeared. We begin with an overview of the historical setting, mentioning some of the people and events which have shaped telecommunications in the past: men of vision and enterprise whose achievements are remarkable by today's standards, let alone those of a far less technically sophisticated age.

A harsh reality in telecommunications was that by 1990 the industry was in a period of rapid change unprecedented in the world economy. This is described briefly, but necessarily in a global setting, illustrated by reference to specific operators, firms, events and the actions of bodies such as the European Commission. The earlier cosy world of national PTTs and their domestic suppliers is changing more or less rapidly in many countries. The winds of change will continue to ensure that this is an age of discontinuity. Another reality, of profound significance, is that for the first time for many centuries the East is challenging the scientific and technological lead of Western countries. This is especially true of Japan, not only in telecommunications but also in basic research. The Kansai Science City located on the borders of Osaka, Kyoto and Nara will be the site of the first multimedia experiments to be held under the auspices of the ministry of posts and telecommunications. The city should be completed in the first decade of the 21st century, with a planned population of some 380 000 people (*Financial Times*, 4 April 1990). The site for the city, on the remnants of the Heijo Palace, is one of the oldest cultural centres in Japan. 'Our definition of culture is art, science and technology together', according to the Planning Manager for the Kansai Research Institute. The advanced telecommunications research institute laboratory, in which NTT is the biggest shareholder, is just one of the centres to be housed in the city at the start of the 1990s. Unlike the Tsukuba technology city, it is intended that the emphasis in Kansai will be on basic science (*Financial Times* Survey 'Kansai', 1 Sept. 1994). The gross regional product of Kansai was approximately equal to that of Canada in 1993 and was forecast to exceed that of the UK by the year 2000. People in the region are noted for their openness to new ideas, their ability to make quick decisions and their talent for spotting new business opportunities. The relative proximity of the region to the Asian continent may help them to capitalise on the huge potential for growth that this continent holds. (The severe earthquake which struck Kobe early in 1995 may well affect existing plans.)

One result of all the changes and rates of change outlined is the increasingly important need for change and faster responses by international bodies such as the CCITT and CCIR; indeed these bodies disappeared in an extensive reorganisation of the ITU. The final section of this chapter has two broad aims: first to explain why there is so much interest worldwide in standardisation, and second to provide an introduction to relevant international organisations and their activities. Both subjects are parts of the context within which those employed in telecommunications and information technology must work.

1.2 A BRIEF HISTORICAL INTRODUCTION

In 1837 an immortally theatrical moment occurred when a young woman of 18 was called in her nightdress from her mother's room in Kensington Palace, London after midnight to be told that she was queen — Queen Victoria. During her long reign of 64 years, much of the theoretical and practical work underlying today's telecommunications was pioneered in Europe and America. In 1833 Carl Friedrich Gauss and Wilhelm Weber, both professors at Göttingen University, Germany, had been the first to establish a system of electromagnetic communication [1]. They used overhead wires to span 3 kilometres. Although they realised that messages could be sent over longer distances they had no money for such a venture. Gauss advised a student of his, Carl August Steinheil of Munich, to develop a system for general use. The first Steinheil telegraph was installed in 1837 between the Munich Royal Academy and the observatory. He discovered that the earth itself was an excellent conductor and could be used in place of a return wire, a discovery of major importance in the development of the telegraph. Another model of a telegraph had been demonstrated in 1835 by Paul Schilling, whose apparatus had been borrowed by a professor from Heidelburg to demonstrate to his students, one of whom had been William Cooke. On his return to London he and Charles Wheatstone set about improving Schilling's system. This they did successfully and were granted the world's first patent for an electric telegraph on 12 June 1837, in London [2].

In 1837 Wheatstone and Cooke installed a demonstration system between Euston station and Camden Town alongside the London and Birmingham railway. It worked so well that they were commissioned by the Great Western Railway to install a 30 km line from Paddington station to Slough. Queen Victoria and Prince Albert consented to become the patrons of the enterprise, and the telegraph link

Figure 1.1 W.F. Cooke

Figure 1.2 C. Wheatstone, 1802–75

was opened, late, in 1844. Then as now, the London public showed little interest in the 'wonder of the age', which could be admired for an admission fee of one shilling at Paddington; but on New Year's Day 1845 something happened which made the telegraph the talk of the town. That day the operator at Paddington had received a sensational telegram, a murder suspect had been seen boarding the train at Slough. The police were informed, and after shadowing the suspect following his arrival in London the two plain clothes detectives arrested him. Subsequently he confessed, was convicted and hanged. The importance of the telegraph was realised.

In 1876 a young Scot named Alexander Graham Bell invented the telephone in America, being granted a patent on his 29th birthday. He secured UK patent rights for his telephone on 9 December 1876. That summer he demonstrated his telephone at the Centennial (of the beginning of the American Revolution) Exposition held at Philadelphia. Sir William Thomson was one of the judges, and along with others in the report of their visit to the exhibition wrote: 'Mr. Alexander Graham Bell exhibits an apparatus by which he has achieved a result of transcendent scientific interest — a transmission of spoken words by electric currents through a telegraph wire. We may confidently expect that Mr. Bell will give us the means of making voice and spoken word audible through the electric wire to an ear hundreds of miles away.' [3]. In September 1876 Thomson addressed the British Association for the Advancement of Science in Glasgow and referred to and exhibited Bell's telephone. As a means of communication it proved to be fast, personal and convenient. Furthermore, it needed no training in the use of codes, and so made electrical communication directly accessible to the general public. No wonder then that the vast bulk of traffic on public networks today is speech. Telephony involves the

Figure 1.3 A.G. Bell, 1847–1922

Figure 1.4 W. Thomson, Lord Kelvin, 1824–1907

transmission of the electrical analogue of acoustic waves, rather than the binary elements of telegraphy. Hence, after about 1880 the developing communication networks were basically designed to handle analogue speech transmission. By 1950 almost all the world's systems were based on analogue transmission. By about 1990 many advanced systems were based on the digital transmission of telephone traffic over optical fibres, verily a dream come true.

Bell was born on 3 March 1847 in Edinburgh, the capital of Scotland, a city that for generations had been a centre of culture second only to London in the British Empire, a city sometimes known as the Athens of the North. The hallmark of the Edinburgh mind was exuberant versatility, and throughout his life as prolific inventor Bell retained a child's joy in the world's diversity [4]. After Aleck's younger and then his older brother had died of tuberculosis, Bell's father emigrated to Canada in 1870, like many Scots before and since, and he persuaded Aleck to go with him. In 1920 Aleck Bell, by then an American citizen revisited Edinburgh to receive the freedom of his native city. In 1990 British Telecom (BT) announced a large investment to provide Edinburgh businesses with the most advanced infrastructure in the UK, in the form of the 'Edinburgh loop' which will carry computer data, graphics and video, as well as speech traffic. It is expected that the facilities will significantly enhance the capital's position as a financial and computer centre second only to London in the UK (*Financial Times*, 22 March 1990).

Boston, Massachusetts, was an American Edinburgh, even claiming the title 'Athens of America'. In the 1870s it was the leading centre of American science and technology, including electricity. Bell's father referred an offer to introduce his visible speech method to a new school in Boston for deaf children to his son, who accepted it eagerly. On Aleck's arrival in Boston someone gave him a copy of John Tyndall's new work on sound, and in Autumn 1872 he heard Tyndall's lecture on the undulatory theory of light propagation. Aleck flourished and in 1873 joined the faculty of Boston University as Professor of Vocal Physiology, but still found time and energy to pursue science and invention. The early days of the telephone began in Boston where Bell and his assistant Thomas A. Watson combined

Figure 1.5 First telephone exchange (from [5]. Reproduced by permission of AT&T. All rights reserved)

their experimental work with lectures and further patent applications — exciting days! [5]. 'Mr Watson, come here, I want to see you', were probably the first words spoken over the telephone by Bell on 10 March 1876. A few years later Watson was to remark 'our transmitter is doing much to develop the American voice and lungs. It is said that all the farmers waiting in a country grocery rush out and hold their horses when they see anyone preparing to use the telephone.' During this time one of their backers, Hubbard, an attorney, over the strong objections of his colleagues made a basic decision that the telephone instruments should be leased. It was a recognition of the true nature of the business as a service industry. The integrity of the service is the customer's prime interest, and in such a business there is no justification for acting otherwise than for the long run.

The first telephone exchange began operating in May 1877, and three months later had installed 700 lines, 90% of the total production up to that time; it is illustrated in Figure 1.5.

At first boys were employed as switchroom operators, but it was soon found that they were too unruly and noisy and too inclined to profanity to make good telephone operators, and they were soon replaced by girls. As a farmer once put it, two boys will do half the work of one boy, and three boys will do no work at all! A very early UK exchange is shown in Figure 1.6.

In August 1877 a newly married Bell sailed for Great Britain and Europe on honeymoon, with a pair of telephone instruments with which 'to start the trouble in that country' as he said. Queen Victoria invited the Bells to Windsor and ordered

Figure 1.6 Glasgow switchboard, 1880–90

a set of telephones which Bell promised to make expressly for her, and he was evidently delighted with the order because the customary demonstration fee of £50 was waived. He spent more than a year promoting and lecturing about the telephone. Professor Bell 'gave a very philosophical and entertaining discourse' to a special meeting on 31 October 1877 of the Society of Telegraph Engineers in London (now the Institution of Electrical Engineers) called to hear his account 'of the nature, history and development of what may be called one of the most interesting discoveries of our age'. The text of this lecture is reproduced in [3]. He was also elected a member that year, joining 957 other members who then formed the Society which had been founded six years earlier by some 66 telegraph engineers. Among the publications to mark the centenary of the telephone, there was an issue of *The Post Office Electrical Engineer's* Journal in Spring 1976 which carried an Editorial, brief biography of Bell and a reprint of an address given to the Institution of Post Office Electrical Engineers in 1928 by Dr. Thomas A. Watson, entitled 'The birth and babyhood of the telephone'. Among the many interesting historical facts presented in this issue was that for one minute during his burial ceremony on 4 August 1922 telephone communication throughout North America was suspended, as a mark of tribute to A.G. Bell.

There are two other well-known names associated with the Victorian era, W.S. Gilbert and A. Sullivan, who wrote the words and music for the hugely successful Savoy Operas, greatly enjoyed then and ever since throughout English-speaking countries. On the day after his 41st birthday (13 May 1883) Sullivan gave a party at his home, and among the guests were the Prince of Wales, the Duke of Edinburgh, some real peers, the painter Millais and of course Gilbert. After the meal and singing by Tosti and Madame Albani, 'Sullivan sprang a surprise on his guests. D'Oyly Carte had assembled on the Savoy stage the cast of Iolanthe, surrounding the telephone, and at exactly 11.15 they began singing while two miles away Sullivan passed an earpiece from ear to astonished ear.' Both men had telephone instruments installed in their homes and another was installed at the side of the stage in the Savoy. At Windsor Castle nine days later Sullivan was knighted by Queen Victoria.

Bell also saw that if the intensity of a light beam in air could be varied according to speech the modulated light could be detected by the photoconductive effect in the element selenium (discovered in 1873). In 1880 he produced his 'photophone' which to the end of his life, he insisted was '... the greatest invention I have ever made; greater than the telephone, ...' Unlike the telephone, it had no commercial value [4].

The entrenched position that the telegraph enjoyed under the British and European Post Offices meant that this new competitor was not regarded with much enthusiasm. However, it was not just Europeans who did not recognise its importance. William Orton, then President of the Western Union Telegraph Company of America, turned down the chance to buy all the rights to the telephone patent, replying scornfully 'What use could this company make of an electrical toy ?' In Britain about a century ago a domestic telephone cost £20 a year to rent (or £720 in 1988 values) and a servant could be employed for about the same amount. Consequently, except in Sweden, home of the well-known firm Ericsson, the telephone never displayed anything like the vigorous growth that characterised the American system. It was only from the late 1980s, when winds of change began sweeping across Britain, Japan and some European countries, that competition started leading to vigorous growth.

1.2.1 From sea to shining sea

One of the boy operators of 1877 was J.J. Carty, who went on through a succession of increasingly responsible posts to become the Chief Engineer of AT&T in 1907. John Carty made many innovations, and is regarded in America as the founder of the profession of telephone engineering. The dream of a transcontinental link became a reality in mid-1914. Kelvin, Heaviside and Rayleigh had provided the basic equations governing transmission, but immense physical obstacles of terrain still lay in the way, but these too were conquered. In 1915, Bell placed the first official transcontinental call from New York to Watson in San Francisco — nearly forty years after their first telephone call. The line consisted of two pairs of 4.23 mm wires with loading which gave 2 two-way channels (circuits) with a third circuit provided by phantoming, as illustrated in Figure 1.7.

Six vacuum tube repeaters were installed, one each at Philadelphia, Pittsburgh, Chicago, Omaha, Denver and Salt Lake City, in a route length of 4800 km to compensate for the 60 dB loss.

This achievement demonstrated two profound lessons which affected the development and character of the Bell System. One was that in an industry so clearly based on science the role of fundamental research to acquire new knowledge and understanding was critical. The other was the experience of planning and carrying out, on schedule, large and difficult operations that took advantage of new knowledge, involved many interrelated factors and required state-of-the-art technologies. The tradition of being able to organise and draw on a great variety of talent was started, and has continued to the present day. A dramatic example of this began about 2.30 p.m. on Monday 15 January 1990, when the most far-reaching service problem ever experienced by AT&T occurred, according to their chairman (*Financial Times* 17 Jan. 1990). At its peak, more than 50% of long distance calls on the public switched network failed to connect. A software 'bug' in a program written in an in-house language which formed part of a new and more sophisticated control system caused a glitch which severely tested hundreds of the best communication and computer experts around the country for more than 8 hours. By 10 p.m. that night the system had started to stabilise, and was back to normal by the following day. At the end of 1988 AT&T had reported the first loss in its 103 year history, due to writing off the remaining analogue equipment in its long-distance network, following accelerated installation of state-of-the-art digital technology to meet its target date of the end of 1990 to carry all US switched traffic in digital form.

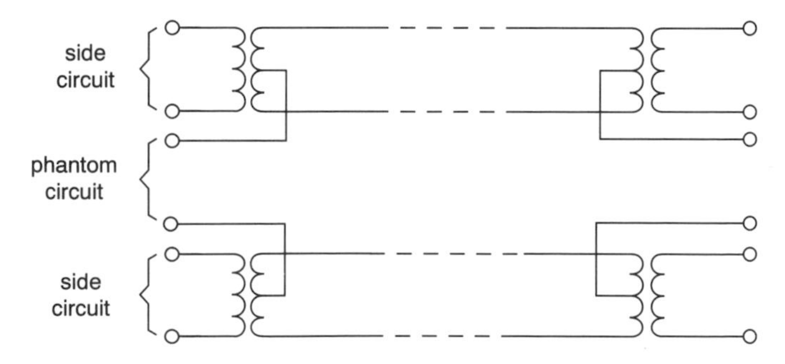

Figure 1.7 First US transcontinental line, 1914 (from [10]. Reproduced by permission of Butterworth-Heinnemann, Oxford)

The outstanding landmark in the AT&T tradition was the setting up of the Bell Telephone Laboratories on 1 January 1925 with its own corporate indentity. Since then a torrent of innovations has come from Bell Labs, seven Nobel prizes have been awarded to serving or past employees, and in the present context the invention of the transistor and the laser stand out. 'That's where they make dreams come true', someone once said, or rather making yesteryear's science fiction into today's science fact. At the end of the 1980s the President of Bell Labs was British-born Ian Ross, who had been with Bell since leaving Cambridge University (UK) in 1952. Dr. Ross is a man credited with a profound understanding of the subtle relationship between research and development. He saw networking as the 'second wave of the information age', following the mainframe computer, in which telecommunications will be exploited on a much larger scale than before. Although a vocal opponent of divestiture, he believes that the laboratories have survived in a very healthy way, and this despite the transfer of five of his research directors and 4000 staff to form Bellcore (Bell Communication Research) (*Financial Times*, 9 Feb. 1987). This new organisation is the cooperative research centre which serves the seven regional operating companies (Baby Bells), and at the end of the 1980s Bellcore was one of the four world leaders in optical fibre systems, the others being AT&T, NTT and British Telecom DuPont (BTD). Bell Laboratories are still arguably the world's finest corporate research and development centre (*Financial Times*, 26 Jan. 1993).

1.3 PAST LIMITATIONS AND INNOVATIONS IN TELECOMMUNICATION NETWORKS

It is interesting and instructive to look briefly at some of the outstanding technical obstacles and their solutions in the evolution of the modern global network. Following Cochrane [6], we divide the elapsed time into eras, beginning with the electric era.

1.3.1 Electric era (1835–1915)

Fundamental limits had appeared by 1915.

(1) Material limit: vast amounts of copper were used and made all services inherently expensive. This physical limitation prompted development of repeaters; it had been calculated that an unrepeatered link across America would require lines of 12–25 mm diameter!

(2) Space limitations: difficulty was experienced in accommodating the necessary open-wire lines and cables; for example, in Broadway, New York, there were telephone poles 90 ft high with up to 50 cross-way arms and 12 wires per arm.

(3) Noise and interference: background thermal noise, crosstalk, talker echo and interference restricted the circuit capacity on long routes to a very low level.

(4) Bandwidth: maintaining bandwidth on long lines was increasingly difficult despite the introduction of loading. Operation at baseband did not utilise the cable capacity efficiently.

(5) Global: transmission limitations on telephony meant there was little hope of achieving a fully interconnected worldwide network.

1.3.2 Electronic era (1915–1965)

This half-century witnessed the widespread introduction of fdm techniques and of early digital systems. These provided full national coverage in the developed countries, and modest international links via submarine cables and satellites. The limits were now principally due to those of discrete electronics and the transmission of traffic in analogue form.

(1) Physical space: difficulties were encountered in exchanges and repeater plant due to bulky technology and problems of heat dissipation.

(2) Noise, gain, power consumption and bandwidth of amplifiers had constrained the upper frequency limit of fdm carrier operation to the range 4–12 MHz in a coaxial cable capable of 12–60 MHz. Crosstalk limited analogue operation on pair-type cables so that no further growth was economically possible using fdm.

1.3.3 Microelectronic era (1965–)

The early years saw the introduction of transistors and integrated circuits, and the domination of the latter has increased steadily to the present. Computer technology and an increasing need to transmit data rather than speech has led to the widespread introduction of low-cost high-complexity integrated and microelectronic circuits into all aspects of the network, facilitating the deployment of digital transmission and switching systems associated with sophisticated terminal equipment. Key realisations flowing from these developments are as follows.

(1) The global network: a fully interconnected world (and national) network of well-defined and controlled standards could be most economically provided by means of digital transmission.

(2) Switching: combining digital switching and transmission results in the most economical network by a factor of at least 50%.

(3) Services: a wholly digital network can provide much more than a plain old telephone service (POTS).

1.4 GENERAL PERFORMANCE COMPARISONS BETWEEN OPTICAL AND TERRESTRIAL MICROWAVE SYSTEMS

In the early 1990s the only competition for optical trunk and junction transmission systems was point-to-point microwave systems. These were deployed extensively to serve commercial customers in urban areas. Since the rate of growth of business traffic is far higher than that of residential customers, it is appropriate to compare these two types of point-to-point system, which with the development of optical systems are likely to fulfil complementary roles.

1.4.1 Introduction

Broad outlines of the major parameters of these two distinct types of systems suffice for our present purpose. The two types share the following features:

rapid progress in technology
extensive introduction into national networks in recent years, e.g. 64-qam systems in the US and provisioning for the next 20 years.

Apart from the long haul applications of microwave links operating in the 4, 6, and 11 GHz bands over typical hop lengths of 50 km, short-haul or local distribution applications at higher frequencies such as 18 GHz are increasingly important for major customers such as banks and large businesses in metropolitan areas. In the UK low-capacity digital radio is used to connect 2 or 8 Mbit/s circuits from an exchange directly into the customer's site, and other traffic applications include PSTN, private circuits and interconnections with cellular radio and other operators. Until the mid-1980s equipment had been cumbersome and time-consuming to install, as well as expensive relative to cable. By the late 1980s new and improved designs of equipment operating in the band centred on 18 GHz were enabling PTTs to meet the growing demand, and provide a service quickly and directly into a customer's premises.

By 1990 many of the techniques developed during the 60-year history of microwave point-to-point systems were being adapted for optical systems. Earlier optical systems were more akin to the early wireless spark transmitters, but with the advent of coherent optical systems and optical amplifiers similar techniques began coming into use. Optical heterodyne techniques and the arrival of intermodulation distortion in optical systems were the first indication of a symbiotic relationship with radio using thousands of electrons/bit and optical systems less than 100 photons/bit [7]. The repeater spacings and bit rates available on fibre had by 1989 begun to far outstrip those obtainable on radio, leading to the very distinct cost advantages illustrated in Figure 1.8 [8].

These facts also led to a realisation that new types of network architectures were becoming available. For instance, networks based upon the concept of an 'optical ether' provide vast bandwidth and are almost totally transparent between nodes.

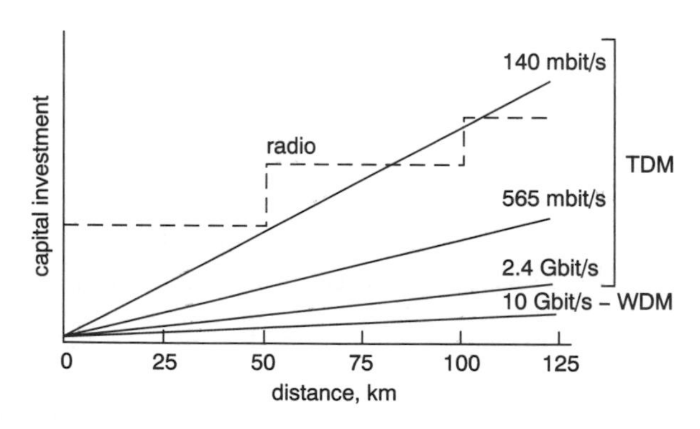

Figure 1.8 Relative UK transmission costs in terms of capital investment versus distance for 140 and 565 Mbit/s and 2.4 Gbit/s tdm and for 10 Gbit/s wdm (from [8]. Reproduced by permission of BT)

Future optical systems are thus likely to resemble radio and satellite systems and networks of 1990.

1.4.2 Trends in optical and radio systems

Radio and optical systems share trends from broadband to narrowband sources, from broadband to narrowband receivers and from simple to complex modulation formats. This last trend and the associated signal processing has not been carried nearly so far in optical systems because fibre (unlike radio) is a guided medium. However, coherent optical systems will probably provide the first steps towards present-day radio signal formats and processing. The development of radio systems has depended critically on the availability of coherent sources to fully exploit the available bandwidth, and Figure 1.9 presents an interesting perspective on the rate of progress across the electromagnetic spectrum since 1832.

1.4.3 Digital radio relay systems (DRRS)

These differ from optical fibre digital line systems in many ways, among which are the following.

(1) DRRS are free-space point-to-point links and are therefore subject to fading. System margins must be allowed for the effects of fading which depend on

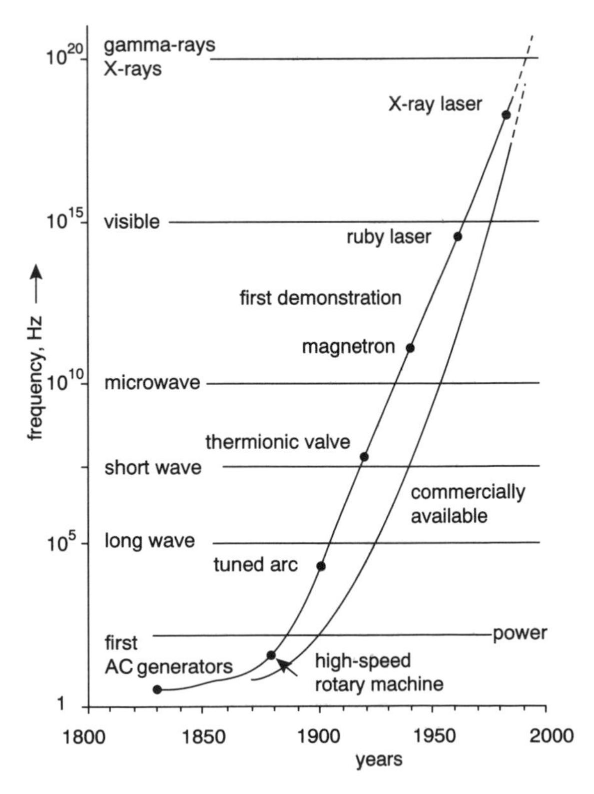

Figure 1.9 The development of coherent waveform generation: carrier frequency versus year (from [8], Figure 3). Reproduced by permission of BT

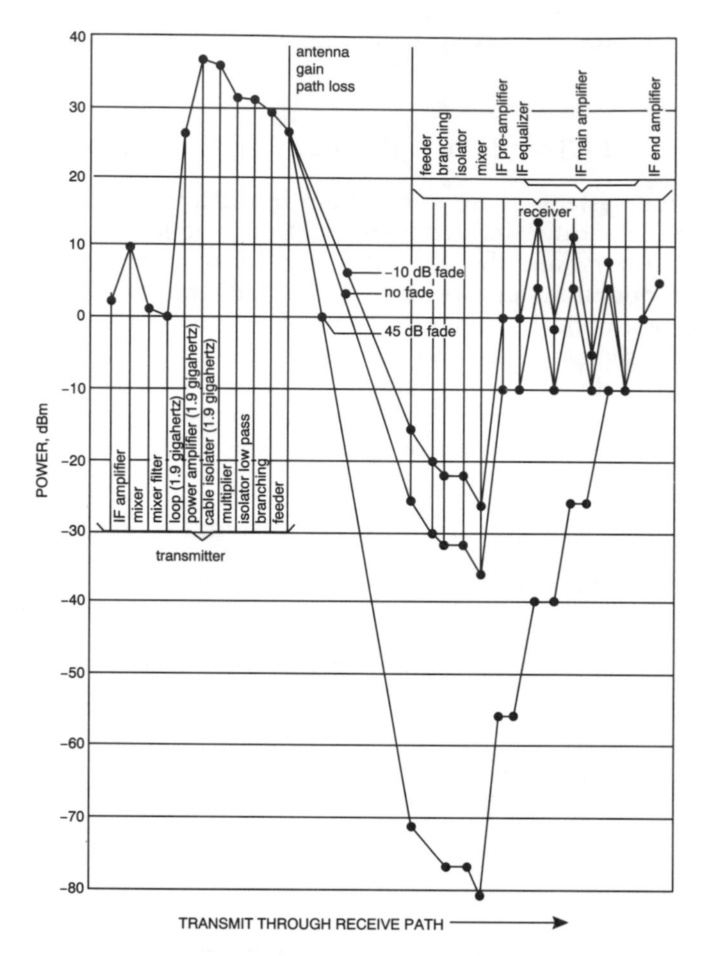

Figure 1.10 Level diagram for digital radio relay system operating in the 7 GHz band ([9], Figure 5). Reproduced by permission of Alcatel

outage times, operating frequency, hop length and other factors. This is illustrated dramatically in a level diagram such as Figure 1.10 [9]. Space diversity reception is usually required.

(2) Unlike direct-detection on-off keying, ofdls radio systems use a variety of modulation schemes applied to carriers which are monochromatic or single-frequency compared with the modulating frequencies. Currently these schemes include 4-psk, 8-psk, and for 140 Mbit/s systems 64, 128 and 256-qam. Spectral efficiency is an important consideration (64-qam gives about 4.5 bit/s/Hz).

(3) Transmitter powers are some 30–40 dB above those used in present-day ofdls. Linearity requirements become more demanding as the number of states increases, and nonlinear distortion must be compensated, e.g. by the use of predistortion in the transmitter.

(4) Unlike on-off keyed (ook) optical systems which use direct detection, coherent detection is employed so carrier regeneration is needed with the attendant complexity. Receiver front-end linearity requirements become more onerous as higher-order modulation schemes are used.

(5) In general, increasingly elaborate equalisation is needed as the number of states increases, leading to the use of dynamically adaptive equalisers for example, whereas equalisation in ofdls is usually very simple. Overall filtering between modulator and demodulator must usually satisfy the Nyquist conditions.

1.4.4 Some comparisons between ofdls and drrs

Following Cochrane [7], we compare: the transmit and receive powers for a given aperture; the ratio of carrier frequency to linewidth; ratios of modulation bandwidth to carrier frequency and linewidth; and radio, satellite and optical fibre technologies.

Comparison of transmitter powers

Optical fibres provide guidance and containment of the energy associated with a modulated carrier, unlike microwave radio; one would therefore expect the transmitter powers to differ substantially. The two situations can be compared by recalling that the transmit and receive powers are proportional to the corresponding apertures, and taking the ratio

$$\frac{P_{\mathrm{R}}}{P_{\mathrm{T}}} = \frac{\Omega_{\mathrm{R}}}{\Omega_{\mathrm{T}}} \tag{1.1}$$

where P_{R} is received power, P_{T} is transmitted power, Ω_{R} is the solid angle subtended at the receiver (dish or fibre plus photodiode), and Ω_{T} is the solid angle subtended at the transmitter (dish or laser)

For simplicity we compare an X-band radio system at 3 cm (10 GHz) and an optical system at 1 μm. Taking a 3 m diameter antenna with a beamwidth θ of about 0.5° in the radio case and an optical beamwidth θ' of 0.0005°

$$\text{ratio of transmitter powers} = \left(\frac{\theta'}{\theta}\right)^2 = 10^{-6}$$

Because of its concentration and guiding of the transmitted power through an extremely low loss path, an optical system requires only milliwatts, where radio requires kilowatts. The influences of per-unit loss and bandwidth of various transmission media are shown in Figure 1.11.

Ratio of carrier centre frequency to carrier width

It is also of interest to compare radio and optical systems with respect to this measure of monochromaticity. Modern drrs transmitters are highly accurate and well stabilised in frequency, so that, for example, for f_{c} in the range 1–10 GHz the widths w_{c} would lie in the range 1–10 Hz, approximately, giving a ratio $f_{\mathrm{c}}/w_{\mathrm{c}}$ around 10^9. For optical systems the typical values pertaining to an operating wavelength of 1300 nm ($f_{\mathrm{c}} = 231$ THz) listed in Table 1.1 show both contrasts and convergence.

The fact that optical carrier source values of $f_{\mathrm{c}}/w_{\mathrm{c}}$ are approaching those for radio is a key turning point in the development of systems, as it will allow for channel packing and spectral densities comparable with microwave systems, and far longer operating spans through the significant improvements in signal-to-noise ratio which become possible.

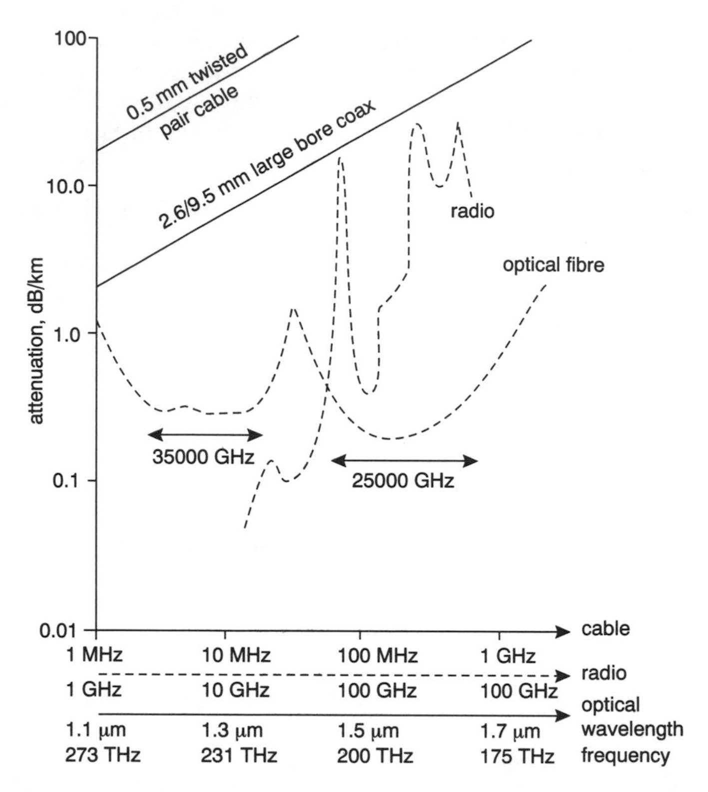

Figure 1.11 Per-unit attenuation versus frequency for coaxial cable, radio and optical fibre ([8], Figure 5). Reproduced by permission of BT

Table 1.1 Comparisons for different electro-optical and radio transmitters [7, p. 10]. Reproduced by permission of IEE, © 1989

Transmitter	Width	Ratio (1)	Ratio (2)	Comments
LED	2000 GHz	10E2	10E-5	Low cost, readily available
F-P laser	500 GHz	10E3	10E-4	Modest cost, readily available
DFB laser	20 MHz	10E7	10E-1	High cost, becoming available
LEC laser	100 kHz	10E9	10E3	Very high cost, scarce
Radio (10 GHz)	10 Hz	10E9	10E7	

Notes:
Ratio (1) f_c/w_c
Ratio (2) B_M/w_c, B_M is modulation bandwidth
LED light-emitting diode
F-P Fabry-Perot
DFB distributed feedback
LEC long external cavity[7, p. 11]

Ratios of modulation bandwidth to carrier frequency and linewidth

The values of the ratio B_M/w_c in Table 1.1 indicate why psk modulation is precluded from optical systems using dfb lasers, whereas fsk modulation is practical; lec (long external cavity) lasers would allow the use of psk.

Table 1.2 A qualitative comparison between radio, satellite and fibre bearers [7, p. 11]. Reproduced by permission of IEE, © 1989

Parameter	Radio	Satellite	Fibre
Bandwidth	Limited	Limited	unlimited
Repeater distance	Line of sight	−10 000 km	100 km installed
Earth surface	(< 100 km)		300 km demonstrated
			1000 km in future
Delay	Not a problem	A severe problem	Not a problem
Interference	<-------A problem	------->	Not applicable
Weather	<-------A problem	------->	Not applicable
Deployment	Rapid	Rapid	Slow
Capital cost	Low	Very high	High
Security	Poor	Poor	Good

Comparisons of radio, satellite and optical fibre technologies

Some of the features of these three are listed in Table 1.2.

Radio is fundamentally capable of rapid deployment and can be quickly adapted to new service requirements on line-of-sight links. Optical fibre takes longer to deploy, even when duct space is available, but in some cases fibre laid in canals, rivers or coastal waters can provide very economical links, as described in Chapter 15.

1.5 THE PUBLIC SWITCHED TELEPHONE NETWORK (PSTN)

The analogue network in most countries has evolved over many decades and some existing plant may be 30 or 40 years old, or even older in countries with less developed telecommunications. The topology or architecture of each PSTN is divided into three broad portions, which in the UK are called the local or subscriber, the junction and the trunk networks. The arrangement and nomenclature used in the largest national network, that of the US, are given in Figure 1.12 [10].

Up to the end of the 1980s optical fibres had been used extensively in the trunk and junction parts. In the UK and in other countries modernisation has been implemented by establishing a digital overlay network on the existing analogue system, and this situation is illustrated in Figure 1.13.

By 1990, the UK network had been automated considerably with the introduction of [11]:

about 400 processor exchanges controlling 3000 remote concentrator units interfacing with customer sites

intelligent network databases for the provision of advanced customer services

dynamic alternative routeing to automatically route customer calls and minimise congestion (call blocking) at peak periods.

1.5.1 The UK local network (1990)

The local area comprises: subscriber apparatus, lines to the local exchange, and local exchanges. In the UK in 1990 there were:

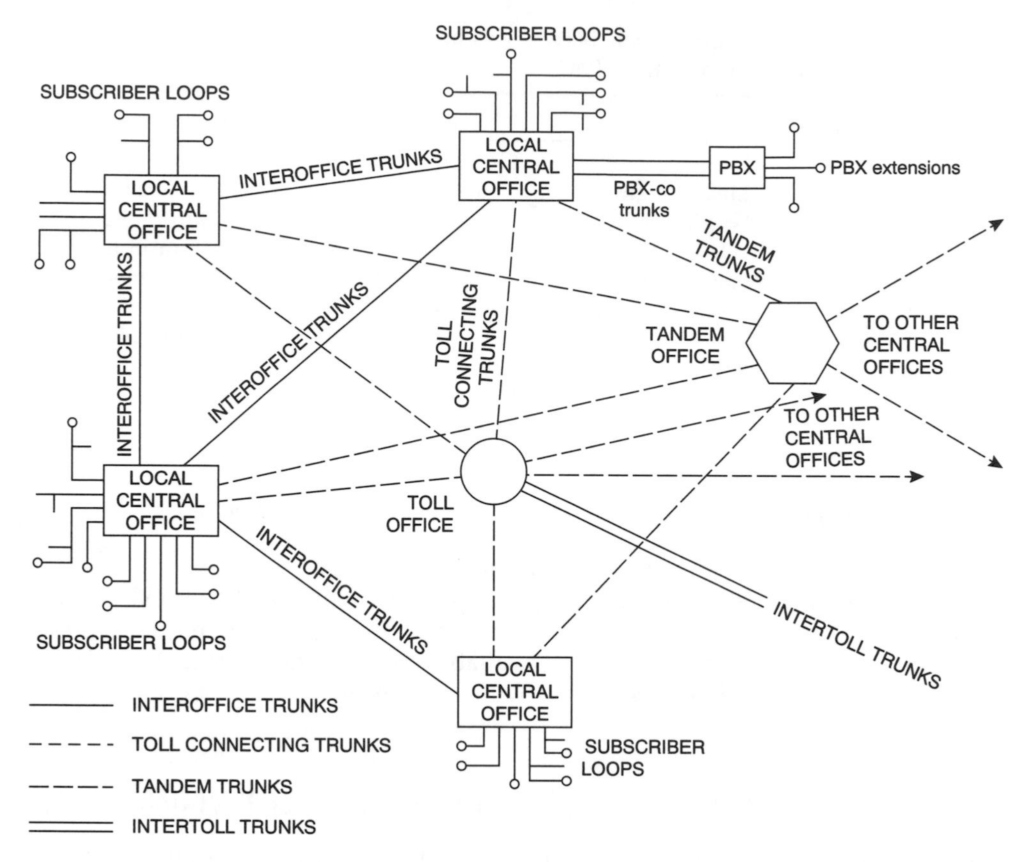

Figure 1.12 Schematic diagram showing architecture of US public switched telephone network ([10], Figure 28). Reproduced by permission of Butterworth-Heinnemann, Oxford

(a) 15 million subscribers
(b) 24 million telephones
(c) 6339 local exchanges.

1.5.2 The UK junction network (1990)

This comprises the lines between a local exchange and another exchange, local or trunk, and there are about 1 500 000 junction lines. Some salient facts about these junction circuits are:

(a) relatively short lengths (1–45 km)
(b) they are economically important
(c) the cost of signalling is an important factor
(d) the older copper cables are similar to those used in the local area but to a tighter specification. Such cables are normally kept under gas pressure, and are provided with automatic monitoring.

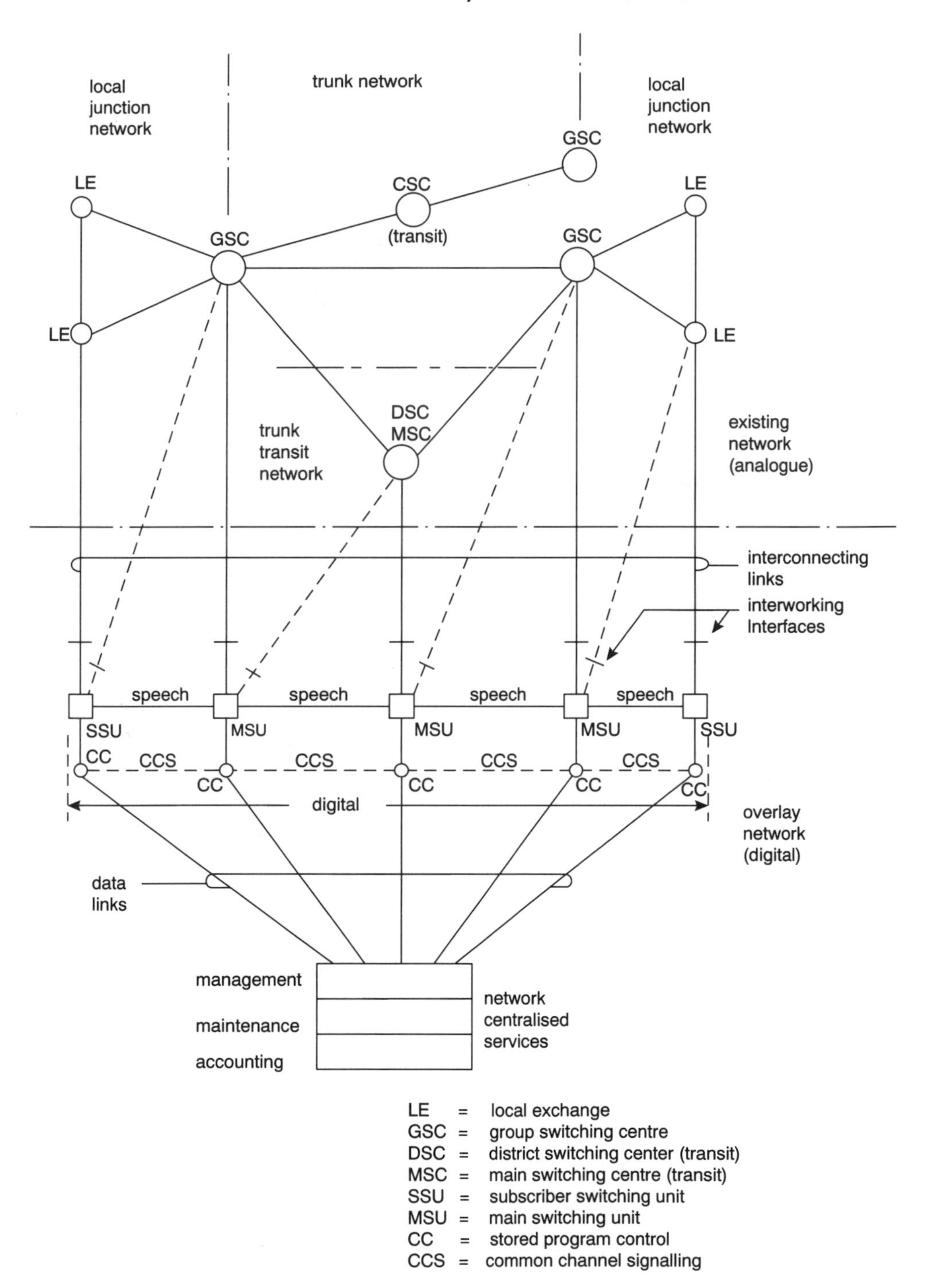

Figure 1.13 UK digital overlay network. Reproduced by permission of BT

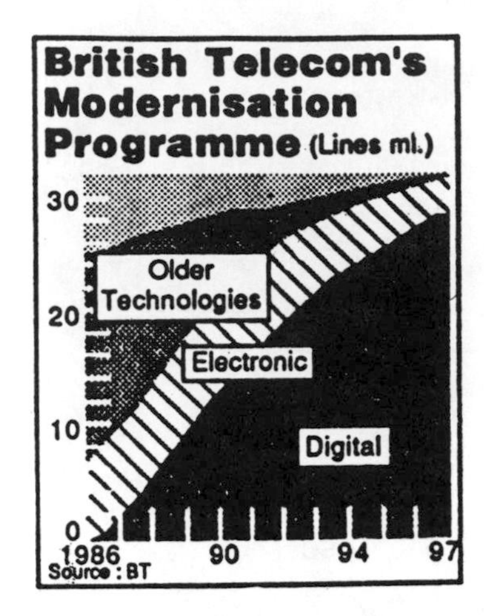

Figure 1.14 British Telecom modernisation programme (1986–1997): planned number of lines and different technologies versus year (*Financial Times*, 24 Mar. 88)

The first point is the most significant because 45 km can be spanned by a repeaterless optical fibre link, thus saving on power feed arrangements, footway boxes, etc.

1.5.3 The UK trunk network (1990)

Turning now to the trunk network, this comprises the trunk exchanges and the interconnecting transmission circuits which carry the heaviest traffic. The trunk exchanges supply the international gateway exchanges, which form part of the international transmission network (ITN) of submarine cables, satellites and high frequency radio circuits. Optical fibres were first installed on trunk routes because of the amount of traffic carried.

As an illustration of a general trend towards better service in public networks resulting from digitalisation, BT's modernisation programme (late 1980s) envisages a steep growth in the number of digital lines, in addition to some growth in the total, as shown in Figure 1.14.

In terms of line faults per annum, these should be reduced from 0.2 to less than 0.05; calls which fail due to network faults should drop from 4.2% to below 0.5%, and the average time taken to connect calls should drop from 12 s to less than 1 s.

1.6 WORLD TELECOMMUNICATIONS: THE WINDS OF CHANGE

The traditional structure of the industry had been strained to breaking point by the flowering of innovative technology and applications in the 1980s. The global equipment market grew from $88 billion in 1986 to $127 billion in 1990, while the corresponding growth in the telecommunications services market was from $175 billion to $279 billion (*Financial Times*, Survey World Telecommunications, 7 Oct. 1991).

1.6.1 Introduction

The 1993 world markets for information communications technology could be broken down by product areas as shown in Figure 1.15.

The top 15 equipment manufacturers ranked by equipment sales in 1992 are listed in Table 1.3.

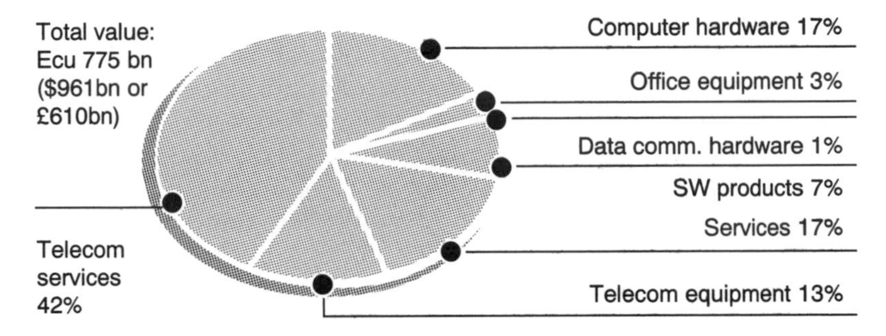

Figure 1.15 World markets for information telecommunications technology, 1993 (*Financial Times,* International Telecommunications, 17 Oct. 1994)

Table 1.3 Top 10 telecommunications equipment vendors by equipment sales (1994) (FT survey, International Telecommunications, section 2, p. 23, 3 Oct. 1995)

Company	Country	US$m	% change 1993–4	% of total company sales	Export sales (%)	Market share (%)
Alcatel	France	20,401	6.8	67.6	72.0	15.8
Motorola	US	14,389	42.9	64.7	44.0	10.8
AT&T	US	14,279	21.2	19.0	9.8	10.7
Siemens	Germany	12,779	−3.1	24.5	58.0	9.6
Ericsson	Sweden	10,699	35.8	100.0	90.0	7.4
NEC	Japan	9,481	0.8	27.1	16.0	7.1
Nortel	Canada	8,223	4.6	92.7	87.0	6.2
Fujitsu	Japan	4,774	3.1	15.5	30.0	3.6
Bosch	Germany	3,413	−29.2	16.1	54.0	2.6
Nokia	Finland	2,531	21.4	43.8	85.0	1.9
Total		100,969	12.3	34.6	40.0	75.7
Others		32,571				24.3
Global total		133,540				100.0

Source: ITU/Company Reports/MarketLine International Database

The renaissance has been brought about by the applications of electronics and optoelectronics, and it opens up prospects which would have seemed extraordinary to well-informed people in the business only 20 years ago, the planning period for distribution cables in the UK. Indeed, reading surveys of trends in line transmission business written some 20 years ago makes one realise how often the best laid predictions of mice and men go astray. In one such report (*circa* 1965), UK growth trend predictions beyond 1985 were very dependent on the use of the viewphone by subscribers. There was of course no mention of optical fibres, and the dramatic changeover from analogue to digital transmission was not forecast.

By the late 1980s users had begun to press for even more rapid changes, anxious to mould a cheaper, uniform communication system out of the national fiefdoms which in many cases still ran the system.

The late 1980s and early 1990s saw the communications industry go through great changes as market forces and a proliferation of information technologies and information service providers enabled convergence between computing, the media and telecommunications to progress. Competition was seen as the dominant process affecting the emergence and evolution of a new communications industry, which was predicted to revolutionise the ways in which companies were structured and run and the ways in which communications technologies would be used [12]. The competitive situation in a number of the major countries in 1993 and that forecast for 1997 is given in Figure 1.31*c*.

'The linking of the world's people to a vast exchange of information and ideas is a dream that technology is set to deliver. It will bring economic progress, strong democracies, better environmental management, improved healthcare and a greater sense of shared stewardship of our small planet' (US Vice-President Al Gore, writing in the *Financial Times*, September, 1994). These noble goals are however still a long way off for the majority of the world's population, who do not even have access to a basic telephone circuit. An ITU study suggested that a $1000 increase in gross domestic product per capita is related to a teledensity (i.e. telephones per 100 population) increase of 2.24%, as shown by the trend line in Figure 1.16*a*.

The foremost global telecommunications challenge is, and will remain, how the enormous investment of money and other resources can best be met, in each individual developing country, to help realise these goals.

The top 20 operators ranked by 1993 revenue are listed in Table 1.4.

In this section we will briefly review some of the events and changes taking place in telecommunications in a number of countries during the early 1990s, when there were four major competitors on the world optical fibre digital line systems scene, Bellcore, AT&T, NTT and BTD (British Telecom DuPont). Then telecommunications and the European Community are mentioned, followed by some data providing a broad picture of the telecommunications marketplace in the early 1990s and a few predictions for some years into the 1990s. The final subsection provides an outline of the situation regarding international bodies and standardisation.

1.6.2 International telephone cartel

In April 1990 the *Financial Times* revealed that consumers across the world were being overcharged by more than $10 billion a year for international calls. The remainder of 1990 saw a series of further articles indicating responses to the revelation, and these led to changes in the early 1990s. For instance, an article headed

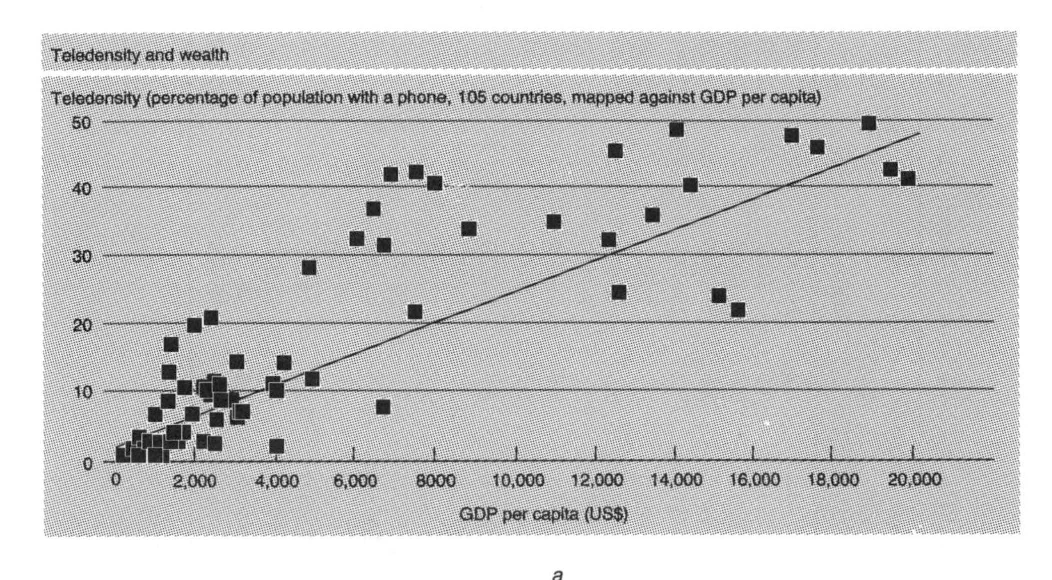

a

The information rich and the informatin poor

Minutes of outgoing International Telephone Traffic (MITT) per inhabitant in selected countries, 1991

OECD		Africa	
Luxembourg	423.7	Botswana	17.7
Switzerland	209.2	Swaziland	14.4
Belgium	82.3	Lesotho	11.1
Austria	81.5	Djibouti	10.0
OECD average	21.1	Africa average	1.6
Spain	18.4	Sierra Leone	0.3
Portugal	14.1	Burkina Faso	0.2
Japan	9.4	Ethiopia	0.2
Turkey	3.4	Tanzania	0.1
World	7.3 Mitt per inhabitant per year	World	7.3 MITT

b

Figure 1.16 (*a*) Teledensity versus gross domestic product per capita (*b*) minutes of outgoing international telephone traffic per inhabitant in selected countries, 1991 (*Financial Times*, International Telecommunications, 17 Oct. 1994)

'Reconnecting charges with costs' (*Financial Times*, 3 April 1990) stated: 'Like OPEC in the 1970s, the international telephone cartel is distorting the development of virtually every branch of industry'.

The biggest carrier among the top 15 is of course AT&T. By mid-1990 it and the other two US long-distance carriers MCI Communications and US Sprint said that they welcomed efforts to cut charges. Proposals to halve the price of international calls were announced by the US telecommunications authorities, and the US led the campaign for lower prices. As an illustration of how the price of calls between

Table 1.4 Some salient facts about the twenty largest telecommunications operators as at 1993 or 1994 (*Financial Times*, 1 Mar. 1995)

Rank 1993	Rank 1992	Company	Country	Telecom service revenue $m	% chge	Pretax profit $m*	As % of revenue	Net profit $m*	As % of revenue	Employees	% chge	Year end**
1	1	NTT	Japan	60 134.9	2.8	1570.3	2.6	743.2	1.2	248 000	−0.4	03/94
2	2	AT&T	US	39 863.0	0.7	−1564.0	−3.9	−3794.0	−9.5	78 387	0.6	12/93
3	3	Deutsche Telekom	Germany	35 679.1	9.3	1934.6	5.4	−1738.2	−4.9	231 000	—	12/93
4	5	France Telecom	France	22 426.4	3.6	3528.9	15.7	848.3	3.8	154 548	−1.5	12/93
5	4	British Telecom	UK	20 539.2	3.3	4139.4	20.2	2711.0	13.2	156 000	−8.6	03/94
6	9	GTE	US	17 266.0	−2.0	1468.0	8.5	900.0	5.2	72 900	−10.1	12/93
7	7	BellSouth	US	15 880.3	4.5	1451.7	9.1	880.1	5.5	95 084	−2.1	12/93
8	6	SIP	Italy	14 872.2	8.6	934.7	6.3	417.7	2.8	86 115	−1.6	12/93
9	8	Nynex	US	13 407.8	1.7	−566.8	−4.2	−394.1	−2.9	76 200	−7.6	12/93
10	10	Bell Atlantic	US	12 534.8	3.7	2195.4	17.5	1403.4	11.2	73 600	3.1	12/93
11	13	MCI	US	11 921.0	12.9	1000.0	8.4	582.0	4.9	36 235	17.0	12/93
12	11	Ameritech	US	11 710.4	5.0	2222.5	19.0	1512.8	12.9	67 192	−5.8	12/93
13	17	Sprint	US	11 367.8	23.2	350.2	3.1	54.9	0.5	50 500	16.4	12/93
14	14	SW Bell	US	10 690.3	6.7	−220.2	−2.1	−845.2	−7.9	58 400	−1.8	12/93
15	16	US West	US	10 293.6	4.8	−2537.0	−24.6	2805.8	−27.3	60 778	−4.6	12/93
16	11	Telefonica	Spain	9587.3	5.7	844.5	8.8	666.6	7.0	74 340	−0.1	12/93
17	15	Pacific Telesis	US	9244.0	−7.0	−1494.0	−16.2	−1504.0	−16.3	14 873	2.2	12/93
18	18	Telstra	Australia	8607.8	3.5	1356.3	15.8	615.0	7.1	72 000	−7.7	06/93
19	21	Telmex	Mexico	7212.0	17.6	2990.1	41.5	2696.8	37.4	48 771	−0.3	12/93
20	25	Telebras	Brazil	6962.1	n/a	n/a	n/a	n/a	n/a	93 574	5.0	12/93

*Profits are for the whole group;
**Dates are for the last day of the month indicated here.

Source: International Telecommunication Union, Switzerland©

The cost of a call

International telephone cost comparison 1994/95 survey

a Local calls (pence per three-minute call)

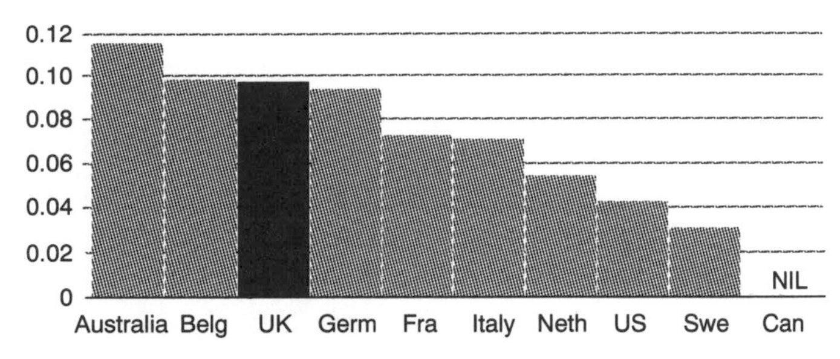

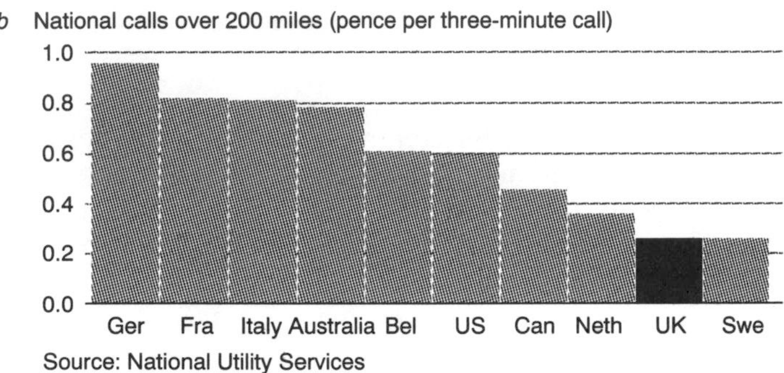

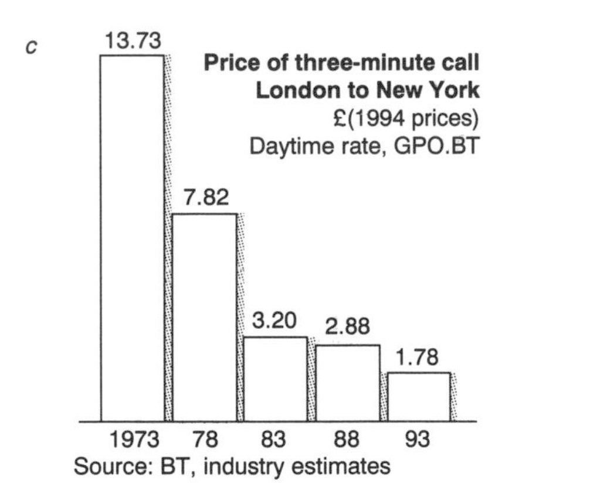

Figure 1.17 Comparisons of (*a*) costs of local calls (pence per 3 minutes) in a number of countries (*b*) national calls over 320 km (pence per 3 minutes) in a number of countries, and (*c*) international calls London to New York at five-year intervals from 1973 to 1993 (£ per 3 minutes at 1994 prices) (*Financial Times*, 15 Mar. 95; *Financial Times*, 17 Sept. 94)

London and New York has changed in the 20 years from 1973 to 1993, consider Figure 1.17*c*, while cost comparisons for international and for trunk calls in some countries are shown in Figures 17*a* and *b*.

The corresponding growth in the number of UK international calls is dramatic, as can be seen from the histogram for the same five years in Figure 1.18.

Large cuts in the cost of calls between the UK and the US were expected to follow after the UK government confirmed in late 1994 that it would allow international

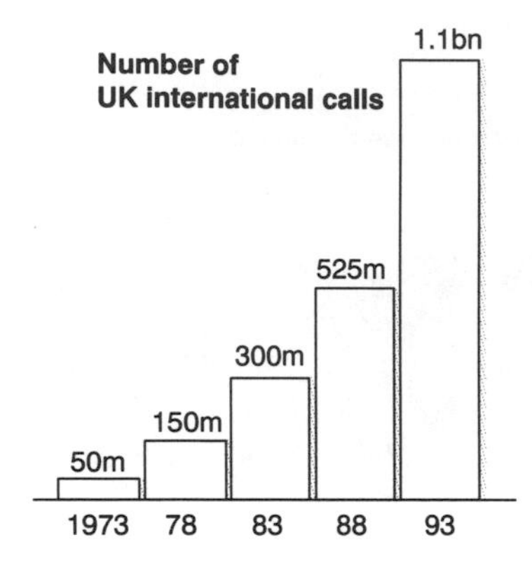

Figure 1.18 Histogram showing number of UK international calls 1973 to 1993 (*Financial Times*, 17/18 Sept. 1994)

simple resale (ISR). That is, allow telecommunications companies to resell capacity, leased from the main transatlantic carriers, to their customers at a discount on existing tarifs. The traffic carried in this way can be connected into the public networks in the US and UK (*Financial Times*, 21 Oct. 94). 19 companies were licensed to do so at that time.

1.6.3 Quality of service

The need for improved quality of service, with its implications for customer responsiveness and availability of services, as well as the need to reduce or at least

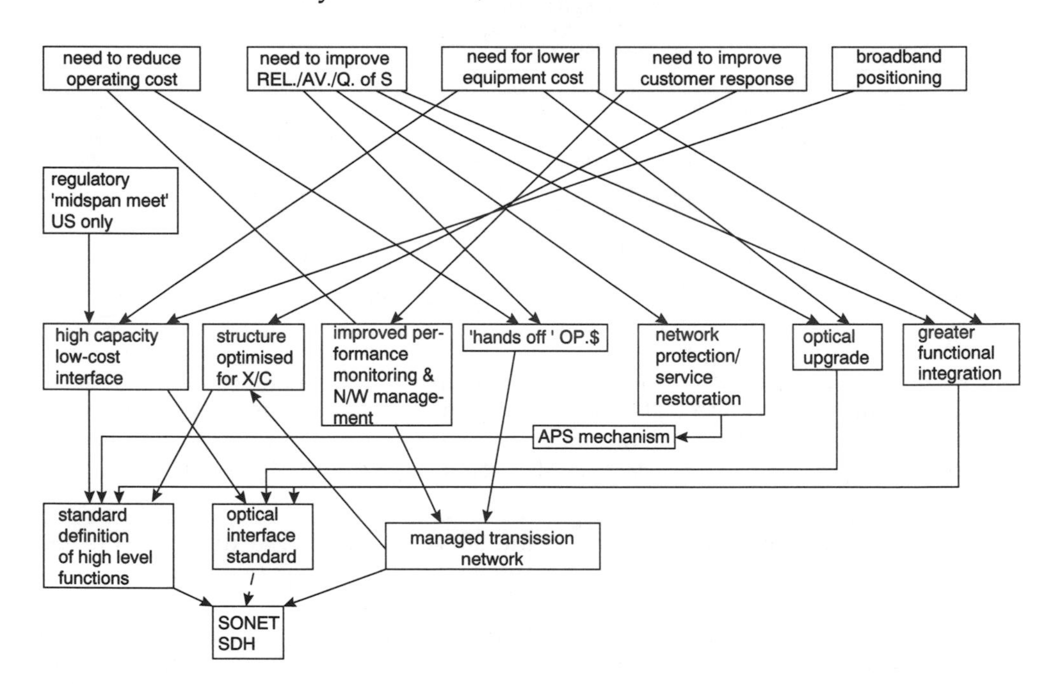

Figure 1.19 Market drivers and corresponding technical responses ([15], Figure 1). Reproduced by permission of BT

contain network operating costs, are driving forces felt to some degree by all modern network operators. The pressures are especially acute in those countries with liberalised regulatory regimes [13]. The new synchronous transmission interface standards, originally developed in the US as SONET, and described in CCITT draft recommendations G707, G708 and G709, are referred to as the synchronous digital hierarchy (SDH), and are described briefly in Chapter 2. These new developments were largely driven by the desire to implement managed optical bit transport networks in response to the main driving forces listed in Figure 1.19, which also illustrates the technical responses.

1.7 REVIEW OF THE TELECOMMUNICATIONS SITUATION IN FOUR MAJOR COUNTRIES

The pressures exerted by these new conditions have led to a number of very significant changes in numerous countries, which in some cases have altered the old world balance. Privatisation of public telecommunications carriers has been one very marked harbinger of changes; Table 1.5 lists material of interest.

We will now look at some of these changes in the UK, USA, Germany and France as examples and to provide the framework within which these national networks (and some others) will operate and change in the 1990s under four broad subheadings: regulatory framework, competitive situation, responses to the competitive marketplace, and external alliances.

1.7.1 United Kingdom

Regulatory framework

Regulation is based upon the Telecommunications Act 1984, in which the Secretary of State and the Director General of Telecommunications (Head of the Office of Telecommunications (Oftel)) have to exercise the functions assigned to them under the Act. In 1990 the Department of Trade and Industry began work on the most far-reaching review of the UK telecommunication market since that Act became law. The framework for the review was laid out in a Green Paper entitled *Competition and Choice: Telecommunications Policy for the 1990s* (HMSO, 1990) published in November 1990, and it was followed by the White Paper in 1991. *Inter alia*, this imposed restrictions on BT in providing or carrying entertainment services until the year 2001, or possibly 1998 if Oftel were to recommend lifting the ban. BT expressed the view that it was difficult to see how the investment of about £15 billion (1994 prices) to install a national optical fibre network over most of the UK could be justified with the ban in force. The UK regulatory framework was, in 1994, the only one in the world to encourage convergence between cable operators and telecom operators by giving cable operators the right to provide both telecommunication services and cable television. The nine main operators dominating the UK cable industry were then all North American in origin. The intention of the Green Paper was to create the world's most liberal telecommunication environment by increasing competition in local, trunk and international telephony (*Financial Times*, 17/18 Nov. 1990). Whatever detailed policies emerge from this, and the most intensive scrutiny since privatisation which began in early 1992, three issues were clear: they will promote

Table 1.5 Pre- and post-1990 privatisations of public telecommunication carriers (*Financial Times*, survey of World Telecommunications, 7 Oct. 1991)

Date	Country	Access lines (000)
	Chile	625
	Hong Kong	2191
Pre-1990)	Japan	51 127
	Spain	10 972
	UK	24 400
Total		**89 315**
	Argentina	3176
	Malaysia	1248
1990	Mexico	4162
	New Zealand	1452
Total		**10 038**
1991	Pakistan	637
	Venezuela	1458
Total		**2095**
	Australia	7420
	Belgium	3525
	Brazil	8434
	Colombia	2065
	Cote d'Ivoire	70
	Czechoslovakia	2125
	Denmark	2792
	Germany	29 840
	Hungary	858
	Ireland	834
	Israel	1472
	Kenya	157
1992 or	Netherlands	6466
later	Nigeria	250
	Poland	2953
	Portugal (CTT)	968
	Portugal (TLP)	1110
	Puerto Rico	872
	Saudi Arabia	1210
	Singapore	939
	South Africa	2866
	South Korea	10 486
	Sweden	5601
	Turkey	4921
	Uruguay	345
Total		**98 579**

Source: BOC

Table 1.6 Some facts relating to British Telecom in 1983/84 and 1993/94 (*Financial Times*, 30/31 July 1994)

	1983/84	1993/94
Turnover	£6.9 bn	£13.7 bn
Pre-tax profit	£990 m	£2.76 m
Fibre in network	Less than 500 000 km	2.6 million km
Employees	241 000	156 000
Payphones	76 500	122 000
Residential lines	16.06 m	20.47 m
Business lines	3.77 m	6.17 m
Proportion of revenue generated in UK	100%	98%
Chairman's salary	£84 198	£663 000

Source: BT

more competition, there will be continuing tough regulation, and the regulatory body Oftel will have to change.

The 85% or so share of the 1994 market held by BT is likely to be undermined as a result. For instance, at the local level customers will be able to connect their telephones to lines supplied by cable television companies, mobile telephone operators, providers of personal communication networks, or indeed anyone else able to provide an appropriate service. Some of the changes which have taken place in BT over the decade since privatisation can be gauged by comparing the items listed in Table 1.6.

Despite the regulatory framework, BT is conspicuously profitable, as can be seen from an international comparison published in 1992, and reproduced in Table 1.7.

The interaction between technology and regulation was described in [14]. Among the topics discussed under competition in the provision of services were: interconnection of public telecommunication operators, restrictions on private networks, value added and data networks. Other topics were competition in the provision of apparatus, quality, provision of services and under future areas of technical interaction progress in competition and liberalisation, the development of private networks, intelligent networks, optical fibres, self-protecting networks and security. In response to the widespread overcharging for international calls, Oftel moved to reduce BT's charges by means of a two-pronged approach: more competition and the introduction of regulation in this sector of the market (*Financial Times*, 2 Oct. 1990).

Competitive situation

Mercury Communications has been British Telecom's main competitor in the fixed trunk network since 1984. By 1994, BT had 130 competitors in the UK, but still retained 85% of the telephony market (*Financial Times*, 30/31 July 1994) with Mercury having 13.5% by revenue. Energis, the third long-distance network (which utilises aerial optical fibre cables on national electricity grid pylons) entered service in autumn 1994, emphasising its distinctiveness as the first information superhighway and claiming to undercut BT. The growth in the number of UK telecommunications operators in the decade 1980–1990, shown in Figure 1.20, illustrates the dramatic increase.

Table 1.7 Telephone company profitability (*Financial Times*, 20 Jan. 1992)

US Regional Bell Operating Companies (RBOCs)	Operating return on net property, plant & equipment (%)	Net income per line ($)	Cash surplus/ (deficit) per line ($)*
Ameritech	16.6	76	7
Bell Atlantic	13.4	75	15
Bell South	12.7	93	4
NYNEX	10.2	62	(3)
Pacific Telesis	13.0	73	19
Southwestern Bell	12.6	91	16
US West	13.4	95	(22)
Average	**13.1**	**81**	**5**
Europe			
BT	**22.6**	**144**	**34**
SIP	6.1	15	(153)
Telefonica	8.2	59	(330)
Japan			
NTT	7.3	37	6

Data for last completed financial year, translated at recent exchange rates (£ = $1.760, $1 = Lira1220, $1 = Peseta102.9, $1 = ¥128

*Cash surplus/(deficit) calculated as net income plus depreciation, less capital expenditure and dividends

Source: Robert Fleming Securities and FT

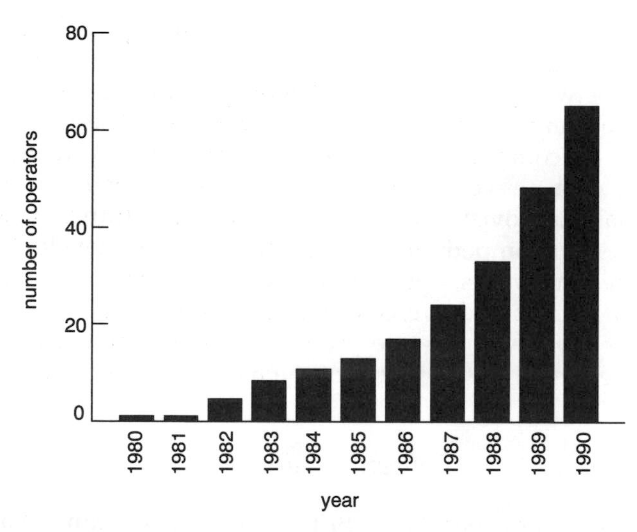

Figure 1.20 Growth of UK telecommunications operators, 1980–1990 ([15], Figure 1). Reproduced by permission of BT

The advent of large numbers of operators means more networks, and a transition from the public utility paradigm to the competitive industry paradigm will bring about many changes in the ways networks are operated and services are provided [15] suggested some of the key differences between these two paradigms, that are listed in Table 1.8.

Table 1.8 Key differences between public utility and competitive industry paradigms ([15] Table 1, p. 30). Reproduced by permission of BT

Public utility Paradigm	Competitive industry paradigm
Monopoly	Large numbers of competitors
Monopoly service provision	Competitive service provision
Monopolist oriented	Customer oriented
Stable pricing regime	Dynamic competitive pricing
One network	Networks of networks
	Competitive interworking issues
'Central Office' philosophy	Pressures to decentralise

Mercury Communications was owned by Cable and Wireless (80%) and Bell Canada Enterprises (20%) in 1993, when its national network consisted of 5690 km of fibre and 2700 km of digital microwave links. The bit rate has increased from the original 140 Mbit/s to 565 Mbit/s and then 2.5 Gbit/s and SDH in 1994. The network was one of the first in Europe to employ two-channel wdm and provides route diversity. A further 40 switching points will be added in 1994. The March 1993/March 1994 results showed 21% growth in turnover, 27% in profit, and that service was being provided for about 30 000 new customers per month. The extent of the Mercury UK trunk network is shown in Figure 1.21, together with current and planned extensions, satellite centres and international submerged cable links as at September 1993. (*source*: Mercury Communications Ltd.)

From April 1994, for five years, Mercury Communications will manage the UK Government's long-distance telecommunication network which, *inter alia*, carries telephony, video-conferencing and mobile communications traffic. This was, in 1993, one of the largest private networks in the UK, serving some 500 offices throughout the country.

Responses to the competitive marketplace

Since about 1983 BT has been engaged in a massive modernisation programme in digital technology, including the installation of several million km of singlemode optical fibre cables. Payphones were another part of the market in which BT's monopoly was broken, and liberalisation was expected to lead to mushrooming demand.

Between 1990 and late 1994, the number of BT employees fell from 232 000 to 148 000 and at the end of this period the target was 100 000. On the positive side, customers in 1994 receive incomparably better service at a lower price and the returns to shareholders have been impressive. On the debit side, effects on employees have been very negative, producing widespread fear of job insecurity and lack of commitment to working for BT (*Financial Times*, 7 Nov. 1994).

Mobile systems and personal communication networks comprise cellular and personal communication networks, and it is in these that BT faces more competition than from Mercury. Cellular phones have in fact proved to be much more successful and so the market in the leading Western countries grew at the start of the 1990s and the growth rate was forecast to accelerate, as can be judged from Figure 1.22.

Manufacturers of equipment have experienced unprecedented growth; for instance, in 1993 Motorola reported 20%, Ericsson 35% and Nokia 55% growth.

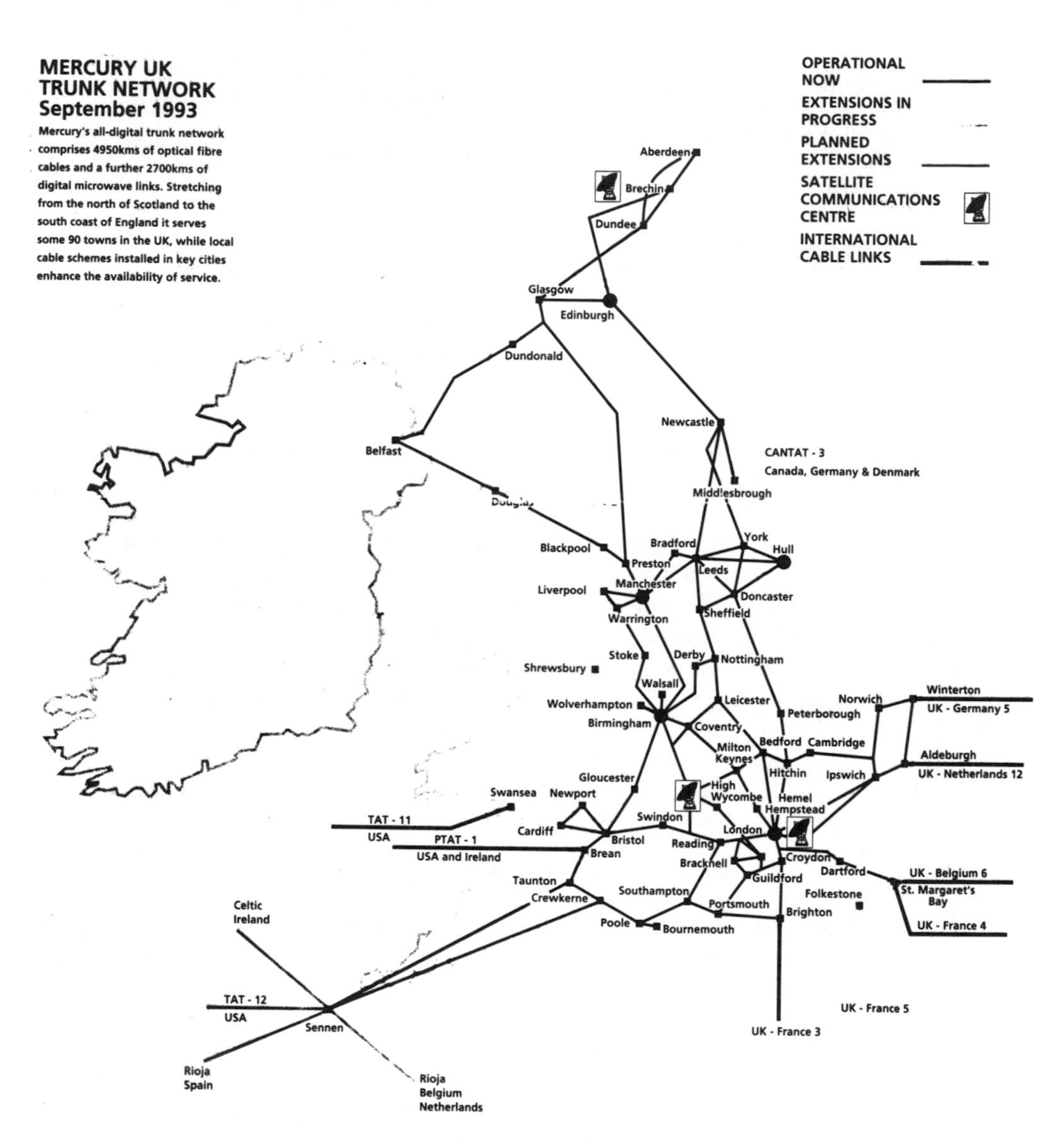

Figure 1.21 Mercury UK trunk network

Yet another aspect of the revolution in the UK since privatisation of BT was the dawn of the era of the personal communication network (PCN) in 1990. Three consortia were licensed in December 1989 to develop and operate such networks; one comprises British Aerospace, Millicom, Pacific Telesis (a Baby Bell), and Matra Communications. A second comprised STC, Telekom, US West and Thorn EMI forming Unitel. A PCN uses a portable telephone that provides the same quality and reliability as the fixed switched network, but is capable of hosting the advanced features of a system designed for the next century. [16]. PCNs are based

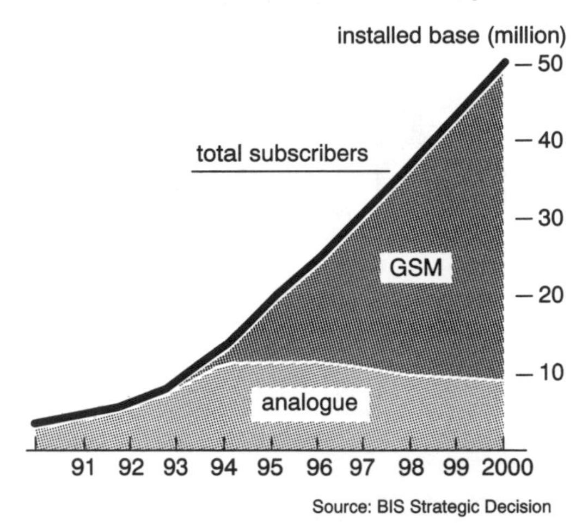

Figure 1.22 Actual and forecast growth in the European cellular market 1990 to 2000 (*The Times*, 1 Mar. 1995)

on pan-European digital cellular telephone technology developed by the Groupe Speciale Mobile (GSM), but operate at higher frequencies and use smaller cells. The creation and evolution of such networks occupied the early 1990s. A personal communicator is a pocket-sized telephone, costing about £200 (1990) and capable of making and receiving calls from almost anywhere in Britain at a cost expected to be little higher than the corresponding charge in the fixed network. They are likely to include a number of useful features such as personal numbers, call barring and diversion and multiple ringing tones to distinguish between types of calls (*Financial Times*, 12 Dec. 1989). By mid-1994 BT was reporting a 12% increase in quarterly revenue, mainly due to activity at Cellnet (cellular joint venture with Securicor) and BT Mobile Division.

External alliances

Convergence between telecommunication and computer or information technologies was one of the dreams of the 1980s, but none became reality, and most proved costly failures. Despite these disappointments, the end of 1990 saw a successful bid by AT&T for NCR, the fifth largest US computer maker (*Financial Times*, 4 Dec. 1990) and BT was then still negotiating with IBM on a partnership to supply a full range of advanced communications and information services to business customers worldwide. Ultimately the goal is to provide a global 'one-stop shop' able to handle everything from simple telephone calls to the management of entire computer operations. Such projects face complex regulatory, political and management hurdles, but if they come to fruition would fulfil long-term aims of the parties concerned (*Financial Times*, 18 Dec. 1990). By early 1995 the pace of forming alliances with the aim of winning business from large multinational companies had quickened, and the situation was described in an article entitled 'Competition down the line' (*Financial Times*, 19 Jan. 1995). The biggest groupings are listed at the top of Figure 1.23.

In mid-1993 BT agreed to take a 20% stake, and set up a joint venture with MCI, the second largest US carrier, in a bold bid for leadership of the

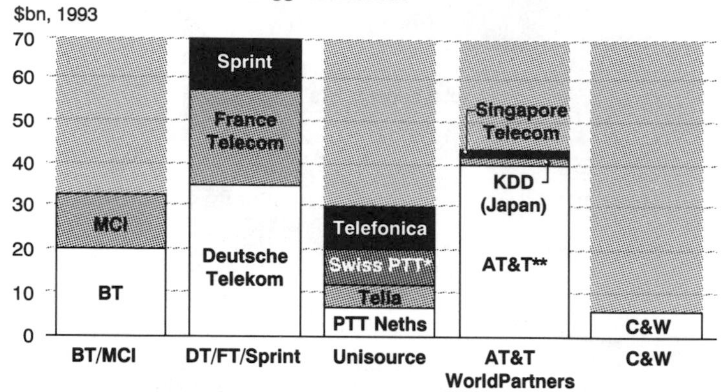

*Includes non-telecoms activity (eg post)
**Excludes equipment sales

telecommunications market by country (m ECU)

	1991	1992	1993	1994	1995	Annual market growth (%) 1991-3	1993-5
France	19,317	20,545	21,859	23,339	24,728	6.4	6.4
Germany	29,802	33,639	36,335	39,275	42,621	10.4	8.3
Italy	15,380	16,556	17,715	18,950	20,230	7.3	6.9
Netherlands	5,385	5,696	6,047	6,421	6,827	6.0	6.2
Spain	9,268	9,415	9,825	10,411	11,006	3.0	5.8
Sweden	5,374	5,648	5,867	6,095	6,388	4.5	4.3
Switzerland	5,495	5,767	6,052	6,348	6,638	4.9	4.7
UK	19,910	20,311	20,932	21,631	22,299	2.5	3.2

ranking by number of main lines

Rank	Company	Country	Lines (m)
1	NTT	Japan	57.3
2	Deutsche Telekom	Germany	35.4
3	France Telecom	France	30.1
4	BT	UK	26.1
5	Telecom Italia	Italy	23.7
6	BellSouth	US	18.7
7	Bell Atlantic	US	18.2
8	Ameritech	US	17.0
9	GTE	US	16.8
10	NYNEX	US	15.7
11	Korea Telecom	Korea	15.6
12	Pacific Telesis	US	14.6
13	Telefonica	Spain	13.8
14	US West	US	13.3
15	Southwestern Bell	US	12.8

AT&T, MCI, Sprint and KDD all have revenues in excess of $1.5bn but do not have lines and are therefore not included in the above table
Source:

Figure 1.23 Telecom alliances involving European partners: upper chart, combined revenues; middle table, market by country and annual growth; bottom table, ranking by number of main lines (*Financial Times*, 19 Jan. 95)

international telecommunications industry (*Financial Times*, 3 June 1993), just one week after AT&T launched its Worldsource venture, a partnership with five Asia-Pacific carriers. Both ventures were aimed at international customers, especially multinational companies. The BT–MCI alliance was cleared by the US Regulatory Authorities in mid-1994 and the joint venture 'Concert' was launched immediately to offer 'one-stop' services to multinational companies.

1.7.2 United States of America

The US industry remained the colossus in the business, and still led the world in the development of high-technology communications switching and transmission equipment and services at the start of the 1990s. By then it had, however, lost the low-technology techniques and products market to the Japanese (*Financial Times*, 17 Aug. 1990).

By early 1993 AT&T was also codeveloper of a technology for high-definition television, competing to become the US national standard, and had forged an alliance with a computer software company (*Financial Times*, 26 Jan. 1993). If the expected convergence of telecommunications, computers and video takes place in the last decade of the 20th century and the first of the 21st century, AT&T will be well placed to lead a single multi-media industry into many new markets.

Regulatory situation

The USA provides the most spectacular example of the interaction of a legal system and a telecommunication operator, following upon the failure of the Federal Communications Commission (FCC) to regulate AT&T. The Congress had enacted the Anti-Trust Laws, the Executive acting through the Attorney-General brought the law suit and in a sense dictated the outcome by agreeing to the decree which broke up AT&T in 1982. The man given the brief to try the case and to enforce the decree is Judge Harold Greene, a man who is deeply involved in the content of what he is enforcing. AT&T was divorced from its 26 local companies, which in turn were formed into seven regional Bell operating companies (RBOCs). These regional companies, or Baby Bells as they are affectionately called, were forbidden to go into three fields: long-distance traffic, information services and equipment manufacture, so long as there is any serious danger that they will be able to engage in anti-competitive activities.

Telecommunication changes began in the USA with legal and regulatory decisions in two areas: *inter alia* that end users should be permitted to own their terminal equipment, a reversal of Hubbard's decision at the beginning of the telephone era; and the so-called divestiture of AT&T summarised earlier. These decisions and their results have led to an unprecedented growth in competition along with a new emphasis on the establishment of international standards for interfaces and network functions to ensure the continued provision of services to the public.

Competitive situation

If new technologies such as PCN came along which would provide effective competition to the fixed networks of the Baby Bells the restrictions could be removed very

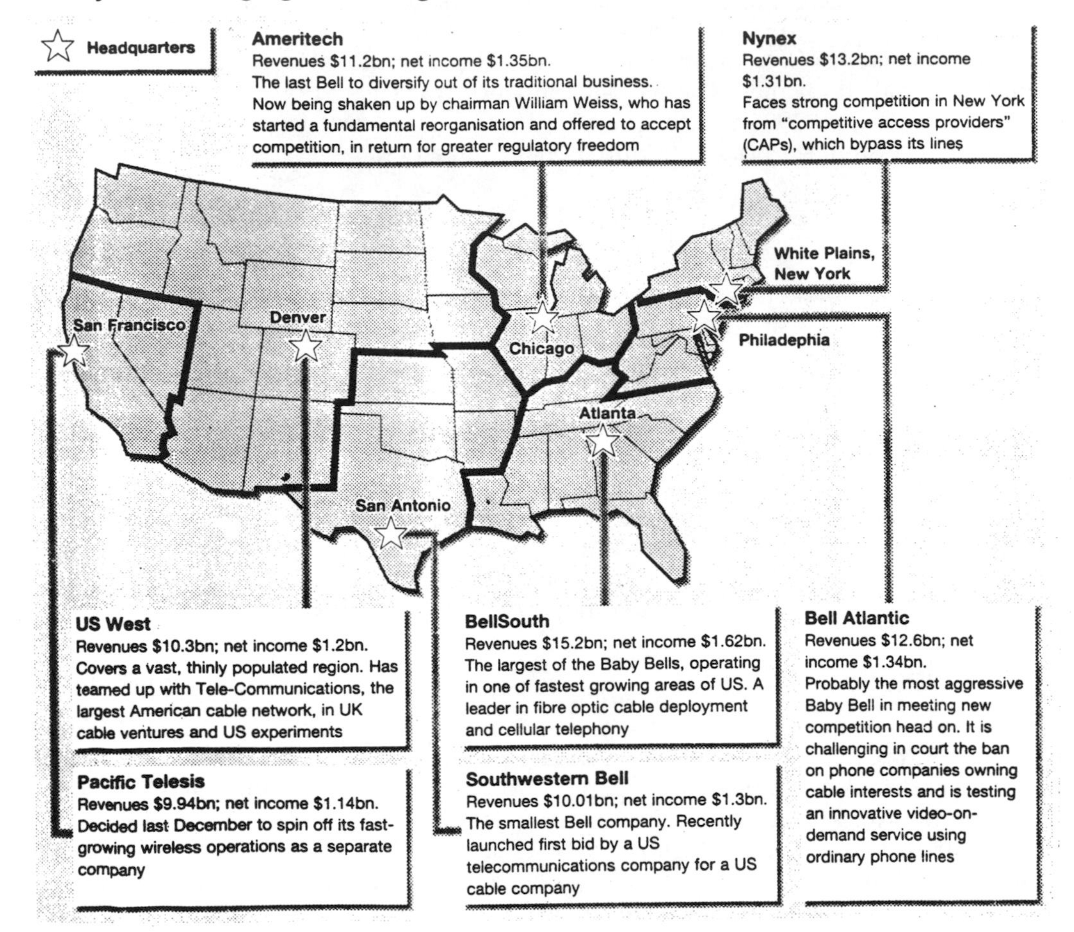

Figure 1.24 Geographical areas covered by Baby Bells and summary of their revenues, net income and policy as at early 1993 (*Financial Times*, 20 Apr. 93)

quickly (*Financial Times*, 25 Sept. 1989). By 1990 the Bells had united to fight in court to remove some of the restrictions and for more coherence in US telecommunication policy. In 1993 Bell Atlantic won a legal case allowing it to provide video services (*Financial Times*, 15 Oct. 1993). The most crucial issue was resolved in 1991 when the seven Baby Bells won a court case giving them the right to enter the cable television field.

The Baby Bells, which cover the states shown on the map in Figure 1.24, faced radical changes in the 1990s, some of which are indicated in this figure. Other changes involving links with cable companies are mentioned in Figure 1.25.

By the mid-1990s the Baby Bells are likely to face competition for business traffic in their own local telephone areas from long-distance carriers such as MCI Communications, and a new, more competitive regulatory framework. In 1994 four of the Baby Bells launched a legal battle to end the operation of the consent decree.

Even after the break-up AT&T is still the largest single entity, for which salient financial figures for 1991 are given in Figure 1.26.

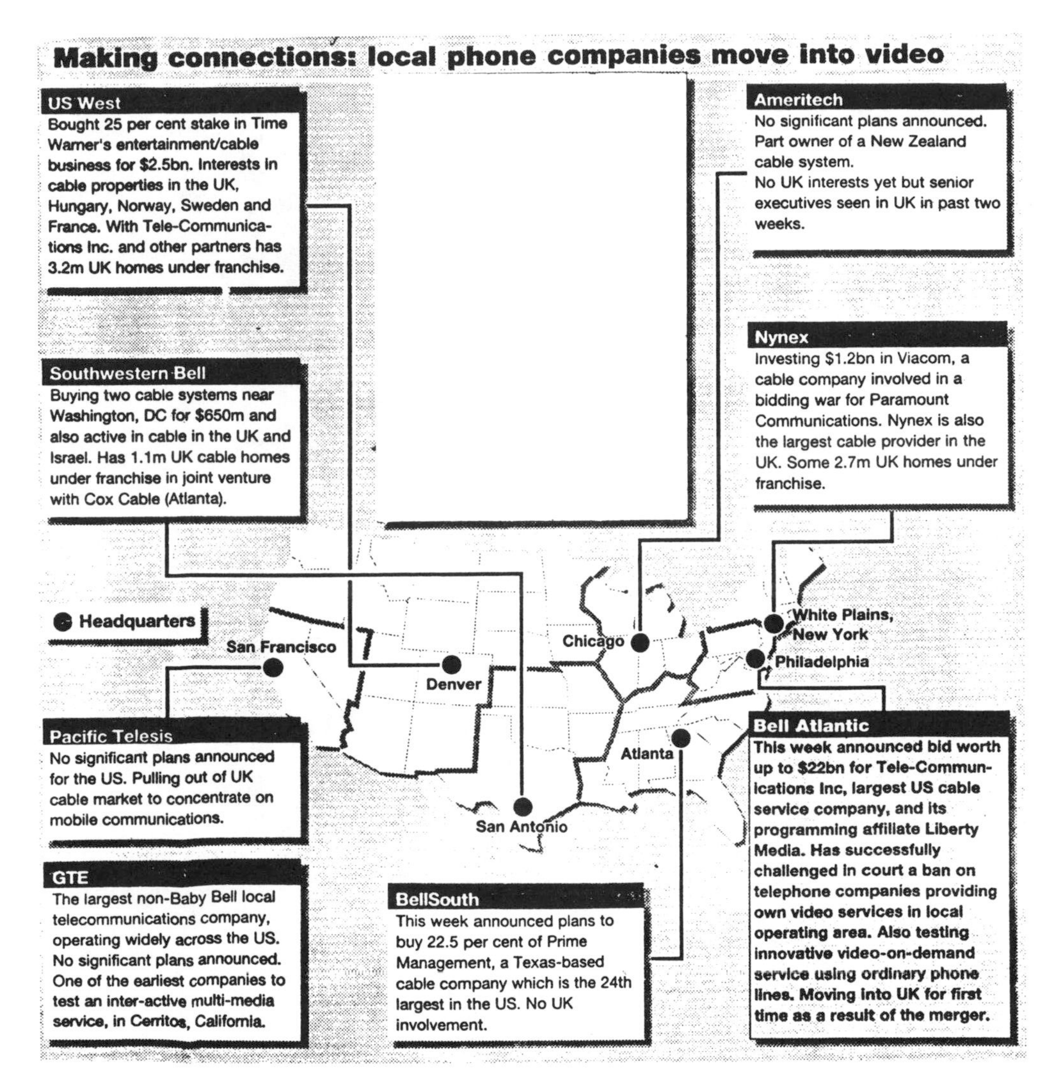

Figure 1.25 RBOCs moves into the US cable television market, 1993 (*Financial Times*, 15 Oct. 1993)

Responses to competition of AT&T and Baby Bells

AT&T responded to the challenge at home with a cost-cutting programme, reorganisation into 20 separate product areas, job losses (some 60 000 over four years to 1993), and investment in new production methods resulting in improving margins. At the same time the company gave every indication that it was prepared to act for the long run in its overseas enterprises (e.g. AT&T was granted a licence in 1994 to provide services throughout the UK, where it was rated better than BT on 3 out of 5 measures of performance by corporate users), just as it has done in America since the early days. For instance, in the first quarter of 1988 it reported that research and development expenditure had grown by 12.3% to $621 million. In 1987 it spent a larger percentage of revenue on R&D than any other public operator, as shown in Figure 1.27, and this is likely to be true for later years. With an ultra-modern

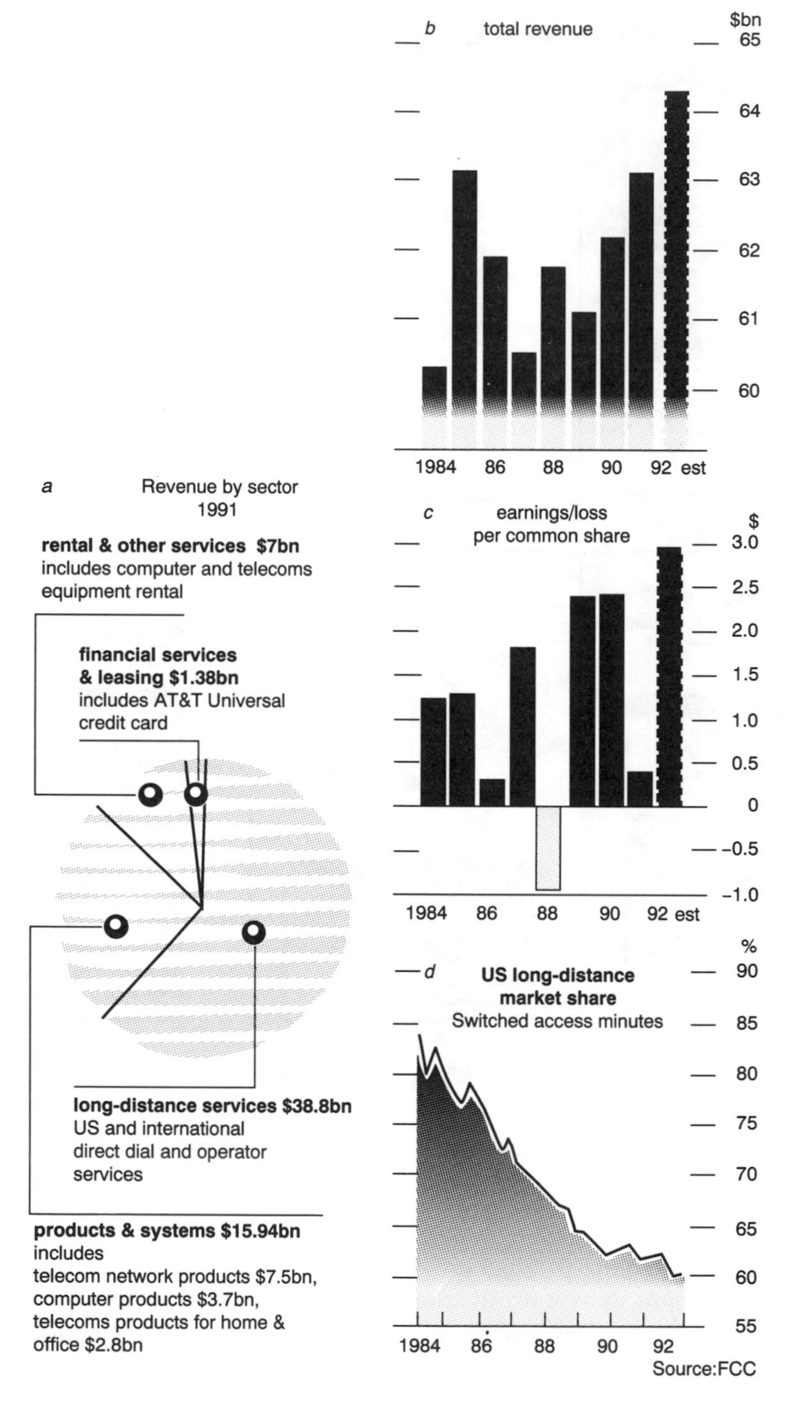

Figure 1.26 (*a*) Revenue by sector in 1991, (*b*) total annual revenues 1984–1992, (*c*) earnings/loss per common share 1984–1992 and (*d*) AT&T share of US long distance market 1984–1992 (*Financial Times*, 26 Jan. 93)

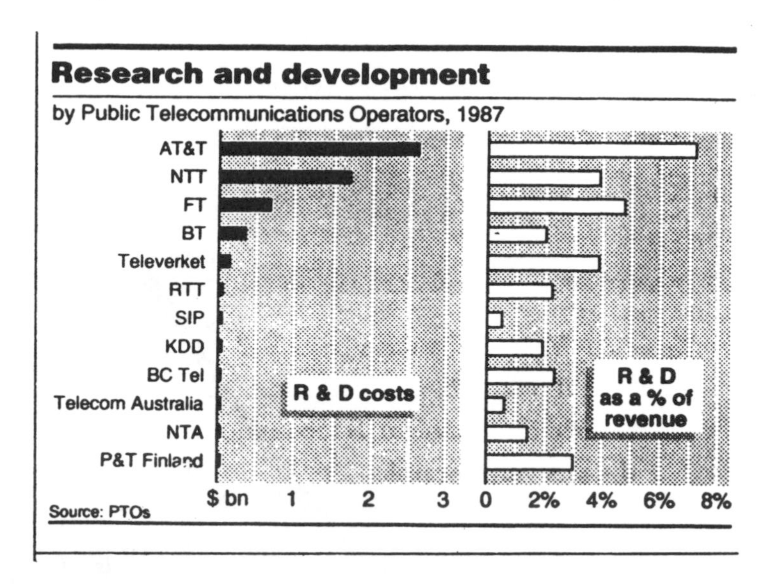

Figure 1.27 R&D costs and R&D as a percentage of revenue for some public telecoms operators, 1987 (*Financial Times*, Survey World Telecommunications, 7 Oct. 1991)

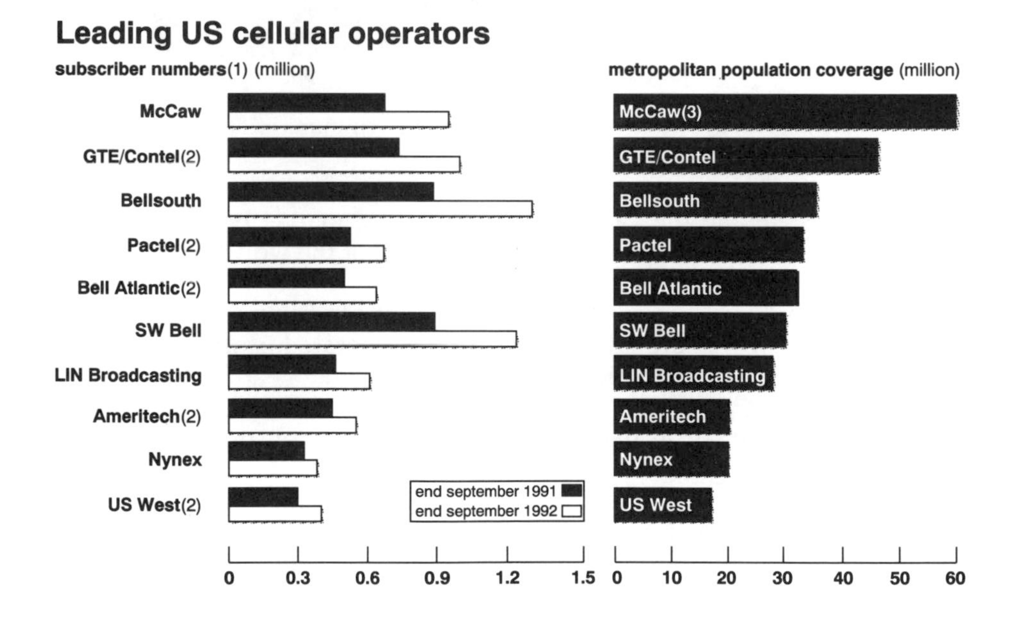

Figure 1.28 (*a*) Numbers of subscribers and (*b*) coverage of metropolitan areas of the leading US cellular operators, 1991/1992 (*Financial Times*, 17 Aug. 1993)

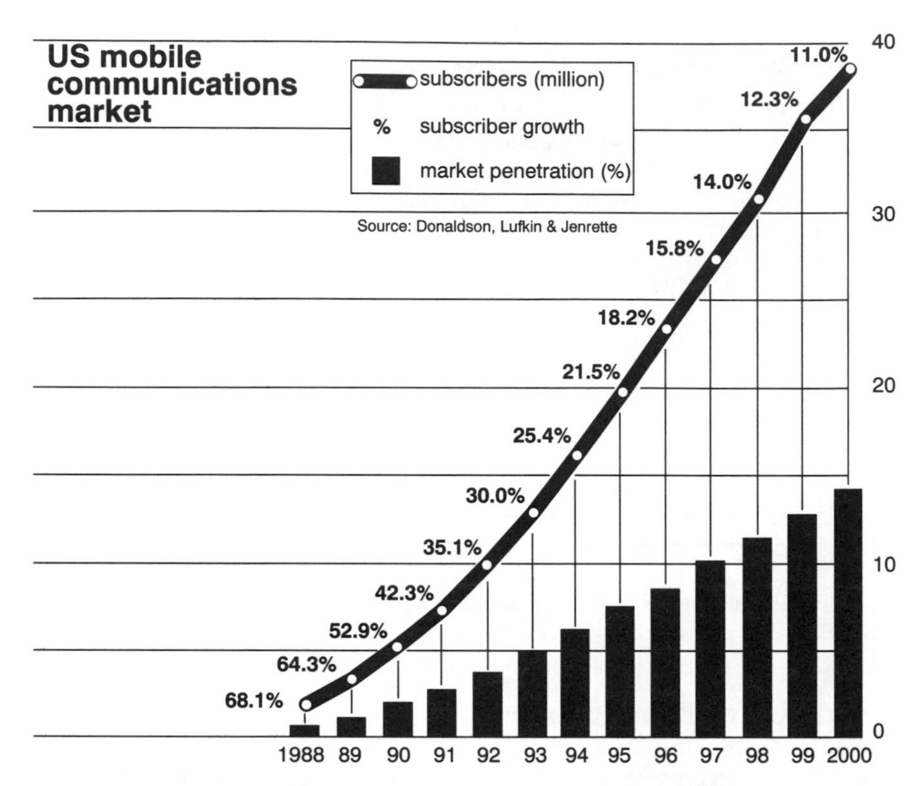

Figure 1.29 Actual and predicted growth in US mobile communications market, 1988-2000 (*Financial Times*, 16 Mar. 1990)

transmission system and the vast potential of the US market behind it, 'there can hardly be a telecommunications company in the rest of the world which would not like to have AT&T's problems.'

In August 1993 AT&T announced the takeover of McCaw Cellular Communications in a bid which was the fifth largest in US corporate history (*Financial Times*, 17 Aug. 1993). The relative positions of the leading operators at the time are shown in Figure 1.28.

A prediction of growth in the US mobile communication market published in 1990 is illustrated in Figure 1.29.

In late 1990 it was announced that AT&T and Zenith Electronics, the sole remaining US-owned television manufacturer, had teamed up to compete for the emerging high-definition television market in the USA. The two companies have jointly developed an all-digital system which will be ready for testing against competing systems, and a decision between competing systems was expected by the FCC in 1993.

External alliances

Denied opportunities in their home market, the Baby Bells are capitalising on their high productivity, modern networks and engineering expertise by scouring the world for business. Most of their overseas activity has been in mobile communications and cable television, but in mid-1990 Bell Atlantic and Ameritech announced the purchase of the Telecom Corporation of New Zealand, the most ambitious and costly purchase to date. US companies won greater access to the Japanese telecommunications market in late 1989, and here too their expertise may give Baby Bells business opportunities. The early 1990s saw a great expansion in their activities overseas.

In 1993 AT&T formed Worldsource with KDD of Japan, Singapore Telecom and several other partners, and in 1994 it selected Unisource (a joint venture between Dutch, Spanish, Swedish and Swiss PTTs) as its European arm to become a truly global company in the market for multinational company business (*Financial Times*, 23 June 1994).

1.7.3 Germany

The Federal German Republic's Deutsche Bundespost (DBP) was Europe's largest business on 1 July 1989, when it underwent the first real reform of the century (*Financial Times*, 30 June 1989). Prior to that date it had been one of the most entrenched telecommunication monopolies with a reputation for conservatism; since that date it has joined British Telecom in the forefront of change in Europe. The Bundespost was founded in 1490 at the behest of the Hapsburg Emperor Maximilian I, ruler of the then German Empire, two years before Columbus rediscovered America; and it celebrated 500 years of postal service in 1990. Figure 1.30 shows

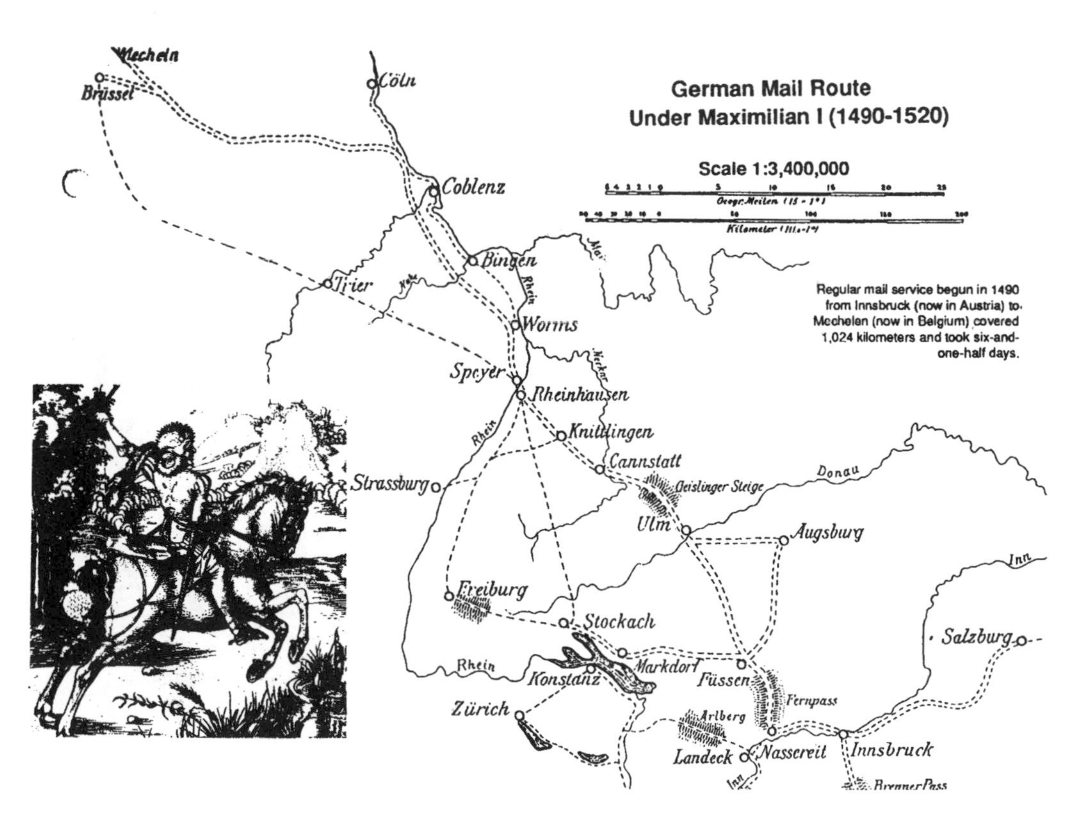

Figure 1.30 German mail routes established under Maximilian I 1490–1520 (Made in Germany, 3/89)

the 1024 km long mail route of the regular service between Innsbruck (Austria) and Mechelen (Belgium) begun in 1490 and later extensions under Maximilian I (1490–1520).

By the mid-1990s it is planned that the DBP will be separated into Telekom, Post-dienst and Postbank, all three of which will become corporations under a holding company controlled by the State.

Regulatory situation

Privatisation was out of the question, as it was banned by the country's Basic Law, an illustration of the impact of legal frameworks on changes in telecommunications. However, by 1993 the government and opposition parties had agreed to change the constitution so that Deutsche Telekom could be privatised, which will allow it to move towards its target of selling 49% of its capital to the private sector, starting in January 1996. The Bundesrat approved postal reform (which enabled the privatisation to go ahead) in July 1994, and Deutsche Telekom became an Aktiengeselleschaft (AG), a joint stock company, on 1 January 1995. Telekom itself has repeatedly warned that it would not be competitive in the liberalised telecommunications market in the EC unless it could have access to global capital markets and restructure itself (*Financial Times*, 2 June 1993). In mid-1994 Deutsche Telekom was valued in the range DM 60–100 billion. Some of the salient facts about Deutsche Telekom are listed in Table 1.9.

Deutsche Telekom has a DM60 billion investment programme in Eastern Germany, for which the actual and predicted growth in number of lines and investment is shown in Figure 1.31.

Table 1.9 Turnover and balance sheet over period 1989–93 and customer connections (*Financial Times*, 6 July 1994)

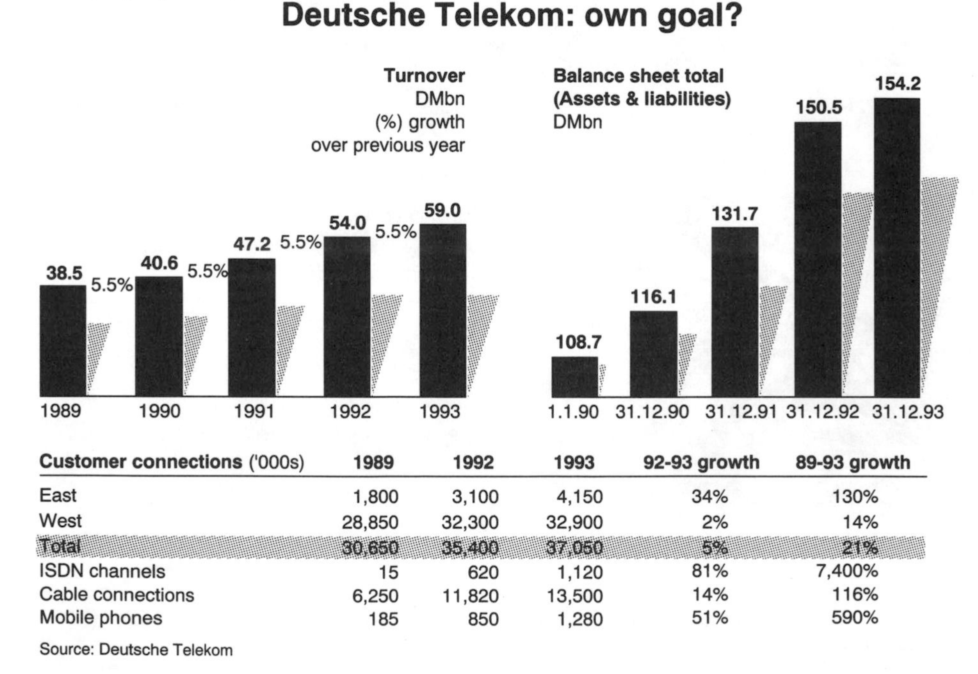

Customer connections ('000s)	1989	1992	1993	92-93 growth	89-93 growth
East	1,800	3,100	4,150	34%	130%
West	28,850	32,300	32,900	2%	14%
Total	30,650	35,400	37,050	5%	21%
ISDN channels	15	620	1,120	81%	7,400%
Cable connections	6,250	11,820	13,500	14%	116%
Mobile phones	185	850	1,280	51%	590%

Source: Deutsche Telekom

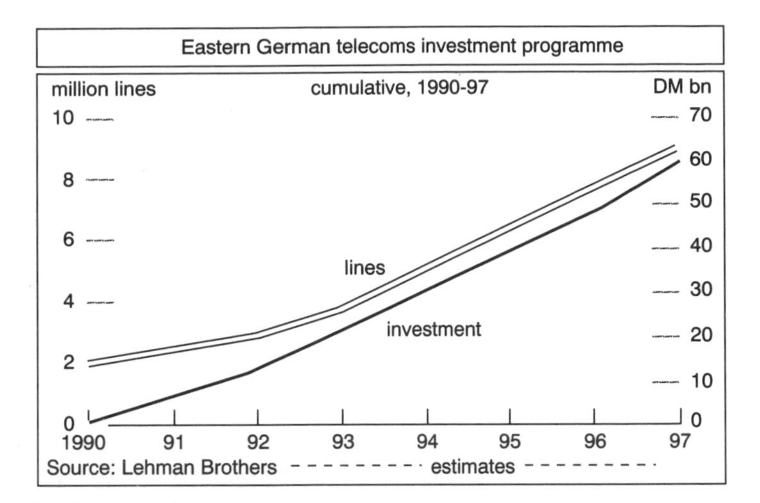

Figure 1.31 Deutsche Telekom investment programme in Eastern Germany 1990–1997 and corresponding growth in number of lines (*Financial Times*, 26 Oct. 93)

Guidelines for the regulation of the industry after the dismantling of the monopoly were also announced in early 1995, and a new telecommunications law was to be drafted which was expected to be passed by parliament in the summer of 1996.

Competitive situation

Competition will be opened to small and medium-sized companies as well as large telecommunication operators after 1998 (*Financial Times*, 28 Mar. 95). Operating licences for regional as well as national operators would be awarded only to applicants that provided evidence of financial stability, market experience and ability to manage a network with a high level of security and integrity.

Responses to the competitive marketplace

These will emerge in due course.

External alliances

The tremendous changes taking place in Eastern Europe at the start of the 1990s saw Telekom extended across a reunified Germany by 1992. In May 1988 it was announced by the French and German PTTs, France Telecom and the Bundespost that they would join forces to supply the growing market for value-added network services, or vans. A jointly controlled commercial company was to be formed to compete with IBM and other computer manufacturers.

By early 1995 a number of alliances had been formed between German companies bidding for Germany's alternative telecommunication networks and external operators, e.g. Viag Interkom (Viag and BT), Daimler Benz Aerospace and Northern Telecom, in Europe's largest telecommunication market (*Financial Times*, 12 Jan.

95). A target date is 1 January 1998, when the European Commission has decided that the telecommunication monopoly in member states will end. There is pressure for an earlier date from US telecommunication companies interested in this lucrative market.

In March 1990 Alcatel's West German subsidiary SEL-Alcatel and VEB Kombinat Nachrichtenelektronik, a Berlin-based company, formed a joint-venture company to modernise large parts of East Germany's telecommunication network, by far the most ambitious operation of its kind by a Western telecommunication company until then (*Financial Times*, 10 Mar. 1990). An initial 3000 km overlay network had been installed by 1991. By 1994 the number of telephones had risen from 10 per 100 to 24.2 before unification, compared with 49.9 in former West Germany, and the target for 1997 was 46.8 (7.2 million new connections) compared with 52.7 (*Financial Times*, 4 May 1994).

The modernisation of Russia's network by installing a digital overlay network connecting 50 cities to Western Europe, involving a consortium of Western telecommunications groups led by Deutsche Telekom, was agreed with Intertelekom, the Russian state-owned long-distance operator in July 1993.

1.7.4 France

The French corporation Alcatel became the world's largest communications company at the end of the 1980s.

By 1990 France Telecom was one of the most technologically advanced and commercially successful networks in Europe, despite its status as a Department of the Civil Service, and was pioneering large-scale implementation of synchronous digital transmission in its long-distance network (*Financial Times*, World telecommunications survey, 7 Oct. 1991). Its phone charges for business and domestic customers were among the lowest in the world. In the financial year which ended in 1991, the network topped 28 million lines, customers could expect only one fault in 7 years, and only 0.14% of customers complained about the accuracy of their bills.

In 1994 France Telecom had the world's most digitised public telephone system, the largest packet switching data transmission network, the world's most extensive and successful videotext service (Minitel, with more than 5.7 m terminals in use by mid-1991 accessing more than 15 000 information services), and the first fully operational nationwide ISDN known as NUMERIS which carries speech, data and video and was expected to serve half a million users by 1995.

Regulatory framework

In March 1990 the Government tabled sensitive plans to release France Télécom from direct ministerial controls (which have existed since 1837 when Louis Philippe was attempting to regulate the very early telegraph service) and turn it into an independent state-owned company (*Financial Times*, 22 Mar. 1990) as a result of a new law taking effect in 1991; but as elsewhere there are likely to be battles ahead with the unions and perhaps the European Commission before a reformed France Télécom is able to invest and plan for the next century. Until late 1990

France Télécom was a staunch defender of the public service ethic and was always in the vanguard of opposition to liberalisation. A government supported five-year development joint programme for high-definition television was confirmed in mid-1990, with the aim of having commercial sets available in 1995 (*Financial Times*, 24 July 1990). If successful, this will generate an enormous amount of traffic on any line systems used.

Competitive situation

The independent Société Francaise du Radiotelephone (SFR) offers competition in the cellular sector.

Responses to the competitive marketplace

France Telecom scheduled the initial opening of its GSM network for end 1991 and the commercial marketing for 1992, and forecast that digital GSM would be the main French cellular technology after 1994. In the fixed network the large-scale implementation of SDH offers the operating benefits outlined in [10] and elsewhere. In addition to building up its internal position against possible competition, France Telecom has taken major shareholdings in some overseas telephone companies.

External alliances

In December 1993 the French and German state operators (France Telecom and Deutsche Telekom) announced a strategic alliance to establish a joint venture to provide data and other advanced services to multinational companies. The joint venture had to be cleared by the European Commission. In 1994 both Deutsche Telekom and France Telecom agreed to invest in the US company Sprint which operates in long-distance, local and cellular telephony. France Telecom Network Services are part of the European service platform based on Transpac's integrated data transmission network and offer single supplier and single billing for international traffic in the UK.

1.7.5 European Union

As mentioned earlier when discussing Germany, the European Commission has set 1 January 1998 as the date for ending telecommunication telephony monopolies in member states (with five extra years of grace for Spain, Ireland, Greece and Portugal). By the year 2000 telecommunications could be Europe's biggest equity sector (*Financial Times*, 11 Oct. 1993). The number of exchange lines, productivity and revenues of the twelve members of the Union in 1992 is presented in the form of bar charts in Figure 1.32 (*Financial Times*, 11 Oct. 1993). Information on cellular subscribers is presented in Table 1.10 (*Financial Times*, 8 Sept. 1993).

Telecommunication alliances involving European partners, and the market by country for the years 1991–1995, can be found in Figure 1.33.

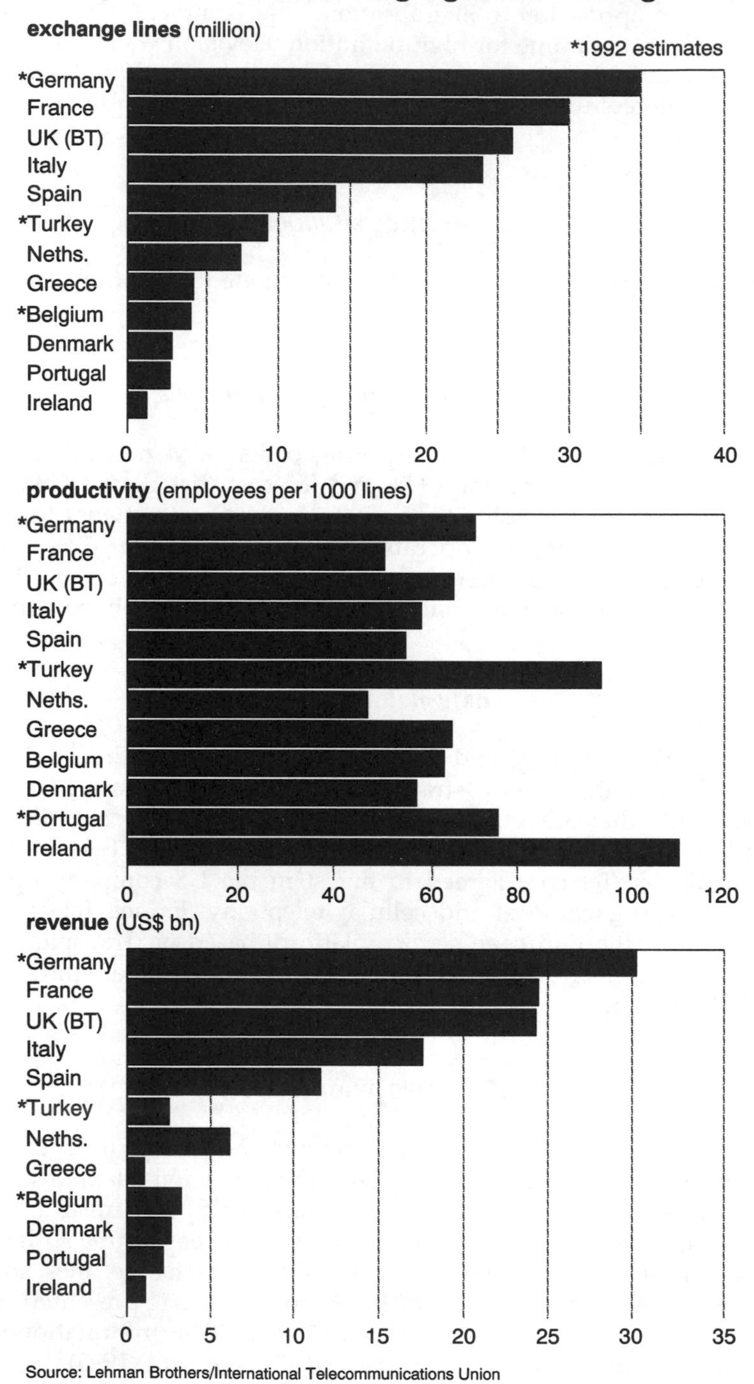

Figure 1.32 Upper chart, number of exchange lines in millions; middle chart, productivity in employees per 1000 lines; bottom chart, (annual) revenue in US$ billion for European telecom operators (*Financial Times*, 11 Oct. 1992)

Table 1.10 European cellular telephone subscribers (mid-1991)

Country	Operator	System	Launch	Subscribers July 1, 1993	Subscribers July 1, 1992	Yearly growth (%)	Penetration July 1, 1993‡
Andorra	PTT	NMT-450	Jul 1990	700	240*	191.67	14.00
Austria	PTT	NMT-450	Nov 1984	55 858	61 459	−9.11	25.55
Austria	PTT	TACS-900	Jul 1990	141 156	81 449	73.31	—
Belgium	Belgacom	NMT-450	Apr 1987	62 995	55 814*	12.87	6.40
Cyprus	Cyprus Telecoms	NMT-900	Dec 1988	11 920	7208	65.37	17.03
Denmark	Tele Denmark Mobil	NMT-450	Jan 1982	46 608	50 281	−7.30	47.03
Denmark	Tele Denmark Mobil	NMT-900	Dec 1986	179 150	143 078	25.21	—
Denmark	Tele Denmark Mobil	GSM	Jul 1992	8000*	n/a	n/a	—
Denmark	Sonofon†	GSM	Jul 1992	8000*	n/a	n/a	—
Faroe Islands	PTT	NMT-450	Jan 1989	1594*	1489	7.05	35.65
Faroe Islands	PTT	NMT-900	Jun 1992	188*	10	1780.00	—
Finland	Telecom Finland	NMT-450	Mar 1982	167 739	157 483	6.51	82.56
Finland	Telecom Finland	NMT-450	Dec 1986	231 718	164 719	40.67	—
Finland	Telecom Finland	GSM	Jul 1992	6500*	n/a	n/a	—
Finland	Radiolinja†	GSM	Jul 1992	6500*	n/a	n/a	—
France	France Telecom	RC2000	Nov 1985	328 000	316 000	3.80	8.42
France	France Telecom	GSM	Jul 1992	15 000	n/a	n/a	—
France	SFR†	NMT-450	Aug 1989	125 000	99 400	25.75	—
France	SFR†	GSM	Dec 1992	7000	n/a	n/a	—
Germany	Deutsche Telekom	C-450	Sep 1985	806 512	677 936	18.97	15.60
Germany	Deutsche Telekom	GSM	Jul 1992	190 000	n/a	n/a	—
Germany	Mannesmann†	GSM	Jun 1992	250 000*	n/a	n/a	—
Iceland	PTT	NMT-450	Jul 1986	16 454	13 989	17.62	65.82
Ireland	Eircell	TACS-900	Dec 1985	47 471	37 000	28.30	13.56
Italy	Sip	RTMS	Sep 1985	33 100	57 800	−42.73	15.76
Italy	Sip	TACS-900	Apr 1990	874 200	633 000	38.10	—
Italy	Sip	GSM	Oct 1992	1200	n/a	n/a	—
Luxembourg	PTT	NMT-450	Jun 1985	873	927	−583	2.30
Malta	Telecell†	ETACS	Jul 1990	5118*	3557*	43.89	14.62
Netherlands	PTT Telecom	NMT-450	Jan 1985	26 419	26 199	0.84	12.79
Netherlands	PTT Telecom	NMT-900	Jan 1989	164 726	116 041	41.95	—
Norway	Tele-Mobil	NMT-450	Nov 1981	158 872	149 860	6.01	75.64
Norway	Tele-Mobil	NMT-900	Dec 1986	160 476	102 392	56.73	—
Norway	Tele-Mobil	GSM	May 1993	1354	n/a	n/a	—
Portugal	TMN	C-450	Jan 1989	28 105	19 535	43.87	6.14
Portugal	TMN	GSM	Dec 1992	16 500	n/a	n/a	—
Portugal	Telecel†	GSM	Oct 1992	20 000	n/a	n/a	
Spain	Telefonica	NMT-450	Jun 1982	57 556*	67 038	−14.14	5.71
Spain	Telefonica	TACS-900	Apr 1990	164 844*	81 165	103.10	—
Sweden	Telia	NMT-450	Oct 1981	256 920	247 495	3.81	86.24
Sweden	Telia	NMT-900	Dec 1986	451 788	368 885	22.47	—
Sweden	Telia	GSM	Nov 1992	3000*	n/a	n/a	—
Sweden	Comvik†	NMT-450	Aug 1981	21 000	21 000#	0.00	
Sweden	Comvik†	GSM	Sep 1992	4000*	n/a	n/a	—
Sweden	Nordic Tel†	GSM	Sep 1992	1500	n/a	n/a	—
Switzerland	PTT Telecom	NMT-900	Sep 1987	240 852	198 404	21.39	36.19
Switzerland	PTT Telecom	GSM	Mar 1993	2000*	n/a	n/a	—
UK	Cellnet†	TACS-900	Jan 1985	710 300	569 800*	24.66	27.99
UK	Vodafone†	TACS-900	Jan 1985	895 141	731 800	22.32	—
UK	Vodafone†	GSM	Jul 1992	1450*	n/a	n/a	—
TOTAL				7 014 857	5 262 453	33.30	19.05

*Mobile Communications estimates.

†Systems operated by private companies. All other systems are operated by national telecommunications operators

‡Figure for each country per 1000 of population, as at July 1 1993.

#Sales have stopped due to lack of spectrum.

Source: *Financial Times Mobile Communications*

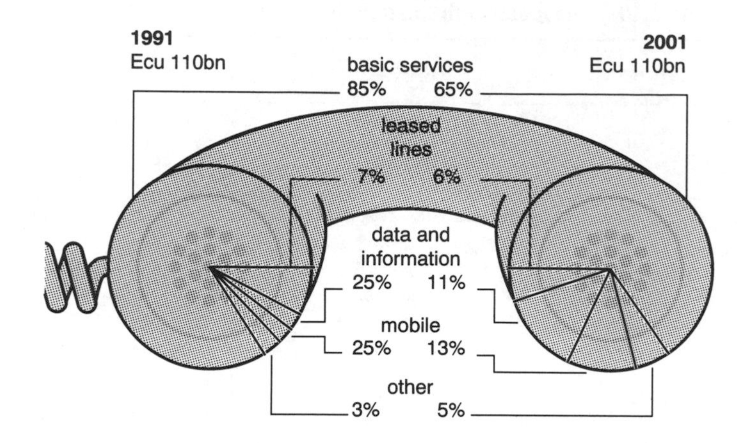

b

Quoted companies	State-owned corporations	Autonomous entities within state structure
UK	Sweden (from July 93)	Germany
Spain	Denmark	Belgium
Italy*	Greece	France
	Ireland	Luxembourg
	Holland	Norway
	Finland**	*Controlling stake held by govt
	Portugal	**Legislation pending to change status

c

	1993		1997*	
	% of revenues exposed to competition	Share held by competitors	% of revenues exposed to competition	Share held by competitors (all services)
UK (BT)	100	10%	100	22%
Germany	6–8	negl.	30	3%
France	6–8	negl.	30	4%
Belgium	6–8	negl.	30	1%
Sweden	80	5%	100	20%
Japan (KDD)	100	17%	100	25%
Japan (NTT)	50	7%	50	15%
US (AT&T long distance)	100	38%	100	45%

Source: Daiwa

*Best estimate

Figure 1.33 (*a*) Telecommunications growth potential within Europe. (*b*) EC/Scandinavian telecommunications operators. (*c*) Telecommunications competition: progress compared (*Financial Times*, 10 May, 1993)

1.8 INTERNATIONAL BODIES AND STANDARDISATION IN TELECOMMUNICATIONS

To provide successful and satisfactory interconnection of all types of communications equipment requires a common understanding among the international standards-setting bodies. Full definition of every aspect relevant to the establishment of suitable unique standards comes about by the participation of many international and national bodies. A complex situation is becoming even more intricate with the convergence of telecommunications and information processing. The broad aims of this section are to:

(1) explain why there is so much interest worldwide in international standardisation
(2) provide an introduction to those international organisations which are participating in these activities.

In Section 1.7 the reader has seen that the world telecommunication industry is in the throes of the most dramatic and unpredicted upheaval to affect any major sector of the economy this century. The ground rules and distinctions between computing and telecommunications are crumbling, and the future of the industry is being radically reshaped.

1.8.1 The European Union

In Europe technological advances, pressure from powerful non-EU operators, the advent of the Single Market, the influence of privatisation and other measures to separate telecommunication monopolies from governments had, by 1993, enabled the European Commission to put forward plans to open up the market. The target date proposed for full liberalisation, including international and domestic calls, is 1 January 1998. This involved discussions with 130 companies, regulators, governments and users (*Financial Times*, 10 May 1993). Between 1994 and 1996 the Commission wants to:

(a) accelerate adoption of existing technical proposals, which would liberalise satellite communications and lay the groundwork for licensing and granting fair access to new telecommunications operators;
(b) study how alternative networks, such as cable television, railways or electricity grids could be used for telecommunications services;
(c) study the thorny issue of allowing companies to establish their own networks to compete with the existing infrastructure.

A 1993 survey by Daiwa Institute of Research predicted that telecommunications would be 'Europe's fastest growing major industry in the 1990s'. Growth potential, EC/Scandinavian operators and competition in Europe are presented in Figure 1.33.

The Commission's campaign for greater harmonisation of telecommunications and information technology policies and markets had made steady progress in the late 1980s. The Commission had set out a list of fairly broad goals with the notable objective of achieving more technical standards. The initiative for greater product harmonisation has come from manufacturers of information equipment, faced with the realisation that international standards were necessary if the goals of interconnection and interworking of products in an open systems environment were to be achieved.

A 1985 list of the organisations involved is given in Table 1.11 [17]. The interactions between these organisations are illustrated in Figure 1.34.

Apart from the increased participation of scientific or industrial groups in the work of the CCITT, this body has realised the crucial need to take a lead in defining

Table 1.11 International organisations [17]. Reproduced by permission of IEE, © 1985

The organisations	
ISO	International Standards Organisation
ECMA	European Computer Manufacturers Association
ITU	International Telecommunication Union
CCITT	International Telegraph & Telephone Consultative Committee
CCIR	International Radio Consultative Commission
CEPT	European Conference of Postal & Telecommunication Administrations
IEC	International Electrotechnical Commission
EEC	European Economic Community
SPAG	Standards Promotion Application Group
CENELEC	European Committee of Electrotechnical Standards
CEN	European Committee for Standardisation
EUCATEL	European Conference of Associations of Telecommunication Industries
ANSI	American National Standards Institute
ECSA	Exchange Carriers Standards Association

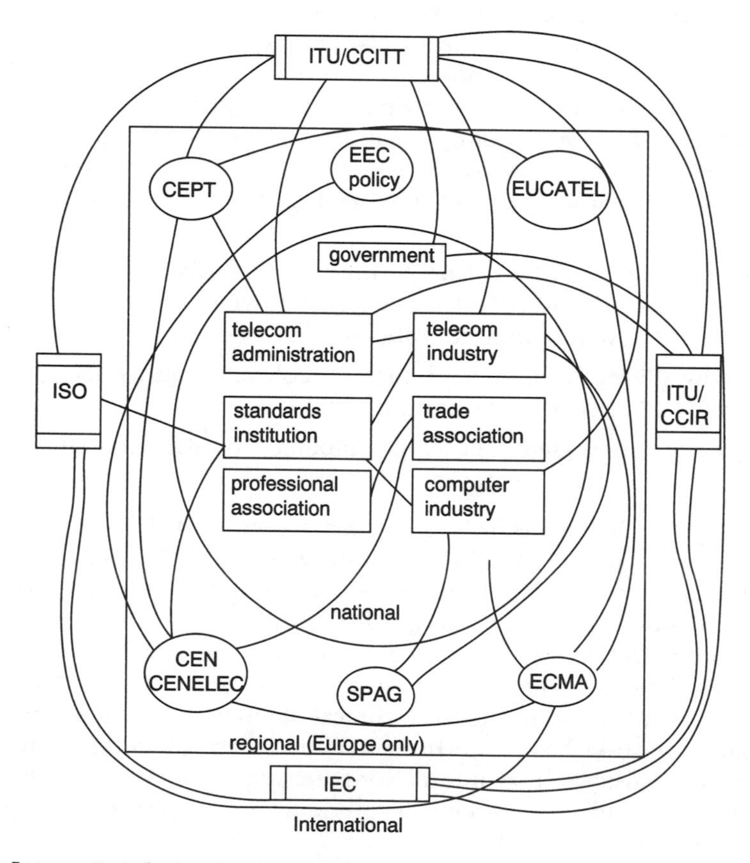

Figure 1.34 International standards organisations and their interactions ([17], Figure 3). Reproduced by permission of IEE, © 1985

and formulating standards, and to do so much more quickly than hitherto. The economic and commercial imperatives pointed clearly to the development of an increasingly international business and the need for standardisation to provide the following key benefits:

(a) the achievement of economies of scale
(b) facilitation of access to new markets
(c) more compatible products and services
(d) facilitation of the satisfaction of important user demands
(e) reduction of product development costs
(f) reduction of product and service lead-in times
(g) timely introduction of new standards

International standardisation is now regarded by most operating administrations and manufacturers as an essential ingredient in the planning, evolution and success of any practical communication system which aims to offer worldwide information processing and transfer capabilities. The past lack of a coordinated effort and commitment by governments, administrations, industry, etc., which led to a fragmented approach, may at last be overcome by a new sense of urgency and direction, with clear strategic objectives to be met.

1.8.2 The International Telecommunication Union (ITU)

The most important organisation in communications is the International Telecommunication Union (ITU), a specialised agency of the United Nations and successor to the International Telegraph Union established in 1865. The bases of the ITU, its conventions, have the status of a treaty between sovereign states in recognition of the importance of its role international relations. The ITU is the oldest of the existing inter-governmental bodies. Three Geneva-based institutions, the ITU, the International Electrotechnical Commission (IEC) and the International Organisation for Standardisation (ISO), between them publish over 96% of all international standards. Broadly speaking, the ITU handles telecommunications and broadcasting, the IEC the electrotechnical sector including electronics, and the ISO everything else (*Financial Times*, 14 Oct. 1994).

As a result of the deficiencies of the pre-1990 working of the CCITT and CCIR, and others brought about by the forces mentioned earlier in this chapter, a restructuring of the ITU was agreed by the plenipotentiaries at the ITU Additional Plenipotentiary Conference held in Geneva in December 1992 [18]. Among the changes were the disappearance of the CCITT and CCIR in their present form. The new Constitution and Convention came into force between members who ratified, adopted or acceded to them on 1 July 1994. In view of the need for swift implementation of the new structure and working methods, the Conference adopted a Resolution which provisionally applied all the provisions of the Constitution and Convention relating to the new structure and working methods from 1 March 1993.

As recently as 1990 the Telecommunications Development Board and the Centre for Telecommunications Development were set up to aid the developing countries. The role and responsibility of the former covered the improvement of telecommunications equipment and systems by means of the dissemination of information and advice, the transfer of know-how, the establishment of training institutes and the enhancement of technical self-reliance among the least developed countries.

Since the last major restructuring in 1947 there had been unprecedented growth in telecommunications, so that by 1990 it was a key technology underlying the whole global information economy and society, with important implications for developments in the political, economic, social and cultural fields. The major developments which brought about this situation were listed as:

(a) increased product and service innovations
(b) increased competition
(c) convergence of information systems and telecommunication networks with development such as broadband networks and universal personal communication networks
(d) globalisation of operations, and with this networks and services
(e) the emergence of related international, regional and national bodies, producing standards which have a major impact on global markets and trade issues
(f) a widening gap between the haves and the have-nots on our planet.

In 1989 a committee was established to draw up a series of wide-ranging recommendations to improve the functioning of the ITU. In the course of their work the committee identified the following needs.

(a) Respond more rapidly and more appropriately to the changing nature of the telecommunications environment and to the priorities of the international telecommunications sector and user community.

(b) Deal more cost-effectively with the constant growth and complexity of the tasks facing the ITU, and reassess the way in which it is financed.

(c) Provide good strategic planning and management at each and every level of the organisation.

(d) Pay greater attention to the needs of the developing countries to close the telecommunications gap.

(e) Give a larger role in the decision-making and strategic planning processes of the ITU to telecommunications manufacturers, service providers and users.

(f) Harmonise activities with those of the regional standards bodies in a more integrated and systematic way.

(g) Cooperate closely with regional telecommunications organisations.

The committee's recommendations will result in a new structure which is shown in Figure 1.35*b* along with the old structure, Figure 1.35*a*.

The standards-setting activities of the CCITT and CCIR have been consolidated into a Telecommunications Standardisation Sector. The remaining CCIR activities have been integrated into a new Radiocommunication Sector, which will include the work of the International Frequency Registration Board (to be replaced by a part-time Radio Regulation Board).

The new Standardisation Sector will study technical, operating and tariff questions, and issue recommendations on them for the purpose of standardising telecommunications on a worldwide basis. Recommendations will also cover interconnectivity of radio systems in public telecommunications networks and

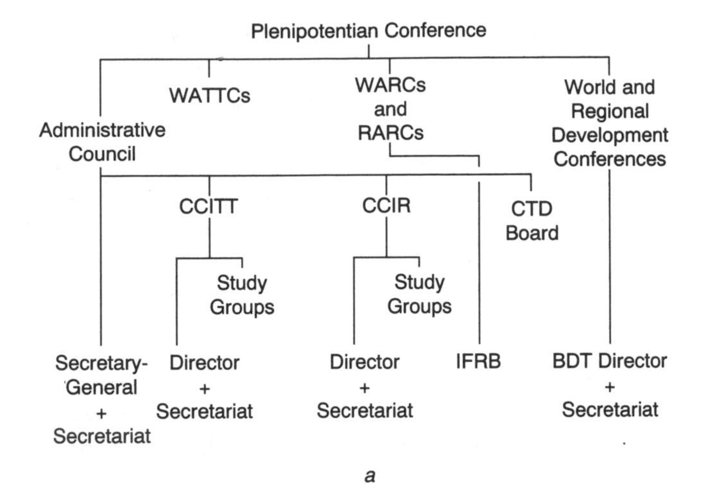

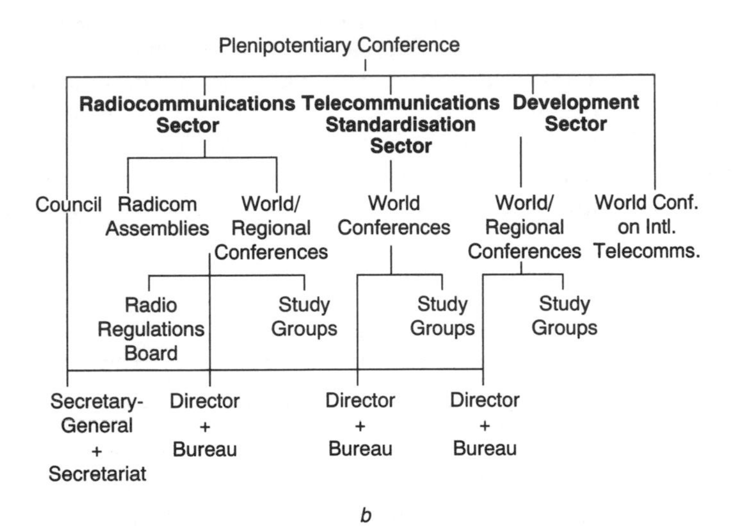

Figure 1.35 (*a*) Old, and (*b*) new ITU structures (*IEE Electronics & Commun. Engineering J.*, 5, no. 1, Feb. 1993)

on the performance required for these interconnections. This sector will operate through World Telecommunications Standardisation Conferences to be held every four years, and through Study Groups and a Bureau headed by an elected Director. At these conferences the work will be similar to that of the present Consultative Committees, namely to consider the Reports of the Study Groups, and approve, modify or reject the draft Recommendations. In addition, conferences will contain and approve programmes of work of the Sector.

The Radiocommunications Sector will operate through World and Regional Radiocommunication Conferences and Assemblies, a Radiocommunication Bureau headed by an elected Director and a Radio Regulation Board. The main functions of conferences will be to review, and revise as necessary, the Radio Regulations. Unlike the Radiocommunications Assembly and the conferences of other sectors, which are open to all members of the corresponding sector, participation in Radiocommunications Conferences will be restricted.

1.8.3 European organisations: CEPT and ETSI

The European Conference of Postal and Telecommunication Administrations (CEPT) was founded in 1959 to facilitate co-operation between 16 Western European PTTs. Its aims are to promote co-operation in resolving administrative, technical and operational questions, to study new techniques and services, to obtain agreement on tariffs, operational procedures and policies within Europe. CEPT is initiating [19] two projects aimed at revolutionising the provision of international data networks for major European companies. The first involves managed data networks, the second the upgrading of international links between their public data networks. It also seeks to coordinate views on matters to be discussed in the wider framework of the ITU or other bodies. The organisation of CEPT is shown in Table 1.12.

In 1987 CEPT welcomed the proposals made by the Commission of the European Communities to establish an institute with a flexible and modern organisation facilitating the elaboration of standards within the field of telecommunications.

This newer body is the European Telecommunications Standards Institute (ETSI), headquartered in the new town of Sophia Antipolis in the Cote d'Azur region of France. Named after the wife of the Paris-based engineer who conceived the original idea, and the Greek word for nearby Antibes, the town is a showcase for French telecommunications (*Financial Times*, 4 Apr. 1990). The organisation, role and membership of ETSI were summarised in [19]. At a plenary meeting in Germany in January 1991 the cellular radio committee of the Groupe Speciale Mobile (GSM)

Table 1.12 CEPT telecommunications commission ([19]). Reproduced by permission of BT

committees	
Satellite Communication	(CCTS)
Harmonisation (Technical)	(CCH)
Harmonisation (Commercial)	(CAC)
Transatlantic Telecommunications	(CLTA)
working groups	
Radio Communication	(R)
Television & Sound Transmission	(TTVS)
Services & Facilities	(SF)
Long Term Studies	(ELT)
Telephony	(Tph)
Telegraph & Telematics	(Tg)
Tariffication	(PGT)
Data Communications	(CD)
Switching & Signalling	(CS)
Transmission	(TR)
Specification & Approval of Terminals	(SSA)
special groups	
IDSN	(GSI)
Mobile Radio	(GSM)
Wideband	(GSLB)

completed and approved the specification of DCS 1800 as the pan-European standard for developing PCNs. The acceptance of this specification will provide both the plan for a network infrastructure and a blueprint for the design and production of products in the early 1990s. ETSI Technical Committees deal with a wide range of issues from digital audio and video to the business case of corporate networks, and reports on the status of selected technologies were published by the International Herald Tribune on World Standards Day, 14 October 1993.

The European Committee for Standardisation (CEN) publishes standards, more than 40% of which are international, and more than 90% of those issued by CENELEC (European Committee for Electrotechnical Standardisation) are international.

The EC Second Stage Telecommunications Terminal Equipment Directive came into force in November 1992. The background to the telecommunication regulatory environment in Europe as at 1994 was explained in the Telecommunications Terminal Equipment (TTE) Handbook published by BSI Standards. The main sections include historical background and current directive, Standards Bodies, Committees and Telecom Standards, Analysis of Telecom Standards, other Technical Requirements, Harmonized Routes to Approval Currently Available, and European Telecom Standards and a great deal of relevant information. A unique updating service forms part of the TTE Handbook.

1.8.4 The International Electrotechnical Commission (IEC)

The International Electrotechnical Commission (IEC) was founded in 1906 as a result of efforts by Lord Kelvin and C. Steinmetz among others, and it operates through voluntary arrangements and participation. The IEC is concerned with the formulation of world standards for electrical and electronic engineering. These standards cover the whole range of these disciplines, and aim to secure unambiguous communication of engineering information, reliability and safety of equipment, interchangeability and mutual compatibility of equipment made by different manufacturers worldwide, and the elimination of unnecessary diversity of components. This body, too, is in the throes of change and facing serious questions about its role into the next century.

1.8.5 International Organisation for Standards (ISO)

Founded in 1947, this body has been concerned with developing standards and operates by consensus among its members in 107 different countries. It is housed on the same site as the IEC and works closely with it where appropriate, but it deals with more everyday matters such as sizes of paper and familiar symbols used on the dashboards of cars to indicate lights, horn, etc. In the ten years from 1983 to 1993 the number of standards published by the ISO increased from 5273 to 9178 (*Financial Times*, 14 Oct. 1994)

The first European technical standard was a Roman regulation which laid down that lead pipes should not be cast shorter than 10 feet long; that was some 2000 years ago when Rome ruled much of what is now Europe ('Made in Germany', 2/89, p. 35), and Roman engineers were good at solving problems. Germany was the first country to establish a standards institute, the Deutsches Institute für Normung, located in Berlin, in 1917. By 1989 there were some 20 000 technical standards which applied in Germany, and from the Deutsche Industrie-Norm the phrase 'Das ist Norm' ('That is standard') emerged in the late 1940s, now abbreviated to DIN.

For governments, operating administrations, manufacturers and others to spend ever increasing sums of money in pursuit of standards in telecommunications and information-processing is clear evidence of real need. Endeavours to provide the marketplace with new and innovative products and services, better to meet human needs, requires up-to-the-minute understanding of the status of international standardisation. Moreover, the radical reshaping of these two areas is bringing about an expansion in standardisation activities in recognition of the fact that worldwide multivendor operation is the new reality in the marketplace.

1.9 REFERENCES

1 LARSON, E.: '*Telecommunications — a history*'. Frederick Muller Ltd., London, 1977.

2 JOHANNESSEN, N.: 'The telegraph — how it all started'. *British Telecom J.*, **8**, no. 1, 1987, pp. 34-37.

3 PRESCOTT, G.B.: '*Bell's electric speaking telephone: its invention, construction, application, modification and history*'. D. Appleton and Company, New York, 1884.

4 BRUCE, R.V.: 'Alexander Graham Bell'. *National Geographic Magazine*, Centennial edition, Sept. 1988.

5 DOHERTY, W.H.: '*The Bell System and the People who Built it*'. Special Publication SP82, Bell Telephone Laboratories, 1968.

6 COCHRANE, P.: '*Future trends in telecommunication transmission — a personal view*'.

7 COCHRANE, P.: 'The future symbiosis of optical fibre and microwave radio systems'. *19th European Microwave Conference*, Wembley, London, Sept. 1989.

8 COCHRANE, P.: 'Future directions in long haul fibre optic systems'. *British Telecom Technol. J.*, **8**, no. 2, Apr. 1990, pp. 5-17.

9 DEBOIS, P.G., LIEKENS, A., and QUAGHEBEUR, G.: 'Broad band solid state microwave link system RR 6-7 for 6 to 7 Gigahertz'. *Elec. Comm.*, **48**, no. 1/2, 1973, pp. 155-162.

10 PIERCE, J.R.: '*Waves and Messages*', 1967, Heinemann Educational Books Ltd., London.

11 HAWKER, I.: 'Evolution of digital optical transmission networks'. *British Telecom Technol. J.*, **9**, no. 4, Oct. 1991, pp. 43-56.

12 HAWKER, I.: 'Future trends in digital telecommunication transmission networks'. *IEE Electronics & Commun. Engineering J.*, **2**, no. 4, Aug. 1990.

13 SEXTON, M.: 'Synchronous networking — SONET and the SDH'. IEE Colloquium '*The changing face of telecommunications transmission*'. 16 Jan. 1989, Digest 1989/5.

14 HORROCKS, R.J.: 'The interaction of technology and regulation in telecommunications'. IEE Colloquium '*The changing face of telecommunications*', 16 Jan. 1989, Digest 1989/5.

15 ADJALI, I., FERNANDEZ-VILLACANAS, J.L., and GELL, M.A.: 'Markets inside telecommunications'. *British Telecom Technol. J.* **12**, no. 2, Apr. 1994.

16 RAMSDALE, P.: 'Personal communications. PCN implementation in the UK'. *IEE Rev.*, **37**, no. 2, 22 Feb. 1991, pp. 75-79.

17 KEARSEY, B.N. JONES, W.T.: 'International standardisation in telecommunications and information processing'. *Electronics and Power*, Sept. 1985, pp. 643-651.

18 'New ITU Structure'. *Electronic and Communication Engineering J.*, **5**, no. 1, Feb. 1993, pp. 22-24.

19 *British Telecommunications Engineering*, **8**, Apr. 1989, pp. 41-41.

2

IMPACT OF OPTICAL FIBRE TECHNOLOGY ON NETWORKS

I saw Eternity the other night, Like a great Ring of pure and endless light, All calm, as it was bright.

Henry Vaughan (1622–95)

2.1 INTRODUCTION

Aspects of the evolution of telecommunication systems and networks and of their limitations and innovations were considered in the preceding chapter. In the 1990s digital radio relay systems and optical systems are complementary, each offering its own peculiar advantages. Both classes of fixed systems are developing in an environment of accelerating rates of change, as are the operators in many of the major countries. The situation in some of the leading nations has been reviewed as evidence of a very fluid background in which some technologies appear very attractive, such as coherent systems and semiconductor laser amplifiers, only to lose appeal through the advent of other innovations such as erbium-doped fibre amplifiers. A brief overview was provided of world and regional markets and of international bodies whose work is directly relevant, especially their standardisation activities. In this chapter we focus on the impact of optical fibre technology and of the Synchronous Digital Hierarchy on networks, ending with consideration of some types of non-voice traffic which may be carried over public networks in the future.

2.2 TECHNOLOGY BACKGROUND

The transmission and switching limitations of twisted copper pairs have dictated the form of today's public switched telephone networks (PSTNs) in which telephone exchanges (offices) are placed in the centres of population, with links to customers' premises. In the UK the lengths of these links lie in the range 3–10 km; the distributions of local line lengths for other countries differ to a greater or

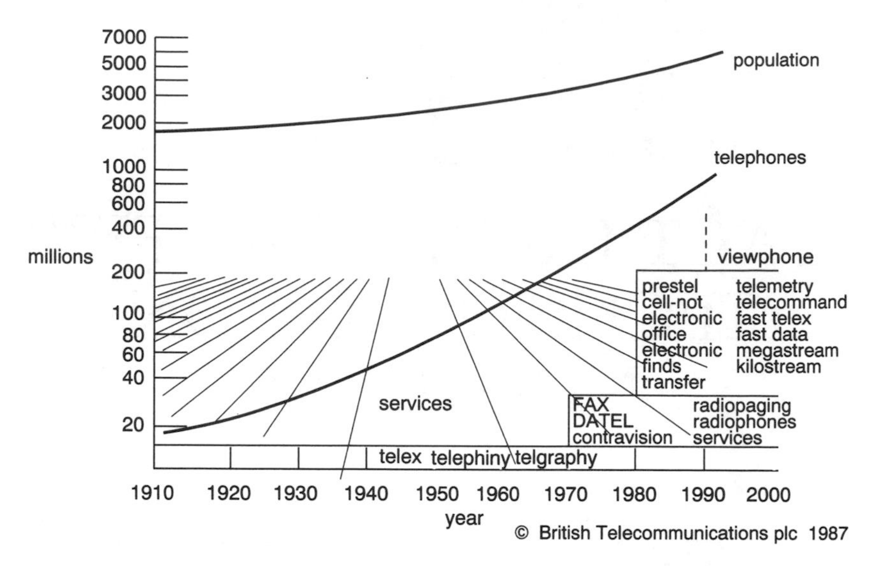

Figure 2.1 Global growth of population, number of telephones and services, 1910–1990 ([1], Figure 4). Reproduced by permission of BT

lesser extent. Similarly, high-capacity coaxial cables in the trunk network required repeaters every few km so that power feed stations had to be located about 30 km apart. These restrictions resulted in some 7600 switching machines, virtually one for every town, village and hamlet to serve some 24 million customers. The traffic was almost entirely telephony, which was transmitted in analogue format.

Telecommunications traffic has been growing exponentially for the past 80 years and this growth shows no signs of slowing, as can be seen from Figure 2.1.

In 1990 about 85% of some 650 million telephones were held by the 15% of the world's population which lived in the so-called developed countries [1]. Just a modest amount of industrialisation in a few of the other countries would increase global use sharply. At the end of the 1980s domestic telephony was growing at a global average figure of 10% annually, international calls at 14%, but the latter were expected to rise to 20% in the next decade. In the developed world there was a rapid rise in the provision of data and wideband services, amounting to some 30% of the total in 1990, and this type of traffic was forecast to exceed telephony sometime between the years 1995 and 2005. The most dramatic and most recent growth was in mobile communications, especially in the Scandinavian countries where in some instances growth rates of up to 50% per annum were achieved, and in the UK to a lesser extent. Bandwidth constraints in the PSTN were becoming a serious limitation under such conditions. A further area of rapid growth was machine-to-machine communication in the financial and commercial sectors, so that, for example, banks were among the first organisations to connect their distributed computer sites via direct 2 and 8 Mbit/s links.

To maintain growth at these rates using radio, satellite and metallic cables would be technically difficult if not impossible, and the costs would tend to be prohibitive. By 1990 optical fibre technology offered the only practical scope for an almost unlimited expansion in traffic and services, because the raw material and bandwidth available were, for all practical purposes, unlimited and at very low cost. In the UK it is predicted that about 90% of all traffic will be transported over fibre cables by 1995. Not only must the evolving network cope with the total transmission capacity required, but another emerging fundamental requirement is the need for a very resilient and adaptable network able to assimilate new services, signal formats and modes of operation with a minimum of disruption.

2.2.1 Quality of service

Until about 1970 all radio and television equipment was based upon thermionic valves and hence were relatively unreliable, as was the case with domestic appliances and the relatively rudimentary office automation equipment then available. By contrast, the telephone network appeared to be very reliable. By 1990 the converse was true [2]. Microelectronics has been widely used in domestic appliances and resulted in lifetimes of about seven or more years in both white and brown goods. One measure of reliability is FITS (10^9 hours), and this quantity is shown for a number of components and systems of interest in Figure 2.2.

In many countries the corresponding mean time between failures (mtbf) of the telephone network was less than 1 to 3 years, values largely dictated by the local line copper link from customer to switching machine. Customers have come to expect high levels of reliability in the home and workplace, and thus to expect a similar or better level in telecommunications. The provision of optical fibre systems, computer-controlled digital switching machines and modern terminal equipment has produced some improvement in quality but the fundamental limit is set by the line plant.

Network topology and reliability

In most telephone networks of any vintage, cables, ducts and overhead lines have been organised in a 'tree and branch' topology. The introduction of synchronous

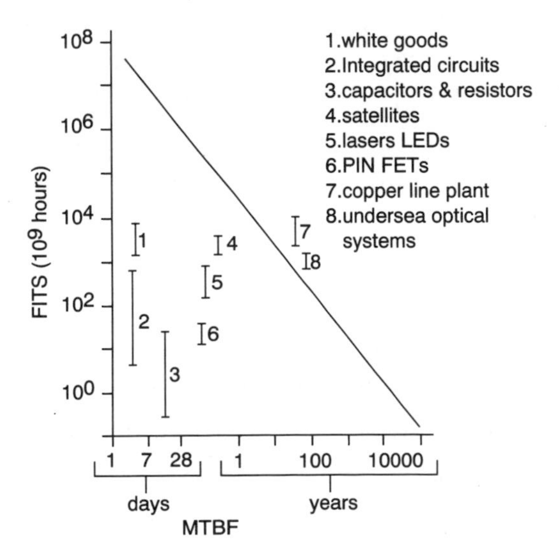

Figure 2.2 Illustrative ranges of component and system reliability ([2], Figure 2). Reproduced by permission of IEE

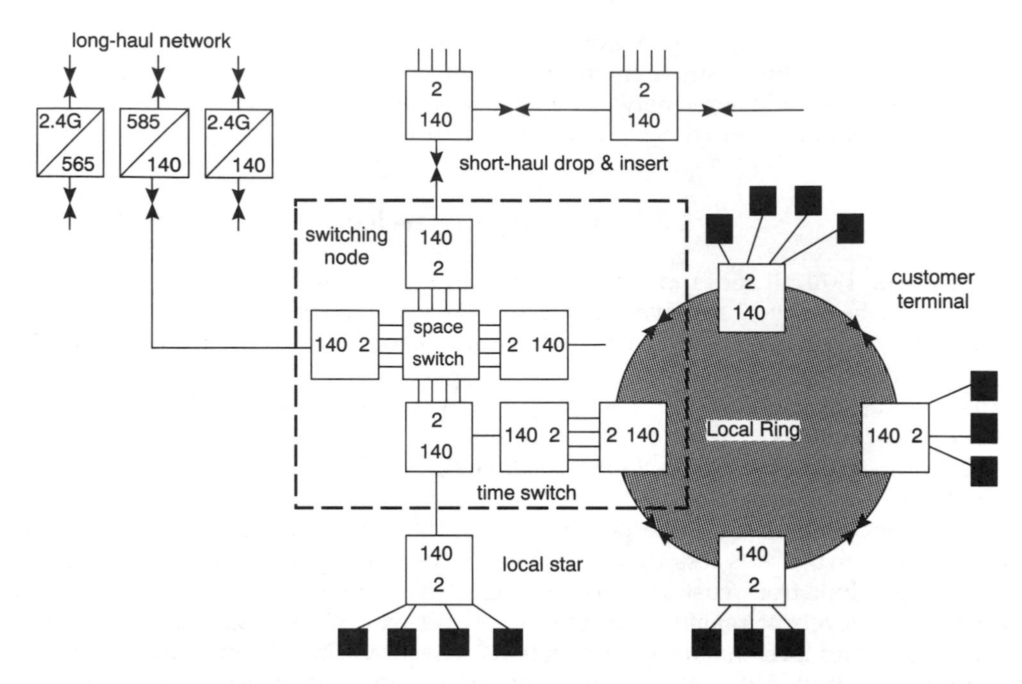

Figure 2.3 Transwitching network ([2], Figure 6). Reproduced by permission of IEE

multiplexing will make it possible to realise star, daisy-chain and ring/star networks of low cost and improved reliability. At sites where such systems converge time-division switching can be installed to interconnect them, and so avoid large quantities of multiplex equipment and interface cabling. Using a suitable system controller over a signal channel also makes it possible to use any time slot to connect any two points, as shown in Figure 2.3.

Connections can be reconfigured to meet traffic demands and, with the ring configuration, to bypass faults and hence significantly enhance network reliability and availability. These advantages can be demonstrated by a simple numerical example for which we assume the reliability figures listed in Table 2.1.

Associated with these values is a mean time to repair (mttr) of five hours for all faults. For the two simple configurations shown in Figures 2.4 and 2.5 and values listed in Table 2.1 it can be shown that the MTBF for 2 Mbit/s access of the synchronous network is more than three times better.

The most fault-prone item in present local networks is the cable that links customer and switch. Far fewer faults occur in the terminal electronics. Examining

Table 2.1 Illustrative reliability figures. Reproduced by permission of IEE

Item of equipment	mtbf (years)
Repeater	20
Synchronous terminal	10
Asynchronous terminal	2
Cable per 100 km	100

Note: mtbf = mean time between failures

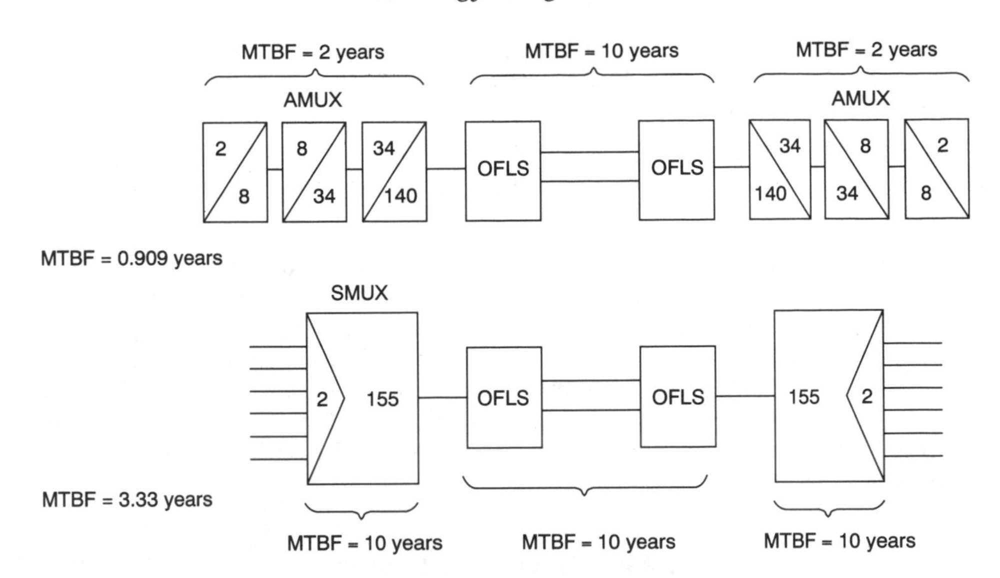

Figure 2.4 Illustrative mtbf values for PDH and SDH point-to-point links ([2], Figure 7). Reproduced by permission of IEE

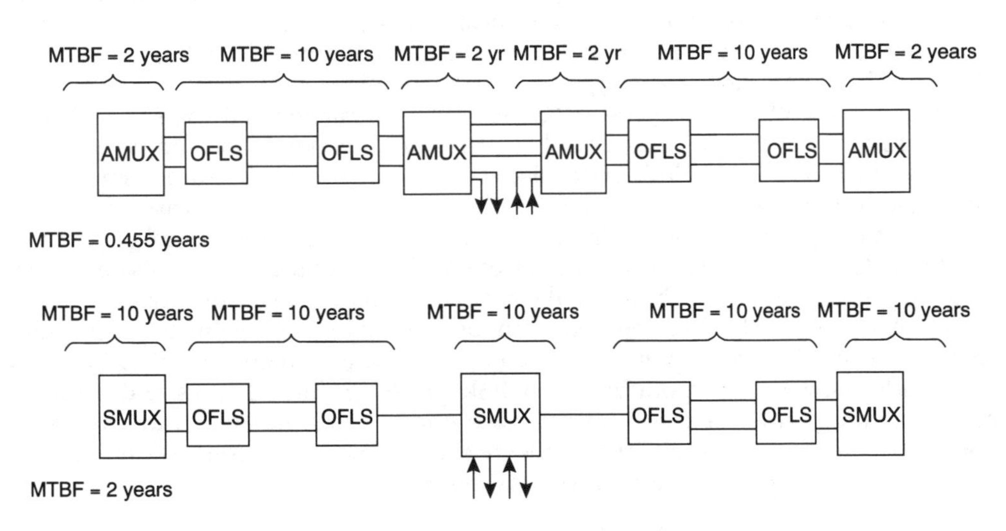

Figure 2.5 Illustrative mtbf values for PDH and SDH drop-and-insert operations ([2], Figure 8). Reproduced by permission of IEE

fault statistics for existing copper networks in various countries revealed that mtbf figures in the range 3–5 years are not uncommon. About half of the failures were due to corrosion or were staff-induced, with some 50% in the last cable section between distribution point and customer.

Factors affecting reliability in the change from copper to glass are:

(1) glass fibre does not corrode and needs far fewer joints
(2) optical cables do not suffer insulation breakdown due to moisture
(3) repeaters are not required

(4) an electronic/optical multiplexer or passive power splitter will reduce the need for physical intervention, thus allowing networks to be reconfigured remotely, which should reduce intervention by staff
(5) equipment voltages, currents and power levels will be reduced
(6) optical and copper cables are equally susceptible to the mechanical digger or spade

Some of these factors clearly tend to improve reliability, but some of the improvement will be offset by the introduction of remote equipment and longer spans, this last because of the 'linear with length digger/spade factor' [2] compared three different elementary options in an analysis which is too lengthy to be included here. His results established that mtbf and reliability are strongly linked, and that the mttr is a critical factor in defining network performance. The key conclusion from the study was that designing for a good mtbf and rapid repair are essential network requirements.

2.2.2 Service protection

The grade of service protection provided through circuit or network switching and duplication of equipment must be appropriate to whichever of the four applications areas (international, trunk, junction and local/access) are of interest. The extent of this protection is driven purely by an economic Quality of Service (QOS) demanded of the application in an increasingly competitive environment.

A comparison of technologies: protection may be categorised in terms of path availability, response time and control complexity. In general, fast simple methods such as automatic $N+1$ link protection have been preferred since they reduce unavailability by about tenfold at cost. Ring protection is a special case of $1+1$ in which traffic is routed in both directions around the ring. More complex schemes involve networking. Network protection offers the advantage of by-passing intermediate failed nodes, and should make best use of redundant capacity.

Differences in protection methods can be quantified by making a computer analysis of an elementary network to compare link and network protection. An eight-node network incorporating $1+1$ link protection on all links and network protection on standby links via intermediate nodes was simulated, and a range of nominal values for node mtbf and mttr were chosen to produce comparative results. The main features of these results were as follows.

(a) Link and network protection significantly improve path availability.

(b) Network protection is more effective than link protection in all cases, although the difference between them can be made marginal for given mtbf and mttr.

(c) Network protection is significantly better for links with mtbfs below 1 year.

(d) Path availabilities increase strongly with source/sink node mtbf for both link and network protection.

(e) If node mtbfs are extended to exceed 30 years (by hardware application) then available performance with link protection approaches that of network protection.

So far it has been assumed that traffic-carrying and protection paths were diverse, whereas in practice ducts or cables may be shared. The effects on path availability were dramatic and demonstrated that, to obtain $\geq$ 99.99% availability, protection routes generally required both cable and duct diversity in addition to full duplication of hardware.

Protection in the UK network at the end of the 1980s

Different topology and traffic-carrying requirements of the trunk (long haul), junction (short haul) and local (access) portions of the network have led to the following protection strategies.

Long haul: this has the best physical protection with cables in earthenware ducts buried between 1 and 2 m underground. All terminals are in surface buildings. Repeaters have been eradicated. The network combines 34/140/565 Mbit/s optical, coaxial and radio point-to-point systems in a meshed topology suited to automatic link and network protection.

Short haul: this distributes traffic between local offices and remote concentrator units (rcu) at 2/8/34/140 and 565 Mbit/s rates, and interfaces with both the trunk and local network. Again, plant is relatively well protected, with cable housed in earthenware ducts organised in a star topology.

Local (Access) network: this distributes 64 kbit/s, 2 Mbit/s and 8 Mbit/s to a customer from an office or an rcu. The majority of this plant consists of overhead or directly buried cables, and it is therefore unprotected. The local network is also organised in a star topology and thus provides little opportunity for diverse routing.

The calculated distribution of unavailability for a 2 Mbit/s circuit of length 1000 km between customer terminals for different types of protection is illustrated in Figure 2.6, based on typical performance values.

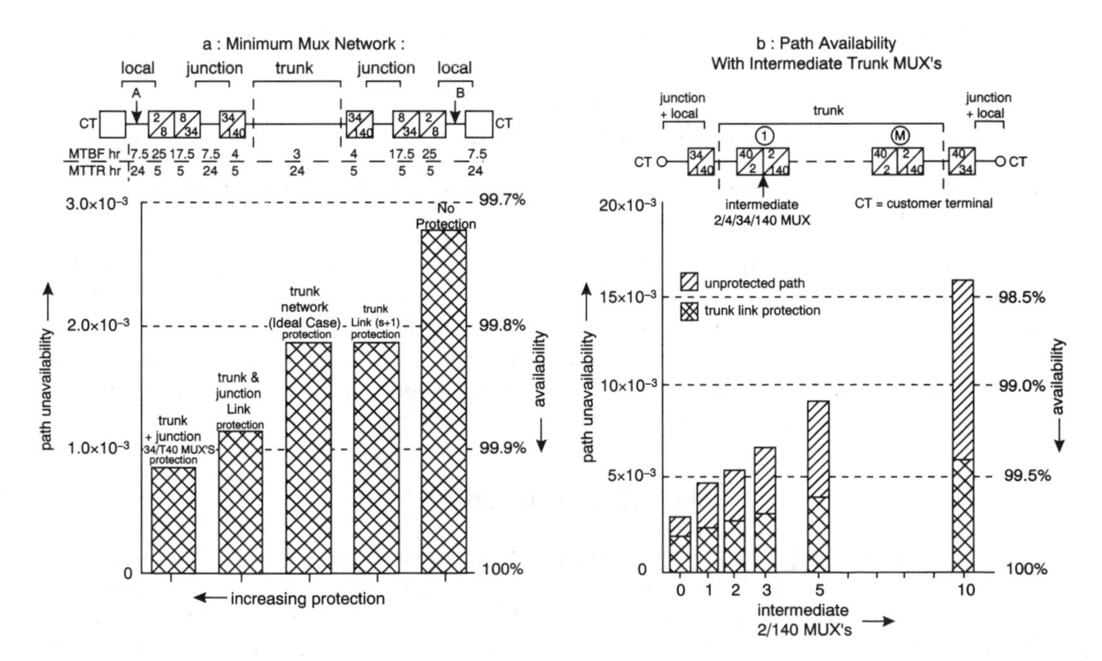

Figure 2.6 Variation of end-to-end availability for intermediate multiplexing and protection ([2], Figure 18). Reproduced by permission of IEE

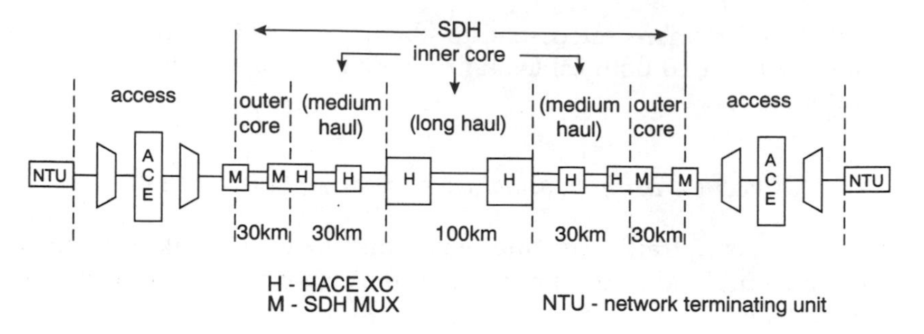

Figure 2.7 Leased line SDH reference circuit ([2], Figure 21). Reproduced by permission of IEE

This shows that the path unavailability depends strongly on the number of intermediate mux stages.

These will be described in some detail later, here we merely compare their availability with the present plesiochronous network. Such systems allow the integration of bit transport, management and protection by means of comprehensive and accessible overhead bytes, combined with the flexibility of automatic digital multiplexing and cross-connects. High path availability will be achieved through extensive redundancy within each layer of the network (leading to simplified software) and hardware duplication. Path availability will be set by remotely allocating redundancy and diversity to meet service objectives. Fast protection response times can be pre-programmed. Such systems possess resilience against major catastrophies.

The dependence of path unavailability on diversity and restoration methods for SDH networks has been studied, and the following key conclusions emerged.

(a) The best availability is obtained with total network protection over diverse paths.

(b) Separate ducts for the main and standby paths greatly improves path availability and makes it independent of duct mtbf.

(c) Cable diversity ensures that path unavailability is independent of cable mtbf and mttr.

A customer-to-customer reference circuit is shown in Figure 2.7, and the detail will be covered later. For the present we concentrate on the path unavailability, which may be as low as 0.1 hour/year under conditions of maximum diversity, and with end-to-end availabilities limited by the Access network. Reliability of digital paths on SDH transmission networks is expected to improve by about tenfold compared with existing PSTNs.

2.2.3 Growth

As in a number of aspects of telecommunications, growth has been exponential. The relatively slow start can be seen if we look at the increasing number of telephones and facilities which became available in the UK during the century from 1876 to 1976, as shown in Figure 2.8 [3]. The article emphasised the early years when the telephone companies and the Post Office were competing to supply the telephone service.

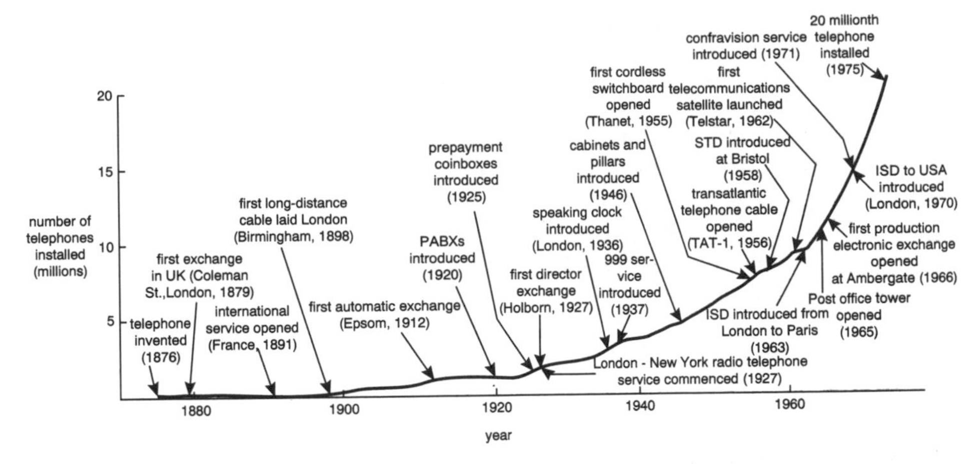

Figure 2.8 Growth of services in the UK during the century 1876–1976 ([3], Figure 7). Reproduced by permission of BT

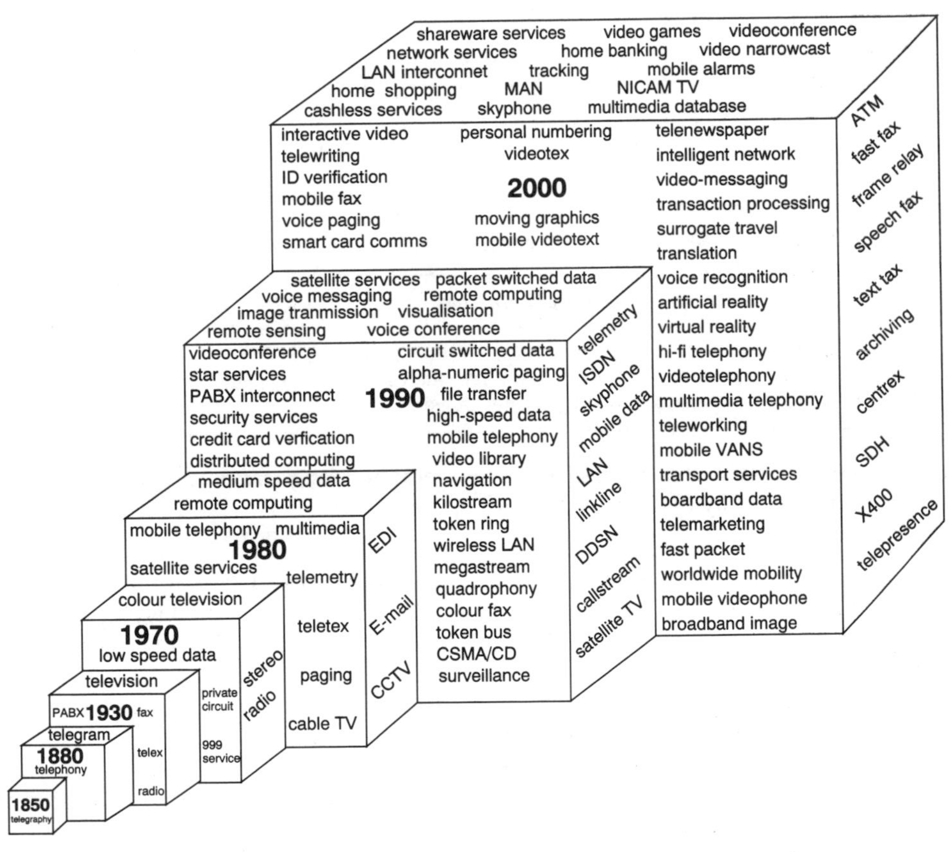

Figure 2.9 Actual and forecast possible growth in services, 1850–2000 ([4], Figure 14). Reproduced by permission of IEE

Table 2.2 Evolution to broadband services (AT&T 1991 forecast). Reproduced by permission of AT&T

Near term (1990–92)	
	Narrow band switching, Integrated services digital network, SS7, Facsimile servers, Premises imaging
Mid-term (1992–94)	
	Broadband switching, SONET, LAN/MAN interconnectivity, Advanced intelligent terminals, Image nodes/servers
Long term (1994 onwards)	
	Broadband ISDN, ATM technology, Photonic switching, Multi-media high speed imaging

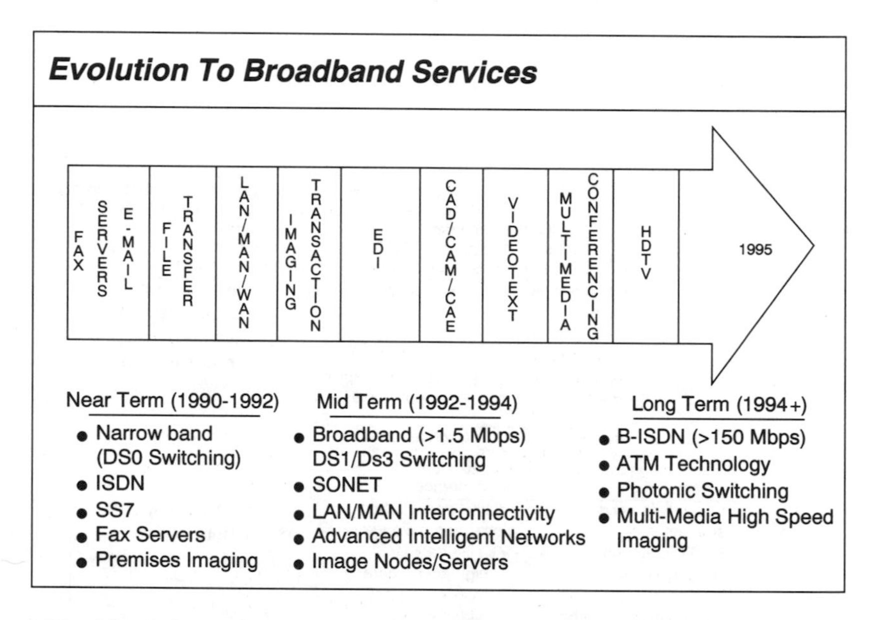

Figure 2.10 Chart above shows near-term, mid-term and long-term evolution to broadband services. Reproduced by permission of AT&T

By 1993 the actual and forecast growth in services could be summarised in the form shown in Figure 2.9 and used by the Managing Director, Development and Procurement, BT, in his inaugural address as President of the IEE for the year 1993–94 [4].

Only some of the potential multimedia services are listed in Figure 2.9, and the article had a section devoted to multimedia applications.

The evolution to broadband services as envisaged by AT&T was outlined by J.S. Mayo, Senior Vice President, Bell Laboratories in a 1991 special issue of *Scientific American* [4a] and is summarised in Figure 2.10.

2.3 SDH-BASED NETWORKS

The role of the Synchronous Optical Network (SONET) and its CCITT twin the Synchronous Digital Hierarchy (SDH) is to provide an international interface standard and frame format for digital transmission over optical fibres. These should lead to easier networking between telecommunication operators, reduced equipment cost and integrated management and protection systems [5]. Both SONET and SDH are developments which started in the mid-1980s, and detailed recommendations covering all aspects of the standards are currently the subject of study in various ANSI and CCITT working parties. By 1990 many Administrations were studying the deployment of SDH-based systems in their national networks, and the advantages and possible disadvantages in the context of the new British Telecom network were described in [5].

The root cause of the problem with the PDH stems from the difficulties associated with the seemingly simple operation known as drop and insert (see Figure 2.5*a*). For example, consider what is necessary to recover the bits that constitute a given 2 Mbit/s channel from the 8 Mbit/s channel in which it is buried. First the exact bit rate at which the 8 Mbit/s channel was assembled must be determined. However, this cannot be done until the corresponding operation has been performed on the associated 34 Mbit/s stream, which must in turn be extracted from the 140 Mbit/s stream. In practice the matter is much more complicated [6a,b]. A synchronous multiplex, by contrast, requires that all tributaries operate at exactly the same digit rate, in which case justification circuits are not needed. Because of the large investment in PDH multiplexing and transmission equipment it is necessary for any viable new synchronous system to be designed to cater, in the first instance, for plesiochronous signals from the current hierarchies. Another reason favouring the evolution to the SDH is the progressive introduction of digital switches that are interfaced and synchronised at 2.048 Mbit/s. Two other references of interest are [7] and [8]. A 1994 tutorial paper which reviewed the features of SDH, outlined its frame structure, reviewed network design and applications and finally summarised the advantages, is also of interest [9].

Despite the falling cost of silicon integrated circuits and the use of new manufacturing techniques such as surface mounted components which have enabled very cost-effective multiplexers for all levels from 64 kbit/s up to 565 Mbit/s to be manufactured, and despite the widespread deployment of optical fibres, every major network operator in the world was, by 1990, actively pursuing a successor to the PDH: the synchronous digital hierarchy (SDH). The major reason for this was the growth of non-telephony traffic resulting from the success of the PDH in reducing the costs of transmission. Most of this new traffic originated from business and commerce, some sectors of which came to depend very heavily on telecommunications. These new customers also demanded other features, such as error-free data transmission, high availability and flexibility of interconnection. In those businesses dependent for their very survival on such services the cost of any interruption could be very serious.

The problems associated with the PDH were common to network operators all over the world, but were particularly acute in the US in the operational complexities arising from the breakup of the Bell System mentioned in Chapter 1. The creation of the regional Bell operating companies (RBOCs) was accompanied by the legal requirement on each to provide standard interfaces at its boundaries with neighbouring RBOCs to facilitate interconnection. There was no standard agreement on optical interfaces (e.g. what optical parameters to use), so wherever an optical transmission link crossed a boundary the optical signal had to be demultiplexed to standard CCITT electrical signals at considerable cost and inconvenience. The lack of a standard optical interface had the further undesirable effect that equipment vendors naturally produced their own proprietary standards. Each RBOC was thus forced to rely on a single vendor for every link in its transmission network.

The problems arising from the lack of a standard optical interface were further exacerbated by the introduction of network management systems to provide operations, administration and maintenance functions. These allowed operators to locate faults within equipment and provided some basic performance information on individual transmission links. Whenever a trunk call could be routed through several RBOCs there was an urgent need to locate quickly any failures in the end-to-end connection. However, once again there were no agreed standards and each vendor developed proprietary systems. Moreover because the PDH has no explicitly defined capacity for management channels, each equipment manufacturer devised a different method of providing such channels. The particular problems of the RBOCs, together with the universal limitations of the PDH, led in the mid-1980s to the American optical interface standard SONET. Developed by the American National Standards Institute, SONET was based on a method of digital transmission called the synchronous transfer mode (STM).

2.3.1 The British Telecom approach

Before discussing some aspects of SONET and the associated synchronous digital hierarchy we will describe briefly the UK approach as representative of one of the most liberalised telecommunication administrations at the start of the 1990s. Like most digital networks the BT network is divided into Core and Access regions. Access incorporates distribution from residential or business users to the local digital exchange. Core regions contain both transmission and 64 kbit/s switching; in the case of BT Core regions are further subdivided into an Outer Core of Processor (main) exchanges (some 300) and their remote concentrator units (about 5000 RCUs), and an Inner Core of about 55 fully interconnected trunk exchanges linked by 140/565 Mbit/s digital line systems [10].

The drive towards the adoption of the Synchronous Digital Hierarchy (SDH) in the Access network comes from the following factors [5,10]:

(a) end-to-end control of transmission networks across Core/Access regions
(b) transparency between Access and SDH Core network
(c) international standards
(d) cheap SDH hardware subsets for Access applications
(e) in-station high bit rate transmission systems under local control, e.g. studio television at 622 Mbit/s.

The relations between the terms Core and Access and the older, better known terms such as trunk and junction are as follows:

Trunk → Inner Core long haul and Inner Core medium haul
Junction → Outer Core
Local → Access

In the UK the planned major milestones of the introduction of the SDH into the main transmission network are shown below

1992 introduce SDH mux rings (Outer Core); begin PC off-load
1993 introduce 622 Mbit/s/2.4 Gbit/s optical line systems (Inner Core)

1994 introduce HACE (Inner Core); introduce SDH to access network
1996 link SDH element managers to a network management centre
1997 introduce broadband switched service
1999 SDH/MTN extended internationally.

Networks in which the first four stages have been completed were designated Managed Transmission Networks (MTN) in the UK. The planned timescales for SDH/SONET roll out over a broad range of networks and operators is not envisaged to be significant until mid to late 1993. By then it is also likely that optical wdm technology will become viable, thus offering further network options and opportunities.

The shape of a possible future MTN/SDH hierarchy is represented schematically in Figure 2.11, and it consists of four levels, the topmost of which is at the highest transmission rate. Most of the terms used in the figure are probably familiar to the reader, with the possible exception of ACE (Automatic Crossconnect Equipment, 64 kbit/s) HACE (High Speed ACE, now SDH Crossconnect Equipment, switching at 2 Gbit/s).

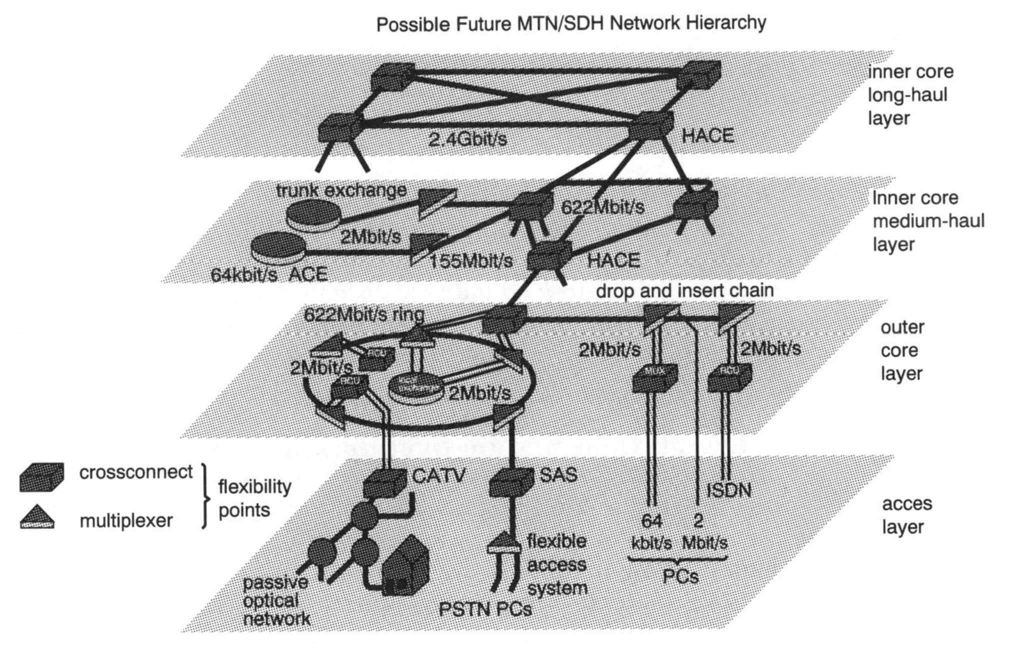

Figure 2.11 Possible future MTN/SDH network hierarchy for the UK ([10], Figure 8)

Table 2.3 Planned evolution of UK network structure during 1990s

Existing plesiochronous network	→	Managed transmission network
Trunk network	→	Inner Core Long Haul
(fully digital by end 1990)	→	Inner Core Medium Haul
Junction network	→	Outer Core
Local network	→	Access network.

Note: by mid-1991 half of the local network exchanges, serving some 13 million customers, were digital.

The existing plesiochronous networks in use throughout the world have a number of limitations which make them increasingly unsuited to modern traffic and operator requirements, for reasons which are well known.

Objectives of the new Core network can be summarised [5] as:

(1) control and management leading to more responsiveness to customer demands, network growth

(2) better availability and quality of service due to reduced quantities of hardware, automatic protection

(3) reduced costs due to optical transmission, SDH/SONET hardware

(4) future proofing because of unlimited bandwidth, software enhancements.

Planned network structure in UK

It is planned to change the existing structure of trunk, junction and local networks described in Chapter 1 to the following:

These changes are illustrated in Figure 2.11.

2.4 PRINCIPLES OF DIGITAL TRANSMISSION OF BROADCAST SERVICES

There are three worlds of digital transmission 'the academic, exotiç space and military applications' and the plain old telephone service (POTS). In academia typical research topics of interest have been theoretical limitations and complicated schemes for modulation and encoding. In space and military systems some applications have involved an extravagant use of bandwidth in order to receive a few bits per second. One of the best known texts covering this field is that of Bylanski and Ingram [11].

2.4.1 Digital transmission over optical fibres

In modern networks at the start of the 1990s it was common to find that more than 30% of the traffic carried was not PSTN, and by the end of the millenium these networks may be dominated by non-PSTN traffic, requiring a far higher grade of service than telephony. In a matter of only 20 years optical fibre technology emerged from the laboratory to become the dominant bearer globally. The penetration of optical fibres from long haul to medium and short haul was well advanced by the end of the 1980s, when the biggest challenge was the local or access network. Optical fibres are now deployed in four distinctly different areas, each of which requires different grades of service and protection. These areas are: intercontinental, internal long haul (trunk), medium/short haul (junction), access or local.

The challenge of the access area will now be discussed briefly. Single-mode fibres began to be installed in the UK for business customers by the late 1980s at a lower cost than their transverse-screen twisted-pair copper predecessors. These were necessarily point-to-point between customer and local exchange, and they only covered some 10% of all customers. Such fibre links are future-proof in that the fibre capacity is effectively unlimited and only terminal equipment need be changed to upgrade capacity. Replacing copper pairs by optical pairs on a one-for-one basis is neither a logical nor a practical option for the future, as we now demonstrate. Some 650 million telephones worldwide are connected by about 1500 million km of copper pairs to their exchanges. The global capacity for the production of optical fibres has been estimated (1990) as some 4 million km per year. Putting these two estimates together and assuming that each copper pair is replaced by one fibre (thus implying duplex operation) would take 375 years! In the UK the figures are 50 million km of copper and an annual production capacity of 350 thousand km, so the corresponding time is about 140 years! Maintenance and repair problems associated with the thousands of cables also militate against this solution, but perhaps the most nonsensical part is the provision of a 50 000 GHz bandwidth link to each customer. Sharing fibres between customers would allow complete provision within the UK over 18 years and globally over 28 years, at 1990 levels of manufacturing capabilities. In the case of residential customers there are much more attractive solutions.

As mentioned earlier, some 70% of the traffic in 1990 was still telephony or voice; consequently this book is concerned largely with the everyday world of the public switched telephone network, or PSTN, in which digital transmission possesses many advantages over analogue transmission. Among these well-known advantages are:

(a) necessity for simple forms of digital modulation in optical fibre systems

(b) use of regenerative repeaters to avoid the cumulative effects of noise in long systems

(c) use of solid-state digital circuitry is much more cost-effective than analogue circuitry

(d) ability to transmit all types of signals of interest: telegraphy and data which are by nature digital, as well as speech, music and television which are analogue in nature

(e) compatability with digital exchanges, which simplifies the network and leads to improved quality and much increased variety of services.

2.4.2 Digital transmission of sound programme signals

In the UK the BBC has used pcm for many years for the transmission of sound programmes [16]. The first system to be developed was sound-in-sync, where the television sound in digital form is carried in the line synchronising pulses. The current system is known as NICAM 3 (near instantaneous companded audio multiplex). In this method, each audio input is pre-emphasised, bandlimited to 15 kHz and variable emphasis limited, before being sampled at 32 kHz. Each sample is then encoded as a 14 bit word, only 10 of which are sent. Depending on the maximum level of the audio signal a range is chosen to determine which 10 bits are transmitted, and there are 5 such ranges. Each range code applies to 32 consecutive samples, hence the description near instantaneous. Tests have indicated that this automatic ranging coding method yields similar performance to 13 bit linear coding and better than A-law instantaneous companding. 13 bit coding allows for $2^{13} = 8192$ different sample amplitudes, which is sufficient to permit the original sound programme signal to be re-constituted with negligible distortion. Recent research has shown that a very high quality sub-band adpcm coding scheme for high fidelity music bandlimited to 15 kHz and sampled at 32 kHz to be equivalent to only 4 bits using pcm. The coder has a total bit rate of only 128 kBd, and is therefore applicable to the integrated services digital network (ISDN). Subjective tests have established that the recovered music is indistinguishable from the original.

2.4.3 Digital transmission of television signals

The rapid conversion of national networks from analogue to digital working also affects television operators. For instance, the BBC alone leased more than 10 000 km of television quality coaxial links within the UK in the mid-1980s. Moreover, there has been for some time in broadcast engineering a swing from analogue to digital techniques in the studio and in other parts of the system. These topics fall outside the scope of this treatment, but the interested readers may wish to consult one or more of references: [12–15].

For television, digital rather than analogue transmission offers the possibilities of lower attenuation and phase distortions when transmitted over long distances. Television is an interesting case because for both monochrome and PAL the video signal is periodic, and the spectrum of the monochrome signal is shown schematically in Figure 2.12.

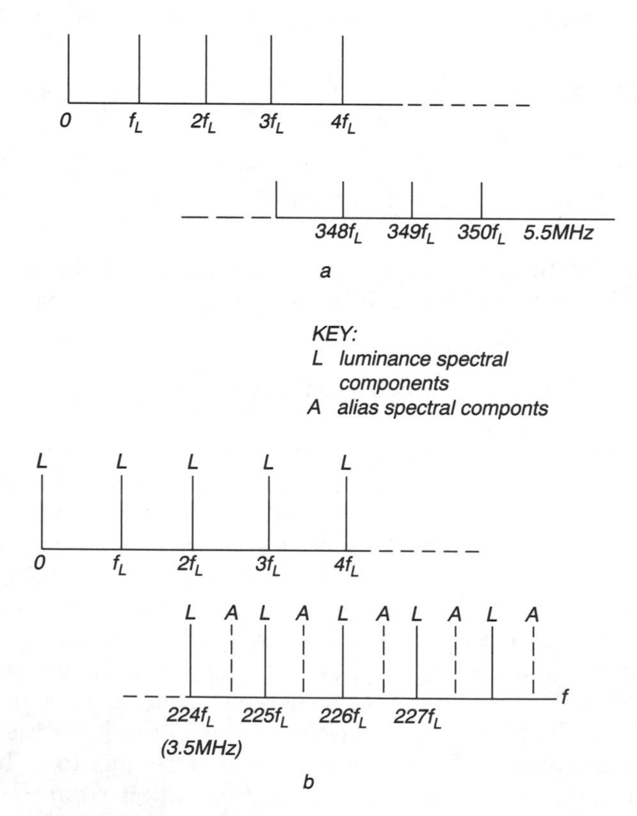

Figure 2.12 (*a*) Monochrome-signal frequency spectrum; (*b*) sub-Nyquist sampling at twice colour subcarrier frequency produces alias components (A) which interleave the luminance (L) components ([16], Figure 10). Reproduced by permission of IEE

It is thus possible to sample at below the Nyquist rate (this is often called sub-Nyquist sampling) in such a way as to ensure that the aliasing components fall between multiples of the line frequency. The advantage of so doing is a reduction in the bit rate. The sampling frequency f_{SA} must be an odd multiple of the line frequency f_L; Figure 2.12*b* shows the result when f_{SA}, is taken as $1135f_L/2$. The video signal can be recovered by means of a comb filter. Digital television has several inherent advantages; for instance, the control of group delay over long connections is no longer a problem, and studio operations such as mixing and fading can be controlled digitally. The hrc (hypothetical reference circuit) for television is simpler than for telephony because it is carried on a dedicated rather than a switched network. CCITT Recommendation J62 applies, but does not specify the national part of the connection. In a small country such as the UK a total of eight sections could be required, thus eight video interconnecting points and the same number of codec operations. Codec requirements are mainly determined by noise and visible interference criteria, but many other parameters must be considered.

The BBC also carried out the world's first transmission of Y, U, V component signals sampled in accordance with CCIR Rec. 601, but sent at a reduced rate over a 140 Mbit/s link between London and Birmingham, in 1983. Component video signals coded according to Rec. 601 require a basic bit rate of 216 Mbit/s. This arises from sampling the luminance component Y at 13.5 MHz, and each of the chrominance components U, V at 6.75 MHz, then using 8 bit quantising. An ultimate target was to achieve efficient and technically adequate transmission at 34 Mbit/s. In the mean time, a vision signal and two associated sound channels, audio programmes, telephony, control and data can be combined onto a single 68 Mbit/s stream. The vision signal is first sampled then quantised using 8 bits; the resulting rate is too high and is reduced to 53 Mbit/s by sub-Nyquist sampling, comb filtering and the use of differential pcm. The dpcm is optimised to maintain transmitted picture quality. Special arrangements are made to handle teletext signals.

2.4.4 Digital transmission of high-definition television (HDTV)

By 1990 extensive tests were in progress in a number of countries on various types of HDTV systems; unfortunately the chances of achieving a world standard appeared to be receding as time went by and many countries were afflicted by recession. High-definition television refers to new television systems which are capable of improvements in performance and definition such that they create 'a new viewing experience'. The proposals involve pictures with aspect ratios like modern cinema screens and with twice the vertical and horizontal detail resolution of existing systems. Thus the network and transmission considerations to send these wideband signals into customers' premises are of crucial importance to service providers and network operators. By 1990 HDTV had been under development in Japan for some 20 years (half a working lifetime) and rather less time in Europe and the US.

Ideally, the picture parameters for such systems should match the most sensitive powers of human visual resolution and perception, but in the late 1980s this presented difficulties to manufacturers of source, transmission and display equipment. However, the increasing penetration of single-mode optical fibres into communication networks provides the bandwidths necessary.

Some of the fundamental picture parameters of HDTV were described in [17]. It was shown how these could be derived from considerations of the performance of viewers' visual systems. Bandwidth requirements of various TV and HDTV systems were compared and a number of optical fibre transmission techniques were considered. The author concluded that the large bandwidth demands of uncompressed HDTV could be met by ofdls (optical fibre digital line system) technology emerging in 1987, in particular, by wavelength division multiplexing. Deployment of sm (single-mode) fibre into the local area and customers' premises will remove the bandwidth constraint.

The HDTV system can be considered as an information source of picture samples arranged, according to the system parameter specifications, in both spatial (horizontal and vertical) and temporal dimensions. Subjective effects play an important role in determining acceptable compromises affecting the demands on both source and display equipment, e.g. those associated with electronic line scanning and line interlace. In the future, electronic beam scanning may not be employed and the enormous bandwidth of optical fibres may make bandwidth compression unnecessary.

2.4.5 Bandwidth requirements for electronically scanned HDTV

The maximum video frequency and hence the required luminance channel bandwidth for balanced horizontal and vertical resolution can be obtained by halving the total number of video samples per second, using the formula:

$$\text{bandwidth} = 0.5 \times K \times I \times \frac{N}{V} \times \frac{N}{H} \times A \times f, \quad \text{Hz} \tag{2.1}$$

where K is the Kell factor (0.7), I is the interlace effect (0.65), V is the vertical active picture line fraction (0.95), H is the horizontal active picture line fraction (0.8), N is the number of active lines, A is the aspect ratio (5.33:3), and f is the picture frequency.

The factors H and V make allowance for the horizontal and vertical scan flyback times required for the source and display process. The numbers in brackets are the values assumed for calculation.

Consideration of the performance of the human visual system, together with the assumptions of a minimum viewing distance of $3h$, where h is the picture height and f at least 60 Hz, lead to $N = 806$ lines per picture for sequential scanning and $N = 1240$ for 2:1 interlace. To ensure total elimination of flicker f would need to be 75 Hz for sequential scanning and 60 Hz for interlace, then for the same spatial resolution and overall flicker performance in the luminance channel the corresponding bandwidths are 40 MHz and 49 MHz for sequential and interlace scanning, respectively. A number of HDTV and standard schemes are compared in terms of their basic parameters in Table 2.4.

2.4.6 Transmission aspects of HDTV

Much of the early experimental work concentrated on satellite or radio systems, but the widespread installation of single-mode fibres offers almost unlimited bandwidth, large power budgets, a wide range of components and extensive multiplexing capabilities. At the network level, the fibre bandwidth could be fully exploited by extensive multiplexing of HDTV channels. In the case of digital transmission, serial bit rates of about 1.8 Gbit/s for RGB tristimulus

Table 2.4 Comparison of HDTV and standard television systems ([17], Table I). Reproduced by permission of BT

System Type	HDTV	HDTV	HDTV	Standard TV	Standard TV
System Name	'Division'	RCA proposal [7]	Eureka	PAL	NTSC
No. of active lines/picture	1035	720	1151	575	485
Interlace ratio	2:1	1:1	2:1	2:1	2:1
Picture Frequency (Hz)	30	30	25	25	30
Luminance Bandwidth Requirement (MHz)	17	25	18	5	4

and 1.2 Gbit/s for YC_1/C_2 components are needed. Multiple channels can be multiplexed by either tdm or wdm techniques, and their performances are compared in Figure 2.13.

The graph balances the power budget equation as a function of the number of terminals/customers served and the bit rate delivered to each terminal. The operating parameters assumed for this illustration are as follows: distance between exchange and terminal 10 km, cable loss 0.7 dB/km (this is a very conservative figure), system margin 5 dB, launched optical power 0 dBm, and bit error rate: 10^{-9}.

As an illustration of the performance available experimentally towards the end of the 1980s, we consider the simultaneous transmission of data and high-definition television traffic using subcarrier modulation (scm) video distribution [18]. Two major concerns for network operators in future applications of the scm technique are the transport of very wideband HDTV signals, and the compatibility of the

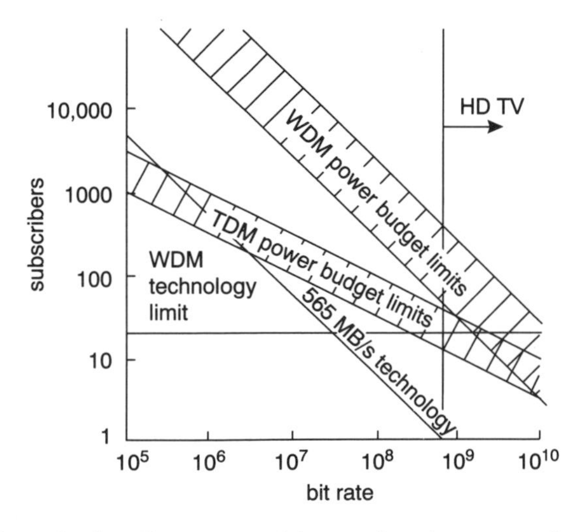

Figure 2.13 Number of subscribers versus bit rate showing comparisons between tdm and wdm multiplexing techniques ([17], Figure 8). Reproduced by permission of BT

technique with baseband digital signals. This MTT-S contribution from two Bellcore authors reported the experimental results obtained over a 2 km length of sm fibre carrying three frequency-modulated NTSC high-definition television channels and one OC-12 (622 Mbit/s) data signal. The unweighted signal-to-noice (snr) of each component HDTV signal was better than 38 dB and with −10 dBm optical power received the ber of the data multiplex signal was less than 10^{-9}. Although it is intended to provide transport for HDTV within SONET, this was not possible in 1989; instead the authors proposed an interim solution in which subcarrier HDTV signals and a baseband B-ISDN signal are transmitted simultaneously over a sm fibre link. The subcarrier frequencies used for the NSTC video signals were 1.3, 1.55 and 1.8 GHz. No pre-emphasis/de-emphasis was used with the fm signals, and the measured system frequency response of each of the fm channels exceeded 20 MHz, the value of the source signal bandwidth. Moreover, the differential delay between the received video components was less than 20 ns and it caused no spatial colour misregistration in the received picture.

The experimental arrangement is shown in Figure 2.14.

The optical transmitter was a dfb laser operating at 1560 nm, and the receiver was a pin photodiode with built-in 50 Ω matching resistor, followed by a low-noise wideband 50 Ω preamplifier. The fm modulator and demodulator were constructed with standard commercially available microwave components. The R, G, B component signals were carried one on each subcarrier. The optical modulation index *m* in each channel was 0.13, where *m* is defined in terms of the peak and average received powers as $P_P/P - 1$, evaluated with all the other channels inactive.

Results: the fm deviation ratio or modulation index, i.e. the peak carrier frequency deviation divided by highest frequency (20 MHz) in the modulating signal, was about 2.1. The corresponding rf bandwidth occupied was taken as that given by one of the standard formulae for fm signals, namely $2(m+2)f_M$. The carrier-to-noise ratio in fm systems is related to the unweighted snr by another standard formula. The measured results are presented in Figure 2.15, and they indicate that the ber performance of the OC-12 signal was little affected by the presence of the

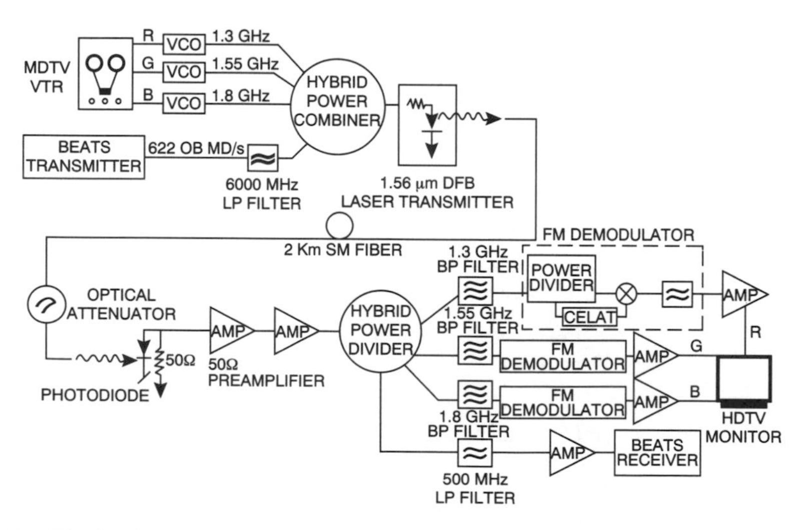

Figure 2.14 Block schematic of experimental arrangement for the simultaneous transmission of fm hdtv and data ([18], Figure 1). Reproduced by permission of IEEE, © 1989

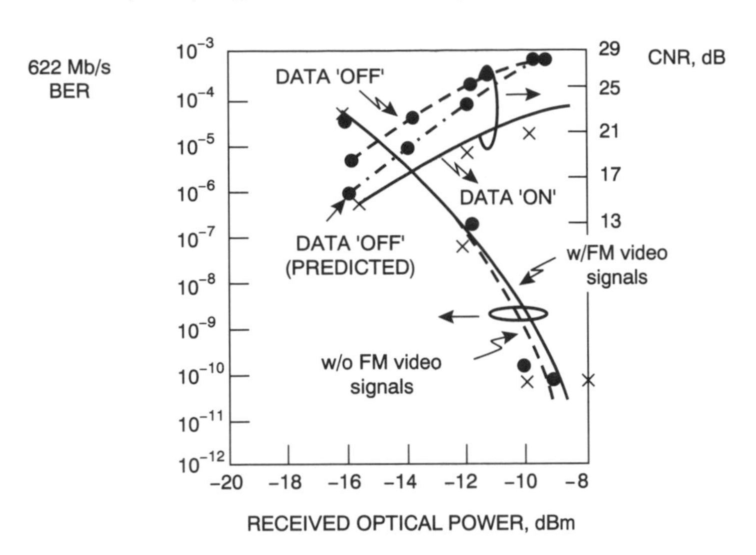

Figure 2.15 Measured values of ber of 622 Mbit/s data and cnr of video signal versus received optical power ([18], Figure 2). Reproduced by permission of IEEE, © 1989

Table 2.5 Measured and calculated values of received hdtv component signals without and with data present ([18], Figure 3). Reproduced by permission of IEEE, © 1989

	Parameter	Channel R	Channel G	Channel B
Data off	Expected CNR		28.0 dB	
	Measured CNR	28.2 dB	26.7 dB	27.3 dB
Data on	Measured CNR	21.0 dB	20.6 dB	22.5 dB
	Unweighted SNR(1)	37.7 dB	37.1 dB	39.6 dB
	Unweighted SNR(2)	38.3 dB	37.9 dB	39.8 dB
	Channel BW	26.0 MHz	32.0 MHz	22.0 MHz
	Diff delay		<20.0 ns	

Notes: (1) measured,
(2) calculated from measured values

video signals, but the converse is not true, for reasons stated. The behaviour is also evident from the measured results summarised in Table 2.5.

Conclusion: the experiment demonstrated the simultaneous transport of a three-component HDTV analogue signal using three microwave frequency subcarriers and a baseband data signal at 622 Mbit/s over 2 km of fibre using standard optical and microwave components, despite which the quality of the received picture was excellent. The authors showed, in principle, an overlay architecture in which there is compatibility between a subcarrier modulation scheme for video transmission and a B-ISDN optical fibre transmission system. This hybrid architecture possesses the advantages of compatibility with existing systems, it provides integrated transport of telephony, data and various video services including HDTV, and can be implemented with commercially available microwave components. It is therefore a possible near-term solution, pending the arrival of the all-digital network.

2.5 REFERENCES

1 COCHRANE, P.: 'The evolution of fibre optic networks'. *Elektrotechnik und Informationstechnik*, 107 Jahrgang, heft 6, 1990.

2 COCHRANE, P.: 'Optical networks', *IEE Vacation School on Optical Fibre Communications*, UCNW, Bangor, Sept. 1990.

3 LINFORD, N.S.: 'The evolution of the United Kingdom telephone service'. *Post Office Electrical Engineering J.*, **69**, Apr. 1976, pp. 12-18.

4 RUDGE, A.W.: 'I'll be seeing you: multimedia communications in the 21st century'. *Electronics & Commun. Engineering J.*, **5**, no. 5, Oct. 1993.

5 WHITT, S., HAWKER, I., and CALLAGHAN, J.: 'The role of SONET-based networks in British Telecom'. *Int. Commun. Conf. ICC*, 1990.

6 a, b. MATTHEWS, M., and NEWCOMBE, P.: 'The synchronous digital hierarchy, part 1: the origin of the species'. *IEE Rev.*, **37**, no. 5, May 1991, pp. 185-89. 'The synchronous digital hierarchy, part 2: survival of the fittest'. *ibid.*, June 1991, pp. 229-33.

7 *'Synchronous Transmission Systems'*, ed. C. Newall, Northern Telecom Europe Ltd., Doc. GH-9, issue 3, 1992.

8 BALCER, R.: *'Managed Transmission Networks: An introduction to the new synchronous digital hierarchy'*. UKC/NSET2.3.2, British Telecommunications plc, 1988.

9 FERGUSON, S.P.: 'Implications of SONET and SDH'. *Electronics & Commun. Engineering J.*, **6**, no. 3, June 1994.

10 HAWKER, I.: 'The role of SDH in the BT core transmission network', 1990.

11 BYLANSKI, P., and INGRAM, D.G.W.: *'Digital Transmission Systems'*, revised second edition, 1980.

12 MORALEE, D.: 'Paving the way to all-digital broadcasting'. *IEE Electronics & Power*, June 1985, pp. 435-440.

13 BARNES, P.: 'Cable TV in the UK'. *IEE Electronics & Power*, June 1985, pp. 441-444.

14 GLEDHILL, S.J.: 'Spectrum analysis of television waveforms'. *IEE Electronics & Power*, June 1985, pp. 445-448.

15 OSBORNE, D.W.: 'Sound and data on DBS'. *IEE Electronics & Power*, June 1985, pp. 449-454.

16 BEARDS, P.H.: 'Pulse-code modulation for broadcast signal distribution'. *Electronics & Power*, June 1979, pp. 425-430.

17 CRAWFORD, D.I.: 'High definition television — parameters and transmission'. *Br. Telecom. Technol. J.*, **5**, no. 4, Oct. 1987, pp. 76-83.

18 LO, C.N., and SMOOT, L.S.: 'Integrated fiber optic transmission of fm hdtv and 622 Mbit/s data'. 1989, *IEEE MTT-S Digest*, pp. 703-704.

FURTHER READING

19 MAYO, J.S.: 'The telecommunications revolution of the 1990s'. Scientific American, special issue, 1991.

3

SINGLE-MODE FIBRE LIGHTGUIDE THEORY

What made this (Maxwell's) theory appear revolutionary was the transition from action at a distance to fields as fundamental variables. The incorporation of optics into electromagnetism, with its relation of the speed of light to the electric and magnetic... system of units... was like a revelation.

Albert Einstein, 'Autobiographical Notes', 1946.

3.1 INTRODUCTION

The first two chapters have been mainly descriptive; in this one we look at the mathematical description of the propagation of light in single-mode optical fibres. Much was written about light transmission in the 1970s but in the main the analysis was devoted to multimode fibres with step or graded refractive index profiles, for which geometrical optics was the appropriate tool. Such fibres were all but rendered obsolete by the development of sm fibres in the early 1980s. The mathematical analysis of these lightguides is considerably more complex, but an approach has been chosen which it is hoped will be more appealing to engineers. Long experience of research and development in industry has suggested that there are relatively few engineers who have the interest or endurance to master long and complex mathematical analysis and arguments, so more than the normal amount of explanation accompanies the analysis given later in the hope that understanding will be more easily gained thereby.

3.1.1 The electromagnetic field

James Clerk Maxwell (1831–1879) was a Scot, and like Alexander Graham Bell born and bred in Edinburgh. Maxwell was a pupil at Edinburgh Academy then a student at Edinburgh University between 1847 and 1850. When a schoolboy of 14 he published his first scientific paper, and two more were published while a student at Edinburgh. Michael Faraday (1791–1867) invented, and Maxwell developed, the concept of field as a region in which electric and magnetic forces are continuous functions of position and of time. In 1864 Maxwell proposed that energy can be transported through dielectrics, including empty space, at a finite velocity by the electric and magnetic fields investigated earlier by Faraday and others. The theory effectively destroyed the concept of instantaneous action at a distance which had been prevalent in physics, especially in continental Europe, until the late 1800s. These fields were orthogonal and travelled together in a direction perpendicular to the fields. Maxwell's concepts led him to a new set of partial differential equations which connected the spatial and temporal differential coefficients of the electric and magnetic fields. These were of vastly greater scope than any reached before; indeed, the scope of his equations was sensational. According to Einstein, the field theory of Faraday and Maxwell represented probably the most profound transformation of physics since Newton's time. They gave a finite velocity for the propagation of the linked electric and magnetic disturbances and identified electromagnetic disturbances as light.

Maxwell predicted (1862) the propagation of electromagnetic waves in air on the basis of his theory, but did not live to see his theory validated by experiments. This was done by Heinrich Rudolph Hertz (1857–1894) in a masterly series of experiments completed in May 1888. In 1879 H. von Helmholtz with H. Hertz in mind, set as a prize problem of the Prussian Academy of Sciences the experimental test of certain implications of Maxwell's theory. Hertz decided that this was too big a risk for his doctoral dissertation, but returned to it in 1884. Hertz embodied an ideal of a physicist, one with equal mastery of mathematical and of laboratory work, and after two years he succeeded in demonstrating the existence of electromagnetic waves. The announcement of his discovery of electromagnetic waves was made in 1888.

Maxwell's equations are a manifestation of the simplest and most elegant known local gauge symmetry that is consistent with the principles of special relativity. Gauge symmetries are profound features of nature, and the electromagnetic field must exist in order to preserve local gauge invariance with respect to changes in electric potential (loosely voltage) from place to place. The field compensates for the place-to-place variations of voltage in exactly the correct way so as to maintain the symmetry of the laws of physics. At the quantum level the concept of gauge change must be extended to the phase of the wave inextricably associated with a charged particle.

3.1.2 Quantum electrodynamics

All of the basic phenomena of the physical world, except those due to gravitational and nuclear forces, are described by quantum electrodynamics (QED). A very readable account of QED was written by Richard P. Feynman [1]. This book also describes reflection and refraction in QED terms and *inter alia* accounts for partial reflection as the absorption of an incident photon by an electron and the subsequent emission by that electron of a new (reflected) photon, i.e. the phenomenon of scattering which occurs throughout an optical fibre. The reference does not give a model for either electron or photon, but attempts have been made to do so by others.

Light consists of quanta (particles) called photons, and only three basic actions are needed to produce all of the phenomena associated with light and electrons, i.e. for the phenomena covered by QED, namely:

(1) a photon goes from place to place
(2) an electron goes from place to place
(3) an electron emits or absorbs a photon.

Each of these actions has a probability of occurrence, and these probabilities are influenced by the state-of-polarisation (sop) of the photons or electrons involved in the action. Let us consider the first two actions. Both photons and electrons have four possible conditions related to the geometry of space–time; in the case of photons these are polarisation states in the one time and three space coordinates. At distances large compared with the wavelength, as in semiconductor laser oscillators, amplifiers and optical fibres, and when travelling at velocity c, the time-like polarisation cancels exactly the polarisation in the direction of travel, leaving two polarisation states. For example, a photon travelling along the z-axis can be polarised in either the x or y direction. Electrons also have four components, but have only two sop available and these are the physical basis of the Pauli exclusion principle, namely that no two electrons can co-exist in the same quantum state at the same point in space–time (whereas photons can).

3.1.3 The QED basis of the electromagnetic field

Let us now turn to the third action, an electron emits or absorbs a photon which, when large numbers of electrons and photons interact over relatively large regions of space–time, is of direct interest in optical sources and detectors. When very large numbers of electrons are moving in unison, and very large numbers of photons are being exchanged, the probability of absorption occurring is independent of whether an individual electron, or any other electron, has absorbed a photon earlier and depends only on the position of the electron in space–time. In such cases the electron is regarded as moving in an external field which is characterised by continuous quantities that depend on space–time coordinates. There are four components in the field, and these correspond to the probabilities of an electron absorbing each of the four different kinds of photons. Those classical field components corresponding to photons polarised in any one of the three spatial dimensions are called the vector potentials, and that corresponding to the time dimension is known as the scalar potential. The more convenient and familiar electric and magnetic field components, which will be used frequently throughout this book, are derived from combinations of these potentials.

Classical optics successfully describe the propagation of laser light inside a transparent medium, because for coherent states of radiation the electric and magnetic fields may be written in terms of their expectation values, and thus their propagation may be treated through the macroscopic Maxwell equations. As is well known, these equations permit the calculation (in principle) of both the spatial progression and temporal evolution of a propagating electromagnetic (em) field, and treat the interaction of the field with a medium phenomenologically through the induced polarisation.

Nonclassical light (i.e. squeezed light and antibunching) requires a different treatment.

3.2 SOME BASIC PROPERTIES OF LIGHT

In practice plane em waves with a single wavelength and constant amplitude are not used because in transmission systems the optical source is modulated by an electric current corresponding to the traffic. The statistics of this modulating signal are usually well approximated by a pseudo-random binary sequence. All installed

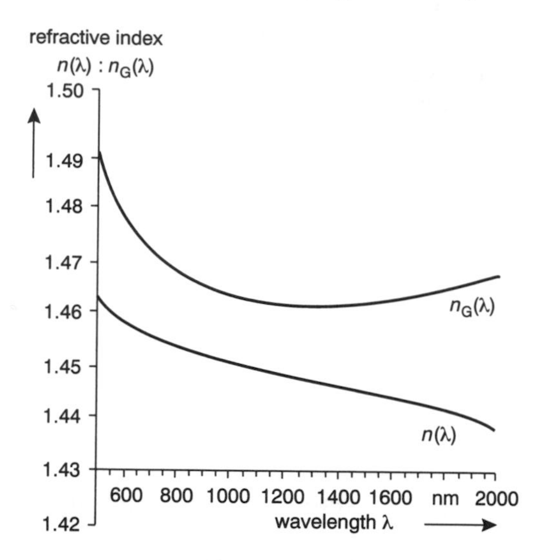

Figure 3.1 Refractive index $n(\lambda)$ and group refractive index $n_G(\lambda)$ for 100% silica fibre ([2], Figure 2.6)

optical fibre systems in service up to the early 1990s use intensity modulation (on–off keying), i.e. bursts of carrier in which the information is conveyed in the envelope. The carrier itself is not nearly as monochromatic as at radio frequencies. The shape of the envelope waveform corresponding to a mark (or one) is often taken to be rectangular or Gaussian, the corresponding spectral representations of single pulses or a periodic train of pulses are well known. Consequently a spectrum of wavelengths centred on a carrier of wavelength λ, or equivalently photons with the corresponding spread in frequency, hence in energy. The speed of propagation of such pulses can be called the group velocity (analogous to group delay at carrier frequencies), and an associated group refractive index can be defined as:

$$n_G = n - \lambda \frac{dn}{d\lambda} \tag{3.1}$$

Curves of n and of n_G for pure fused silica glass as a function of λ are given in Figure 3.1 [2].

Note that the curve of n_G reaches a minimum value around 1300 nm, which is one of the operating wavelengths of interest mentioned above. More accurate values are needed for calculations, and these are supplied in Table 3.1.

When calculating transmission times of optical pulses, the value of n_G should be used and not n itself.

3.2.1 Basic structure of optical fibre waveguides

As is well known, the basic structure of an optical fibre consists of a cylindrical core surrounded by a coaxial cladding. By convention, the maximum value of the refractive index of the core is denoted by n_1 and that of the cladding by n_2, and in all cases n_1 is greater than n_2.

In general, two types of ray paths need to be considered: those which pass through the axis periodically, and those which do not. The former are called meridional and the latter skew rays.

Table 3.1 Refractive index $n(\lambda)$ and group refractive index $n_G(\lambda)$ for 100% silica fibre ([2], Table 2.4)

Wavelength λ (nm)	Refractive index n	Group refractive index n_G
600	1.4580	1.4780
700	1.4553	1.4712
800	1.4533	1.4671
900	1.4518	1.4646
1000	1.4504	1.4630
1100	1.4492	1.4621
1200	1.4481	1.4617
1300	1.4469	1.4616
1400	1.4458	1.4618
1500	1.4446	1.4623
1600	1.4434	1.4629
1700	1.4422	1.4638
1800	1.4409	1.4648

A meridional ray path in the core at an angle α to the normal to the core–cladding interface may undergo both reflection and refraction as just described, however there will be a unique angle of incidence for which the reflected ray lies at 90° to the normal, i.e. along the interface. This particular angle is the critical angle α_c, and is related to the refractive indices of core and cladding as before by:

$$\sin\alpha_c = \frac{n_2}{n_1} \tag{3.2}$$

For example, using the same 1.5 value for glass as before, the critical angle for an air–glass interface is about 42°. Total internal reflection can only occur at an interface where light travels from an optically denser medium to a less dense one, never in the reverse direction.

3.2.2 Numerical aperture

It follows from equation (3.2) that all meridional rays that do not diverge from the axis by more than $90° - \alpha_c$ will be reflected back into the core and so will be guided. In order to satisfy this condition, light must be launched into the fibre at suitable angles to the axis. The theoretical maximum value of this angle can be determined from the law of refraction, giving:

$$\frac{\sin\theta}{\sin(90-\alpha_c)} = \frac{n_1}{1} \tag{3.3}$$

hence

$$\sin\theta = n_1\cos\alpha_c = \sqrt{[n_1^2 - n_2^2]}$$

by making use of equation (3.2). Measured values are usually smaller. The maximum allowable value, say θ_M is known as the acceptance angle for meridional rays, and it clearly depends only on the two values of n, and the sine of this

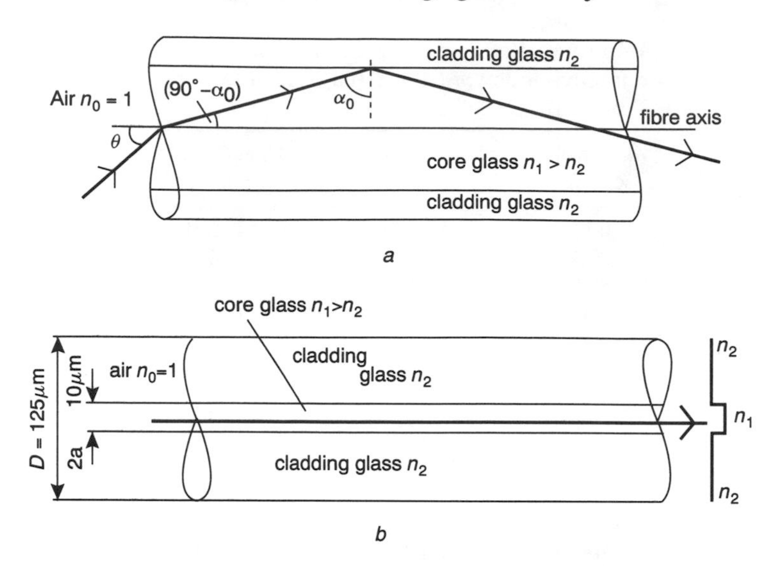

Figure 3.2 Guidance of light in (*a*) a multimode fibre, (*b*) a single-mode fibre ([2], Figure 2.8)

angle is called the numerical aperture, often abbreviated to na. This is an important quantity in launching power from a laser or led into a fibre. Evidently the numerical aperture is:

$$\sin\theta_M = \text{na} = \sqrt{[n_1^2 - n_2^2]} \tag{3.4}$$

3.2.3 Theoretical studies of propagation in an optical fibre lightguide

Theoretical studies of propagation can be based on two approaches: geometrical optics or em modes. The former is of interest because it is simpler and gives more immediate physical insight, and can provide good approximations in the case of multimode fibres. On the other hand, the electromagnetic approach must be used in all problems involving coherence or interference, and when dealing with fibres which support only one or a few modes (the so-called single-mode fibres) and the only ones of technical interest in line systems since the late 1980s. Unfortunately, this rigorous method does not often lead to analytical solutions, and numerical solutions may not be cost-effective. When such is the case good results can be obtained from a quasi-classical approach, which provides a correction to geometrical theory and includes effects to the first order in wavelength, thus allowing undulatory effects to be treated.

3.2.4 Refractive index profiles

Optical fibres are manufactured to give one of a small number of types of variation of the refractive index n as a function of the radius r in the core, and the term 'refractive index' profile is used to denote this variation. The index n can thus be written as $n(r)$, and as one would expect the propagation of rays depends on the variation of refractive index with radial distance from the axis, i.e. on the shape of

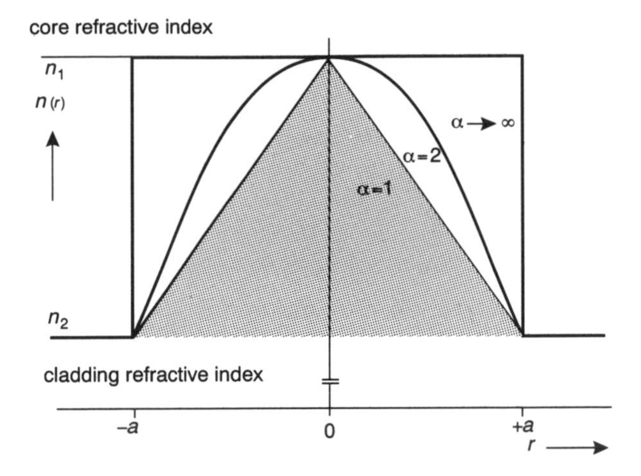

Figure 3.3 Three commonly used idealised refractive index profiles ([2], Figure 4.1)

$n(r)$. Three simple shapes are of practical importance, and they are illustrated in Figure 3.3 [2].

The three simple profiles correspond to particular values of the exponent α:

$\alpha = 1$ triangular profile

$\alpha = 2$ parabolic profile

$\alpha = \infty$ rectangular profile

It is evident that there is only one value of α for which the refractive index in the core is constant. For all finite values n varies with r and these are called 'graded index fibres'. In practice the parabolic and the triangular profiles are used because of their good guiding properties, but other more elaborate profile shapes are also employed, as will be seen later.

These are special cases of the power law index profiles which are described by the equations for the core and cladding, respectively:

$$n^2(r) = n_1^2\left[1 - 2\Delta\left(\frac{r}{a}\right)^{\alpha}\right], \quad \text{for } r < a \tag{3.5}$$

and

$$n^2(r) = n_2^2, \quad \text{for } r > a \tag{3.6}$$

where n_1 is the refractive index along the fibre axis ($r = 0$), Δ is the normalised refractive index difference, r is the radial distance from the axis in μm, a is the core radius in μm, α is the profile exponent, and n_2 is the refractive index of the cladding.

From equations (3.5) and (3.6) the quantity Δ is related to the numerical aperture na as:

$$\Delta = \frac{(\mathrm{na})^2}{2n_1^2} \tag{3.7}$$

where the numerical aperture is defined as:

$$(\mathrm{na})^2 = n_1^2 - n_2^2 \tag{3.4}$$

from which we obtain:

$$2\Delta = 1 - \left(\frac{n_2}{n_1}\right)^2 \tag{3.8}$$

All practical fibres have Δ very much less than one and thus satisfy the condition for weak guidance of the propagating modes.

3.2.5 Single-mode fibres

Geometric optic theory can be applied to multimode glass fibres of the types used in early optical telecommunication systems for which four simplifying assumptions [3] can be made.

(1) The index profile is circularly symmetric.

(2) The core diameter $2a$ is very much larger than the wavelength of the guided light (more than 100 times greater), so that hundreds of modes can propagate simultaneously, giving multipath transmission.

(3) The total change in refractive index within the guiding region is restricted to a few percent, so that the propagating modes can be treated as transverse electromagnetic waves.

(4) Variations of refractive index over lengths of the order of a wavelength are negligible, and the conditions are such that geometrical optics (or the zeroth order of the WKB approximation) is valid.

The variation of refractive index can be more general than the simple types mentioned above, for instance, profiles with a central notch (as often the case in practice) and those with one or more concentric annuli of different index as in dispersion-flattened fibre. Evidently in the case of single-mode fibres assumption (2) is no longer valid.

It is feasible to decrease the number of modes so that only one mode can propagate. As one would expect from metallic circular waveguide theory, this is done by reducing the critical dimension sufficiently, in fact by reducing the core diameter to around 9 μm. More precisely, by reducing the mode field diameter (the size of the radial field amplitude of the fundamental mode) so that only the fundamental mode is guided.

[2] gave the following example of typical dimensions of a step-index sm fibre: mode field diameter 10 μm, cladding diameter 125 μm, core refractive index 1.46 and Δ of 0.003. For these values $(\text{na})^2 = n_1^2 - n_2^2 = n_1^2 \times 2\Delta$ gives na as $1.46 \times \sqrt{(0.006)} \simeq 0.113$, and the acceptance angle is $\arcsin(\text{na}) \simeq 6.5^\circ$.

Normalised frequency: an important property of optical fibres is described by means of the normalised frequency or v-value, which is dependent on the core radius a, the numerical aperture na and the wavelength λ of the light in the core. The v-value is a pure number given by:

$$v = \frac{2\pi \times a \times (\text{na})}{\lambda} \tag{3.9}$$

The number of modes N which are guided in a fibre is a function of v given approximately by:

$$N \simeq \frac{v^2 \alpha}{2(\alpha + 2)}, \tag{3.10}$$

for the general power law profile. The values for the step and parabolic profiles are readily found from this expression. As a practical example of a multimode lightguide, consider a fibre with core diameter $2a = 50$ μm, na $= 0.2$ and $\lambda = 1.3$ μm. The v-number is found from equation (3.9) as $v = 2\pi \times 25 \times 0.2/1.3 = 24.2$. The corresponding number of guided modes is approximately $N \simeq v^2/4 = 146$.

Single-mode fibres often have step index profiles, and in such cases it can easily be shown that the v-value becomes 2.405, and the corresponding fundamental mode is denoted as LP_{01}, where LP stands for linearly polarised. The number 2.405 is in fact the first zero of the Bessel function $J_0(x)$, and this and higher-order Bessel functions are particularly appropriate for the analysis of cylindrically symmetrical systems such as coaxial cables, hollow waveguides and optical fibres.

Cut-off wavelength: the value of λ above which only the fundamental mode is guided is known as the cut-off wavelength, and it corresponds to $v = 2.405$. This v-value enables the cut-off wavelength of a given lightguide to be calculated; for example, consider a sm fibre with $2a = 8.5$ μm, a step index profile and na $= 0.113$, then $\lambda_{co} = 2\pi \times a \times (\text{na})/v$ is $\pi \times 8.5 \times 0.113 \div 2.405 \simeq 1255$ nm.

The cut-off wavelength is the shortest wavelength at which only the fundamental mode is capable of propagating in a given fibre; thus in sm fibres the cut-off wavelength must be less than the shorter wavelength edge of the transmission band of interest. In the example, the fibre will permit single-mode propagation at wavelengths in the 1300 and 1550 nm bands. As the operating wavelength increases above the cut-off value the fundamental mode radial distribution extends further into the cladding.

The term 'single mode' is not strictly correct, since the fundamental mode contains photons which can have either of two mutually perpendicular polarisation states, e.g. along x-axis and the y-axis. Moreover, sop fluctuations occur along the fibre because of imperfections. The presence of two states rather than one per mode does not affect direct detection systems, but must be considered in systems employing semiconductor optical amplifiers or other components which are polarisation-sensitive, and coherent systems.

3.3 PROPAGATION IN SINGLE-MODE FIBRES

The treatment is based mainly on that given in [4], and is aimed at practising engineers.

Before plunging into the mathematical detail, it may be helpful to the reader to outline the general steps [5] which are appropriate to optical fibres. As in most practical applications, we begin with Maxwell's vector equations, then:

(1) consider these equations for linear, isotropic material in the absence of conducting currents and charges

(2) perform the operation of curl on the equations for curl $\boldsymbol{E}$ and curl $\boldsymbol{H}$, apply a vector identity, and reduce the two to a scalar equation in which the change in the dielectric constant over the distance corresponding to a wavelength is taken to be small

(3) choose a cylindrical coordinate system and transform the curl curl equations to these coordinates; this gives two sets of three equations for the components of $\boldsymbol{E}$ and $\boldsymbol{H}$ in terms of one another

(4) solve the resulting equations for the transverse components in terms of the axial ones by assuming solutions which are harmonic in time and the axial direction

(5) the axial components of $\boldsymbol{E}$ and $\boldsymbol{H}$ must now be found from partial differential equations subject to the appropriate boundary conditions.

3.3.1 Vector and scalar wave equations for optical fibres

We now consider the vector and scalar wave equations derived from Maxwell's equations and which are applicable to the problem in hand. For a medium of dielectric constant ε (or alternatively refractive index n) which is a function of the Cartesian coordinates x, y, z one can obtain a pair of vector wave equations:

$$\nabla^2 \boldsymbol{E} + \nabla\left(\boldsymbol{E} \cdot \frac{\nabla\varepsilon}{\varepsilon}\right) = \frac{\varepsilon}{c^2}\frac{\partial^2 \boldsymbol{E}}{\partial t^2} \tag{3.11}$$

$$\nabla^2 \boldsymbol{H} + \frac{[(\nabla\varepsilon) \times (\nabla \times \boldsymbol{H})]}{\varepsilon} = \frac{\varepsilon}{c^2}\frac{\partial^2 \boldsymbol{H}}{\partial t^2} \tag{3.12}$$

$\boldsymbol{E}$ and $\boldsymbol{H}$ are the electric and magnetic field vectors as usual. The general solution for a dielectric cylinder has non-zero components of these vectors in each of the three directions. The presence of the term $\nabla\varepsilon$ in the equations prevents them being separated into their scalar components, thus complicating their solution. Fortunately, in optical fibres the variations of dielectric constant and hence of refractive index (since for glass $n = \sqrt{\varepsilon}$) over distances comparable with the wavelength are small and can be neglected. By taking the curl of these equations and applying a vector identity, the above equations can be reduced to a set of six identical scalar wave equations of the form:

$$\nabla^2 \Psi = \frac{\varepsilon}{c^2}\frac{\partial^2 \Psi}{\partial t^2}, \tag{3.13}$$

where Ψ stands for any Cartesian component of the em field.

As usual, a cylindrical coordinate system r (radial), ϕ (azimuthal), and z (axial) is now chosen, where z lies along the axis of the fibre and is positive in the direction of propagation. Transforming the curl equations into these coordinates yields two sets of three equations for the components of $\boldsymbol{E}$ and of $\boldsymbol{H}$ in terms of one another. These can be solved for the transverse (two r and two ϕ) components in terms of the two axial (z) components. From the physics of the problem we expect solutions to be of the form:

$$\Psi = \Psi(r, \phi)\exp[-\mathrm{j}(\omega t - \beta z)] \tag{3.14}$$

which represents waves propagating in the positive z-direction with propagation constant β. Equation (3.14) may be rewritten as:

$$\nabla_T^2 \Psi + (k^2 n^2 - \beta^2)\Psi = 0 \tag{3.15}$$

where the subscript T denotes transverse, and k is the free space or vacuum wave number $2\pi n/\lambda$, with $n = 1$. This is an eigenvalue equation which has distinct field solutions (eigenfunctions) for specific values of β (eigenvalues); these solutions correspond to the discrete propagating modes which the lightguide supports. Equation (3.15) can be solved for β and any one of the six field components. The remaining five components are then found by substitution into Maxwell's equations.

By assuming solutions of the form of equation (3.14) one can obtain the four equations giving the r and ϕ components in terms of E_z and H_z, etc. To complete the solution, the scalar wave equations (3.13) must now be solved for E_z and H_z. Equation (3.13) is first transformed to cylindrical coordinates and the variables are separated by assuming that E_z and H_z are proportional to $F(r)\exp(jm\phi)$, where F is an as yet unknown function of r, and m must be an integer to ensure azimuthal periodicity. The differential equation for $F(r)$ is then:

$$\frac{\partial^2 F}{\partial r^2} + \frac{1}{r}\frac{\partial F}{\partial r} + \left(k^2 - \beta^2 - \frac{m^2}{r^2}\right)F = 0 \tag{3.16}$$

which can be solved for F and for β in specific lightguiding structures, subject to the appropriate boundary conditions. We will now consider the specific case of sm fibres with step index profiles.

3.3.2 Solution for a step-index single-mode fibre

One of the few cases for which the equation (3.16) can be solved in closed form is for the case of a homogeneous core of constant refractive index n_1 surrounded by a cladding of infinite extent and index n_2. Then the solutions are Bessel functions chosen to ensure that $F(r)$ is finite at $r = 0$, and that $F(r)$ tends to zero as r tends to infinity. As before, we adopt cylindrical polar coordinates for the components of the field, and look for solutions of the components in the z-direction; thus we write:

$$\Psi_z = F(r)\Phi(\phi) \tag{3.17}$$

This equation is now applied to the core and to the cladding regions separately, and in both regions the refractive index is a constant, so that we obtain two equations of the form:

$$\frac{r^2}{F}\frac{\mathrm{d}^2 F}{\mathrm{d}r^2} + \frac{r}{F}\frac{\mathrm{d}F}{\mathrm{d}r} + r^2(k^2n^2 - \beta^2) = 0 \tag{3.18a}$$

and

$$\frac{1}{\Phi}\frac{\mathrm{d}^2\Phi}{\mathrm{d}\phi^2} = 0 \tag{3.18b}$$

For a valid solution the radial and azimuthal parts must each be constant, say m^2; then the azimuthal part, equation (3.18*b*), represents a simple harmonic oscillator which has the well known solution:

$$\Phi = A\cos(m\phi + B) \tag{3.19}$$

where A and B are arbitrary constants. The quantity m is the azimuthal mode number, and for circular periodicity it must take only integer values. Taking the radial part of equations (3.18) given by the terms in F and r gives a Bessel equation, which has the solutions:

$$F(r) = C_1 J_m\frac{ur}{a} + D_1 Y_m\frac{ur}{a}, \text{ in the core} \tag{3.20}$$

and the modified Bessel or Hankel function

$$F(r) = C_2 K_m\frac{wr}{a} + D_2 I_m\frac{wr}{a}, \text{ in the cladding} \tag{3.21}$$

where the C and D are arbitrary constants, while the parameters u and w are defined by:

$$u^2 = a^2(k^2 n_1^2 - \beta^2) \tag{3.22}$$

$$w^2 = a^2(\beta^2 - k^2 n_2^2) \tag{3.23}$$

in order that the mode field can be expressed by Bessel functions in the core and by Hankel functions in the cladding region. On physical grounds the mode field must be localised close to the core and be bound, and this requirement is satisfied by setting the constants D to zero in equations (3.20) and (3.21). This leads to the general solution for a step index fibre given by:

$$\Psi_z = A_1 J_m \frac{ur}{a} \cos(m\phi + B), \text{ in the core} \tag{3.24}$$

$$\Psi_z = A_2 K_m \frac{wr}{a} \cos(m\phi + B), \text{ in the cladding} \tag{3.25}$$

we will return to these equations in Section 3.3.4. So far the relations between A_1 and A_2, and between u, w and β, have not been found, but they will be determined later from boundary conditions.

3.3.3 Normalised frequency and cut-off

In the meantime we introduce the normalised frequency or v-value denoted by v, which it will be recalled is defined by

$$v^2 = a^2k^2(n_1^2 - n_2^2) = u^2 + w^2 \tag{3.26}$$

where the relation with u and w follows from equations (3.22) and (3.23). The v-value is a constant of the lightguide and it gives much information concerning its

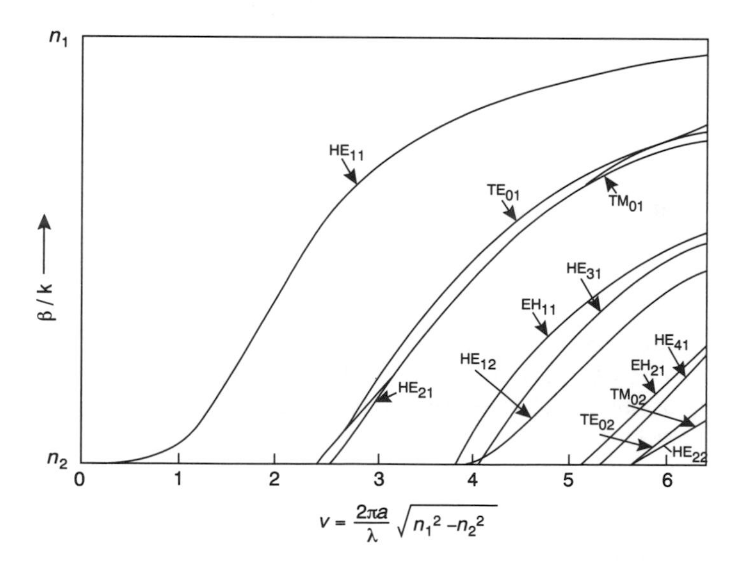

Figure 3.4 Normalised propagation constant versus v-value for a few of the lowest-order modes in a step index lightguide ([5], Figure A.4)

properties. It can be seen from equations (3.22) and (3.23) that acceptable solutions require that the effective index seen by a given mode must lie between n_1 and n_2, so that we have $kn_1 > \beta > kn_2$. For large v-values the modes are tightly bound to the core region where the refractive index is near n_1, whereas lower values correspond to modes in regions further from the core. In the limit the effective index becomes n_2 and w vanishes, the fibre ceases to guide that particular mode and the power is radiated away. This condition is known as cut-off. All modes apart from the fundamental have a cut-off frequency. In the case of the fundamental mode the field still spreads out as the v-value is reduced, but w only approaches 0 in the limit $v = 0$, i.e. at infinite wavelength or zero frequency. Some of these properties are illustrated in Figure 3.4, which presents information for a few of the lowest order modes.

3.3.4 Approximate solutions

The weakly guiding linearly polarised approximation: exact equations for the longitudinal field components are provided by equations (3.24) and (3.25), which can be solved by satisfying the boundary condition that the tangential components of $\boldsymbol{E}$ and of $\boldsymbol{H}$ are continuous at the core–cladding interface. This condition gives four homogeneous equations in the unknown constants A, B, C and D. A solution exists only if the determinant of the coefficients vanishes. This process of matching the fields at the interface, deriving the relations between $u(v)$ or $w(v)$, and finding the transverse components, results in solutions which are too complicated for practical use. Instead, we make use of the fact that the difference between n_1 and n_2 is of the order of 1%; consequently the fibre provides only weak guidance to the transmitted modes. This property enables certain modes to be combined into a set whose transverse fields are essentially polarised in one direction, because the longitudinal components are negligible by comparison. This set of modes is known as linearly polarised or LP modes, and their use simplifies the theory considerably. LP modes should not, in general, be mistaken for the natural modes of the fibre, since they comprise sets of natural or true modes which only become degenerate in the limit $\Delta = 0$. The one exception to this statement is the fundamental mode whose LP solution is a natural mode. The polarisation characteristics of the fundamental mode were discussed in [6]. Concise analytical expressions were stated for the fields corresponding to both low and high values of the normalised frequency v. The results provided the first accurate picture of polarisation of the fundamental mode, and corrected a longstanding error in the literature. In practical fibres the quantity Δ is less than 0.01, and the electric and magnetic fields are virtually orthogonal and plane-polarised everywhere in the fibre cross-section; hence the description weakly guiding.

3.3.5 LP solution for a step-index fibre

We assume two dominant transverse fields, say E_x and H_y, which both satisfy equation (3.13) with solutions identical to (3.22) and (3.23). The boundary conditions at the core–cladding interface are found by matching the transverse field components in the limit $\Delta = 0$, yielding:

$$\Psi(r = a_-) = \Psi(r = a_+) \tag{3.27}$$

where Ψ stands for E_x or H_y. The remaining part of the boundary conditions is to match the longitudinal components; to do so, use is made of the two curl equations of Maxwell, resulting in:

$$\frac{\partial\Psi(r = a_-)}{\partial r} = \frac{\partial\Psi(r = a_+)}{\partial r} \tag{3.28}$$

It follows that the field and its first derivative are continuous across the core–cladding interface. Applying this result to the derived fields (cf. equations (3.24) and (3.25)) shows that the following characteristic equation must be satisfied by every LP mode in a step index fibre:

$$\frac{uJ'_m(u)}{J_m(u)} = \frac{wK'_m(w)}{K_m(w)} \tag{3.29}$$

where the prime denotes the derivative of the function with respect to its argument. The derivatives can be eliminated by making use of the recurrence relations for Bessel functions, resulting in either of the two forms:

$$\frac{uJ_{m+1}(u)}{J_m(u)} = \frac{wK_{m+1}(w)}{K_m(w)} \tag{3.30}$$

$$\frac{uJ_{m-1}(u)}{J_m(u)} = -\frac{wK_{m-1}(w)}{K_m(w)} \tag{3.31}$$

The characteristic equation (3.29) and equation (3.26) taken together define a unique $u(v)$ or $w(v)$ curve for each LP mode, from which all the properties, e.g. field shape, propagation constant or dispersion, follow. A pictorial representation of the characteristic equation is given in Figure 3.5, and it is reminiscent of a Foster reactance curve in network theory. This figure is obtained by plotting the right- and left-hand sides of equations (3.30) separately for the case of $m = 0$. The intercepts give the value of LP_{01}, LP_{02}, and so on. A similar family of curves is obtained for the first

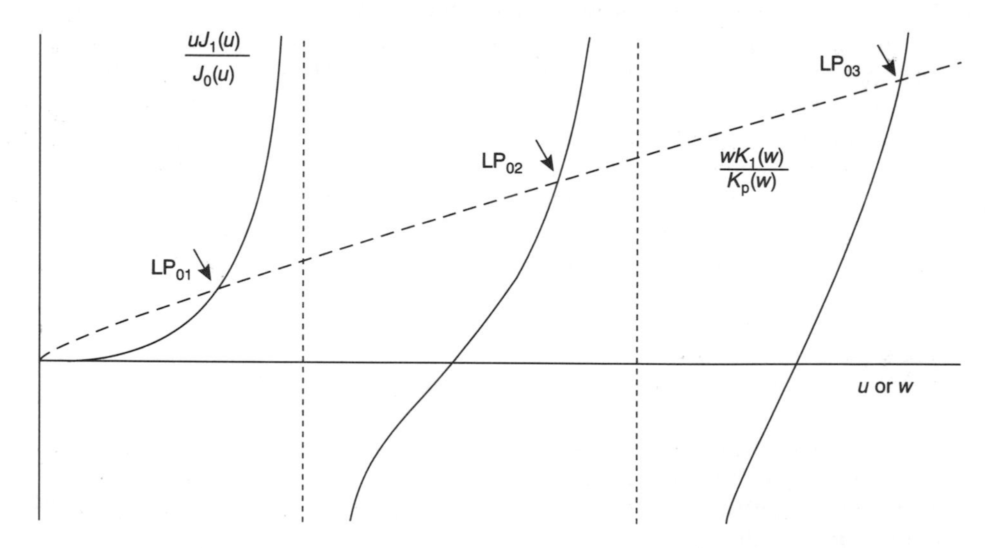

Figure 3.5 Graphical solutions of the characteristic equation (3.30) for $m = 0$ ([4], Figure 6). Reproduced by permission of IEE

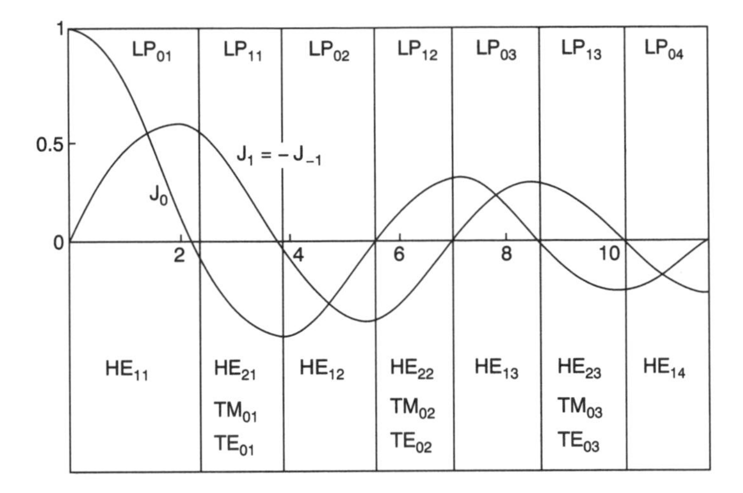

Figure 3.6 Diagram showing the first two Bessel functions $J_0(u)$ and $J_1(u)$ versus u (see Equation 3.30) and relations between the LP and true modes of the electromagnetic field. Reproduced by permission of Opt. Soc. America

few integer values of m, giving LP_{12}, etc. The position of the origin and the scales depend on the chosen v-value and u and w are then constrained by equation (3.26).

Let us now examine this diagram to obtain information about the modes. It is apparent that valid solutions are confined to segments of the abscissa between a zero and the next pole to its right. In reactance terms, the allowable reactances are non-zero, positive and finite. This is confirmed by equation (3.31). Both w and the r.h.s. tend to zero at cut-off; thus u must be a zero of $J_{-1}(u) = 0$. For large v-values both w and the r.h.s. tend to infinity requiring u to be a zero of $J_m(u) = 0$. Now consider a mode LP_{mp}, where m denotes the azimuthal and p the radial mode number, which are measures of the degree of structure present in the transverse field function in the corresponding directions. With m positive u must lie between the mth zero of $J_{m-1}(u)$ and $J_m(u)$, while for $m = 0$ u lies between the $(p-1)$ th zero of $J_{-1}(u)$ and the pth zero of $J_o(u)$. Consider the LP_{01} mode: in this case $0 < u < 2.405$, since this is the first zero of $J_o(u)$; furthermore, this must be the fundamental mode, since u and therefore v both vanish at cut-off. The next propagating mode is LP_{11}, corresponding to the range $2.405 < u < 3.832$, where 3.832 is the first zero of $J_1(u)$. It is evident that v must be kept below 2.405 if single-mode operation is desired. To round off this consideration of the characteristic equation and its solutions for step-index fibres, we note that LP_{01} corresponds to the natural mode HE_{11}, whereas LP_{11} corresponds to the TE_{01}, TM_{01} and HE_{11} natural modes. The relations between the lower-order LP modes and the true modes of the electromagnetic field and the corresponding intervals on the u-axis are illustrated in Figure 3.6 [7].

3.3.6 Field shape and the Gaussian approximation

For the fundamental mode the field shape is bell-like, and is illustrated for two v-values in Figure 3.7.

In the core region the shape is that of a Bessel function of type J, and this is matched to a modified Bessel K-type shape in the cladding. It can be seen that the

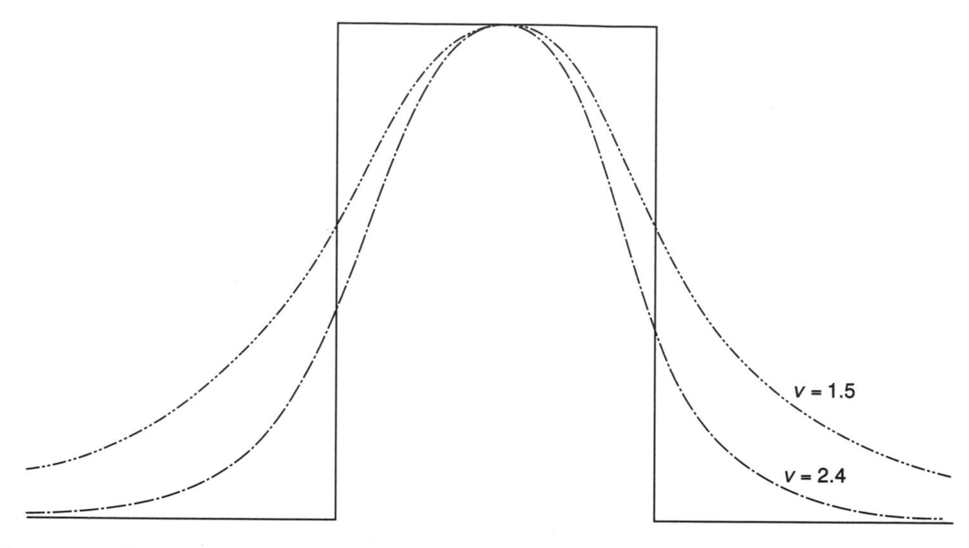

Figure 3.7 Field distribution of fundamental mode versus radial distance from fibre axis for two v-values ([4], Figure 7). Reproduced by permission of IEE

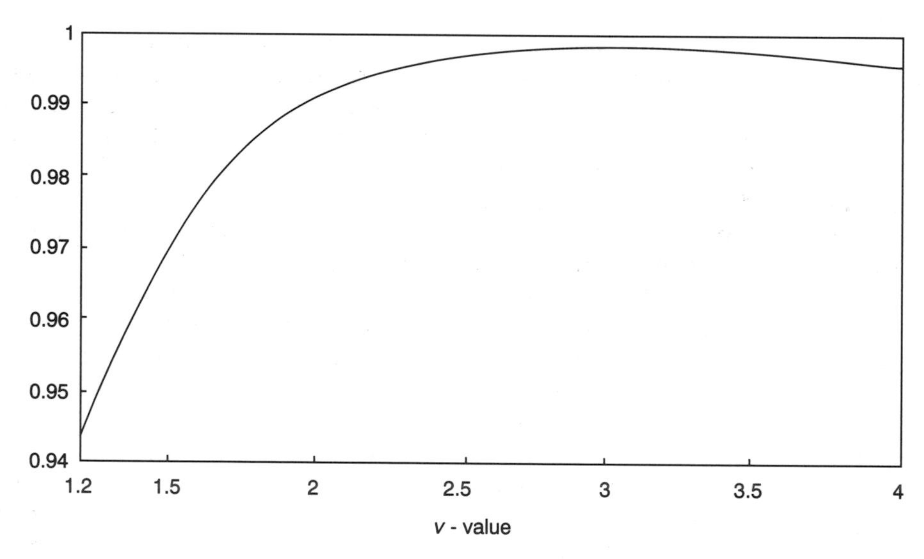

Figure 3.8 Optimum launching efficiency versus v-value of a Gaussian beam into a step index fibre ([4], Figure 8). Reproduced by permission of IEE

field extends some way into the cladding, and that even at cut-off ($v = 2.5$) only some 80% of the power propagates in the core. For lower values of v such as 1.5, the fraction is less and the field extends farther into the cladding. The overall shape looks somewhat Gaussian, and it is possible to launch a Gaussian beam into a step index fibre very efficiently (up to 99.65%), so that the fundamental mode shape is often approximated by the particular Gaussian beam which maximises the coupling efficiency, as shown in Figure 3.8.

A 1981 paper in the *Bell System Technical Journal* [8] gave a detailed treatment of the electromagnetic fields, field confinement and energy flow in dispersionless single-mode lightguides with graded-index profiles.

3.4 REFERENCES

1 FEYNMAN, R.P.: '*QED. The strange theory of light and matter*'. Penguin Books, 1990.
2 MAHLKE, G., and GÖSSING, P.: '*Fiber Optic Cables: Fundamentals, Cable Engineering, Systems Planning 2nd Edition*'. Siemens, Munich, Germany, 1993
3 GLOGE, D., and MERCATILI, E.A.J.: 'Multimode theory of graded-core fibers'. *Bell System Technical Journal*, **52**, 1973, pp. 1563–1578.
4 WRIGHT, J.V.: 'Propagation in optical fibres'. *IEE Vacation School on Optical Fibre Telecommunications*, 1986.
5 KECK, D.B.: 'Optical fiber waveguides' in '*Fundamentals of Optical Fiber Communications*'. ed M.K. Barnoski Academic Press Inc., N.Y., 1976.
6 ZHENG, X.-H., HENRY, W.M., and SYNDER, A.W.: 'Polarisation characteristics of the fundamental mode of optical fibres'. *IEEE/OSA Journal of Lightwave Technology*, **6**, no. 8, Aug. 1988, pp. 1300–1305.
7 GLOGE, D.: 'Weakly guiding fibers'. *Applied Optics*, **10**, 1971, pp. 2252–2258.
8 PAEK, U.C., PETERSON, G.E., and CARNEVALE, A.: 'Electromagnetic fields, field confinement and energy flow in dispersionless single-mode lightguides with graded-index profiles'. *Bell System Technical Journal*, **60**, no. 8, Oct. 1981, pp. 1727–1743.

4

SINGLE-MODE FIBRE AND CABLE CHARACTERISTICS

4.1 INTRODUCTION

The preceding chapter was very much concerned with the physical and mathematical aspects of signal propagation in single-mode fibre waveguides. The emphasis in this and the next chapter changes to practical aspects of fibres by themselves and when cabled. A short description is given of fibre materials, fabrication and drawing with reference to four very widely used fabrication techniques. This is followed by sections dealing with reliability, refractive index profiles and optimised fibre designs. Advances in fibre design and manufacture continued to be reported during the early 1990s. The penultimate section deals with the topic of microbending loss in some detail because of its important bearing on system loss budgets, again based on publications from the early 1990s. The last section covers macrobending and its system implications.

Systems engineers are of course concerned not with the performance of the best available fibres, but with that obtained routinely from production runs. The performance of production fibres must conform to the appropriate specifications, which will cover all the necessary points. As an example we consider the points covered for a single-mode fibre to a British Telecom specification (CW1505E).
The specification covers: optical, geometrical and mechanical properties. Notes are then given on the definitions of these properties and the corresponding test methods.

Additional specifications and performance data cover fibre materials, coating materials, optical characteristics, mechanical characteristics, environmental exposure and hydrogen performance. Full information on the performance of sm fibres in hydrogen is covered in an Application Note.

These are followed by a diagram relating mode-field diameter to cut-off wavelength.

Until 1988, at least, fibres were almost exclusively fabricated from silicate glasses, the vast majority by the vapour deposition of silica, germania and related oxides, usually in a sequence of layers within a tube of pure silica. This tube is then collapsed, pulled and coated, processes which will be described in some detail later in this chapter. By a coat is meant the layer applied directly to the outside of the cladding layer during production. This coating must be removable for coupling light into or out of the fibre, and also for splicing. It may be composed of several layers of plastic, and must be applied evenly over the entire length without bare patches or variations in thickness. Details will be described later.

4.2 MATERIALS, FABRICATION AND FIBRE DRAWING

It is widely known that silicon is a very abundant material, whereas copper is not. In fact, the solid outer crust of the Earth consists of 50% oxygen and 25% silicon by weight. The great abundance of these two elements, numbers 8 and 14, respectively, in the periodic table, arises from the composition of the crust, which is mainly quartz, and its compounds with the metallic oxides, the silicates.

4.2.1 Materials

Rock crystal, the purest form of quartz, is anisotropic in its optical and mechanical properties. In contrast with this, fused silica glass is an amorphous or non-crystalline glassy solidified melt of silicon dioxide, which only appears to be a solid because of its high viscosity. It therefore has no melting point but only becomes softer at higher temperatures, and vaporises directly from this state without going through the liquid state.

The viscosity is an essential property for the entire production and forming process of glass [2] of Chapter 3. This property is used to describe the internal friction or resistance to flow. The units commonly used are decipascal seconds, and the property is denoted by η, and so:

$$1 \text{ decipascalsecond} = 1 \text{ g/cm s}$$

In fused silica glass the viscosity decreases monotonically with rising temperature, as shown in Figure 4.1.

Although there are no obvious points of note on the curve, for technical reasons several temperatures have been identified with the help of the logarithm of the viscosity. These particular values are listed in Table 4.1.

The upper and lower annealing points are at the boundaries of the transformation range from the viscoelastic condition of a melt to the brittle state of fused silica glass. Within the range of the softening point the shape of the fused silica glass body changes under its own weight.

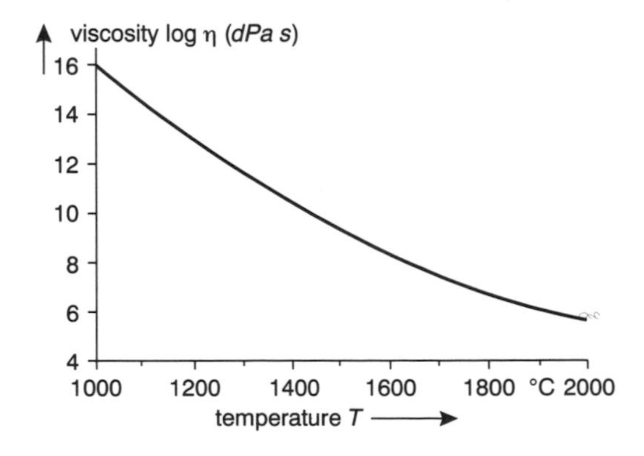

Figure 4.1 Viscosity of fused silica glass versus temperature ([3.2], Figure 3.1)

Table 4.1 Description of temperature values for fused silica glass in terms of log η ([3.2], Table 3.2)

Viscosity log η	Description	Temperature in fused silica glass (°C)
7.6	Softening point	1730
13	Annealing point	1180
14.5	Strain point	1075

Many preparation techniques for optical fibres have been investigated, but they can be grouped into two categories, broadly speaking. The majority of installed fibre manufacturing facilities use vapour phase reactions to deposit pure silica or silica doped with various additives, e.g. GeO_2, P_2O_5, B_2O_3 or F. The other category involves liquid phase techniques for melting and converting low-temperature glasses such as sodium calcium silicates or sodium borosilicates into fibres. These glass systems offer great flexibility in the choice of optical, mechanical and chemical properties. However, because they are melted in crucibles from powders, it is difficult to avoid some contamination, and thus loss. The breakthrough in the production of very low loss fibres was only achieved when the liquid phase methods were replaced by vapour phase techniques, the first of which was pioneered by Corning Glass Works of America in 1970. Liquid phase methods will therefore not be described.

4.2.2 Fabrication techniques

The basis for all vapour phase techniques for the fabrication of optical fibres is a gas phase oxidation reaction on to a substrate to produce a solid preform, which is later drawn down into a fibre. The systems used in these techniques are based on pure silica doped with a suitable additive to modify the refractive index, and confer other desirable properties. The effects of various dopants on the refractive index are shown in Figure 4.2.

A major advantage of vapour phase techniques over the older liquid phase methods is that the starting materials are volatile liquids which can readily be distilled to reduce the concentration of most transition metal impurities to below 1 part in 10^9. Differences between the various techniques can be subtle, but they can be clarified by considering the means by which energy is supplied for the reaction; these can be a flame, a heated tube or various types of plasma.

The production of fibres usually involves several steps, making it possible to selectively optimise mechanical, geometric and optical properties. In addition, this multistep process allows fast and economic mass production, as now required to meet the demand for fibres.

All of the techniques in use by the mid-1980s began with the production of a preform. This is a glass rod consisting of a core surrounded by a cladding, and is a very much scaled up version of the final fibre, with the same geometric proportions and refractive index profile. The final fibre is produced from the preform by heating one end to a higher temperature than the end from which the fibre is drawn. At the same time a coating is applied as a protective sleeve. After describing these methods in outline we will look at some technologies introduced in the late

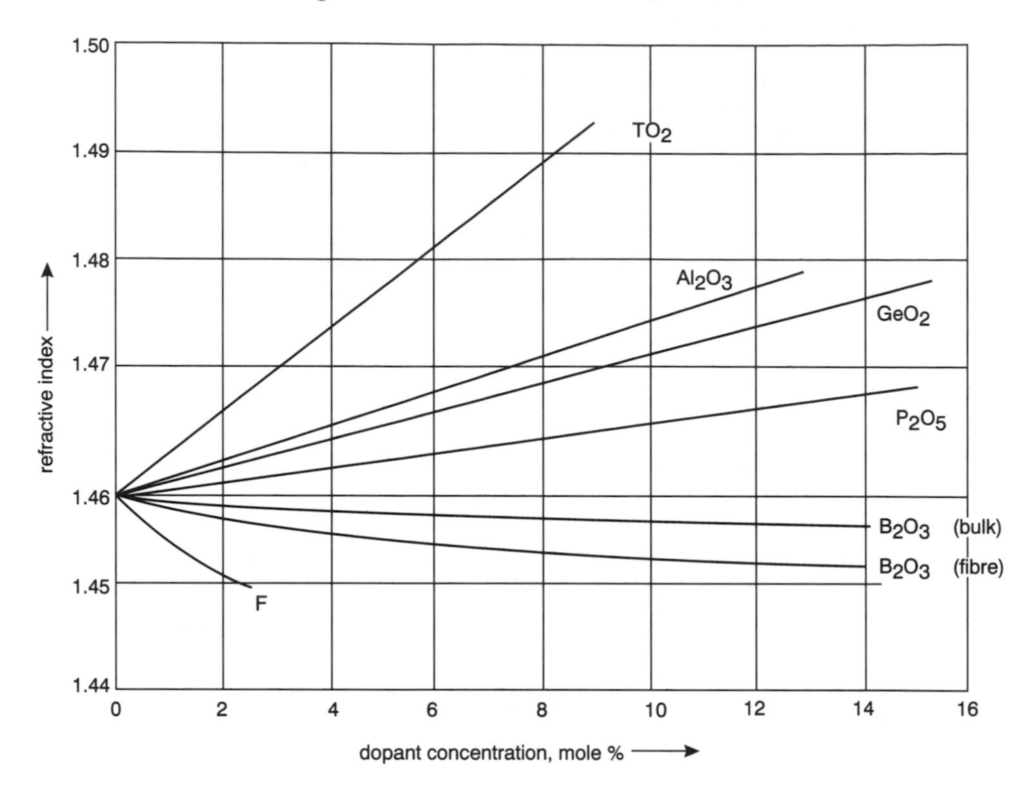

Figure 4.2 Refractive index versus dopant concentration in silica glass ([1], Figure 11). Reproduced by permission of IEE

1980s. A few typical refractive index profiles obtained using an inside tube technique and chemical vapour deposition are shown in Figure 4.3 and are reproduced from a late 1988 publication [2]. This publication also contained a section headed 'Comparative analysis', giving some quantative data current in 1988. For instance, for the four common methods to be described in the next section, the average losses ranged from 0.35 to 0.38 dB/km at 1300 nm, and at 1550 nm the range was 0.19–0.21 dB/km. Extending the comparison to other parameters such as process flexibility, fibre tensile strength and geometrical variations may be useful. In the future the authors suggest that research will be increasingly targetted on reducing production costs without at the same time degrading performance. Key factors were listed as production yields, price of raw materials, material yields, labour costs (preform size, production rate, etc.) and costs of machinery. The paper goes on to describe attempts to improve both inside-a-tube and tubeless techniques. Ericsson, too, have developed a new process called the intrinsic microwave-heated chemical vapour deposition in which a microwave cavity replaces the torch. According to an article in *Lightwave* for August 1989, the yield of a 50 cm preform in the new method is about 50 km, compared with only 15 km for the same size preform used with the current *modified chemical vapour deposition* process.

Another method for improved yields at lower cost was described in late 1988 by authors from Philips Research Laboratories in Aachen [3] in which a hybrid technology was used for the preparation of standard and dispersion-shifted sm fibres. Instead of the whole fibre being fabricated by the plasma-activated chemical

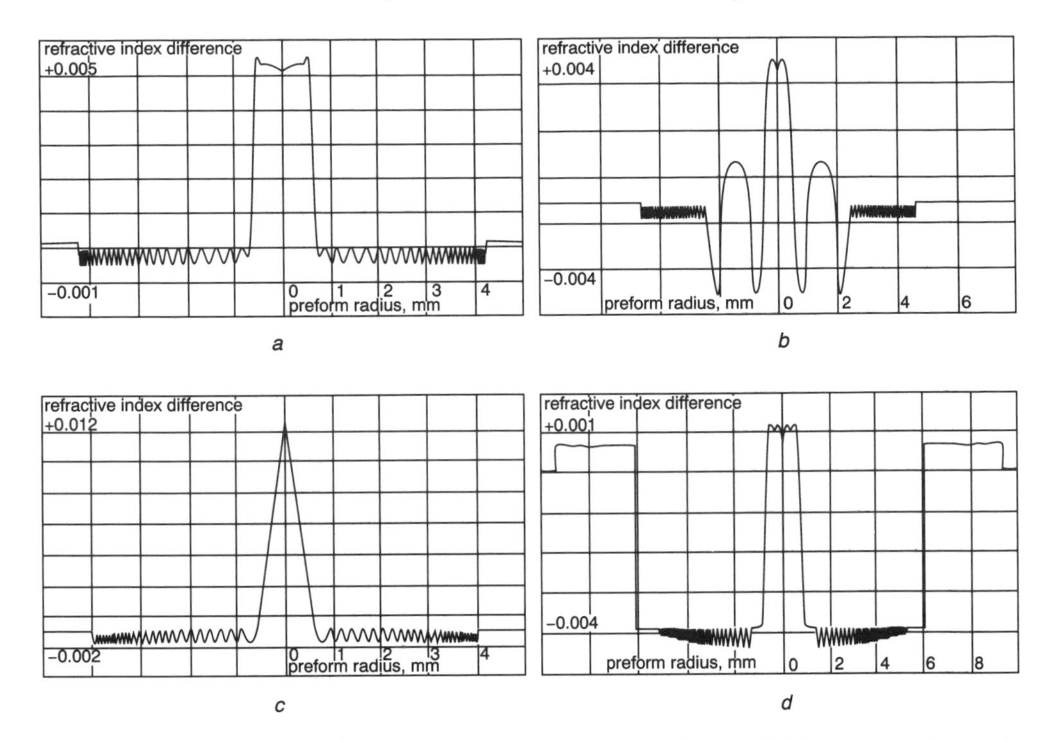

Figure 4.3 MCVD tailored profiles: (*a*) matched cladding, (*b*) dispersion flattened, (*c*) dispersion shifted, (*d*) 'non-doped' core ([2], Figure 3). Reproduced by permission of Alcatel

vapour deposition process, abbreviated to pcvd, only the core region is so fabricated. The cladding is added by substrate and jacket tubes are obtained from low-cost shaping and sintering techniques. Owing to *in situ* plasma etching, no disturbance is caused by the interfaces. Low loss is achieved with less than 5% of the fibre volume made by the pcvd technique. This hybrid technique allows economic mass production of preforms, yielding more than 300 km of fibre, and so makes the introduction of fibres to the local area network more competitive with the existing copper. The concept of the hybrid technique was described first, and tabular data on the parameters of conventional pcvd process and of the hybrid were included. Comparisons were extended in a second table and they showed that the newer technique allowed a convenient choice of the major process variables needed to prepare large preforms. The hybrid method rests on the availability of high purity silica tubes, the ability to smooth and clean the optically sensitive surfaces and the possibility of preparing a central preform rod without bulk contamination from the collapse phase. The hybrid technology is appropriate for large preforms containing several hundreds of km of fibre, and is equally suited to all types of dispersion-modified and step-index single-mode fibres.

The methods in use to make preforms by melting glass fall into four main categories, as given below:

(1) outside vapour deposition (ovd)
(2) vapour axial deposition (vad)

(3) modified chemical vapour deposition (mcvd)
(4) plasma chemical vapour deposition (pcvd)

The first two are very similar processes and so are the last two, as will be seen later. Let us look at each of these in turn, following the treatment given in [1] and [4].

Outside vapour deposition method

Preform production takes place in two stages. In the first stage a target rod of fused silica glass, Al_2O_3 or graphite is rotated about its longitudinal axis in a lathe, and a narrow zone heated on the outside by means of an oxy-hydrogen or propane gas burner, as illustrated in Figure 4.4. [4].

A carrier gas of oxygen saturated with the desired dopants is fed through the traversing burner and transformed into the corresponding oxides. Typically the reactants would comprise one or more of $SiCl_4$, $GeCl_4$, BCl_5 or PCl_3. The metal halides react with excess oxygen at the very high temperature in the flame to produce a stream of very fine particles of soot (about 0.1 μm diameter) which are deposited on the rotating rod. The rod itself is also traversed axially so that a porous glass preform is built up layer by layer. Each layer can be doped differently from the previous layers, for example, by adding a certain amount of impurity material to the basic SiO_2. This allows the desired index profile to be built up. For instance, graded index profiles are obtained by continuously reducing the doping of the core with GeO_2 from the first layer outwards, until pure SiO_2 is deposited for the cladding. Other profiles are obtained by obvious modifications. As soon as enough layers have been deposited to form the core and the cladding, the process is stopped, and the cylindrical tube made of soot particles is removed from the target rod.

The second stage involves heating the preform in segments over its entire length to the melting point. At temperatures between 1400 and 1600°C the preform collapses or shrinks to a solid, bubble-free transparent glass rod in which there is no longer a central hole; this rod is called a blank. It is then sintered and during this stage the preform is continually flooded with gaseous chlorine as a drying agent to remove all traces of water from the glass. The ovd process allows great

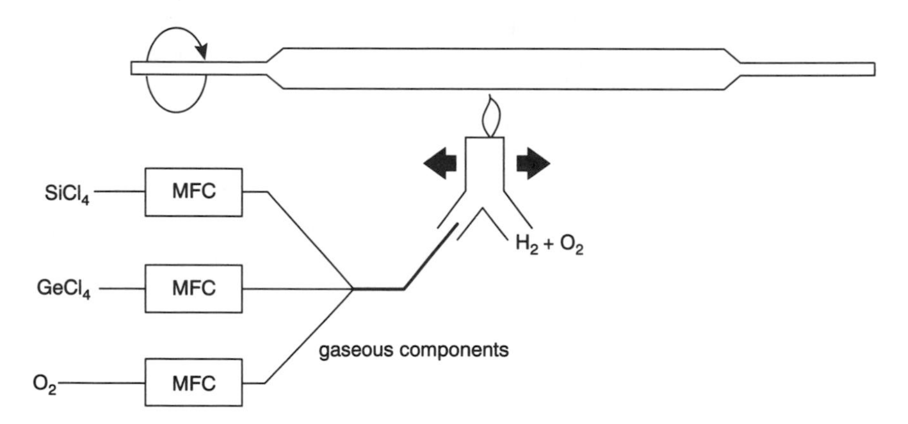

Figure 4.4 Schematic arrangement of outside vapour deposition (ovd) method ([4], Figure 1). MFC = mass flow controller. Reproduced by permission of IEE

flexibility in glass composition for both core and cladding. The ovd process yields preforms with capacities exceeding 100 fibre-km.

Vapour axial deposition method

This particular technique (vad) was pioneered by the Japanese. In the vad method the soot particles created in the oxy-hydrogen burner(s) are deposited on the end of a rotating fused silica glass rod as shown in Figure 4.5.

The particles are formed from vaporised reactants such as $SiCl_4$ or $GeCl_4$, or others shown in the sketch. The porous preform so created is drawn upwards at a rate which maintains a constant distance between the burners and the preform which is growing axially. Several burners are used simultaneously to produce the correct refractive index in core and cladding. Different profiles can be produced by altering the distance, the temperature during deposition and the construction of the burners. As in the ovd process, following deposition the porous preform is collapsed to a transparent blank in a ring shaped furnace, and dried as before. Fibres with a thicker cladding can be manufactured by sliding a fused silica glass tube as a cladding over the preform, in a similar manner to the rod-in-tube technique. This process also yields preforms with capacities exceeding 100 fibre-km.

The development and performance of fully fluorine-doped sm fibres using the vad process were described by workers from the Furukawa Electric Company in [5]. A high yield of fibre having a per-unit loss below 0.20 dB/km at 1550 nm was achieved in standard fibre in which fluorine doping was used in the cladding and a small amount was added to the core. A similar approach was applied successfully to dispersion-shifted fibres with a dual shape core index-profile and gave encouraging results which indicated that the transmission loss might be reduced to as low

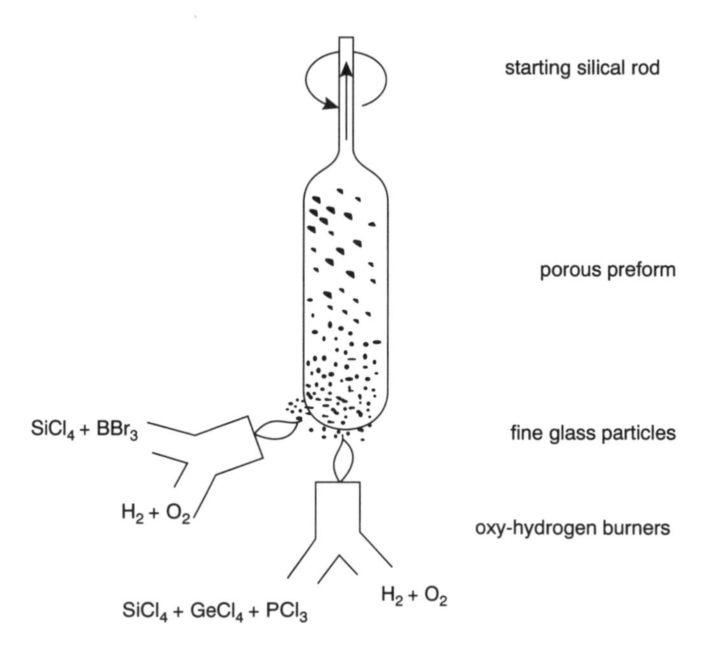

Figure 4.5 Schematic arrangement of vapour axial deposition (vad) method ([4], Figure 2). Reproduced by permission of IEE

as 0.20 dB/km in this type of fibre as well as the standard type. Fluorine doping throughout the entire cross-section of the fibre was shown to be a key requirement. A newly developed vad-based process allowed doping down to 0.6%. The paper described the dopant design and the performance for standard fibres, before discussing the design of the refractive-index profile in dual-shape-cores and the fabrication and dispersion control in dispersion-shifted fibres, and the performance of this type. The characteristics of the fluorine-doped fibres with respect to hydrogen diffusion and reaction and breaking strength were treated and results presented as for the earlier parts of the paper. In conclusion, the improved process was found to minimise transmission loss and to offer high protection against changes induced by hydrogen, and it therefore looked promising for various applications in terrestrial and underseas systems.

Modified chemical vapour deposition method

The mcvd method is the preferred technology at AT&T and at Alcatel; a paper describing the manufacture of such fibres at Alcatel was published in 1989 [6]. Preforms are produced in two steps, in the first of which a fused silica glass tube is rotated around its longitudinal axis in a lathe while a narrow zone is heated by means of a oxy-hydrogen burner. The burner traverses over a length of the preform, and the basic arrangement is sketched in Figure 4.6.

Meanwhile, oxygen and the gaseous halide compounds such as $SiCl_4$, $GeCl_4$, PCl_4 needed as dopants flow through the tube. The halides in the tube oxidise with excess oxygen to form a fine doped silica soot which adheres to the inner wall of the preform in clear glassy layers at the operating temperature of 1600°C. Reactant flow rate and mix is adjusted to create the desired profile. The tube itself forms the outer part of the cladding. The vapour phase deposition on the inside surface forms the inner layers of the cladding and all of the core.

The second stage of the mcvd method is to heat the preform to 2000°C in segments along its length, causing it to shrink to a rod. If the reactants have been kept free of hydrogen no special drying process is needed because the hydrogen-rich gases used for heating only come in contact with the outside of the tube. Any other environmental contaminants also have no effect. This process yields 50 cm preforms with capacities around 15 fibre-km. In the late 1980s Ericsson of Sweden developed an improved process called intrinsic microwave-heated chemical vapour deposition, or imcvd, which yields some 50 km of fibre from a 50 cm preform.

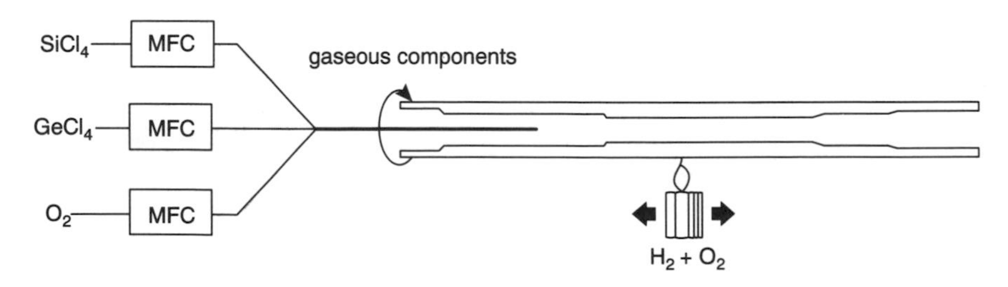

Figure 4.6 Schematic arrangement of inside vapour deposition method ([4], Figure 3). MFC = mass flow controller. Reproduced by permission of IEE

Plasma chemical vapour deposition method

This process was developed by Philips, and it has many features in common with the mcvd process described above. The difference lies in the reaction technique employed. The oxy-hydrogen burner is replaced by a low pressure (about 10 mbar) plasma formed when the gas is excited by RF energy at microwave frequencies, and the gas becomes ionised. When these charge carriers recombine considerable energy is released as heat, which is used to melt the high melting point materials. The local plasma is moved along the quartz or silica tube to oxidise the vapours. The halides react with the oxygen to form SiO_2, and the resulting soot particles are deposited directly as layers of glass at around 1000°C. Because the plasma region can be moved very quickly back and forth along the tube, more than a thousand very thin layers can be deposited, and this clearly allows a better approximation to an ideal profile to be obtained. The essential parts of the apparatus are shown in Figure 4.7.

After deposition has been completed, the tube is collapsed as in the mcvd method, and it is then protected with additional glass material. The preform capacity of this process is also around 30 fibre-km.

The first two methods produce a blank of soot (glass particles) containing hydroxyl impurities produced by the burners. These impurities must be removed by drying, followed by sintering to form a solid void-free glass preform.

The porous mass of soot material is placed in a consolidation furnace where the preform is heated to temperatures ranging from 900 to 1600°C while drying gases are circulated around and through the preform. The commonly used agent for this purpose is chlorine gas. After drying, the temperature is raised still higher, the soot material sinters and the central hole left by the withdrawal of the target rod closes. In the ovd process the entire cross-section of the fibre is vapour-deposited and sintered. In the vad method the sintered core preform is usually jacketed with a silica tube to adjust the cladding to core ratio of the fibre to the desired value. When consolidation is complete the result is a solid transparent preform with only a few parts in 10^9 of hydroxyl contaminant left in the silica-germania matrix. A diagram of this process is shown in Figure 4.8.

The next stage is fibre drawing.

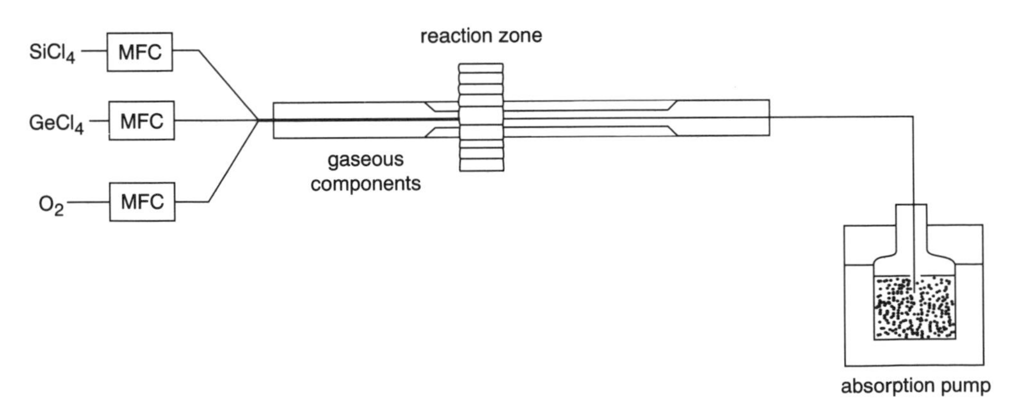

Figure 4.7 Schematic arrangement of plasma chemical vapour deposition (pcvd) method ([4], Figure 4). MFC = mass flow controller. Reproduced by permission of IEE

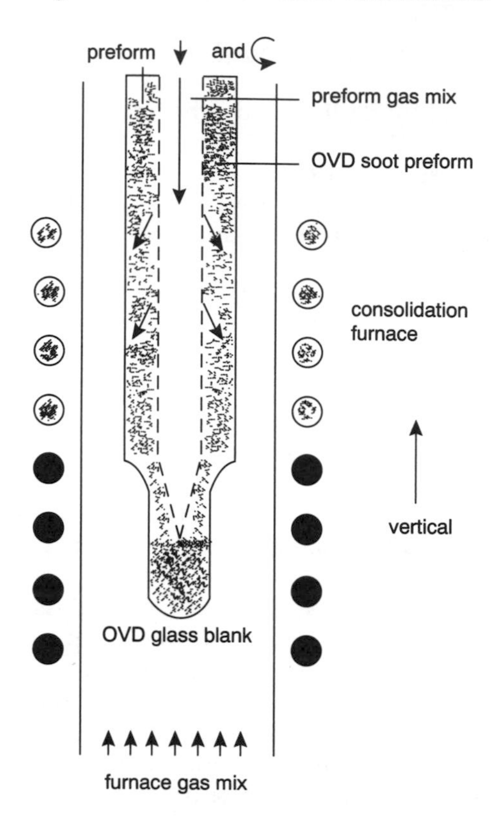

Figure 4.8 Schematic arrangement for drying and sintering ([4], Figure 5). Reproduced by permission of IEE

4.2.3 Fibre drawing

The sintered preform is placed inside a high-temperature furnace and the tip is heated to 2100–2200°C, then free drawn to a hair-thin fibre with outside diameter 125 μm for use in telecommunications. A representation of the process is given in Figure 4.9.

Below the furnace there is a fibre diameter monitor followed by the application of one or two layers of polymeric coating materials. The coated fibre then passes through the capstan drive and on to the take-up drum. The capstan pulling apparatus is used to draw the fibre. The diameter monitor controls the speed of the capstan to maintain a constant diameter. Both drawing speed (typically 200 m/min) and the feed mechanism are precisely adjustable by means of an automatic control system. During drawing from the relatively large diameter preform the core and cladding glasses keep their relative geometries, despite the reduction in radial dimensions of perhaps 300 to 1. Thus, the core to cladding ratio and the refractive index profile built up in the preform are faithfully reproduced in the fibre itself. When the coating has been hardened by heating or by uv curing, the tensile strength of the fibre is checked as it passes through an in-line test monitor. This is done by running the fibre over an array of rollers which apply a precisely measured tensile force; the fibre must withstand this minimum stress before it is wound on to a drum.

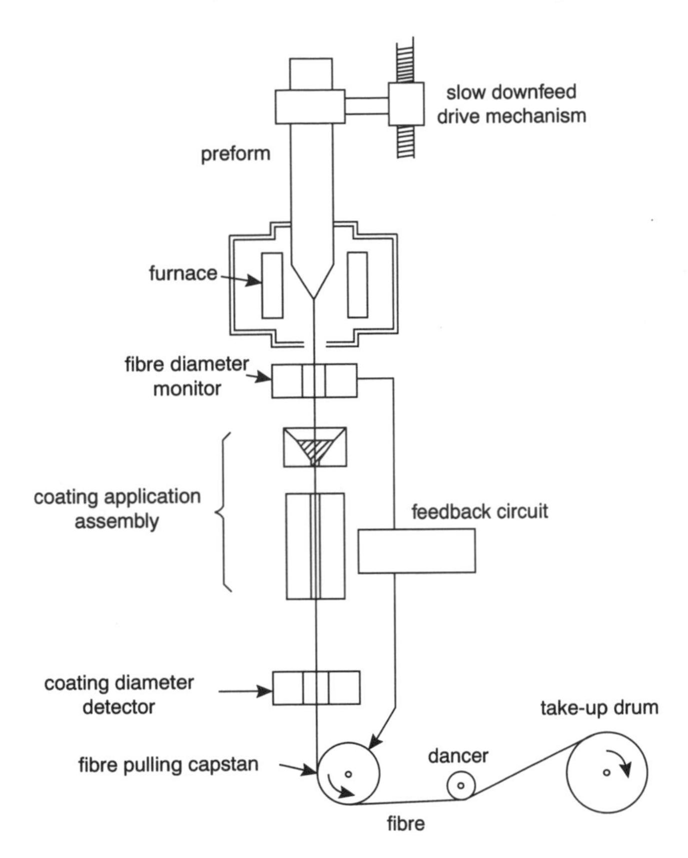

Figure 4.9 Schematic arrangement for fibre drawing ([4], Figure 6). Reproduced by permission of IEE

During the fibre drawing process specific components may be forced to occupy non-equilibrium states at low temperatures owing to rapid quenching from high temperatures. These states may give rise to so-called drawing-induced absorption bands [7]. Other parameters such as index profile and cut-off wavelength do not vary appreciably for drawing temperatures in the range 1940–2200°C. Adjustments of drawing speed and temperature have been found to improve the fibre properties by avoiding undesired effects.

The influence of the drawing process on the lifetime of optical fibres was the subject of a 1989 paper from Philips Research Laboratories [8]. Strength and lifetime measurements were made on fibres with different residual tensile stresses at the surface, these stresses were measured optically, and the results were included in a theory for fibre strength and lifetime. The data from the strength measurements were compared with those obtained from measurements on pure silica fibres without any residual stress. In the prediction of lifetime the residual stress in the outer region of the fibre has to be included, but experiment revealed that the decrease in the intrinsic strength was much larger than the increase in residual stress in the outer region, and it depended on the drawing conditions. The phenomenon was not well understood.

Service lifetimes of optical fibres are usually extrapolated from time-to-failure measurements carried out under laboratory conditions, but in order to calculate the correct time, the stress in service as well as the residual stress must be taken into account. Many types of sm fibres have a tensile stress in their surface regions. The aim of the authors [9] was to check quantitatively the consistency of experimental results obtained in static and dynamic fatigue tests and the theory by including measured values of the residual stress. The work demonstrated that the drawing force and the fibre structure are of vital importance for the internal stress in a fibre. Neglect of the internal stress in commonly used single-mode fibres led to predictions of lifetime which were ten times too high. It was also demonstrated that the intrinsic strength depended on the drawing force, but was independent of fibre structure if the residual stress in the outer region of the fibre was taken into account.

4.2.4 Coatings for optical fibres

A coating forms an integral part of a lightguide, and serves a number of purposes, which will be outlined shortly. The intrinsic strength and the optical properties of the fibre must be preserved; this requires that the coating be applied immediately after the fibre leaves the furnace and monitor, as shown in Figure 4.9. At this stage in the process the fibre is still in a clean environment so that the risk of surface contamination is minimised. It is, of course, necessary to apply the coating concentrically and homogeneously without damage to the pristine surface of the fibre, and for the coating to become solid before the newly coated fibre reaches the take-up capstan.

At the end of the 1980s the most common technique was to apply a liquid polymer with either an open bath withdrawal, or a die, system. Solidification takes place on curing either by heat or by uv radiation as indicated earlier.

Broadly stated, the contributions which the coating must make to the performance of the fibre in use are as follows.

Abrasion protection: the coating must provide enough abrasion resistance to protect the fibre from subsequent handling and cabling without loss of strength.

Microbending loss protection: the attenuation of a fibre can be increased significantly by the presence of bends of small radius which cause light to leak into the coating. The periodicity of bending at which loss becomes significant is around 1 mm. External forces that deform the fibre axis can give rise to a variation of refractive index along the fibre. Such deformations can induce mode coupling in multimode fibres and loss through radiation in both types.

Static fatigue protection: the surface of the glass must be protected from external particles which could damage it by initiating flaws which weaken the glass dramatically, and the polymer itself must be free from injurious agents. The coating must also provide chemical protection against liquid or gas ingress which might enhance any stress corrosion.

Experience to 1990 suggested that no one coating can satisfy all three requirements simultaneously, indeed some requirements were in conflict with others. For example, to isolate the fibre mechanically from its environment and thus avoid microbending loss, suggests a soft and compliant covering material which cushions the fibre. Such a material would not afford protection from abrasion. A compromise is made by using a composite structure consisting of a soft inner layer surrounded by a hard outer shell. The coating must be applied in such a manner as to avoid non-uniformly distributed stresses when the fibre is subjected to thermal cycling and ageing. Defects such as voids, lumps and uncoated areas must also be avoided, as they contribute to microbending loss, and much work has been done to produce concentric, bubble-free coatings.

As an illustration, the system adopted by one manufacturer uses an epoxy acrylate uv-cured coating applied in two concentric layers. The inner layer has a low Young's modulus, and serves to isolate the fibre from external compressive stresses,

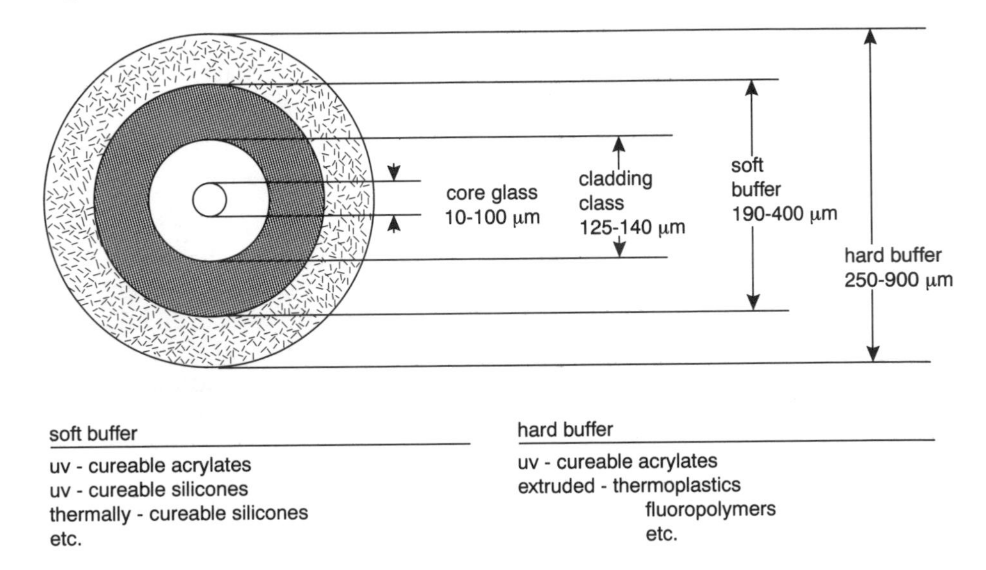

Figure 4.10 Schematic arrangement of composite buffered coatings ([4], Figure 7). Reproduced by permission of IEE

Table 4.2 Properties of coatings and their system implications ([4], Table 1). Reproduced by permission of IEE

Properties of Coating	Impact
Refractive Index	Mode stripping, concentricity monitor
Viscosity	Processability
Surface tension	Processability
Particle size/count	Processability/strength
Shelf life	Stability
Gaseous evolution	Processability, hydrogen evolution
Glass transition temperature	Temperature dependance of attenuation
Elastic modulus	Compressibility, abrasion resistance
Coefficient of thermal expansion	Temperature dependance of attenuation
Moisture uptake	Ageing
Chemical resistance	Ageing
High temperature resistance	Processability, ageing
Adhesion to glass	Strippability, strength
Cure speed	Economics
Coefficient of friction	Compatibility with subsequent jacketing layer

Table 4.3 Properties of fused silica glass ([3.2], Table 3.3)

Property	Unit	Value
Density γ	$\frac{g}{cm^3}$	2.20
Young's modulus E	$\frac{N}{mm^2}$	72 900
Shear modulus G	$\frac{N}{mm^2}$	33 000
Linear thermal expansion coefficient	K^{-1}	5.5×10^{-7}

to improve lateral load resistance and to ease splicing by possessing low adhesion to the surface of the cladding so that it can be stripped off mechanically without difficulty. The outer coating layer has a high modulus to provide sufficient rigidity to prevent buckling induced by shrinkage, to give abrasion resistance and flexure. A diagram of composite buffered coatings is given in Figure 4.10.

A summary of the important attributes of coating properties, and their effects on the processability and performance of optical fibres, is presented in Table 4.2.

Material properties of fused silica glass

Fused silica glass is an isotropic medium so that its physical properties are identical in all directions and are well known, in particular its behaviour when subjected to rapid changes of temperature. Due to the extremely small linear thermal coefficient of expansion, this type of glass is extraordinarily resistant to temperature changes. This coefficient and others related to the physical properties are listed in Table 4.3. Numerical examples for the values involved were given in [2] of Chapter 3.

4.3 FIBRE RELIABILITY

After drawing and coating the fibre is ready to be proof-tested. The mechanical and elastic properties and failure mechanisms differ from those for copper or steel wires. In particular, optical glass fibres remain within the elastic limit for a few percent strain, above which they fail in brittle tension. The strength of fibres is very dependent on the quality of the surface, since it is mainly governed by the size of flaws present. Hence the stress laid earlier on the need to keep the surface as pristine as possible. It is impossible to avoid them completely, and they can grow under the influence of stress, leading to a weakening of the fibre. This effect is accelerated if moisture is present. Studies of flow growth mechanisms and accelerated ageing experiments have shown that in order to achieve fibre lifetimes in the range 25–40 years, such as are required for telecommunication plant, the residual stresses should not exceed 25–30% of the proof stress for each fibre.

The reliability of fibres is governed by the manufacture and the subsequent handling. Clearly, a fibre must be strong enough to survive the short-term stresses occurring during fabrication and cabling, and on installation, as well as the long-term stresses that can be present during the working life of 25–40 years. The short-term strength of fibre is limited by the brittle nature of glass, and the long-term strength by the process of static fatigue. Let us explore these in a little more depth. Brittle behaviour leads to a strength dependence on flaws present in the glass; since their size and location are random throughout the material, the strength of an optical fibre is a statistical property. On the other hand, static fatigue results from ageing, i.e. the dependence of strength on time, possibly assisted by the gradual growth of surface flaws due to moisture.

To ensure that fibres exceed a minimum strength the entire length is proof-tested. This consists of subjecting a fibre to a tensile stress greater that any values than can be anticipated in manufacture, storage or installation. The proof stress must be sufficiently high to ensure an adequate service life in adverse environments such as humid atmospheres. Any fibres which break during testing are discarded to ensure that all fibres which are used satisfy the test.

In service a fibre cable is subject to a number of environmental factors which must be considered at the design stage in order to guarantee survivability over the planned lifetime: 25 years in the case of submerged systems. Factors which may limit the life of an optical fibre include mechanisms which cause the attenuation to change and failure due to static fatigue, and these were treated in [9].

Changes in fibre loss may result from the absorption of hydrogen, fibre stress induced by mechanical deformation or thermal effects and ionising radiation. We will now remark briefly on each in turn.

Considerable research into the loss mechanism and into sources of hydrogen in cables followed the first reporting in 1983 that this gas can increase attenuation. Hydrogen can diffuse fairly readily into silica fibres, but hermetic fibres coated with a carbon layer some 50 nm thick can increase the diffusion time beyond the lifetime of even an undersea cable. This effect will be considered in a later chapter.

Fibre loss also increases with mechanical deformation, in particular with bending, and although the coatings are designed to protect the fibre it is not practical to eliminate bending completely. In particular, thermal expansion of the materials used may well induce a continuous series of small bends along the entire length of the cable, an effect called microbending. This too will be described in detail in a later chapter.

The third cause mentioned is ionising radiation. Naturally occurring and man-made radionuclides, such as those found in seabed sediments, can emit gamma radiation which ionises the atoms in optical fibres. Some of the resulting free electrons will migrate and become trapped in defect sites where they will cause increased loss at operating wavelengths. Calculations predict that the corresponding increase will be less than 0.2 dB for 100 km of cable over a 25 year life.

4.3.1 Fibre fatigue

Silica fibre will fracture under stress but at a level which varies from sample to sample, because each fibre contains a number of randomly positioned flaws which redistribute the stress. Small cracks on the silica surface grow at a rate dependent on the level of stress, the humidity, the temperature and the pH of the environment. A crack will grow relatively slowly until it reaches a critical size, whereupon it becomes unstable and the growth suddenly accelerates to give catastrophic failure. The point at which this occurs is a material property and is adequately described by the critical stress intensity factor. Tensile tests on fibres have shown that there are generally two or more distributions of crack size.

Deep water cables installed in the late 1980s have an operational cable safety factor approaching 2, but it may be possible to reduce this safety factor while still maintaining the required fibre reliability.

4.4 REFRACTIVE INDEX PROFILES

The reader will have noticed that most practical profiles are irregular and they differ, sometimes considerably, from ideal shapes. The influence of irregular index profiles on lightguide dispersion, cut-off wavelength of the fundamental mode and modal field diameter was investigated systematically, and the results were reported in [10]. The paper discussed how smooth profile transitions between core and cladding affect the fundamentals of propagation such as waveguide dispersion, cut-off wavelength and modal field diameter, and how the very common axial dips in the profile affect the same properties. The authors pointed out the possibility of obtaining a negative waveguide dispersion over a substantial fraction of the v-value range of interest in sm fibres, thus facilitating the design of dispersion-shifted and dispersion-flattened fibres.

The theoretical dependence of the effective cut-off wavelength λ_{eco} in depressed inner cladding sm fibre on index profile has been studied and was reported in [11]. The cut-off wavelength of the first higher order mode is a critical parameter for characterising the performance of a sm fibre, hence its position is very important. Too high a value may result in excessive modal noise especially for splices in the 1300 nm window; too low a value may result in excessive bending loss in the 1550 nm band. The measurement of the effective cut-off wavelength λ_{eco} is well-defined in the CCITT Recommendation G.652. In general, the theoretical cut-off λ_{eco} for depressed cladding fibres is greater than the effective value. An empirical relation was used between these two wavelengths in terms of the ratios b/a and $\Delta n'/\Delta n$ as a mathematical model, where these four quantities are defined in Figure 4.11.

The behaviour of λ_{eco} as a function of each ratio showed good agreement between predicted and measured values of this wavelength. The relation between values of λ_{eco} calculated from the model using refractive index data measured on the preform to values of λ_{eco} measured for a family of fibres shows excellent correlation, as the reader can judge from Figure 4.12.

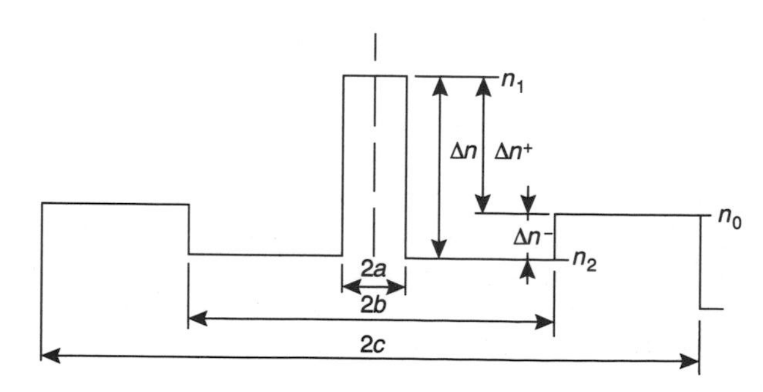

Figure 4.11 Refractive index profile of depressed cladding fibre ([11], Figure 1). Reproduced by permission of IEE

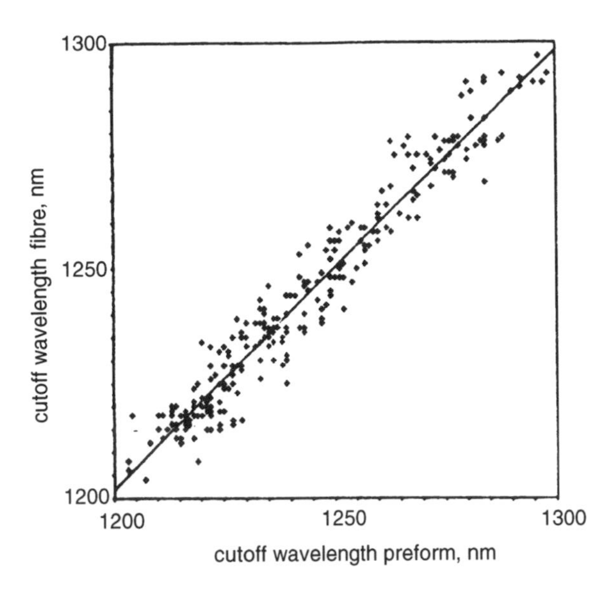

Figure 4.12 Correlation between effective cut-off wavelength in fibre to that predicted in preform from model. $p^2 = 0.96$ ([11], Figure 4). Reproduced by permission of IEE

4.5 OPTIMISED FIBRE DESIGN

By 1990 any realistic single-mode fibre was based on an oxide glass fabricated by one of the methods described earlier to deposit pure silica and/or silica doped with a number of additives such as GeO_2, P_2O_5, B_2O_3 or F_2. Mechanisms which may prevent the ideal low loss being attained can arise from:

(1) absorption or scatter intrinsic to the fibre material
(2) impurity absorption
(3) absorption and scatter consequent on the processing required in manufacture.

When the primary coated fibre is stored or is cabled and jointed to form part of a link, additional losses will arise from:

(4) splices
(5) loss of guidance resulting from bending or micro bending, for example on a storage reel, in a joint housing or in a cable.

Measurements and estimates of the dominant contributions to the loss of a realistic GeO_2, SiO_2 system fibre, prior to cabling, are shown in Figure 4.13 as a guide to the values which could be expected. It shows the measured and estimated losses of a low loss fibre fabricated by the mcvd process [12] in 1978. This fibre had $\Delta = 0.19\%$, a core diameter of 9.4 μm and outer diameter of 125 μm, and was fabricated with a GeO_2 core and SiO_2 cladding. The cut-off wavelength was 1100 nm.

The origin of the individual curves will be considered in this section. From a system standpoint, it is the lower extremities of the total loss curve which are of interest. Figure 4.13 clearly shows the locations and values of the minima of loss around 1300 and 1550 nm which provided the impetus to develop sources and

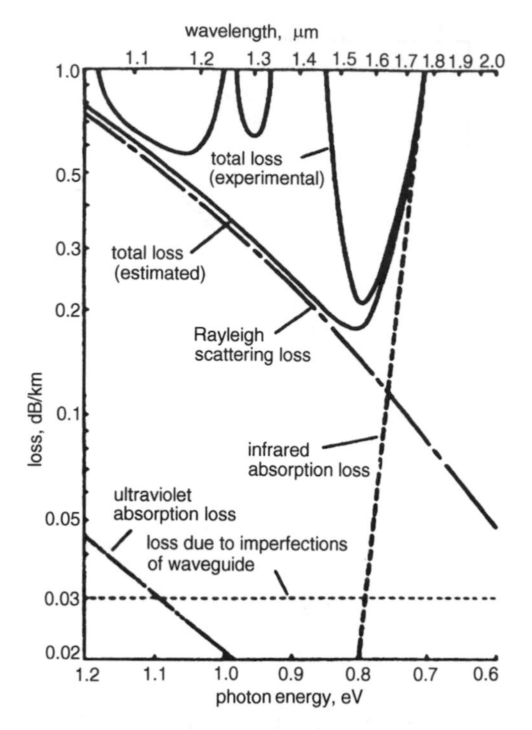

Figure 4.13 Component parts and total of the theoretical per-unit transmission loss versus wavelength (upper scale) and photon energy (lower scale), experimental curves of total dB/km ([12], Figure 3). Reproduced by permission of IEE

detectors to operate in one or both of these bands. To establish the mechanisms which contribute to this low-loss region with any accuracy, it is essential to make spectral measurements over the widest possible wavelength range. The range 700–1700 nm was measured by a grating monochromator with a halogen lamp as source.

4.5.1 Loss mechanisms

Optical attenuation is due to the sum of the bulk material attenuation and attenuation due to imperfections in the structure of the lightguide. Bulk material attenuation comprises absorption losses (intrinsic ultra-violet, intrinsic infra-red, impurity) and scattering. Thus there are many sources of loss in glass and fibres, depending on glass composition, methods of preparation and fibre design, but they fall into two broad categories, absorption and radiation. These will now be considered in detail.

Absorption loss

The absorption of light can be intrinsic (caused by interaction with one or more of the major glass components) or extrinsic (caused by impurities). In the infrared or far infrared (5–50 μm) there are strong absorption bands due to oscillations of the structural units such as Si–O (9.2 μm), P–O (8.1 μm), B–O (7.2 μm) and Ge–O (11.0 μm) within the glass. At the short wavelength side, in the ultraviolet (UV) region, there are electronic transitions which cause strong absorption bands as indicated earlier. Between these extremes most common oxide glasses have transmission windows separated by absorption bands arising mainly from the presence of hydroxyl groups, and trace impurities which are largely comprised of transition metal ions.

The impurity absorption due to the hydroxyl (OH ions, written as OH^{++}) groups which are bonded into the structure of the glass is particularly important for systems operating at 1300 nm and at 1550 nm. These groups have fundamental stretching vibrations which occur between 2.7 and 4.2 μm, depending on the position of the hydroxyl group in the glass. The fundamental vibrations give rise to overtone and combination bands extending down to visible wavelengths. In most glasses the water content is sufficient to attenuate these absorption peaks substantially. Another absorption mechanism can be caused by colour centres which can be produced in glasses and fibres by exposure to ionising radiation. These induced absorptions are usually observed in the uv and visible but they can have tails which extend into the near infrared.

Ultraviolet absorption

Extensive study has been devoted to the causes and effects of UV absorption, and a formula has been proposed [4.7] as a model for the UV edge; it is:

$$\alpha_{uv}(E) = \alpha_{uvo} \exp \frac{hf}{E_0}, \text{ dB/km/w \% Ge} \tag{4.1}$$

where α_{uv} is the uv absorption coefficient, α_{uvo} is the material-dependent constant, h' is the photon energy, E_0 is the material-dependent constant, and w is the percentage by weight (of Ge).

In practice, it is difficult to verify such an equation, and direct absorption measurements tend to show losses perhaps five times those given by equation (4.1). A modified form of this equation can be used to give better agreement with measured results. As an illustration, the results for a GeO_2–SiO_2 core single-mode fibre with $\Delta = 0.27\%$ are plotted in [7]. It can be seen from this figure that although the measured absorption remains the minor contributor, it is not negligible and, ignoring OH effects, it adds about 22% to the loss above 1 μm.

Infrared absorption

The loss at the longer wavelengths arises from the silicon–oxygen and cation–oxygen vibrational bands. The effects depend on whether one is dealing with pure fused silica or with doped silica, where, as one might expect, they are more complicated, and dependent on the dopant. An exponential tail extends back into the 1300–1600 nm region, as indicated in Figure 4.13. Despite discrepancies in the literature regarding the precise location and relative strengths of IR absorption bands in doped silica, general conclusions are in reasonable accord with the observed spectral loss properties between 1 and 2 μm. As in the case of the ultraviolet edge, the infrared edge can be expressed by means of a formula, which for GeO_2–SiO_2 core single-mode fibres is:

$$\alpha_{ir} \simeq 7.81 \times 10^{11} \exp\left(\frac{-48.48}{\lambda}\right), \text{ dB/km} \tag{4.2}$$

where λ is in μm. This function was plotted in [7].

Impurity absorption

A variety of transition metal contaminants and residual hydroxyl groups introduce bands of loss between the two edges just described. Loss from the metals can be made very small by using high-purity starting materials as remarked previously. At this point it is worth recalling that silica-based fibres exhibit losses of 0.35–0.4 dB/km about 1300 nm and 0.2–0.25 dB/km around 1550 nm, so that hundredths of a dB are significant.

The OH absorption is a more difficult problem to contain, but some spectral responses of fibres have shown no peaks at the expected wavelengths. However, in some production fibres the effects could be seen, and the fibre acted as a filter with several passbands. The two major consequences of this impurity in fibres intended for telecommunication applications are the increases in loss in the second window due to tails of nearby peaks, and an increase in loss in both the 1300 and the 1500 nm windows due to the joint action of P and OH.

4.5.2 Scattering loss

It is generally accepted that between, say, 600 and 1600 nm Rayleigh scatter is the dominant intrinsic source of loss in doped silica fibres. It is caused by scattering from a large number of randomly distributed inhomogeneities smaller than the wavelength of light. This process can therefore be described by the Rayleigh scattering law, and is the most fundamental scattering mechanism in glasses. The inhomogeneities arise from density and compositional fluctuations which are frozen into the glass on cooling. The magnitude of the component of loss due to scattering, α_{sc}, due to density fluctuations is given by [1]:

$$\alpha_{sc} = \frac{8\pi^3 n^8 P^2 \beta T_G}{3\lambda^4} \tag{4.3}$$

where P is the average photoelastic constant, β is the isothermal compressibility, and T_G is the glass transition temperature, and the other symbols have their usual meanings.

The density fluctuations increase with increasing T_G; consequently the loss is higher in pure silica glass than in glass with a lower value of T_G. Multicomponent glasses have an additional scattering loss due to local fluctuations in composition. An expression which has been derived for this loss is:

$$\langle \Delta C^2 \rangle = \frac{\rho_0 kT / C_0 V}{\partial\mu / \partial C} \tag{4.4}$$

where C is the mole fraction of one of the components, μ is the chemical potential difference between the minor and major constituents, and V is the molar volume, and the partial derivative is to be evaluated at C_0, T. Equations (4.3) and (4.4) can be used as guides in fibre design, but this aspect will not be considered.

The most important fact embodied in equation (4.3) is that this component of loss, which accounts for 0.7 dB/km at 1 μm, is inversely proportional to λ^4. It is this fact which has spurred efforts to achieve operation at wavelengths longer than the original 850 nm of the first generation systems. Comparing the scattering loss at 850, 1300 and 1550 nm relative to that at 850 nm gives values of 100%, 18% and 9%, respectively [4]. Some indication of the values to be expected are given in Table 4.4.

Table 4.4 Rayleigh scatter values for germania-silicate glasses (Midwinter, Table 1)
(Data from Garrett & Todd (1982). The incremental attenuation over pure silica for phosphorous doped glasses of the same index difference is approximately one half that for Germania doping.)

	Rayleigh scattering in decibels per kilometre at		
material	850 nm	1300 nm	1550 nm
pure silica	1.20	0.22	0.11
0.18%	1.59	0.29	0.14
0.27%	1.80	0.33	0.16
0.39%	2.01	0.38	0.19

A useful way to show the importance of Rayleigh scattering in the overall loss of a fibre is to plot the per-unit loss as a function of λ^{-4} using a linear abscissa scale and dB/km ordinate scale. On such a plot the straight line proportional to λ^{-4} intersects the vertical axis above the zero loss level at a value which corresponds to the loss due to structural imperfections. One such plot is illustrated in Figure 4.14.

The general theory of scattering of radiation of wavelength λ from small spherical particles of radius r, where the ratio of the circumference to the wavelength $2\pi r/\lambda$ is very much less than one, has been known for many years. The condition that $2\pi r$ is to be very much smaller than λ is known as the Rayleigh scattering limit. A perfect crystal will have no Rayleigh scatter, but in optical fibres there are many scattering centres. Olshansky has written a concise review of Rayleigh scatter theory applied to fibre materials, but when one is concerned with multicomponent glass fibres the Rayleigh scatter cannot be predicted accurately from first principles. A detailed treatment can be found in [7].

To provide the reader with some feel for the numbers involved in these loss mechanisms, let us consider the case of a GeO_2–SiO_2 core sm fibre with Δ less than 0.4%. An approximate but self-consistent set of data has been established and the calculated losses are given in Table 4.5.

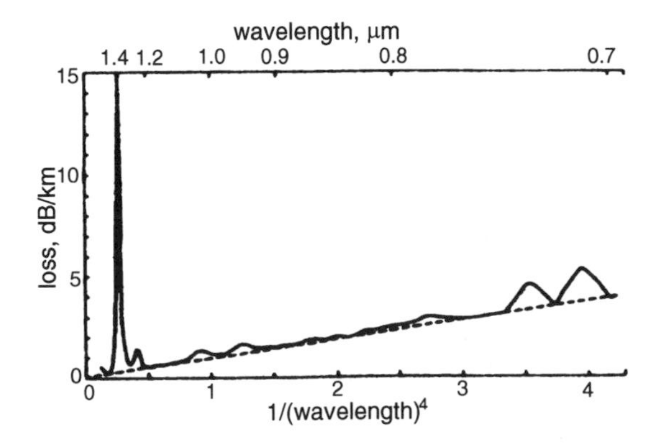

Figure 4.14 Per-unit loss versus $1/\lambda^4$ (lower scale) and wavelength (upper scale): full line curve measured; intersection of dash line with vertical axis gives value due to structural imperfections ([12], Figure 2). Reproduced by permission of IEE

Table 4.5 Self-consistent minimum losses for germania-silicate core single-mode fibre ([7], Table 2). Reproduced by permission of IEE

Δ%*	Losses at 1.3 μm (dB/km)			Losses at 1.55 μm (dB/km)		
	Rayleigh α_{Rs}	Absorption α_{abs}	Total α_T	Rayleigh α_{Rs}	Absorption α_{abs}	Total+ α_T
0	0.22	—	0.22	0.11	—	0.13
0.18	0.29	0.06	0.35	0.145	0.035	0.20
0.27	0.33	0.09	0.42	0.16	0.05	0.23
0.39	0.38	0.13	0.51	0.19	0.07	0.28

Within the experimental error, these results are in agreement with those reported by others in the early 1980s. The author remarked that much work remained to be done before the loss mechanisms in this type of fibre were fully understood.

4.5.3 Splice loss

Among the factors which have contributed to the remarkable spread of single-mode systems from the mid-1980s has been the ability to achieve routinely low-loss fusion splices in the field. The factors which contribute to power loss in splices have been studied extensively. It is very desirable to maximise the mechanical strength by making a fully fused splice each time, and the surface tension forces present in such splices will align the outer circumferences of the fibres in the absence of impurities. Unless the two fibres happen to be identical, their fields will be mismatched to a greater or lesser extent when spliced together. The effects of mismatches of profile, of concentricity and other factors have been quantified.

The loss of a well-made splice in single-mode fibre is comparable to that in a multimode fibre, despite the large difference in core diameters. However, because of the lower per-unit loss of the former, splice loss is a more significant fraction in the overall power budget for single-mode systems. Splice loss can arise from a number of causes, as listed below:

(1) mismatch between the profiles of the fibres at the joint
(2) misalignment of the cores due to core concentricity error
(3) poor mechanical alignment, e.g. due to lateral offset, separation or tilt
(4) core deformation due to the fusion technique employed.

Some of these contributions are dependent on the fibre production method used, while others depend on the design of the splicing apparatus. In the design of fibres one aims to minimise the loss due to the above causes, but not all can be minimised at the same time. For instance, a large spot size is needed to reduce loss due to core offset, whereas a small spot size reduces loss due to tilt. With a conventional step index single-mode fibre for use in 1300 nm systems, perhaps the most important of these loss mechanisms is offset, because this is proportional to the square of the ratio of offset to spot size. This loss could amount to 1 dB for an offset of 2.3 μm. By taking all the contributions together it is possible to keep fusion splice loss in conventional step index single-mode fibres with a mean concentricity error of 0.5 μm to a mean value of about 0.2 dB. The subject of splices is also treated in Section 16.4.

4.6 MICROBENDING LOSS

Any random bending of the fibre axis will cause loss due to microbending. When a primary coated fibre is in mechanical contact with a microscopically rough surface, e.g. secondary coating or a drum, it can suffer repeated microbends, subject to the mechanical rigidity of the fibre. Microbending can be a very important source of loss, perhaps amounting to tenths of decibels per km. There have been a number of studies of the effects on the propagation of the fundamental HE_{11} or equivalent LP_{01} mode, but the effect is quite complex for real fibres. An approximate formula derived from an analysis of a particular microbending loss model due to Petermann [13] gives:

$$\alpha_M \simeq \frac{C W_G{}^{4P+2}}{\lambda^{2P+2}} \tag{4.5}$$

where C is a constant related to the curvature statistics, P is a constant also related to the statistics, and W_G is a weighted field width of Gaussian approximation to LP_{01} field.

The problem with any theory of this effect is to judge the validity of any numerical predictions based on it. Some work was reported in 1982 [14]. The authors investigated the applicability of Petermann's model. Twenty lengths of sm fibre were packaged in a tube of oriented polypropylene which formed a loose tube for the fibres in their package. A slight back tension amounting to about 0.07% strain was applied to the fibres during packaging, to ensure that the fibres pressed against the inner wall of the loose tube as it is reeled at the far end of the packaging line. The surface roughness of the wall together with irregular bends in the tube were thought to be responsible for the microbending. Each fibre was measured for total loss before and after the packaging process over the spectral range from 1100 to 1750 nm. The differences of the two measurements at each wavelength gave the microbending loss; an example of the results is shown in Figure 4.15.

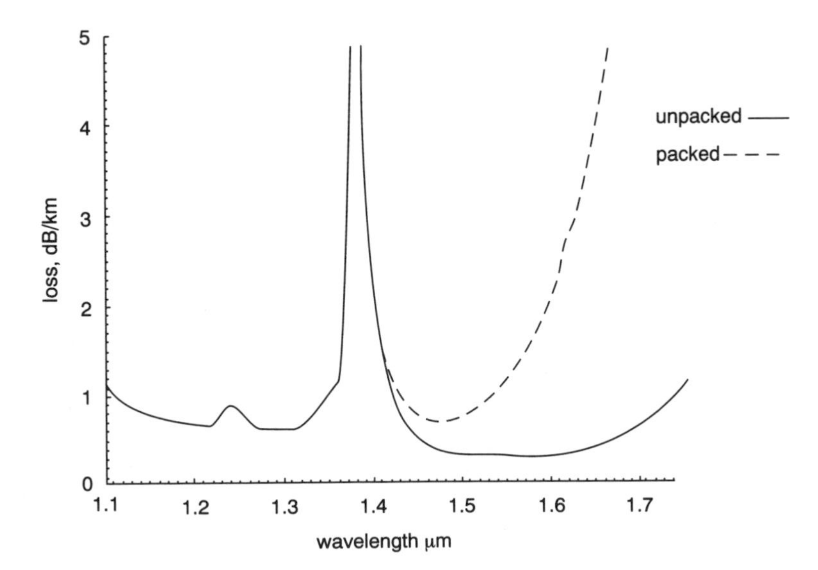

Figure 4.15 Spectral loss versus wavelength of sm fibre sensitive to microbending, before and after packaging ([14], Figure 1)

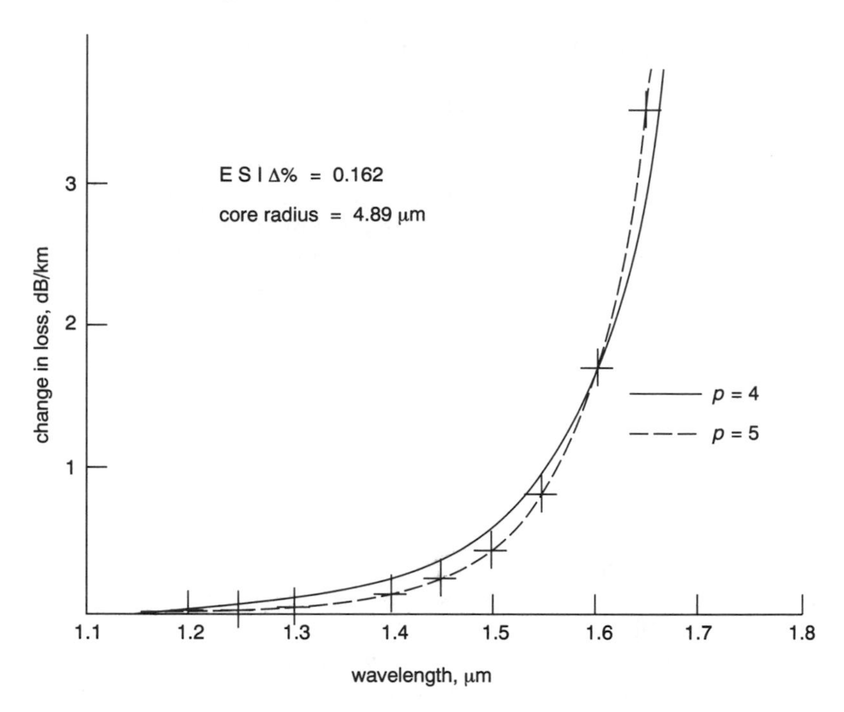

Figure 4.16 Measured incremental loss versus wavelength for fibre in Figure 4.15. Curves show calculated microbending loss for $\rho = 4$ and 5. Both curves are normalised to the experimental value at 1600 nm ([14], Figure 2)

It is immediately evident from the drawing that the effect in the third window at around 1550 nm is severe; indeed it is known as the 'microbending catastrophy', and is attributed to the loss of power from the guided mode, principally by the mechanism of mode coupling. On the basis of Petermann's formula, the authors were able by a suitable assumption of spectrum and choice of parameter to achieve the fit to measured data illustrated in Figure 4.16.

The agreement is quite satisfactory. The loss is also a strong function of the value of Δ and this is shown in Figure 4.17.

The curve corresponds to a fixed value of normalised wavelength, or v-value of 1.7, and the experimental points are shown with error bars. The diagram clearly shows that the value of Δ cannot be chosen too small, otherwise the additional loss due to microbending will become unacceptable. The Petermann model in its simplest form, applied to a series of sm fibres, was found to give good agreement with the results of measurements over a spectral range for individual fibres, and over a range of fibres for a fixed v-value. The value of one parameter needed to be taken as some three times the value previously suggested to obtain the good agreement achieved.

The solution of Petermann's auxiliary function for microbending loss was discussed in [15]. The spot sizes and auxiliary function that determine the microbending sensitivity of monomode fibres were shown to have analytic solutions for step index fibres. Although a numerical solution is necessary in the core region when the profile is inhomogeneous, the derivation is simplified by matching to the analytic solution within the cladding region.

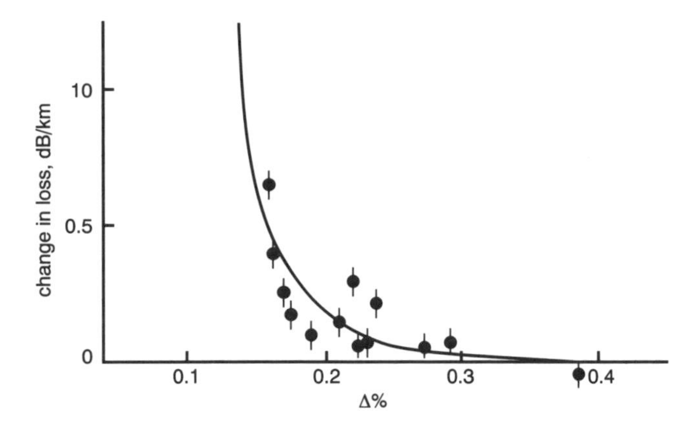

Figure 4.17 Incremental per-unit loss versus Δ% at $v = 1.7$ for 14 different fibres; curve represents least-squares fit using equation (5), giving $\rho = 4.26$ ([14], Figure 3)

4.6.1 Theory of microbending loss

Several papers appeared in the late 1980s which dealt wholly or in part with microbending loss, and three of these will be considered to round out the paragraphs above. The theory of microbending was verified experimentally by workers at the Electromagnetics Institute in Denmark [16]. The existence of discrete cladding modes in sm fibres was demonstrated by inducing microbends periodically along the axis of the fibre then the spectrum of the resulting additional losses was measured. In order to explain the results of the measurements it was necessary to use a microbending theory which took account of coupling between the guided mode and a number of discrete cladding modes. Very good agreement between the measured results and this theory was achieved. The authors then discussed the consequences of the existence of discrete cladding modes on the correct choice of artificial microbending spectrum to be used for the characterisation of the sensitivity of the fibre to microbending.

The introduction discussed existing methods to obtain the microbending sensitivity and their limitations, and pointed out that to discriminate between the different methods of calculation a periodic distribution of deformations along the fibre axis was appropriate. When such a distribution was employed, the existence of discrete cladding modes could be determined from simple spectral microbending measurements. Section 2 considered the evaluation of the microbending loss coefficient from the power coupling between the guided mode and these cladding modes, and further details. The determination of the spectral loss from a standard attenuation measurement arrangement was discussed in the next section and the results were compared with theoretical curves. Finally, the proper choice of microbending deformation device to use in the characterisation of fibres was treated.

4.6.2 Power spectrum of microbends

The power spectrum of microbends in sm fibres had been the subject of an earlier letter [17]. A new model was proposed in this contribution from BTRL, and it was based upon the idea that microbends were extended rather than point effects, as had been assumed previously. As in [17], the model was fitted to the experimental data and it yielded values for the wavelength and correlation length of the disturbances which caused the microbends. In this letter the mechanism of microbending loss was attributed to coupling between the propagating mode and radiation modes caused by bends in the axis of the fibre, but the mechanical origin of these bends remained a mystery. The use of the simple model rather than an abstract spectrum provided some insight into the possible nature of the imperfections which cause microbending loss, and enabled a coherent search for the origin of this loss to begin.

4.6.3 Coatings which possess very low microbending sensitivity

To prevent microbending losses from occurring over a wide temperature range, a new buffer coating and a new top coating were developed and described in [18]. The buffer coating had a low Young's modulus over a wide temperature range, whereas the modulus of the top coating was high compared with those normally applied. The resulting fibres possessed very low microbending sensitivity except at low temperatures where added loss occurred as a result of a change in the radial stress in the buffer coating from compression to tension. This effect can be avoided if the thickness of this layer does not exceed a critical value. In addition to its favourable microbending properties the new coatings provided good mechanical protection. To provide the reader with some numerical data, the results of a sandpaper test on a drum and the temperature test carried out on single-mode fibres are presented in Table 4.6.

The entries in the table show that the fibres with the new coatings possess very low microbending sensitivities compared with conventional coatings, and that to avoid unacceptable increase in loss at −60°C the buffer layer must be less than 24 μm in thickness.

Resistance to microbending (and to macrobending) may form parts of the specification for an optical fibre; for instance, for sm fibres to a British Telecom specification mentioned at the beginning of this chapter: microbending resistance: a typical attenuation change of 1 dB/m for 1 kgf/m of fibre tested on 100 micron grit sandpaper between flat plates.

Two general approaches have been adopted to minimise microbending loss, and they are usually known as the loose or tight tube packaging methods. These will be treated in Chapter 5.

By 1994 production standard fibres were approaching the theoretical limits of material attenuation due to Rayleigh scattering and absorption. Any additional

Table 4.6 Increase in per-unit loss for a sample of sm fibres ([18], Table 2). Reproduced by permission of IEEE, © 1989

Thickness		Sand-paper test		Temperature test	
Buffer layer	1300 nm	1550 nm	1700 nm	−60°C	80°C
32(1)	1.3	2.7	5.5	<0.05	0.0
0(2)	1.3	2.0	3.1	0.0	0.0
12	0.1	0.1	0.6	0.0	0.0
20	0.1	0.1	0.6	0.0	0.0
24	—	—	—	0.8	0.0
27	0.1	0.1	0.6	1.4	0.0
33	<0.05	<0.05	0.1	>4	0.0

(1) coated with reference coatings BC_2 and TC_2
(2) only top coating TC_1 was applied

losses due to cable manufacturing (bending and microbending) and joints (splices and connectors) thus represented a considerable contribution to the total attenuation of the fibre link. Whereas macrobends and joints were then well understood and documented, serious difficulties still arose from the practical determination and experimental simulation of the irregular curvature of fibre in a cable and characterisation of its sensitivity to microbending. The application of the mode-coupling theory of Petermann allowed the theoretical dependence of excess loss due to microbending on mode field diameter (mfd), cut-off wavelength (λ_{co}) and mac-value (= mfd/λ_{co}) to be established [19]. The theoretical considerations were confirmed by systematic experiments using two different methods. The investigation of microbending effects in buffer tubes and cabled ribbon stacks verified that the methods simulated the realistic conditions. In conclusion, the two authors from Siemens AG stated that a small sensitivity to microbending could be achieved by selecting fibres with low mac-values and as thick a coating as possible with proper combinations of materials.

4.6.4 Relationship of the microbending loss to the other fibre parameters

The relationship of the microbending loss to the other fibre parameters can be shown graphically in the so-called fibre design diagram, an example for a step index fibre being shown in Figure 4.18.

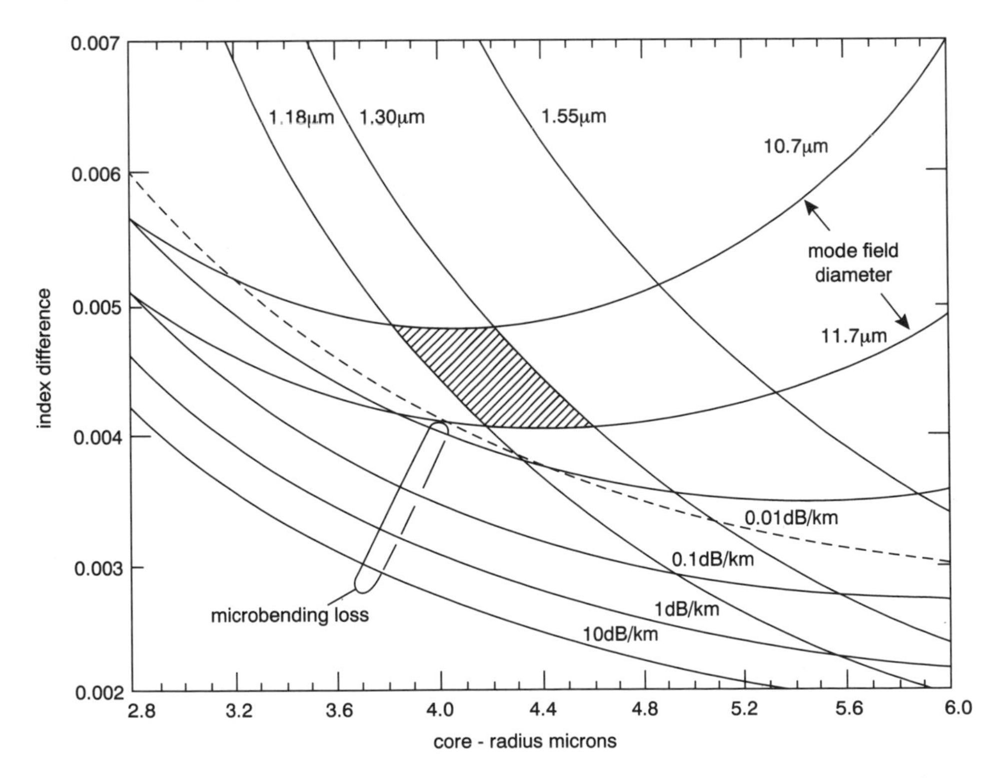

Figure 4.18 Fibre design diagram showing index difference versus core radius with wavelength, mode field diameter and per-unit loss as parameters ([9], Figure 3). Reproduced by permission of SPIE

The dash contour was obtained from an expression for the pure bend loss of a step index fibre and corresponds to 1 dB/m attenuation; the upper part of the diagram thus corresponds to the strongly guided regime in which bend loss will not be a problem. Typical contours of microbending loss calculated from Petermann's model are shown as solid curves in the lower half of the figure. The curve for 0.01 dB/km lies close to the dash line, so that the upper half of the diagram also corresponds to good microbending performance. The shaded region designates a suitable fibre specification having a cut-off wavelength between 1180 and 1300 nm and a mode field diameter between 10.7 and 11.7 μm. Fibre designs in this region are expected to have a good tolerance to microbending and are not expected to exhibit the problems experienced with designs current in 1990.

4.7 MACROBENDING LOSS

In addition to losses caused by microbending, further losses may arise from macrobending, i.e. when a fibre is bent. Thus the loss of a fibre, loosely wound on a drum or mandrel, increases when the radius R decreases, varying as exp $(-R)$. The effect may be visualised by considering the wavefront of a mode as it travels round a bend. It extends some distance into the cladding and at some outer radius its phase velocity must exceed a limit imposed by the speed of light. Under this condition power will leak from the bound mode into the radiation field at a rate determined by the radius of curvature and the amplitude of the field at this point. This is described as pure bend loss and the resulting attenuation for a step-index fibre has been modelled by an equation. Another paper to discuss the topic appeared in 1986 [20]. The particular fibre used in the investigation was of the dispersion flattened type with a refractive index profile shown in idealised form in Figure 4.19.

The phase velocity of the fundamental mode (HE_{11}) mode is smaller than the velocity of uniform plane waves in the outer cladding, so that if the fibre is bent the phase velocity is forced to increase in proportion to the distance along the radius of curvature. At the radius itself the phase velocity exceeds that of the plane wave in the outer cladding and the power is radiated out of the cladding, a condition which has been analysed by a number of workers and for which a mathematical model exists. An expression for a curvature loss coefficient valid for a small amount of uniform curvature, and hence a small amount of loss, was given in [20]. As a result of the investigation the authors found that for triple-clad sm fibres of the dispersion-compensated types discussed the details of the profile had little influence on the macrobending loss. This loss could only been kept low if the difference between the effective refractive index of the fundamental mode and the outer cladding index

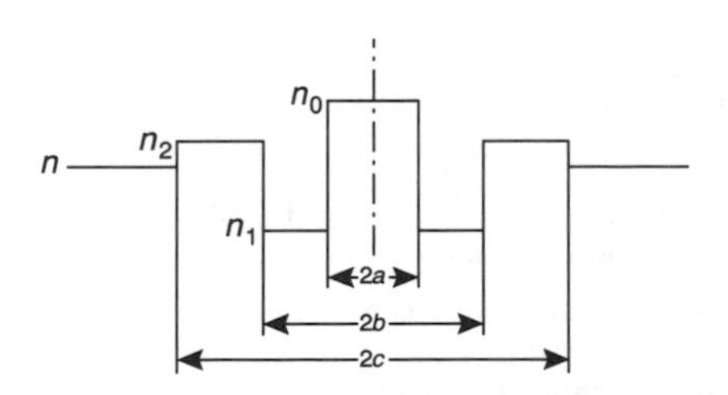

Figure 4.19 Refractive index distribution of triple clad fibre ([20], Figure 1). Reproduced by permission of Prof. Dr.-Ing. H.G. Unger

was made sufficiently large; a value exceeding 0.3% was quoted. In addition the core radius must be dimensioned so that the normalised frequency parameter v must be near 2.3 at 1300 nm.

As in the case of microbending there are specifications for fibres in this respect, which for single-mode fibres for British Telecom is as follows. Macrobending resistance: less than 1.0 dB (0.05 dB/m) increase in attenuation at 1550 nm when 100 turns of fibre are wound loosely around a mandrel of 60 mm diameter.

4.8 REFERENCES

1 BEALES, K.J.: 'Fibre materials and fabrication'. *IEE Vacation School*, 1983.

2 DORN, R., and Le SERGENT, C.: 'Preform technologies for optical fibres'. *Electrical Commun.*, **62**, no. 3/4, 1988, pp. 235-241.

3 GEITTNER, P., HAGEMANN, H.-J., LYDTIN, H., and WARNIER, J.: 'Hybrid technology for large sm fiber preforms'. *IEEE/OSA J. Lightwave Technol.*, **6**, no. 10, Oct. 1988, pp. 1451-1454.

4 DENTSCHUK, P.: 'Optical waveguide manufacturing techniques and optical cable requirements', *IEE Vacation School on Optical Fibre Telecommunications*, 1986.

5 OGAI, M., IINO, A., MATSUBARA, K., TAMURA, J., KOGUCHI, M., NAKAMURA, S., and KINOSHITA, E.: 'Development and performance of fully fluorine-doped single-mode fibers'. *IEEE/OSA J. Lightwave Technol.*, **6**, no. 10, Oct. 1988, pp. 1455-1461.

6 KIESSEWETTER, W.: 'Manufacturing optical fibres for telecommunication systems'. *Elec. Commun.*, **63**, no. 2, 1989, pp. 176-182.

7 TODD, C.J.: 'Losses in monomode fibre'. *IEE Vacation School*, 1982.

8 BOUTEN, P.C.P., HERMANN, W., JOCHEM, C.M.G., and WIECHERT, D.U.: 'Drawing influence on the lifetime of optical fibers'. *IEEE/OSA J. Lightwave Technol.*, **7**, no. 3, Mar. 1989, pp. 555-559.

9 WRIGHT, J.V., and SIKORA, E.S.R., SCOTT, J.M., and PYCOCK, S.J.: 'Considerations for the lifetime of submarine optical cables'. *Fibre Optics '90*, SPIE, **1314**, 24-26 Apr. 1990.

10 IRRERA, F., and SOMEDA, C.G.: 'Monomode fibers with smooth core-cladding transitions and index dips: an analytical investigation'. *J. Lightwave Technol.*, **6**, no. 7, July 1988, pp. 1177-1184.

11 YANG, R., HOFFART, M., SCHÜRMANN, H., and SOMMER, R.G.: 'Profile dependence with empirical model for effective cut-off in depressed inner cladding fibre'. *Electronics Lett.*, **25**, no. 16, 3 Aug. 1989, pp. 1070-1071.

12 MIYA, T., TERUNUMA, Y., HOSAKA, T., and MIYASHITA, T.: 'Ultimate low-loss single-mode fibre at 1.55 μm'. *Electronics Lett.*, **15**, 1979, pp. 115-116.

13 PETERMANN, K.: 'Theory of microbending loss in monomode fibres with arbitrary refractive index profiles', *A.E.U.*, **30**, 1976, pp. 337-342.

14 HORNUNG, S., DORAN, N., and ALLEN, R.: 'Monomode fibre microbending loss measurements and their interpretation'. *Opt. Quantum Electronics*, **14**, 1982, pp. 359-362.

15 WRIGHT, J.V.: 'Microbending loss in monomode fibres: solution of Petermann's auxiliary function'. *Electronics Lett.*, **19**, no. 25, 8 Dec. 1983, pp. 1067-1068.

16 PROBST, C.B., BIARKLEV, A., and ANDREASEN, S.B.: 'Experimental verification of microbending theory using mode coupling to discrete modes'. *IEEE/OSA J. Lightwave Technol.*, **7**, no. 1, Jan. 1989, pp. 55-61.

17 BLOW, K.J., DORAN, N.J., and HORNUNG, S.: 'Power spectrum of microbends in monomode optical fibres'. *Electronics Lett.*, **18**, no. 11, 27 May, 1982, pp. 448-450.

18 BOUTEN, P.C.P., BROER, D.J., JOCHEM, C.M.J., MEEUWSEN, T.P.M., and TIMMERMANN, H.J.M.: 'Doubly coated optical fibers with a low sensitivity to temperature and microbending'. *IEEE/OSA J. Lightwave Technol.*, **7**, no. 4, Apr. 1989, pp. 680-686.

19 UNGER, C., and STÖCKLEIN, W.: 'Investigation of the microbending sensitivity of fibers'. *IEEE/OSA J. Lightwave Technol.*, **12**, no. 4, Apr. 1994, pp. 591–596.
20 FRANTSESSON, A.V., YANG, R., and UNGER H.-G.: 'Macro-bending loss of triple-clad single-mode fibres'. *A.E.U.*, **40**, no. 1, 1986, pp. 132–133.

FURTHER READING

21 LAPP, J.C., BHAGVATULA, V.A., and MORROW, A.J.: 'Segmented-core single-mode fiber optimized for bending performance'. *J. Lightwave Technol.*, **6**, no. 10, Oct. 1988, pp. 1462–1465.
22 ANDREASEN, S.B.: 'New bending loss formula explaining bends on loss curve'. *Electronics Lett.* **23**, no. 21, 8 Oct. 1987, pp. 1138–1139.
23 MIDWINTER, J.E., The Clifford Paterson Lecture 1983. Int. J. Electronics, 1983, **55**, no. 2.

5

DESIGN AND SYSTEM ASPECTS OF FIBRES AND CABLES

5.1 INTRODUCTION

This is the third and last chapter to be devoted to the bearer of light from point to point. At the end of the preceding chapter the losses in cabled fibres due to microbending and macrobending were examined in terms of their impact on system performance. Design aspects are taken further here for standard fibres and for dispersion-shifted and dispersion-flattened types which have been deployed in operational systems. The important topic of pulse dispersion and its effects on performance are considered in some detail, not only for single fibres but for concatenated fibres, which necessitates the use of statistics. Consideration in the context of real systems is continued in the next topic which is the effect of production tolerances on dispersion, followed by an example of the temperature dependence of chromatic dispersion. The chapter ends on the subject of group delay in fibres.

5.2 ASPECTS OF SINGLE-MODE FIBRE DESIGN

A key factor in the development of inland single-mode transmission systems is the performance of the final cabled and jointed link. The careful balancing of the various loss mechanisms outlined earlier has led to a small number of designs which have found wide acceptance.

5.2.1 Design diagrams

These have appeared in a number of published papers, e.g. [1]. With either step index or other profiles it is possible to construct a design diagram based on the four main interdependent fibre parameters, namely:

(1) core radius a

(2) index difference Δ

(3) Petermann mode spot size ω (radius)

(4) cut-off wavelength λ_{co}

Item (4) refers to the LP_{11} mode. The diagrams are constructed by choosing any two of the four as axes, and plotting contours of the remaining two parameters. One must, however, always bear in mind that such diagrams refer to ideal (and not real) profiles. They serve to highlight the overall design concepts. The final optimum design must be derived empirically. This is done by considering a practical profile and calculating the propagation characteristics by the finite element method.

Practical type-T fibres for operation in 1550 nm band

Several examples will be given later, but a typical example is shown in Figure 5.1 to illustrate some of the points which arise.

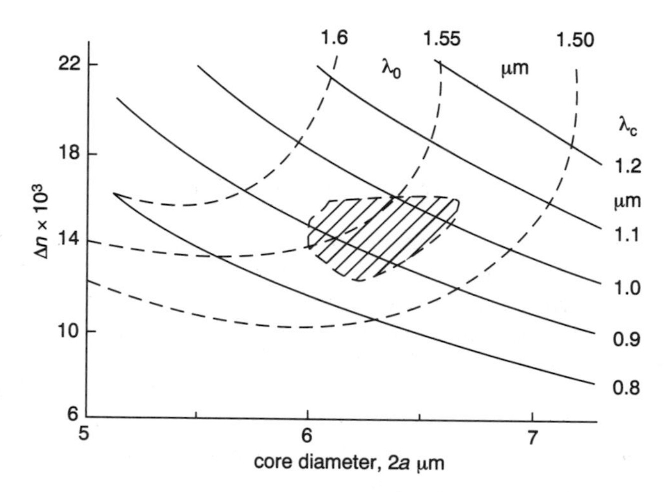

Figure 5.1 Design diagram showing Δn versus core diameter with cut-off and zero dispersion wavelengths as parameters for practical T-type fibre ([1], Figure 8). Reproduced by permission of BT

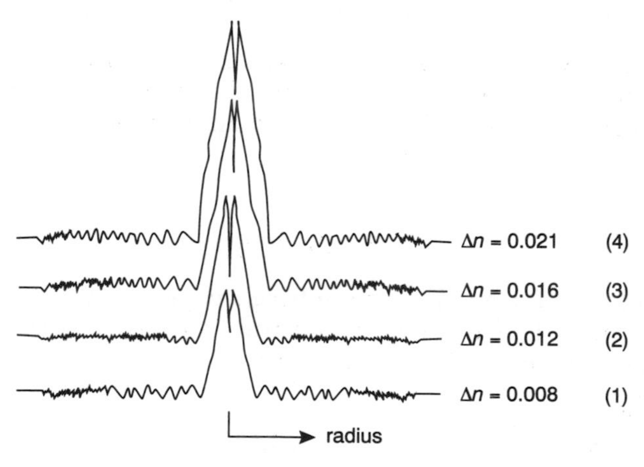

Figure 5.2 Preform refractive index profiles of four graded-index type T fibres for use at 1550 nm with corresponding values of Δn ([1], Figure 1(*c*)). Reproduced by permission of BT

Table 5.1 Parameters of graded-index type T fibres for use at 1550 nm ([1], Table 1). Reproduced by permission of BT

Fibre	Peak Δn	Core ϕ, μm	λ_c μm (1m)	λ_0 μm	Field radius, μm	Loss at 1.55 μm, dB/km
1	0.008	7.5	0.99	1.42	5.0	0.24
		6.5	0.79	1.5	5	10
2	0.012	6.1	0.85	1.54	4.4	0.25
3	0.016	6.4	1.0	1.54	4.1	0.28
4	0.021	6.0	1.23	1.59	2.9	0.37
		6.4	1.24	1.52	—	0.41

The figure shows contours of constant zero dispersion wavelength and of constant cut-off wavelength, with core diameter and normalised index difference as axes. Practical constraints restrict the ranges of parameters to a much smaller region than is covered by the values shown, and this is indicated by the shaded target area. The upper and lower limits for Δ and for $2a$ are determined from the losses measured in fibres with low values of Δ and spot size for good joining. Calculations of the mode field as a function of core radius also give some insight into the probable bending performance of different designs. The parameters of the four graded-index type T fibres for operation at 1550 nm, whose profiles are given in Figure 5.2, are listed in Table 5.1.

For instance, the parameter values for fibre 3 locate a point near the upper edge of the target region. The dependence of the loss at 1550 nm on Δ (column 7) should be noted. Another, more recent choice of axes is shown in Figure 5.3 taken from the manufacturer's data sheet for sm fibre SM-02-R, with cut-off wavelength as abscissa and mode field diameter as ordinate. This diagram enables the performance of the fibre to be assessed in terms of dispersion and bend characteristics.

To explore the lowest attenuation limit of type T fibre profiles, the best fibre was chosen and its loss was plotted so that the calculated Rayleigh scattering loss becomes a straight line with positive slope through the origin. The result is shown in Figure 5.4.

Between 1100 and 1200 nm the loss approaches the Rayleigh limit obtained by assuming that a fixed fraction of the total power is guided in the core, and using Olshanksky's data for scattering coefficients in GeO_2–SiO_2 glasses. At shorter wavelengths the loss increases above the limit owing to increased drawing induced loss. Around 1300 nm some of the excess loss may be attributable to the OH contribution. Finally, the excess of about 0.04 dB/km in the 1550 nm window suggest that, perhaps, losses below 0.2 dB/km are achievable in type T fibres. Values around 0.22 dB/km achieved in 1987 seem to indicate such losses are not obtained in production fibres. An important aspect of many engineering designs, for example in networks or amplifiers, is the sensitivity of the performance to changes in the main parameters. In the present context, two of the sensitivities of the type T design can be illustrated by plotting the variation in wavelength of zero dispersion against core diameter $2a$ and Δ for fixed values of the other parameter, as shown in Figure 5.5.

From this graph the reader can see that the core diameter must be restricted to a tolerance of $\pm 4.7\%$ if $1540 < \lambda_0 < 1560$ nm, while for the same range of wavelengths the tolerance on Δ is $\pm 8\%$. Both of these tolerances can readily be achieved in practice.

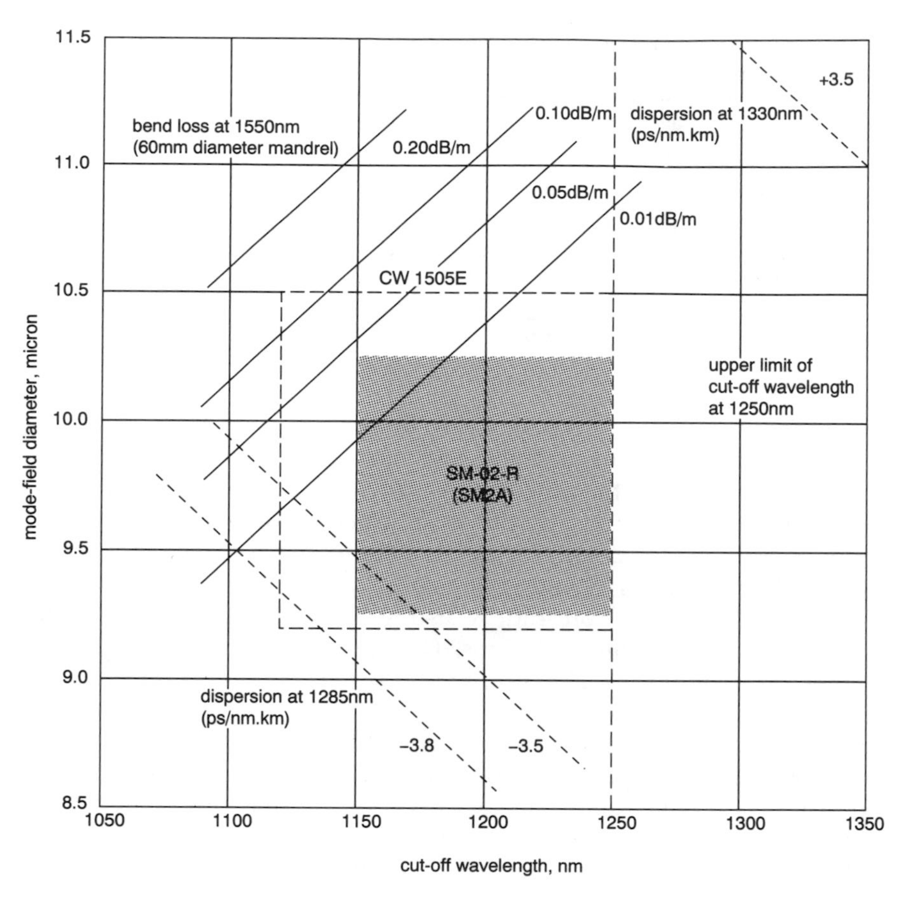

Figure 5.3 Mode-field diameter versus cut-off wavelength with dispersion and bend loss as parameters for SM2A fibre compared with specification CW1505E (SM2A fibre spec., CW1505E spec.)

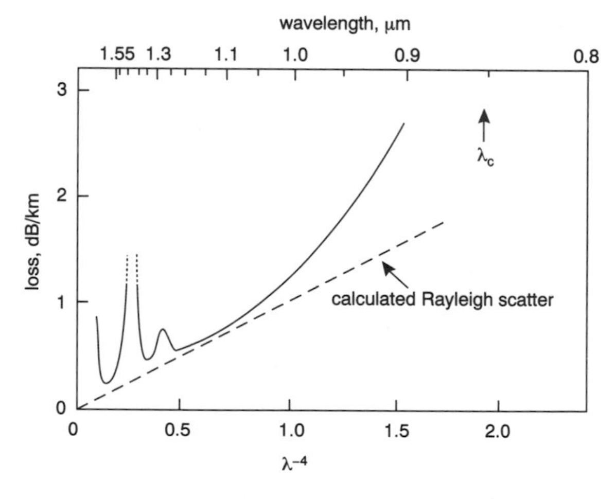

Figure 5.4 Rayleigh scattering plot of type T fibre with loss of 0.22 dB/km at 1600 nm ([1], Figure 7). Reproduced by permission of BT

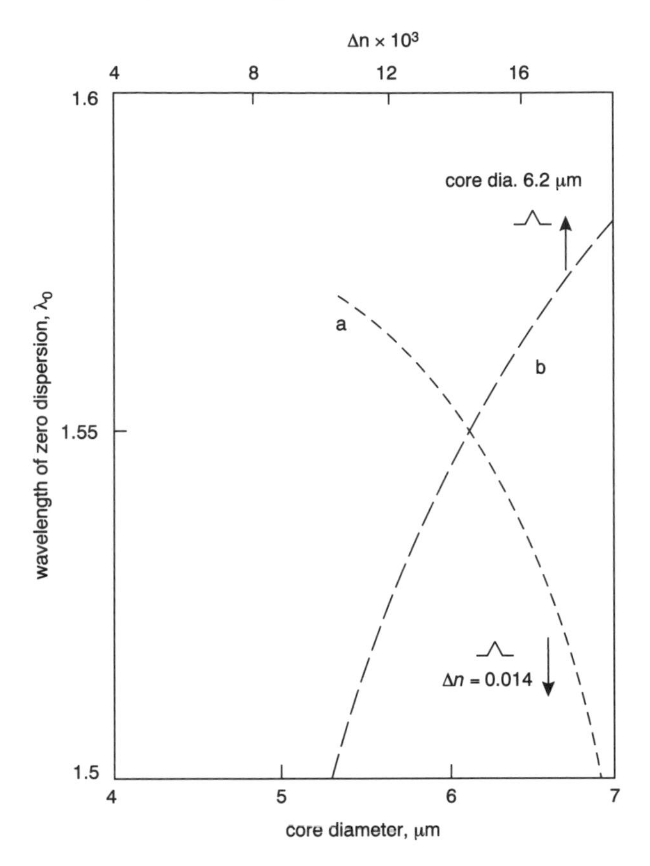

Figure 5.5 Wavelength of zero dispersion: (*a*) versus core diameter showing sensitivity of cut-off wavelength to changes in d when $\Delta n = 0.014$, and (*b*) versus Δn showing sensitivity of cut-off wavelength to changes in Δn when $d = 6.2$ μm ([1], Figure 10). Reproduced by permission of BT

Some typical preform profiles

Before considering some other fibre designs in detail, let us look at a few typical actual preform profiles dating from 1988 [4.2]. We recall that the idealised step-index profile and electric field distribution are as shown in Figure 5.6.

By the late 1980s the large central dips and ripples in the refractive index in the core region, which were obvious in many fibres fabricated early in the decade, had been virtually eliminated, [2] of Chapter 4 showed typical examples of the four most useful profiles and Figure 4.3 is repeated below as Figure 5.7 for convenience.

Dopants for control of parameters

The choice of dopants is restricted by the fact that considerable power is carried in the cladding; therefore the materials used must introduce the smallest possible additional loss consistent with other requirements. The relative powers in the LP_{01} mode carried at various radii as a function of the normalised frequency are given in Figure 5.8 [2].

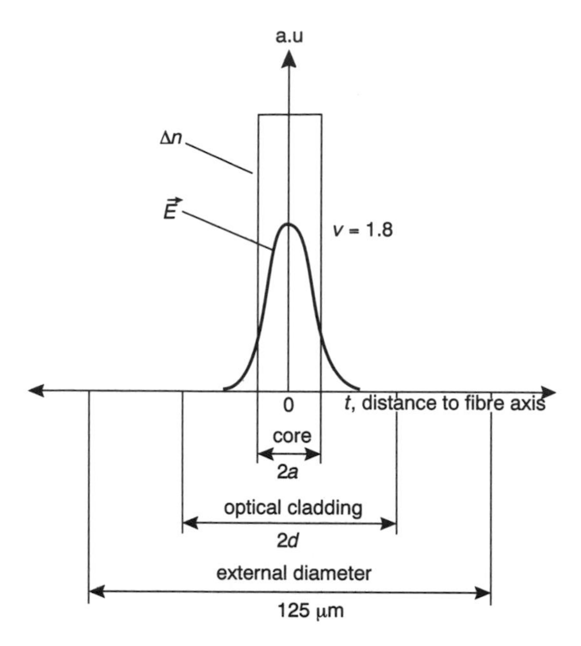

Figure 5.6 Idealised step index profile and corresponding electric field disribution in a single-mode fibre ([2] of Chapter 4, Figure 1). Reproduced by permission of Alcatel

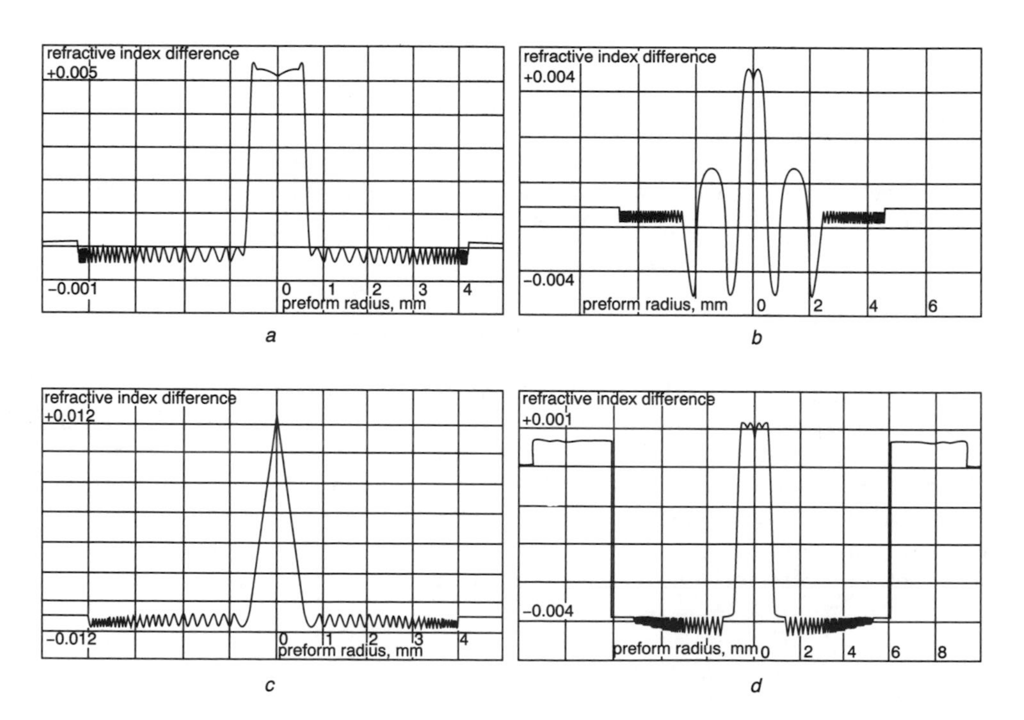

Figure 5.7 MCVD tailored sm profiles: (*a*) matched cladding, (*b*) dispersion-flattened, (*c*) dispersion shifted and, (*d*) non-doped core ([2] of Chapter 4, Figure 3). Reproduced by permission of Alcatel

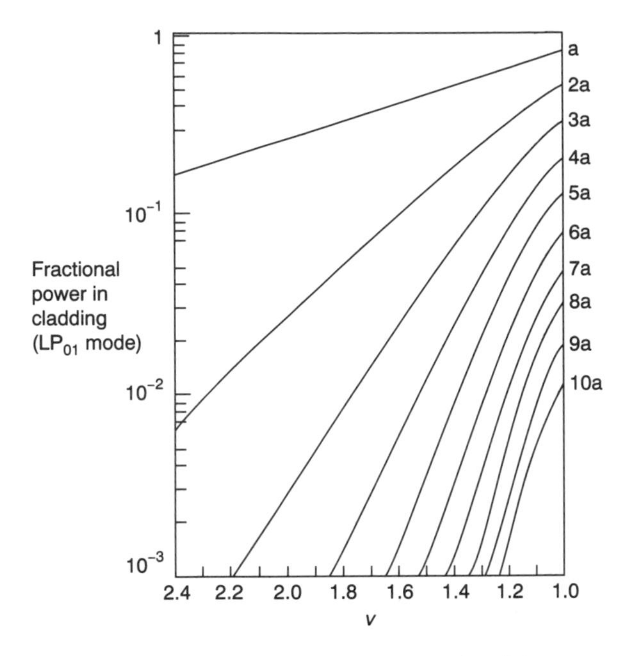

Figure 5.8 Fractional power of LP_{01} mode versus normalised frequency (v-number) with radius as parameter. Note that only the LP_{01} mode propagates for $v \leq 2.4$ ([2], Figure 2). Reproduced by permission of IEEE, © 1982

v-values of 1.6 or so represent the practical limit of operation before bending losses become appreciable, and the reader can see from Figure 5.8 that about 10^{-3} of the power travels in cladding regions more than $5a$ from the axis. It is for this reason that the radius of the cladding is taken to exceed five times the radius of the core.

In 1982 there were only four dopants of practical interest, and these could be divided into two groups [3]. Those containing boron and fluorine decrease the refractive index, while those containing germanium or phosphorous increase it, as we have already seen. Of these, B should be avoided wherever possible, as it introduces significant losses in both second and third generation windows, even when present in very low concentrations. When OH is present, P introduces an additional loss centred on 1600 nm, with a tail which can effect the loss at 1300 nm. Even without OH, too much P will move the ir edge into the window around 1550 nm. Turning to fluorine F, this is used to accurately control the refractive index of silica up to a maximum Δ_F of about -0.2%. Two dopants, Ge and F, are thus available for adjusting the index of silica; P is reserved to the subsidiary but important role of reducing the cladding deposition and sintering temperature to ease the deposition process and minimise tube distortion, thereby achieving lower splice losses through improved geometrical tolerances. With these ingredients one can construct a standardised quasi-step index mcvd fabricated single-mode fibre, as shown in Figure 5.9.

In this matched cladding structure F is used to offset the increase in n due to the P dopant, and to accurately match the cladding indices to that of silica. The amount of P is tapered to zero through the inner cladding where the power is appreciable. A selection of matched cladding designs are presented in Table 5.2.

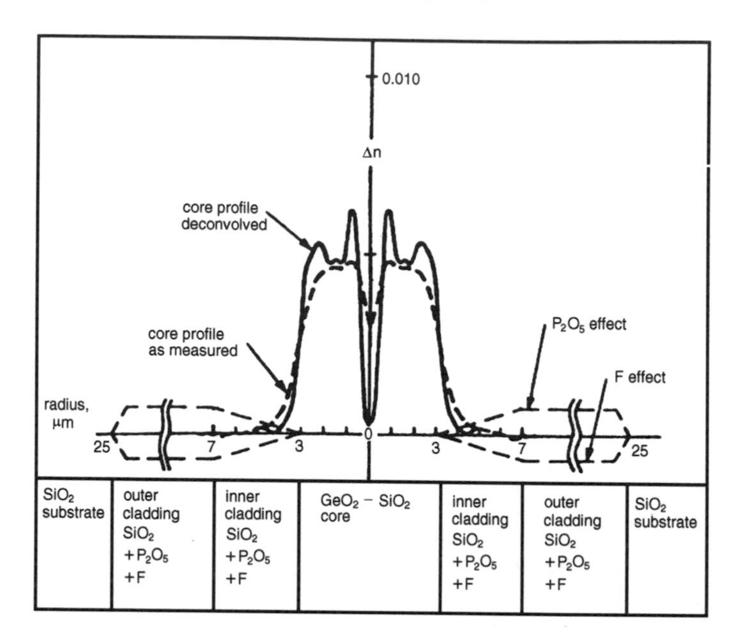

Figure 5.9 Quasi-step index mcvd sm deposition structure for a matched cladding design (support tube to 100 μm not shown). The deconvolved core profile makes allowance for the finite probe size in the refracted near-field technique. The higher phosphorous level eases deposition of the outer cladding, and this level is tapered to zero at the inner cladding to avoid P-OH and IR edge contributing to the spectral loss. Fluorine is used to match the cladding index to that of the silica support tube ([3] Figure 19). Reproduced by permission of Chapman & Hall

Table 5.2 Four main classes of single-mode matched cladding fibre designs ([3], Table 4). Reproduced by permission of Chapman & Hall

Fibre type	Δ	$2a$	λ_0	Dispersion D (ps $km^{-1}nm^{-1}$)	
	(%)	(μm)	(μm)	1.3 μm	1.55 μm
Weak-guidance monomode (WGM)	<0.2	>9	$\simeq$ 1.30	<1	$\simeq$ 18
Standard monomode (SM)	0.20–0.39	8–9	1.31–1.325	1–3	16–18
Intermediate Shifted monomode (ISM)	0.48–0.60	$\simeq$ 6	1.38–1.42	7–11	7–11
Fully shifted monomode (FSM)	0.65–0.90	$\simeq$ 4.5	1.50–1.55	15–20	<3

This table contains an intermediate and a fully dispersion zero-shifted design. The former supports transmission at 650 MBd in both bands, whereas the fully shifted design allows 300 MBd at 1300 nm and more than 1.2 GBd in the 1550 nm band (roughly inverting the performance of the standard design). The four matched cladding designs can be categorised on the basis of fibre loss and the results are shown in Figure 5.10.

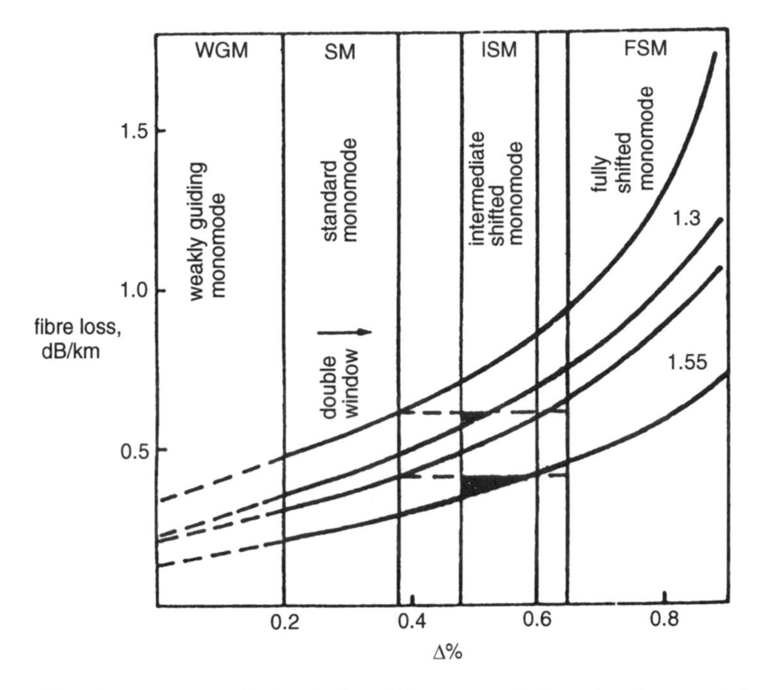

Figure 5.10 Fibre loss versus relative index difference Δ% showing four matched cladding design classes. The arrow emphasises the restriction on Δ if dual window fibre is required ([3], Figure 20). Reproduced by permission of Chapman & Hall

5.2.2 Optimised designs of fibres

Design diagrams can be constructed for fibres which will enable optimum designs to be achieved subject to the appropriate constraints as illustrated earlier. In particular, the following factors must be considered:

(a) cut-off wavelength
(b) zero dispersion wavelength
(c) microbending
(d) splice loss.

These are best examined on a design diagram, one form of which is shown in Figure 5.11 [4] of Chapter 3.

The factors listed above can be examined, and two cases were considered: a step index design for 1300 nm, and a shifted-zero one for use at 1550 nm. The spot size limits are shown by the solid curves, a pair for each design. Above the upper curves the region corresponds to small spot size, hence splice loss dominates; below the lower curves lies the region of large spot size, hence microbending loss dominates. The cut-off wavelengths for each design are shown by the dashed lines labelled λ_{co}; for single-mode operation one must work in the regions to the left of these lines. The final constraint is imposed by the two dotted curves which correspond to contours of zero dispersion. The designer must choose operating conditions on or close to these dotted lines if dispersion penalties are to be avoided. A study of the diagram will reveal that a fibre cannot be designed for zero dispersion at 1300 nm

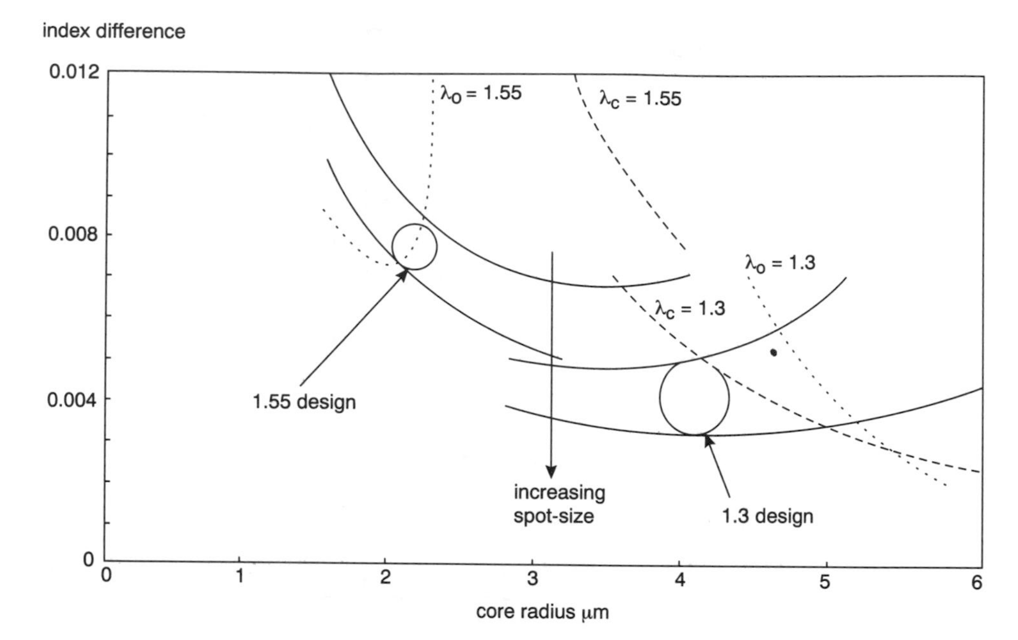

Figure 5.11 Fibre design diagram based on relative index difference versus core radius for operation at 1300 and at 1550 nm; wavelength of zero dispersion indicated by dotted curves and cut-off wavelength by dash curves ([4] of Chapter 3 Figure 13). Reproduced by permission of IEE

because of the constraint imposed by λ_{co}. Taking all the constraints into account, one is left with an area on the diagram within which optimum fibre performance can be expected. For the 1300 nm design this is centred on a core radius of 4 μm and an index difference of 0.004. In the case of the zero-shifted design, the corresponding values are just over 2 μm and 0.008.

Another form of design diagram is given in Figure 5.12 and was constructed from experimental data on a loose-tube cabled standard single-mode fibre.

Looking in detail at this fibre, Figure 5.12 shows that the loss performance of the primary coated fibre is acceptable, but clearly it is the performance when cabled and spliced that matters to the system designer. The lower boundary to the value of Δ will be determined by the onset of the cabled fibre 'microbending catastrophy' (see Section 4.6), and the boundary needs to be located for the particular structure under consideration. Referring to Figure 5.12, the two sloping lines contain the error bars on the data specifying the required values of Δ as a function of the v-value to ensure a microbending loss of about 0.05 dB/km. If the only performance requirement is operation at 1300 nm over a span of 30 km, then the cut-off of the LP_{11} mode must lie below 1300 nm and the relative index difference Δ is less than or equal to 0.39%. Δ is based on the peak core refractive index and is referenced to silica; it is presumed that the OH peak at 1380 nm is less than 2 dB/km. From Figure 5.12 there appears to be quite a large design area in which the microbending loss lies below 0.05 dB/km for the standard fibre. In practice, it is preferable to restrict the cut-off wavelength and the relative index rather more; appropriate ranges might lie between 1 and 1.25 μm and between 0.28 and 0.38% for λ_{co} and Δ, respectively. Expressed in terms of the tolerances on these quantities the design requires the

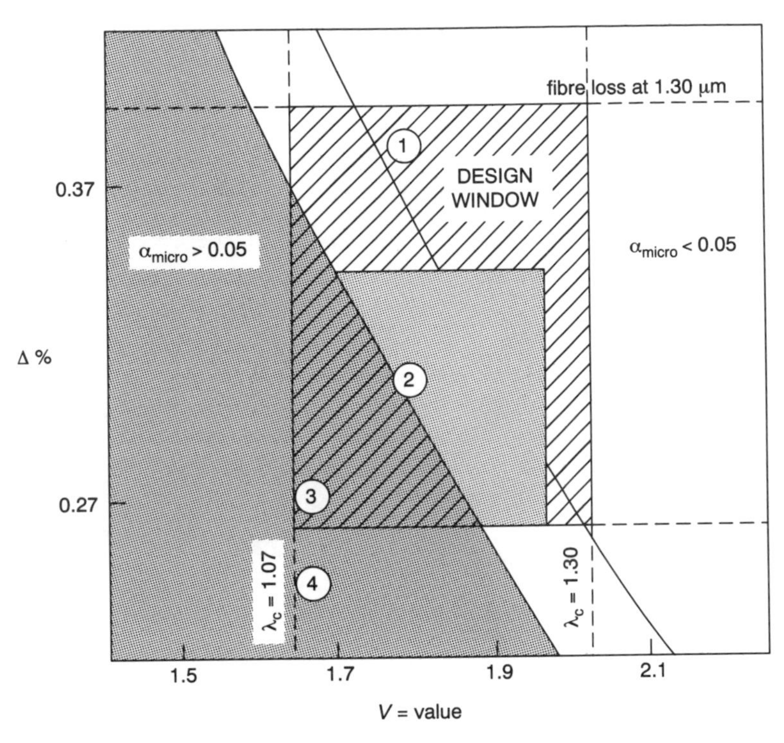

Figure 5.12 Fibre design diagram based on relative index difference versus normalised frequencies ([3], Figure 21*a*). Reproduced by permission of Chapman & Hall

values $\lambda_{co} = 1140 \pm 140$ nm and $\Delta = 0.33 \pm 0.05\%$; these tolerances might need to be tightened as the splice loss dependence on fibre profile parameters becomes better understood.

Dual-window fibre

The conditions for keeping 'open' the window at 1550 nm as well as at 1300 nm are more critical. Referring to the figure, the rectangular areas (larger striped, smaller shaded) effectively cover the largest ranges of λ_{co} and Δ to give adequate performance in a loose tube cable in both windows, given a reasonably uniform spread of fibre parameters throughout the design rectangles. For this two-window design using standard fibre, the abscissa scale should be taken as v-values at 1550 nm, assuming the step index theory. The right-hand boundary defines the largest v-value at 1550 nm to allow single-mode operation at 1300 nm, assuming that λ_{co} is defined by means of measurements on short lengths of fibre. The left hand and lower boundaries are less well defined. The closeness with which they can be approached will depend on the level of incremental loss that can be tolerated at 1550 nm. Too low a value of Δ makes the fibre difficult to handle and measure. To obtain the lowest losses at 1550 nm, and keep the window at 1300 nm open, may require the design area to shrink to the inner rectangle in Figure 5.12. In such cases the tolerances become approximately $\lambda_{co} = 1220 \pm 60$ nm and $\Delta = 0.31 \pm 0.03\%$. Comparing these values with the previous example suggests that the dual window fibre will be about twice as difficult to realise on profile grounds.

5.2.3 Design of dispersion-shifted fibres

These fibres became of greater technical interest when reliable commercial lasers operating at 1550 nm became available. We now examine dispersion-shifted fibre designs in rather more detail. If a step index fibre is zero-shifted the result has a fairly large index difference, and moreover it suffers from a dopant-dependent loss, as shown in Figure 5.13.

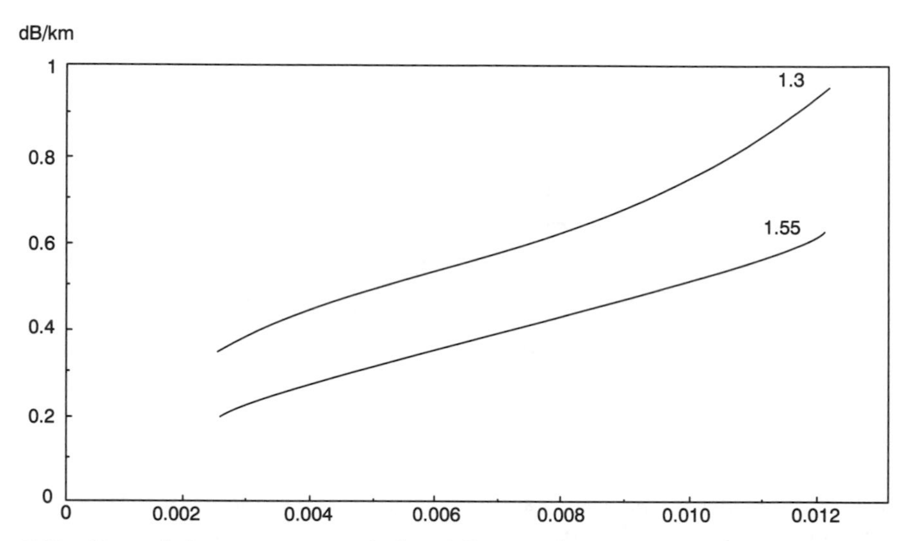

Figure 5.13 Per-unit loss versus step index difference for operation in the 1300 band and the 1550 nm band ([4] of Chapter 3, Figure 12). Reproduced by permission of IEE

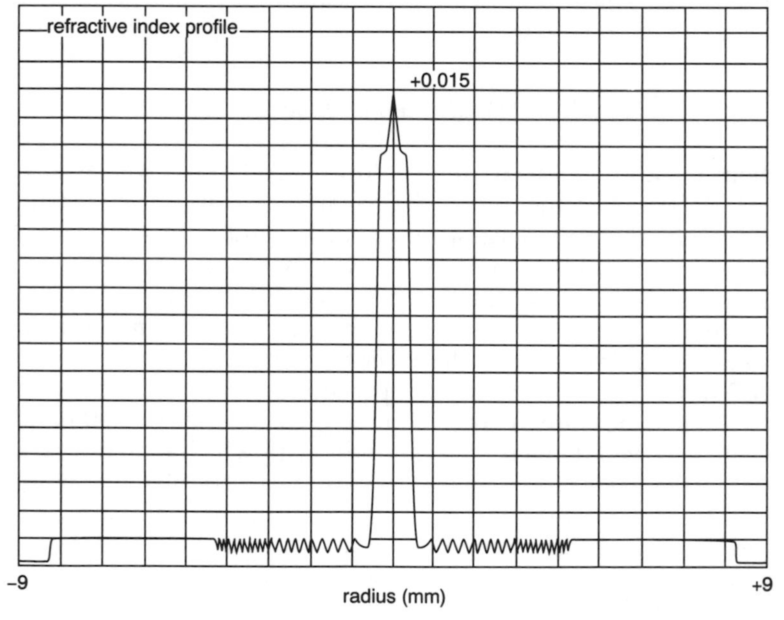

Figure 5.14 Typical refractive index profile of T-type fibre ([4], Figure 1). Reproduced by permission of Alcatel

In fact the loss at 1550 nm can be as high as the loss of a conventional design at 1300 nm, so that no advantage can be taken of the lower loss window. The careful balancing of these and the other relevant factors has led to a small number of designs that have gained widespread acceptance. The triangular profile, or type T fibre is one of the most successful, and it corresponds to the profile index α equal to one, as mentioned in Chapter 3. A typical equivalent profile realised in a design for submerged system use and published in 1987 [3] is shown in Figure 5.14.

Spread of per-unit loss and of wavelength of zero dispersion

The absence of a notch centred on the axis, which was characteristic of earlier designs of this type of profile, should be noted. The reproducibility of the optical and geometrical parameters of a fibre in manufacture are very important, especially in an undersea system. The performance of the present fibre in these respects is illustrated in the next two figures, starting with reproducibility of the per-unit loss of the fibre shown in Figure 5.15.

Careful study of this distribution provides a realistic indication of what can be achieved with good quality control. The corresponding cumulative distribution curve for the wavelength of zero dispersion is shown in Figure 5.16, and again the reader is invited to study the curve carefully. Further information based on [4] appears in Chapter 15.

Use of equivalent triangular profile in design

Although the T-type (or triangular) design reduces the dopant-dependent loss, the fibre has to be operated far from cut-off and so becomes more sensitive to

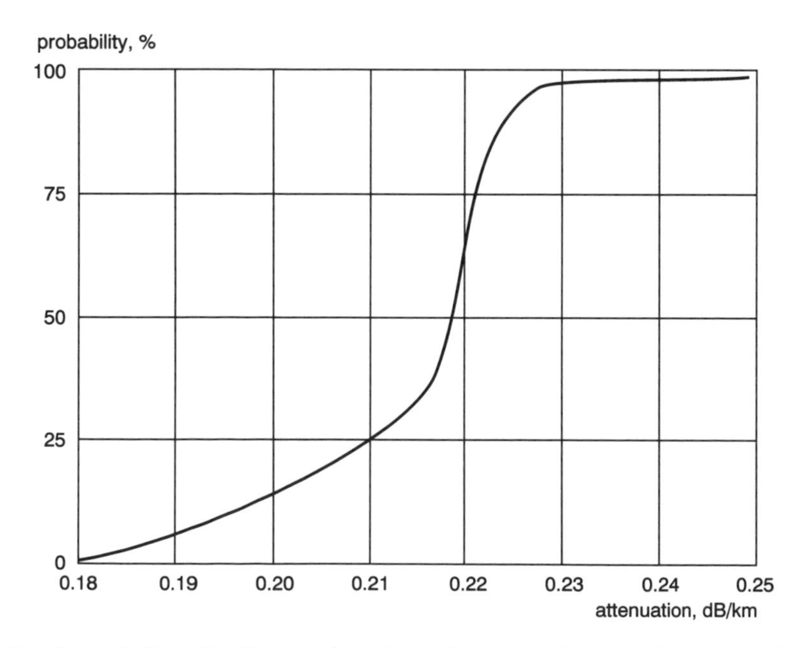

Figure 5.15 Cumulative distribution function of per-unit loss of dispersion-shifted fibre showing reproducibility ([4], Figure 2). Reproduced by permission of Alcatel

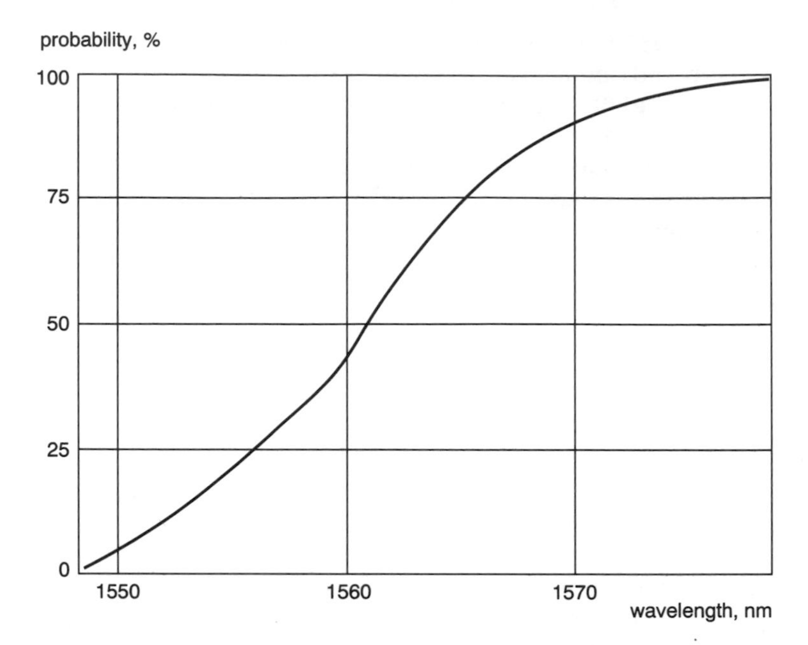

Figure 5.16 Cumulative distribution function of wavelength of zero dispersion of dispersion-shifted fibre showing reproducibility ([4], Figure 3). Reproduced by permission of Alcatel

bending. This was taken into account in the recent design of T-type fibre with matched cladding described in [4]. For design purposes a model was developed which enabled the designer to derive an equivalent triangular profile from the actual measured profile. The equivalent triangular profile fibre had the same transmission characteristics as the actual fibre. The equivalent refractive index difference Δ_E is 1.06%, and the typical equivalent profile obtained was shown in Figure 5.14. The average transmission characteristics of this design of matched cladding fibre in the waveband 1550–1570 nm were:

$$\alpha_1 = 0.22 \text{ dB/km}$$

$$\lambda_0 = 1560 \pm 10 \text{ nm}$$

$$2\omega_{FF} \leq 8.2 \text{ μm}$$

where λ_0 is the wavelength of zero dispersion, and $2\omega_{FF}$ is the mode field diameter in far field.

This particular type of fibre was manufactured by the mcvd process. Optical and geometric characteristics of the preform were measured after the deposition and collapse stages. The most suitable sleeving tube for the preform was then chosen to adjust the value of zero dispersion wavelength in the 125 μm diameter fibre. Excellent reproducibility of the deposition process was obtained by using a multiprocessor unit to control the functioning of the equipment used to produce the gases, the burner and the mcvd lathe. The deposition of the layers and the collapse stage were fully automatic. The average splice loss obtained by means of new splicing equipment was less than 0.1 dB.

Example of design of fibre for submerged systems

A further example of design is taken from a 1987 publication [5], which described aspects of the choice of optimum parameters for use in submerged systems at 1550 nm. Design considerations were discussed for sm fibre with the dispersion zero at 1300 nm, but for use at 1550 nm. For the purpose of determining the optimum design, the design requirements were listed as follows.

(a) The Rayleigh scattering loss and structural imperfection loss should be reduced significantly over existing designs.

(b) To suppress incremental loss due to OH formation, lattice defects near the core axis must be avoided, as they concentrate the EM field distribution.

(c) The bending sensitivity at 1550 nm is to be cut to at least half that of conventional 1300 nm optimised fibres, given as 0.01 dB/km.

(d) Single-mode operation must be achieved in the cabled fibre.

(e) The splice loss is to be about half that of an optimised 1300 nm design, namely some 0.06 dB/km.

(f) The choice of the most suitable parameters is to be specified.

From the first two requirements a pure silica core, or one with slight Ge doping, with F depressed cladding in either case, would be desirable. The pure silica with F depressed cladding combination was expected to possess lower transmission loss than the other.

Next consider items (c) and (d) from the list. As λ_{co} and Δ become larger the bending sensitivity decreases, but the intrinsic loss tends to increase above $\Delta = 0.45\%$, thus setting an upper limit to the relative index difference. In addition, the upper limit of λ_{co} is determined by the need to ensure single-mode operation.

Leaving aside (e) for the moment, the important parameters are the mode field diameter 2ω and the effective cut-off wavelength. These are taken as the abscissa and ordinate, respectively, of a field on which the dependent parameters of equi-bending loss α_B, dispersion D and Δ, are displayed, as shown in Figure 5.17.

In this Figure the area within the parallelogram corresponds to 1300 nm single-mode optimised designs as recommended by the CCITT, and the dot labelled 6 to the values $2\omega = 10 \pm 0.1$ μm and $1100 \leq \lambda_{co} \leq 1280$ nm.

From the three straight lines to the right of the diagram it can be seen that this depressed cladding design has significant bending loss at the longer wavelengths. Cabling loss due to microbending can be considered as equivalent to the loss resulting from uniform bending of radius R_E. The incremental loss due to this cause was about 0.015 dB/km at 1550 nm for a fibre with v-value 1.7 and R_E of 46 mm. The theoretical microbending loss was calculated with R_E equal to 45 mm. To achieve a microbending loss of less than 0.005 dB/km the design diagram shows that the upper limit of mode field diameter $2w$ is about 11.5 μm, whereas the lower limit of cut-off wavelength is close to 1350 nm. The parameters of optimised fibres lie inside the rectangular area shown in Figure 5.17. A number of possible designs are indicated by the dots labelled 1–5 and the corresponding parameters and performance characteristics are listed in Table 5.3.

Experimental values of the bending loss of these six designs of fibres are shown in Figure 5.18. This figure shows that the bending losses of the 1550 nm optimised fibres at this wavelength are smaller than those of the typical CCITT optimised

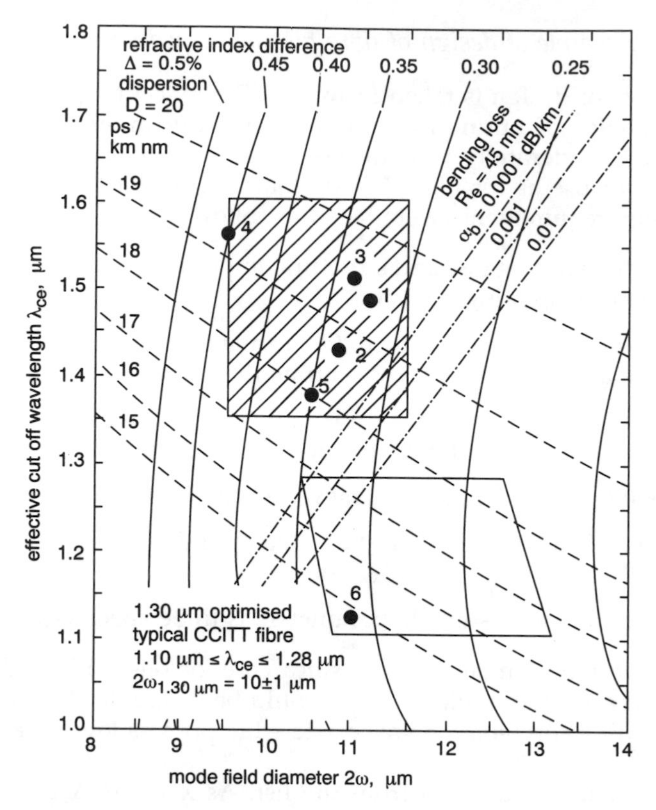

Figure 5.17 Design diagram showing effective cut-off wavelength versus mode-field diameter with dispersion D and index difference Δn at 1550 nm as parameters ([5], Figure 1). Reproduced by permission of IEE

Table 5.3 Parameters for 1550 nm loss-optimised test sm optical fibres ([5], Table 1). Reproduced by permission of IEE

Fibre	MFD at 1.55 μm	λ_{ce}	Δ	Loss at 1.55 μm (1.3 μm)	Dispersion at 1.55 μm
	μm	μm	%	dB/km	ps/km nm
1	11.1	1.48	0.33	0.176 (0.320)	19.27
2	10.8	1.43	0.34	0.184 (0.323)	18.69
3	10.9	1.51	0.35	0.181 (0.348)	18.92
4	9.50	1.56	0.45	0.189 (0.334)	18.99
5	10.1	1.38	0.35	0.197 (0.351)	17.67
Min~ max	9.50~ 11.1	1.38~ 1.56	0.33~ 0.45	0.176~ 0.197 (0.320~ 0.351)	17.67~ 19.27

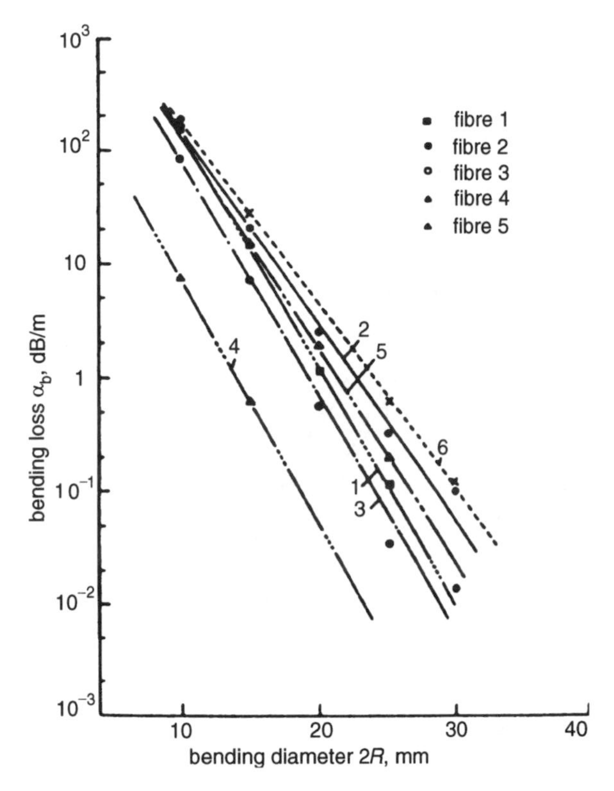

Figure 5.18 Per-unit bending loss versus bending diameter of test sm fibres at 1550 nm ([5], Figure 2). Reproduced by permission of IEE

1300 nm fibre shown by the line of dashes. The results of the work showed that the optimum ranges of mode field diameter and of cut-off wavelength are 10.5 ± 1 μm and 1350–1600 nm, respectively. The average loss and dispersion measured on test lengths are about 0.187 dB/km and 18.7 ps $km^{-1}nm^{-1}$.

Long-term stability of dispersion-shifted fibres exposed to hydrogen

The last topic to be discussed under the heading of dispersion-shifted fibres is their long-term stability when exposed to hydrogen. Although such fibres had been installed in trunk systems in public networks by the late 1980s, their long-term stability under hydrogen exposure was first evaluated and reported in [6]. The letter reported estimates of hydrogen-induced losses due to the reaction at 4°C over the 25 years nominal service life of undersea systems. The estimates were based on chemical kinetics in dispersion-shifted and standard fibres, and the loss increases at both 1300 and 1550 nm were negligibly small, despite the assumed exposure to hydrogen at 1 atm. Both types of fibre were fabricated by the vad process described earlier, and fluorine doping was used during the consolidation of the soot blanks. Details of the fibres used, the two main mechanisms which account for most of the induced losses caused by the reaction in silica-based fibres, and the results, were given. The initial rate of increase in the losses due to the short-wavelength edge and to hydroxyl group (OH) absorption was found by the theory of chemical kinetics

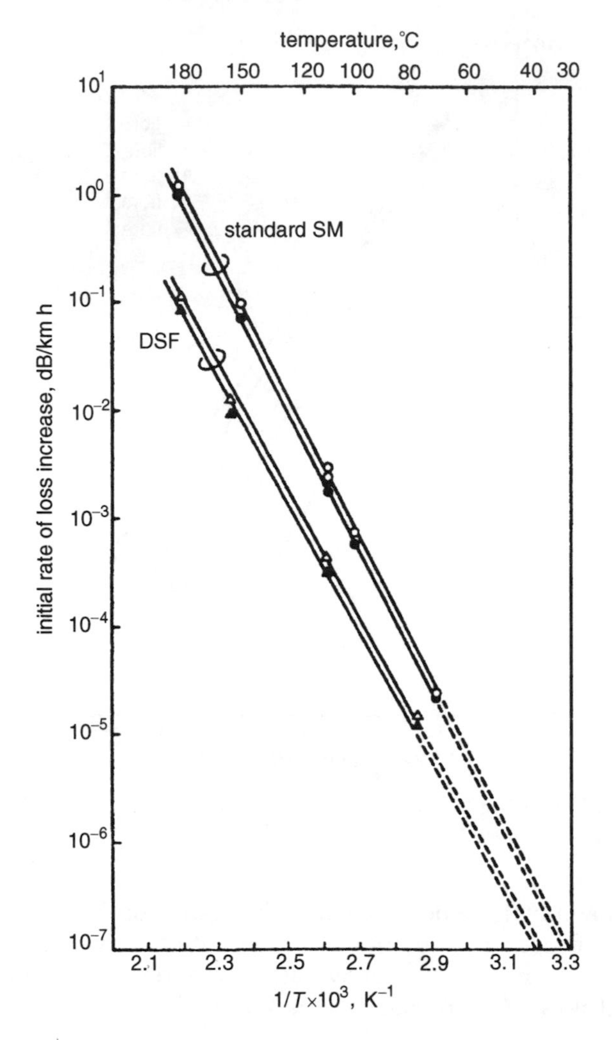

Figure 5.19 Arrhenius plot of initial rate of increase of loss of standard and dispersion-shifted sm fibre ([6], Figure 2). Reproduced by permission of IEE

Table 5.4 Loss increases due to hydrogen reaction at 1 atm and 4°C ([6], Table 2). Reproduced by permission of IEE

Type	Wavelength	Initial rate[1]	Loss Increase after 25 yr[2]
	μm	dB/km h	dB/km
DSF	1.3	4.1×10^{-10}	9.0×10^{-5}
	1.55	3.1×10^{-10}	6.8×10^{-5}
Standard SM	1.3	8.4×10^{-10}	1.8×10^{-4}
	1.55	6.5×10^{-10}	1.4×10^{-4}

to be a simple exponential law, and this was confirmed experimentally. The results obtained are presented in Figure 5.19.

There is a difference between the two types of fibre, but it is not very large. Extrapolating the straight lines on the Arrhenius plot provides the estimates presented in Table 5.4. Even over the very long lengths of some submarine systems, particularly those in the Pacific basin, the values in the last column would still be negligible.

5.2.4 Design of dispersion-flattened (segmented-core) fibres

Instead of shifting the wavelength of the zero of dispersion it is possible to design fibres which will have a low value of dispersion in both windows, and research on such fibres at a number of laboratories took place throughout the late 1980s. The desirable goal of low dispersion in both windows is achieved by balancing the waveguide and material components of the total dispersion. More complicated index profiles are required in order to increase the number of available degrees of freedom, and these are usually in the form of segmented core profiles; for example, as shown later in Figure 5.29. As one would expect, the design is more complicated, more difficult to make, and it suffers from excess attenuation in one or both windows. Nonetheless, Philips of Holland had begun production of their dispersion-flattened single-mode fibres (dff) by 1988 and had proposed the new fibre to the CCITT and the IEC as a standard. Such fibres have also been fabricated at Corning Glass Works R&D Division using the segmented-core approach, which gave improved macrobending and microbending performance without a sacrificial increase in either the cut-off or zero-dispersion wavelengths [4.21]. A related paper described the design and experimental results on a segmented-core fibre optimised for use in long-haul and in local-area applications [5.7]. This new fibre was compatible with installed fibres above 1300 nm, and possessed a bandwidth × distance product of 2–4 GHz × km over the 800–900 nm band where several modes can propagate. Such a fibre has potential application to future local-area optical fibre networks.

Earlier, in connection with Table 5.2, the performance of a 30 km span of intermediate shifted fibre designed to equalise the dispersion in both bands was given.

5.3 SINGLE-MODE FIBRE DESIGNS USED BY AT&T IN THE LATE 1980s

AT&T was the dominant long-distance carrier in the largest national market and was very much at the leading edge of optical fibre technology. It is therefore appropriate to describe the salient features of the types of fibres that AT&T Bell Laboratories designed. It is worth noting that almost all of the fibre produced for 1300 nm systems in the late 1980s was of the depressed-cladding kind.

The design principles and performance of the four types of fibres were described in [8]. We first consider the type designed for 1300 nm systems for which more than a million km had been shipped by the end of 1986, and nearly 16000 km of optical fibre cables had been installed. This depressed-cladding fibre was designed to provide low intrinsic loss, high bandwidth, low splice loss, and a high resistance to macrobending- and microbending-induced loss. During 1988 and 1989 about the same length of ocean cables using this fibre was to be laid. The fibre was developed to meet the diverse transmission needs and environmental considerations

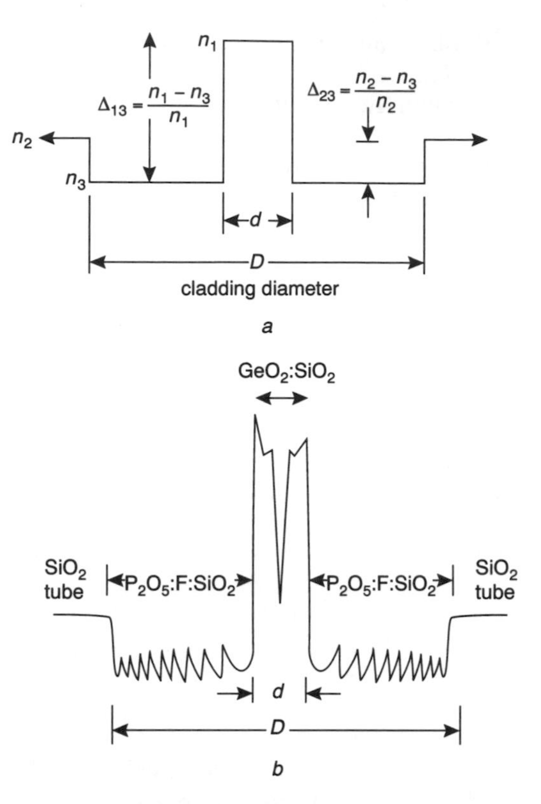

Figure 5.20 (*a*) Idealised profile of depressed-cladding fibre; (*b*) measured profile of preform ([8], Figures 1*b*, 4). Reproduced by permission of AT&T

from the subscriber loop to the deep ocean. Primary transmission requirements are, of course, low intrinsic loss and high bandwidth or low pulse dispersion at 1310 nm. The chosen ideal index profile is shown in Figure 5.20*a*, and a corresponding measured preform index profile in Figure 5.20*b*. The departures from the ideal profile are a consequence of the mcvd process, are reproducible and do not affect fibre performance.

Depressed-cladding (type W) fibre

Clearly this is a depressed-cladding type (sometimes called the W-type from the shape of the profile) and Figure 5.20*b* shows the dopants used in the AT&T mcvd fabrication process. As we have seen earlier, matched cladding profiles can be described by two variables, Δ and core radius a (typically $\leq 0.3\%$ and $2a \geq 8.7$ μm for matched cladding designs). The depressed-cladding design has two additional variables Δ_{23} and b which allow more freedom to optimise the design; for instance, the waveguide dispersion can be tuned by adjusting Δ_{13} through choice of Δ_{23}. More importantly, the high attenuation of the first higher-order mode is dominated by a leaky mode mechanism in depressed-cladding designs which depend on both the core and cladding indices. The loss due to the leaky mode lowers the cut-off wavelength by as much as 100 nm from the value in a matched cladding fibre with identical parameters. Alternatively, the profile parameters can be changed and selected so that the depressed-cladding fibre is still single-mode for v-values greater than 2.405. This has the advantage that the fundamental mode is more closely bound to the core, and this effect is reinforced by choosing the proper combination of Δ_{23} and b, resulting in a smaller spot size, hence more power near the axis. The values chosen for the AT&T design were:

$$\Delta_{13} = 0.37\%$$

$$2a = 8.3\ \mu\text{m}$$

$$\Delta_{23} = 0.12\%$$

$$b/a = 6.5$$

$$2\omega = 8.8\ \mu\text{m}$$

5.3.1 Transmission performance in 1300 and in 1550 nm bands

Details of the per-unit loss and the dispersion versus wavelength of a typical production fibre are sketched in Figure 5.21*a* with details in the 1300 nm band shown in Figure 5.21*b*.

The median loss is 0.35 dB/km at 1310 nm, whereas the loss at 1550 nm for the typical fibre above is 0.2 dB/km. In the mcvd process chlorine is used during the collapse stage to produce a typical total O–H absorption peak at 1390 nm of less than 1.5 dB/km; a low value being important to ensure a relatively flat 'passband' for the fibre. For example, the typical fibre characteristic shown in Figure 5.21*b* has about 0.03 dB/km variation between 1285 and 1330 nm, thanks to the low absorption peak.

The slope of the dispersion curves in the neighbourhood of the zero is approximately 0.09 ps $\text{nm}^{-2}\text{km}^{-1}$, and is about the same for both matched and depressed-cladding types of profiles. The value is relatively insensitive to the normal fibre-to-fibre variations in composition and profile. The dominant contribution to the slope is the material component because of the low concentration of germanium in the fibre, and the small slope of the waveguide dispersion. It follows that the maximum dispersion over a range, e.g. the passband above, is controlled by variations in λ_{co} resulting from fibre-to-fibre variations in profiles which alter the waveguide dispersion. The dispersion specification limits shown by the top and bottom of the box in Figure 5.21*b* of less than or equal 3.2 ps $\text{nm}^{-1}\text{km}^{-1}$ relate to a cut-off wavelength of 1310 ± 10 nm.

Bandwidth × distance product in 1300 nm band

The value of this product is of interest in dispersion-limited second generation trunk transmission systems. In the late 1980s the bit rate used in such system fibres would be 140 Mbit/s or the equivalent. The limitation arises from the combination of fibre dispersion and source linewidth so the information is presented with source linewidth $\delta\lambda$ as parameter in Figure 5.22.

The use of the dispersion-flattened fibre design with a 2 nm source linewidth increases the 50 GHz × km bandwidth from that shown in the figure to the band from 1310 to 1600 nm.

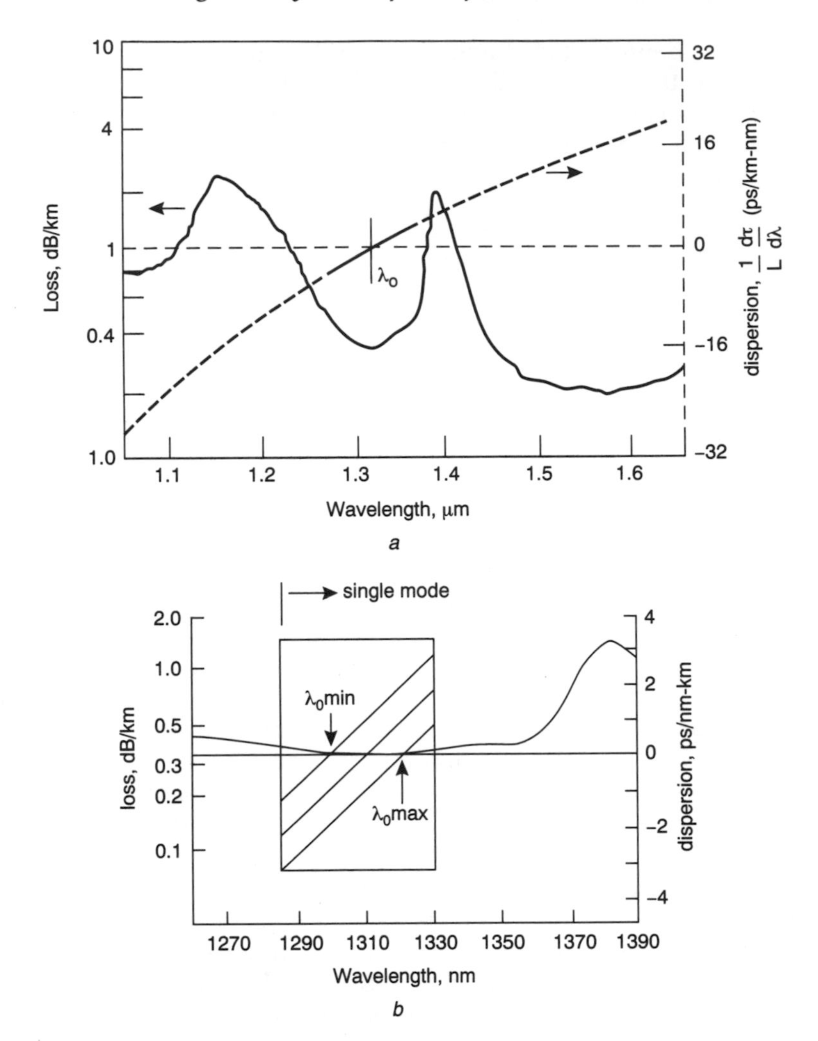

Figure 5.21 (*a*) Per-unit loss and dispersion versus wavelength; (*b*) in-band loss and dispersion spread versus wavelength of AT&T design of depressed-cladding fibre ([8], Figures 2*a*, 5). Reproduced by permission of AT&T

5.3.2 Choice of cut-off wavelength

The basic requirement on the value of λ_{co} in sm systems is that higher-order modes are attenuated sufficiently over a short distance in the fibre to avoid modal noise and an increase in pulse dispersion. In high bit-rate systems modal noise is the more serious limitation on performance. In practice, λ_{co} is determined from measurements on test lengths of fibre, and is found to be length-dependent as illustrated for several types of fibre in Figure 5.23.

It is evident from the figure that over the test lengths shown there is no sign of an asymptote; i.e. the attenuation of the higher-order modes is greater than for the fundamental, even to wavelengths well below 1310 nm. Conversely, at any wavelength there will be a length of fibre beyond which the power in higher-order modes becomes negligible; thus modal noise and pulse broadening become

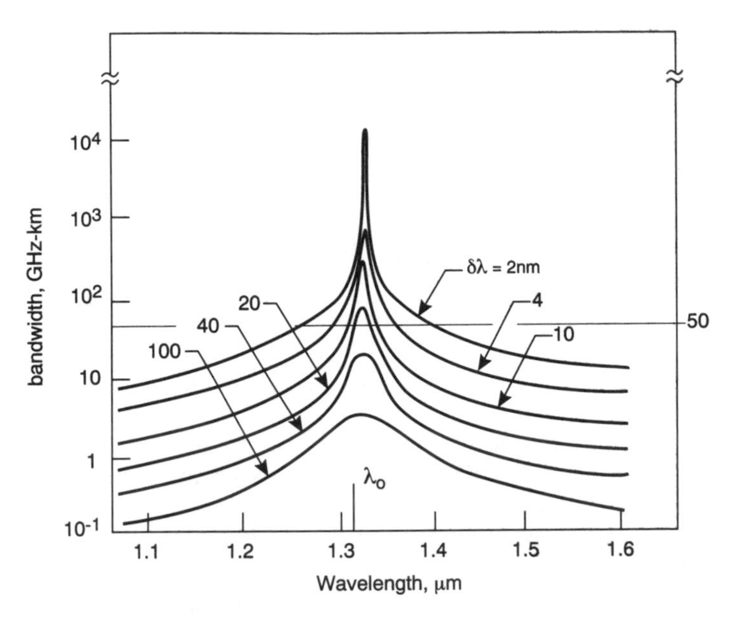

Figure 5.22 Calculated bandwidth × distance product versus wavelength with source linewidth as parameter of AT&T design of depressed-cladding fibre ([8], Figure 2b). Reproduced by permission of AT&T

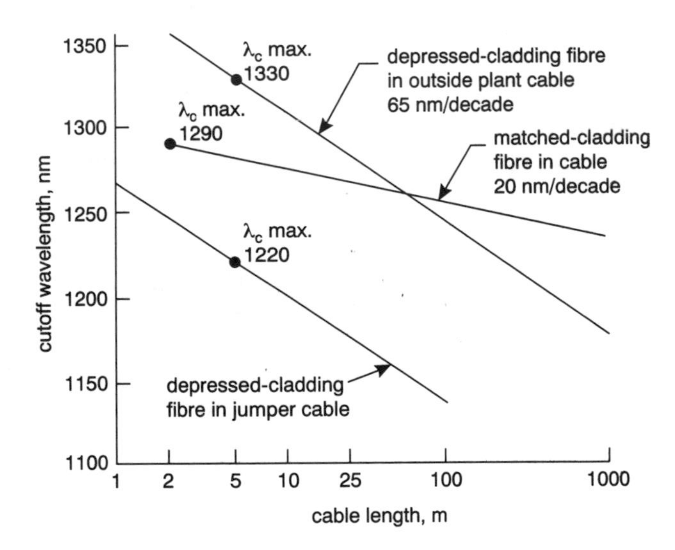

Figure 5.23 Cut-off wavelength versus sample length for two types of cabled depressed-cladding fibre and for cabled matched-cladding fibre ([8], Figure 6). Reproduced by permission of AT&T

negligible too. It will be seen from Figure 5.23 that two different slopes are present; these reflect the fact that two different loss mechanisms for the higher modes apply. Figure 5.23 shows that the same requirement for cut-off wavelength cannot be used for both types of profile. The dots are the maximum values of λ_{co} normally specified for each fibre. The value of 1330 nm for the depressed-cladding fibre was established by experiments and modelling of modal noise as a function of laser characteristics, fibre length between splices and splice losses. It gives a very conservative bound for avoiding modal noise in systems operating at 1300 nm. The authors quoted an example in explanation, and stated that no noise penalty would occur at 1285 nm for any cable length greater than 10 m between two high loss splices when the cut-off wavelength measured in a 5 m length is 1330 nm. In AT&T practice, outside cable plant lengths always exceed 10 m, even for short length repairs, so that this value of λ_{co} causes no problem. Short-length fibres for use as jumpers or with components (fibre tails) are selected from those having cut-off values below 1220 nm in a 5 m length in order to avoid modal noise.

5.3.3 Mode field diameter

The effects on splice loss, on macrobending and on microbending losses of the spot size or mode field diameter has already been discussed in Section 4.5. These effects will be illustrated for the particular AT&T design under consideration.

Firstly, the effect on the per-unit loss versus wavelength characteristic which is shown in Figure 5.24.

The most striking feature is the steep increase in loss due to macrobending, and the catastrophic effect on the window at 1550 nm. This loss edge moves to shorter wavelengths as the bend radius decreases. On the other hand, the microbending loss spectral dependence can have a downwards, zero or upwards slope depending upon the bend spectrum, although experiments on cables to induce microbending loss always show a positive slope.

Secondly, the effects on splice loss, macrobending and microbending losses shown in Figure 5.25 enable the reader to compare the effects for depressed-cladding and for matched cladding fibres.

This refers to 1310 nm and constant v-values based on the designs described earlier. The curves for splice loss are derived from data on arc fusion splices on

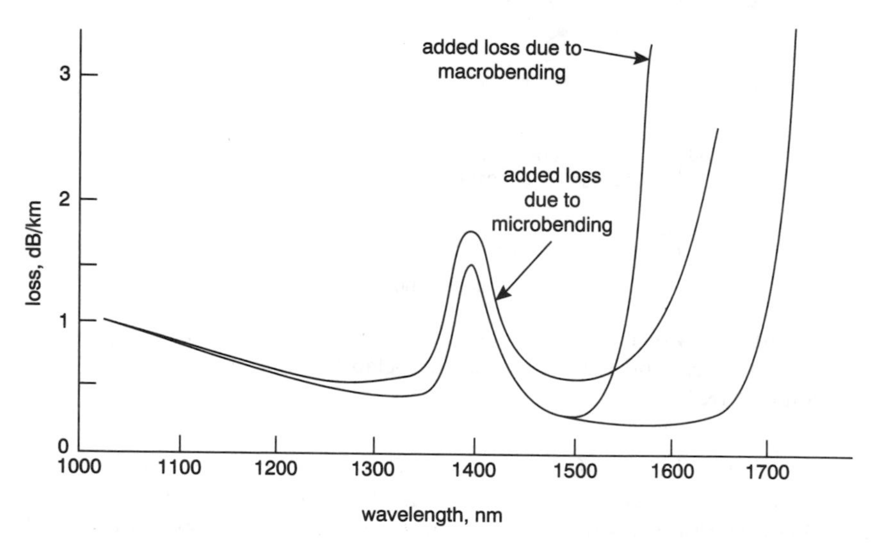

Figure 5.24 Per-unit loss curves versus wavelength for depressed-cladding fibre with added macrobending loss and added microbending loss superimposed ([8], Figure 7). Reproduced by permission of AT&T

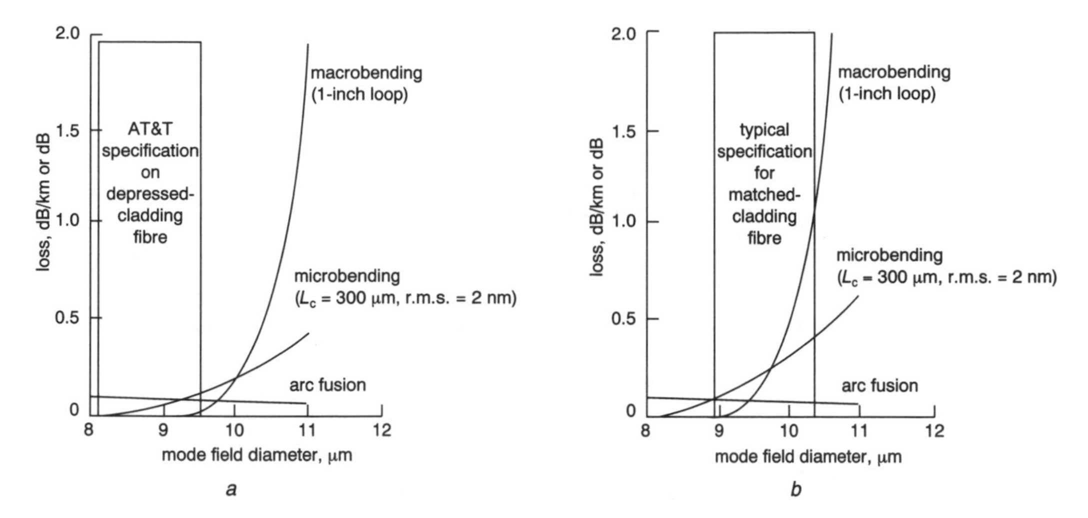

Figure 5.25 Calculated per-unit loss curves versus mode-field diameter for (*a*) depressed-cladding fibre and specification, and (*b*) matched-cladding fibre ([8], Figure 8). Reproduced by permission of AT&T

cables in the field, for which a median value of 0.08 dB loss was found. It is apparent that the splice losses are only very slightly dependent on spot size. In sharp contrast, the calculated values of the macrobending loss in particular are strongly dependent on mode field diameter. One can ask what would be the effect of changing either of the designs, for example, decreasing the v-value (or operating at 1550 nm which has the effect of reducing this value). The answer would be that the curves would be displaced to the left, consequently the fundamental mode would be less closely bound to the core resulting in a higher loss for a given spot size. The loop diameter for the macrobending calculations was chosen somewhat arbitrarily for the purpose of illustration; the effect of increasing this value is to slide the macrobending curve to the right. Calculations of microbending loss were based on a correlation length of 300 nm and a 2 nm r.m.s. deformation amplitude as a result of cable tests in the laboratory.

5.3.4 Dispersion-shifted fibres (dsf)

Typical total dispersion characteristics of matched- and depressed-cladding fibres are shown by curve (1), dispersion-shifted fibre by curve (2) and dispersion-flattened fibre by curve (3) in Figure 5.26*a*.

The slope of the total dispersion curve (1) for matched-cladding and depressed cladding (types (a) and (b) in Figure 5.28) at the wavelength of zero dispersion is about 0.09 ps $nm^{-2}km^{-1}$, as seen in Figure 5.21*b*, [8] quoted specification limits of ≤ 3.2 ps $nm^{-1}km^{-1}$ on dispersion; hence for the given slope the corresponding specification on the wavelength of zero dispersion is 1310 ± 10 nm.

The total dispersion shown in this figure is in fact made up of two effects; material dispersion and waveguide dispersion. Material dispersion is a bulk effect due to the nonlinear dependence of refractive index on wavelength which results in the group velocity of each propagating mode being different. Waveguide dispersion depends upon the waveguide structure (e.g. geometry, index profile) which

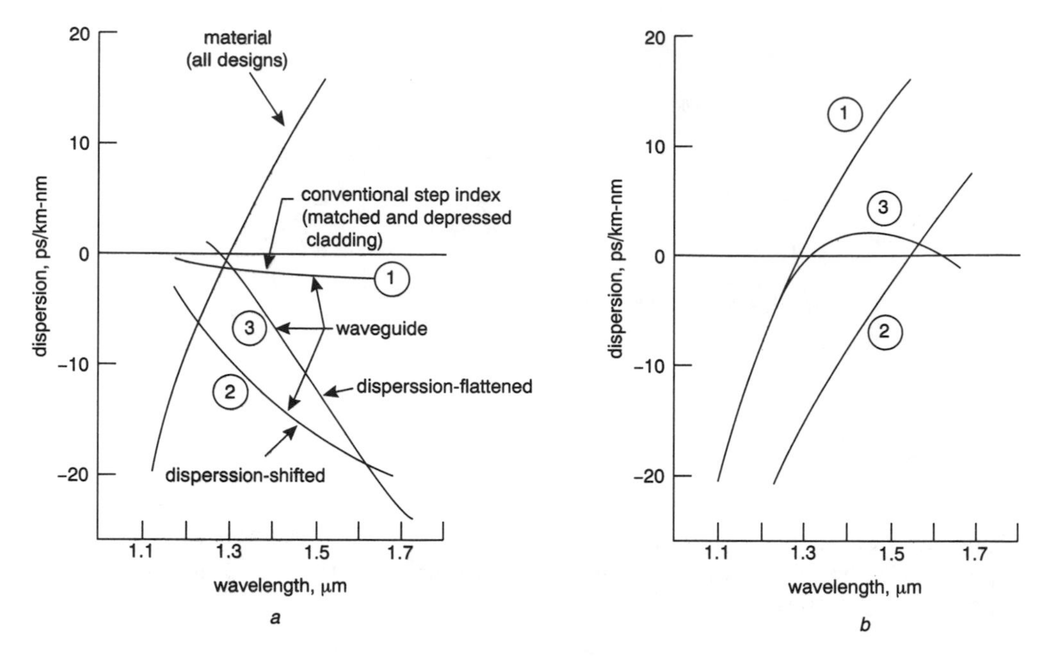

Figure 5.26 Dispersion versus wavelength characteristics for all four kinds of fibres: (*a*) curves of total dispersion, (*b*) corresponding material and waveguide components ([8], Figure 3). Reproduced by permission of AT&T

results in the group velocity also being a nonlinear function of frequency. Typical curves of these components for the four kinds of fibres of interest are illustrated in Figure 5.26*b*.

Note that all four types have the same material dispersion, which is positive in both 1300 and 1550 nm bands, and that all waveguide components are negative in these bands. The combination of the material dispersion and the appropriate waveguide dispersion curve produce the curves shown in Figure 5.26*b*.

The dispersion versus wavelength of dsf are shown in curve 2 of Figure 5.26*a*, with curve 2 in Figure 5.26*b* showing the waveguide and material components separately.

The profile type can be one of several, with triangular core being common and one which has been discussed already. The distribution of power in the core and in the cladding for step index and for triangular profiles is shown in Figure 5.27.

It can be seen that the triangular profile 'focuses' the transmitted power better than the step index, and so produces a smaller spot size, and a larger waveguide dispersion. Clearly, the increase in dispersion can be traded off against increase in core size, leading to the same value of dispersion as a step index fibre of smaller core size. The wavelength of zero dispersion is a function of core size and the relations for the two profiles of interest are shown by the families of curves in Figure 5.28.

In both cases the curves go through maxima, so the designer can choose the ratio of the index at the apex of the triangle to that at the base to minimise the requirements on the dimensions. If the desired zero wavelength is 1550 nm, then Δ is chosen so that the corresponding peak just reaches this value; the value of Δ is then the minimum which will satisfy the requirement. However, this conflicts with

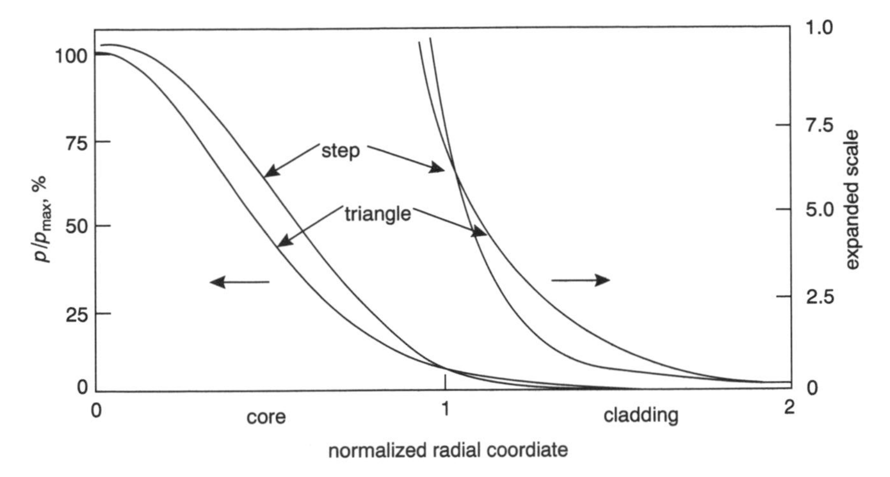

Figure 5.27 Relative power distribution versus normalised radial distance for step and triangular index profile dispersion-shifted fibres ([8], Figure 9*a*). Reproduced by permission of AT&T

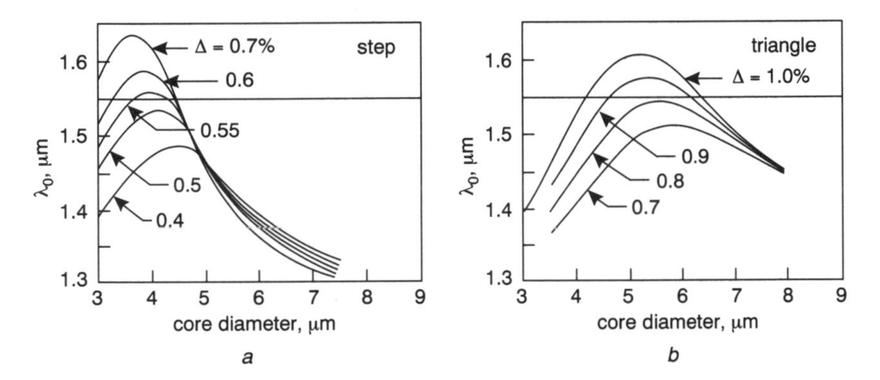

Figure 5.28 Cut-off wavelength versus core diameter: (*a*) for step index profile, and (*b*) triangular profile, with Δ as parameter ([8], Figure 9*b*). Reproduced by permission of AT&T

the need to operate at as high a v-value as possible to keep bending effects small. It follows that rather than operate at the peak, it is preferable to work with values from the right of the peak. From Figure 5.28*b* the best operational choice for the T-type profile is to take:

$$0.8 < \Delta < 1.0\% \quad \text{and} \quad 0.5 < \frac{\lambda_{co}}{\lambda} < 0.6$$

Dispersion-shifted fibres fabricated around the beginning of 1987 had per-unit losses below 0.25 dB/km at 1550 nm, with some fibres exhibiting as low as 0.20 dB/km. The fact that λ_{co} is sensitive to variations in fibre parameters does not cause unacceptable performance in systems comprising links each containing many jointed fibres, because the overall dispersion is a length-weighted average of the individual lengths of fibre between splices.

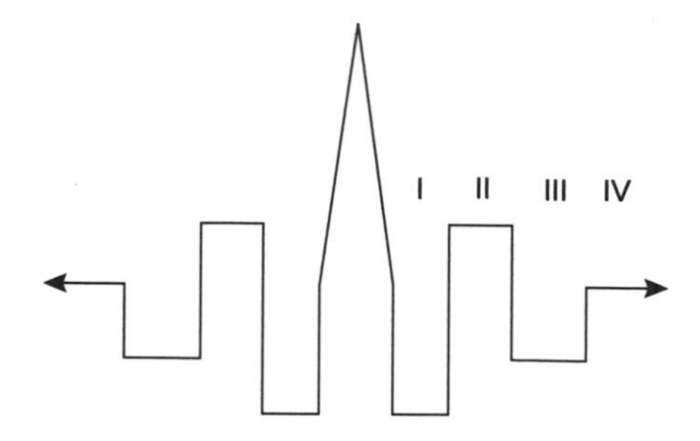

Figure 5.29 Refractive index profile of dispersion-flattened fibre with four cladding annuli ([8], Figure 1*d*). Reproduced by permission of AT&T

5.3.5 Dispersion-flattened fibres (dff)

Lightguide structures with more than two claddings have been designed which overcome the radial leakage of light through single well profiles; one such is the so-called Q-clad (quadrature) profile illustrated in Figure 5.29.

This has two wells separated by an annular ring guide in the second cladding which serves as a barrier to further radial leakage. As one would expect, the more complex structure provides more design parameters 'eight independent radial and index parameters are at the designer's disposal' to be optimised to achieve low loss and dispersion. The curves labelled 3 in Figure 5.26 refer to this kind of fibre. The corresponding bandwidth-distance products are most impressive, greater than 1000 GHz km near the dispersion minimum, and higher than 50 over the entire waveband from 1310 to around 1600 nm even for a source of linewidth of 2 nm, typical of a multimode laser at these wavelengths. As the reader can see from the figure, the dispersion is less than 2 ps $km^{-1}nm^{-1}$ in the same band. This performance is of course paid for by an increased sensitivity to deviations from the ideal profile, rather like a high-order reactance filter is more sensitive than a simpler filter, [8] quoted a minimum loss value of about 0.3 dB/km near 1520 nm for this dff design, but suggested that improved designs were reducing this figure. More recent designs will be mentioned elsewhere.

5.4 PULSE DISPERSION AND ITS IMPACT ON SYSTEM PERFORMANCE

Dispersion in optical fibres manifests itself in the broadening of light pulses which can cause intersymbol interference when transmitted through long lengths of fibre. It is often the most important limiting factor on the section length between repeaters in high bit rate systems. The term is used to cover the variation of the refractive index with wavelength, or more generally the variation of the propagation constant with wavelength. It may also be used to cover other mechanisms in fibres which cause distortion of these pulses. In discussing propagation in sm fibres it is important to distinguish between dispersion arising from material effects and that which is due to the waveguide properties. The more important of these causes will now be treated. In single-mode fibres the broadening is due only to distortion of the lowest order propagating mode rather than to some hundreds of propagating modes in multimode fibres; nevertheless, the effects degrade system performance.

5.4.1 Theory of dispersion in sm optical fibres

Chromatic dispersion

This is a combination of two related effects which arise because optical signal carriers, unlike carriers at radio frequencies, are not monochromatic (single tone). The two effects are material dispersion and waveguide dispersion. The first is due to the wavelength dependence of the refractive index of the medium and is given by $(\lambda/c)\mathrm{d}^2n/\mathrm{d}\lambda^2$. The waveguide dispersion is due to the ratio a/λ being frequency-dependent so that the field distributions and group velocities of the modes are wavelength-dependent.

The field pattern of each mode travels through the fibre with a phase velocity given by V_P, where

$$V_P = \frac{\omega}{\beta} = \frac{ck}{\beta} = \frac{c}{n_{EP}} \tag{5.1}$$

where n_{EP} is the effective phase index, and $n_2 < n_{EP} < n_1$. The corresponding speed of propagation of modulated signals is the group velocity V_g given by:

$$V_G = \frac{\mathrm{d}\omega}{\mathrm{d}\beta} = c\frac{\mathrm{d}k}{\mathrm{d}\beta} = \frac{c}{n_{EG}} \tag{5.2}$$

where n_{EG} is the effective group index. The reciprocal of this is the per-unit group delay τ_1 and is of more technical interest. The derivative of τ_1 with respect to the wavelength i.e. $\mathrm{d}\tau_1/\mathrm{d}\lambda$, a quantity known as the chromatic (or wavelength) dispersion and is also important, and will be considered shortly. Let us now apply these concepts to the step index fibre.

Dispersion in a step-index fibre

It is convenient to introduce the dimensionless quantity b as:

$$b = \frac{w^2}{v^2}, \quad 0 < b < 1 \tag{5.3}$$

which can be regarded as a normalised propagation constant. At cut-off it is zero because w equals 0, and at large v-values it tends to unity; this behaviour is shown in Figure 5.30.

Since the difference between the core and cladding indices is small, the propagation constant itself can be written as:

$$\beta = kn_2(1 + b\Delta) \tag{5.4}$$

where $2\Delta = 1 - (n_2/n_1)^2$. Provided that Δ does not vary with λ for the range of interest, the result of differentiating equation. (5.4) is:

$$\frac{\mathrm{d}\beta}{\mathrm{d}k} = N_2 + \Delta N_1 \frac{\mathrm{d}(bv)}{\mathrm{d}v} \tag{5.5}$$

where

$$N = \frac{\mathrm{d}(kn)}{\mathrm{d}k} \tag{5.6}$$

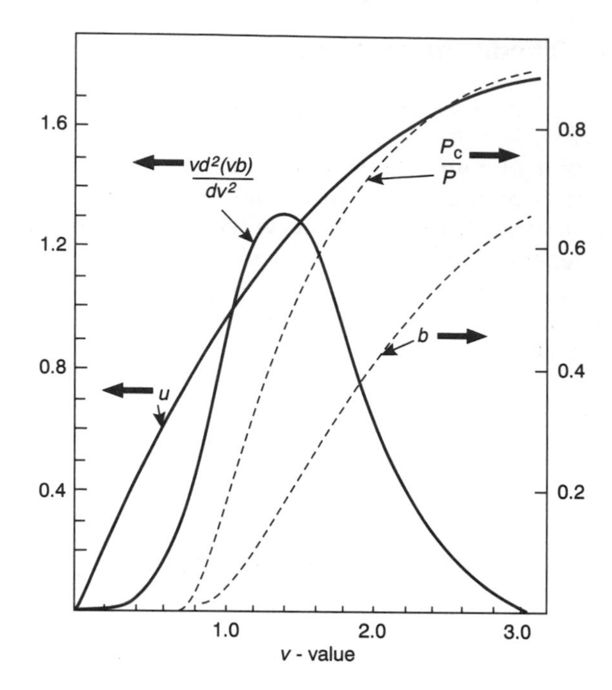

Figure 5.30 Universal lightguide parameters b, u and $v[\mathrm{d}^2(\mathrm{d}v)/\mathrm{d}v^2]$ versus v-value, relative power in core versus v-value ([3], Figure 11)

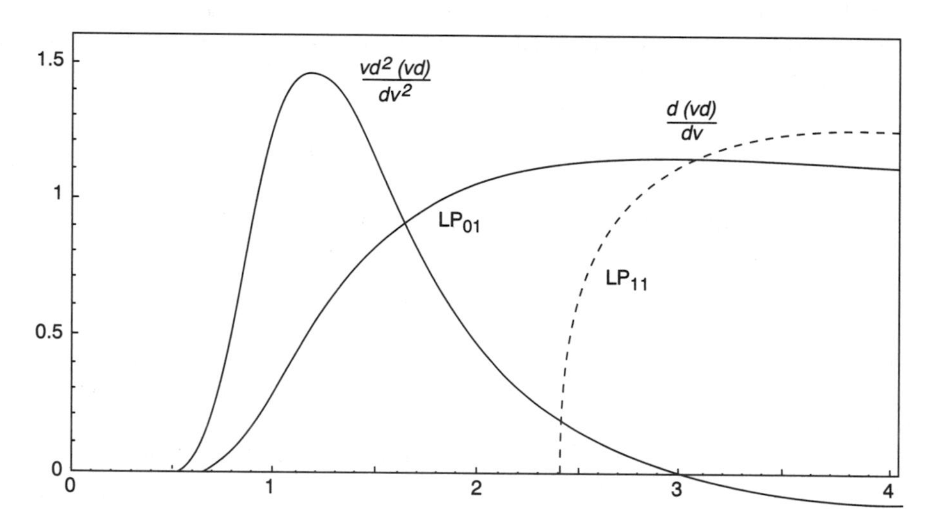

Figure 5.31 Dispersion parameters versus v-values ([4] of Chapter 3, Figure 10). Reproduced by permission of IEE

The terms N_1 and N_2 are the group indices of the core and cladding materials, and they measure the delay of plane waves travelling through a homogeneous volume of the bulk material. Clearly they are parameters of the materials; they are the same for all modes. The lightguide or modal aspect is expressed by $\mathrm{d}(bv)/\mathrm{d}v$, and this can be regarded as a normalised delay. The behaviour of this quantity for the first two modes is sketched in Figure 5.31.

The curves of $\mathrm{d}(bv)/\mathrm{d}v$ are zero at cut-off when v equals 0 or 1, and approaches 1 at large v-values. As each mode approaches cut-off its delay tends to N_2/c, since this is the delay of a plane wave in the bulk cladding material. On the other hand, far from cut-off its delay tends to N_1/c for the corresponding reason. Note that when the second mode just begins to propagate the difference between the delays of the first two modes is of the order of $N_1\Delta/c$. This is as large as the maximum delay spread in a step-index multimode fibre and illustrates once more the necessity to operate below with $v < 2.405$. To find the dispersion we differentiate equation (5.5) which yields:

$$c\frac{\mathrm{d}\tau}{\mathrm{d}\lambda} = \frac{\mathrm{d}N_2}{\mathrm{d}\lambda} \quad \text{(material dispersion term)}$$

$$- \frac{\Delta N_1}{\lambda}\left[v\frac{\mathrm{d}^2(vb)}{\mathrm{d}v^2}\right] \quad \text{(waveguide dispersion)} \tag{5.7}$$

Let us look at these in turn.

Material dispersion

This is the dispersion associated with the propagation of a non-monochromatic signal through a fibre with refractive index which is wavelength-dependent (or, equivalently, of the velocity of propagation within the fibre). For the material dispersion component of the dispersion $\mathrm{d}\tau/\mathrm{d}\lambda$ the material dispersion parameter $M(\lambda)$ is defined as:

$$M(\lambda) = \frac{1}{c}\frac{\mathrm{d}N_2}{\mathrm{d}\lambda} = \frac{\lambda}{-c}\frac{\mathrm{d}^2n}{\mathrm{d}\lambda^2} \tag{5.8}$$

and the usual unit for M is ns $\mathrm{nm^{-1}km^{-1}}$. For pure silica this contribution to the dispersion changes sign at 1270 nm; indeed, for most optical materials of interest it becomes zero in the vicinity of 1300 nm. Below this zero wavelength λ_0 it is negative; above it is positive, and reaches about 20 ps $\mathrm{km^{-1}nm^{-1}}$ at 1550 nm. The sign is chosen to make M negative below the zero dispersion wavelength and positive above it. The behaviour of this quantity over the wavelength range 1000–2000 nm can be seen from Figure 5.32.

Waveguide dispersion

This is the dispersion associated with the use of a non-monochromatic source due to the fact that the ratio a/λ (a is core radius) and consequently the field distributions and group velocities of the propagating modes are wavelength-dependent. In practice, waveguide and material dispersion always occur together and the combined effect is known as chromatic dispersion. The waveguide dispersion is given by $v\mathrm{d}^2(vb)/\mathrm{d}v^2$ in equation (5.7) and is shown in Figure 5.30 for the fundamental mode. It can be seen that it attains a maximum value at around $v = 1.2$ and changes sign from positive to negative close to $v = 3$, which lies above cut-off. It follows that the waveguide dispersion will be negative for single-mode systems, since these operate with $v < 2.4$. Because this is the case, the two components of dispersion can be made to cancel at wavelengths above 1300 nm or so, where they do for the ideal step index and depressed-cladding profiles. Dispersion-shifted fibres can also be designed which shift the zero dispersion wavelength into the 1550 nm band. A

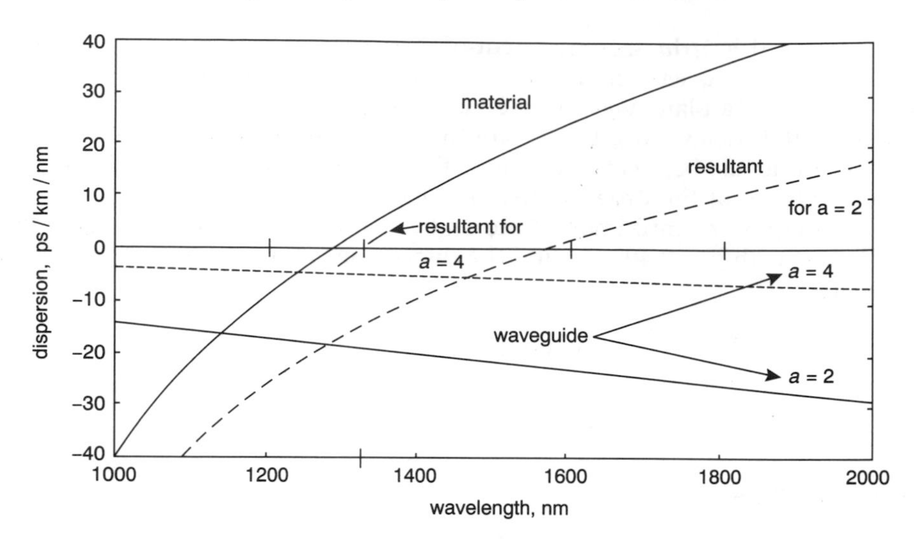

Figure 5.32 Curves of dispersion versus wavelength, a = core radius in μm ([4] of Chapter 3, Figure 11). Reproduced by permission of IEE

quantitative estimate of how far the zero can be shifted can be found by means of the useful approximation due to Rudolph and Neumann, namely:

$$w = 1.143v - 0.996 \tag{5.9}$$

which is valid for $1.3 < v < 3$, a range which includes the values of interest in single-mode systems. Using this in conjunction with equation (5.8) leads to a value of $-\lambda/[cn(2\pi a)^2]$ for the waveguide dispersion, from which the importance of the core radius can be seen. The strong influence of core radius a on waveguide dispersion is apparent from Figure 5.32. In particular, halving the value of a from 4 to 2 μm results in a shift of the zero dispersion wavelength from about 1310 to 1570 nm according to the figure.

Polarisation mode dispersion (pmd)

The fundamental mode consists of two independent and orthogonal polarisations which are degenerate when the waveguide has perfect circular symmetry. Strain in the lightguide or geometric imperfections will remove this degeneracy and cause birefringence. The two polarisations then have different group delays, resulting in pulse-broadening irrespective of the source linewidth.

PMD sets an ultimate limit to the bit rate at which information can be transmitted over long-haul systems.

In the preceding analysis the variation of Δ with λ has been ignored, so that the profile dispersion term $d\Delta/d\lambda$ has been taken as constant. This is not strictly true, and though smaller than the other components of dispersion it is not negligible.

Power law profiles and their generalisation

In terms of the maximum refractive indices n_1 and n_2 in the core and cladding, respectively, equations power law profiles can be expressed in the simple form repeated here from equations (3.5) and (3.6):

$$n^2(r) = n_1^2 - (na)^2 \left(\frac{r}{a}\right)^\alpha, \text{ for } r < a \tag{5.10a}$$

$$n^2(r) = n_2^2, \text{ for } r \geq a \tag{5.10b}$$

The step, parabolic and triangular profiles are obtained by putting $\alpha = \infty$, 2 and 1, respectively. This pair of equations can be generalised by replacing $(r/a)^\alpha$ by $f(r)$, giving:

$$n^2(r) = n_1^2 - [n_1^2 - n_2^2]f(r), \text{ for } r \leq a \tag{5.11}$$

$$n^2(r) = n_2^2, \text{ when } r > a \tag{5.12}$$

Both indices are wavelength-dependent, but assuming linear profile dispersion $f(r)$ is independent of wavelength. The modal group delay is defined as:

$$-2\tau(\lambda) = \pi c \times \lambda^2 \times \frac{\mathrm{d}\beta}{\mathrm{d}\lambda} \tag{5.13}$$

where $\mathrm{d}\beta/\mathrm{d}\lambda$ can be obtained by differentiating the stationary expression for β^2, yielding:

$$\frac{\lambda^2 \mathrm{d}\beta}{\mathrm{d}\lambda} = \text{(a)} \times \text{(b)} \times \text{(c)} \tag{5.14}$$

where

$$\text{(a)} = -\lambda\beta\left(1 + \frac{2}{\omega_\mathrm{P}^2\beta^2}\right) \tag{5.15a}$$

$$\text{(b)} = \frac{2\pi^2}{\beta} \times \frac{\mathrm{d}n_2^2}{\mathrm{d}\lambda} \tag{5.15b}$$

$$\text{(c)} = \left\{\left(\beta^2 + \frac{2}{\omega_\mathrm{P}^2}\right) - \frac{4\pi^2 n_2^2}{\lambda^2}\right\}\frac{\mathrm{d}P}{\mathrm{d}\lambda} \tag{5.15c}$$

(a) represents the pure waveguide contribution obtained when $\mathrm{d}n_1^2/\mathrm{d}\lambda = 0 = \mathrm{d}n_2^2/\mathrm{d}\lambda$. (b) is the pure material contribution corresponding to the case when the two derivatives are equal but nonzero. (c) is the linear profile contribution.

In the expressions the letter p stands for $\ln(n_1^2 - n_2^2)$ and ω_P denotes the Petermann spot size for which a defining expression was given. The authors showed that equation (5.14) could be integrated explicitly and exactly without ignoring any of the terms. The resulting equation corresponding to (5.14) is then:

$$b'(v) - b'(v_0) = \int \frac{4\mathrm{d}v}{v^3\omega^2} \tag{5.16}$$

where v denotes the normalised frequency defined in equation (3.9) and $b' = a^2 b$ with b given by equation (5.3). The upper and lower limits of integration are v and v_0, respectively. The quantity ω is ω_P/a and v_0 is the normalised frequency corresponding to a reference wavelength λ_0 not equal to λ. The significance of the reference wavelength is that $b'(\lambda_0)$ can be found accurately using bend loss measurements and combined with a knowledge of the behaviour of the Petermann spot size between this wavelength and λ itself to yield a value for $b(\lambda)$. Differentiating equation (5.16) with respect to v gives an important expression for the quantity ω_P/a, namely:

$$\left(\frac{\omega_P}{a}\right)^2 = \frac{4}{v^2(b_1' - b')} \tag{5.17}$$

This result leads to a known relationship between the fractional power in the core region η, the degree of guidance Ω and the spot size, q.v.:

$$\left(\frac{\omega_P}{a}\right)^2 = \frac{4}{v^2(\eta - b')\Omega} \tag{5.18}$$

5.4.2 Dispersion statistics in concatenated single-mode fibres

Dispersion is of particular interest in the case of long-haul systems in which there may be a large number of concatenated single-mode fibres or at high bit-rates. Two important parameters are the wavelength of zero dispersion λ_0, and the slope of the chromatic (i.e. the sum of the material and the waveguide) dispersion characteristic at this wavelength. The reason for their importance lies in the possibility of chromatic dispersion limiting the maximum repeater spans attainable, especially when multi-longitudinal mode lasers are used. Since measurements in the field are often time-consuming and therefore expensive, it is desirable to deduce relations between the performance of concatenated fibres and the statistics of the zero wavelength and the slope of individual lengths. In particular, the effects of the variances on repeater section length dispersion budgetting have been clarified for matched or depressed cladding fibres for operation at 1300 nm [9].

Dispersion measurements were made in the factory on two sets of 112-fibre and 153-fibre cable lengths, and the group delays of individual fibres were also measured. A three-term Sellmeier curve was fitted to these delay data. Assuming that the delay of a link of many jointed fibres is the sum of the individual delays, a Monte Carlo technique is used to determine histograms of slope and λ_0 for 6×10^4 concatenated links. The analysis was done for links with up to 9, 16 and 25 cable lengths, and under two different scenarios: these attempt to cover the following situations. The cable route contains only fibres made by the same process and under similar conditions; the cable contains fibres from two different manufacturers; or the cable fibres come from the same manufacturer but were fabricated by two different processes. Results of the analysis indicated that the variances of the slope at the wavelength of zero dispersion, σ_s and of the zero dispersion wavelength itself σ_0 of the dispersion in jointed fibres decrease as $1/N$, where N is the number of concatenated fibres in the cable route. Furthermore, the overall λ_0 is almost independent of the individual slopes, regardless of actual measured values, if random values uniformly distributed in the range 0.08–0.1 ps $\text{km}^{-1}\text{nm}^{-2}$ were used instead, since the values of the average and standard deviation of λ_0 changed by less than 0.005% and 1%, respectively. The main conclusion of the study suggested that dispersion measurements in the field may be unnecessary, if statistical data on slope and zero dispersion wavelength obtained in the factory are available. More information can be found by consulting the reference.

5.4.3 Chromatic dispersion measurements on zero-shifted concatenated sm fibres

The results of measurements in the 1550 nm band were reported in late 1986 [10]. The letter described measurements on a 123 km length of jointed fibres using the semiconductor laser phase shift method.

Experimental arrangement: this consisted of four sinusoidally modulated lasers, an apd detector and a phase detector. The optical carrier wavelengths were at 1496, 1520, 1560 and 1590 nm. With sinusoidal modulation the modulated light travels along the guide with a phase shift given by:

$$\phi(\lambda) = 2\pi f_M \times L \times \tau(\lambda) \tag{5.19}$$

where f_M is the modulating frequency (Hz), L is the overall length (km), τ per-unit time delay (ps/km).

By detecting phase differences among lasers, and using the least-mean-square fitting approximation for the Sellmeier equation:

$$\tau(\lambda) = a\lambda^2 + b + c\lambda^{-2} \tag{5.20}$$

the wavelength-dependent relative time delay can be calculated. Differentiating this delay yields the chromatic dispersion.

For the experiment, twelve zero-shifted fibres with lengths from 8 to 17 km were spliced into a 123 km link, and some details of individual fibres are given in Table 5.5.

Column 4 lists the dispersion of the separate fibres and the reader can see that there is quite a variation in the values. Again from the table, columns 5 and 6, it can be seen that the agreement between calculated and measured values is very good. The average per-unit loss of the fibres was 0.25 dB/km, and the average splice loss was 0.125 dB, both referred to 1550 nm.

Table 5.5 Dispersion values of constituent fibres ([10], Table 1). Reproduced by permission of IEE

Fibre	Length, km		Dispersion (ps/km/nm) at 1.55 μm		
no.	Individual	Total	Individual	Calculated	Measured
1	8.3	8.3	0.07	0.07	0.07
2	8.3	16.6	2.57	1.32	1.36
3	8.3	24.9	3.41	2.02	2.03
4	8.3	33.2	3.32	2.34	2.35
5	8.3	41.5	3.12	2.50	2.50
6	8.3	49.8	6.24	3.12	3.22
7	12.6	62.4	1.39	2.77	2.87
8	17.1	79.5	1.04	2.40	2.61
9	8.7	88.2	2.93	2.45	2.48
10	8.7	96.9	2.31	2.44	2.53
11	8.7	105.6	1.43	2.36	2.45
12	17.0	122.6	1.33	2.21	2.23

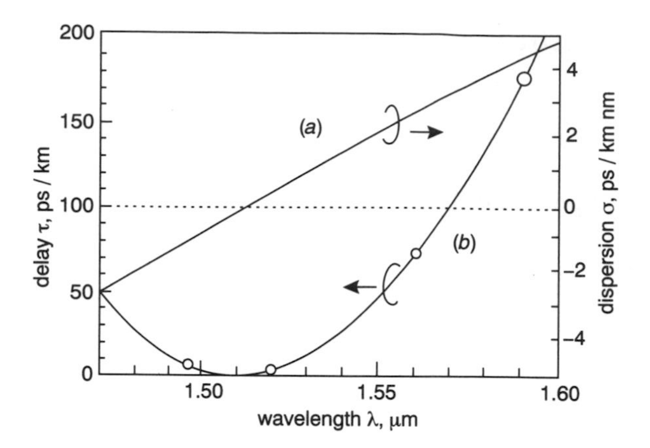

Figure 5.33 (*a*) Measured chromatic dispersion of 123 km of dsf, (*b*) corresponding group delay ([10], Figure 1). Reproduced by permission of IEE

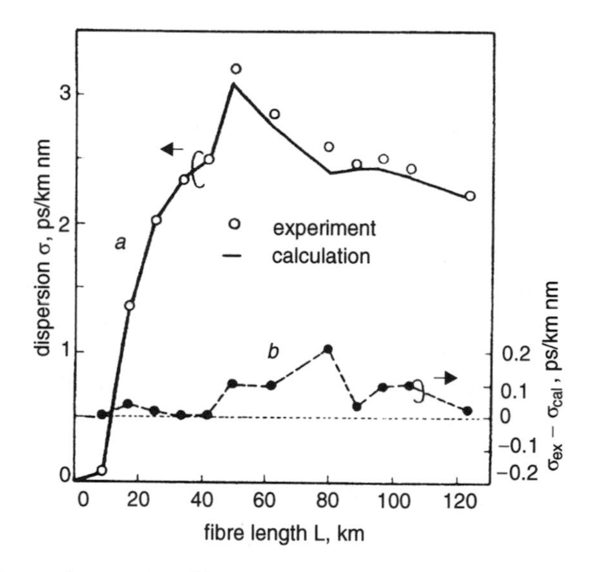

Figure 5.34 (*a*) Dispersion versus fibre length, and (*b*) difference between measured and calculated dispersion versus fibre length ([10], Figure 2). Reproduced by permission of IEE

Results: the measured dispersion values on the link are given in Figure 5.33 together with the per-unit delay. Experiments gave the wavelength of zero dispersion as 1510 nm, while the dispersion at 1550 nm was 2.33 ps $km^{-1}nm^{-1}$. The dynamic range of the measuring equipment is about 32 dB, and this would allow lengths up to 150 km to be measured if the per-unit loss was less than 0.21 dB/km. The measured dependence on length of the dispersion is presented in Figure 5.34 where it is indicated by open circles.

The solid lines in this figure correspond to the calculated values given in Table 5.5. For further details the reader is referred to the original. In conclusion, the authors claim to have measured chromatic dispersion over this length with a standard deviation of 0.01 ps $km^{-1}nm^{-1}$, and to have confirmed the addition rule for the dispersion of a long length of concatenated dispersion-shifted fibres, i.e. the measured value agreed well with that estimated from the sum of the dispersion values of all the constituent fibres.

5.4.4 Impact of production tolerances on dispersion

This important topic was discussed in 1987 in a paper from Denmark [11]. The author showed that optimum values of the slope of the dispersion, evaluated at the wavelength of zero dispersion, can be found as functions of the production tolerances of single-mode fibres. These were based on expressions which were independent of index profile. The optimum value minimises the spread in dispersion, or maximises the available wavelength interval of the fibre. The maximum allowable tolerances for dispersion-flattened fibres are derived.

The most important variances can be represented by the relative standard deviations of the two main fibre parameters, core radius a and Δ, which can be written as σ_a/a and σ_Δ/Δ, respectively. Based on a first-order expansion of the scalar wave equation a simple relation can be derived between the slope of the dispersion at the zero wavelength, $\mathrm{d}D/\mathrm{d}\lambda$, and the relative dispersion sensitivities $a\mathrm{d}D/\mathrm{d}a$ and $\Delta\mathrm{d}D/\mathrm{d}\Delta$. The simple relation is then used to find optimum slopes for given standard deviations and maximum allowable dispersion of the laser source to be used in the system. The theory can be used for dispersion-shifted fibres, and for dispersion-flattened designs to give the maximum values of the standard deviations.

Theory: based on a first-order expansion of the scalar wave equation, and assuming weakly doped fibres, it can be shown that the sensitivities of the total dispersion to the two production variables a and Δ may be written:

$$a\frac{\mathrm{d}D}{\mathrm{d}a} = \frac{\mathrm{d}[\lambda(M + P - D)]}{\mathrm{d}\lambda} \tag{5.21}$$

$$\Delta\frac{\mathrm{d}D}{\mathrm{d}\Delta} = 0.5a\frac{\mathrm{d}D}{\mathrm{d}a} + D - M \tag{5.22}$$

where M is the material dispersion, P is the profile dispersion, and W is the waveguide dispersion, D is the sum of all three components.

For the fibres being considered, P is negligible and M is almost equal to that of pure silica, so that these equations can be approximated at the zero dispersion wavelength by:

$$a\frac{\mathrm{d}D}{\mathrm{d}a} \simeq M_1 - \lambda S \tag{5.23}$$

$$\Delta\frac{\mathrm{d}D}{\mathrm{d}\Delta} \simeq M_2 - \frac{\lambda S}{2} \tag{5.24}$$

with the two parameters M dependent only on λ, and where S equals $\mathrm{d}D/\mathrm{d}\lambda$. Thus, at the wavelength of zero dispersion, the sensitivities depend solely on the slope of the dispersion there. The author then considers a type T fibre with $\lambda_0 = 1550$ nm and $\omega_0 = 3.6$ µm, and gives curves showing the errors in the M, arising from the approximations of the profile and the material dispersion. In all the calculations, Sellmeier's coefficients are used for pure silica, and Kobayashi's for silica with 7.9 mole % GeO_2. If the two standard deviations (r.m.s. values σ_a and α_Δ) are uncorrelated, then the mean square deviation $\sigma_{2D}{}^2$ on the dispersion at λ_0 is the sum of their squares. Let the maximum allowable value of dispersion at λ_0 be D_M; then the usable waveband is given by:

$$2\delta\lambda = \lambda_0 \pm \delta\lambda = \frac{D_M - \sigma_D}{S} \tag{5.25}$$

For given values of the relative standard deviations one can find two values of S: one which minimises the r.m.s. value of the dispersion, say S_1; a second maximises the value of $\delta\lambda$, denoted by S_2. The value of S_1 is shown to be a function only of the ratio $(\sigma_\Delta/\Delta)/(\sigma_a/a)$ for a given λ_0. In a numerical example the range of this quantity was given as:

$$0.052 < S_1 < 0.082 \text{ ps km}^{-1}\text{nm}^{-2}$$

Conclusion: the two relative deviations have been shown to define optimum dispersion slopes at the zero wavelength independent of fibre design. To minimise variations of dispersion at this nominal wavelength for dispersion shifted fibres, the slope of $D(\lambda)$ should lie in the range quoted, with the lower value chosen when Δ is the more critical, and the upper value when the core radius is hard to control. For the largest waveband, the slope should be less than or equal to that which minimises the r.m.s. value of the total dispersion. Turning to dispersion-flattened designs, the author suggests that the quantities σ_a/a and σ_Δ/Δ should be less than 1.6 and 5%, respectively, and goes on to suggest that fibres with two well-controlled zeros at 1300 and 1550 nm and a more or less well-controlled point in between may be easier to manufacture than completely flattened designs.

5.4.5 Temperature dependence of chromatic dispersion in single-mode vad fibres

Highly accurate measurements of chromatic dispersion over the range −60 to +250°C were reported in [12]. These measurements demonstrated a linear relationship for $\lambda_0(\theta)$ for temperatures below 150°C with a slope of 0.025 nm/°C, but with an increasing dependence above 150°C. Both below and above this value the results were independent of fibre coating and structure. All fibres were manufactured by the vad process, and further details of fibre structure and fabrication can be found in [12]. The preparation and measurements were also given prior to figures illustrating the results. Of the two major contributions to chromatic dispersion, the waveguide component was virtually independent of temperature, leaving the material part to account for the observed dependence.

5.5 GROUP DELAY IN SINGLE-MODE FIBRES

The definition of group delay was given in equation (5.2) as the reciprocal of the group velocity. As at much lower frequencies, such as radio or microwave, the exact interpretation of group delay is sometimes difficult to establish. Methods for measuring this quantity for optical systems have been devised.

5.5.1 Polynomial approximations to the per-unit group delay

The accuracy of chromatic dispersion measurements on dispersion-shifted fibres was examined experimentally for four proposed polynomial functions, and it was

Table 5.6 Comparison of four group delay fitting equations. Reproduced by permission of IEE

Fitting equation	Measuring wavelength array	DSF-A Fitting error	DSF-A λ_0	DSF-A S_0 $\times 10^{-2}$	DSF-B Fitting error	DSF-B λ_0	DSF-B S_0 $\times 10^{-2}$	DSF-C Fitting error	DSF-C λ_0	DSF-C S_0 $\times 10^{-2}$
		ps/km	nm	ps/km nm^2	ps/km	nm	ps/km nm^2	ps/km	nm	ps/km nm^2
$\tau_1 = A_1\lambda^2 + A_2$	① Narrow	0.80	1531.79	5.952	1.32	1544.78	8.686	0.81	1553.00	7.468
$+ A_3\lambda^{-2}$	② Broad	16.60	1537.81	5.180	36.62	1559.42	7.053	34.48	1577.73	5.327
	|①-②|	15.80	6.02	0.772	35.30	14.64	1.633	33.67	24.73	2.141
$\tau_2 = A_1\lambda^4 + A_2\lambda^2$	① Narrow	0.19	1532.02	5.555	0.59	1545.29	8.806	0.08	1553.67	7.706
$+ A_3 + A_4\lambda^{-2}$	② Broad	4.71	1529.31	6.146	7.21	1544.13	8.628	1.11	1552.05	7.821
$+A_5\lambda^{-4}$	|①-②|	4.52	2.71	0.591	6.62	1.16	0.178	1.03	1.62	0.115
$\tau_3 = A_1\lambda^3 + A_2$	① Narrow	0.57	1532.12	5.878	0.83	1544.81	8.766	0.38	1553.23	7.607
$+A_3\lambda^{-1}$	② Broad	6.87	1526.55	6.581	11.75	1545.03	9.082	9.80	1557.59	7.134
	|①-②|	6.30	5.57	0.703	10.92	0.22	0.316	9.42	4.36	0.473
$\tau_4 = A_1\lambda + A_2$	① Narrow	0.58	1532.11	5.884	0.78	1544.87	8.764	0.39	1553.21	7.619
$+A_3\lambda$ In λ	② Broad	11.80	1525.44	6.744	11.09	1542.78	9.392	4.15	1554.33	7.424
	|①-②|	11.22	6.67	0.860	10.31	2.09	0.628	3.76	1.12	0.195

found that the five-term Sellmeier polynomial provided the best fit to dispersion-shifted fibres with various index profiles [13]. The authors also used the laser phase shift method of measurement at six different wavelengths between 1490 and 1590 nm to demonstrate that the maximum difference in the zero of dispersion between the four equations was less than 0.47 nm.

Results: a comparison of the results from the use of the four polynomials is given in Table 5.6.

The headings DSF-A, B, C refer to three different typical profiles (illustrated in [5.13]) of the dsf used in the measurements. A resolution of better than 0.2 ps/km was achieved for the group delay, and the standard deviation for the measured dispersion values was less than about 0.01 ps km^{-1}nm^{-1}, whereas that for the zero wavelength was less than about 0.05 nm. The combinations of semiconductor lasers used in the experiment are given in Table 5.7 below.

As can be seen, two arrays were used, and it was found that the values obtained with the narrowband array were more accurate than those with the broadband array, regardless of which approximating polynomial was chosen. As an illustration, the results of the dispersion measurements on fibre A are shown in Figure 5.35.

Table 5.7 Measured wavelength array ([13], Table 2). Reproduced by permission of IEE

Wavelength	① Narrowband array	② Broadband array
	nm	nm
λ_1	1492.9	1205.4
λ_2	1518.1	1270.8
λ_3	1528.3	1309.4
λ_4	1548.8	1358.9
λ_5	1569.5	1528.3
λ_6	1587.8	1587.8

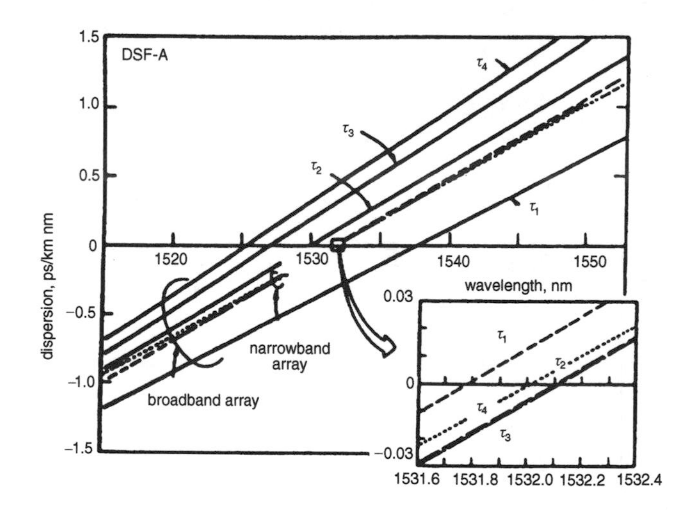

Figure 5.35 Examples of measured dispersion versus wavelength for dsf with profile type A ([13], Figure 2). Reproduced by permission of IEE

Study of the inset shows that the difference between equations is less than 0.33 nm for this fibre; for fibre B the corresponding value was 0.51 nm and for fibre C it was 0.67 nm. Larger differences were found with the broadband array, and overall the best fit is obtained with the five-term polynomial, as the reader can see from Table 5.7. In addition, the differences between the zero wavelengths and between the slopes of the dispersion at the zeros were less in the five-term equation than in any of the others. The authors concluded that a five-term Sellmeier polynomial fit to the group delay is most appropriate for use with dispersion-shifted fibres of arbitrary profile.

5.5.2 Measurements of group delay

Highly accurate measurements of group delay have been reported, e.g. [14]. These measurements were made in long lengths of fibre over the band 1200–1600 nm by means of the phase shift of a modulated optical carrier. The method allowed measurements of group delay to be made over the range 1200–1600 nm with a resolution of 0.93 ps. A 33 dB optical dynamic range was available at the wavelength of peak emission. An experimental arrangement with a fully remote light source was described for measurements on installed links.

The high modulation frequency was shifted down to the kHz range using a second optical carrier at the fibre output, thus preserving the resolution obtained by use of the high-frequency modulation while obtaining the ultrasensitive detection and phase measurement possible at the lower frequency. Among the results quoted using two optically coupled leds and fibres several km in length were those for (a) a standard step-index sm fibre and (b) a dispersion-flattened design shown in Figure 5.36, which shows the excellent fits achieved with the Sellmeier polynomials (full lines) and the excellent temporal resolution obtained. The stability and reproducibility of the measurements were tested by calculating the wavelength of zero dispersion from 12 independent measured group delay versus wavelength characteristics. A histogram of the resulting distribution is given in Figure 5.37.

The standard deviation on λ_0 was 0.09 nm.

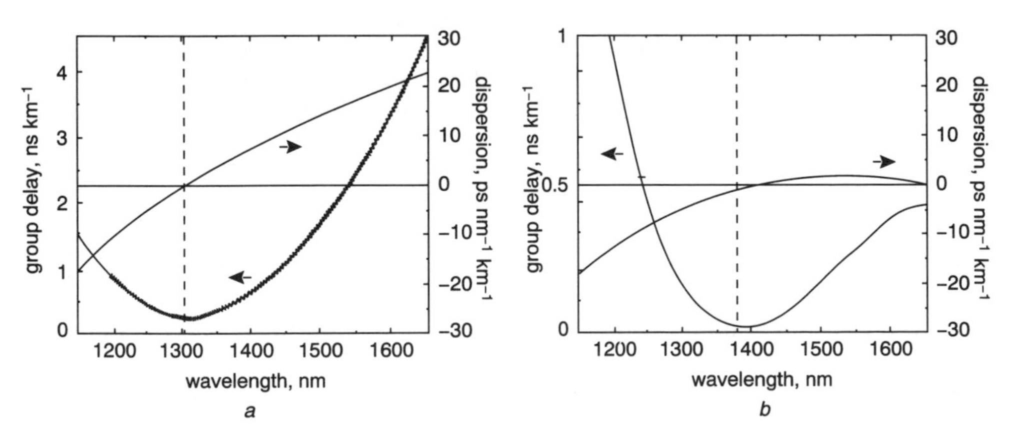

Figure 5.36 Group delay versus wavelength measured using two optically coupled LEDS and (*a*) standard step-index fibre, (*b*) dispersion-flattened fibre, together with fitted Sellmeier polynomials and chromatic dispersion ([14], Figure 5). Reproduced by permission of IEEE, © 1988

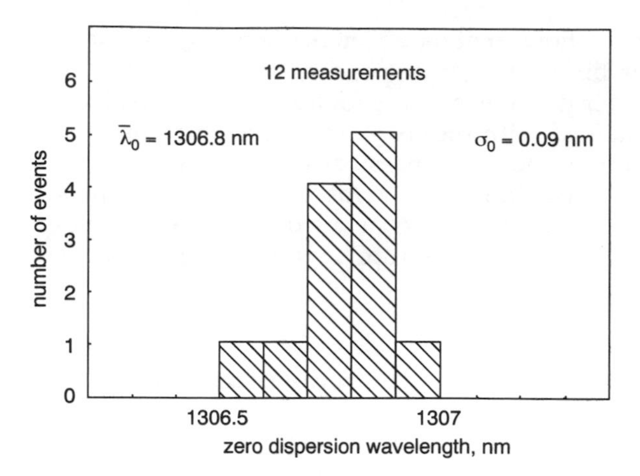

Figure 5.37 Histogram of the distribution of wavelength of zero dispersion calculated from independent measurements of the same fibre ([14], Figure 6). Reproduced by permission of IEEE © 1988

5.5.3 Temperature stability of transit time in loose-tube cabled fibre

The transit time of a single-mode fibre in a loose tube cable have been measured as a function of temperature and the results are quoted in [15]. The temperature range was −50 to +60°C, and the results indicated a positive linear slope of 6.7 ppm/°C for the transit time, a value which was the lowest reported as at 1983. Loose-tube single-mode fibres exhibit high stability of transit times against variations of temperature; consequently there is a very low cable-induced clock phase jitter and drift. The equation for the dependence of transit time on temperature is given as:

$$c\frac{\delta\tau}{\delta\theta} = L\frac{\delta n}{\delta\theta} + n\frac{\delta L}{\delta\theta} \tag{5.26}$$

where τ is the transit time, θ is the temperature, L is the length of fibre, n is the refractive index of core, and c is the speed of light in vacuum.

The first term on the right-hand side represents the temperature dependence of refractive index, the second the dependence on length. In turn, the value of $n(\theta)$ is a function of bulk density, electron polarisability and stress-optic effects. For doped silica fibres, typical values due to these effects are −1 ppm/°C, 1–20 ppm/°C and a two to tenfold increase in jacket induced stress-optic effects, respectively. The two latter contributions are usually dominant, leading to overall values of 16–40 ppm/°C for plastic jacketed fibres and 16–150 ppm °C for solid cables with strength members. Measurements were made on a 200 m length of bare doped silica sm fibre, and on a 1058 m length of loose tube cable containing the same type of fibre. In the results a high degree of correlation was found between the bare fibre and the cabled fibre, as can be seen in the Figure 5.38.

Over the wide temperature range covered by the measurements the relative phase changes were almost linear, and had a slope of about 6.7 ppm/°C. The slight hysteresis effect found with the cable was due to the measurement and not to the cable itself. This value was the best reported until that date for single-mode loose tube cabled fibre, and it compares with a value of about 10 ppm/°C for the best coaxial cable using a foam dielectric, and some 16 ppm/°C for a solid cabled

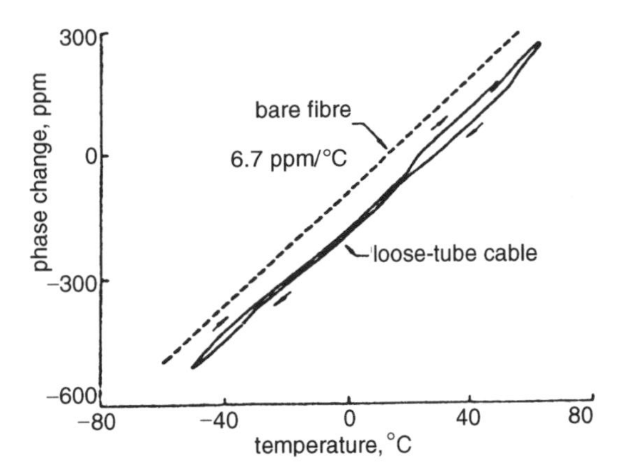

Figure 5.38 Relative phase change (transit time) versus temperature for bare and loose tube cabled sm fibres ([15], Figure 2). Reproduced by permission of IEE

fibre. The authors draw the conclusion that their results suggest that the loose tube structure is one of the more viable candidates for Gbit/s systems.

5.6 REFERENCES

1 AINSLIE, B.J., BEALES, K.J., COOPER, D.M., and DAY, C.R.: 'Monomode optical fibres with graded-index cores for low dispersion at 1.55 μm'. *British Telecom Technol. J.*, **2**, no. 2, Apr. 1984, pp. 25-33.

2 AINSLIE, B.J., BEALES, K.J., DAY, C.R., and RUSH, J.D.: 'The design and fabrication of monomode optical fiber'. *IEEE J. Quantum Electronics*, **QE-18**, no. 4, Apr. 1982, pp. 514-523.

3 GARRETT, I., and TODD, C.J.: 'Components and systems for long-wavelength monomode fibre transmission'. *Opt. and Quantum Electronics*, **14**, 1982, pp. 95-143.

4 CARRATT, M., REINAUDO, C., JOCTEUR, R., and TREZEGUET J.-P.: 'Dispersion shifted fiber for long unrepeatered submarine systems'. *Elect. Communication*, **61**, no. 4, 1987, pp. 384-388.

5 NAMIHIRA, Y., HORIUCHI, Y., KUWAZURU, M., NUNOKAWA, M., and IWAMOTO, Y.: 'Optimum fibre parameters of low-loss single-mode optical fibres for use in 1.55 μm wavelength region'. *Electronics Lett.*, **23**, no. 18, 27 Aug. 1987, pp. 963-964.

6 IINO, A., TAMURA, J., SENTSUI, S., and NAMIHIRA, Y.: 'Long-term loss stability of dispersion-shifted single-mode fibres exposed to hydrogen'. *Electronics Lett.*, **25**, no. 14, 6 July. 1989, pp. 926-927.

7 BHAGAVATULA, V.A., LAPP, J.C., MORROW, A.J., and RITTER, J.E.: 'Segmented-core fiber for long-haul and local-area network applications'. (See 4.22), pp. 1466-1469.

8 KALISH, D., and COHEN, L.G.: 'Single-mode fiber: from research and development to manufacturing'. *AT&T Tech. J.*, **66**, no. 1, Jan./Feb. 1987, pp. 19-32.

8A SHARMA, A., SHARMA, E.K.: 'Exact relationships between field and dispersion of single-mode fibres in presence of material and linear profile dispersion.' *Electronics Lett.*, **24**, no. 14, 7 Jul. 1988, pp. 873-874.

9 DE LA IGLESIA, R.D., and AZPITARTE, E.T.: 'Dispersion statistics in concatenated single-mode fibers'. *IEEE J. Lightwave Technol.*, **LT-5**, no. 12, Dec. 1987, pp. 1768-1772.

10 MIYAJIMA, Y., OHNISHI, M., and NEGISHI, Y.: 'Chromatic dispersion measurement over a 120 km dispersion-shifted single-mode fibre in the 1.5 μm wavelength region'. *Electronics Lett.*, **22**, no. 22, 23 Oct. 1986, pp. 1185-1186.

11 ANDREASEN, S.B.: 'Dispersion limited production tolerances of single-mode fibres'. *IEEE/OSA J. Lightwave Technol.*, **LT-5**, no. 9, Sept. 1987, pp. 1183–1187.
12 HATTON, W.H., and NISHIMURA, M.: 'Temperature dependence of chromatic dispersion in single-mode fibers'. *IEEE J. Lightwave Technol.*, **LT-4**, no. 10, Oct. 1986, pp. 1552–1555.
13 MIYAJIMA, Y., KAWATA, T., and NEGISHI, Y.: 'Considerations on group delay fitting equations for dispersion shifted fibres'. *Electronics Lett.*, **23**, no. 25, 3 Dec. 1987, pp. 1381–1382.
14 THEVENAZ, L., and PELLAUX, J.-P.: 'Group delay measurement in single-mode fibers with true picosecond resolution using double optical modulation'. *IEEE/OSA J. Lightwave Technol.*, **6**, no.10, Oct. 1988, pp. 1470–1475.
15 BERGMAN, L.A., ENG S.T., and JOHNSTON, A.R.: 'Temperature stability of transit time delay for a single-mode fibre in a loose cable', *Electronics Lett.*, **19**, no. 21, 13 Oct. 1983, pp. 865–866.

FURTHER READING

16 WALKER, G.R., and WALKER, N.G.: 'Alignment of polarisation-maintaining fibres by temperature modulation'. *Br. Telecom Technol. j.*, **6.** no. 1, Jan. 1988, pp. 60–69
17 NAMITHTRA, Y., HORIUCHI, Y., RYU, S., MOCHIZUKI, K., and WAKABAYASHI, H.: 'Dynamic polarisation fluctuation characteristics of optical fiber submarine cables under various environmental conditions'. *IEE/OSA J. Lightwave technol.*, **6**, no. 5, May 1988, pp. 728–738
18 PETERMANN, K.: 'Nonlinear transmission behaviour of a single-mode fiber transmission-line due to polarisation coupling'. *J. Optical Commun.*, **2**, no. 2, 1981, pp.59–64
19 SASAKI, Y.: 'Long-length low-loss polarisation-maintaining fibers'. *IEEE J. Lightwave Technol.*, **LT-5**, no. 9, Sept. 1987, pp. 1139–1146
20 SASAKI, Y., TAJIMA, K., and SEIKAI, S.: 'Low-loss dispersion-shifted polarisatin-maintaining fibres'. *Electronics Lett.*, **23**, no. 6, 12 Mar. 1987, pp. 279–280
21 IMAI, M., TERASAWA, Y., and OHTSUKA, Y.: Polarisation fluctuation characteristics of a highly birefrigent fiber system under forced vibration'. *IEEE/OSA J. Lightwave Technol.*, **6**, no. 5, May, 1988, pp.1366–1375.
22 IMAI, T., and MATSUMOTO, T.: 'Polarisation fluctuations in a single-mode optical fiber'. *OSA/IEEE J. Lightwave Technol.*, **6**, no. 9, Sept. 1988, pp. 1366–1375.
23 PETERMANN, K.: 'Constraints for fundamental-mode spot size for broadband dispersion-compensated single-mode fibres'. *Electronics Lett.*, **19**, no. 18, 1 Sept. 1983, pp. 712–714.

6

SYSTEMS ASPECTS OF SEMICONDUCTOR LASER DIODES

6.1 INTRODUCTION

This is the first of three chapters devoted to aspects of semiconductor laser diodes which are of most concern as components in ofdls. Throughout these chapters there is a very deliberate policy of emphasising some of the details which system designers have a need to know. This approach was strongly influenced by the author's experience of working in the company where a great deal of the pioneering work on fibres and lasers in the UK was done, and with his serving on two of the Working Parties on Standard Optical Systems for British Telecom in the early 1980s. In the very early days yields were very low and the few devices which did survive had lifetimes measured in hours. The present chapter takes up this approach in terms of the inverted ridge waveguide (IRW) laser about which much has been published and which was, at one time, the type of device whose reliability pedigree was best known. Lasers of this type are in service on both inland networks, and in submerged systems where its proven reliability is an important part of the 25 year system life. In addition to details of device characteristics the topics of laser to fibre coupling and of laser packaging are treated. The final section discusses transmitters.

The development of efficient and reliable sources suitable for use with optical fibre communication systems has been a long, arduous and costly business, lasting over half a working lifetime. In 1958 C.H. Townes and A.L. Schawlow at Bell Labs, published a paper [1] outlining the general conditions under which stimulated emission of light could be amplified. This followed some years after Townes and his graduate students had invented the maser (microwave amplification by stimulation of radiation) and was a practical embodiment of an idea pointed out by Einstein in 1916, the year his first book was published (it was on general relativity). Towards the end of that same year Einstein had two papers published, followed by a third in early 1917, on quantum theory. These contained the coefficients of spontaneous and induced emission and absorption. The basic concepts of the stimulated emission process, and the possibility of coherent 'negative absorption' from atoms in the upper level of an atomic transition were clearly outlined by Einstein in the 1917 paper 'On the quantum theory of radiation'.

In the UK much of the work of laser development from the earliest days was carried out at the Standard Telecommunication Laboratories of STC Ltd. (where the inventor of pcm A.H. Reeves had worked), in association with British Telecom Research Laboratories. From the beginning the potential for semiconductor lasers in telecommunications was recognised because of their size ('it appears as a dark speck of dust to the human eye') and ease of modulation with information-bearing signals. Despite its size a laser chip embodies considerable optical, electrical and material physics and chemistry knowledge and engineering skills. The difficulties of turning this potential into practical devices for use in systems was vastly underestimated throughout the industry worldwide. Unlike optical fibre technology, which has seen continuous development of one basic material, silica, over the past 20 or so years, sources and detectors have seen considerable changes from early 850 nm devices for multimode systems to those for single-mode systems at 1300 and 1550 nm. Semiconductor technologists have had to cope with changing from gallium arsenide

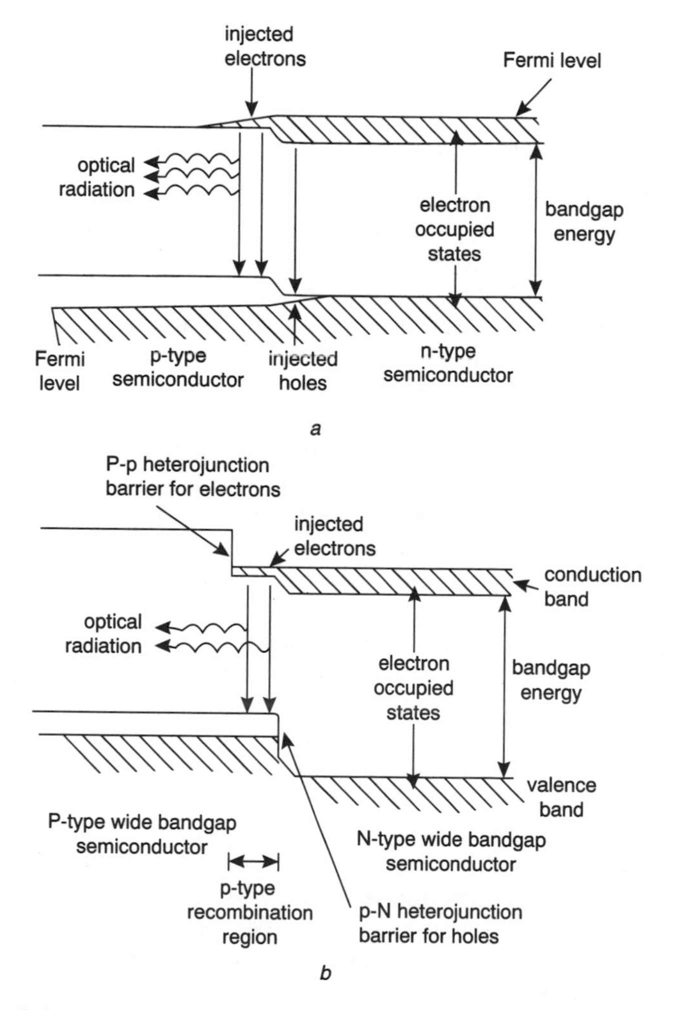

Figure 6.1 Band diagrams for (*a*) injection across a p–n homojunction and (*b*) injection and confinement in a double heterostructure laser ([2], Figure 2). Reproduced by permission of IEE

and silicon-based technologies to that of indium phosphide. A personal review by two well-known British Telecom engineers of the factors which have been important in the progress of sources and detectors in the first 20 years appeared in 1987, and provides suitable introductory material [2].

Probably the most significant advance in bulk laser technology occurred in the late 1960s with the introduction of heterostructures, in which additional semiconductor layers with wider bandgaps and lower refractive index could be included in existing laser structures to provide better confinement of the electrical charge carriers and of the photons. The reasons for the importance of a viable heterostructure technology lay in its ability to provide this confinement simultaneously for both electrical and optical carriers. Figure 6.1*a* shows how radiative emission is achieved by electron and hole injection across a heavily doped forward-biased pn (homo) junction.

The emission takes place over a volume determined by minority carrier diffusion lengths, which are several μm in GaAs. Figure 6.1*b* illustrates how the imposition of heterojunctions, rather than homojunctions, of suitably wider bandgap can be used to confine the carriers in a much smaller volume in the double heterostructure (dh). An additional property and advantage of the wider bandgap confinement layers in the dh laser is that their refractive indices are less than that of the narrow bandgap recombination (or active layer) at the emission wavelength. This structure forms a strongly guiding slab waveguide for the emitted radiation which effectively confines the light to the plane of the pn junction, even when the active layer is very thin, of the order of 0.1 μm. Typical bulk laser chip dimensions are 250 μm in length, 500 μm in width and 100 μm in height [3]. All significant optical and electrical activity takes place in a centrally located stripe region having typical dimensions of 250 μm by 1 μm wide by 0.1 μm deep which forms a cavity. This region exhibits optical gain due to recombination of electrical carriers. The remainder of the chip consists of materials which mechanically, thermally, electrically and optically support the active region. These dimensions can be compared with those for the IRW laser in Figure 6.2.

Another very significant advance in technology was achieved with the development of quantum well lasers for use in the 1300 and 1550 nm bands. These were

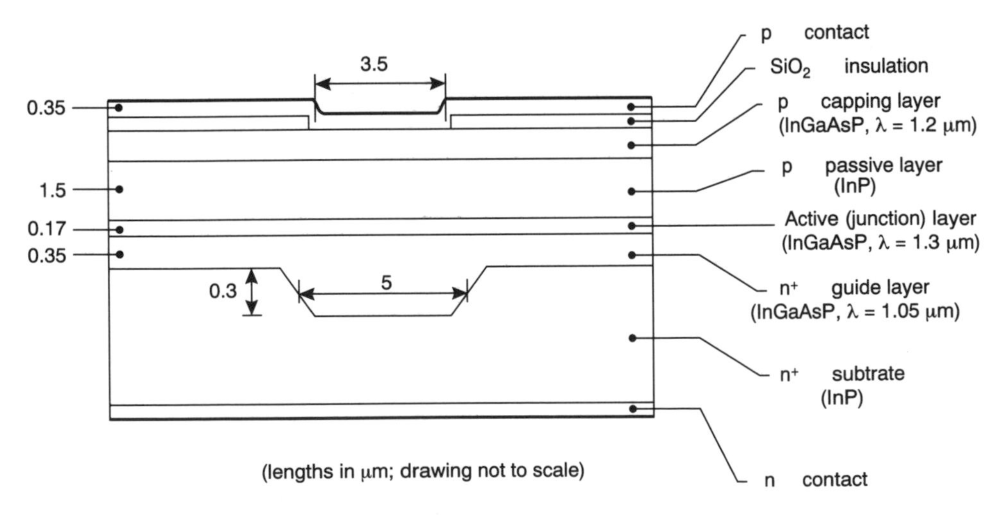

Figure 6.2 Schematic diagram showing layered structure and some dimensions of IRW laser ([4], Figure 1). Reproduced by permission of IEE

first developed by R.W. Dingle and C.H. Henry at AT&T Bell Labs in 1975 [3] but only became a practical reality in the early 1990s, when the ability to grow very thin uniform layers (1-10 nm) of $In_{1-X}Ga_XAs_YP_{1-Y}$ on InP reproducibly and when greater understanding of the electrical and optical properties of the quantum well lasers and the necessary material systems were achieved.

Only sources for use in the second and third windows will be treated in this book. GaAs is not a suitable material at 1300 and 1550 nm; instead, it had been shown that a range of InGaAsP alloys could be grown lattice-matched to InP. In the late 1970s the first devices using these newer materials gave encouraging results, especially in that their operating lifetime at room temperature was in excess of 1000 hours, whereas the early GaAs devices lased only for a few hours. A number of the failure mechanisms (e.g. facet burn-off, erosion) found in the GaAs lasers did not appear to the same extent in the newer materials. Partly as a result of the apparent absence of such internal degradation mechanisms, and partly as a result of all the accumulated experience on the physics of lasers and on growth and processing techniques for GaAs, the whole development cycle for these sources has been significantly compressed compared with the earlier technology. Thus, less than four years after decisions by British Telecom to install 850 nm systems in the trunk network, systems at 1300 nm using lasers were being installed in the UK and elsewhere. Furthermore, decisions were also made by international telecommunication operators to install submerged systems both on short routes and across the oceans, as described in Chapter 15.

Two broad categories of electro-optical semiconductor sources have emerged for use in systems. These are the light-emitting diode (LED) or the laser diode, or laser for brevity. In some situations a system designer may have a choice of either LED or laser as transmitter; if so, the following comparison between system aspects of both types of sources may be of interest.

Comparing the properties of lasers and LEDs, the light-emitting diodes have:

(1) wider operating temperature range
(2) higher reliability (in some cases)
(3) incoherent emission
(4) no interference effects from optical feedback
(5) a spectrum which is unaffected by direct modulation
(6) no need for optical power monitors
(7) no need for power control circuitry
(8) lower launched powers into sm fibre
(9) smaller modulation bandwidths
(10) slower response speeds
(11) broader spectrum of emission
(12) potentially better linearity
(13) lower cost

For inland junction and trunk applications and for submerged systems the preferred source is almost always a laser, because of its advantages over the alternative LED, the most important of which are perhaps: (a) very high modulation rates, in excess of 10 Gbit/s, and (b) the power which can be launched into standard single-mode fibres.

Until the early 1990s all lasers used in systems were types based upon bulk active laser diodes operating in the 1300 nm band. Research and development into lasers which utilise the special two-dimensional physical properties of very thin semiconductor layers forming their active regions has resulted in quantum well devices which outperform bulk active devices [3]. This very readable reference looked in depth at the bulk active laser diode as an introduction to qw diode lasers, described the advantages of the latter when used as transmitters in terms of modulation bandwidth, launched power and chirping. Other topics treated included the benefits of qw lasers as pumps for erbium-doped fibre amplifiers, design, fabrication and future directions. Commercial versions of quantum well devices with appropriate reliability pedigree are preferred for applications as transmitters and as pump lasers for fibre amplifiers in 1550 nm systems, such as those which have entered service from 1994 onwards.

6.2 ILLUSTRATIVE CHARACTERISTICS OF AN IRW LASER

This is a bulk active device in which the bulk or uniform composition active laser behaves electrically like a pn junction diode. It was chosen to provide the reader with as much quantitative data as possible on many of the characteristics of importance in ofdls applications because a considerable amount of data is available. This type, the so-called inverted rib waveguide (IRW) had been in service in both landline and in undersea routes since the mid-1980s. It is an InGaAsP device for use at 1300 μm and the structure is shown in Figure 6.2 [4]. The characteristics are detailed below.

6.2.1 Absolute maximum ratings (to IEC 134 at 25°C unless otherwise stated)

Storage temperature −20 to +65°C

Case operating temperature −10 to +55°C (limited by cessation of lasing)

Thermal impedance 40°C/W

Optical output power 2–4 mW, subject to coupling efficiency to fibre

Forward current, dc, 250 mA, but subject to the prior condition above; drive current above threshold max. 60 mA; peak reverse current max. 1.0 μA

Maximum reverse voltage 1.0 V

Forward current, pulse: pulse width dependent (another manufacturer quotes 300 mA, 100 ns duration pulses with 1 kHz repetition rate)

Photodiode ratings: bias voltage min. 0, max. −10 V; forward current max. 10 mA

Mechanical ratings: fibre pull force max. 5 N, fibre bend radius (storage) min. 35 mm; fibre bend radius (operation) min. 35 mm; radius distance from body min. 45 mm

6.2.2 Electrical and optical characteristics at 25°C

The characteristics listed in Table 6.1 refer to a case operating temperature of 25 ± 1°C and a monitor photodiode bias voltage of −5 V, unless stated otherwise.

Table 6.1 Operating characteristics of IRW laser

Quantity	min.	typ.	max.	units
laser threshold current		85	130	mA
mean threshold current temperature dependence over range 10–35°C			3	%/°C
drive current above threshold for P = 0.5 mW		40	50	mA
laser forward voltage at P = 0.5 mW		1.3	1.7	V
peak spectral output at P = 0.5 mW	1275		1325	nm
temperature dependence of peak spectral output		+0.5		nm/°C
monitor dark current			750	nA
monitor photocurrent at $I_{TH} + 50$ mA	150		2000	μA
rise time of monitor diode, 10–90%		7		ns

CW light output versus drive current with temperature as a parameter. A family of curves measured on a laser chip are shown in Figure 6.3. In broad terms, other commercially available contemporary devices possessed similar static characteristics, but with higher or lower threshold currents at the same temperature.

Devices were characterised at the 5 mW level on the cw output power curve for which the minimum value of power launched into a butt-coupled fibre was 0.5 mW, giving a coupling loss of −10 dB.

The slope (also known as external quantum or incremental) efficiency at a given na is found from the static output characteristics as:

$$\frac{P_2 - P_1}{I_2 - I_1} \tag{6.1}$$

and as can be seen from Figure 6.3 it decreases with increasing temperature. The efficiency is expected to decrease with time in a similar manner, with the end-of-life slope being similar to that at the absolute maximum operating temperature. The value of the slope is the same for both front and rear facets, and a typical value is 36% up to 5 mW and 25°C, falling to 30% at 50°C.

Maximum threshold current, $I_{TH,M}$, with temperature and time as parameters

On day 1 the start of use in a system, the typical value is 80 mA with a maximum of 130 mA at 40°C. Another manufacturer quotes 100 mA at 50°C, increasing to 200 mA during the operating life of the laser. Further information on this topic will be given later.

The relationship between forward current and voltage in the laser diode is shown in Figure 6.4.

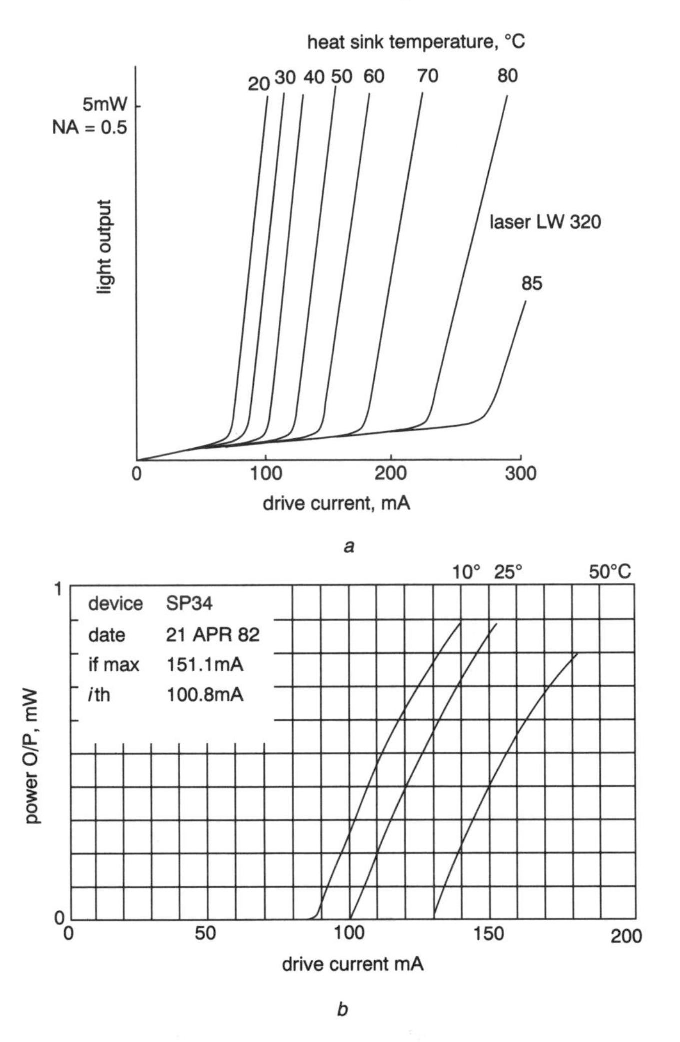

Figure 6.3 (*a*) CW light output versus drive current characteristics of typical IRW laser with temperature as parameter; (*b*) light output versus drive current for device SP34 at 10°, 25° and 50°C (a [4], Figure 2). Reproduced by permission of a IEE, b Nortel

Monitoring output

Type: internal photodiode monitoring the output from the rear facet.

Ratio of monitor current to optical output power: the current tracks the power with a nominal 1 mA/mW slope, which at the start of life may range from 0.5 to 2 mA/mW. The behaviour at three temperatures of 1982 devices is illustrated in Figure 6.5.

Type of diode and leakage current: a pin InGaAs/GaAs mesa photodiode is used. Typical and maximum values of leakage current are < 0.1 μA and 0.75 μA, respectively, at 25°C.

Diode speed, rise and fall times: the photodiode speed is sufficiently fast for use as a mean power monitor at 320 MBd (TAT-8 line rate), with rise and fall times below 2.5 ns, and capacitance less than 10 pF.

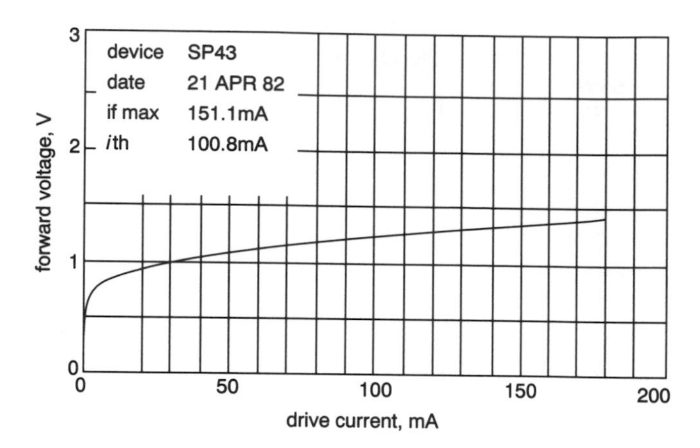

Figure 6.4 Forward voltage versus drive current characteristic of an IRW laser diode SP34. Reproduced by permission of Nortel

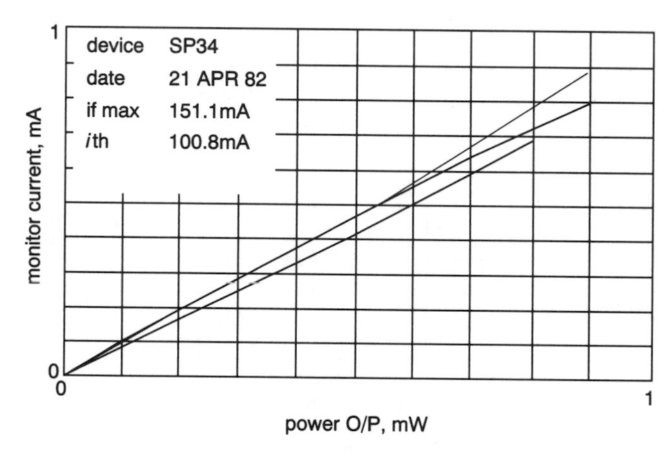

Figure 6.5 Monitor current versus output power at three different temperatures for device SP34 showing tracking ratio. Reproduced by permission of Nortel

Spectral characteristics

Lasing peak wavelength λ_P: the values at 25°C are 1290 (min.), 1310 (typ.), and 1330 nm (max.). These values increase by some 0.5 nm/°C over the temperature range of interest. Similar values were quoted by other manufacturers.

Spectral behaviour: the laser operates solely in the fundamental transverse mode, as with most other practical devices, and the zero-order longitudinal mode in terms of the time-averaged spectrum. At around 20 mW on the output characteristic of Figure 6.3, extended and significant first-order mode propagation appears.

Spectral width: in single-mode operation this quantity refers to the single 'line' rather than to the envelope of a number of 'lines'. The single longitudinal mode has a Lorentzian spectral shape down to at least −10 dB relative to the peak. The fwhm width δf is typically 10–100 MHz for $I/I_{TH} \simeq 1.1$. By comparison, reported results on GaAlAs lasers indicate that $\delta f \simeq 1/(I/I_{TH} - 1)$ for current ratios between 1.1 and 1.4.

Transient spectral behaviour: so far, time-averaged spectra have been considered which give no information on the spectral content of individual pulses nor on the variations from one pulse to the next. However, the spectrum of a laser carrying traffic is a dynamic and not a static quantity, and the evolution of the spectrum during a pulse is useful for predicting system performance. When pulsed from below threshold the IRW operates for 2 to 3 ns with several modes and consequently a broadened spectrum, in a manner similar to that shown in Figure 6.6.

The theoretical expression for the spacing $\delta\lambda$ between adjacent modes is:

$$\delta\lambda = \frac{\lambda^2}{2LN_G}, \tag{6.2}$$

where λ is the operating wavelength (nominally 1310 nm), L is the length of the laser cavity (200 μm), and N_G effective refractive index ($\simeq 4.5$)

Coherence characteristics: these are hard to obtain but are important in multimode systems because of modal noise. Since most current and future systems are and will be single-mode, this topic will not be further considered here.

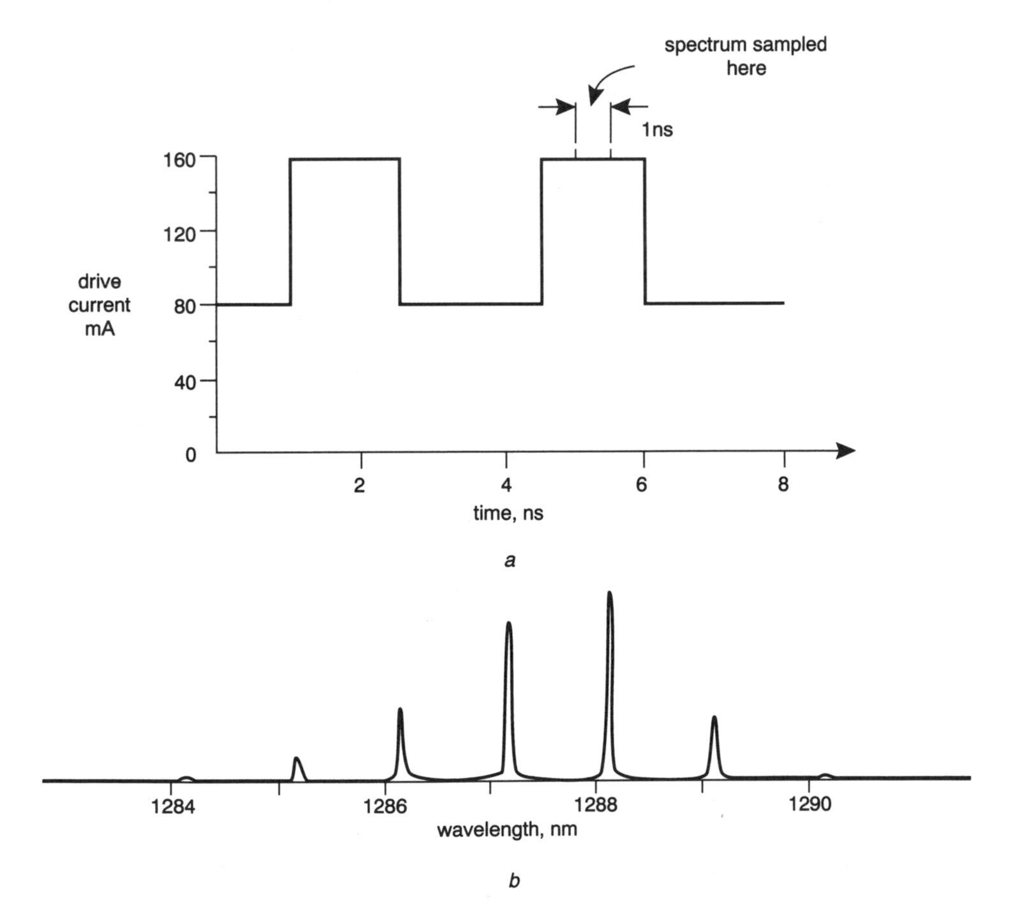

Figure 6.6 An example of the time resolved spectrum of an IRW laser under simulated 320 MBd ask modulation: (*a*) applied signal, (*b*) resulting spectrum with 1 ns resolution. Reproduced by permission of BT

Signal-to-noise ratio: this number exceeds 40 dB at any output level up to the maximum rated output power, when measured in the appropriate system bandwidth. In a laser comparable with the IRW a range of 45–50 dB at $I/I_{TH} = 1.3$, a modulating frequency of 100 MHz and a measuring bandwidth of 175 ± 5 MHz were quoted.

Sensitivity to reflections

Reflection noise figures were not available at the time, and are difficult to measure because they depend on many factors. Sometimes reflections are divided into near end (several cm from front facet) and far end (many cm to km). This topic will be discussed elsewhere in the chapter. Good package design, and index matched joints in the system fibre were expected to make reflection effects insignificant. This was borne out in experimental tests carried out in 1983, and the measured bit error rate curves for maximum and minimum reflections at an adjustable joint between the tail fibre and the test fibre gave the results presented in Figure 6.7.

Modulation characteristics

Extinction ratio: with a 0 at threshold and a 1 at a minimum mean launched power of 750 μW (−1.25 dBm) the typical value of the ratio P_0/P_1 for the IRW is 1:40 or 2.5%. The ratio varies with modulation current, being lower for larger values and larger as the modulating current tends to zero. Minimum and maximum values had not been established at the time of the investigation, but it can be seen from Figure 6.3

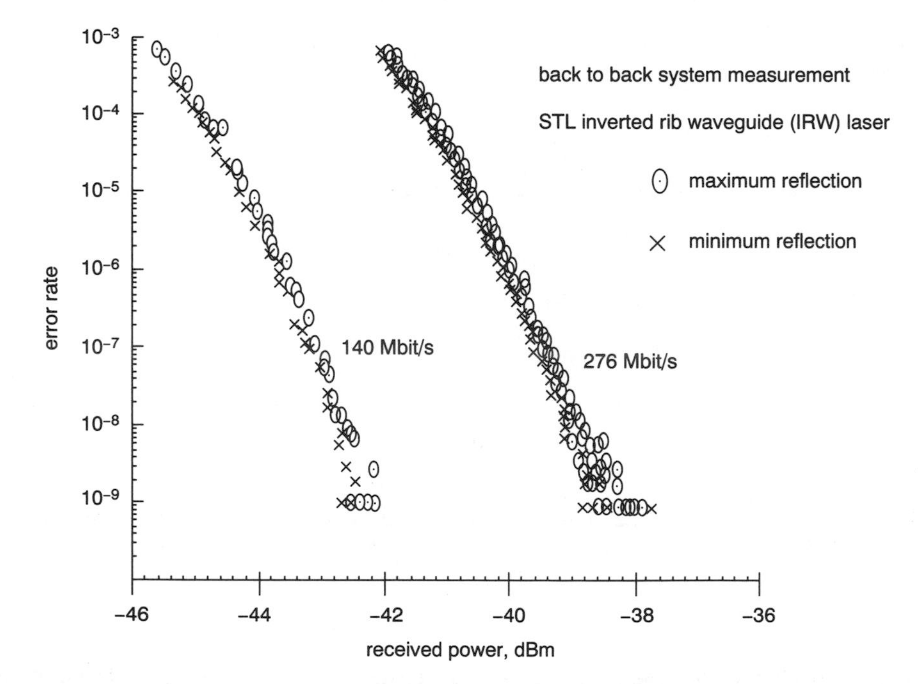

Figure 6.7 Measured ber versus received power for an IRW laser at 140 and at 276 Mbit/s under maximum and minimum reflection conditions. Reproduced by permission of BT

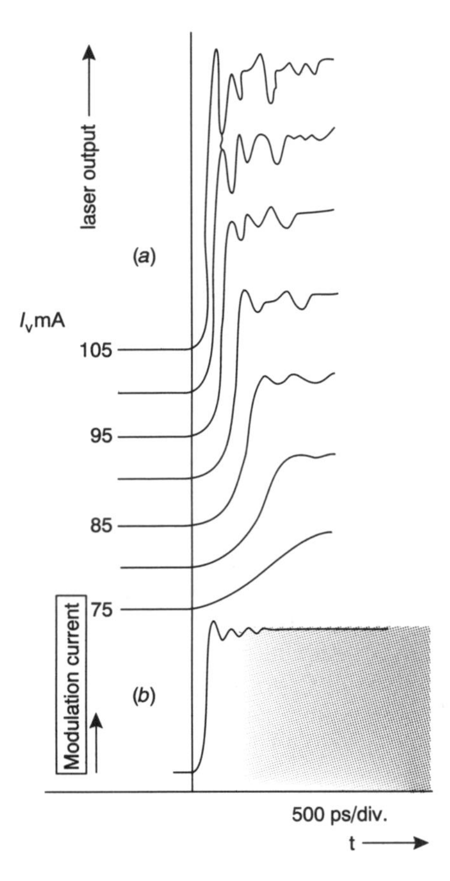

Figure 6.8 Variation of turn-on delay and rise time of optical power output response to current steps of amplitude from 75 to 105 mA in steps of 5 mA

that the extinction ratio is sufficiently small because of the very small light output due to spontaneous emission below threshold, thus causing a negligible system penalty.

Turn-on delay: for a laser biased at the threshold, increasing the modulation current reduce this delay. For the IRW the delay is typically 1 ns at low power levels, falling to 0.2 ns at maximum power output. Optical rise and fall times are quoted as less than 0.5 and 0.8 ns, respectively, which should be compared to line symbol intervals of approximately 7 ns and 3.5 ns for 140 and 274 Mbit/s traffic, respectively. The rise time depends upon the amplitude of the modulating current, and this is shown for a different type of laser in Figure 6.8.

Small-signal electrical or electro-optical equivalent circuits are seldom supplied by a manufacturer.

6.2.3 Life and reliability

A comprehensive programme of tests on the long-term reliability of the IRW laser chip was in progress in 1984, by which time a total of 1.24 million device hours had been accumulated [5]. The programme itself was reported in detail in [6]. As reported in 1984, the results from the tests on packaged IRW lasers indicated that this combination was then the most reliable laser source worldwide. More detailed and recent information was published in September 1988 [9].

End-of-life is defined by a 50% increase in the operating threshold current above the value at the start of life in the system.

6.2.4 Package and cooler

A hermetically sealed block with drive pins and fibre tail forms the package. Glass compression seals are used for the pins and the lid is welded in position. All other seals, including that for the fibre tail, are formed with solders. No organic materials are employed within the package.

The package is designed for optimum heat transfer, and for the fitting of a cooler. Some other contemporary lasers were available with integral coolers. The reliability of the cooler must, of course, also be thoroughly tested.

The optical monitor is a pin diode mounted behind the laser.

If a cooler is needed, an optional package with insulation can be used. This allows operation up to 60°C ambient temperature, while maintaining the package at about 10°C with less than 1 W of cooler dissipation.

The nominal package (laser heat sink) temperature when operating with a cooler lies somewhere in the range 5–20°C, with longer life expectancy at the lower value.

The pull-out strength of the fibre tail is such that the device performance is unaffected by a 100% proof stress of 1 kg m^2.

Type of tail fibre

A single-mode 9/125 μm fibre with an extended coating of silicon resin and Nylon 12, giving an outer diameter of 0.85 mm and a minimum bend radius of 35 mm.

Alternatively, the same package can be fitted with a 50/125 μm graded index fibre of na 0.2 with extruded primary and secondary coatings of silicone rubber and Nylon 12, respectively. The outer diameter is 1 mm and the minimum bend radius is 25 mm.

6.2.5 Optical safety

GaInAsP/InP lasers emit intense radiation in excess of 1 mW continuous power with or without a coupled optical fibre. According to BS 4803 as amended, they are classed as 3B lasers and appropriate precautions must be taken.

Details of how the IRW laser was designed for use in both landline and submerged sm systems by minimising known degradation hazards are described in Section 6.4.

System performance of the IRW laser

System studies have confirmed that the laser package with an suitable external modulator and thermoelectric cooler performs satisfactorily up to 700 Mbit/s, although originally designed for 320 MBd operation. The excellent modulation characteristics were shown in eye diagrams [5] for modulation rates of 320 and 560 Mbit/s and 1 Gbit/s.

The landline version of the package was used in the first European 140 Mbit/s single-mode commercial link, that installed between Luton and Milton Keynes, which became operational in early 1984.

6.3 LASER TO FIBRE COUPLING

There are three essential challenges in coupling lasers to sm fibres:

(1) achieving maximum efficiency
(2) reducing back reflections to a minimum
(3) combining both these desiderata in a stable package.

By the mid-1980s the most widely used coupling was some kind of microlens termination on the fibre pigtail; these were very reproducible and achieved coupling efficiencies of around 50%. Although reflections from this lens back into the laser are extremely low, feedback of optical power from discontinuities downstream of this lens, particularly the first connector which interfaces the transmitter to the system fibre, can degrade performance. Isolators provide one obvious solution, but they also introduce complications such as additional cost and loss and a sensitivity to wavelength and to polarisation. Experiments had demonstrated severe degradation and ber floors in Gbit/s systems due to the interferometric conversion of laser phase noise to intensity noise by multiple reflections from connectors.

A new approach to minimising reflections from the first and subsequent interface connectors was described in [7]. The new connectors were fabricated with beam expanding fibre tapers with bevelled end faces (a bevel angle of about 1° was sufficient to reduce reflections to practically unmeasurable levels). For the joint between the transmitter and the first system fibre, the input pigtail to this connector was terminated with a microlens fabricated by a new technique, which was combined with the unitary and self-aligning nature of the new design. Calculated back-reflections from tapers of 55 and 100 μm mode diameters versus bevel angle were presented, and they indicated a value of −80 dB for a 1° angle in a 55 μm mode diameter taper.

The bevelled taper design could also be used without the microlens in an expanded beam connector, thus eliminating problems associated with multiple point reflections. It was straightforward to add the bevel to a perpendicular end-face expanded beam connector and an insertion loss of 0.1 dB was measured in such a case. The microlensed pigtail version was fully compatible with existing laser packages, and so could replace conventional lens tapers. The bevelled connector end could be retrofitted into a biconic connector. As a connector component in a system, the taper pigtail would be either fusion or rotary-spliced to a system fibre. The bevelled ends could be prepared under factory conditions, thus eliminating the delicate and time-consuming procedure of polishing fibre ends in the field.

The coupling efficiency from laser to single-mode fibre can be increased significantly over that obtainable from butt coupling by means of a high index microlens grown on the cleaved end of the suitably tapered fibre as shown below in Figure 6.9. The method belongs to one category of coupling techniques based on the fabrication of a microlens on the fibre itself. The other category comprises methods which use separate micro-optical components. Each category has its advantages and disadvantages.

The fabrication of the short taper is simple and reproducible. The performance of a short taper combined with a high-index lens is better than a long taper with hemispherical end. The performance of these two approaches have been compared using a 1300 nm bh laser and a 6.8/125 μm sm fibre with an index difference of 0.36%. An optical glass with a refractive index of 1.9 was used for the microlens. The measured results are shown in Figure 6.9.

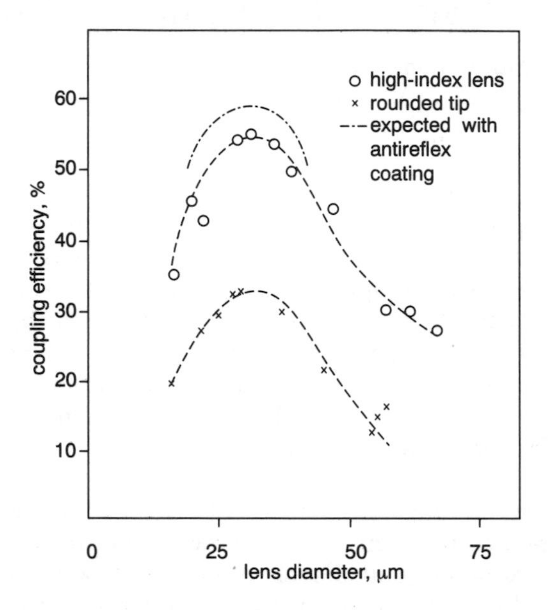

Figure 6.9 Measured coupling efficiency versus lens diameter for high-index lens and for long-taper rounded tip ends, and predicted values after anti-reflection coating ([8], Figure 3). Reproduced by permission of IEE

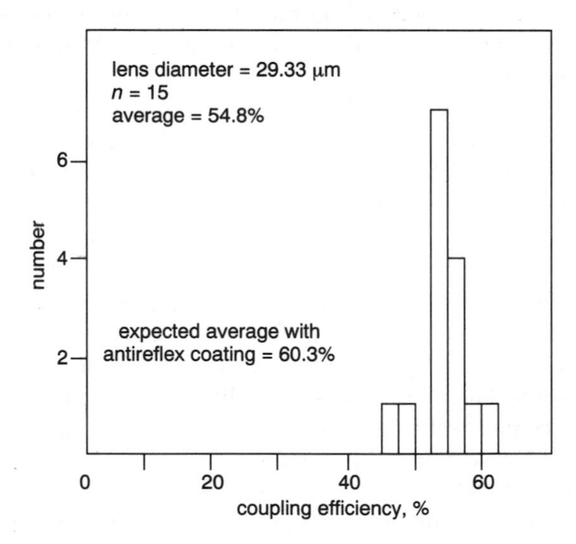

Figure 6.10 Histogram of measured coupling efficiencies between bh lasers and sm fibre obtained using a high-index uncoated lens of near optimum size at the tip of a short taper ([8], Figure 4). Reproduced by permission of IEE

The dash-dot line assumes a reflection loss at the air-to-lens interface boundary reduced to a few percent by an antireflection coating, which gives about a 10% improvement. It is clear from the figure that a lens diameter of around 30 μm is best. The measured coupling efficiencies are reported in Figure 6.10, and they give the reader some idea of the spreads obtained. The introduction of high index ended short tapers, which are both simple to fabricate and reproducible, achieves coupling efficiencies of around 50% between bh (buried heterostructure) lasers and sm fibres. The addition of antireflection coating to the lens is expected to improve this value by perhaps 5% or so.

6.4 LASER PACKAGING

An important factor in the widespread and successful installation of optical fibre systems has been the availability of packaged laser as components which system designers can use with ease and confidence. This is especially important in submerged applications where system design lifetimes are 25 years, so that reliability is of primary concern. The following details are given to provide the reader with an indication of the technology needed to achieve a reliable product.

6.4.1 Design considerations

The design and realisation of a high-reliability laser for single-mode systems was described in [9]. The minimisation of known reliability hazards in the design of the InGaAsP inverted rib laser for single-mode systems at 1300 nm is described. The results of an extensive reliability assurance programme are summarised, and lead to the conclusion, based on 7 million lifetest hours, that this design of laser possesses an extremely good reliability performance. These devices are now being used extensively in landline systems, and for the UK–Belgium submarine link. They were also selected for other submerged routes, UK–Denmark, the UK segment of TAT-8 (the first transatlantic optical fibre system which began working in October 1988, and entered service in December 1988), PTAT-1 and the STC segment of NPC (see Chapter 15).

The prime target was a failure of less than 1 device in 60 over the 25 year operating life of submerged systems. Critical factors in the design are mode control, metallisation, junction formation and packaging, and these factors are listed in Table 6.2, together with a summary of the known hazards and techniques for control of degradations. These factors are fully treated in the reference, therefore only a very brief summary will be given here.

Spatial mode control: operation in the zero-order spatial mode only is necessary to avoid problems arising from the nonlinear $P(I)$ characteristics, poor coupling efficiencies and spectral behaviour. A cross-sectional view of the structure of an IRW laser was shown in Figure 6.2.

The mesa bh and the channel or rib guide both produce real dielectric guides in the plane of the junction.

Metallisation: electrical contacts and bonds to the heat sink must be stable under high currents throughout the life, and two types of contact were evaluated, alloyed and ohmic Schottky contacts, and the latter was found to be better for high reliability. The solders used for bonding were carefully chosen for compatibility with

Table 6.2 Critical design factors in InGaAsP lasers and the control of degradation hazards ([9], Table 1). Reproduced by permission of IEEE, © 1988

Factor	Degradation hazard	Method of degradation control
Mode control	Simple buried heterostructures, regrown interface contact aging	Use planar or regrown structures which do not intersect junctions
Metallisation	Alloyed contacts introduce dark regions in electro-luminescence soft solder migration producing voids	Use ohmic Schottky contacts near active region. Use hard solders
Packaging	Movement and chip contamination	Epoxy-free and hermetic
Junction formation	Dark lines and cross-hatching in electroluminescence	Lattice matching, uniform LPE

the diamond heat sinks, and the need to avoid the hazards of electro-migration and whisker formation.

Facet damage due to optical power density has not proved to be the hazard with InGaAsP lasers that it was with GaAlAs devices, so that facet passivation is unnecessary at low power levels.

Heterojunction formation: the lattice dimension of the alloys and the InP substrate are carefully matched to achieve efficient luminescent material, and the layers are closely lattice-matched too. The uniformity and perfection of the interfaces is also important. Dark line defects, which were the main cause of degradation in early GaAlAs/GaAs lasers, are normally absent in InGaAsP/InP devices.

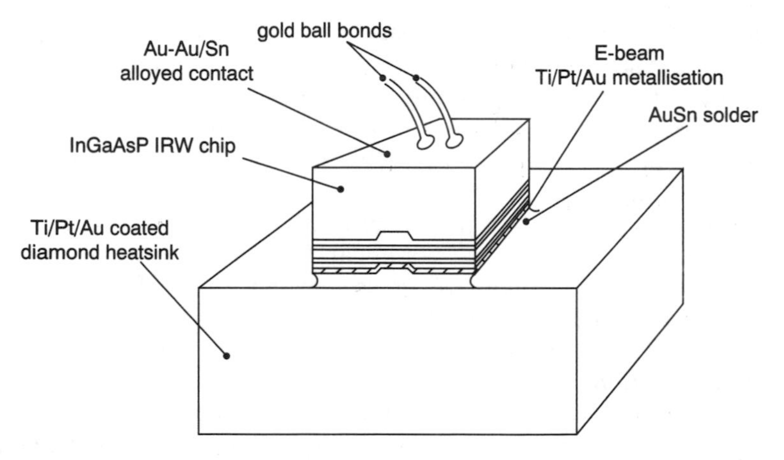

Figure 6.11 Schematic diagram of high reliability laser and mount ([9], Figure 3). Reproduced by permission of IEEE, © 1988

Laser package design: the minimisation of reliability hazards and other aspects of package design are treated in the 1984 paper, which is summarised later.

IRW chip and heat sink structure: the assembly adopted after all the reliability considerations had been satisfied is sketched in Figure 6.11, and it embodies the methods of controlling degradation hazards detailed in Table 6.2.

Typical output characteristics

The light versus drive current curves of an IRW chip and heat sink before and after 23 775 hours of operation under standard conditions of 5 mW per facet, and heat sink temperature of 50°C, are shown in Figure 6.12.

The modest increases in threshold currents and decreases in slope, happened largely in the initial period of the lifetest, thereafter the curves are virtually unchanged.

6.4.2 Reliability assurance

This was intended to give users of the lasers confidence that all important device parameters could be predicted to stay within specified limits during the design life (25 years) of an submerged system. A three-part programme of tests to achieve this aim comprised:

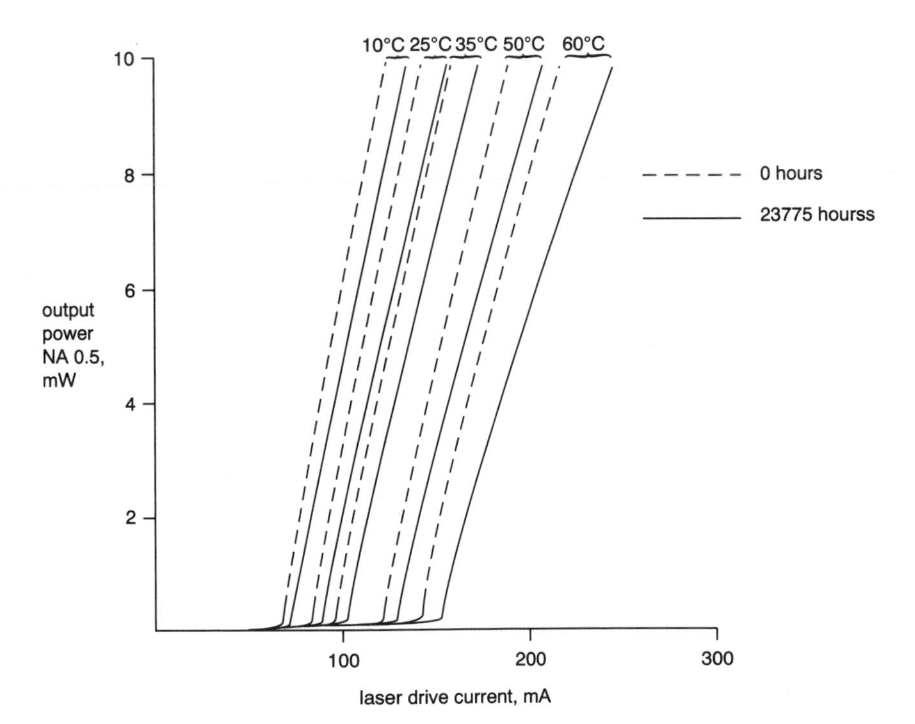

Figure 6.12 Sample family of light output versus drive current curves of an IRW laser before and after lifetest at 5 mW and 50°C ([9], Figure 4). Reproduced by permission of IEEE, © 1988

(1) lifetests under standard conditions, including study of the drift of optical and electrical parameters with time; these validation tests were used to determine which wafers were suitable for packaging; validation testing consisting of pre-testing, endurance testing at 50°C and re-testing, and covering (a) drive and threshold current drift, (b) optical and thermal parameter drift, (c) conclusions from standard lifetests

(2) lifetests under overstress conditions covering (a) temperature overstress and (b) current overstress

(3) package design and lifetests.

The design and performance of an extremely reliable 1300 nm laser package for single-mode high bit rate operation was first described in a paper published in 1984 [5]. The package was developed originally for submerged applications at 320 MBd, and thus required very careful selection of materials as well as a design that permitted components to be screened or 'burned-in' before insertion into the package. As a means of optimising the package performance lens coupling was adopted, and the preferred choice was a tapered hemispherical lens formed directly on the end of the fibre. This technique maximised launched power and minimised the effects of optical feedback on the spectral and noise performance of the laser, without demanding excessively tight alignment tolerances. Nevertheless, the long-term positional stability of components required was still 1 μm, which placed severe demands on the types of solder in critical situations. In addition, differential movement of components with changes in temperature were minimised by the expansion matching of materials, component design and assembly techniques. Nd/YAG laser welding, combined with a purpose-designed termination carrier, was chosen as the method for providing final anchorage of the optical fibre termination. The localised spot welds allow the termination to be anchored very close to the laser chip without reheating pre-screened components. An extremely stable launch configuration with a high degree of immunity to shock and vibration was provided by this means. The final major factor in achieving a package which could be driven at high bit rates was the need to minimise the inductance between the laser chip and the external drive circuitry. Very low inductance values were obtained by using two laser drive pins in parallel coupled to a stripline inside the package. Many of the topics mentioned in this paragraph are directly related to those treated in an earlier section, e.g. coupling, optical feedback.

Package design

The internal details of the submarine laser package are given in Figure 6.13, and the main components are described briefly in the following paragraphs, beginning with the package body itself.

Package body: this piece part is machined from aircraft grade selected steel, and the surface is protected by nickel and gold plating. The circular flange is a feature of the submerged application and is absent from the landline version. AgCu eutectic is used for flange and termination feed-through brazing, while pins are held in place by glass-to-metal compression seals. After insertion of the components and final alignment, the package is outgassed in vacuum, then hermetically sealed in a dry He/N_2 mixture either by projection welding or by seam sealing. The sealed package is finally tested for leaks down to about 10^{-10} atmosphere cm^3s^{-1}.

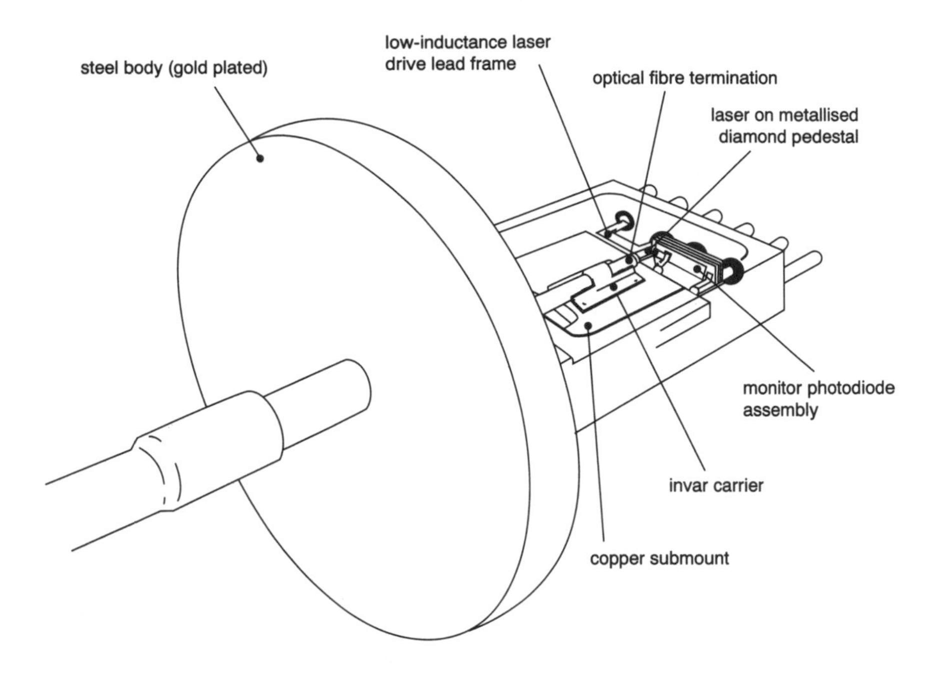

Figure 6.13 Sketch of single-mode submerged laser package ([5], Figure 1). Reproduced by permission of IEE

Laser submount: the base is formed from copper sheet shaped to aid accurate component location, and to allow access for forming a hermetic seal between the fibre feed-through spigot and the optical fibre termination by soldering. The monitor photodiode is mounted on part of the near face of the base, which has been machined at an angle of 20° to the vertical, to avoid optical feedback from the photodiode surface. The submount is nickel and gold plated before a metallised diamond pedestal and the laser chip are attached. Final electrical connection to the laser chip is made pulse-tip thermo-compression ball bonding a pair of gold wires between the top p-contact surface of the chip, and the laser drive LED frame.

Details of the monitor photodiode unit materials, mechanical construction, and other aspects are given by the authors. The photodiode itself is a GaInAs pin diode with an active area of approximately 350 × 200 μm. Typical values of dark current lie in the range 10–100 nA.

Optical fibre tail

This consists of a 5 m length of Halar coated sm fibre terminated at both ends. The system fibre end is simply a cladding mode stripping termination that ensures only power launched into the fibre core is measured during electro-optical testing. This termination is eventually removed before the tail is spliced to the system fibre. At the input end, the termination has been specifically designed for entry and fixing in the package body. The fibre tail must be hermetically sealed to its termination; this is aided by sputter coating a region adjacent to the lens end of the tail with a suitable metallisation that ensures excellent adhesion to the fibre during the subsequent soldering operation to form the seal.

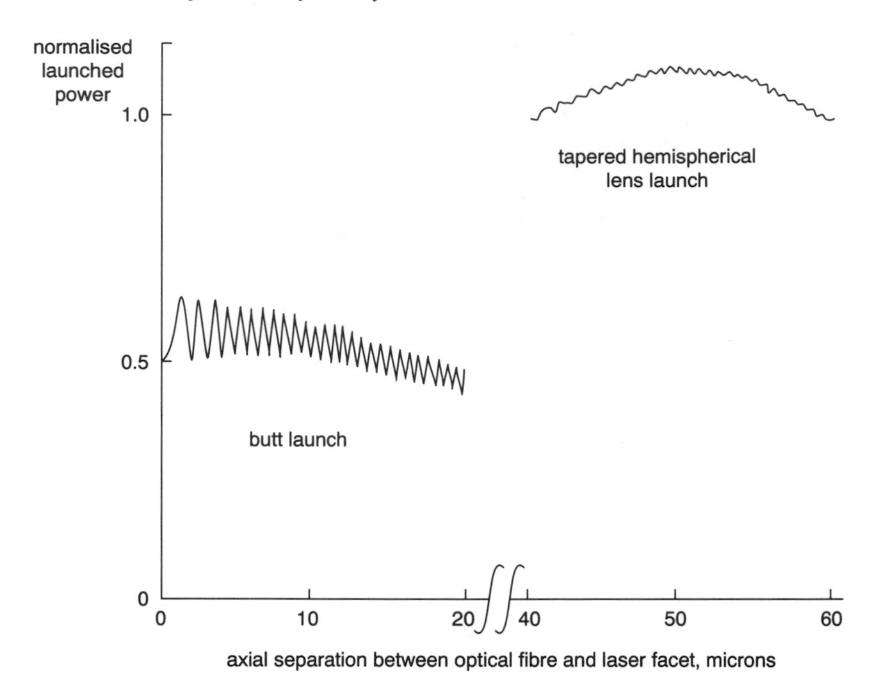

Figure 6.14 Normalised launch power versus axial separation between optical fibre and laser facet for tapered hemispherical lens launch and for butt launch ([5], Figure 3). Reproduced by permission of IEE

Lens formation and performance: a tapered hemispherical lens with a tip diameter of about 40–50 μm is formed directly on the fibre by a combined drawing and remelting technique, which had been proved to be the best method for use with the IRW laser. The direct formation of the lens on the fibre minimises the number of optical interfaces and so contributes to the low reflections achieved. The curvature of the lens itself contributes significantly to the reduction of optical feedback. As we have seen, changes in temperature cause changes in the length dimension of any external cavity formed between the front facet of the laser and the reflecting surface. In the specific case of this package the effects are shown in Figure 6.14, which compares butt launch with that adopted for the package.

Three points are worthy of note; the first is that the variations in power with variations in length are much more pronounced for the butt launch than for the lens case, the second is the 3 dB improvement obtained with the latter. The large changes in the butt launch case occur at intervals of $\lambda/4$ (0.33 μm) due to constructive and destructive interference of the forward and return waves. The third point is the advantage in the case of the lens launch, of the spacing between laser and fibre tip which becomes about 50 μm, thus facilitating final cleaning and inspection. Typically, 24% efficiency is achieved with the lens launch, compared with 10 or 12% for the butt launch.

Termination carrier

A small but extremely important component in the package design is the Invar termination carrier that is designed to provide positional stability between the terminated optical fibre and the active region of the IRW laser after alignment. The Invar carrier and the diamond pedestal mentioned earlier are closely matched in terms of height and thermal expansion coefficients; these properties and the configuration used provide launch conditions which are insensitive to changes in temperature over the full operating range of the equipment.

Package assembly

Mechanical and solder details are supplied by the authors for the various parts which go to make up the package; again, reliability considerations play a large part in the design approach and techniques adopted.

Packaged laser characteristics

A typical packaged laser characteristic is shown in Figure 6.15, which shows plots of light output from the tail, P_T, and monitor photodiode current, I_M, versus drive current. Typically, 1.5 mW output is obtained at a drive level 50 mA above the threshold current, and I_M is 1 mA with the photodiode biased at −3 V. This diode forms part of a mean power feedback control loop for the optical output power. It is therefore essential that the relationship between P_T and I_M does not vary significantly over the operating temperature range of the package. As can be seen from Figure 6.16, the current accurately tracks the output; the effect of temperature variations on the tracking ratio P_T/I_M is illustrated in Figure 6.16, from which one can see that changes of less than 6% in the ratio are achieved throughout the range 20–50°C. The tracking ratio is very sensitive to relative movement between the laser chip,

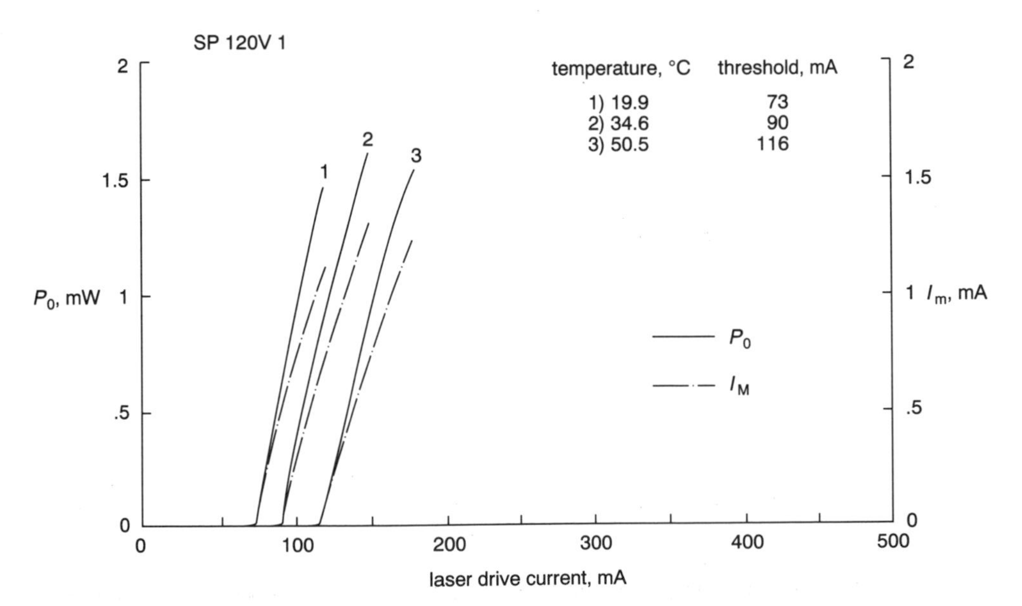

Figure 6.15 Light power versus drive current and threshold current versus drive current of packaged IRW laser SP120V 1 with temperature as parameter ([5], Figure 5). Reproduced by permission of IEE

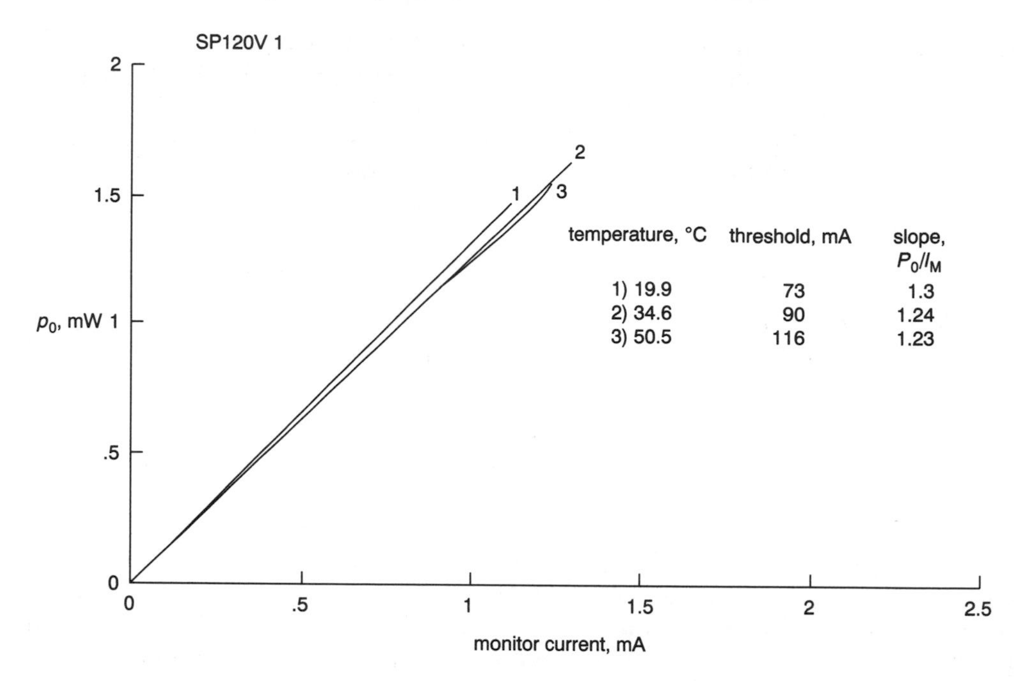

Figure 6.16 Package tracking ratio showing light output versus monitor current of IRW laser SP120V 1 with temperature as parameter ([5], Figure 6). Reproduced by permission of IEE

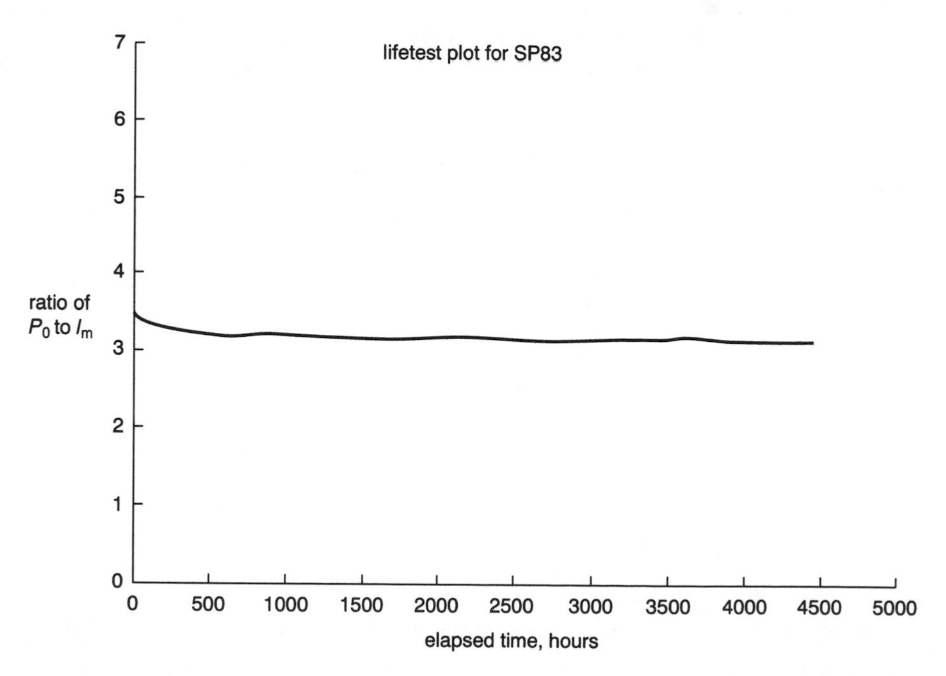

Figure 6.17 Tracking ratio drift showing ratio of output power over monitor current versus elapsed time for device SP83 ([5], Figure 7). Reproduced by permission of IEE

the optical termination and the monitor photodiode, as one would expect; so the constancy of tracking achieved augered well for the continued excellent stability provided by the launch mechanics. Variations of this ratio with time have been investigated, and a plot covering 5000 hours duration is shown in Figure 6.17.

Predictions of tracking ratio drift made from such data indicate typical changes of less than 10% over a 25 year operating life.

6.5 ELECTRO-OPTICAL TRANSMITTERS

These are required in the transmit side of the main transmission path in terminal transmission equipment and in the main transmission paths of any electronic regenerator pairs with which a system is equipped. Repeatered systems using optical fibre amplifiers only require them at terminals. A laser transmitter is taken as a subsystem which requires only an electrical modulating signal and dc power inputs to provide a modulated light output.

6.5.1 Introduction

The basic functions performed in regenerators are: pre-amplification, agc and equalisation, clock recovery, regeneration, laser modulation, laser biasing, supervision. Only modulation and biasing will be considered here. A digital regenerative repeater requires a much more complex circuit than its analogue counterpart, so that for reasons of space, power consumption and reliability it is desirable to integrate as much of the circuitry as possible. By 1988, papers had been published on transmitter designs for rates up to 1.7 Gbit/s, then the highest rate in use in the AT&T digital network.

The principal requirements of the circuits which realise these two functions are: laser modulation (speed, dynamic range, purity); and laser biasing (stability with time, voltage and temperature; power; dynamic range).

Generally speaking, commercially available technologies are not able to satisfy all the analogue and digital requirements in regenerators, so that proprietary design and manufacture to guarantee the necessary system performance is common [9].

6.5.2 Laser modulation circuits

1980s high-capacity long-haul transmission systems required milliwatt laser launched power levels, and therefore drive currents approaching 100 mA above threshold [10]. The drive pulses were generated by either modulated current sources, or by resistively matched transmission line coupled voltage sources. In practice, the latter was conveniently realised with 50 Ω output impedance amplifiers and transmission line resistively matched to the dynamic driving-point impedance of the laser.

Performance of modulated current source laser drivers is limited by available device speeds at high currents, and by the effects of parasitic inductance and capacitance. The former arises from bond wires on the drive transistors and on the laser chip, and from interconnections, whereas capacitance limitations are dependent on device sizes. Parallel bipolar transistor current sources suffer from a straight trade-off between current capability and available bandwidth restriction due to parasitic

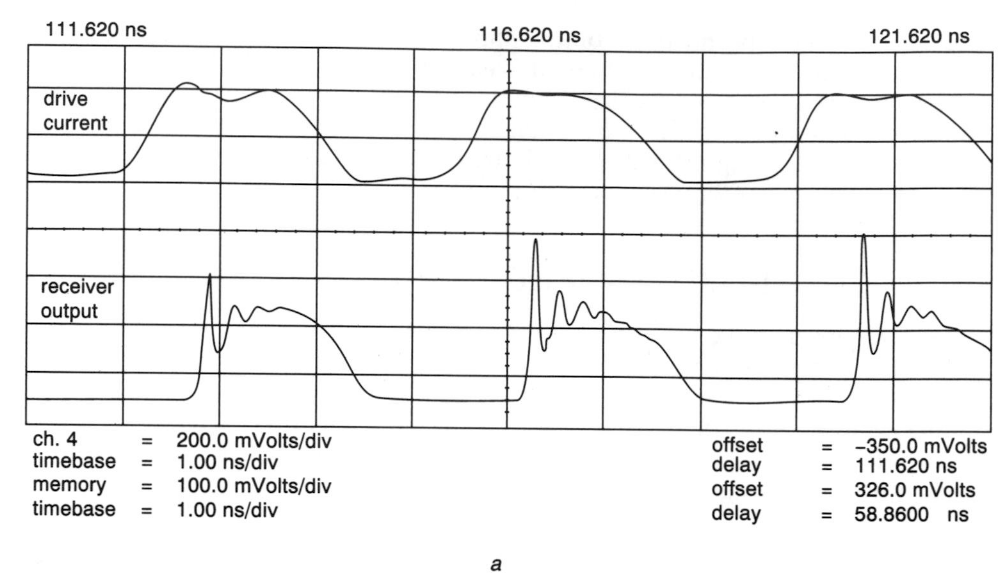

a

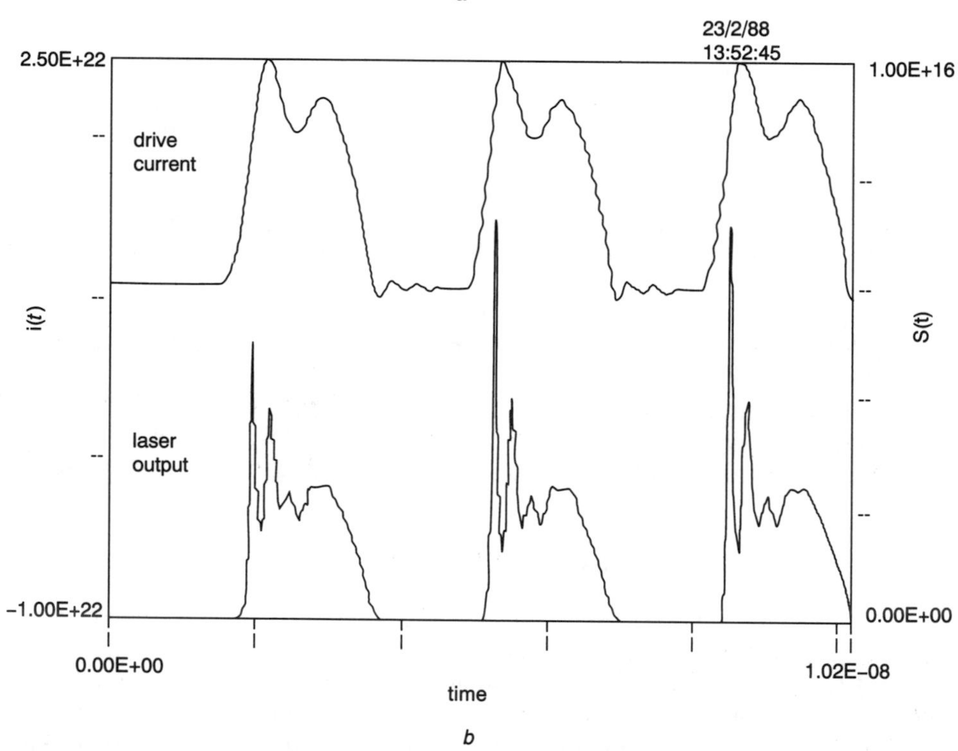

b

Figure 6.18 (*a*) Measured results and (*b*) simulated results of a 600 Mbit/s current switch drive circuit showing the output current pulse oscillations superimposed on the optical output, which includes the relaxation oscillations of the laser ([12], Figure 2). Reproduced by permission of IEE

capacitance. The combination of these parasitics has a direct effect on the optical output through the severe transient oscillations they cause in the laser drive signal when a current source is employed. The effect is illustrated in Figure 6.18, which is a trace of the 600 Mbit/s output from a 1550 nm dfb laser driven by a IC current source comprising eight large-area transistors in parallel, and associated buffers producing 40 mA modulation current above the threshold value. These measured waveforms can be compared with those computed from SPICE modelling of the driver circuit and the laser and shown in Figure 6.18*b*. The laser model for use in SPICE 2 implements the rate equations for a single mode of oscillation.

Significant improvement in transient behaviour is possible if the current source is replaced by an approximation to a voltage source driver. This is shown in Figure 6.19, which refers to the same laser and receiver, but with a 50 Ω impedance-matched driver stage.

The authors state that further improvement over the current drive could be achieved if a more accurate model of the laser was available for incorporation in an optimisation package.

The 1985 paper [13] described the experimental evaluation of a semi-custom array for use as a regenerator at 140 Mbit/s. The advantages seen in this approach were: commercial availability in production quantities, a low non-recurring development charge, a fast turn-round time, and low pro rata costs. In the UK and some other European countries 140 Mbit/s systems use 5B6B or 7B8B line codes, hence the line rates are higher (168 and 160 MBd, respectively). Thus the regenerator IC must operate at these speeds with adequate margins for production spreads in parameters, temperature variations and ageing. It is also desirable that the chip should operate at lower order rates for local area systems.

By 1990 systems operating at 1.7 or 2.4 Gbit/s were in service. An increasing research effort was being directed at circuits capable of operating at 10 Gbit/s and beyond. Typical components such as multiplexers, demultiplexers, amplifiers and decision circuits embodying bipolar devices or field-effect transistors (FETS) had been demonstrated. The first successful design and testing of a bipolar laser driver for use at such speeds was reported in [14]. The driver was fabricated as a single

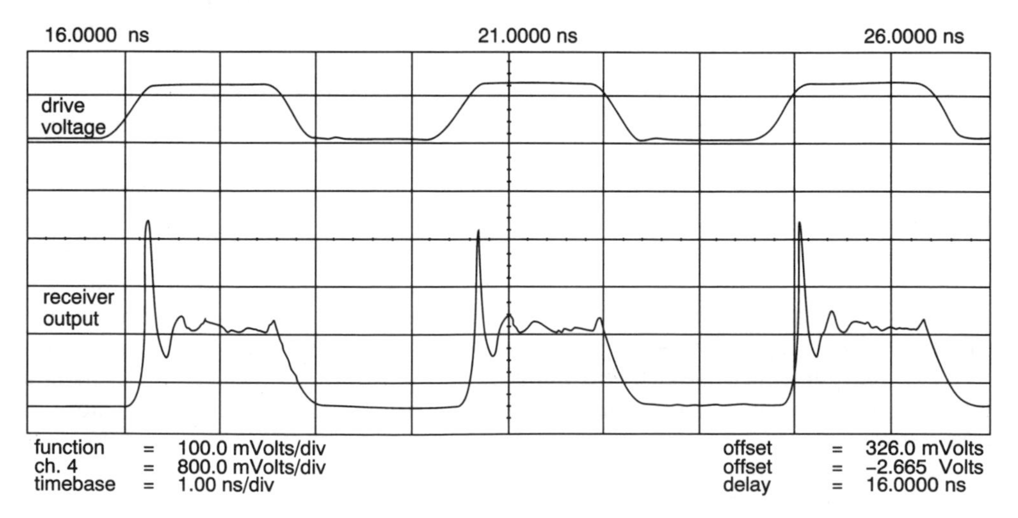

Figure 6.19 Measured results for laser driven by a resistively matched 50 Ω voltage drive circuit, note relaxation oscillations of laser ([12], Figure 3). Reproduced by permission of IEE

chip in InP/InGaAs HBT technology, and typically delivered a modulating current of 100 mA and 50 mA dc bias with less than 1 W power dissipation. In tests, clean output eye patterns were obtained and were attributed to the small amount of jitter together with fast rise and fall times.

Laser drivers are crucial circuits in traditional transmitters and although their design is simple in concept there are stringent requirements which make their implementation challenging. Heterojunction bipolar transistors (HBTs) are naturally suited to this application because of their very wide bandwidth and high current densities, hence high gain. This latter property permits fewer stages to obtain required output levels, thus reducing the dissipation and timing jitter (phase noise). A block diagram was given and the operation described, followed by a figure showing the circuit the details of which were also described. The driver chip was RF-on-wafer probed, and the performance was evaluated by applying a 10 Gbit/s prbs signal at the input and a high-frequency sampling oscilloscope to monitor the output. Measurement-limited rise and fall times of 35 and 40 ps, respectively, were obtained. Cascade-connected amplifiers on the same wafer had 3 dB bandwidths of 12.3 GHz and dc gains of 17.3 dB, after probe and cable effects had been taken out in the course of the small-signal measurements.

As mentioned earlier, optical laser sources and photodiode detectors are now so fast that system performance is limited by the speed of the associated electronics. Although the operating speed of silicon integrated circuits is being extended into microwave (>1 GHz) frequencies, the newer materials such as gallium arsenide are preferred in microwave engineering applications, and increasingly in very high capacity optical line systems. A state-of-the-art transmitter has been described by authors from AT&T Bell Laboratories in a 1988 paper [15]. A block schematic of the 1.7 Gbit/s transmitter is given in Figure 6.20.

The electronics circuitry consists of a high-speed GaAs laser driver chip which performs the functions shown in the left-hand box in the figure, together with a low-speed Si chip for the feedback control of the laser output power and other functions listed in the right-hand box. Let us briefly look at these boxes in turn.

The GaAs chip contains the laser driver which converts the incoming modulated signal of peak-to-peak amplitude 0.5 V to modulation currents adjustable between 50 and 100 mA, depending on which of the three chip variants is chosen, for driving the laser. As shown in the diagram, the GaAs chip also provides a current $i_c(t)$ to the feedback control circuitry, and the complement $\bar{i}_c(t)$ to the duty-cycle tracking bias circuit for duty-cycle independent control of the laser output power. The time-averaged value $\langle i_c(t)\rangle$, is proportional to the duty cycle of the input signal, whereas $\langle \bar{i}_c(t)\rangle$ is proportional to (1 − duty cycle). If the duty cycle of the input is constant, the input to the feedback control circuit may be switched from $i_c(t)$ to a dc reference value I_R, shown at the right-hand side of Figure 6.20. The laser driver comprises three cascaded stages of differential pairs, to take advantage of their functional properties such as symmetry, and relative independence of individual FET threshold voltages. The symmetry is expected to have the advantage of insensitivity to temperature changes, and the pair configuration minimises the transient switching noise in the power supply nodes. This integrated circuit is fabricated in GaAs enhancement/depletion mode, etched gate, n-channel MESFET technology. Feature sizes are nominal gate lengths of 1 μm, minimum line separations of 8 μm and a minimum line width of 4 μm.

The silicon chip, CBIC, provides two inputs to the laser driver chip; these are the currents i_{RD} and i_{RL} and they control the operation of the driver chip through current

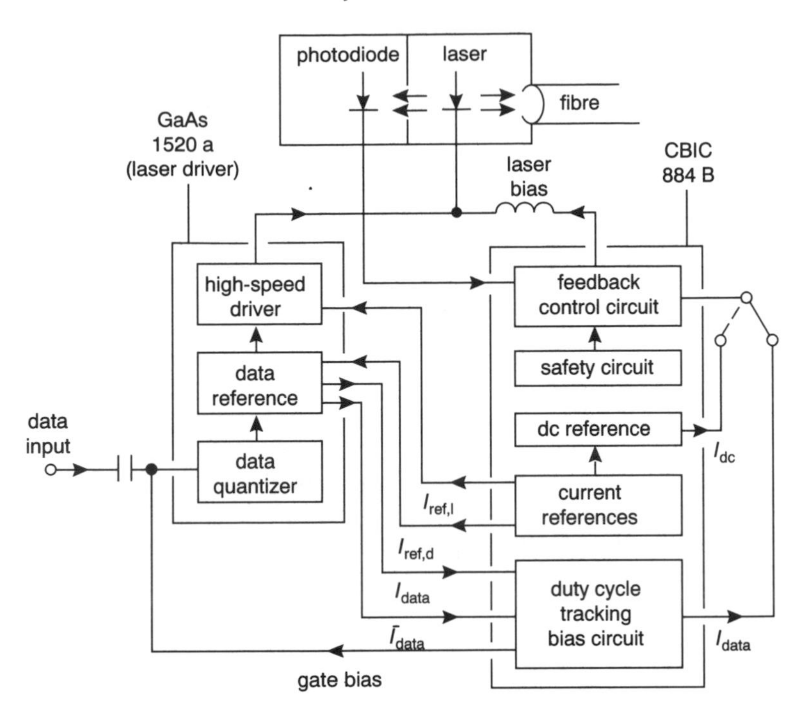

Figure 6.20 Block schematic diagram of transmitter for 1.7 Gbit/s system ([15], Figure 1). Reproduced by permission of IEEE, © 1988

mirrors. Input i_{RD} controls both i_c and $\bar{i}_c$, and can also be adjusted to optimise the rise and fall times of the modulation current. The other input i_{RL} can be adjusted to vary the magnitude of the modulation current.

The performance of the GaAs laser driver integrated circuits is detailed in the paper, and only the highlights will be given here. Eye diagrams of the modulation current at 1.7 Gbit/s and 30 mA peak-to-peak measured over the temperature range 0–70°C showed very little change in the eye crossings or the extreme levels of 0 and −30 mA. The sensitivity of the same eye diagram to variations of the input signal between 0.35 and 0.71 V was measured, and the rise and fall times of typically less than 200 ps shorten as the level increases between 0.35 and 0.5 V input, thereafter staying the same because the circuit saturates. No significant differences in the rise and fall times of the output from this chip were found when the corresponding input quantities varied from 180 to 300 ps, thus confirming that the pulse sharpening takes place within the chip. Now consider the control of the laser output power by means of the two control currents. The eye crossings can be set to the mid-value of the modulation current by a voltage adjustment at any given duty cycle, but this voltage must be optimised for varying duty cycles because ac coupling removes the mean value of the modulation current. For operation independent of duty cycle the high and low states of the modulation input should remain at fixed levels, and this is done by varying the input voltage as a linear function of duty cycle. This optimisation is carried out by means of a feedback loop, and results in the two control currents being linearly proportional to the duty cycle over the range 1:8 to 7:8. Variations of power supply voltage of ±5% cause no perceptible degradation in performance.

An interesting approach was proposed in 1987 [16]. This new laser modulator is based upon current subtraction to produce the laser injection current. The main advantages claimed for the novel circuit are:

(a) simplicity of operation as an analogue modulation current source
(b) ease with which the bias and modulation currents can be controlled
(c) dc coupling
(d) capability for shaping the eye diagram
(e) reduced number of components giving cost and reliability advantages.

In addition, fast switching times, easy regulation of the optical power and of the extinction ratio with an automatic feedback network are features of the design.

The operating principle of the circuit is illustrated by the ideal scheme portrayed in Figure 6.21.

From the figure the laser injection current $i_L(t)$ is given by the difference between the total dc current I_T and the modulation current $i_M(t)$ responsible for the switching the laser between 1′s and 0′s. The operation is clearly visualised with the help of Figure 6.22.

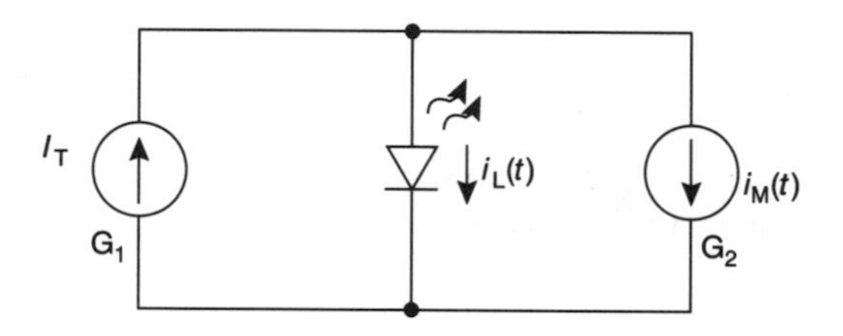

Figure 6.21 Circuit diagram of idealised drive circuit to illustrate operating principle ([16], Figure 1). Reproduced by permission of Alta Frequenza

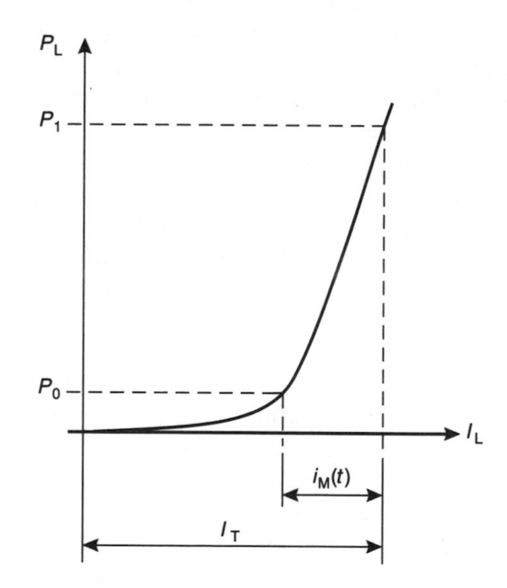

Figure 6.22 Graph of output power versus drive current showing two specific values of drive current and corresponding power levels ([16], Figure 2). Reproduced by permission of Alta Frequenza

Let the independent generator supplying the modulating current switch ideally between a value I_M during the 'on' state and zero in the 'off' state, while the other generator provides the steady current I_T. When i_M is zero, all the current available from the dc generator must flow into the laser, thus determining the high logic level P_1, and therefore we have

$$I_T = I_{TH} + I_1 \tag{6.3}$$

On the other and, when the modulation generator is in its 'on' state, it supplies a current I_M in opposition to the other two components, giving a lower injection current, which holds the laser near threshold at P_0, corresponding to a zero. Broadly speaking, the enhanced speed of operation results from the complementary way in which the driving transistor and the laser behave during every transition. A laboratory prototype of the proposed circuit was realised using the Fujitsu type FLD 130 D4BK device and BFT 99A and 93 bipolar transistors. At 620 MBd, the measured rise and fall times were about 320 and 560 ps, respectively. Eye diagrams at 620 MBd and at 140 Mbit/s show the 'bit rate scaling' capability of the circuit. Laboratory tests confirmed the results of simulations and strengthened interest in the principle as a laser driver capable of covering a wide range of applications from Gbit/s to low Mbit/s at low cost.

The design of laser driver circuits for use in the Gbit/s range can be assisted by suitable computer simulation software. A 1987 letter [17] described a subcircuit three-port model comprising the laser chip, the electrical parasitics of the package, and incoherent optical reflections from the end of the fibre pigtail implemented in the SPICE program. The use of one or other variant of SPICE, with all their limitations, has been widespread in the design of some subsystems such as transmitters or receivers, but in the opinion of the author much more realistic simulation packages have been available for some years. There is a very widespread reluctance to learn to use new tools among many engineers. Computed results were compared with measured results from dfb and FP lasers. The chosen representation of the module is shown in Figure 6.23, in which node 1 represents the electrical input and node 3 the optical output P. The voltage at node 2 is a measure of the carrier concentration n that can be related to the laser wavelength λ in order to model the chirping behaviour. Internally, the three-port model is divided into three boxes A, B, and C, where box A is the simple circuit model given in Figure 6.22*b* for the chip and parasitics.

Bondwire inductances, stray and chip capacitances and losses are all included. Box B is described by two single-mode laser normalised rate equations, taken as:

$$\frac{\mathrm{d}s}{\mathrm{d}t} = \left[K_1(n-1)(1-K_2) - \frac{1}{\tau}\right] s(t) + K_3 n + s_E(t) \tag{6.4}$$

$$\frac{\mathrm{d}n}{\mathrm{d}t} = i(t) - n - (n-1)(1-Ks(t))s(t) \tag{6.5}$$

where $s(t)$ is the photon concentration, s_E is the external photon injection, $n(t)$ is the carrier concentration, $i(t)$ is the injection current, K_1 is the constant related to optical gain, K_2 is the constant related to gain compression, and K_3 is the constant related to spontaneous emission.

The pigtail is modelled by a transmission line as shown in Figure 6.24*c*, is driven by an e.m.f. $E(s)$, and loaded by an impedance $Z \geq Z_0$ chosen to give the correct value of reflection coefficient $\rho = (Z - Z_0)(Z + Z_0)$. The reflected signal, and hence

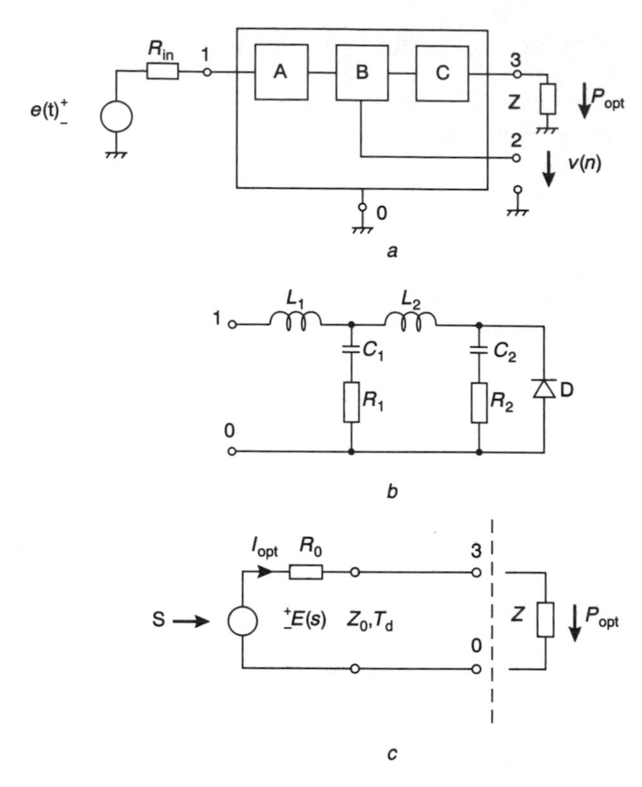

Figure 6.23 Subcircuits representation of laser module for use with SPICE 2G.5: (*a*) three-port black box; (*b*) subcircuit A for electrical behaviour; (*c*) subcircuit C for fibre pigtail. ([17], Figure 1). Reproduced by permission of IEE

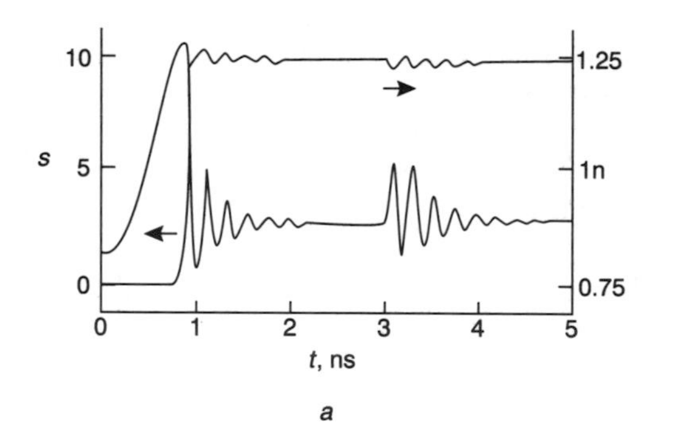

Figure 6.24 Response to step voltage input with laser biased below threshold: (*a*) numerical results: upper trace wavelength chirp $\Delta\lambda(t + \tau_D) \simeq n(t)$; lower trace $p(t + \tau_D) = s(t) \times 0.125$ mW; (*b*) waveform of $s(t)$ ([17], Figure 2). Reproduced by permission of IEE

s_E, are found from the relation:

$$s_E = K_4(0.5E - Z_0 I_{opt}) \tag{6.6}$$

With these choices of models, the three-port model can be implemented as a subcircuit in SPICE, version 2G.5. The parameter values used in a typical calculation are:

$$\tau = 2.91 \times 10^{-3}$$
$$K_1 = 1.420 \times 10^3$$
$$K_2 = 3 \times 10^{-3}$$
$$K_3 = 1$$
$$K_s = 15$$

where $K_s = s_E/s = 0.5\rho K_4 e/s$. For an input voltage step (see Figure 6.25*a*) to the laser biased below threshold, the computed optical response $p(t)$ is sketched in the lower curve of Figure 6.24 in terms of $s(t) = p(t + \tau_D)$.

The characteristic overshoot and ringing are present, but in addition, after an interval of about $2\tau_D$, more oscillations occur, this time due to the optical feedback. This second burst of damped oscillations causes oscillations in λ, i.e. additional chirping.

In the experiments, lasers were driven from below threshold by a periodic train of pulses, where the period was adjusted to slightly exceed one round-trip delay $2\tau_D$, by 2 ns in fact, to cause overlapping of a given launched pulse and the reflected pulse from the immediately preceding launched pulse. The response of a 1300 nm FP laser optical output is illustrated in Figure 6.24*b*, and it shows the same qualitative behaviour as the computed one. Measurements on dfb lasers gave similar results. Comparison of the computed and measured results show good agreement for the turn-on overshoot and for the ringing effects, as well as for the frequency

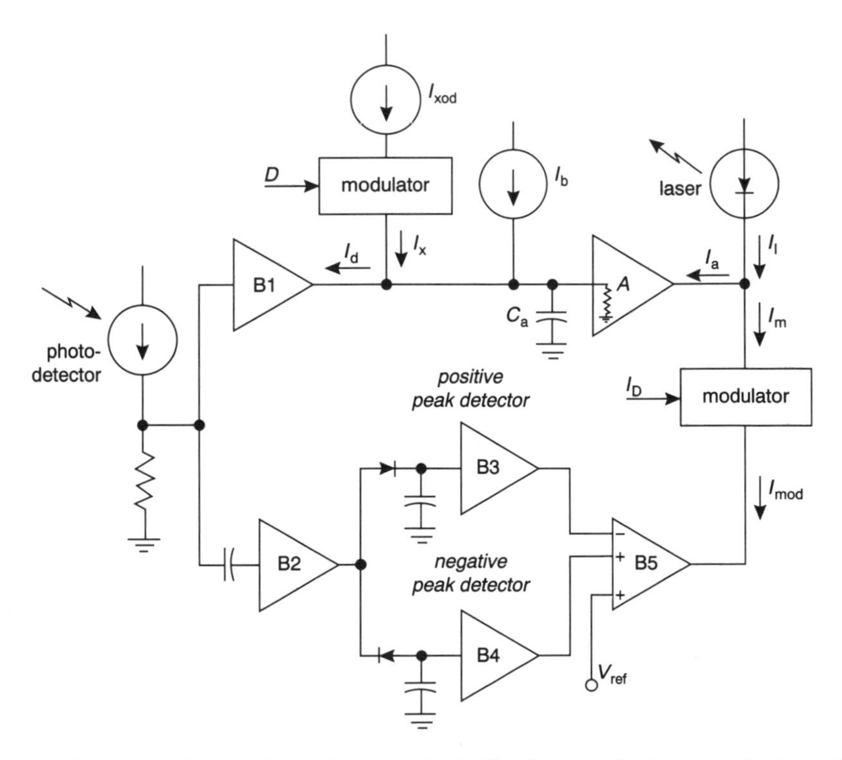

Figure 6.25 Improved laser driver circuit with feedback control of bias and of modulating currents, and with output level settings independent of laser parameters ([17], Figure 3). Reproduced by permission of AT&T

of the feedback-induced oscillations, but the damping of these oscillations does not agree. The author concludes that the *ad hoc* addition of the photon injection term into equation (6.4) gives qualitative agreement with measured results, and is therefore a useful first approximation and a powerful tool for the design of laser drivers in Gbit/s systems.

As bit rates increase and laser operation is no longer confined to the small-signal regime, more accurate large-signal models become necessary for use in the computer-aided design (CAD) of optical fibre microwave links and other applications. The first known application of the harmonic balance method to the characterisation of laser diodes appeared in 1989 in a letter from well-known workers in the microwave field [18]. The authors claimed that the harmonic balance technique had many advantages over time-domain solutions such as those used in SPICE, as it can be applied to highly nonlinear Gbit/s lasers to determine the large-signal intensity and frequency modulation responses, and is applicable to MESFETs and other microwave transistors, and to lumped or distributed networks containing many elements. The method does not rely on numerical integration, and unlike SPICE 2 it can also be used to analyse the modified circuit which accounts for coherent reflected waves from the laser mirror facet. The nonlinear single-mode rate equations were given as the mathematical model of the device from which a circuit representation of the active region, parasitics, stimulated emission current, photon loss and storage, and spontaneous recombination was derived.

6.5.3 Biasing of lasers

A paper [19] on the design of biasing and control circuits addressed three major considerations:

(a) whether to bias above or below threshold
(b) how to stabilise the optical output levels independently of variations in the mean output power
(c) to what extent can the output levels be stabilised relative to various circuit and device parameters.

The results of this study indicated that in order to eliminate any dependence of the optical output on a specific laser's characteristics, or on the mean value of the drive current, feedback control of both bias and modulation currents is necessary. The paper presented an analysis of a generalised method of negative feedback stabilisation of both these currents, an evaluation of the effectiveness of this technique, and the determination of the critical parameters of the feedback loop. A simplified block schematic of an improved laser driver is given in Figure 6.25.

The operation of this scheme can be found in the reference, but here we simply note that the modulating current $i_M(t)$ is proportional to the difference between the optical outputs corresponding to 1 and 0, and that the secondary negative feedback loop controlling this current will therefore maintain this difference at a constant value.

6.5.4 Design of laser transmitters

The design of laser transmitters for optical systems was described in a 1983 paper by two authors from British Telecom [20]. Restricting our attention to the modulation parts of the paper, we note that the theoretical performance can be determined from a numerical solution of the rate equations. This shows a damped relaxation oscillatory response to a driving current step. The frequency of this oscillation, f_R, is a determining factor limiting the maximum modulation rate. An approximate expression for f_R is:

$$(2\pi f_R)^2 \simeq \frac{(I_F/I_{TH}) - 1}{\tau\tau'} \tag{6.7}$$

where I_F is the total drive current, I_{TH} is the threshold current, τ is the photon lifetime, and τ' is the carrier lifetime

The authors mention the range of 0.8–3 GHz for this frequency in practice in 1983 devices; ten years later values of around 8 GHz were achieved. Relaxation oscillations do not occur in lasers with a deliberate lateral waveguide structure, or in some other designs.

Another theoretical relation is that for the time delay between the leading edge of a current step and the onset of stimulated emission if the laser is driven from below threshold, and is:

$$\tau_{TO} = -\tau' \ln\left(1 - \frac{I_{TH} - I_0}{I_P}\right) \tag{6.8}$$

where τ_{TO} is the turn-on delay, and I_P is the peak-to-peak amplitude of the modulation pulse. In practice, lasers are usually pre-biased close to threshold in order to reduce this delay.

As already described, the output characteristic is temperature-dependent, and it also degrades with time; therefore it is necessary to use automatic level control (alc) of output power, and to eliminate the possibility of supply current transients. In some circumstances, especially for devices operating at 850 or at 1300 nm, it may also be necessary to meet eye safety requirements. Since photodiodes have stable transfer characteristics, they are used as the feedback or return element in a control loop, providing that the tracking ratio is predictable. In digital systems, low values of the extinction ratio P_0/P_1 are needed, as are short turn-on delays; therefore it is usual to control one or more of the mean power, peak power, or pedestal position. Some details are given in the paper.

Control of the emitted spectrum is also important and is discussed by the authors, who cover injection locking and transmitters for coherent systems.

Further information on laser transmitter design can be found in [23].

The performance of a component-level transmitter design which operates up to 3 Gbit/s, nrz, has been used in a systems experiment at 2.4 Gbit/s over 35.2 km of dispersion-shifted fibre, and it achieved a sensitivity of −33.3 dBm at a ber of 3×10^{-11} [22]. The details give state-of-the-art figures for NMOS technology as at 1988, and are therefore of interest. The integrated transmitter is housed in a hermetic ceramic package which includes the laser driver, a 1550 nm dfb laser, automatic power control, and a thermo-electric heat pump controlled by a proportional heat/cool circuit. The output sm fibre pigtail is connected to a rotary bevelled splice of return loss exceeding 50 dB, then to an in-line optical isolator which provides more than 35 dB isolation between the laser output and the system fibre. The data signal is connected to the laser driver by a 50 Ω microstrip having more than 10 dB return loss at 4 GHz. The NMOS driver IC requires only 3.3 V, which is supplied from a regulator connected to a single −5 V power supply. The driver is interfaced to the laser by means of a 25 Ω microstrip, and the laser itself has a 3 dB bandwidth of 7–11 GHz, and a single-mode suppression ratio (smsr) better than 36 dB; both of which properties contribute to its suitability for multi-gigabit rate systems.

A block schematic of the driver appears in Figure 6.26, and it was designed in 0.75 μm silicon NMOS technology, with f_T in the range 8–12 GHz, to provide current pulses up to 60 mA into the 25 Ω load. The buffers amplify the signal to appropriate NMOS switching levels, and the shaping circuit minimises the over-swings at the output of the differential stage, as well as maintaining a stable cross-over point for the rising and falling edges of the current eye diagram, irrespective of operating and processing conditions. The IC also provides a data reference current proportional to the mark density of the input signal, which is used for optical output power control independent of the mark density.

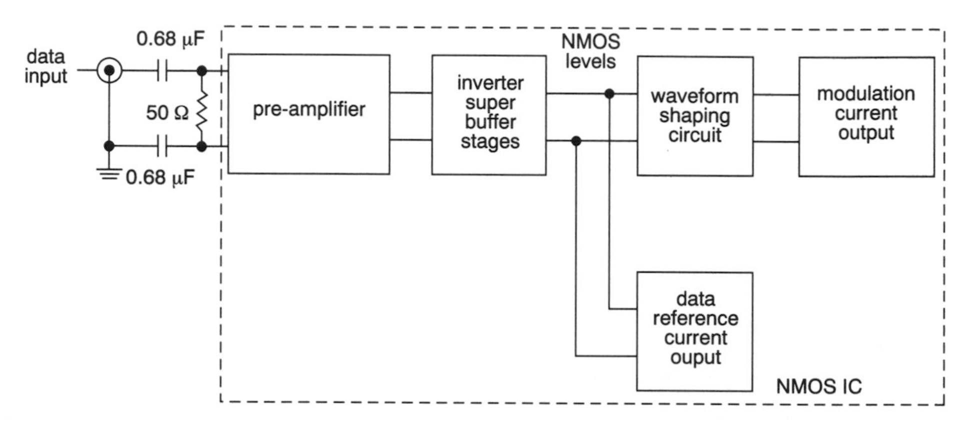

Figure 6.26 Block schematic of NMOS laser driver integrated circuit ([22], Figure 2). Reproduced by permission of IEE

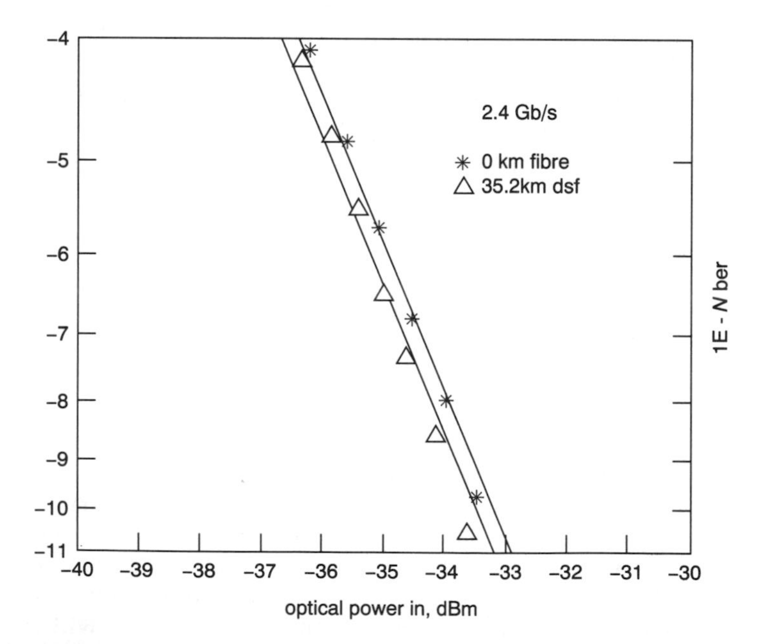

Figure 6.27 BER versus received optical power for 1550 nm laser transmitter ([22], Figure 4). Reproduced by permission of IEE

The quality of an optical transmitter is assessed most realistically in a systems environment, and this has been done, with the following results. The error rate curves after 3 km of sm fibre and after an additional 35.2 km of dispersion-shifted fibre are shown in Figure 6.27.

The ds fibre was used to minimise the system degradations due to chirping and mode partition noise of the laser, so that the performance of the transmitter itself could be studied. With the laser biased at threshold and 3 km of fibre the sensitivity was −33 dBm at a ber of 3×10^{-11}, and when 35.2 km of the ds fibre was added the sensitivity only changed to −33.3 dBm, a change within the reproducibility of the test set.

6.6 REFERENCES

1 TOWNES, C.H., and SCHAWLOW, A.I.: 'Infrared and optical masers'. *Phys. Rev.*, **112**, 15 Dec. 1958, pp. 1940-1949.

2 NEWMAN, D.H., and RITCHIE, S.: 'Sources and detectors for optical fibre communications applications: the first 20 years'. *IEE Proc.*, **133**, Pt. J, no. 3, June 1987, pp. 213-229.

3 GOUSIC, J.E., HARTMAN, R.L., KOREN, U., TSANG, W.-T., and WILT, D.P.: 'Quantum well lasers in telecommunications'. *AT&T Tech. J.*, **71**, no. 1, Jan./Feb. 1992, pp. 75-83.

4 TURLEY, S.E.H., HENSHALL, G.D., GREENE, P.D., KNIGHT, V.P., MOULE, D.M., and WHEELER, S.A.: 'Properties of inverted rib-waveguide lasers operating at 1.3 μm wavelength'. *Electronics Lett.*, **17**, no. 23, 12 Nov. 1981, pp. 868-870.

5 EALES, B.A., BRICHENO, T., ASHTON, J.E.U., and JANSSEN, A.P.: 'A high reliability 1.3 μm single mode laser package'. *Proc. Electro-optics and Lasers Int. '84*, Brighton, England, Mar. 1984, pp. 101-114.

6 ROSIEWICZ, A., KIRKBY, P.A., and JANSSEN, A.P.: 'The reliability of IRW lasers for use in 1.3 μm monomode systems'. *9th ECOC*, Geneva, 23-26 Oct. 1983.

7 PRESBY, H.M., and BENNER, A.: 'Bevelled-microlensed taper connectors for laser and fibre coupling with minimal back reflections'. *Electronics Lett.*, **24**, no. 18, 1 Sept. 1988, pp. 1162-1163.

8 KHOE, G.D., POULISSEN, J., and DE VRIEZE, H.M.: 'Efficient coupling of laser diodes to tapered monomode fibres with high index ends'. *Electronics Lett.*, **19**, 17 Nov. 1983, pp. 305-306.

9 GOODWIN, A.R., DAVIES, I.G.A., GIBB, R.M., and MURPHY, R.H.: 'The design and realisation of a high reliability semiconductor laser for single-mode fiber-optical communication links'. *IEEE J. of Lightwave Technol.*, **6**, no. 9, Sept. 1988, pp. 1424-1434.

10 JANSSEN, A.P., COX, A.A., GOODWIN, A.R., and DAVIS, I.G.A.: 'High reliability lasers for submarine use'. *Suboptic 86, Int. Conf. Optical Fiber Submarine Telecommunication Systems*, Feb. 1986, Paris, France, pp. 193-198.

11 BRIERRE, J., CADENE, J.-C., FOURRIER, J.-Y., THILLIEZ, J.-M., and TRIBET, D.: 'Reliable high speed bipolar i.cs. for long distance transmission'. *Int. Conf. Optical Fiber Submarine Telecommunication Systems*, Feb. 1986, Paris, France, pp. 396-403.

12 ABERNETHY, T.W., and WALKER, S.D.: 'Computer-aided laser transmitter optimisation techniques for high-capacity submarine optical transmission systems'. *IEE Coll. 'Submarine Optical Transmission Systems'*, 24 Mar. 1988, London.

13 STEVENSON, A., and FAULKNER, D.W.: 'Proprietary semiconductor regenerator i.c. for 140 Mbit/s optical transmission systems'. *IEE Proc.*, **132**, Pts E and I, no. 2, Mar./Apr. 1985, pp. 62-67.

14 BANU, M., JALALI, B., NOTTENBURG, R., HUMPTREY, D.A., MONTGOMERY, R.K., HAMM, R.A., and PANISH, M.B.: '10 Gbit/s bipolar laser driver'. *Electronics Lett.*, **27**, no. 3, 31 Jan. 1991, pp. 278-280.

15 CHEN, F.S., and BOSCH, F.: 'GaAs mesfet laser-driver i.c. for 1.7 Gbit/s lightwave transmitter'. *IEEE J. Lightwave Technol.*, **6**, no. 3, Mar. 1988, pp. 475–479.
16 BOTTACHI, S., and OSELLADORE, M.: 'A new current subtraction semiconductor laser modulator operating above 565 Mbit/s'. *Alta Frequenza*, **LXI**, no. 4. June 1987, pp. 235–240.
17 WEDDING, R.: 'Spice simulation of laser diode modules'. *Electronics Lett.*, **23**, no. 8, 9 Apr. 1987, pp. 383–384.
18 SNOWDEN, C.M., HOWES, M.J., and IEZEKIEL, S.: 'Harmonic balance model of laser diode'. *Electronics Lett.*, **25**, no. 8, 13 Apr. 1989, pp. 529–530.
19 SWARTZ, R.G., and WOOLEY, R.A.: 'Stabilised biasing of semiconductor lasers'. *Bell Syst. Tech. J.*, **62**, no. 7, Pt 1, Sept. 1983, pp. 1923–1936.
20 SMITH, D.W., and MATTHEWS, M.R.: 'Laser transmitter design for optical fiber systems'. *IEEE J. Selected Areas in Commun.*, **SAC-1**, no. 3, Apr. 1983, pp. 515–523.
21 CHEN, F.S.: 'Simultaneous feedback control of bias and modulation currents for injection lasers'. *Electronics Lett.*, **16**, no. 1, 3 Jan. 1980, pp. 7–8.
22 YANUSHEFSKI, K.A., YANUSHEFSKI, M.J., HOKANSON, J.L., SHASTRI, K.L., SMELTZ, P.D., WIAND, G.T., and RUNGE, K.: 'An integrated lightwave transmitter operating up to 3 Gbit/s, n.r.z. using a d.f.b. laser modulated with a sub-micron silicon n.m.o.s. driver'. *ECOC 88*, pp. 1–4.

FURTHER READING

23 WANG, Z.G., BERROTH, M., NOWOTNY, U., GOTZEINA, W., HOFMANN, P., HÜLSMANN, A., KAUFEL, K., RAYNOR, B., and SCHNEIDER, J.: '15 Gbit/s integrated laser diode driver, using 0.3 μm gate length quantum well transistors'. *Electronics Lett.*, **28**, no. 3, 30 Jan. 1992, pp. 222–224.
24 MONTGOMERY, R.K., REN, F., ABERNATHY, C.R., FULLOWAN, T.R., KOPF, R.F., SMITH, P.R., PEARTON, S.J., WISK, P., LOTHIAN, J., NOTTENBURG, R.N., NGUYEN, T.V., and BOSCH, F.: '10 Gbit/s AlGaAs/GaAs hbt driver for lasers or lightwave modulators'. *Electronics Lett.*, **27**, no. 20, 26 Sept. 1991, pp. 1827–1829.

7

SOME CHARACTERISTICS OF LASER DIODES

Hop, hop, hop
Chirp, chirp, chirp
Jitter, jitter, jitter

(Hewlett Packard advertisement for frequency and time interval analyser, 1988)

7.1 INTRODUCTION

Chapter 6 dealt with an illustrative specification for an IRW laser to provide a specific example of a well-characterised device, then proceeded to consider laser-to-fibre coupling and laser packaging, before ending with aspects of transmitters. Now we broaden the treatment of lasers to cover other subjects of importance in system design.

For some years all the information published in graphical or text form on the spectral characteristics of intensity-modulated lasers referred to time-averaged spectra, without necessarily making this clear. Such information was of limited use in assessments of system performance, because of variations in spectral behaviour within individual pulses and between adjacent pulses. Because of its importance a method was developed at BTRL to examine the time-resolved spectra of lasers, and this work revealed the essentially statistical nature of the phenomenon. A second very important topic is that of chirp in lasers under on–off modulation.

The third section deals with aspects of the effects of temperature variations on threshold current, output characteristics and spectrum, and discusses the use of thermoelectric coolers for temperature control. Fine control of temperature is sometimes used to adjust the oscillating wavelength to a desired value. The operation and performance of semiconductor lasers is dependent on internal feedback and is strongly affected by the presence of external feedback. In the last two sections the effects on laser properties, including noise behaviour, are described.

7.2 SPECTRAL CHARACTERISTICS OF BULK LASERS

Fabry–Perot and distributed feedback cavities have been employed widely, and are shown schematically in Figure 7.1 [1].

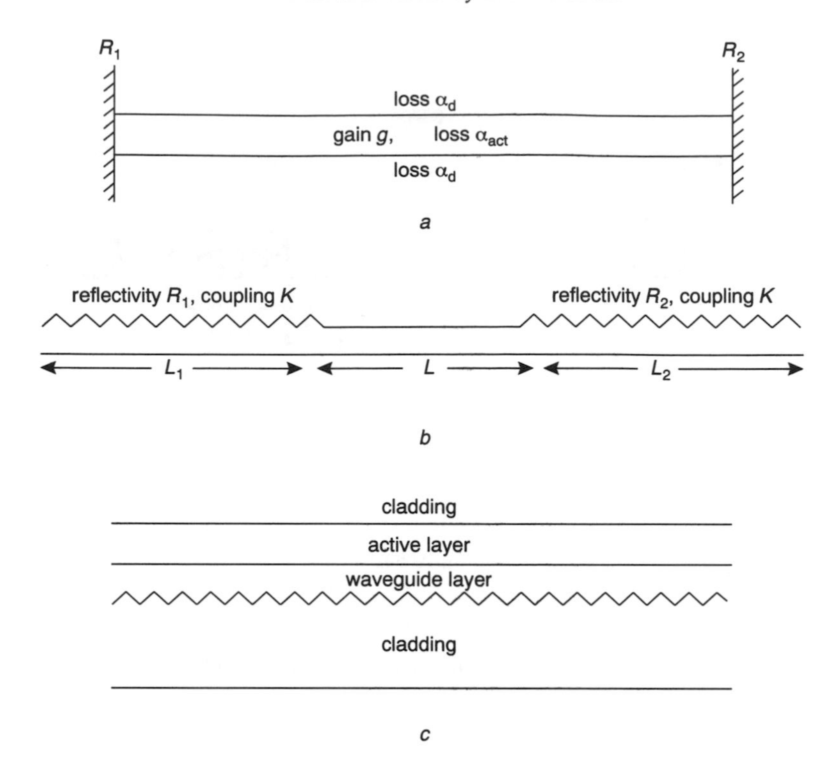

Figure 7.1 Schematic diagrams of (*a*) Fabry–Perot, (*b*) distributed Bragg, and (*c*) distributed feedback cavities ([1], Figure 3.12). Reproduced by permission of IEE

In these sketches the symbols have the following meanings: R_1 (R_2) is the power reflection coefficients at facets 1 (2); α_A and α_C are the loss per unit length in active, cladding region; g is the material gain per unit length; L is the length of laser cavity; and K is the grating coupling constant.

For bulk lasers operating in the 1300 nm band, typical 'linewidths' were given in Table 1.7, part of which is repeated here for convenience.

Fabry–Perot lasers	500 GHz
DFB lasers	20 MHz
LEC lasers	100 kHz

In double heterostructure configurations (Figure 6.1*b*) a symmetric slab waveguide is formed, which for an InGaAsP lattice matched to an InP cladding has a core region with refractive index around 3.54 at 1300 nm and 3.57 at 1550 nm. The cladding indices are 3.20 and 3.17, respectively. An important parameter of this waveguide is the confinement factor Γ which is defined as the ratio of the optical power confined to the core to the total cross-sectional power. The value of Γ is a function of the thickness of the active layer, the operating wavelength and the refractive indices of core and cladding regions. Values approaching unity correspond to tight confinement, whereas values tending towards zero denote poor confinement. For the numerical values quoted and for single transverse mode operation, the required thickness of the active layer becomes 0.43 μm at 1300 nm and 0.47 μm at 1550 nm. For F–P lasers operated just below the threshold, longitudinal modes corresponding to wavelengths at which there are an integral number of half-wavelengths in L occur on either side of the operating wavelength at intervals of $\lambda^2/2Ln_G$, where n_G is the group effective index (about 4.5); for $L = 300$ μm the separations $\delta\lambda$ are:

$$\delta\lambda = 0.7 \text{ nm at } 1300 \text{ nm}$$
$$\delta\lambda = 1.0 \text{ nm at } 1550 \text{ nm}$$

The spectral behaviour of lasers is important because fluctuations combine with fibre dispersion to degrade the received signal, and so give rise to system penalties. Changes in the output spectrum become relative delays at the end of a long dispersive fibre, resulting in jitter and partition noise (i.e. fluctuating division of power between two or more modes with time). Timing extraction may also be degraded, for instance if the circuitry has stabilised while the laser is producing a stable output. A small temperature change could then produce a rapid shift of timing phase as the spectral output changes, possibily resulting in a burst of errors while the timing extraction circuitry readjusts. Mode selection effects may permit satisfactory operation only at specific temperatures.

The spectrum of a laser is dynamic, not static, especially under modulation conditions. The spectrum of the optical response at a particular instant on the current waveform can usually be separated into constituent longitudinal modes, perhaps five in number, with spacings of about 1 nm. During a drive pulse the total energy in the optical response fluctuates between these modes, resulting in a continously evolving spectrum, although the total mean power P_1 is nearly constant throughout the duration of the first level. If the high-speed modulating waveform is a stream of pulses, the spectra are different again, due to the patterning effect causing the time-resolved spectra to exhibit in-filling of the pulses corresponding to individual modes. The effect observed is of the light output at a given instant being dependent on the signal during the preceding few symbol intervals, and it is this 'memory' which is called patterning. It occurs when the line rate is so high that the carrier density cannot reach equilibrium during one symbol interval. Very short relaxation times and low parasitics are called for if packaged lasers are to handle rates in the GBd range.

Packaged lasers are used in systems; unfortunately this carries the risk of optical feedback occurring within the package. In one type of device results suggested an external cavity length of about 60 μm, which is consistent with the distance from the rear facet of the laser chip to the monitor photodiode. The observed effects are caused by movements in both the gain peak and the external cavity length with changes in temperature.

7.2.1 True time-resolved spectra of modulated lasers

There is a new technique of avoiding pulse-to-pulse (time) averaging by a technique developed at BTRL in 1982 which separates the emission in each longitudinal mode during a given pulse [2]. Ideally, the outputs of a streak camera placed on the exit slits of a monochromator could simultaneously monitor the distribution of power in many modes during modulation, but cameras sensitive at the required wavelengths did not exist. The required information can however also be recorded using a transient digitiser coupled with time-of-flight mode separation. The experimental arrangement used was given in [2]. The separate simultaneously launched modes were collected by an array of fibres of different lengths which permitted the emission from each longitudinal mode to be separated in time and displayed on a

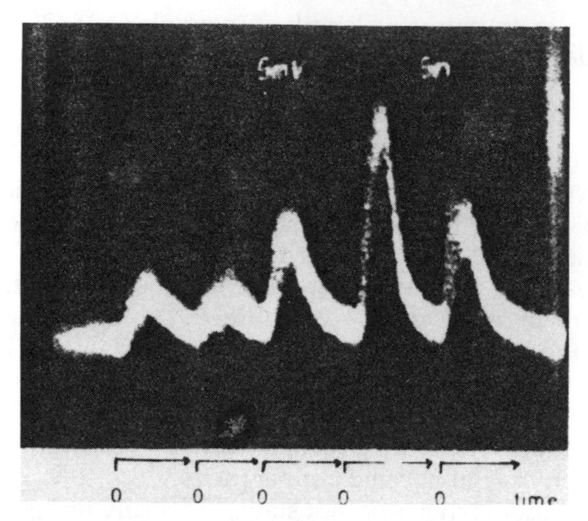

Figure 7.2 Sampling oscilloscope display of modes from fibre array ([2], Figure 2). Reproduced by permission of IEE

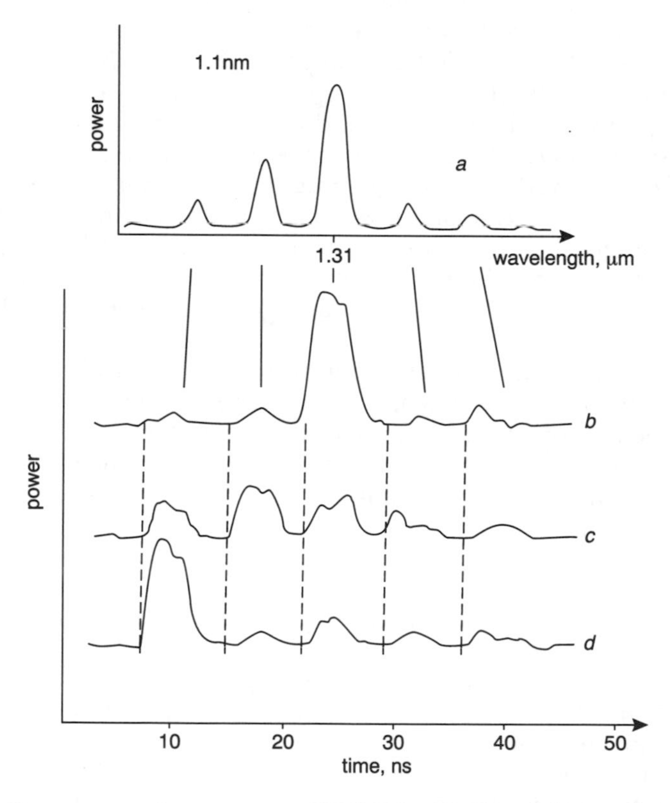

Figure 7.3 (*a*) Time-averaged spectrum at 320 MBd with 57 mA bias and ±12.5 mA pulse; (*b*)–(*d*) single shot transient spectra with 45 mA bias and 25 mA pulse ([2], Figure 4). Reproduced by permission of IEE

sampling oscilloscope. The technique is applicable generally to all laser wavelengths of interest and to any type of structure.

A typical output from the fibre array is illustrated in the oscillogram of Figure 7.2 for a 1300 nm laser biased at threshold, and it clearly shows in-filling of the pulses due to patterning. The separate arrival of the five dominant modes can be seen clearly.

A great deal of spectral variation on a pulse-to-pulse basis was also observed, and examples of the five strongest modes are shown in Figure 7.3 together with the time-averaged spectrum at 320 MBd (a line rate used in some medium and long-haul submerged systems), reproduced from [2]. Although the power in each mode fluctuated considerably, the total output from all the modes was found to vary by less than 5% from pulse to pulse, indicating that more than 90% of the available power was in these five modes. The oscillations superimposed on the signals from each mode were caused by the resonance behaviour of the laser. Similar effects were also observed on a channel substrate 1550 nm device.

The evolution of a true time-resolved spectrum throughout the duration of a pulse is illustrated in Figure 7.4, which was obtained for a 1300 nm channelled substrate buried-crescent laser.

In this diagram the line shape for each mode has been artificially generated to produce spectra of conventional appearance.

The spectra obtained from pulse-to-pulse measurements vary considerably, and so are best treated statistically. The quantities of interest are the average mode position, and the variance of the spectrum at any particular instant on the drive current waveform. Expressions for the mean and the variance were defined in terms of mode intensities at the chosen instant, and by recording a large number of spectra the corresponding PDFs can be generated. Probability density functions for the mean and variance from 2600 events at bias currents of 80 and 100% of the threshold value were obtained by recording the spectra at the middle of the 3 ns symbol interval (equivalent to a single 'one' in a 320 MBd NRZ system) [3], and are shown in Figure 7.5.

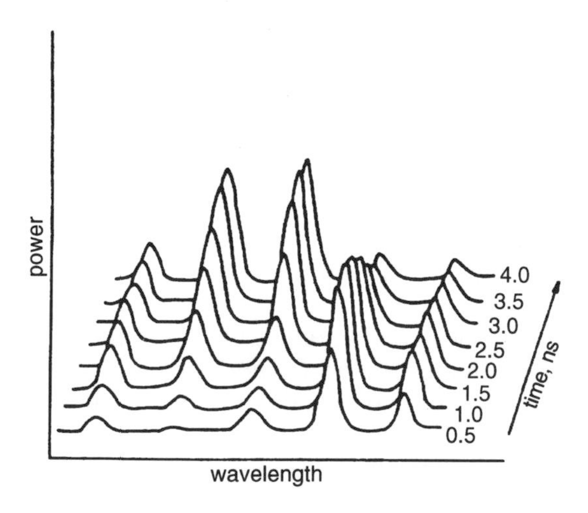

Figure 7.4 True time-resolved spectrum throughout the pulse duration, 1300 nm cs-bc laser ([3], Figure 2). Reproduced by permission of IEE

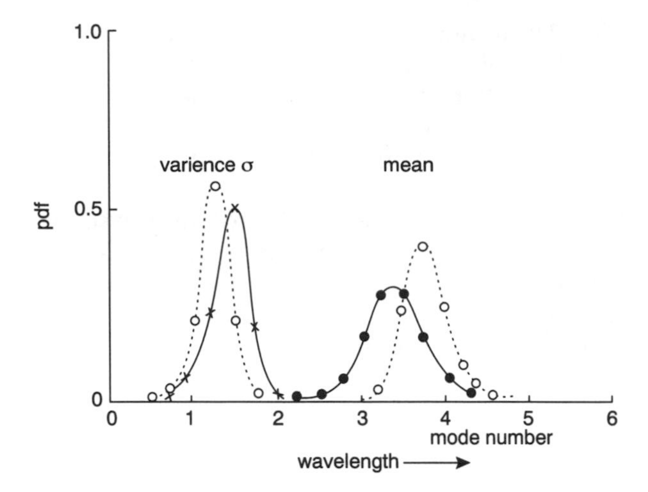

Figure 7.5 PDFs of mean and variance produced after sampling 2600 events ([3], Figure 3). Reproduced by permission of IEE

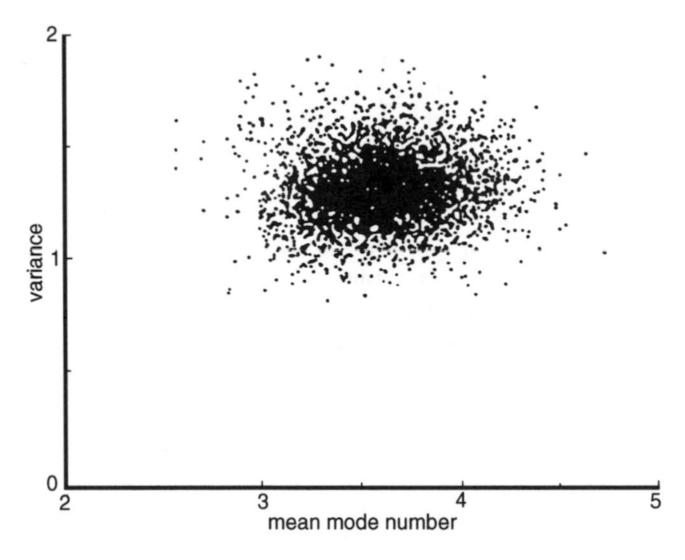

Figure 7.6 Correlation between variance and mean mode number for a 1300 nm IRW laser ([3], Figure 4). Reproduced by permission of IEE

Changing the sampling instant also changes the PDFs; for instance, at the beginning of the pulse much broader distributions were found, including zero variance events, i.e. single-mode operation.

Correlation between the mean and variance was examined for an IRW laser. Little correlation between the mean mode number and the variance is to be expected from partitioning, and this was confirmed experimentally, as can be seen from the scatter diagram in Figure 7.6.

To simulate reflections such as might occur in a packaged laser fitted with a fibre tail, an experimental arrangement was devised, and the above sampling process was repeated. The effect of reflections on the scatter diagram is shown clearly in the following Figure 7.7.

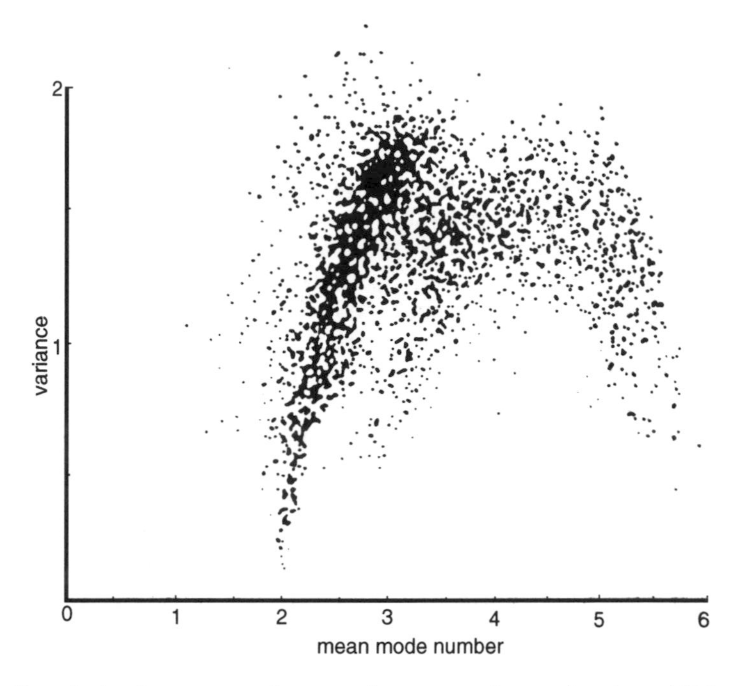

Figure 7.7 Correlation between variance and mean mode number for a 1300 nm IRW laser with feedback ([3], Figure 5). Reproduced by permission of IEE

From this scatter diagram it can be seen that any one of the four modes may predominate. The measurements also indicate that the laser was locked into a single mode for periods of about 1 ns. Similar effects were seen with a different type of packaged laser.

Considerable effort has been devoted to establishing a satisfactory measurement method for the changes in wavelength of lasers under modulation, and over long operating periods. Experience suggests that the shift of individual lines is not so important as shifts in the 'centre of gravity' of the whole evolving spectrum, as system impairment seems to arise from a combination of the latter with the chromatic dispersion of system fibres.

This work by the same two authors from BTRL was amplified in [4], and is summarised below.

The mean and variance were obtained from spectral measurements on a 280 μm long device emitting at 1520 nm, which when modulated at 320 MBd had seven dominant lasing modes. The experimental arrangement enabled the spectral evolution to be observed during a 3 ns current pulse of about 30 mA amplitude, corresponding to an output of some 4 mW peak power. PDFs were obtained by sampling at the centre of the pulse for 1000 events with the laser biased just below threshold. The curve of mean values spanned some three mode spacings and the variance two.

The spectral statistics for 100, 210 and 280 μm cavity lengths of the same type of device were also reported in the form of graphs of PDF versus r.m.s. width for the variances and PDF versus mean wavelength about the nominal value. The r.m.s. widths became narrower as the length increased, causing the mode separation to decrease. The PDF curves showed that the device with a 100 μm cavity had a much narrower distribution about the nominal mean wavelength; calculations suggested that a 70 μm cavity length and reflection coated rear facet would support only one lasing mode.

Correlation measurements showing scatter of variance against mean values for a large number of pulses were very similar for both 210 and 280 μm devices, but these differed greatly from the diagram for the 100 μm cavity laser, again when biased just below threshold. The scatter diagram for the 100 μm device at 19°C showed nearly single-mode operation, whereas at 26°C the mean spectrum contained two main modes.

Tests were also performed on a packaged 300 μm laser oscillating at 1300 nm and butt-coupled to a length of sm fibre. Variance versus mean diagrams at 20.4°C, 20.9°C, 24.2°C and 27.5°C illustrated how strongly mode selection depends on case temperature. This was attributed to the presence of feedback within the package resulting from an external cavity supporting modes that lay between the internal modes 0 and 1, and between 3 and 4. This corresponded to an external cavity length of about 60 μm, consistent with the distance between the rear facet and monitor photodiode. Short-term correlation measurements were also made using a 101 pattern of 1.5 ns duration pulses (the fastest useful operating speed of the digitiser). The diagram of mean spectral position in pulse 2 versus mean in pulse 1 showed no significant correlation.

All of the results reported were obtained with lasers biased just below threshold, because when biased above threshold the greatly increased noise on the zero level prevented meaningful measurements being taken [3] ended with a discussion which mentioned the significance of the results in systems and in the prediction of system penalties.

7.2.2 Chirp in intensity-modulated lasers

Intensity-modulated lasers exhibit dynamic wavelength shift (chirping) arising from gain-induced variations in refractive index. The combination of linewidth broadening and fibre dispersion can set a limit to system performance and the subject has therefore received much attention. A rigorous investigation of the effect of chirping on bit rate appeared in 1989 [5], when a methodolgy and numerical results were presented. Models of chirped signals and of sm fibres were used to evaluate system ber by taking account of injection current extinction and rise time and in the presence of isi due to the combined effect of chirping and finite receiver bandwidth, shot noise, apd multiplication noise and preamplifier circuit noise.

The drive current was modelled as a sequence of equiprobable marks and spaces, the laser output and chirping by large-signal coupled first-order rate equations for a noise-free laser biased above threshold and operating in a single longitudinal mode. Patterning effects were included. The optical receiver was modelled as an ideal photodetector with additive noise, a linear pulse shaping filter and a sampling device.

A given set of parameter values and 2.4 Gbit/s were used to obtain numerical results. These showed predicted light output curves, without and with chirp for a given input current nrz pulse, ber versus received optical power for two values of extinction ratio, receiver sensitivities versus extinction ratio for two values of gain compression factor with 100 km of sm fibre, receiver sensitivities versus linewidth enhancement factor for two values of gain compression factor with 100 km of sm fibre, receiver sensitivities at 1.7 Gbit/s versus extinction ratio for two values of gain compression factor with 100 km of sm fibre. In the summary the authors stated that the usefulness of their method for assessing the implications of important system and device parameters on system performance could be extended to rz and pre-equalised drive pulses and signal processing in the receiver.

To understand the relation between chirp and intensity modulation, the rate equations can be studied [6]. The authors show that the optical frequency shift can be expressed by the relation:

$$\delta f(t) = \frac{\alpha}{4\pi} \frac{\mathrm{d}(\ln p)}{\mathrm{d}t} \tag{7.1}$$

where α is a parameter relating the refractive index and gain, and $p(t)$ is the emitted optical power. In order to achieve chirp-free behaviour, it is necessary for δf to be constant during the symbol interval, and for this condition equation (7.1) can be solved, giving:

$$p(t) = P_0 \exp \gamma t \tag{7.2}$$

yielding a constant optical carrier frequency throughout the pulse duration given by:

$$f = f_0 + \frac{\gamma\alpha}{4\pi} = f_0 + \delta f \tag{7.3}$$

Hence, if the optical power $p(t)$ satisfies equation (7.2) in the symbol interval and is zero elsewhere, then no chirp occurs. A number of ideal current pulse shapes, including the rectangular pulse for which the highest bit rates are obtained, would result in optical pulses of the same shape under ideal conditions. The ideal rectangular shape can be modified by considering finite rise and fall times, resulting in a twofold increase in line rate over Gaussian shaped pulses, if the modulation bandwidth is at least 2.5 times the line rate. Simulation showed that a 4 Gbit/s stream could be transmitted over 100 km of sm fibre, but this entails severe requirements on bias level and drive current pulse shaping.

Spectral studies of dfb lasers directly modulated at high rates have shown the chirp to depend on bias level, modulation level and frequency, and device structure. The importance of the chirp-induced dispersion penalty on system performance makes it very desirable to measure laser chirp characteristics quickly and cheaply. Measurements of time-resolved chirp in Gbit/s nrz pulses from a gain switched laser were reported in [7]. The results demonstrated the utility of the low-cost, compact and easy-to-use arrangement for resolving the spectrum of high-speed and gain-switched dfb laser pulses with ps durations.

The laser oscillated at 1530 nm when biased at 58 mA (threshold current 48 mA) and was modulated with 52 mA nrz pulses from a pattern generator at a 2 GHz repetition rate. The evolution of the power in the various transmitted wavelengths selected by an F–P etalon (resolution 0.05 nm) for resolving these pulses is plotted in Figure 7.8.

The curves show the 'blue' shift at the leading edge and the more pronounced 'red' shift at the trailing edge of the pulse. The small peaks occurring later than 1200 ps were attributed to relaxations in the lower level.

The chirp of gain-switched pulses is of interest for transmission of Gbit/s wdm signals over dispersion-shifted fibres, so the spectral behaviour of the device biased at 97 mA when gain-switched by a 150 mA peak-to-peak sinewave at 2 GHz was examined using a second F–P etalon (resolution 0.5 nm) to resolve these pulses of 35 ps fwhm duration. The peaks of the time responses were located and the wavelengths within the main part of the pulses versus time are shown in Figure 7.9.

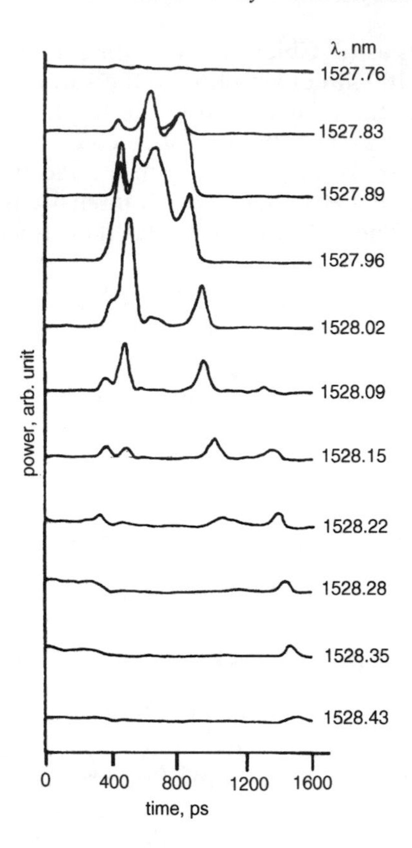

Figure 7.8 Time-resolved chirp of isolated 2 Gbit/s nrz pulses from a dfb laser ([7], Figure 2). Reproduced by permission of IEE

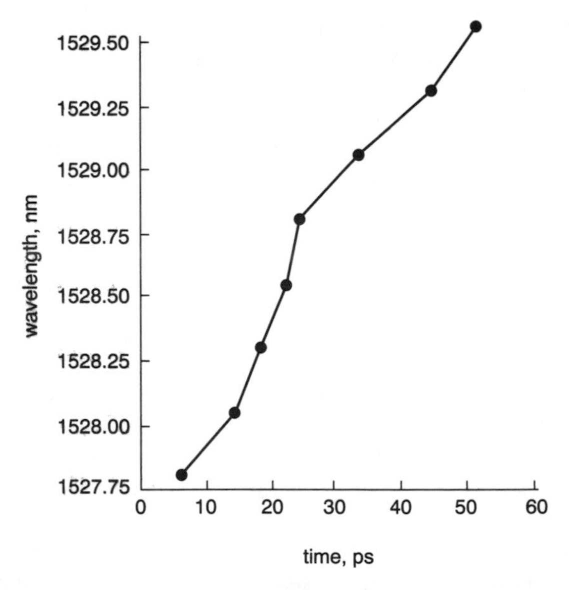

Figure 7.9 Time-resolved chirp of 35 ps pulse from laser gain-switched at 2 GHz ([7], Figure 3). Reproduced by permission of IEE

The figure shows the expected red shift chirp and that the chirp rate is not constant and is lower under the trailing edge.

The authors emphasised that the sampling of chirped time-varying signals in the wavelength domain results in a narrowing of the angular optical spectrum $\Delta\omega$, together with a shortening of the pulse duration Δt. Heisenberg's uncertainty relation sets a limit to the accuracy with which the position in time of a narrowband signal can be determined, i.e. the transform limitation. Observed values of $\Delta\omega\Delta t$ lay in the range 3–5.

The transmission-line laser model (TLLM) has been used to predict the dynamic performance of 1520 nm mode-locked lasers, which is in good agreement with experimental results. The TLLM includes both dispersive bandwidth-limiting elements and a refractive index dependent on carrier concentration which governs chirping. The predicted time-resolved spectrum of a 10 ps pulse from a dispersive grating-controlled actively mode-locked laser showed a rapid red shift during the first half of the pulse duration, followed by a small slow blue shift [8]. This behaviour was thought likely to affect pulse dispersion, especially in highly dispersive fibres used for soliton generation and transmission. Graphs of computed values of time-resolved frequency shift, power output and carrier density versus time were given, and a second figure showed the power spectral density over four pulses versus frequency shift. The external cavity modes were visible but not resolved. The centre of the spectrun was moved by 13 GHz to the red as expected. A spectral width of 35 GHz gave a time × bandwidth value of 0.35, which was close to the value of 0.32 for sech^2 pulses, but increased by chirping.

A method for measuring the evolution of the mean wavelength in a modulated device was described in [9]. The results show a high correlation between chirp and regenerator dispersion penalty when dispersive fibres are used. The chirp and intensity waveforms of a modulated dfb device are shown in Figure 7.10.

The chirp, or change in the mean wavelength (defined as the ratio of the first moment of the time-resolved spectrum divided by the intensity), is plotted for the intervals when the intensity rises above 20% of its maximum value. The curves

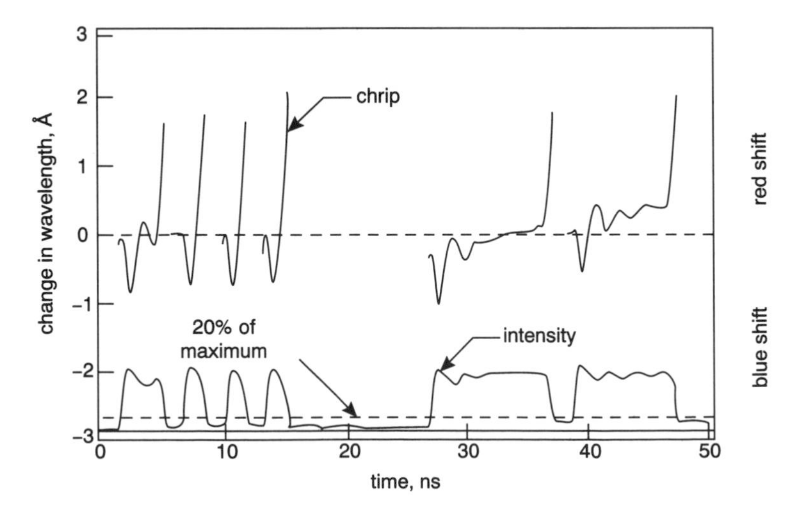

Figure 7.10 Change in mean wavelength versus time of dfb laser modulated by nrz pulses from a 600 Mbit/s $2^7 - 1$ prbs pattern generator ([9], Figure 3). Reproduced by permission of IEE

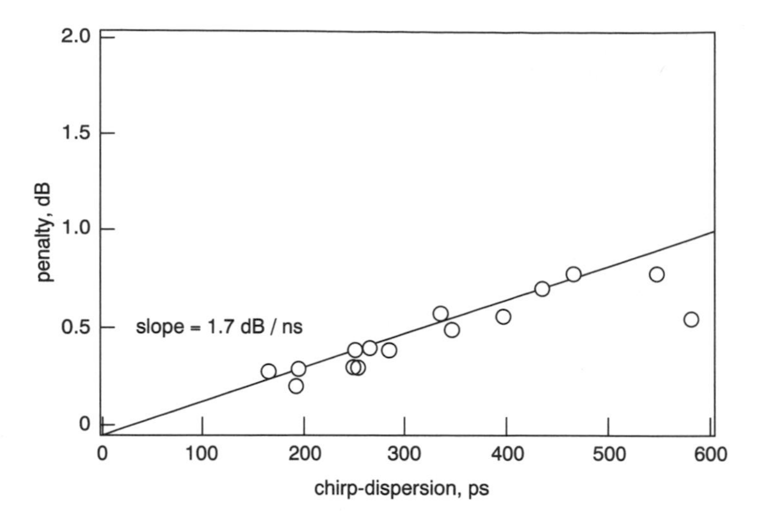

Figure 7.11 Measured dispersion penalty (o) versus chirp × dispersion product for ber = 10^{-9} at 600 Mbit/s, 1500 nm dfb lasers, 92 km sm fibre ([9], Figure 4). Reproduced by permission of IEE

show the characteristic blue shift on rising edges, and red shift on falling edges. This particular laser exhibits a shift of about 0.3 nm. At the right of the figure, the reader can see that for long drive pulses the chirp has a damped oscillatory behaviour which mimics the overshoots in the current. The drift towards longer wavelengths on longer pulses does not recover when the drive is momentarily interrupted, and the drift may arise from heating of the laser junction which has a time constant several times longer than the bit period used. A consequence of this drift is that the leading edges of contiguous pulses can occur at different wavelengths, which, if associated with a large amount of fibre dispersion, could lead to higher jitter. It would also suggest that line codes which do not permit repeated ones to occur should be used in such circumstances.

The transmission performance of a number of dfb lasers was examined at 600 Mbit/s over 92 km of step-index sm fibre with approximately 1.5 ns/nm of chromatic dispersion and the chirp measured with the same arrangement as before. The resulting regenerator sensitivity penalty at 10^{-9} ber, as a function of the chirp-dispersion product, is shown in Figure 7.11.

The closeness of these points to the straight line illustrates the high degree of correlation.

7.3 EFFECTS OF TEMPERATURE ON LASER CHARACTERISTICS

When 1300 nm lasers were sufficiently developed and engineered for production and deployment in single-mode operational systems in the early 1980s it became essential to obtain and record more detailed information on their properties, especially in the case of long-lived lasers. Automated acquisition and recording systems for production laser characteristics were designed and used by leading manufacturers. One such system stored measured values of light output, drive current and forward bias voltages for both ternary (880 nm) and quaternary (1300 nm) lasers for a range of temperatures, and it was described in [10]. The results were obtained for planar oxide-insulated stripe geometry devices, and it was stated that the thermal behaviour of IRW lasers was similar.

Temperature changes have three main effects on 1300 nm devices:

(1) an increase in threshold current and in forward bias threshold voltage with increasing temperature
(2) a decrease in slope of the output light versus drive current with increases in temperature
(3) a spectral shift or mode hopping.

7.3.1 Temperature dependence of threshold current and bias voltage

Families of cw light output L as a function of drive current I are very well known; much less known are those for L as a function of forward bias voltage V and [10] gave the first published $L(V)$ curves for 1300 nm cw planar oxide-insulated stripe-geometry laser diodes. A knowledge of both is desirable for a full understanding of those laser properties which impinge directly on system performance [10] appeared in late 1982. This study reported examples of $L(I)$ and $L(V)$ families of curves over the temperature range 10–45°C, at intervals of 5°C, of $V/V(15°\mathrm{C})$ at 4 mW output versus temperature, and of driving current versus forward voltage at 10 and 45°C for ternary and quaternary devices. Forward voltage at threshold and threshold current versus temperature characteristics were presented for the same quaternary laser. The thermal behaviour of ternary (GaAl) (As) and quaternary (GaIn) (AsP) cw planar oxide-insulated stripe-geometry devices was studied. Results for the latter are similar to those for the IRW type of laser. Standard cw quaternary $L(I)$ characteristics are illustrated in Figure 7.12*a*, and they correspond to a characteristic temperature θ_C of 62 K (a range of 50–80 K for GaInAsP lasers has been quoted, the range for IRW lasers is given later). The corresponding $L(V)$ family are shown in Figure 7.12*b* over the same range 10–45°C. $V(\theta)$ is shown in Figure 7.13 for the same device LW71 and for a ternary laser.

Curves were also given for the threshold current and threshold voltage versus temperature characteristics of laser LW71, and finally for the I(V) curves at 10 and 45°C, which show the usual diode shape.

7.3.2 Effect of temperature on slope of output characteristics

At higher temperatures the slope dL/dI is also a function of temperature and tends to decrease as temperature increases, especially above $\theta = 70°\mathrm{C}$, giving rise to a reduction in modulation depth even if mean power feedback is used. For example, over the temperature range 20–70°C the variation in slope of a particular family of output characteristics was sufficient to halve the modulation depth. Again, either temperature control or electronic control techniques can be used to overcome this effect. In one design the cooler maintained the output power to within ±0.13 mW of the mean level of 0.95 mW at 23°C over the −40 to +65°C range [11]. The corresponding changes in bias current and in the cooler current are shown in Figure 7.14.

The cw output characteristics $L(I)$ commonly have slopes above threshold which lie in the range 0.1–0.4 mW/mA. For an IRW laser with an incremental efficiency

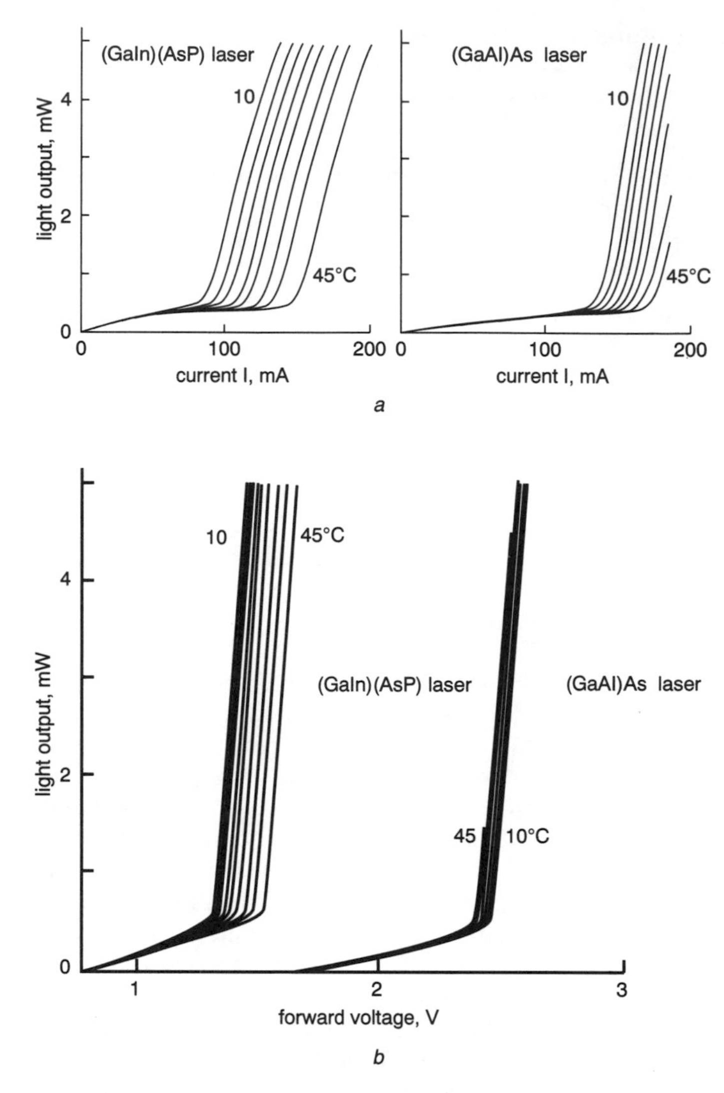

Figure 7.12 (*a*) Light output versus drive current at 5°C intervals; (*b*) light output versus forward bias voltage at 5°C intervals ([10], Figures 1*a*, 2*a*). Reproduced by permission of IEE

(slope) of 0.36 up to 5 mW into 0.5 na at room temperature, a rise to 50°C reduces the slope to 0.30. The change in slope with temperature can also be fitted by an expression by an appropriate choice of θ_C, as described in the next paragraph.

Threshold current increases with temperature according to an empirical relation of the form:

$$I_{TH}(\theta_2) = I_{TH}(\theta_1) \exp \frac{\theta_2 - \theta_1}{\theta_C} \tag{7.4}$$

where θ_1, θ_2 and θ_C are the room, package (heat sink) and characteristic temperatures, respectively, in °C.

As a particular example consider the cw output characteristic of an IRW laser with temperature as the parameter of interest. At 20°C the threshold current may be as low as 69 mA and as high as 90 mA for chips from the same slice. The

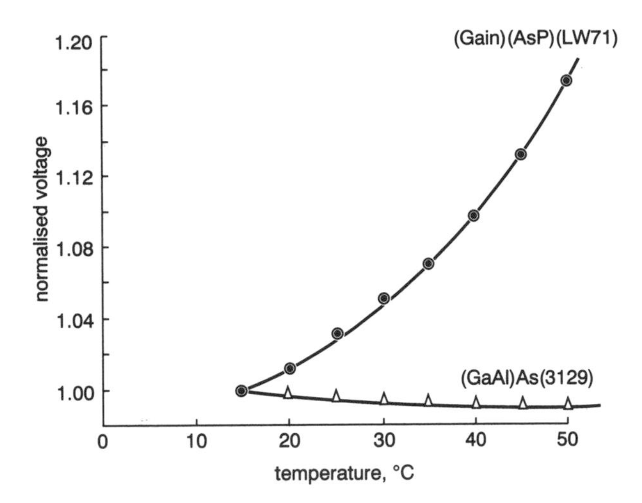

Figure 7.13 Normalised voltage ($V(T)/V(15°C)$) versus temperature at constant $L = 4$ mW for 1300 nm and 880 nm devices ([10], Figure 3). Reproduced by permission of IEE

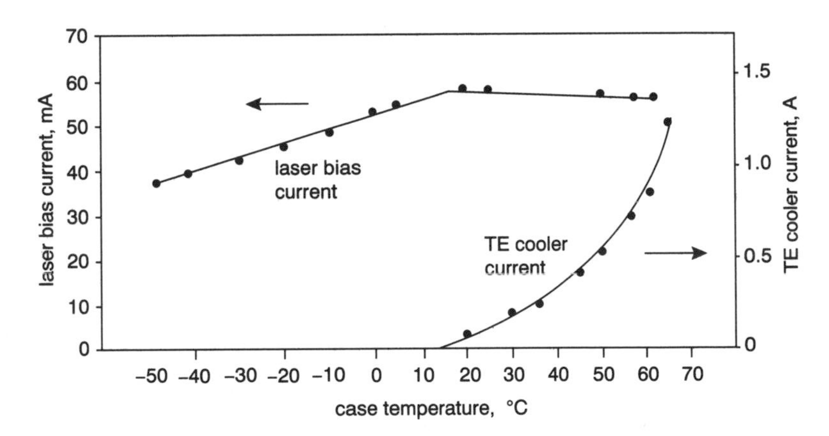

Figure 7.14 Laser bias current versus case temperature and thermo-electric cooler current versus case temperature ([11], Figure 7)

value of θ_C is found to be 62 K up to 50°C, but lower at higher temperatures when considering cw thresholds; for pulsed operation θ_C remains at 62 K to at least 90°C, where the device still lases but is close to thermal runaway. For this type of laser a maximum value of 2.5 mA/°C has been quoted for the temperature dependence of I_{TH} between 20 and 50°C.

It should be noted that the temperature dependence of I_{TH} in cw and in pulse operation differ at higher temperatures. An example of this for buried crescent lasers with separate optical confinement is shown in Figure 7.15 [12].

7.3.3 Temperature sensitivity of laser spectrum and mode hopping

Changes in temperature cause shifts in wavelength of about 0.55 nm/°C in 1300 nm lasers, which is more than twice that in GaAlAs devices at 850 nm and this heightened sensitivity was one of the penalties of operating at the longer wavelengths.

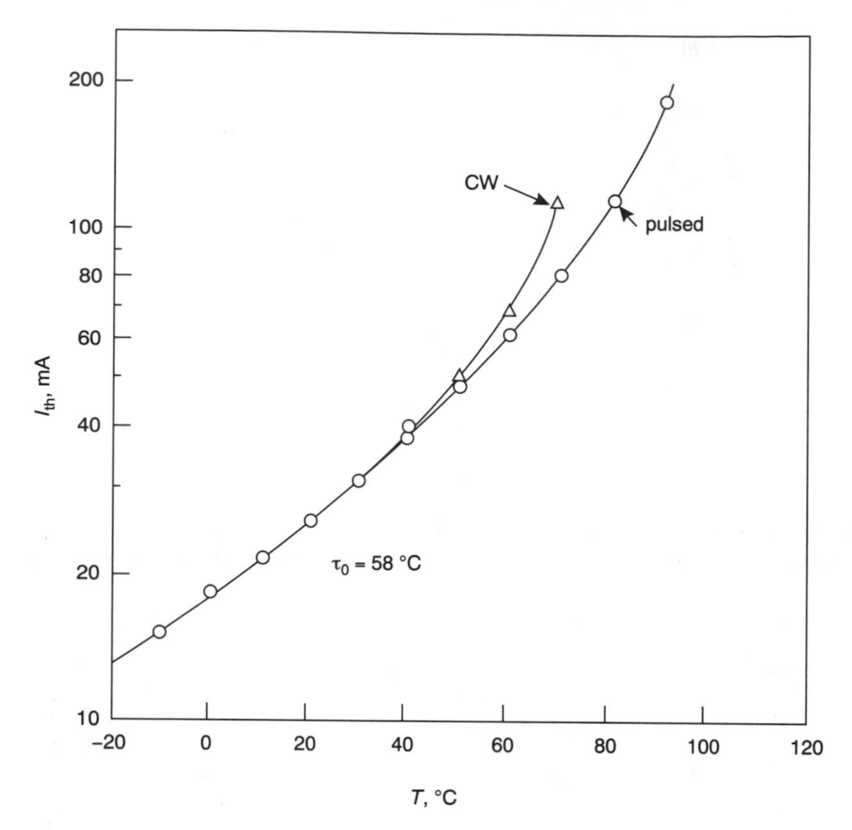

Figure 7.15 Threshold current versus temperature characteristics of a buried crescent laser under cw and pulsed pump conditions ([12], Figure 2). Reproduced by permission of IEE

An informative illustration of the continuous changes in wavelengths due to temperature changes was published in 1981 [13]. The curves were based on those for oxide stripe devices and are illustrated in Figure 7.16.

This diagram shows four properties of multi-longitudinal mode spectra: (a) mode spacing, (b) the spectral width, (c) the slope of the (envelope) peak wavelength, and (d) the slope of individual mode wavelength.

System implications of mode hopping

For an IRW laser of cavity length 300 μm the wavelength spacing $\delta\lambda$ between adjacent longitudinal modes is approximately 1 nm, and the fwhm of the near field distribution is 2.8 μm. Under normal conditions the device operates solely in the zero-order mode, but hopping to an adjacent mode can occur. A mode jump has to be considered as a timing jitter effect. The system implication of a mode hop of 1 nm over a 30 km span of standard sm fibre in a system carrying 140 Mbit/s traffic can be found by evaluating the increment in propagation delay $\delta\tau$ given by:

$$\text{per-unit dispersion} \times \text{fibre length} \times \text{mode hop}$$

The maximum dispersion depends upon the spread in peak wavelengths λ_P of the laser, and on the fibre dispersion as a function of wavelength. For instance, a given

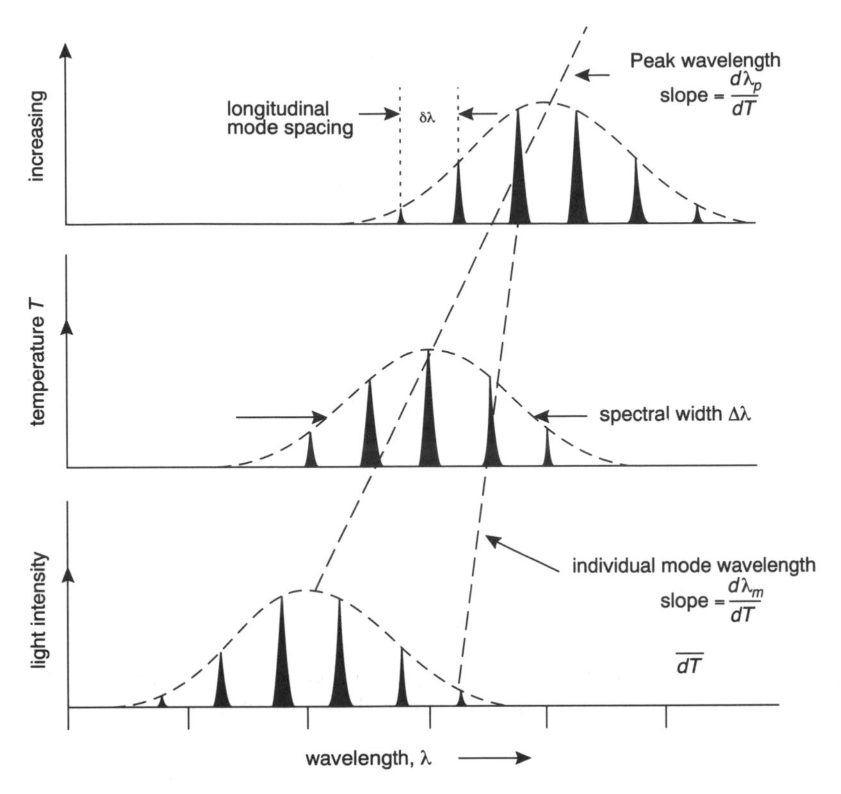

Figure 7.16 Diagram to show the effect of temperature on a multimode spectrum: (*a*) mode spacing, (*b*) the spectral width, (*c*) the slope of the (envelope) peak wavelength, and (*d*) the slope of individual mode wavelength ([13], Figure 10. Reproduced by permission of IEE

laser may have the following values of λ_P: 1280 (min.), 1300 (typ.) and 1320 nm (max.). At room temperatures the fibre dispersion as a function of wavelength can be found from measured values of per-unit group delay τ_1 ns/km by fitting a fifth-order polynomial of the form:

$$A\lambda^{-4} + B\lambda^{-2} + C + D\lambda^2 + E\lambda^4 \tag{7.9}$$

to the measured points, then differentiating the expression to yield values for $d\tau_1/d\lambda$. Results for a single-mode fibre with $\lambda_0 = 1370$ nm are shown in Figure 7.17 [14].

The curve of τ_1 is roughly parabolic, hence the first-order dispersion is approximately linear with slope $d^2\tau_1/d\lambda^2 \simeq 0.1$ ps km^{-1}nm^{-2} over the above range of room temperature values of λ_P. Therefore, specifying a value of chromatic dispersion less than 6 ps km^{-1}nm^{-1} between 1270, and 1330 nm requires that the zero dispersion wavelength λ_0 must lie within the range 1270–1340 nm. Worst case (maximum) dispersion would result if λ_0 were to lie at the upper limit and simultaneously λ_P were to be at its minimum value, resulting in a separation of 60 nm which, multiplied by the slope, gives about 6 ps km^{-1}nm^{-1}. The corresponding delay over 30 km for a 1 nm mode jump is then $6 \times 30 \times 1 = 180$ ps. This figure is around 1.5% of the bit interval for 140 Mbit/s of $10^{-6}/140 \simeq 7$ ns. Several repeater sections with such a delay could possibly give rise to sufficient jitter accumulation to cause

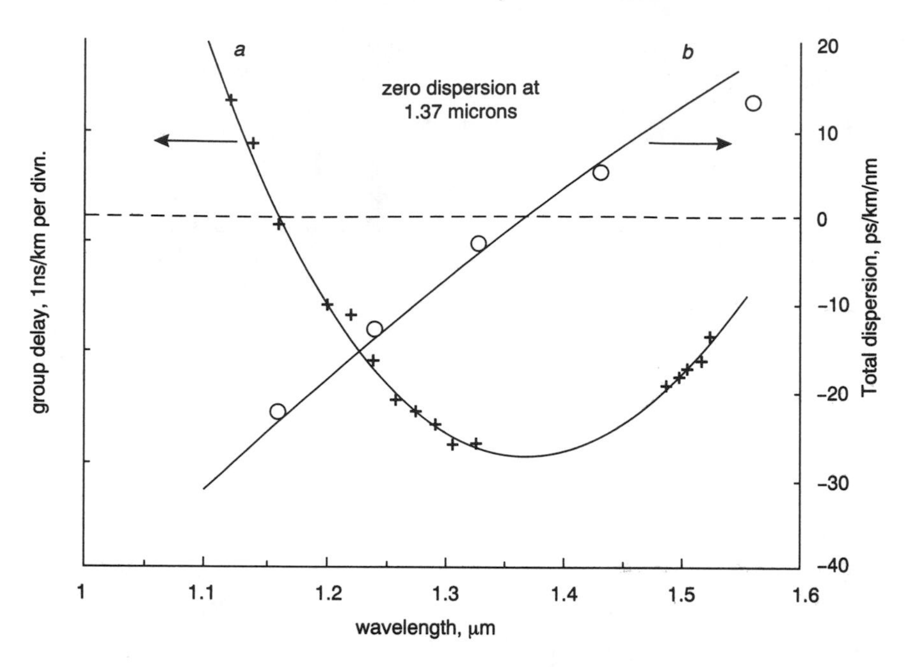

Figure 7.17 (*a*) Measured group delay and polynomial versus wavelength, and (*b*) derived dispersion versus wavelength ([14], Figure 5). Reproduced by permission of IEE

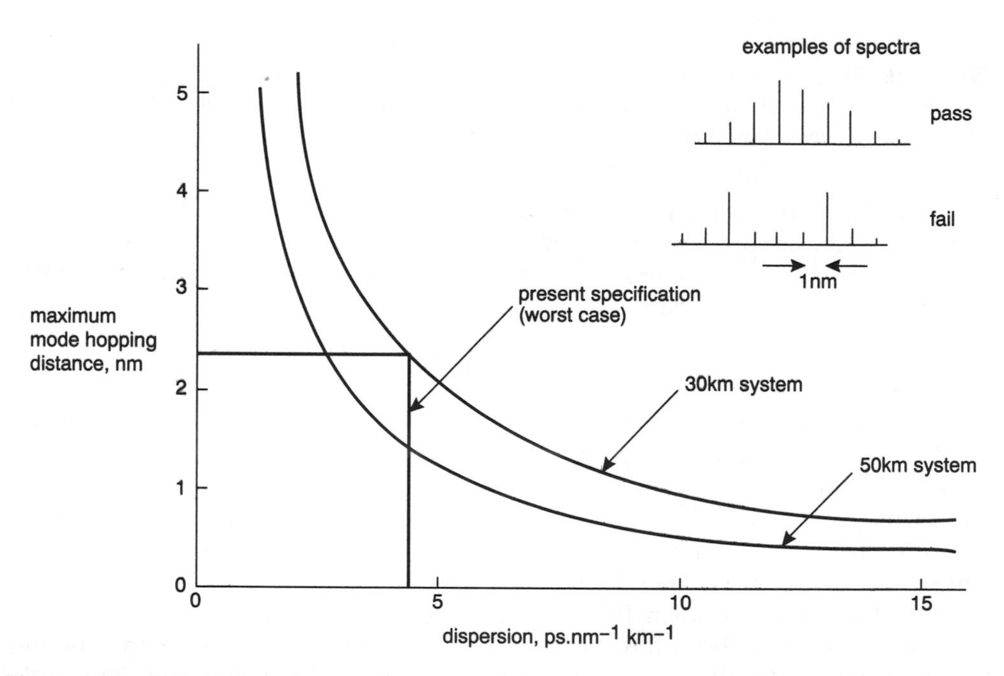

Figure 7.18 Sketch showing maximum mode hop versus per-unit dispersion for 30 and 50 km spans in a 280 Mbit/s system assuming worst-case jitter of 300 ps, together with examples of spectra. Reproduced by permission of Nortel

a burst of errors, while the phase of the repeater clock achieves synchronisation. It should be noted that mode jumps can occur in temperature-controlled lasers if the nominal temperature and variations combine adversely.

The general form of relation between maximum allowable mode jump, dispersion and repeater span, assuming a worst-case jitter of 300 ps, can be seen from the sketch in Figure 7.18 which refers to a 280 Mbit/s system (note: actual repeater spans were 45 km in early 280 Mbit/s submerged systems and 55 km in later ones).

It is clear from these curves that mode jumping may be an important constraint on system performance. This has been confirmed in later work.

7.3.4 Temperature control using thermo-electric cooler

Temperature control can be used or a mean power feedback controller to adjust the laser bias current for changes in I_{TH}. Peltier coolers are often convenient for this purpose; they control by balancing the heat pumped away, the heat generated by the laser and the heat leakage due to convection from surfaces and conduction along wires. Convection can be reduced by lagging and conduction by using thin leads. A method of calculation is given below.

The heat leakage Q through the walls of an insulated box is given by:

$$Q = A\left(\frac{\Delta\theta}{(\Delta x/k + 1/h)}\right) \tag{7.5}$$

where A is the surface area, $\Delta\theta$ is the difference between surface and ambient temperature, Δx is the thickness of insulation, k is the average thermal conductivity of insulation, and h is the average value of coefficient of heat transfer (typically 28.5 $Wm^{-2}K^{-1}$).

Similarly, when the leads are connected to a good heat sink, the heat conducted along a lead wire is given by:

$$Q = kA\frac{\Delta\theta}{\lambda} \tag{7.6}$$

where k is the thermal conductivity of wire, A is the cross-sectional area of wire, l is the length of wire, and $\Delta\theta$ is the temperature difference between the ends of wire.

The mass of the laser package does not enter either of the above equations, but does determine the time taken to change the temperature by a given amount, and this quantity can be derived from the following equation:

$$t = mc_P\frac{\Delta\theta}{Q} \tag{7.7}$$

where t is the time taken, m is the mass of package, c_P is the specific heat capacity of package, $\Delta\theta$ is the change of temperature required, and Q is the heat transferred by cooler.

The range of ambient temperature over which equipment should operate is specified by the telecom operator concerned, for instance −5°C to +40°C for line terminating equipment in the UK. By assuming a shelf temperature of 15°C above ambient, one arrives at a range of −5°C to +55°C over which the cooler is to maintain the laser temperature constant. The thermal resistance of the heat sink required is found from:

$$\frac{(\theta_{HS} - \theta_A)}{Q_{HS}} \tag{7.8}$$

Table 7.1 The values given in Table 7.1 refer to a lagged packaged laser having dimensions 25 × 21 × 5 mm with four leads each 15 mm long by 0.1 mm diameter. The thermo-electric cooler is a Melchior, type FC 0.6-32-06L.

Heat generated by laser	0.3W	0.3W
Surface leakage	161mW	161mW
Conduction leakage	80mW	80mW
Max/Min ambient temp.	+55°C	−5°C
Max/Min heat sink temp.	+60°C	−10°C
$T_{HS} - T_A$	5°C	5°C
Laser temp.	25°C	25°C
ΔT across cooler	35°C	35°C
Heat to be pumped	0.54W	—
Cooler current	0.45A	—
Cooler voltage	1.8V	—
Elec. power	0.83W	—
Heat to/from heat sink	1.37W	—
Heat sink thermal resis.	3.6°C/W	—
Mass of package	25 gms	
Time to equiliibrium (laser off)	317 secs	

where θ_{HS} is the heat sink temperature, θ_A is the ambient temperature, and Q_{HS} is the total heat flow into the heat sink.

Illustrative numerical values are given in Table 7.1 which refers to a lagged packaged laser, with dimensions 25 mm × 21 mm × 5 mm, four leads each 15 mm long by 0.1 mm diameter. The thermo-electric cooler is a Melchior type FC 0.6-32-06L.

7.4 EFFECTS OF EXTERNAL OPTICAL FEEDBACK ON LASER PERFORMANCE

Many different effects of feedback (which always has magnitude and phase) on the operation of lasers have been observed and studied, with a view to understanding them and avoiding or minimising resulting degradations in performance. The earliest effect was the increased level of intensity noise when feedback in the range −30 to −10 dB relative to the output power was present; extreme linewidth broadening was also present. Broadening or narrowing of the emission line have been observed at very low feedback levels (< −50 dB). For sufficiently large amounts of feedback the spectrum becomes a multivalued function of the phase of the feedback. Still higher levels lead to intensity noise and line broadening. The use of antireflection coatings to increase the amount of feedback to make the linewidth narrower, is well known.

Other experimentally observed phenomena induced by feedback include the introduction of ripples into the laser cw characteristics above threshold, possibly accompanied by hysteresis effects, and periodic changes in the output spectrum with distance to the reflecting surface or with drive current. In addition, the noise spectrum and transient are also effected by feedback. These and other effects will be treated in this section.

Reflections can also arise from backscattering in the system fibre. Feedback from far end reflections and from backscattering can be reduced by inserting an optical isolator between laser and system fibre, and this is commonly done in practice. Even with an optical isolator, near-end reflections can still occur, e.g. in packaged lasers. These can however be reduced by means of an anti-reflection coating on the near-end of the fibre. Feedback is strongly affected by the coupling technique between chip and fibre tail, for instance, and a variety of designs have evolved which can reduce reflections very significantly.

7.4.1 Feedback regimes in dfb and F–P lasers

The effect of feedback on the spectrum of dfb lasers oscillating in the 1550 nm band was investigated and reported in [15]. The work covered the range in the magnitude of the feedback F from -80 dB to -8 dB of dfb lasers with an external reflector. The feedback factor F is a ratio with magnitude $|F|$ and phase $\arg F$ in general. Five distinct regimes were discovered; these are listed in order of increasing feedback:

Narrowing or broadening of linewidth, depending on the phase of F ($\arg F$): 30% changes in linewidth were observable at -80 dB when an external reflector was located $L_E = 400$ mm from the output facet; this corresponds to 10 pW of power fed back for 1 mW output power.

At a value of F which depends on the distance to an external reflector, linewidth broadening, which occurs at the lowest values of $|F|$ for out-of-phase feedback, changes to an apparent splitting of the emission line arising from rapid mode hopping. The amount of splitting depends on $|F|$ and the distance L_E to the external reflector.

Increasing values of $|F|$ in the range -45 to -39 dB, but independent of the value of L_E, lead to suppression of mode hopping and single-mode operation results. In this regime the behaviour is very sensitive to any other reflections of comparable magnitude.

Independently of L_E and of $\arg F$, and at about $|F|$ of -40 dB, side modes separated from the main mode by the relaxation frequency appear: The size of these side modes increases as $|F|$ increases, and the linewidth broadens to about 50 GHz. This phenomenon is known as 'coherence collapse' because of the drastic reduction in the coherence length of the laser.

When $|F|$ is larger than about -10 dB, the laser operates as a long external cavity device with a short active region and, provided that the external cavity is sufficiently frequency selective, the laser will operate in a single longitudinal mode with narrow linewidth for all values of $\arg F$. Operation in this regime is relatively insensitive to other external feedback effects.

Essentially the same effects have been seen at about the same levels in F–P lasers. The boundaries between the regimes are unexpectedly sharp and are displayed in Figure 7.19.

The analysis proceeded from the Van der Pol equation for the electric field $e(t)$ within the laser, plus an additional source term corresponding to a delayed reinjection of the field and neglecting the noise. By assuming $e(t)$ of the form $E_0 \exp(-j\omega t)$ the change in steady-state gain is given by $-2\kappa \cos(\omega\tau_E)$, where κ is the feedback fraction and τ_E is the round-trip transit time in the external cavity. An expression

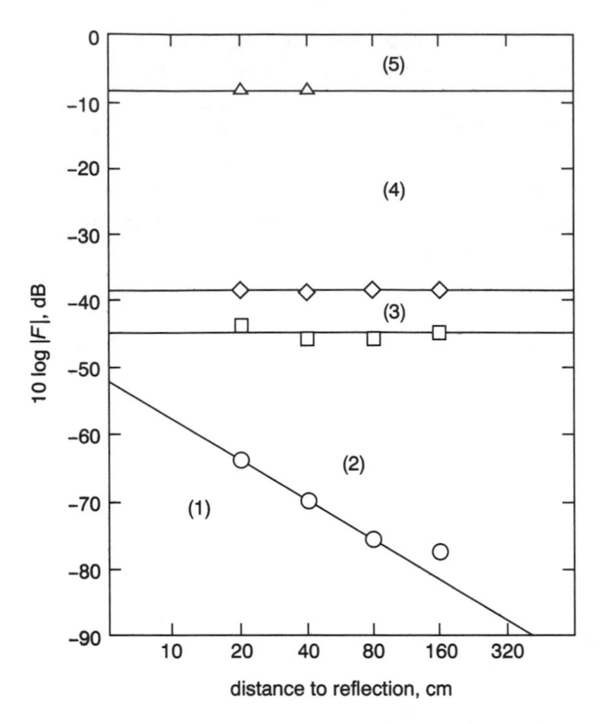

Figure 7.19 $|F|$ versus distance to reflection illustrating the five distinct regimes when external feedback is present. Lines are drawn through data points for emphasis ([15], Figure 8). Reproduced by permission of IEEE, © 1986

was also given for the change in linewidth valid when τ_E is less than the coherence time of $e(t)$.

7.4.2 Theory of Fabry–Perot laser diodes with feedback

A number of attempts have been made to model this situation analytically, which resulted in expressions which were not in good agreement with numerical or experimental results. Another expression for the stable operating range of lasers with optical feedback was proposed in [16], which showed very good agreement with experimental and numerical results. The rate equations for a single-mode laser with optical feedback were evaluated in the small-signal domain to predict the behaviour of the laser under modulation up to microwave frequencies. The transfer function relating a modulating current (input) to the modulated output of the mode of minimum linewidth (the most stable external cavity mode) is very similar to the well-known expression for an amplifier with a single external feedback loop, and was given as:

$$H(jf_M) = \frac{[1 - \beta(jf_M)] \times \mu(jf_M)}{[1 - \mu(jf_M)\beta(jf_M)]} \tag{7.10}$$

where f_M is the modulating frequency, H is the transfer function with feedback, μ is the forward path gain (loosely gain without feedback, $\beta = 0$), and β is the feedback fraction.

The quantity μ was given as the reciprocal of a second-order polynomial, so we can write:

$$\mu^{-1} = \left[\frac{jf_M}{f_R}\right]^2 + \frac{jf_M}{f_D} + 1 \tag{7.11}$$

where f_R is the relaxation resonance frequency and f_D is the damping frequency.

With weak feedback and Henry's linewidth enhancement factor $\alpha > 1$, the quantity β for a F–P laser can be written as:

$$\beta = j\frac{\kappa_E\sqrt{1+\alpha^2}}{f_M}[1 - \exp(-jf_M\tau_E)] \tag{7.12}$$

$$\kappa_E = \frac{2[(1-R_2)/\sqrt{R_2}] \times \sqrt{[(P_R/P)]}}{\tau_L} \tag{7.13}$$

where τ_E is the round trip delay in an external feedback loop, τ_L is the round trip delay in an internal cavity of length L, P_R is the reflected power returned to the output facet, and P is the power emitted from the output facet.

The authors showed that the minimum level of feedback which makes the denominator of 7.10 vanish corresponds to the onset of coherence collapse. They assumed that $f_R < f_D$ so that $\mu(jf_M)$ has its maximum value when $f_M \simeq f_R$, i.e. at the pole of equation (7.10). By further assuming a long external cavity so that $f_R\tau_E \gg 1$, β is a maximum when the exponential term becomes -1, i.e. at odd multiples of 180°. The corresponding expressions for κ_E and of P_R/P yield the critical values of these quantities.

Comparison with numerical results followed. The first figure showed the rin in the band 5–500 MHz versus P_R/P when $\alpha = 6$ and $P = 1$, 2, 5 and 10 mW. The next figure presented numerical results derived from the rate equations for the same ordinate and abscissa and α, a fixed $P = 5$ mW and laser lengths of 150, 300, 450, 600, 900, 1500 and 2000 μm. In both figures the predicted critical values lay at or close to the 'knee' of the curves, showing the good agreement achieved. The third figure showed the relative intensity noise (rin) in the band 5–500 MHz versus P_R/P for $P = 5$ mW and $\alpha = 6$, 4, 2, and 0, and the fourth showed the rin in the band 5–500 MHz versus P_R/P for $\alpha = 6$ and $P = 5$ mW for several values of the nonlinear gain coefficient. Again the predicted critical values displayed good agreement between numerical and theoretical results. The final figure showed the critical values of P_R/P versus α and the much improved agreement obtained by the present analysis. For bulk InGaAsP–InP dfb lasers f_D has been found to be about 20 GHz, and a similar value for InP-based mqw devices. Some typical values were given for bulk (mqw) lasers as follows: $f_R = 6(10)$ GHz, $\alpha = 6(3)$, hence the critical value of P_R/P is 5×10^{-5} (1.7×10^{-3}); these values indicate that mqw lasers tolerate some 14 times more feedback than bulk devices.

7.4.3 Theory of dfb laser diodes with feedback

The effect of external feedback on resonant frequency, threshold gain and spectral linewidth of dfb lasers was analysed in [17] for any type of laser cavity formed by a corrugated waveguide limited by partially reflecting facets. Sensitivity to feedback on a facet was shown to be closely related to the power emitted. Numerical results were given for this sensitivity and for wavelength selectivity for conventional dfb lasers with one antireflection (ar) coated facet and for $\lambda/4$-shifted devices with ar coatings on both facets. Both laser types were found to be more sensitive than F–P lasers to small values of feedback on their ar coated facet but that $\lambda/4$-shifted lasers were relatively insensitive at larger values of feedback. The analysis is considerably more complicated than in [16], as indeed is the whole paper. The abscissa of most graphs was the coupling constant taken as κL, where the coupling coefficient κ (cm^{-1}) is given by [18]:

$$\kappa = \frac{\pi n_G \Delta\lambda}{(\lambda_B)^2} \tag{7.14}$$

where n_G is the group index of laser structure, $\Delta\lambda$ is the stop band width, and λ_B is the centre (Bragg) wavelength.

Dependence of mode stability and linewidth on R and on κ

Experimental data for ar coated 1550 nm dfb lasers was reported in [18], which demonstrated a correlation between mode stability and coupling coefficient parameter κ. This parameter is important in predicting which devices will maintain the same single mode of operation after coating reasonable agreement between theory and experiment on the increase of linewidth was also reported.

The lasers were fabricated with second-order continuous gratings and cleaved to give $L = 250$ μm long cavities with random phase changes at the two facets. The reflectivity of the output facet was reduced from 32 to 5% by an ar coating, which resulted in the linewidth increasing by 1.6 times, compared with 2.3 from theory.

Mode stability: the dominant emission modes before and after coating were correlated using the stop band as an internal mode reference. 15 lasers from two wafers were assessed and for most devices the change in stop band width was 0.2 nm or less compared with a mode spacing of 1.4 nm. Values of κ ranged from 67 to about 195 cm^{-1}, with an average of 120 cm^{-1}. 37% of devices with κ less than this average exhibited mode jumps or multimode dfb spectra after coating, compared with 67% of those with κ above 120 cm^{-1}, as expected, because of the stronger backwards Bragg scattering and hence feedback implied by the larger values of κ. The average value of the coupling constant κL was 3.

Linewidth broadening: an equation was given for the linewidth Δf which was proportional to the facet loss and to the spontaneous emission factor, both of which are influenced by the coating. After presenting the theory, the results of measurements were given. 14 lasers from the same wafer, which all operated in a single mode with a minimum side-mode suppression-ratio (smsr) of 25, were measured. The minimum linewidths before and after coating one facet with a film with $R = 3$–4% were measured, and the resulting average increase was 17–32 MHz, a factor of 1.8. The power outputs from both facets of each device, measured at the same bias as the linewidths, exhibited an average increase of 23%, giving a power × linewidth average increase of 2.2 as a result of the antireflection coating.

Control of the coupling constant κL

The value of the coupling constant exerts a strong influence on current threshold, slope efficiency, side-mode suppression ratio and other determinants of laser performance. dfb lasers have been fabricated in various cavity lengths and the corresponding optimum value of κ is different for each length. DFB devices with

$\lambda/4$-shifted gratings require the same value of κ to be fabricated on both sides of the shift region. A method for the accurate control of the grating in dfb lasers operating in the 1300 nm band was described in [19]. The accurate control of κL was achieved by a novel mocvd technique and the introduction of a barrier layer. The results of measurements of stop band width on 300 μm devices from four wafers were given from which values of κL were calculated (from equation (7.14)). The spreads of κL in each wafer were small enough to allow optimum values of κ to be obtained in fabrication. Curves of typical far-field patterns showed that despite the asymmetry of the structure the patterns were Gaussian in shape.

By the end of the 1980s extremely narrow linewidth dfb lasers had been fabricated, for instance the 1200 μm long device emitting 20 mW at 1500 nm grown by movpe and described in [20]. Control of the coupling coefficient κ so that $\kappa L \leq 1.0$ was essential to avoid the spatial hole burning which occurs in long cavities due to mode coupling and which causes the linewidth to increase dramatically. The thickness of the active layer was reduced to 70 nm, and improvements in the geometrical uniformity of the active layer were also needed to obtain the desired performance, which resulted in a linewidth of about 1 MHz. This can be compared to the value of 20 MHz quoted earlier for dfb lasers operating at 1300 nm. Among the facts mentioned was the linear relation between linewidth and the reciprocal of the cavity length, suggesting that the internal cavity loss was the dominant factor affecting linewidth. cw threshold currents as low as 20 mA were found for the 1200 μm devices, with an external quantum efficiency of 0.20 mW/mA and single-mode operation to above 20 mW output.

7.5 EFFECTS OF EXTERNAL FEEDBACK ON NOISE BEHAVIOUR OF LASERS

High bit rate directly modulated systems are very attractive because of their low cost and large transmission capacity, and by 1990 bitrate × span products up to 1000 Gbit/s km had been realised in experiments. Such systems are, however, very susceptible to optical feedback effects which can severely degrade the performance. The effects are mainly due to intensity noise and linewidth broadening; consequently a mathematical description of intensity noise in lasers under intensity modulation is needed to analyse such systems.

7.5.1 Analysis of behaviour under cw and pcm modulated conditions

The lack of such a tool was remedied by Petermann and Wang in [21] which analysed the intensity noise and linewidth of pcm modulated single-mode lasers with distant optical feedback and provided experimental confirmation using a 1300 nm (Hitachi HL 1314A) device. The spectral behaviour of a laser operating in the coherence collapse regime (the fourth regime of Section 7.4.1) must be included in the analysis.

After describing the experimental arrangement the authors stated the theory and listed numerical values of the parameters of the dfb laser which were used in the numerical analysis to simulate the rate equations and showed a block diagram of the computer simulation model. Measured and calculated results for cw and sinusoidal

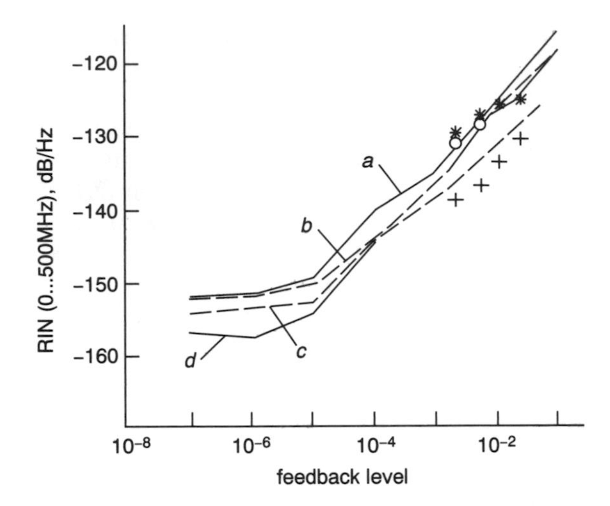

Figure 7.20 Measured ($*$, $+$, O) and calculated (a, b, c, d) values of rin versus feedback fraction averaged over a 500 MHz band ([21], Figure 3). Reproduced by permission of IEE

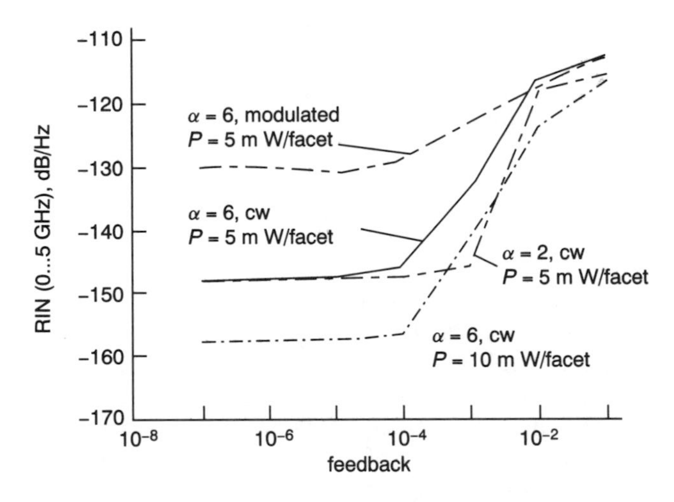

Figure 7.21 Calculated values of rin versus feedback fraction averaged over a 5 GHz band for four different combinations of parameters ([21], Figure 5). Reproduced by permission of IEE

modulation, and measured results using a 1 Gbit/s random pcm modulating signal, were presented for the rin, as shown in Figure 7.20.

The increase in rin as the laser enters the fourth regime is clearly seen. Measurements and simulation both show that modulation does not increase the rin, and in fact counter-phase feedback reduces the rin evenly. The rin under cw conditions can therefore be used as a guide to the values when modulated, and this includes pcm. Simulation results for 10 Gbit/s pcm are given in Figure 7.21.

The clear threshold levels for the onset of coherence collapse seen in the cw curves in Figure 7.21 are not nearly so marked when the device is modulated.

Typical spectra of the rin for a laser without feedback and with and without a 5 GHz modulating tone (approximately equal to the laser's relaxation resonance

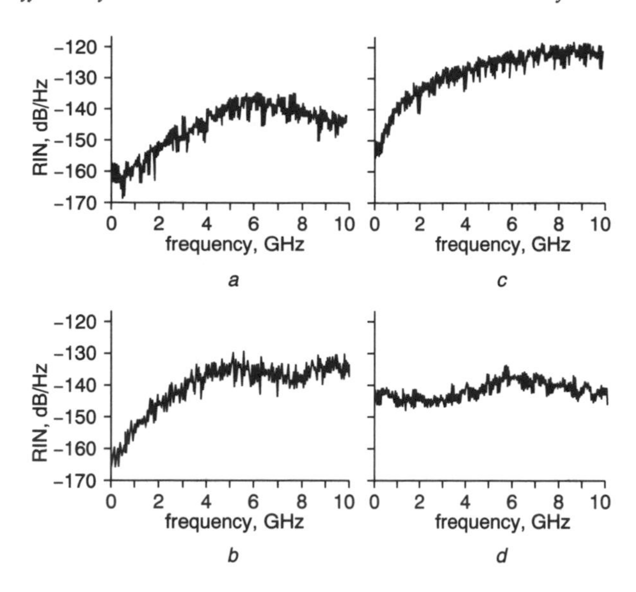

Figure 7.22 rin spectrum of single-mode laser with no external feedback: (*a*) cw, (*b*) 5 GHz sinusoidal modulation, (*c*) 10 Gbit/s random pcm modulation, and (*d*) 5 GHz external modulation ([21], Figure 6). Reproduced by permission of IEE

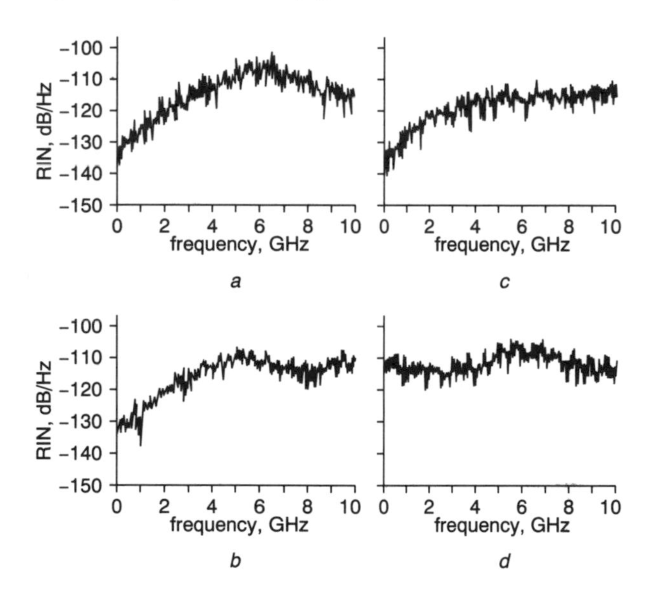

Figure 7.23 rin spectrum of single-mode laser with feedback fraction of 10^{-2}: (*a*) cw, (*b*) 5 GHz sinusoidal modulation, (*c*) 10 Gbit/s random pcm modulation, and (*d*) 5 GHz external modulation ([21], Figure 7). Reproduced by permission of IEE

frequency) are illustrated in Figure 7.22. These results can be compared with those for a feedback fraction of 10^{-2} which are presented in Figure 7.23.

The expected down-conversion of noise to low frequencies when modulated by the drive current is not observed, as can be seen by comparing Figure 7.23.

The above results lead to the conclusion that when a laser is modulated the rin is increased at or below the feedback fraction threshold for coherence collapse, whereas for more feedback the rin is about the same with or without modulation; thus when a large amount of feedback is present the rin under cw conditions can be used in system design, as it gives the worst case.

7.5.2 Relation between rin and ber

In systems the ber is largely related to the rin and to evaluate ber one must consider the pdf of the noise. This renders an analytical approach very difficult; however, if a filter is used in the receiver to reduce the noise level the resulting noise can be treated as Gaussian, at least up to moderate bit rates. With this assumption, ber and rin under 100% modulation (on–off) are related by equation (7.15).

$$\frac{1}{\text{snr}} = \frac{1}{\text{snr}'} + \text{rin} \times B \tag{7.15}$$

where snr is the signal-to-noise ratio of system, snr′ is the signal-to-noise ratio of the receiver itself, B is the equivalent noise bandwidth of receiver, assumed to be $f_S/2$, and f_S is the symbol rate on the line.

For Gaussian statistics and a ber $\geq 10^{-9}$ we have snr $\geq$ 15.6 dB, and assuming a 1 dB penalty at the receiver snr′ $\geq$ 16.6 dB. The corresponding requirement on the product is then rin $\times B < 5.7 \times 10^{-3}$ in good agreement with experimental results at 565 Mbit/s. For a 10 Gbit/s system, B is 5 GHz and we have rin < -119 dB/Hz. Referring to Figure 7.21, the curve corresponding to 10 Gbit/s modulation, this inequality cannot be satisfied for values of feedback fraction exceeding 10^{-2}, so an optical isolator would be needed.

The most important devices parameters which influence the rin for relatively large amounts of feedback were given as f_R the relaxation frequency, τ the round trip transit time in the laser cavity, and κ the feedback coefficient. When the modulating frequency $f_M \ll f_R$, the modulation transfer function is approximately one,

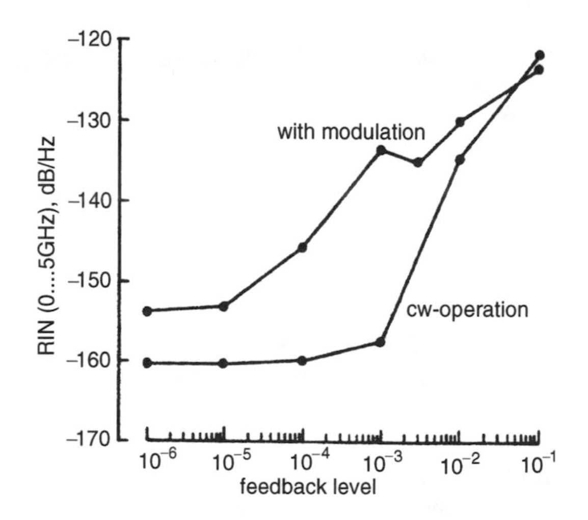

Figure 7.24 Calculated values of rin versus feedback fraction averaged over a 5 GHz band for a quantum well laser with $f_R = 12$ GHz and $\alpha = 2$ in cw operation and when modulated at 10 Gbit/s ([21], Figure 8). Reproduced by permission of IEE

and the rin averaged within the band 0 to f_M can be integrated to yield an expression for this quantity. The most influential parameter is f_R because rin is inversely proportional to the fifth power of the relaxation frequency. At 10 Gbit/s, a value of rin < −120 dB/Hz is needed, and for a bulk laser of length 300 μm, $\tau = 9$ ps, linewidth enhancement factor $\alpha = 6$ and a feedback factor of 0.1, this requires that $f_R > 12$ GHz. The relaxation frequency is generally higher in a quantum well laser than in a bulk device and is an important technical fact in favour of the former. [23] gave an example of a qw laser with $\alpha = 2$ and a gain coefficient 2.5 times that of the bulk laser for which the value of rin satisfied the inequality for all values of feedback. Figure 7.24 shows numerical results for a qw device.

7.5.3 Dependence of linewidth on feedback and modulating frequency

The dependence of linewidth Δf of a Hitachi HL 1314A device operating at 3 mW per facet on feedback factor was illustrated by measured and calculated results, as reproduced in Figure 7.25

The measured maximum of about $\Delta f = 40$ GHz was in agreement with that computed from the theoretical expression:

$$\Delta f = f_R\sqrt{[\ln 4 \times (\alpha^2 + 1)]} \tag{7.16}$$

with $\alpha = 6$ and $f_R = 6$ GHz for the device concerned this gives $\Delta f = 43$ GHz. Linewidth broadening is also caused by chirping, and here the equation quoted was the following:

$$\Delta f = (\alpha + 1)f_M \tag{7.17}$$

The authors stated that for relatively low bit rate systems feedback is the main cause of linewidth broadening, whereas at higher rates chirp is the main contributor.

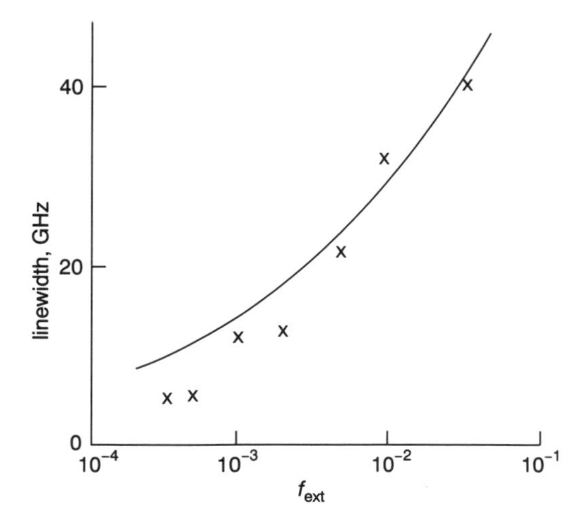

Figure 7.25 Measured values (x) of linewidth versus feedback factor of a Hitachi HL 1314A laser at 3 mW per facet, and the fitted theoretical curve ([21], Figure 9). Reproduced by permission of IEE

Experimental and theoretical studies have shown that the intrinsic intensity noise (rin) and the frequency noise of a laser can be reduced significantly by the feedback resulting from an external cavity of air length below 10 mm. The effects of external cavity resonances can thereby be avoided and the noise spectrum becomes relatively flat up to very high frequencies. However, long external cavities cause feedback which gives rise to sharp peaks in both types of noise. A unified study of intensity and frequency noise, intensity and frequency modulation responses, reduction of chirp, frequency and damping rate of the relaxation oscillations in F–P and in dfb lasers with arbitrary external feedback and arbitrary external cavity length was published in [22] but is too long and mathematical to abstract. In both this and the preceding reference a list of all the relevant laser parameters needed in the analysis was given in tabular form.

7.5.4 Evaluation of power penalties caused by feedback noise in dfb lasers

Such lasers have lower reflectivity facets to restrict Fabry–Perot oscillations compared with other types of devices, and as a result tend to suffer from feedback induced noise power penalties in high bit-rate systems. These penalties were investigated in [23], which reported both experimental and theoretical results covering rates from 500 to 2000 Mbit/s. Power penalties could be suppressed by reducing the relative intensity noise below −135 dB/Hz at 2 Gbit/s, and this could be achieved by using an optical isolator with more than 25 dB isolation.

The observed noise power density for the dfb device to sm fibre combination was found to be almost flat over the band 200–800 MHz and a wide range of coupling losses (5–21 dB), whether the laser was operating cw or with sinusoidal modulation (with corresponding noise peaks and troughs). Therefore, the feedback-induced noise density was taken to be flat in the theoretical analysis, and was treated like Gaussian white noise. The averaged rin values as functions of the coupling loss are shown in Figure 7.26.

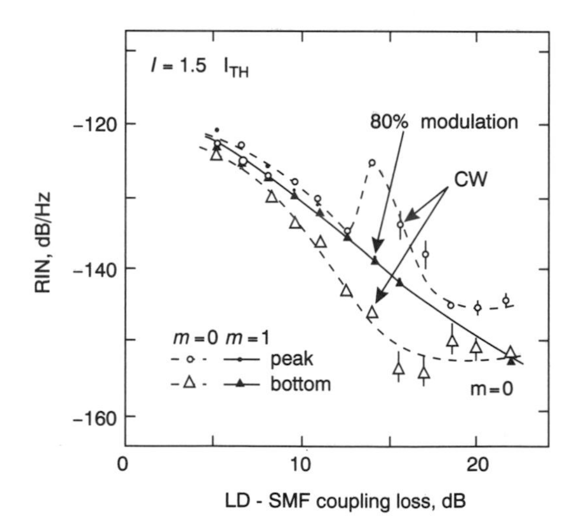

Figure 7.26 rin versus coupling loss between laser output and system fibre for cw operation and for 80% amplitude sinusoidal modulation ([23], Figure 3). Reproduced by permission of IEEE, © 1988

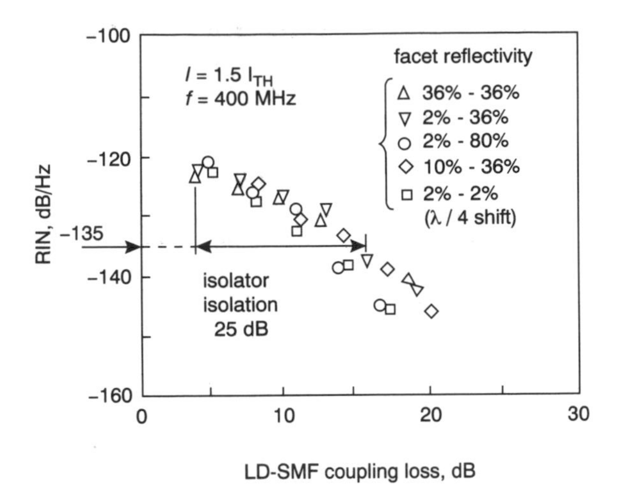

Figure 7.27 rin versus coupling loss between laser output and system fibre for dfb lasers with various facet reflectivities ([23], Figure 4). Reproduced by permission of IEEE, © 1988

The effects of dfb lasers with different facet reflectivities on this same quantity are illustrated in Figure 7.27.

For the results reported in these two figures the values of rin were measured at 400 MHz in a bandwidth of several MHz.

The theoretical estimates of feedback-induced noise power penalties were based on the following expression for the mark-to-noise ratio snr_1:

$$\text{num.} = 4(M \times I_P)^2$$

$$\text{den.} = \left(\sqrt{[2q(I_P + I_D)M^{2+x} + \langle i^2 \rangle + \langle i_F^2 \rangle M^2]} + \sqrt{[2qI_D M^{2+x} + \langle i^2 \rangle]}\right)^2 \times B_N, \tag{7.18}$$

where I_P is the apd rin peak output current, M is the apd rin multiplication factor, x is the apd rin excess noise factor, I_D is the apd rin multiplied dark current, $\langle i^2(t) \rangle$ is the long time average of receiver input noise, $\langle i_F^2(t) \rangle$ is the long time average of feedback-induced noise, q is the electronic charge, and B_N is the equivalent noise bandwidth of the receiver.

As customary, the noise on spaces is neglected as the laser is biased below the threshold in this state. The power penalty is defined as the ratio of apd rin current increment needed for a ber of 10^{-9} when feedback noise is present, to the current when this noise is absent, expressed in dB. Values of this quantity computed from equation (7.18) are presented in Figure 7.28, by the full lines, and measured results are shown by means of the circles.

The experimental results were obtained from measurements at 560 Mbit/s and at 1.8 Gbit/s rates, using nrz format and a pseudo-random sequence of length $2^{15} - 1$. This format was chosen to reduce the effect of feedback delay due to the distance between laser and optical connector. The measured ber performance is presented in Figure 7.29.

At 560 Mbit/s, a rin value of −120 dB/Hz causes less than 0.3 dB penalty, but the situation is much more severe at the higher bit rate. There is good agreement between these measured values and the calculated ones.

	500 Mb/s	2 Gb/s	4 Gb/s	
$\sqrt{\langle i_t^2 \rangle}$	8	15	25	pA/$\sqrt{\text{Hz}}$
M	25	20	15	
I_d		5		nA
B		1 × (bit rate)		Hz

p: receiver sensitivity

a

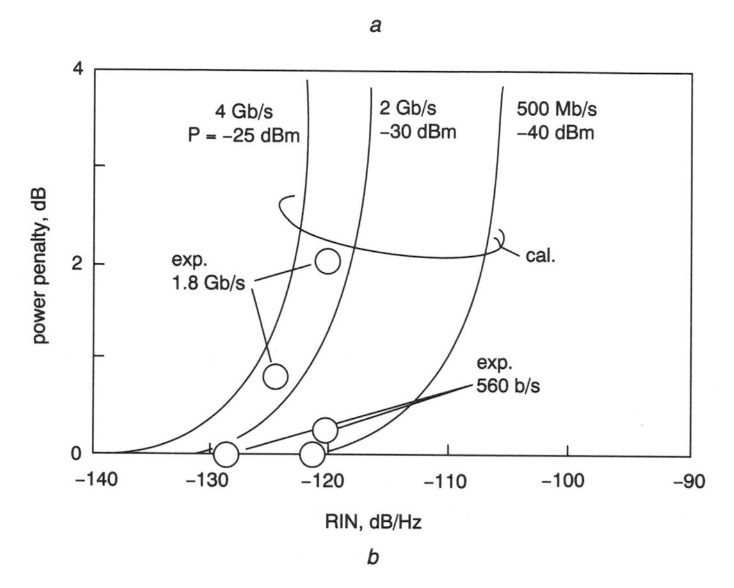

Figure 7.28 Measured and calculated values of power penalty versus rin at different bit rates ([23], Figure 5). Reproduced by permission of IEEE, © 1988

The effects of optical feedback on mode partition events and on chirping were neglected, because even the maximum feedback from the optical connector showed no effect on the output spectrum as measured with a monochromator. The authors also confirmed by means of a test at 560 Mbit/s on 100 km of fibre at 1500 nm that the feedback showed no effect at the receiver end, even when the rin was as large as −120 dB/Hz.

Conclusion: a dfb laser used with commercially available optical isolators with more than 25 dB isolation at 2 Gbit/s (thus reducing the penalty to less than 0.1 dB) will become a reliable source for long-distance high-capacity systems.

7.5.5 Theoretical and experimental results on rin and phase noise (spectral linewidth)

A new theoretical approach to the effects of external optical feedback on the amplitude and phase noise of dfb lasers, together with experimental results on rin and ber obtained by direct detection at 2.3 Gbit/s, was described in [24]. The new theoretical approach to the complex quantity C [17] can be applied to all values of facet reflectivity. Expressions were given for linewidth and for the sensitivity of the output facet. The authors showed that lasers which have high values of the side

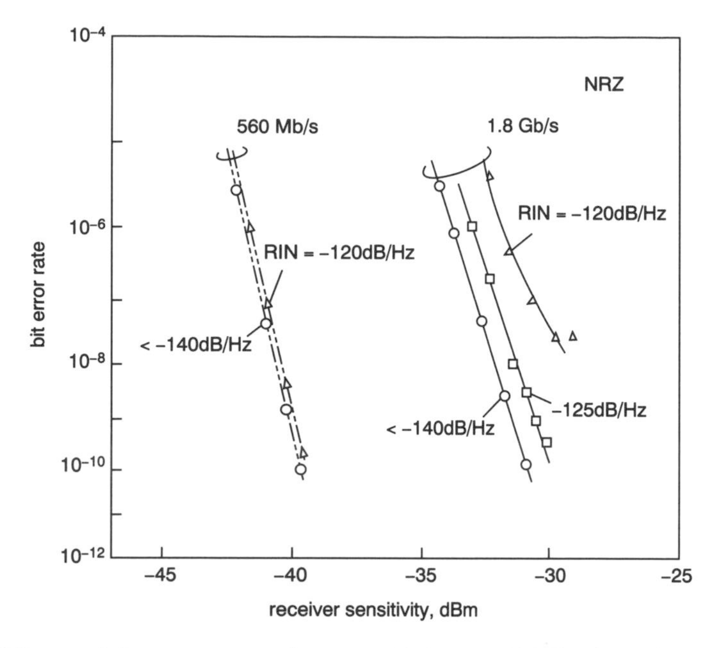

Figure 7.29 Measured ber versus receiver sensitivity at 560 Mbit/s and at 1.8 Gbit/s for different values of rin ([23], Figure 6). Reproduced by permission of IEEE, © 1988

mode suppression ratio (smsr > 30 dB) at output powers around 0.5 mW are less sensitive to the effects of optical feedback. Their experiments indicated that a value of rin below −135 dB at 5 mW emitted power is a good guarantee that there will be no system penalty with 2.3 Gbit/s nrz modulation.

Ten dfb lasers operating in the 1300 nm band were used in the experimental work, and four had ar coatings. Power was coupled from the laser chip into the tapered end of one port of a 50:50 sm coupler. A second port was connected to a reflector of adjustable position which provided the external feedback. A third port was connected to a photometer to measure the feedback, and the fourth to a two-way switch to select either ber or rin measurement.

Figures were presented which showed: the average magnitude of C versus κL for three different pairs of values of left and right facet reflectivities; normalised threshold loss versus wavelength illustrating the effect of feedback; the magnitude of the sensitivity factor of the right-hand facet versus κL for three different pairs of values of left and right facet reflectivities; and rin versus drive current for three values of feedback for two devices exhibiting differing noise behaviour. This showed that laser 1 required an isolator whereas laser 2 did not, even for relatively large values of feedback (e.g. −12 dB) at 2.3 Gbit/s. Figure 6 in that reference showed plots of ber for two or three values of feedback and another parameter for the same two devices, and another displayed measured values of smsr at 0.5 mW versus the sensitivity classification showing that high values of smsr correspond to low values of sensitivity to external feedback.

7.5.6 Terminal electrical noise measurements

Terminal electrical noise (TEN) measurements can be used to study the intrinsic behaviour of lasers, the sensitivity to optical feedback and for advanced

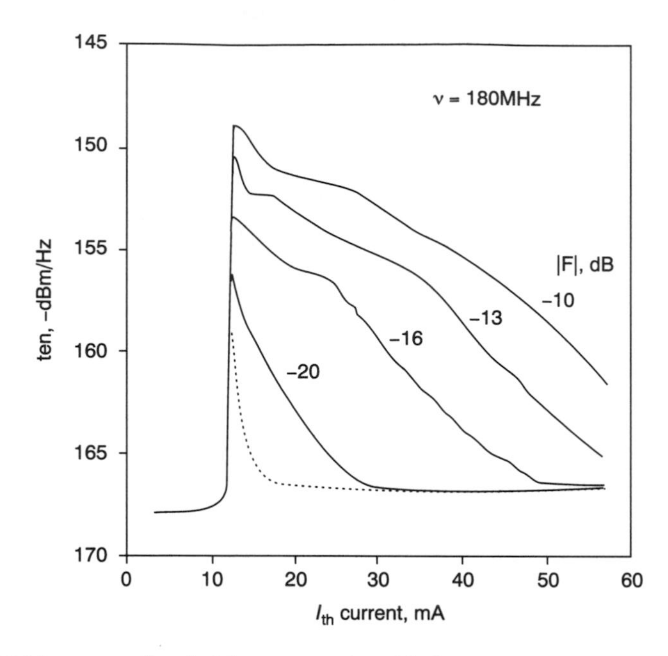

Figure 7.30 TEN versus threshold current of a dfb laser with various values of feedback ratio ([25], Figure 3*a*). Reproduced by permission of IEE

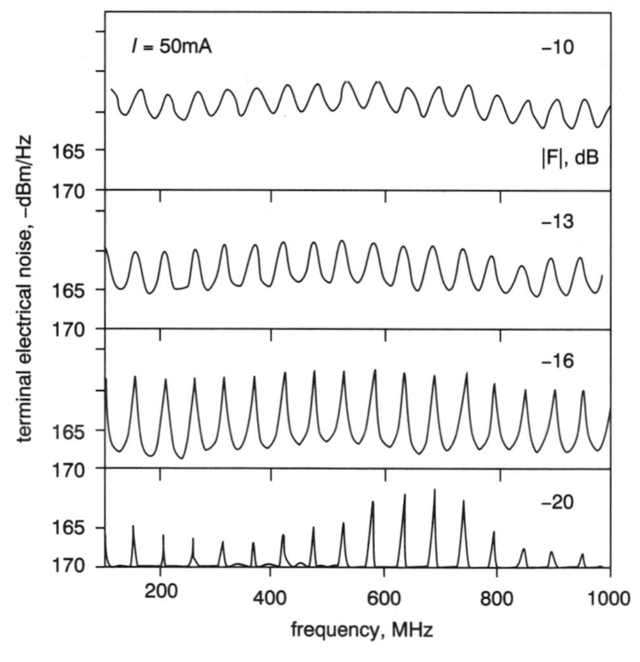

Figure 7.31 TEN versus frequency of a dfb laser with various values of feedback ratio ([25], Figure 3*b*). Reproduced by permission of IEE

characterisation of lasers. They show changes in emission spectra, feedback reflection sensitivity dependent on the κL product of the grating, and sidebands of the field spectrum [25]. The technique was used to measure dfb laser noise without feedback, to relate the results to mode spectra, and to investigate the noise of devices with high side-mode suppression and different κL products under the influence of external optical feedback. For the first time, the sidebands of the field spectrum are demonstrated in TEN spectra. The results of this measurement are used to calculate relaxation frequency, damping coefficient and nonlinear gain. The devices under test were 1300 nm dc-dcpbh structures with cleaved facets, and with second-order gratings with κL products in the range 1.5-4. These lasers have threshold currents around 10 mA, and a 3 dB down frequency of 12 GHz at high injection levels. Since the technique is both new and able to characterise lasers well, it is of interest to reproduce several of the diagrams. First, we consider the terminal electrical noise as a function of drive current with the amount of feedback as a parameter, and we would expect a noise peak in the vicinity of threshold, as we have already seen for the rin; this is indeed the case, as is evident from Figure 7.30.

The curves show that increasing amounts of feedback cause larger values of TEN. If we now turn to frequency as the independent variable, the curves show a characteristic ripple effect with spacing given by the reciprocal of the external round-trip delay, the amplitude and shape of which are dependent on the amount of feedback present, as shown in Figure 7.31.

The authors also demonstrate that the TEN decreases markedly between $\kappa L = 1.5$ of the lasers to which the figures refer and the value 4. The figures are for devices with side-mode suppression ratios better than 30 dB. The much lower sensitivity of dfb lasers to external optical feedback compared with Fabry-Perot structures was also demonstrated, and it is due to the strong dfb coupling.

7.6 REFERENCES

1 ADAMS, M.J., STEVENTON, A.G., DEVLIN, W.J., and HENNING, I.D.: '*Semiconductor lasers for long wavelength optical fibre communications systems*'. Peter Peregrinus Ltd., 1987.

2 HENNING, I.D.: 'Technique for measuring true time resolved spectra of a semiconductor laser'. *Electronics Lett.*, **18**, no. 9, 29 Apr. 1982, pp. 368-369.

3 HENNING, I.D., and FRISCH, D.A.: 'Statistical analysis of instantaneous laser spectra'. *Electronics Lett.*, **18**, no. 11, 27 May 1982, pp. 465-466.

4 HENNING, I.D., and FRISCH, D.A.: 'Real-time measurements of semiconductor laser spectra'. *J. Lightwave Technol.*, **LT-1**, no. 1, Mar. 1983, pp. 202-206.

5 CARTLEDGE, J.C., and BURLEY, G.S.: 'The effect of laser chirping on lightwave system performance'. *J. Lightwave Technol.*, **7**, no. 3, Mar. 1989, pp. 568-573.

6 PETERMANN, K., and KRUGER, U.: 'Chirp reduction in intensity-modulated semiconductor lasers for maximum transmission capacity of single-mode fibres'. *Archiv für Electronik und Übertragungstechnik.*, **40**, no. 5, 1986, pp. 283-288.

7 OLSEN, C.M., IZADPANAH, H., and LIN, C.: 'Time-resolved chirp evaluation of Gbit/s nrz and gain-switched dfb laser pulses using narrowband Fabry-Perot spectrometer'. *Electronics Lett.*, **25**, no. 16, 3 Aug. 1989, pp. 1018-1019.

8 LOWERY, A.J.: 'Time-resolved chirp in mode-locked semiconductor lasers'. *Electronics Lett.*, **26**, no. 13, 21 June 1990, pp. 939-940.

9 BERGANO, N.S.: 'Wavelength discriminator method for measuring dynamic chirp in d.f.b. lasers'. *Electronics Lett.*, **24**, no. 20, 29 Sept. 1988, pp. 1296-1297.

10 THOMAS, B., KAR, A., HENSHALL, G.D., and JONES, K.: 'Forward bias voltage characteristics for (GaA1)As and (GaIn)(AsP) lasers'. *IEE Proc.*, **129**, no. 6, Dec. 1982, pp. 312-315.

11 STOCKTON, T.E., and CASPER, P.W.: 'Injection laser diode transmitter design'. *Fiberoptic Technology*, Oct. 1982, pp. 93–96.
12 LOGAN, R.A., VAN DER ZIEL, J.P., TEMKIN, H., and HENRY, C.H.: 'InGaAsP/InP (1.3 μm) buried-crescent lasers with separate optical confinement'. *Electronics Lett.*, **18**, no. 20, 30 Sept. 1982, pp. 895–896.
13 KIRKBY, P.A.: 'Semiconductor laser sources for optical communications'. *The Radio and Electronic Engineer*, **71**, no. 7/8, July/Aug. 1981, pp. 362–376.
14 WHITE, K.I., HORNUNG, S., WRIGHT, J.V., NELSON, B.P., and BRIERLEY, M.C.: 'Characterisation of single-mode optical fibres'. *The Radio and Electronic Engineer*, **57**, no. 7/8, July/Aug. 1981, pp. 385–391.
15 TKACH, R.W., and CHRAPLYVY, A.R.: 'Regimes of feedback effects in 1.5 μm distributed feedback lasers'. *J. Lightwave Technol.*, **LT-4**, no. 11, Nov. 1986, pp. 1655–1661.
16 HELMS, J., and PETERMANN, K.: 'A simple analytic expression for the stable operation range of laser diodes with optical feedback'. *IEEE J. Quantum Electronics*, **QE-26**, no. 5, May 1990, pp. 833–836.
17 FAVRE, F.: 'Theoretical analysis of external optical feedback on dfb semiconductor lasers'. *IEEE J. Quantum Electronics*, **QE — 23**, no. 1, Jan. 1987, pp. 81–88.
18 EDDOLLS, D.V., PARK, C.A., and BUUS, J.: 'Dependence of mode stability and linewidth of dfb lasers on facet reflectivity and coupling coefficient'. *Electronics Lett.*, **27**, no. 7, 28 Mar. 1991, pp. 590–591.
19 TAKEMOTO, A., OHKURA, Y., KAWAMA, Y., KIMURA, T., YOSHIDA, N., KAKIMOTO, S., and SUSAKI, W.: '1.3 μm distributed feedback laser diode with grating accurately controlled by new fabrication technique'. *Electronics Lett.*, **26**, no. 3, 2 Feb. 1989, pp. 220–221.
20 KONDO, Y., SATO, K., NAKAO, M., FUKUDA, M., and OE, K.: 'Extremely narrow linewidth (1 MHz) and high power dfb laser grown by movpe'. *Electronics Lett.*, **25**, no. 3, 2 Feb. 1989, pp. 175–177.
21 WANG, J., and PETERMANN, K.: 'Noise characteristics of pcm-modulated single-mode semiconductor laser diodes with distant optical feedback'. *IEE Proc.*, **137**, Pt J, no. 6, Dec. 1990, pp. 385–390.
22 FERREIRA, M.F., ROCHA, J.F., and PINTO, J.L.: 'Noise and modulation performance of Fabry-Perot and dfb semiconductor lasers with arbitrary external optical feedback'. *IEE Proc.*, **137**, Pt J, no. 6, Dec. 1990, pp. 361–369.
23 SHIKADA, M., TAKANO, S., FUJITA, I., and MINEMURA, K.: 'Evaluation of power penalties caused by feedback noise of distributed feedback laser diodes'. *IEEE/OSA J. Lightwave Technol.*, **6**, no. 5, May 1988, pp. 655–659.
24 BEYLAT, J.-L., BALLAND, G., and JACQUET, J.: 'External optical feedback effects on dfb semiconductor lasers — theoretical analysis and experimental results'. *ECOC 88*, pp. 392–395.
25 IDLER, W., SCHWEIZER, H., LANG, R.J., KLENK, M., WUNSTEL, K., and MOZER, A. 'Advanced noise investigations on InGaAsP/InP d.f.b. lasers'. *ECOC 88*, pp. 380–383.

FURTHER READING

26 SCHUNK, N., PETERMANN, K.: 'Measured feedback-induced intensity noise for 1.3 μm dfb laser diodes'. *Electronics Lett.*, **25**, no. 1, 5 Jan. 1989, pp. 63–64.

8

MORE ADVANCED SEMICONDUCTOR LASER DIODES

8.1 INTRODUCTION

This is the last of three chapters devoted to aspects of semiconductor lasers which are, or may become, of direct interest to designers of ofdls. The first of the three gave many details of an illustrative specification and associated performance evaluation for a bulk active type of device, which is in widespread service as the transmitter source for both inland and submerged systems.

Chapter 7 opened with the spectral characteristics of lasers, with emphasis on a treatment of time-averaged and true time-resolved spectral characteristics of modulated lasers. This was followed by the topic of chirp in dfb and mode-locked lasers under intensity modulation. The following sections dealt with the effects of external influences on laser performance, beginning with temperature, then considered the effects of optical feedback effects on signal and on noise behaviour.

The present chapter opens with consideration of relative intensity noise and of mode partition (or mode competition) noise, which set an upper limit to the achievable signal-to-noise ratio in a system. The latter type of noise is examined further in the next section. Quantum well laser diodes are assuming increasing importance in the 1990s, both as transmitters and as pump diodes for use with fibre amplifiers, and devices for both applications are discussed. Optical sources for systems operating in the Gbit/s range, either as a single device or in the form of an array, form the subject of the next section, and the chapter concludes with comments on reliability.

8.2 NOISE IN SEMICONDUCTOR LASERS

A complete analysis of a typical ofdls with or without regenerative or optical amplifier repeaters requires that the noise contributions of all the noisy components be considered. By 1992 suitable models had been developed and one such will be mentioned at this point [1]. The transmission-line laser model (TLLM) has demonstrated its usefulness in design and simulation for both semiconductor lasers and optical amplifiers (soa). The model was originally developed to investigate the signal characteristics of both types of component, such as spectral behaviour and modulation response. More recently it was shown that it can provide an accurate model for the optical noise behaviour of both. For example, it is able to simulate the intensity noise of lasers with or without feedback and the turn-on jitter, whereas for soa devices of complex structures it can be used to evaluate the spontaneous emission spectrum. In the present context, the laser TLLM includes random noise generators to represent spontaneous emission into the lasing mode, which degrades the laser performance by:

(a) broadening the linewidth
(b) adding phase and intensity noise
(c) causing random jumps in power distribution between modes
(d) inducing random jitter in the turn-on time of the laser.

All of these effects can contribute to system impairments and hence penalties.

Mode partition noise ((c) above) and reflection-induced noise (part of (b) above) were considered in an invited paper which appeared in 1982 [2]; however, more recent and detailed references can be consulted, e.g. [3].

Intrinsic noise in a laser is observable if the output illuminates a photodiode directly without an intervening fibre. This type of noise is governed by quantum processes inside the laser cavity which include:

(a) shot noise of injection current
(b) spontaneous recombination of carriers within the active layer
(c) light absorption and scattering
(d) stimulated emission.

Unlike reflection noise, quantum noise is unavoidable, and it sets an absolute upper limit for the system signal-to-noise ratio which can be achieved for a given launched power. This limit is of interest when information is transmitted in analogue format, e.g. multichannel television, but it is not a constraint for digital systems.

8.2.1 Intrinsic noise

The intrinsic noise behaviour of a laser may be described by adding a Langevin shot noise term to each of the coupled pair of rate equations for a noise-free device, and repeating this for each lasing longitudinal mode. The large-signal active region model describing the dynamic behaviour of photon and carrier densities within this cavity is given by the rate equations [4]. Choosing notation in which time-varying quantities are represented by lower-case letters, the pair of equations for a single mode is:

$$\frac{\tau \mathrm{d}s}{\mathrm{d}t} = \tau\Gamma g(n - N_0)(1 - \varepsilon s)s - s + \frac{\beta\Gamma N_0}{\tau'}\tau + F \tag{8.1}$$

$$\frac{\tau' \mathrm{d}n}{\mathrm{d}t} = \frac{\tau' I_A}{qV} - n - \tau' g(n - N_0)(1 - \varepsilon s)s + F' \tag{8.2}$$

where τ is the photon lifetime, s is the normalised photon density in mode j, $j = 1, 2, \ldots$, Γ is the mode confinement factor, g is the normalised gain, n is the normalised carrier density, N_0 is the value of n for transparency, ε is the gain compression parameter (zero for small-signal cases), β is the fraction of spontaneous emission entering lasing mode, F is the Langevin noise term in equation for $\mathrm{d}s/\mathrm{d}t$, τ' is the spontaneous carrier lifetime, I_A is the current entering the active region, V is the volume of the active region, q is the electronic charge, and F' is the Langevin noise term in the equation for $\mathrm{d}n/\mathrm{d}t$.

Numerical values for some or all of these quantities are listed in papers dealing with numerical solutions to the rate equations and in papers using models for lasers, e.g. [1]. For more than a single mode there will be one term involving the quantities s, g, β and F for each mode present; see for instance [5]. Note that instantaneous spectral intensity distributions cannot be expressed explicitly by equations (8.1) and (8.2). The fluctuations which occur within a pulse produce mode partition (competition) noise which can affect transmission quality if the pulse width approaches the carrier lifetime of about 2 ns (symbol interval at 500 MBd).

8.2.2 Relative intensity noise in semiconductor lasers

The effect of laser rin on system ber results was reported in [6] for a number of devices which were commercially available in 1989. The 1300 nm lasers were modulated by a 600 Mbit/s prbs and the output transmitted to a pin detector over 10 km of standard sm fibre including several connectors. The rin was calibrated and the effect on ber evaluated by injecting white noise into the modulating current. The corresponding fluctuations in the light output swamped the quantum noise, causing an increase in rin defined as the ratio of mean square value of the fluctuation to the square of the mean value. In turn, these quantities in the ratio were replaced by the corresponding current expressions involving the m.s. value of the white noise current, injected and threshold currents. The m.s. values were then related to corresponding values of rin in dB/Hz. Measurements on the AT&T device indicated that the intensity noise at 1 mW output was equivalent to an added m.s. noise current density of $5 \times 10^{-20}\ \mathrm{A}^2/\mathrm{Hz}$ (or $0.22\ \mathrm{nA}/(\mathrm{Hz})^{1/2}$). ber results on a Thomson laser indicated that for rin less than -128 dB/Hz the effect on transmission quality could be neglected, but that a noise floor began to appear for rin $= -98$ dB/Hz.

The intrinsic noise can be characterised by either the relative intensity noise (rin) or the dc signal-to-noise ratio, defined at the output of the photodetector as follows:

$$\mathrm{rin} = \langle |\mathrm{d}S\,(\mathrm{j}f)|^2 \rangle 2 \frac{\mathrm{d}f}{S^2} \tag{8.3}$$

$$\mathrm{snr} = \frac{1}{\mathrm{rin}} \tag{8.4}$$

$\langle |S\,(\mathrm{j}f)|^2 \rangle$ represents the noise power two-sided spectral density at the frequency f, and $\mathrm{d}f$ is the positive frequency noise bandwidth. The behaviours of the measured snr for lasers of the two types are presented in Figure 8.1.

The left-hand ordinate scale refers to the dc signal-to-noise ratio defined above with $\mathrm{d}f = 10$ MHz; the right-hand ordinate refers to the modulation signal-to-noise ratio with 50% modulation and $\mathrm{d}f = 5$ MHz. The pronounced dips in the neighbourhood of the threshold current are evident, as is the crossover between the gain-guided and the index-guided lasers well above threshold.

A significant noise amplitude is always superimposed on the mean light level, due to the amplification of quantum fluctuations inherent in laser action. The relative noise amplitude as a function of frequency is bias-dependent. Usually the noise power is a maximum near the threshold, and hence the snr is a minimum, as shown

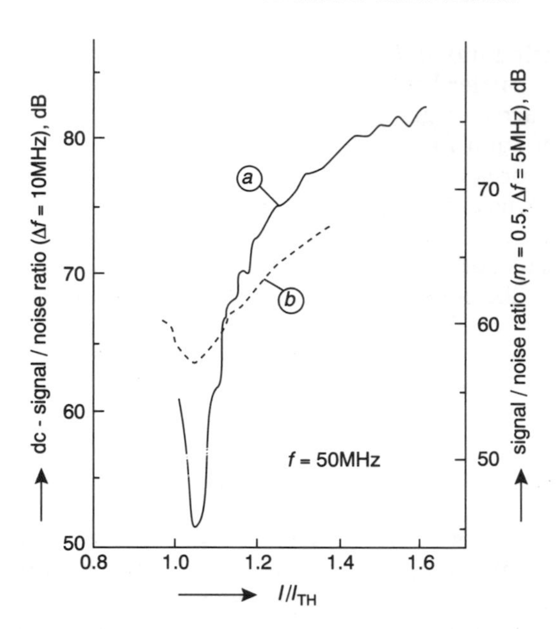

Figure 8.1 dc snr (left-hand ordinates) and snr when modulated (right-hand ordinate) for two early types of bulk lasers versus I/I_{TH} ([2], Figure 5). Reproduced by permission of IEEE, © 1982

in Figure 8.1. When measured at the output of a photodetector under conditions in which laser noise predominates, the peak value was usually less than 5 μW r.m.s. electrical power in a bandwidth of 100 MHz.

Amplitude-modulated quantum noise in a buried heterostructure (bh) laser and in a self-aligned structure 1300 nm InGaAsP laser, both operating in a single longitudinal mode have been measured and calculated, and the results were reported in [7]. The relevant parameters of the two types of lasers used in the investigation are listed in Table 8.1.

Based on these values and the single-mode rate equations, the amplitude modulation (am) quantum noise characteristics were calculated, and the results, together with those obtained by measurements, are presented in Figure 8.2.

The agreement between the calculated (full or dash lines) and the two measured sets of points (noise power measured in a 3 MHz bandwidth centred at 75 MHz) is good, except near the threshold of oscillation (I/I_{TH} tending to 1), where multimode rate equation analysis is needed to explain the behaviour. The relative intensity

Table 8.1 Parameters of 1300 nm lasers used in measurements ([8], Table 1). Reproduced by permission of IEE

	Structural parameters			Material parameters			Noise parameters	
LD (cavity length)	Optical mode volume V_0	Mode confinement factor Γ	Photon lifetime τ_p	Differential gain constant A	Absorption carrier density N_0	Carrier lifetime τ_s	Spontaneous emission coefficient β	Population inversion parameter n_{sp}
µm	m^{-3}		ps	m^3/s	m^{-3}	ns		
BH (300)	3.5×10^{-16}	0.32	2.0	4×10^{-12}	9.6×10^{23}	1.75	2×10^{-5}	3.6
SAS (200)	6.4×10^{-16}	0.32	1.5	4×10^{-12}	9.6×10^{23}	1.64	1×10^{-5}	3.3

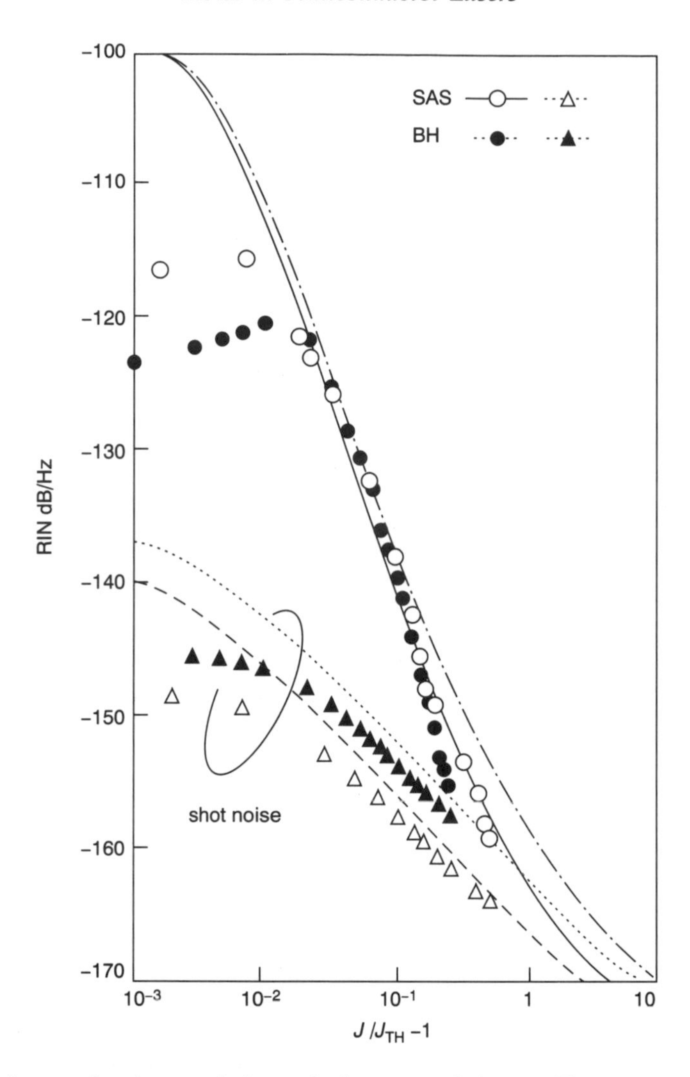

Figure 8.2 Measured points and theoretical curves of rin per Hz versus normalised bias for buried heterostructure and self-aligned 1300 nm InGaAsP lasers operated with a single longitudinal mode ([7], Figure 1). Reproduced by permission of IEE

noise is defined as the noise power generated in unit resistance and in unit bandwidth, and is expressed as a function of the normalised bias level $I/I_{TH} - 1$. The study showed that the am quantum noise level under normal operating conditions is about an order of magnitude above the shot-noise floor, and operation at drive levels at more than twice I_{TH} is required to approach the noise floor.

Measurements have been made of relative intensity noise in a two-section dbr (distributed bragg reflector) laser (based on a ridge waveguide structure and emitting at 1545 nm) at low (<700 MHz) and at high (> 2 GHz) as a function of grating current [8]. Using different grating currents but keeping output power constant, rin was measured at frequencies between 10 MHz and 1 GHz and between 2 and 18 GHz, and the results were compared with frequency noise measurements by others. A large increase in the rin was observed for low frequencies up to about

700 MHz as the grating current was increased to overcome the increased losses in the grating section. The increase in this current was needed to maintain constant output power under wavelength tuning. This was attributed to shot noise generation in the passive grating section of the device. In the band 2–18 GHz the most distinctive feature was the resonance peak at around 4 GHz, where the rin rose to about −145 dB/Hz from −153 dB/Hz at 2 GHz, before declining monotonically to about −160 dB/Hz at 16 GHz. No rin increase was observed with increasing grating current over this band.

8.2.3 Mode partition (or competition) noise

The combination of fluctuations of optical output power between modes revealed in true time-resolved spectra (see Section 7.3.2) and dispersion in a system fibre gives rise to mode partition noise. Instantaneous spectral intensity distributions cannot be expressed explicitly by equations (8.1) and (8.2), and they require a statistical description. The statistical characteristics of the intensity fluctuations in each mode was treated in [5] and expressions for the mean and variance obtained as functions of a 'mode partition constant'. This quantity was found to be independent of active volume, structure, wavelength and spontaneous emission fraction.

Experimental results obtained in the early 1980s [5] with Fabry–perot (index guided) InGaAsP/InP dh lasers confirmed theoretical predictions that this noise was roughly independent of the structure, of wavelengths in the range 800–1600 nm and of bit rates varying from 400 to 1600 Mbit/s. Some of the findings are summarised in Table 8.2 ([5], part of Table 1).

The mode partition constant is zero when the laser response to a current pulse is truly single-mode. Measured results on a laser of type 2 in the table gave values of this constant as 7 and 9.5 when modulated at 400 and 1600 Mbit/s, respectively. The corresponding time-resolved spectra are shown in Figures 8.3 and 8.4.

In both figures, waveforms (*a*) are the responses of very fast photodiodes to optical pulses of the same shape. Waveforms (*b*) show the photodiode responses to successive central longitudinal modes selected from successive spectra by a monochromator. In (*c*) the time-averaged spectra are illustrated. It is evident that

Table 8.2 Parameter values for several different bulk laser structures. Reproduced by permission of IEEE

Item	1	2	3	4	5	6
structure	bc	bh	pcw	pcw	bh	bh
wavelength	1295	1295	1315	1509	1509	1570 nm
threshold curr	42	24	106	58	14.2	124 mA
fwhm envelope	2.4	1.7	4.5	4.0	5.0	5.5 nm
mode partition constant	7.8	7.0	10.3	7.5	8.6	8.5

Notes:
bc buried cresent
bh buried heterostructure
pcw plano-convex waveguide

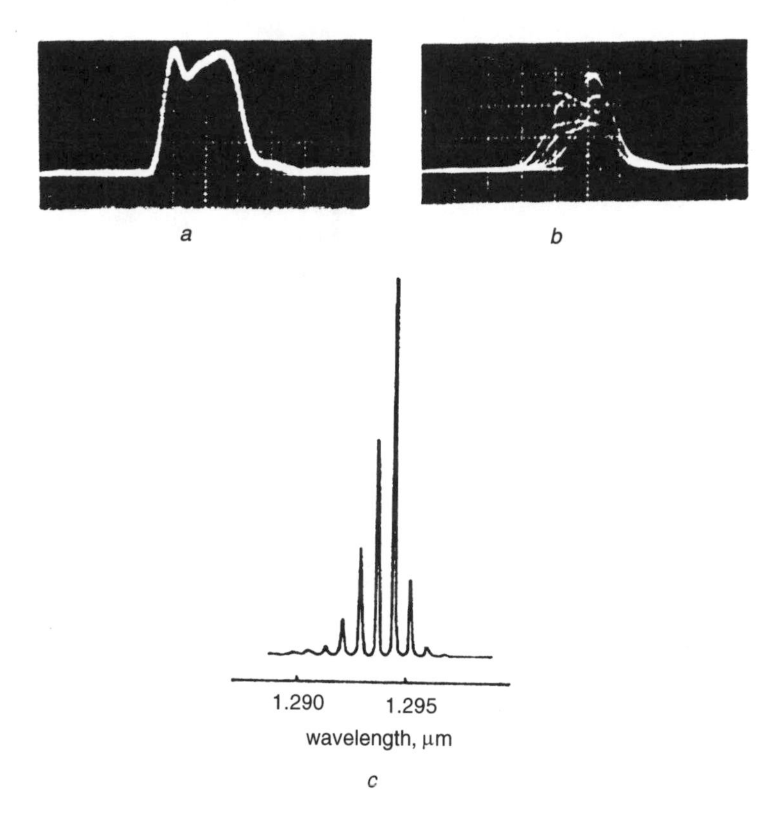

Figure 8.3 (*a*) output waveform; (*b*) waveforms of middle longitudinal mode; (*c*) output time-averaged spectrum when modulated at 400 Mbit/s ([5], Figure 4). Reproduced by permission of IEEE, © 1982

there are considerable variations in (*b*), indicating that the central mode varies from pulse to pulse.

This form of noise is also known as mode competition noise [1] of Chapter 7, and it is most severe when only two longitudinal modes occur. Under certain conditions of current and temperature for which the gain peak lies between the modes, the noise in one mode reaches a peak, as can be seen from the sketch in Figure 8.5.

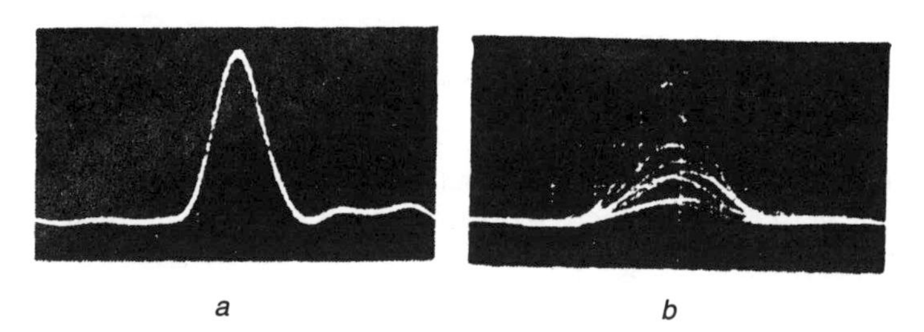

Figure 8.4 (*a*) output waveform; (*b*) waveforms of middle longitudinal mode; (*c*) output time-averaged spectrum when modulated at 1.6 Gbit/s ([5], Figure 5). Reproduced by permission of IEEE, © 1982

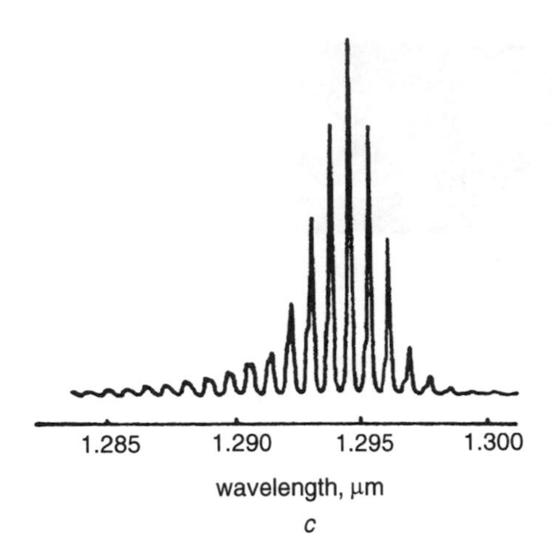

Figure 8.4 (*continued*)

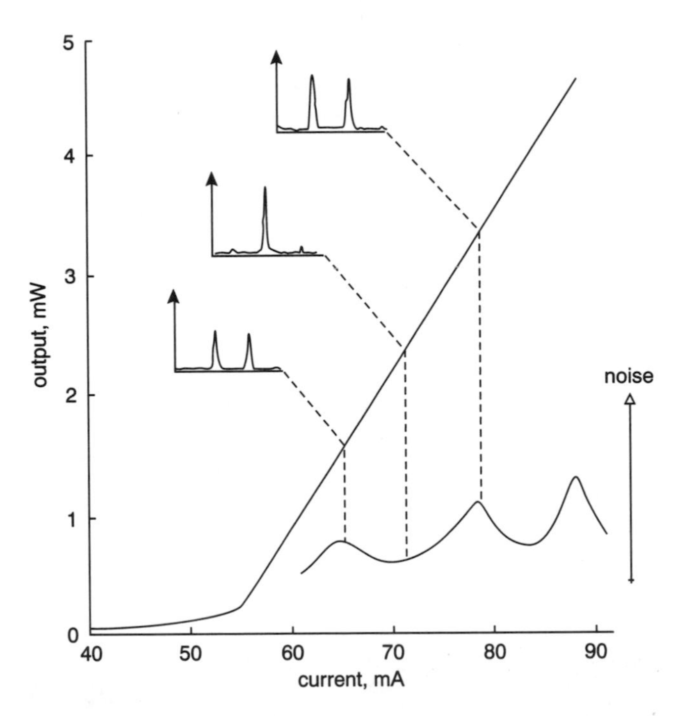

Figure 8.5 Light output versus drive current and noise power output versus drive current at three points on the $P(I)$ curve with corresponding time-averaged longitudinal mode spectra (Adams *et al.*, Figure 7.6). Reproduced by permission of IEE

These peaks correspond to changes of mode, the average spectrum consisting of two dominant modes as shown. 100% intensity modulation occurs when the gain peak lies midway between them. In addition to the amplitude noise, phase and frequency noise is also produced and must be considered in coherent system applications. Some details are given in [1] of Chapter 7.

8.3 MODE PARTITION NOISE IN NEARLY SINGLE-MODE LASERS

For lasers with more than three or four longitudinal modes, the effects of mode partition noise on system performance have required the use of second-order statistics to characterise the fluctuations. The system implications of mode partition noise in almost single-mode lasers was treated more rigorously in a 1988 paper [9]. The reason for undertaking the study lay in the fact that dfb devices can have a vestigial side mode, and that the amount of power in the side mode can increase with age, leading to increasing system penalties. Probability density functions based on recent research are assigned to the total laser intensity and to the side mode intensity. This enables the pdf of the dominant mode, and the joint statistics of both dominant and side modes to be derived. These joint statistics are used in the evaluation of the effects on system performance.

8.3.1 Theory

The numerical results are based on the assumption of a Gaussian pulse of width a at the $1/e$ points propagating in the axial direction (z-axis) in a sm fibre, so that the received pulse can be written as:

$$h(t) = I_M h_M(t) + I_S h_S(t) \tag{8.5}$$

where I denotes the optical intensity, $h(t)$ is the pulse response, and the subscript M stands for main or dominant and S for side modes, respectively. The ratio of the mean intensities or powers $I_S/(I_M + I_S)$ can be written as $\chi/(1 + 2\psi)$. It is assumed that the spectrum of the laser is constant during one symbol interval, but that it varies independently from one interval to the next. The side pulse response includes information on whether it precedes or follows the main pulse. The received pulse responses can also be written as:

$$h_M(t) = P(0) \exp\left(\frac{-t^2}{b^2}\right) \tag{8.6}$$

$$h_S(t) = h_M(t - \tau) \tag{8.7}$$

In these equations we have:

$$b^2 = a^2 + (DL)^2 \left[\left(\frac{\lambda_{MP}{}^2}{2\pi ca}\right)^2 + \sigma_\lambda{}^2\right] \tag{8.8}$$

$$\tau_D = -\delta\lambda DL \tag{8.9}$$

The meanings of the symbols and their numerical values are listed below; note that standard sm fibre is implied by the value of D:

$P(0)$ = launched power, i.e. power at $z = 0$
τ = time interval between arrival of main and side modes
D = fibre dispersion parameter (17 ps km^{-1}nm^{-1})
L = length of fibre
λ_{MP} = wavelength of main mode peak emission (1550 nm)
a = launched pulse half-width at $1/e$ relative amplitude ($0.24T$) (98% of energy in launched pulse is within symbol interval)

σ_K = $1/e$ half-width of Gaussian spectral distribution of individual modes (0.05 nm)
$\delta\lambda$ = longitudinal mode spacing (0.76 nm)

8.3.2 Numerical example

The receiver comprising a photodetector followed by a filter is assumed to possess a Gaussian response of 3 dB bandwidth B equal to $0.47/T$, where T is the symbol interval. The mean square noise power spectral density N_F/B at the output of the filter is assumed to be related to the white noise power N at the receiver input by $N_F = 1.06N$. The numerical results to follow are based on the values above, plus those listed below for transmission at 1.2 Gbit/s (higher rates would require chirp to be included in the model).

γ = extinction ratio (10)
G = photodiode gain (1)
I_D = photodiode dark current (10 nA)
η = photodiode quantum efficiency (0.8)
N = mean square noise power at input to filter (18.13 $(\mathrm{pA})^2\mathrm{s}$)
N_F = mean square noise power at output from filter ($1.06N$)
ψ = constant related to pdf of total intensity (2×10^{-4})
t_S = sampling instants (0 s)

For this particular value of ψ the relations between values of mode power, mode ratio and correlation coefficient are given in Table 8.3.

In calculating the error rate the threshold level has been optimised for each value of the received optical power. In all the numerical results presented the sampling instants t_S have been chosen to maximise the system response to the main mode. Four different values of relative side mode power have been used in the calculations, namely −20 dBr, −17 dBr, −15.2 dBr and −14 dBr relative to the total power.

Figure 8.6 shows the bit error rate for a 75 km length of fibre as a function of the received mean power P for the four values above.

Intersymbol interference is caused by the presence of a delayed or advanced side mode pulse, and the degradation in ber performance increases as the relative side

Table 8.3 Equivalences between side mode power, mode ratio, and correlation coefficient for $\psi = 2 \times 10^{-4}$ ([9], Table 1). Reproduced by permission of IEEE, © 1988

Side mode		Mode	Correlation
dB	Power	Ratio	Coefficient
−20	0.010	99.0	−0.333
−18.2	0.015	65.7	−0.469
−17	0.020	49.0	−0.578
−16.8	0.021	46.6	−0.596
−16	0.025	39.0	−0.662
−15.2	0.030	32.3	−0.728
−14	0.040	24.0	−0.817

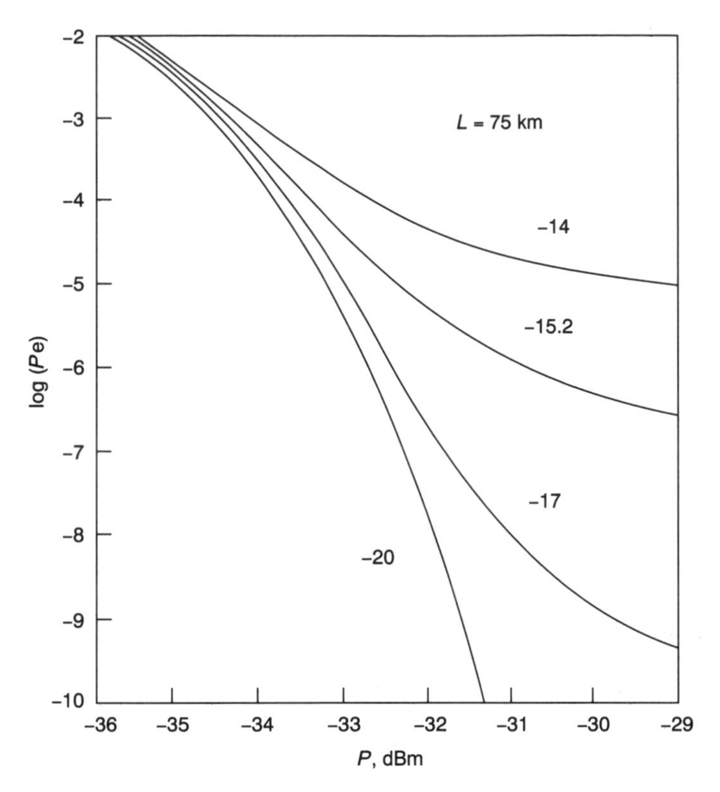

Figure 8.6 Probability of error versus received power for $L = 75$ km and four values of side mode suppression ratio in dBr ([9], Figure 3). Reproduced by permission of IEEE, © 1988

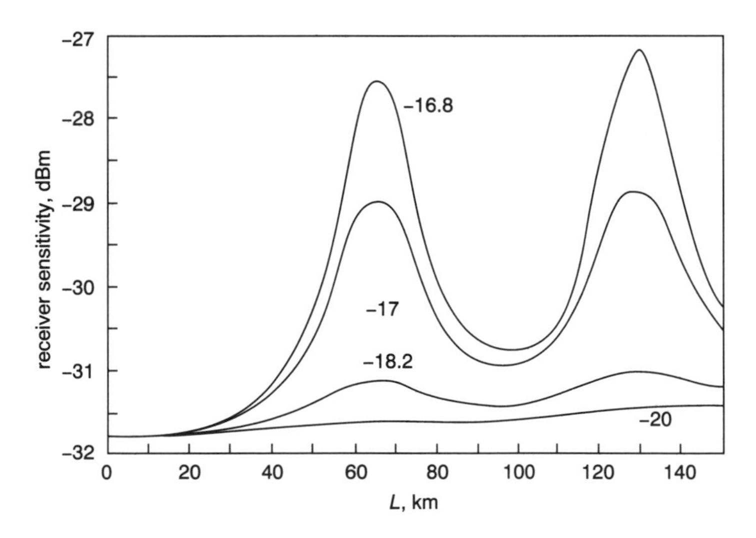

Figure 8.7 Receiver sensitivity at 1.2 Gbit/s versus fibre length for four values of side mode suppression ratio in dBr ([9], Figure 4). Reproduced by permission of IEEE, © 1988

mode power increases. These results suggest that −20 dBr is about the maximum relative power which can be allowed before noise floors start to appear.

Next consider the effect on the receiver sensitivity as a function of fibre length L for four values of relative power, −20 dBr, −18.2 dBr, −17 dBr and −16.8 dBr. The results are shown in Figure 8.7, and they show clearly the combined effects of mode partition noise and dispersion due to linewidth.

The local maxima on each curve correspond to τ taking integer values of $\pm T$, $\pm 2T$. The author states that significant increases of sensitivity above the value at the fibre input ($L = 0$) indicate the onset of error rate floor characteristics.

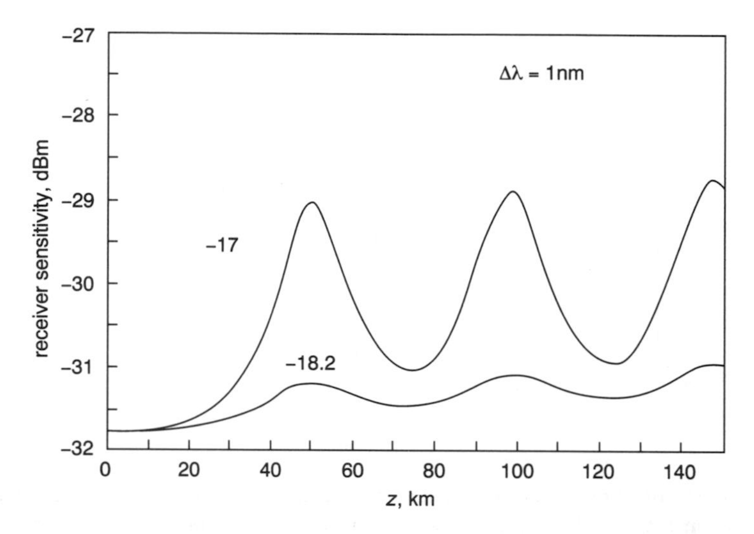

Figure 8.8 Receiver sensitivity at 1.2 Gbit/s versus fibre length for 1 nm mode spacing and two values of side mode suppression ratio in dBr ([9], Figure 5). Reproduced by permission of IEEE, © 1988

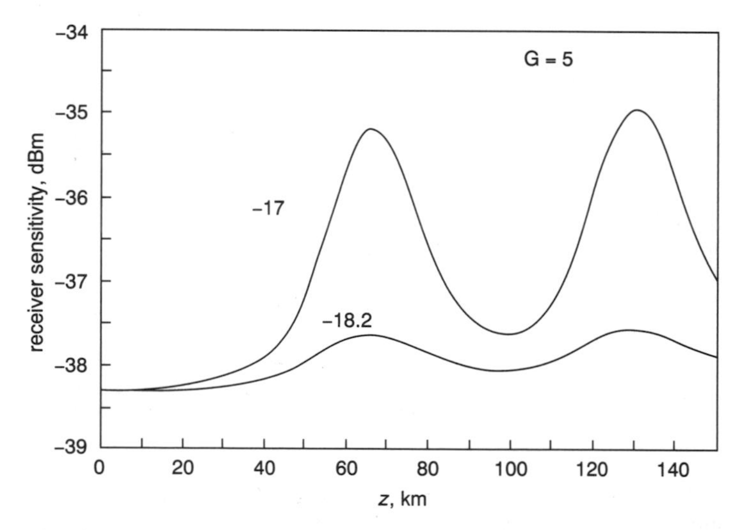

Figure 8.9 Receiver sensitivity at 1.2 Gbit/s versus fibre length using an apd with $G = 5$ and electron ionisation coefficient of 0.3 and two values of side mode suppression ratio in dBr ([9], Figure 6). Reproduced by permission of IEEE, © 1988

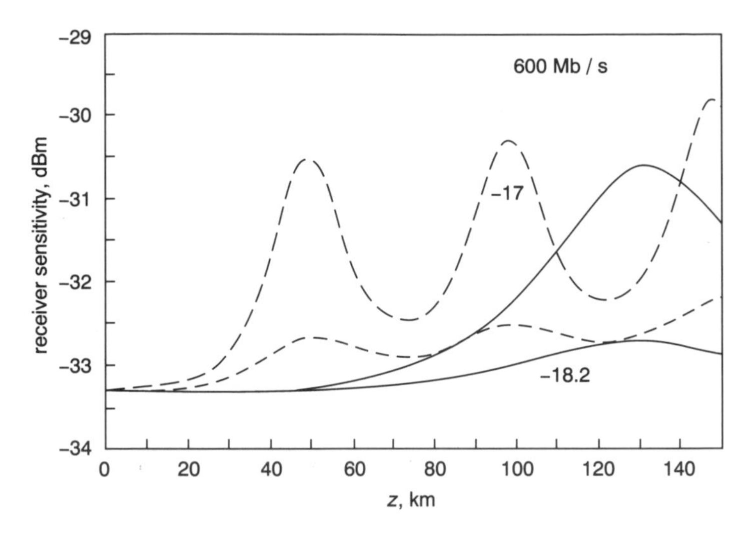

Figure 8.10 Receiver sensitivity at 600 Mbit/s versus fibre length for two different combinations of mode spacing and mode width ([9], Figure 7). Reproduced by permission of IEEE, © 1988

The effect of changing the spacing $\delta\lambda$ to 1 nm on the receiver sensitivity as a function of length for two values of the relative power is presented in Figure 8.8. As the reader can see from this diagram the penalty for this value of spacing differs little from that for the smaller value 0.76 nm.

Turning now to the dependence of sensitivity on fibre length for $G = 5$ and $k_E = 0.3$, Figure 8.9 shows that a penalty is incurred, relative to Figure 8.6 and $z = 0$, when the photodiode has larger than unity gain, i.e. when an avalanche diode is used. This is caused by the known change in slope of the error rate curve as a function of G when intersymbol interference is present. At large values of relative power, and so considerable intersymbol interference the use of an apd does not raise the noise floor above that of a pin photodiode.

Two different data rates were considered next. At 600 Mbit/s the sensitivity as a function of length for two pairs of conditions is illustrated in Figure 8.10. The corresponding results at the lower rate are shown in Figure 8.11.

The effects of the relation between τ and T and of the dispersion due to the linewidth show up in the ripples. When the combination of parameters is such that τ becomes an integer multiple of T at values of fibre length greater than those of practical interest, the penalty due to mode partition fluctuations becomes less severe, because one is working on portions of curves below the first peak.

8.3.3 Comparison between results using present method and Gaussian approximation

Finally, the methodology used in [9] was compared with alternative techniques of evaluating the average probability of error. For this comparison the author chose a 560 Mbit/s rate, $D = 16$ ps km^{-1}nm^{-1}, $\delta\lambda = 1$ nm and $\sigma_K = 0.125$ nm and operation at 1550 nm. The computation of error rate is much simplified if the waveforms of the photocurrents at the sampling instants are assumed to have

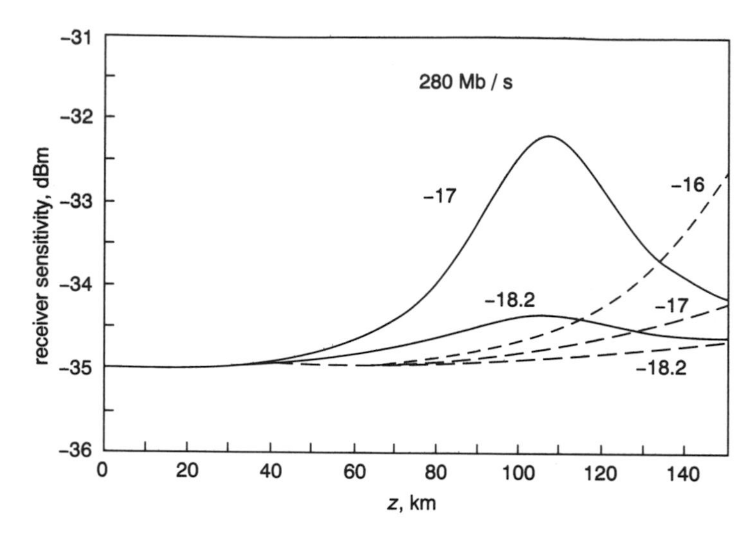

Figure 8.11 Receiver sensitivity at 280 Mbit/s versus fibre length for two values of mode spacing and smsr values as given ([9], Figure 8). Reproduced by permission of IEEE, © 1988

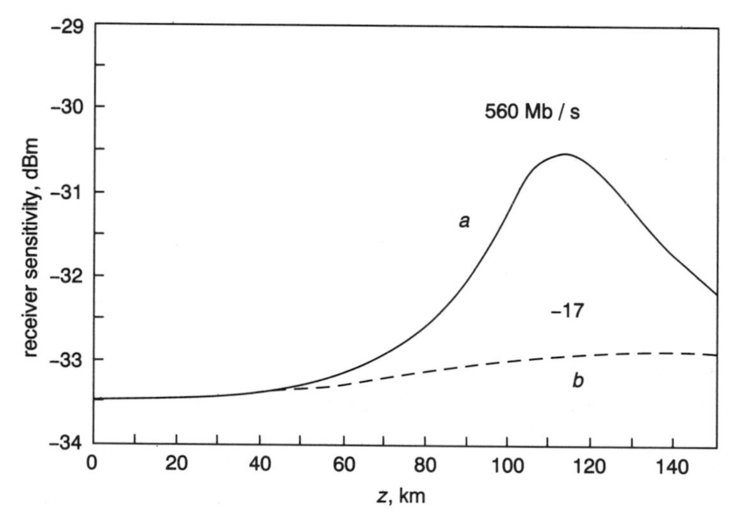

Figure 8.12 Receiver sensitivity at 560 Mbit/s versus fibre length for smsr of 17 dB evaluated: (*a*) by the Gauss quadrature rule, (*b*) using Gaussian approximation ([9], Figure 9). Reproduced by permission of IEEE, © 1988

Gaussian distributions. The receiver sensitivities calculated from the Gaussian quadrature method and from the assumption differ considerably, especially in the range 110–130 km corresponding to actual repeater spacings achieved in submerged systems, as shown in Figure 8.12.

It is evident from the figure that the assumption of Gaussian distribution is very optimistic, and should therefore not be used in system calculations. The discrepancy between results derived from the two methods of calculation was attributed to the effect of isi on the pdfs of the filtered output currents when 1 and when 0 were transmitted. These are the currents at the input to the decision circuit. That Gaussian

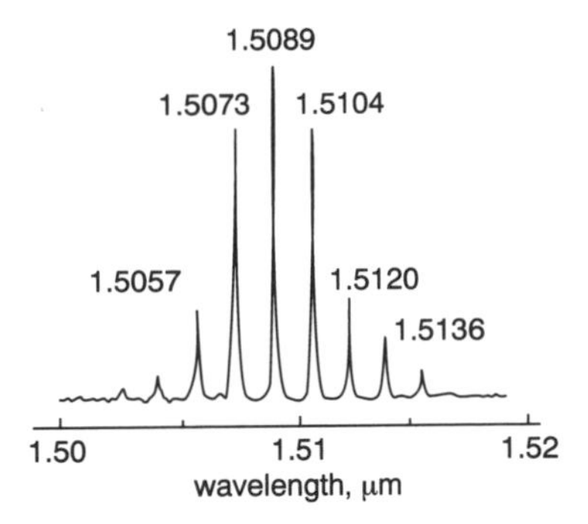

Figure 8.13 Spectrum of bh laser modulated by a 397.2 Mbit/s rz pulse train of pattern ...100... showing six main modes ([5], Figure 1). Reproduced by permission of IEEE, © 1982

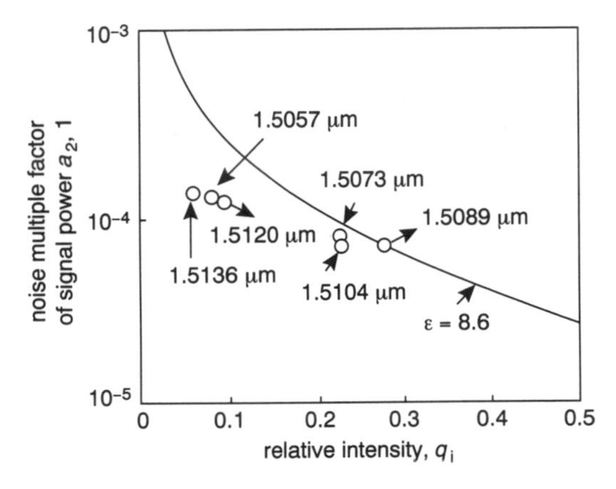

Figure 8.14 Measured mode partition noise coefficients versus fractional power in each mode for laser under same conditions as in Figure 8.13 and calculated curve taking the mode partition constant as 8.6 ([5], Figure 2). Reproduced by permission of IEEE, © 1982

approximations are inappropriate for the purpose is proved by constructing discrete approximations to these pdfs from the Gauss quadrature rule.

The author summarised his results by indicating that the side mode suppression ratio must exceed 18.2 dB if the receiver sensitivity penalty is not to exceed 1 dB for fibre lengths up to 150 km.

8.4 QUANTUM WELL LASER DIODES

Much of this section is based upon a very readable paper by senior members of staff at Murray Hill which appeared in the *AT&T Technical Journal* in 1992 ([3] of Chapter 6) and an earlier invited paper [10], also from Murray Hill. This invited paper described the status of InGaAsP/InP distributed feedback quantum well lasers with either lattice-matched or strained quantum wells forming the active layer. This layer was placed within an optimised graded-index waveguide structure with very low internal loss. BH F–P lasers based on these structures possessed low threshold currents, high power output and quantum efficiencies. DFB lasers based on these structures retained most of these properties, together with smsr as high as

50 dB and linewidth as low as 440 kHz. All of the devices were grown by the atmospheric pressure mopve process. Much of the contents were devoted to details of the mechanical, electrical and optical properties of the structures considered, which will not be described here for lack of space.

In ([3] of Chapter 6) the authors first considered the properties of bulk lasers as an introduction to the basic concept and properties of quantum well structures, emphasising those features which would serve for comparison between the two. The next section introduced the quantum well laser, drawing attention to the fundamental role played by the thickness of the active layer, and illustrated the differences between the energy versus density of carrier states in the two cases. The confining effect of a well was explained with the aid of a figure which showed the decreasing transition energy of the lowest electron–hole recombination with increasing well thickness for 1500 nm InGaAsP devices. Schematic diagrams showing bandgap energy and refractive index profiles versus active layer thickness in a multiple quantum well structure were given.

The next section described some of the advantages of lasers utilising quantum confinement over bulk devices, in terms of bandwidth, launched power, threshold current and chirp, each heading being illustrated by a figure. The benefits in applications other than as transmitter sources came next, including their use as pump lasers for erbium-doped fibre amplifiers. Two sections followed, the first on design considerations and the second on fabrication. In the final section on future directions mention was made of a number of possibilities, including the extension of quantum confinement of the carriers by shrinking two ('quantum wire') or three ('quantum dot') dimensions. In the treatment which follows only those parts of ([3] of Chapter 6) which have a direct bearing on the role of mqw lasers in ofdls will be used.

As mentioned above, more than half a working lifetime was devoted to the development and engineering of reliable bulk lasers in an enormous concentration of effort in many organisations. The goal of reliable mqw devices has occupied, and will continue to demand for some years, large resources of highly skilled staff, money and equipment.

8.4.1 Features of particular interest for system designers

Quantum well active layers discriminate strongly in favour of the TE mode, even when facet reflectivity is suppressed. In strained mqw devices in which the well thickness prevents the electron–light hole transition, the TM component of the gain spectrum was undetectable. The resulting large difference between the TE and TM modes in mqw lasers should result in low mode partition noise.

MQW-DFB laser have unusually pure single longitudinal mode behaviour with smsr as large as 40 dB in 500 μm devices and larger at 1000 μm. No mode jumps were seen on crossing the threshold.

Realistic values of α are about 2 for mqw devices, whereas in InGaAsP conventional bulk structures values lie in the range 4–8. Dynamic chirp Δf and α are related by $(\Delta f)^2 \simeq 1 + \alpha^2$, thus the chirp of mqw lasers should be several times less than that of bulk devices. Experiments on 500 μm lattice-matched devices revealed almost no differences in linewidth in lasers biased at 20 mA with or without modulation by a 40 mA peak-to-peak waveform at 1.7 Gbit/s, and virtually no differences in spectra for 1000 μm cavities. Similar experiments were performed with strained

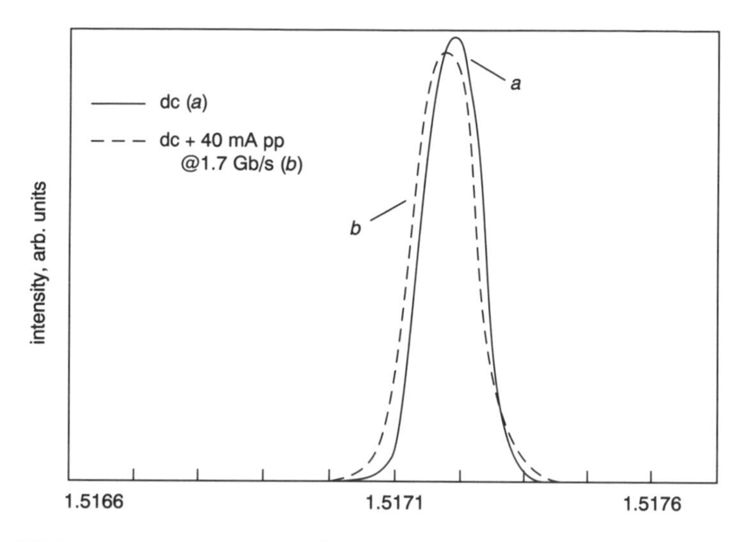

Figure 8.15 High-resolution spectrum of Bragg mode: (*a*) dc bias only, (*b*) bias plus 40 mA peak-to-peak modulation at 1.7 Gbit/s ([10], Figure 9*a*). Reproduced by permission of SPIE

mqw-dfb lasers up to their relaxation oscillation frequency of 10 GHz and showed only minor changes under sinusoidal modulation up to 20 dBm. An example of the Bragg mode spectrum with and without modulation at 1.7 Gbit/s is presented in Figure 8.15.

A transmission experiment to evaluate the chirp penalty was performed using 70 km of standard sm fibre. A 1.7 Gbit/s $2^{15}-1$ prbs modulated a 500 μm lattice-matched laser biased at two slightly different levels and the results are presented in Figure 8.16.

The linear current scale was chosen to enhance the difference between the two conditions. The measured chirp penalty of 0.25 dB was some 8 to 10 times lower than that of a bulk device under identical conditions.

3 dB bandwidths of compressively strained dfb devices were measured at 20 mA above the 20 mA threshold, this combination being chosen because drivers for conventional lasers are designed to supply 40 mA and lasers are modulated from below threshold. These types of laser were expected to be readily usable at 10 Gbit/s. At higher drive currents and thus power level P, the bandwidth increases as $\sqrt{P}$ with slope 5.6 mW/(mW)$^{1/2}$ up to 10.6 GHz, a behaviour similar to that of fast bulk laser diodes. Measured values are shown in Figure 8.17.

8.4.2 Advantages of mqw lasers as transmitters and as pump sources

In this subsection we consider only semiconductor sources for on-off modulated direct detection systems and for pumping optical fibre amplifiers, beginning with transmitter applications.

MQW transmit lasers

Bandwidth: as with all semiconductor devices there are two main determinants of bandwidth in bulk and quantum well lasers: the intrinsic speed of the light response

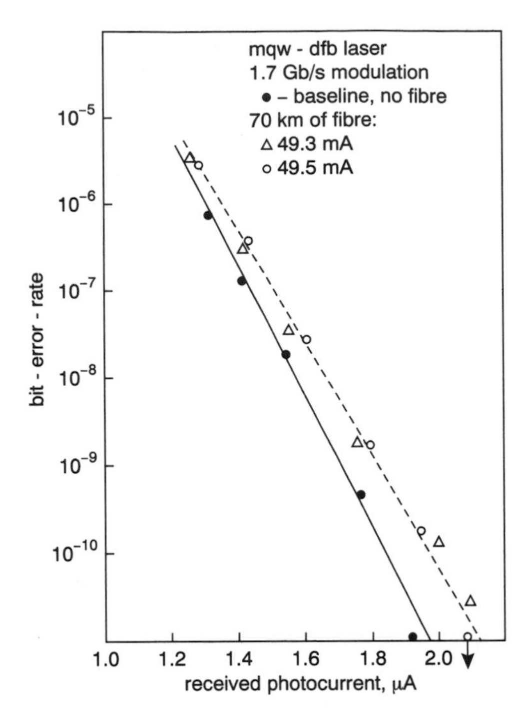

Figure 8.16 ber at 1.7 Gbit/s versus received photocurrent for 1m-mqw-dfb back-to-back and with 70 km of smf at 49.3 and 49.5 mA bias ([10], Figure 9*b*). Reproduced by permission of SPIE

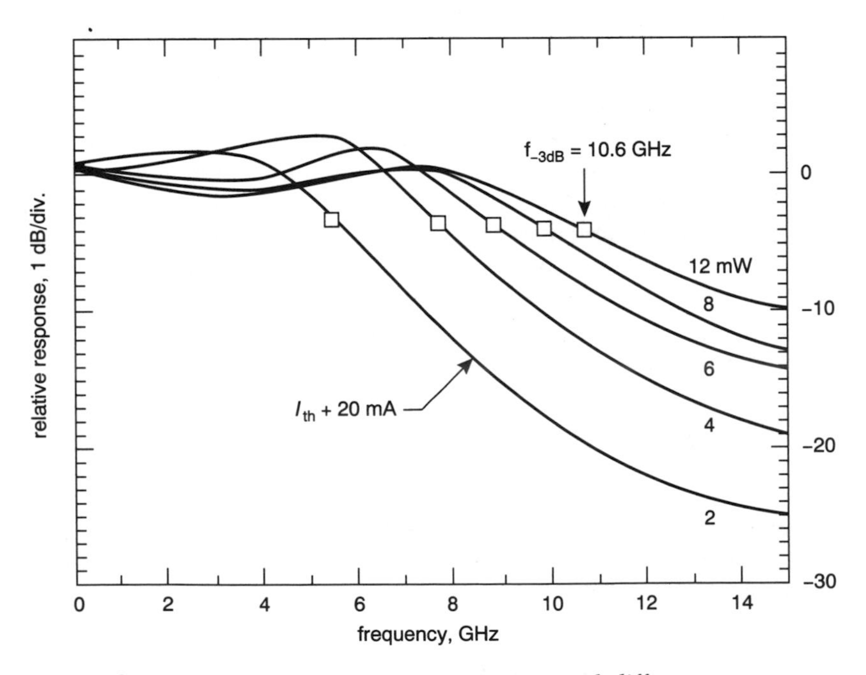

Figure 8.17 Relative responses versus frequency at several different output power levels, note 3 dB bandwidths ([10], Figure 10). Reproduced by permission of SPIE

to drive current within the chip; and parasitic effects. The latter consist of internal frequency band-limiting effects such as capacitance shunting drive current away from the active region, and external capacitances and inductances associated with packaging the chip into a usable component. MQW lasers have roughly 1.7 times higher intrinsic frequency response than bulk devices, owing to their higher optical gain per injected carrier.

Launched power: the two intrinsic features which affect the amount of power launched into a fibre are: the conversion efficiency, and the angular divergence of the light beam seen by the fibre. In both respects mqw devices are superior, and illustrative pairs of $L(I)$ and $\mathrm{d}L/\mathrm{d}I$ curves are given in Figure 8.18.

Coupling to sm fibre: the higher conversion efficiency of mqw lasers enables the very tight coupling between the active region and the optical waveguide mode needed with bulk devices to be relaxed slightly. Looser coupling permits an expanded optical mode, hence a narrower angular divergence, and this can be seen in the far-field profiles for both types of laser in Figure 8.19.

Chirp: this is related to the ratio of the change of refractive index to the change of gain, both resulting from a time-varying drive current. The higher conversion efficiency of mqw devices causes this ratio to be smaller, and hence there is less chirp by a factor of about 1.6. Chirp distributions for both kinds of devices are illustrated in Figure 8.20.

Gain spectra: considering mode partitioning, the higher gain of the TE modes relative to the TM gain in mqw lasers enables TM modes of oscillation to be suppressed as shown in Figure 8.21.

The looser coupling also provides tighter control over higher-order waveguide modes of oscillation. The same mechanism that is responsible for lower chirp when the laser is modulated also leads to narrower linewidths in cw operation.

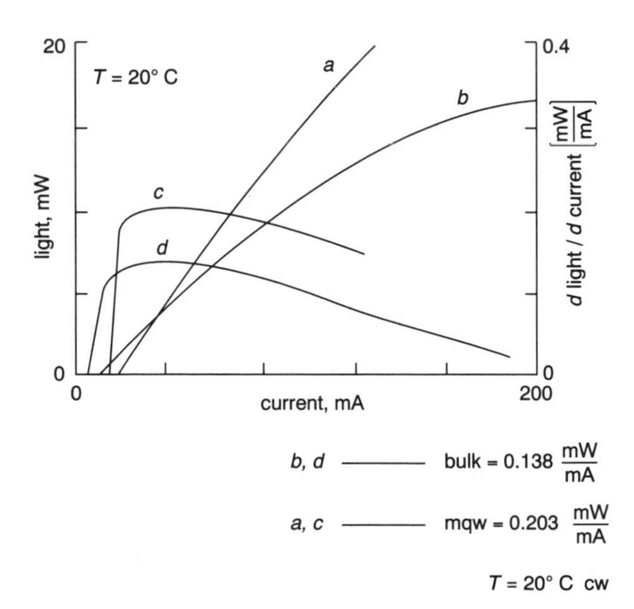

Figure 8.18 Output power versus drive current: (*a*) for a mqw device, (*b*) for a bulk laser, slope of output characteristics, (*c*) for mqw laser, (*d*) for bulk device.

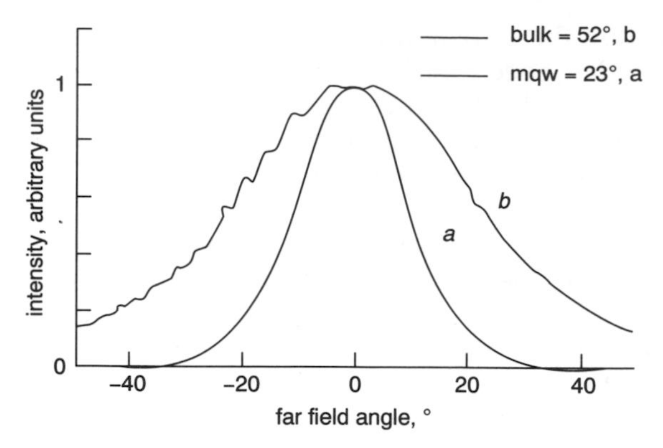

Figure 8.19 Intensity versus far-field angle showing representative far-field profiles: (*a*) for mqw laser, fwhm 23°; (*b*) for bulk device, fwhm 52°. Copyright © 1992 AT&T. All rights reserved. Reprinted with permission

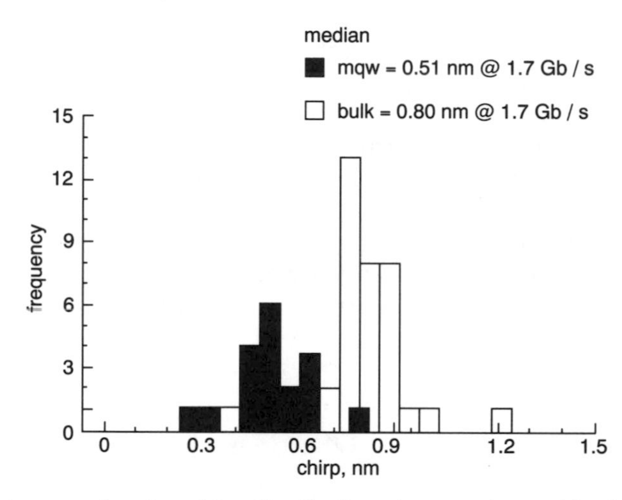

Figure 8.20 Histogram showing chirp distributions for mqw lasers (shaded) and bulk lasers; corresponding median values of chirp at 1.7 Gbit/s were 0.51 and 0.80 nm, respectively. Copyright © 1992 AT&T. All rights reserved. Reprinted with permission

MQW pump laser diodes

For use as a pump laser source in operational systems, high output power, precise control of wavelength and high reliability are mandatory. Both yield and cost of manufacture are also important considerations. By 1994 the reliability was apparently sufficiently high for applications in submerged systems, since predicted initial estimates of lifetimes of about 100 years were reported for certain types of mqw lasers.

Pump lasers for erbium-doped fibre amplifiers may be required to generate more than 100 mW of power in order to produce the pump power levels needed within the amplifier itself at a given operating pump wavelength (e.g. up to 45 mW at 1480 nm or 70 mW at 980 nm). The advantages of mqw devices over bulk laser pump diodes are then as follows.

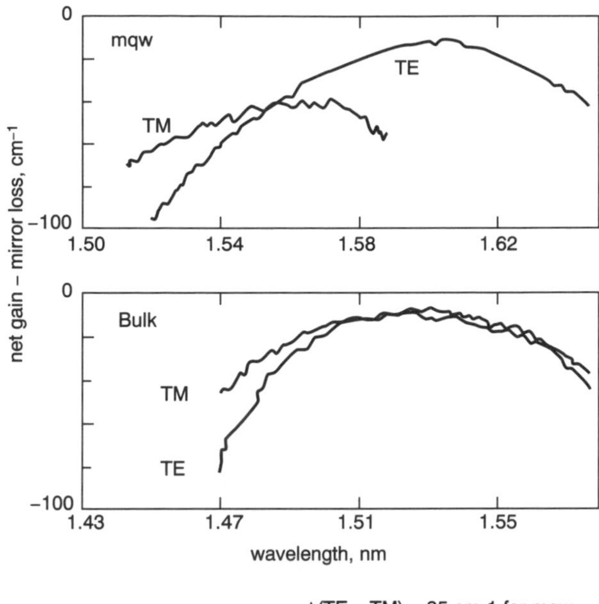

Figure 8.21 Spectra of TE and TM gains for mqw lasers (upper figure) and for bulk lasers (lower figure). Note splitting of the gain peaks in mqw devices and the associated reduction of the TM peak by about 35 cm^{-1} ([3] of Chapter 6, Figure 8). Copyright © 1992 AT&T. All rights reserved. Reprinted with permission

Higher conversion efficiency: this leads to lower drive currents for a given optical power output and consequently easier thermal management.

Higher power capability: although bulk laser diodes were used widely for pumping erbium amplifiers in experimental and demonstration systems, their inferior power capability suggests that mqw devices will be deployed in operational systems by the mid-1990s.

Lower internal losses: almost three times lower than in bulk devices. This allows longer cavities, hence lower injected current densities and therefore less heat generated per unit area. The equivalent series resistance is therefore smaller, which increases the conversion efficiency. The thermal impedance is also lower, leading to reduced temperatures or higher output power levels for the same temperature rise above ambient.

Further information on pump laser diodes for erbium-doped fibre amplifiers can be found in [11].

A practical 1300 nm optical fibre amplifier was not a viable component by 1994, but work on praseodymium-doped fibre amplifiers had been reported from a number of laboratories. A very practical and informative treatment by authors from BTRL, Martlesham, published in early 1993 [12], described the reasons for the then current interest and included information on pump power requirements as well as experimental results. For example, under small-signal operating conditions 15 dB of net gain at 1310 nm required 500 mW of absorbed pump power, whereas for power amplifier applications 100 mW of output required about 1 W of launched pump power. Rapid development and commercial availability of semiconductor pump lasers oscillating at about 1000 nm at output levels around 1 W would provide a practical pump source.

8.4.3 Multiple quantum well structures

When lasing, the active region in bulk devices has three distinguishing features, as follows.

For the range of active layer thicknesses as used in practical devices (about 150–100 nm) the carrier (electrons, light and heavy holes) energy levels form a continuum in the conduction and valence bands. The density of states for each type of carrier is parabolic. Radiative transitions take place from electron states to either type of hole states.

The gain is the same for both TE and TM polarisations.

The presence of many injected carriers with energies near the conduction and valence band edges which do not contribute to the lasing action because the low density of states, is insufficient to permit lasing. These states must, however, be 'filled' before lasing can occur between the higher energy states, and this needs more injected carriers.

When the active layer thickness is reduced to about 30 nm or less the carriers are affected by the presence of the built-in barriers above and below the active region. This causes the energy levels in the conduction and valence bands to break into discrete bands, giving a step density distribution of states. Each step is associated with a different confined quantum state of the carriers. Moreover, where previously the bulk valence band edge for light holes (primarily TM polarised) and heavy holes (primarily TE polarised) was the same, in quantum well structures they are split, thus breaking the symmetry between the behaviour of TE and TM modes. Such active regions are described as quantum wells because of the quantum mechanical

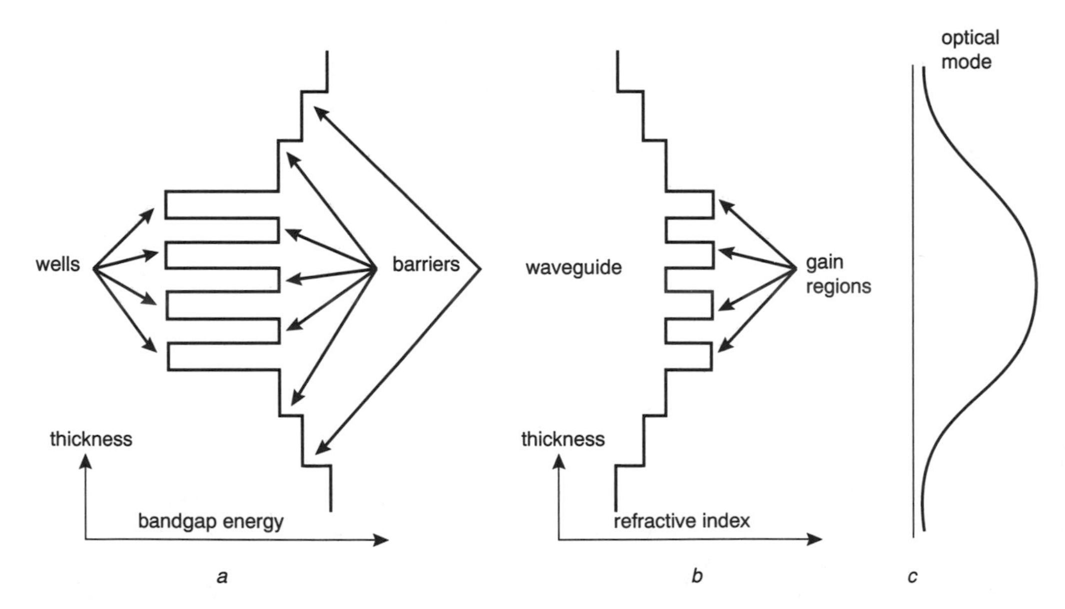

Figure 8.22 Schematic of multiple quantum well active region thickness dimension showing: (*a*) profile of material bandgap energy; (*b*) refractive index profile; and (*c*) optical power distribution ([3] of Chapter 6, Figure 4). Reproduced by permission of AT&T

nature of the energy splitting in the well.

Referring to the features listed for bulk devices, the corresponding statements for quantum well devices are as follows:

The density of states for each type of carrier is step-like, and the bulk valence band edge for heavy and light holes are no longer the same. Radiative transitions take place from electron states to either type of hole states.

The gain is no longer the same for both TE and TM polarisations.

Energy need no longer be pumped into unused carrier states, i.e. those near band edges or those that amplify the 'wrong' polarisation, but only into the wanted states, thus increasing the differential optical gain per injected carrier and so lowering the threshold current required. In addition the device oscillates in the desired polarisation more reliably.

In practice, multiple quantum wells are used to improve the performance; they are shown schematically in Figure 8.22. Figures (*b*) and (*c*) can be compared with corresponding figures for an optical fibre of the dispersion-flattened type (see Section 4.4).

8.5 OPTICAL SOURCES FOR HIGH-CAPACITY OFDLS

In order to provide some examples of 1994 state-of-the-art devices which may be deployed in operational systems in the late 1990s, this section describes several types of source which were developed in the UK in the early 1990s [13]. These sources were: an individually addressable 8-wavelength source with 2 nm spacing for wdm; a high power mqw device with low chirp achieving 2 dB penalty over 100 km of standard sm fibre; a transmitter with external modulator providing 5 mW coupled power and modulation bandwidth exceeding 20 GHz and chirp below 0.3; and a low-cost laser with integral fibre grating external cavity which produces soliton pulses at bit rates between 300 Mbit/s and 8 Gbit/s.

8.5.1 Multichannel source

This provided 8 channels at 2 nm spacing in a single fibre output and operated at 1 Gbit/s. The arrangement of the chip on its carrier in plan view is shown schematically in Figure 8.23.

The eight-channel laser chip itself was rectangular and consisted of two rows of four rectangular 'windows', with a centrally placed strip carrying the eight conductors for the drive currents. The output spectrum of the array when all lasers were driven simultaneously is illustrated in Figure 8.24.

8.5.2 Low-chirp high-speed laser

Structure: the device active region comprised four quantum wells with the geometry and dimensions indicated in Figure 8.25.

Linewidth: this is inversely proportional to the output power, as shown in Figure 8.26.

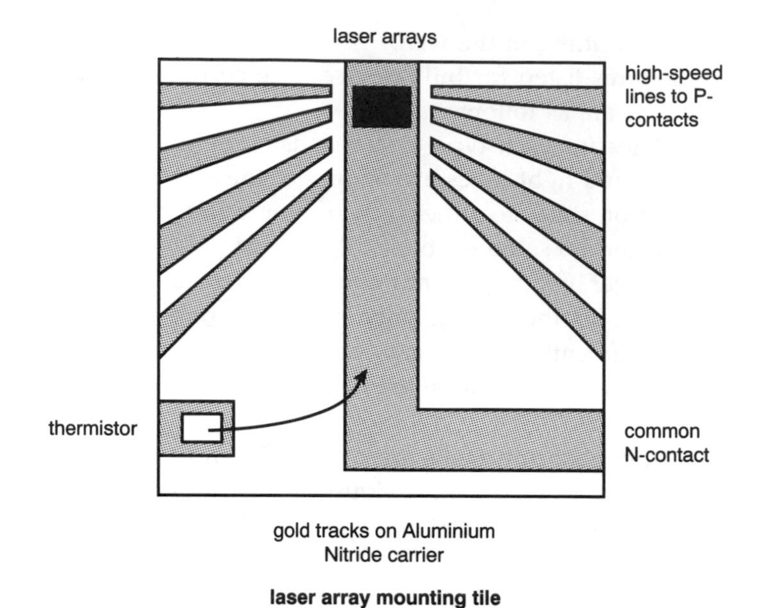

Figure 8.23 Laser array mounting arrangements showing major features ([13]). Reproduced by permission of IEE

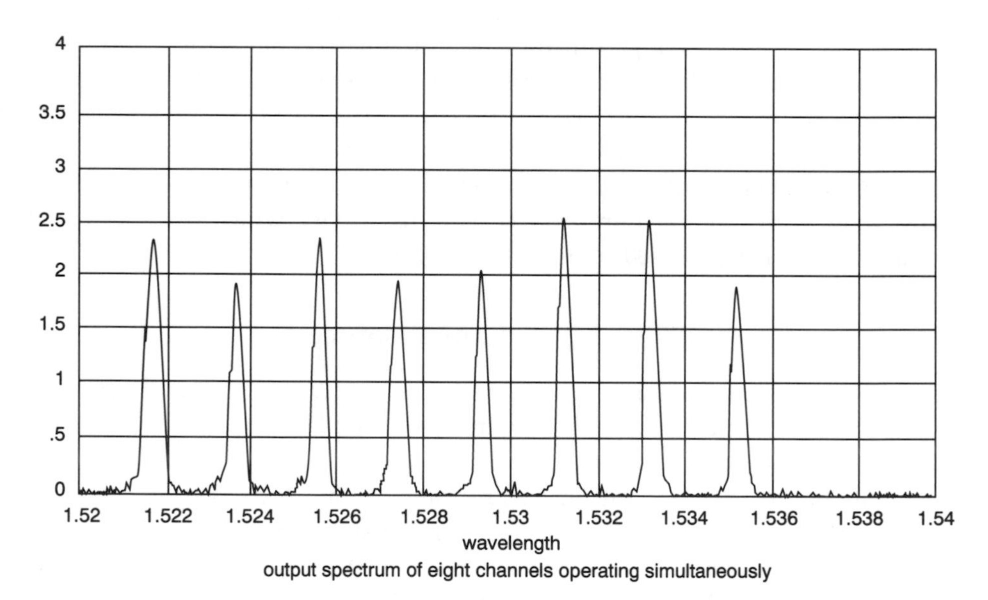

Figure 8.24 Time-averaged spectrum of laser array when driven simultaneously ([13]). Reproduced by permission of IEE

8.5.3 Performance in 2.488 Gbit/s intensity-modulated direct-detection system

Experimental arrangement: measurements of ber at 2.488 Gbit/s were carried out using the arrangement shown in Figure 8.27.

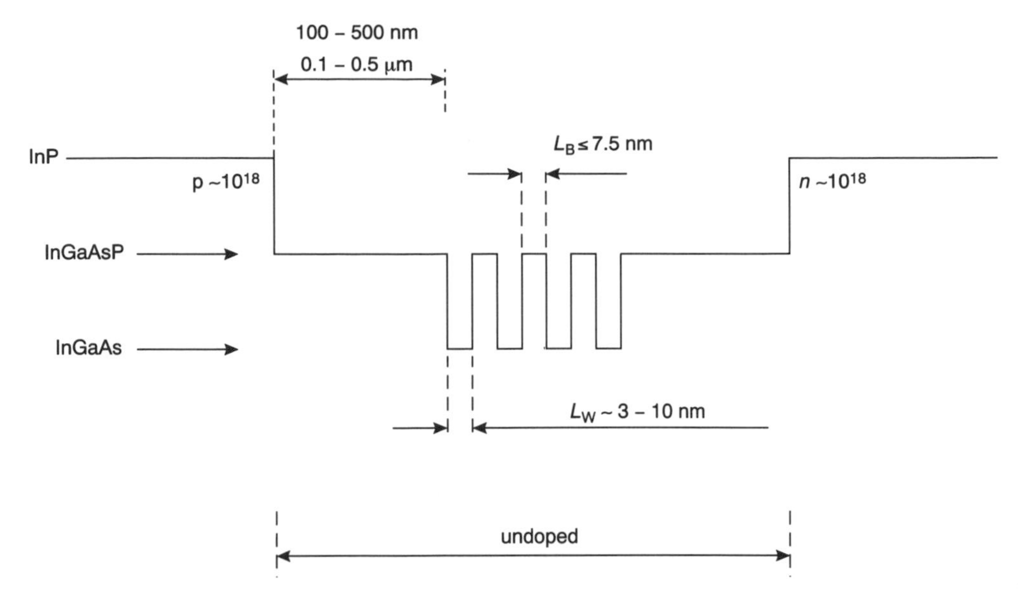

Figure 8.25 Active region structure of mqw laser giving dimensions and showing regions with different compositions: wells are InGaAs, barriers InGaAsP and cladding layers p-InP or n-InP ([13]). Reproduced by permission of IEE

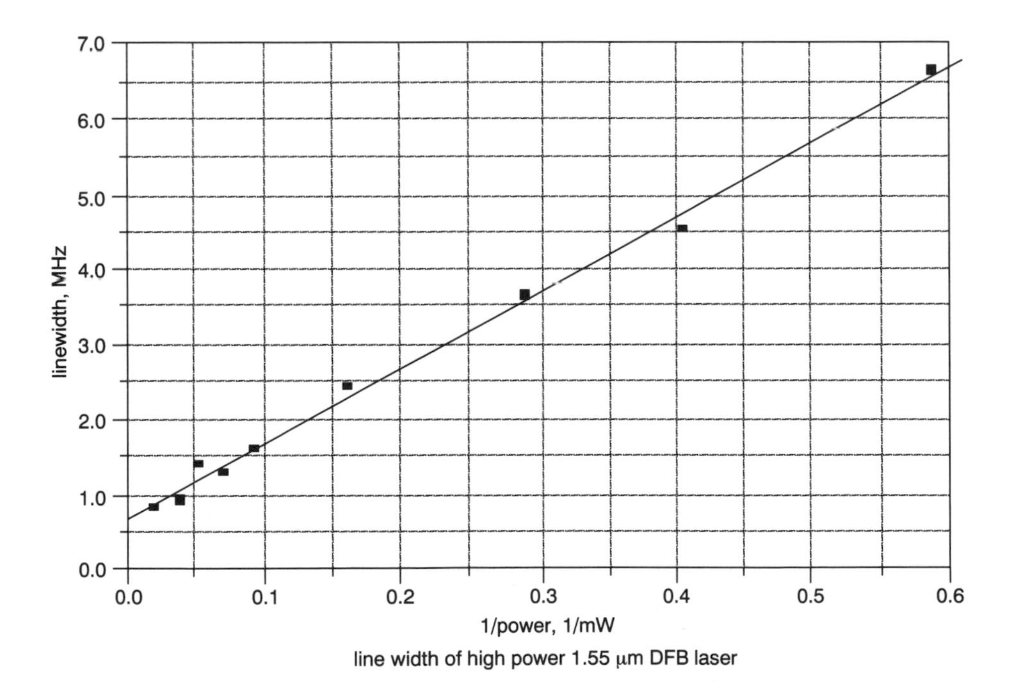

Figure 8.26 Measured values of linewidth (MHz) versus reciprocal of output power (1/mW) and best fit straight line for high power dfb laser operating in 1550 nm band ([13]). Reproduced by permission of IEE

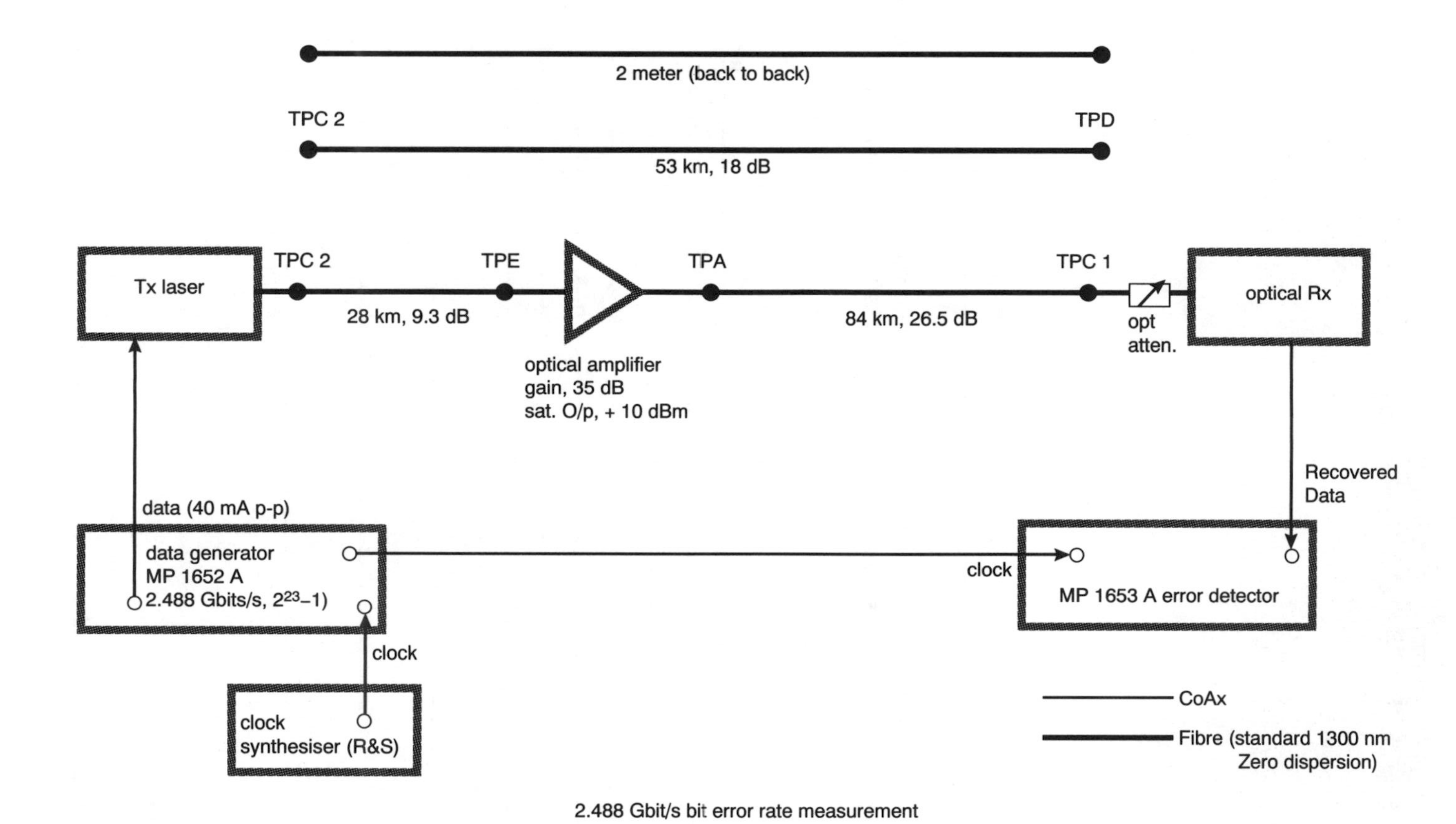

Figure 8.27 Experimental arrangements for measurements of ber using high-power dfb laser transmitter: (*a*) back-to-back, 2 m; (*b*) 53 km span; (*c*) 28 km, edfa and 84 km of standard sm 1300 nm fibre ([13]). Reproduced by permission of IEE

back to back

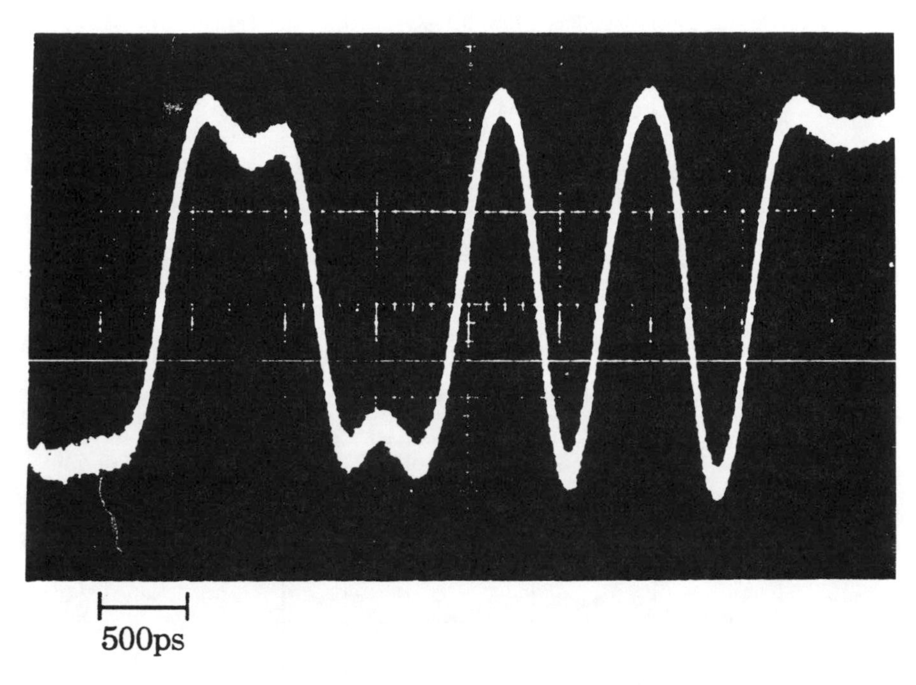

after 112 km 1300 nm zero-dispersion fibre

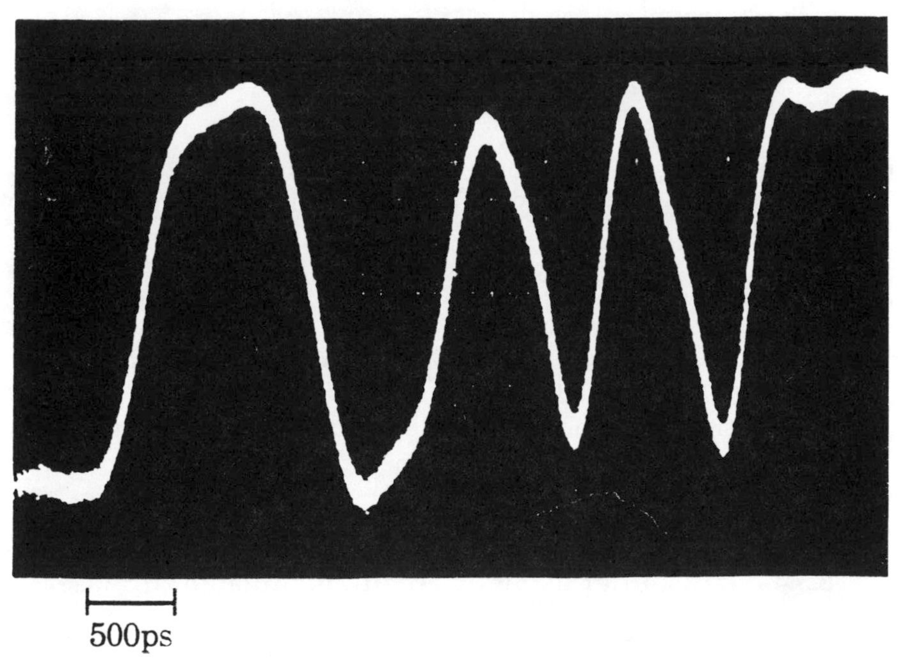

Figure 8.28 Oscillograms showing waveforms of received 2.488 Gbit/s signal for a specific device under stated conditions: (*a*) upper trace back-to-back; (*b*) lower trace after 112 km of standard fibre ([13]). Reproduced by permission of IEE

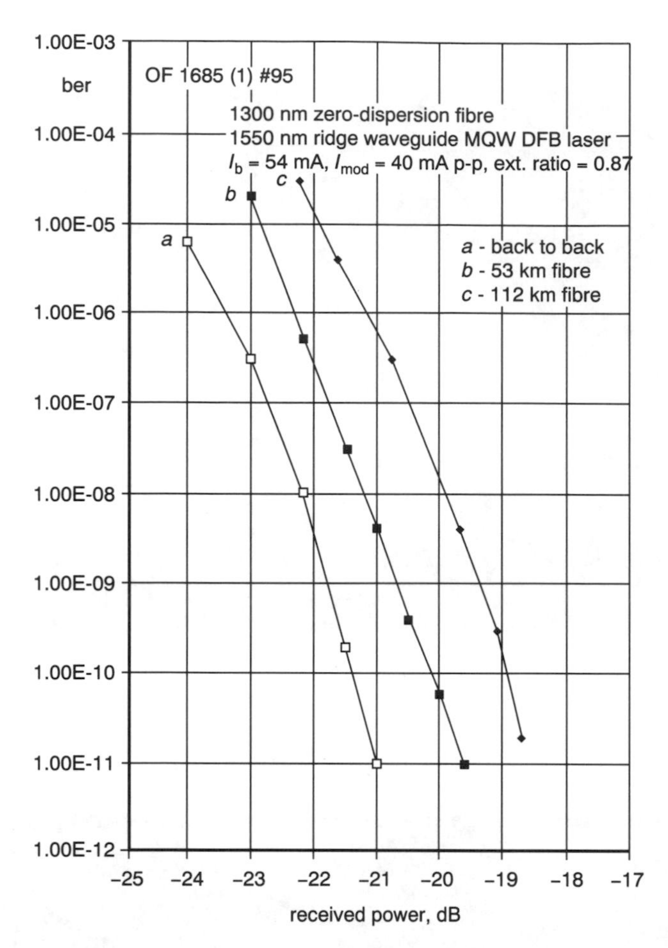

Figure 8.29 ber versus received optical power: (*a*) back-to-back, 2 m; (*b*) 53 km span; (*c*) 28 km, edfa and 84 km of standard sm 1300 nm fibre ([13]). Reproduced by permission of IEE

Corresponding waveforms: those for (*a*) and (*c*) are reproduced in Figure 8.28. ber versus received power: measurements were made for all three cases given as (*a*), (*b*) and (*c*) in Figure 8.27, and the results are showed in Figure 8.29. For case (*c*) the sensitivity at a ber of 10^{-9} is about −19.4 dBm, which compares very unfavourably with a value of −32 dBm quoted for the same bit rate and type of fibre, realised in submerged systems.

8.5.4 Transmitter using externally modulated dfb laser

This class of transmitter is used when very low chirp is needed, such as in coherent systems, and is therefore beyond the scope of this book. Therefore only very brief mention will be made of some salient performance features. A 20 GHz modulation bandwidth was achieved at 5 mW launched into the system fibre, as illustrated in Figure 8.30.

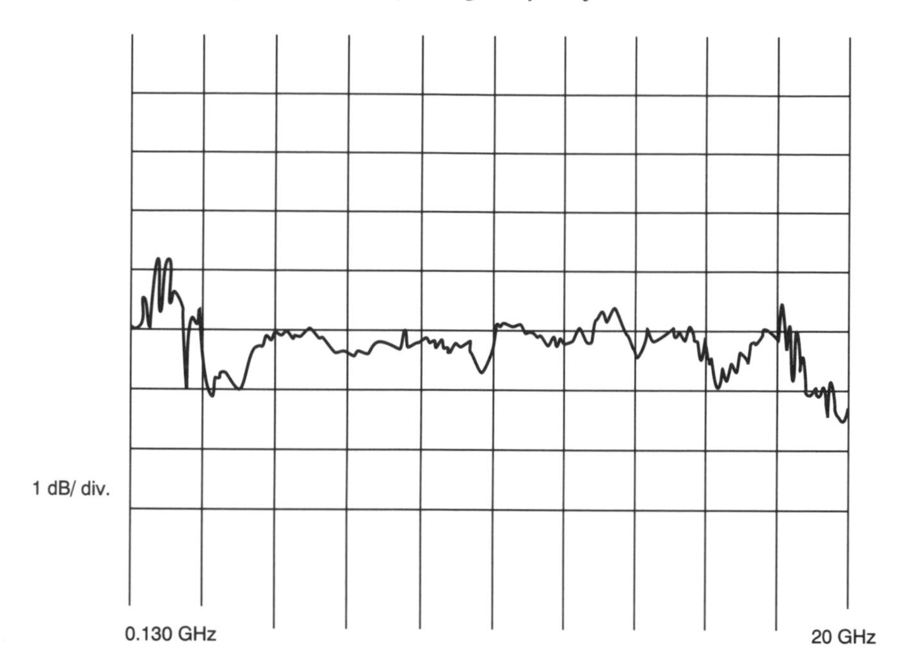

Figure 8.30 Modulation frequency response of externally modulated dfb laser at 5 mW output power ([13]). Reproduced by permission of IEE

8.5.5 Performance of various mqw–dfb lasers at 10 Gbit/s

Strained layer devices

The excellent characteristics of 1300 nm sl-mqw-dfb lasers were reported for the first time in [14]. Spans of 50 km of both standard and ds fibre were achieved without penalty. The device structure was based on a bh containing eight 3 nm InGaAs layer quantum wells and a barrier layer of 10 nm of InGaAsP, and the cavity length was 300 μm. The device oscillated at 1310 nm, had a threshold current of 21 mA and the side-mode suppression ratio exceeded 30 dB.

To measure the linewidth enhancement factor the laser was biased at 1.21 times the threshold and was driven by 40 mA peak-to-peak current pulses to generate 45 ps optical pulses; the results indicated values of α as low as unity, compared with more than 3 for bulk lasers. The 20 dB down chirp width at this bias was 0.35 nm compared with 0.39 nm for the bulk device, and 0 dBm was launched into the system fibre.

Resonance frequencies as high as 90 GHz were predicted for these types of device in 1988. Values achieved in the early 1990s fell far short of this value, as can be seen from Figure 8.31 for some 1550 nm devices [15].

The bandwidth of the sl–mqw laser at 40 mA above threshold was around 16 GHz, and the estimated differential gain constant (dg/dn) at 4.8×10^{-20} m^2 was 2.2 times that for the unstrained device and three times larger than the bulk laser. The corresponding value of the nonlinear damping (K–) factor (proportional to $\varepsilon/(dg/dn)$, where ε is the nonlinear gain parameter) at 0.13 ns was about half that of the unstrained mqw laser. The theoretical value of the intrinsic maximum bandwidth of a device was stated as $2.8\pi/K$, which gives 68 GHz in the present

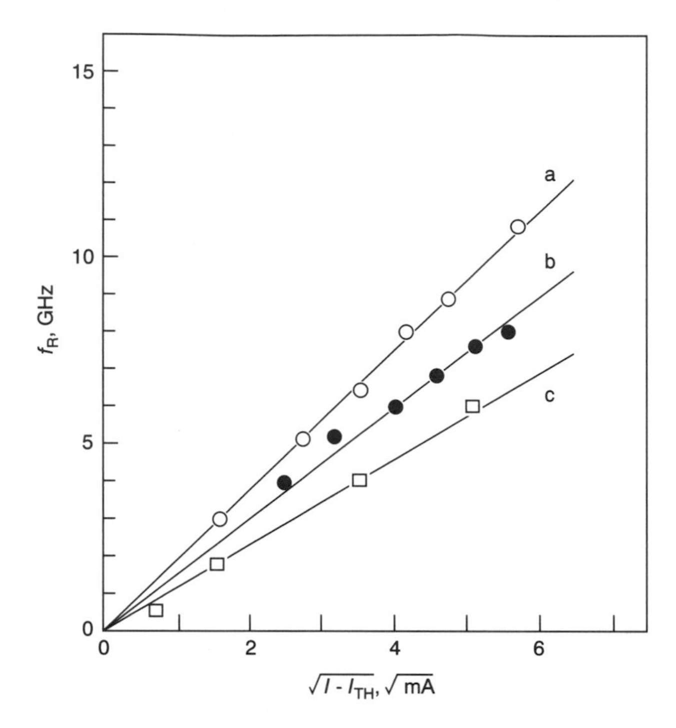

Figure 8.31 Resonance frequency versus square root of drive current above threshold current characteristics for: (*a*) strained layer mqw device; (*b*) unstrained mqw; and (*c*) bulk laser ([15], Figure 2). Reproduced by permission of IEEE, © 1991

case. A later and more detailed treatment of the relation between damping and 3 dB intrinsic bandwidth has been given.

Devices incorporating semi-insulating buried-heterostructures

Lasers with semi-insulating (si), buried heterostructure current blocking layers combined with mqw–dfb structures have also shown very promising results at 10 Gbit/s. The sibh structure offers special advantages with either bulk or mqw active regions; for example, threshold currents as low as 7 mA were achieved in 400 μm devices without facet coating, and output powers of 11 mW per facet for 100 mA drive current [16] in mqw–dfb–sibh devices. Side mode suppression ratios higher than 40 dB under cw conditions and greater than 38 dB with 10 Gbit/s nrz modulation were reported for the mqw laser. The very low parasitics associated with these devices gave 3 dB bandwidths greater than 11 GHz. When driven by trapezoidal current pulses with rise and fall times of 33 ps the corresponding optical pulse times were 36 and 44 ps, the shortest reported at that time. Modulation experiments at 10 Gbit/s under large-signal conditions were performed on both mqw and bulk dfb–sibh devices and open eye diagrams were obtained at an extinction ratio of about 5:1. When modulated by 10 Gbit/s nrz pulses, the wavelength chirp of the mqw devices was 0.47 nm and for bulk lasers it was 1.0 nm.

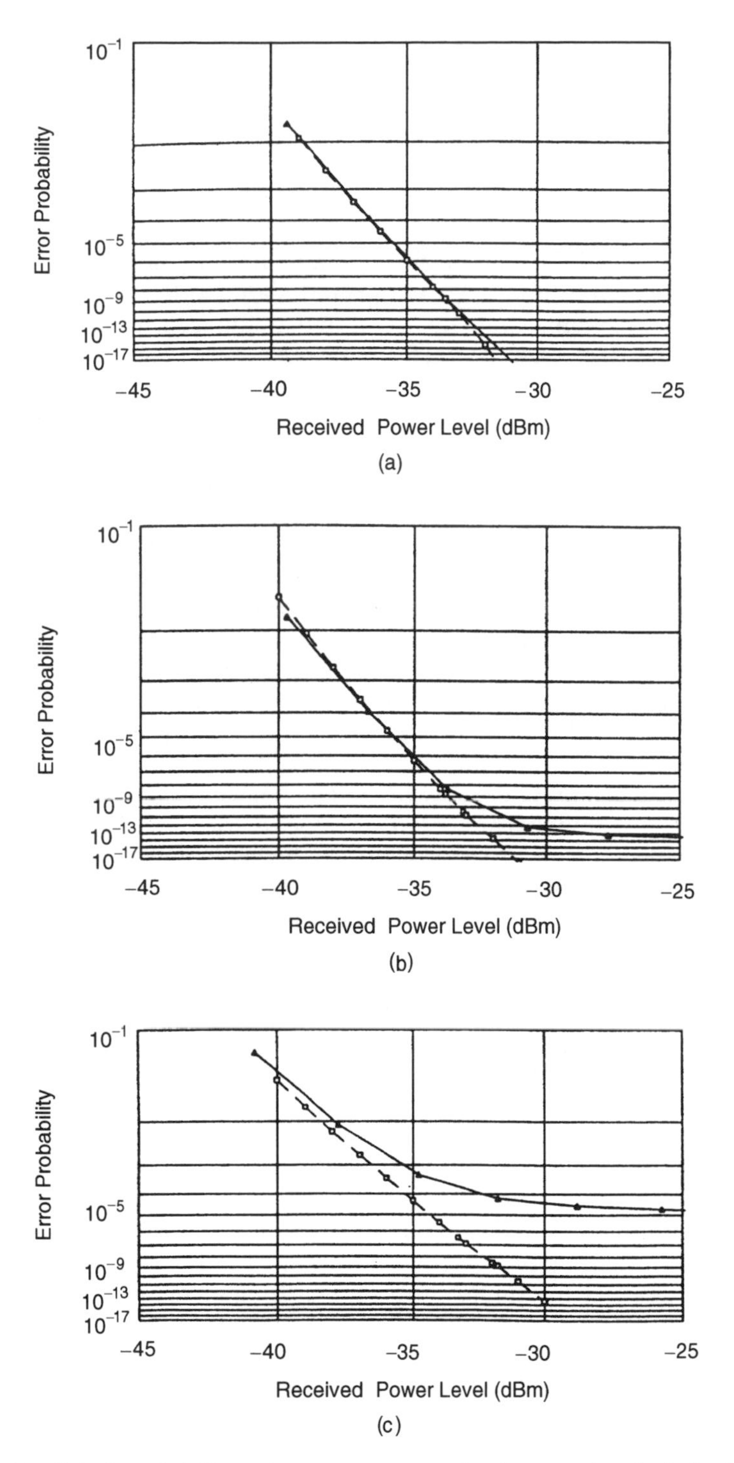

Figure 8.32 Predicted probability of error versus received power level; dashed line curves from deterministic model, full line curves from stochastic model, when biased (*a*) 0.8 mA above threshold (*b*) 0.5 mA below threshold and (*c*) 3 mA below threshold ([17], Figure 11). Reproduced by permission of IEEE, © 1995

8.5.6 Systems effects of laser turn-on jitter and chirp noise

The error rate performance of imdd systems operating at Gbit/s rates is critically dependent on the characteristics of the transmit laser. Up to 1994, the analysis of such systems were based on solving the deterministic laser diode rate equations for given bias and modulation currents. The resulting output waveform was transmitted along the system fibres and the received signal used to evaluate the ber, taking account of intersymbol interference, detector noise and receiver noise by means of numerical techniques, e.g. the Gaussian quadrature numerical integration. Such techniques were extremely useful in providing insights into the combined effects of chirp and dispersion on performance. Attempts were also made to utilise the results to assist in the specification and design of lasers for such systems e.g. ITU(T) Study Group XV in Recommendation G.957 addressed the specification of lasers for 2.5 Gbit/s systems operating in the 1550 nm band over conventional (standard) fibres and 60 km spans ('L16.2 application'). Recommendation G.957 considers several measurements for selecting laser diodes for such applications.

However, in actual systems significant error rate floors have been observed e.g. [14] of Chapter 13 (as large increases in power incident on the photodiode resulting in only marginal improvements in system ber) when the transmit laser was biased near or below threshold, which were not predicted by system simulations based upon models using deterministic rate equations. The cause of these error rate floors was suspected to be the stochastic turn-on behaviour of lasers when biased near to or below threshold. Fluctuations (jitter) in turn-on times and the resulting 'chirp noise' are direct results of the dominance of spontaneous emission photons in the active region of the laser while below threshold. This effect could only be modelled once an analytical technique had been developed for determining the stochastic turn-on behaviour of a laser biased at arbitrary levels. This new approach separates the laser output into two distinct regimes (stochastic and deterministic), depending on the number of photons in the active region. When this number is small, the evolution of the photon density is a random process and is modelled by a set of modified stochastic rate equations. When the number is large, the evolution is governed by the usual deterministic rate equations. The choice of bias level strongly influences which regime the laser operates in and for how long, and when modulated the laser may operate alternately in the deterministic and stochastic regimes.

The new approach was used in [17] to evaluate the error rate floors induced by chirp noise in a complete system. First, the probability density functions of the laser turn-on times were calculated, then the conditional error probability of the entire system for a given turn-on time was computed. The results were combined to determine, for the first time, the error rate floor due to turn-on time jitter and associated turn-on chirp and overall system ber performance. Predicted probability of error curves were presented for three different bias points using the new simulation technique and are reproduced in Figure 8.32.

These plots show that the deterministic model is only adequate when the laser is biased above threshold, and that the farther the bias is below threshold the larger the probability at which the error floor occurs.

The results presented by the authors from Telecom Australia show that failure to include the impact of turn-on jitter and chirp noise in system design will give a misleading relationship between laser bias level and system performance. The model allows system designers to understand the trade-offs between system error performance and factors such as extinction ratio, laser bias, laser dynamic response and lineshape. The selection of transmit lasers and their operating conditions for use in long-haul high-capacity imdd systems has thus been placed on a more rigorous foundation.

8.6 COMMENTS ON RELIABILITY

By 1994 strain compensation had been employed in many mqw structures. By growing alternate layers with opposite senses of strain a large number of strained layers can be realised, thus providing device engineers with an extra degree of freedom in design. For commercial applications, device reliability must not be compromised. The long-term stability of strain-compensated and zero-net-strain structures was, however, questioned in a paper published in 1993. The results of preliminary reliability studies at BT Laboratories, Martlesham, on both the above types of InGaAs(P)/InP heterostructure lasers were reported in [18].

Wafers with 16 zero-net-strain and conventionally strained quantum wells and 8 strain-compensated quantum wells were grown, and a small sample was used for reliability studies, as detailed in the reference. Such devices operate with significantly lower threshold currents than other mqw lasers, and the strained quaternary well lasers exhibited record low threshold currents of only 3.1 mA. The modulation response and temperature sensitivity were, however, much the same.

The conventional definition of lifetime as the time taken for the drive current to increase by 50% from its initial value was used (i.e. current needed to maintain a constant output, here 4 mW). Using extrapolation methods developed for standard bulk active region lasers, and a conservative value of activation energy of 0.35 eV, resulted in a room temperature lifetime due to thermal degradation of around 100 years.

A new reliability hazard: mode hopping due to degradation

Error-free operation of an ofdls using dispersive fibres requires that the laser source emits in a single mode and never hops or changes to another wavelength during its operational life. The deployment of single-frequency devices has introduced a new reliability hazard 'mode hopping due to degradation' and it is expected that there will be some residual chip degradation even in devices for submerged applications. In general, degradation may be uniform or localised in some part of a chip, and where present it will change the injected carrier density. This in turn will perturb the effective refractive index and alter the optical pitch of the grating. For localised degradation in the grating an additional phase shift will be introduced into the original design and if the added phase is large enough it will cause mode hopping.

Confidence in long-term system reliability requires information on the spectral behaviour of degraded lasers, and this information can be obtained experimentally by overstressing devices to cause degradation, then measuring the spectrum, or by simulation. A theoretical model, with deliberate changes made to recombination parameters to simulate degradation, which can be solved for the spectral modes of a dfb laser was described in [19]. The model was applied to a dfb laser with uniform grating and to another with a $\lambda/4$ phase-shifted grating, and both uniform and localised types of degradation were used. The treatment covered the modelling of nonuniform injected dfb lasers, uniform nonradiative recombination, and localised injected current loss. The authors drew the conclusion that uniform degradation was much less likely than localised degradation to cause mode hopping. The simulation confirmed that overall the $\lambda/4$ phase-shifted laser was likely to resist mode hopping better.

The work demonstrates that degradation modelling is a powerful tool to compare the susceptibilities of different dfb laser designs to mode hopping arising from this cause, and it enhances knowledge of degradation hazards in lasers. It can also be used to screen designs to establish which are more resistant to this effect.

8.7 REFERENCES

1 LOWERY, A.J.: 'Transmission-line laser modelling of semiconductor laser amplified optical communication systems'. *IEE Proc. J.* **139**, no. 3, June 1992, pp. 180–188.

2 PETERMANN, K., and ARNOLD, G.: 'Noise and distortion characteristics of semiconductor lasers in optical fiber communication systems'. *IEEE J. Quantum Electronics,* **QE-18**, no. 4, Apr. 1982, pp. 543–555.

3 PETERMANN, K.: *'Laser diode modulation and noise'*, Dordrecht, The Netherlands, Kluwer Academic Publishers, Tokyo, ADOP Advances in Optoelectronics, 1988.

4 IEZEKIEL, S., SNOWDEN, C.M., and HOWES, M.J.: 'Large signal analysis of laser diodes for directly modulated microwave fibre-optic links'. *19th European Microwave Conf.*, London, 4–7 Sept. 1989, pp. 985–990.

5 IWASHITA, K., and NAKAGAWA, K.: 'Mode partition noise characteristics in high speed modulated laser diodes'. *IEEE J. Quantum Electronics,* **QE–18**, no. 12, Dec. 1982, pp. 1000–1004.

6 JOINDOT, I., BOISROBERT, C., and KUHN, G.: 'Laser rin calibration by extra noise injection'. *Electronics Lett.*, **25**, no. 16, 3 Aug. 1989, pp. 1052–1053.

7 MUKAI, T., and YAMAMOTO, Y.: 'AM quantum noise in 1.3 μm InGaAsP lasers'. *Electronics Lett.*, **20**, no. 1, 5 Jan. 1984, pp. 29–30.

8 SUNDARASAN, H., and FLETCHER, N.C.: 'Measurements of low frequency and high frequency relative intensity noise in dbr lasers'. *Electronics Lett.*, **27**, no. 17, 15 Aug. 1991, pp. 1524–1526.

9 CARTLEDGE, J.C.: 'Performance implications of mode partition fluctuations in nearly single longitudinal mode lasers'. *IEEE J. lightwave Technol.*, **6**. no. 5, May 1988, pp. 626–635.

10 TEMKIN, H., LOGAN, R.A., and TANBUN-EK, T.: 'InGaAsP/InP distributed feedback quantum well lasers'. *Proc. SPIE. The Intern. Soc. for Opt. Engineering,* 'Laser diode technology and applications III, **1418**, 23–25 Jan. 1991, CA, USA.

11 DESURVIRE, E.: *'Erbium-Doped Fiber Amplifier — Principles and Applications'*. John Wiley & Sons, Inc. New York, 1994.

12 WHITLEY, T.J., WYATT, R., SZEBESTA, D., and DAVEY, S.T.: 'Towards a practical 1.3 μm optical fibre amplifier'. *BT Technol. J.*, **11**, no. 2, Apr. 1993, pp. 115–126.

13 FARRIES, M.C., WALE, M., DUTHIE, P.J., FELL, P., and CARTER, A.C.: 'Optical sources for high capacity optical communications' *IEE Coll.* 1994.

14 MIYAMOTO, Y., HAGIMOTO, K., ICHIKAWA, F., YAMAMOTO, M., and KAGAWA, T.: '10 Gbit/s, 50 km dispersive fibre transmission experiment using strained multiquantum well dfb laser diode'. *Electronics Lett.*, **27**, no. 10, May 1991, pp. 853–854.

15 HIRAYAMA, Y., MORINGA, M., SUZUKI, N., and NAKAMURA, M.: 'Extremely reduced nonlinear K-factor in high speed strained layer multiquantum well dfb lasers' *Electronics Lett.* **27**, no. 10, 9 May, 1991, pp. 875–876.

16 SPEIER, P., BOUAYAD-AMINE, J., CEBULLA, U., DÜTTING, K., KLENK, M., LAUBE, G., MAYER, H.P., WEINMANN, R., WÜNSTEL, K., ZIELINSKI, E., and HILDEBRAND, O.: '10 Gbit/s mqw-dfb-sibh lasers entirely grown by lpmovpe'. *Electronics Lett.* **27**, no. 10, 9 May, 1991, pp. 863–865.

17 STEPHENS, T., HINTON. K., NADERSON, T., CLARKE, B.: 'Laser turn-on delay and chirp noise effects in Gb/s intensity-modulated direct-detection systems'. *J. Lightwave Technol.*, **13**, no. 4, Apr. 1995, pp. 666-674.

18 SELTZER, C.P., PERRIN, S.D., HARLOW, M.J., STUDD, R., and SPURDENS, P.C.: 'Long-term reliability of strain compensated InGaAs(P)/InP mqw bh lasers'. *Electronics Lett.*, **30**, no. 3, 3 Feb. 1994, pp. 227-229.

19 GOODWIN, A.R., WHITEAWAY, J.E.A., and MURPHY, R.H.: 'Modelling effects of degradation on spectral stability of distributed feedback lasers'. *Electronics Lett.*, **25**, no. 2, 19 Jan. 1989, pp. 164-165.

20 WENKE, G., and HENNING, I.D.: 'Spectral behavior of InGaAsP/InP 1.3 μm lasers and implications on the transmission performance of broadband Gbit/s signals'. *J. Optical Commun.*, **3**, 1982, pp. 122-128.

21 OGAWA, K., and VODHANDEL, R.S.: 'Measurements of mode partition noise of laser diodes'. *IEEE J. Quantum Electronics*, **QE-18**, no. 7, July 1982, pp. 1090-1093.

22 KIKUSHIMA, K., HIROTA, O., SHINDO, M., STOYKOV, V., and SEUMATSU, Y.: 'Properties of harmonic distortion of laser diodes with reflected waves'. *J. Optical Commun.*, **3**, 1982, pp. 129-132.

9

ASPECTS OF OPTICAL RECEIVERS FOR IMDD SYSTEMS

9.1 INTRODUCTION

Optical receivers are critical components in such systems. Their performance plays a dominant role in determining repeater spacings and in the flexibility of the system in terms of its tolerance to component ageing and permissible variations in section length.

At the receive terminal and at each electronic regenerator in a repeatered system the received signal must be converted into an electrical signal, amplified and processed to give an estimate of the transmitted signal. These three functions are somewhat similar to those in a digital radio relay system, but the ways in which they are performed differ. Conversion of variations of the envelope of the optical carrier to corresponding variations in the electrical signal is done in a semiconductor photodiode. Two basic classes of photodiode have been deployed in operational systems, the so-called pin diode and the avalanche photodiode (apd); system applications of both will be given. Both classes are available as small, lightweight, cheap and reliable devices with wideband linear responses and high quantum efficiency in either of the 1300 or 1550 nm bands using appropriate material systems. Their operating principles and details are given in Chapter 10.

Each new generation of system is built upon the experience gained with previous ones but new and unexpected phenomena have continued to emerge requiring repeater designers to maintain due prudence in their approach [1]. For instance, the change from coaxial to digital transmission using optical fibre was a major discontinuity in technology and techniques, and presented new sets of requirements for designers. Factors which have to be considered by designers of regenerative or optical amplifier repeaters can be grouped under the headings:

(1) transmission performance
(2) supervisory subsystems
(3) power supply
(4) mechanical construction
(5) reliability.

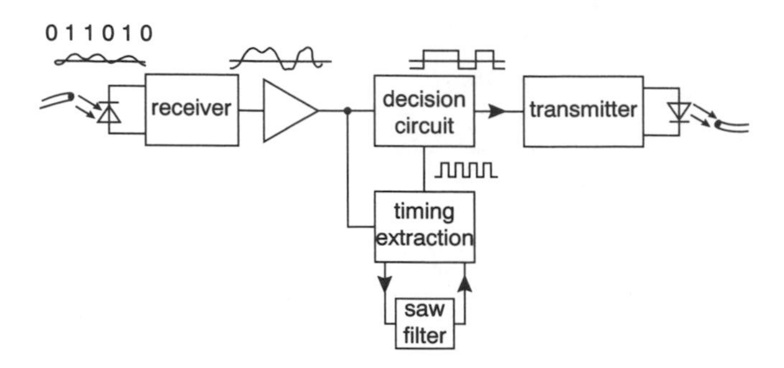

Figure 9.1 Simplified block schematic of regenerative repeater ([1], Figure 1). Reproduced by permission of BT

(1) included the topics of error ratio, design penalty and system margins, jitter and timing circuit design, alignment jitter, channel equalisation, spectral behaviour of lasers.

(2) included purpose and general requirements, characteristics, facilities (loop-back, transmitter health, error events, received light level or margin, remote switching), telemetry channels. Most of these will be considered in the three chapters dealing with receivers.

Transmission performance: a simplified block diagram of a dependent repeater which reshapes, regenerates and retimes the line signal (3R regenerator) such as used until the mid-1990s in regenerative repeatered systems is shown in Figure 9.1.

Repeater design involves compromise between conflicting requirements, and for regenerative repeaters there is a conflict between spacing and error ratio. The target is to maximise spacing while ensuring as far as possible that the end-of-life error ratio is acceptable. The two major items in the system power budget are the transmitter power launched into the fibre tail and the receiver sensitivity at the bit rate(s) of interest. The difference between these two quantities (in dB) is then reduced by an amount corresponding to tolerances and known penalties to give the net path loss capability. This is allocated to fibre loss (including splices), repair allowance, ageing allowance and the operating margin at end-of-life. These lead to unmeasurably low error ratios at the start of life due to the steepness of the curve of error ratio versus snr. For example, an individual submerged repeater with a 6 dB electrical (3 dB optical) margin against an error ratio of 10^{-12} at the end of life (corresponding to an overall value of 10^{-9} in a long-haul system with some 200 repeaters) would have a ratio of 10^{-33} on day 1 (a regenerator operating at 300 Mbit/s transmits about 10^{18} bits during a 25 year lifespan).

Factors which can cause the noise performance to be degraded during the system lifetime include changes in decision threshold, change of sampling epoch, ageing of laser transmitter resulting in reduced power or extinction ratio, and (of major concern in the next three chapters) changes in the photodiode detector.

9.2 SYSTEM SPECIFICATIONS FOR OPTICAL RECEIVERS

The performance and availability of optical detectors and receivers for 1300 nm systems had been extensively investigated during the 1980s in the UK and in other countries leading the world in the application of fibre optics to telecommunication networks. The amount of data gathered, and the interactions with component manufacturers contributed to providing a firm base from which components could be chosen with confidence as to their performance and reliability. The two PDH levels of most interest in the UK in the late 1980s were 34 and 140 Mbit/s, and the information in this section will therefore concentrate on devices for these two bit rates. Nevertheless, much of the material will be applicable to higher rate systems as well. The market for electronic and opto-electronic components of the quality levels demanded for terrestrial and undersea optical systems is international, and one of the difficulties facing designers in their choices is that of assessing the reliability programmes of potential suppliers whose manufacturing base lies across the oceans, perhaps on the other side of the world.

Before discussing specifications and characterisation, it is desirable to provide the reader with a circuit diagram of a real pinfet receiver and the associated bias circuitry, which was commercially available for 1300 nm system use from about 1982. The circuit, although simple, has a considerable number of elements in the circuit diagram shown in Figure 9.2a.

Stray and parasitic capacitances and inductances are not shown, but do influence circuit performance including the stability of the module under normal operating conditions.

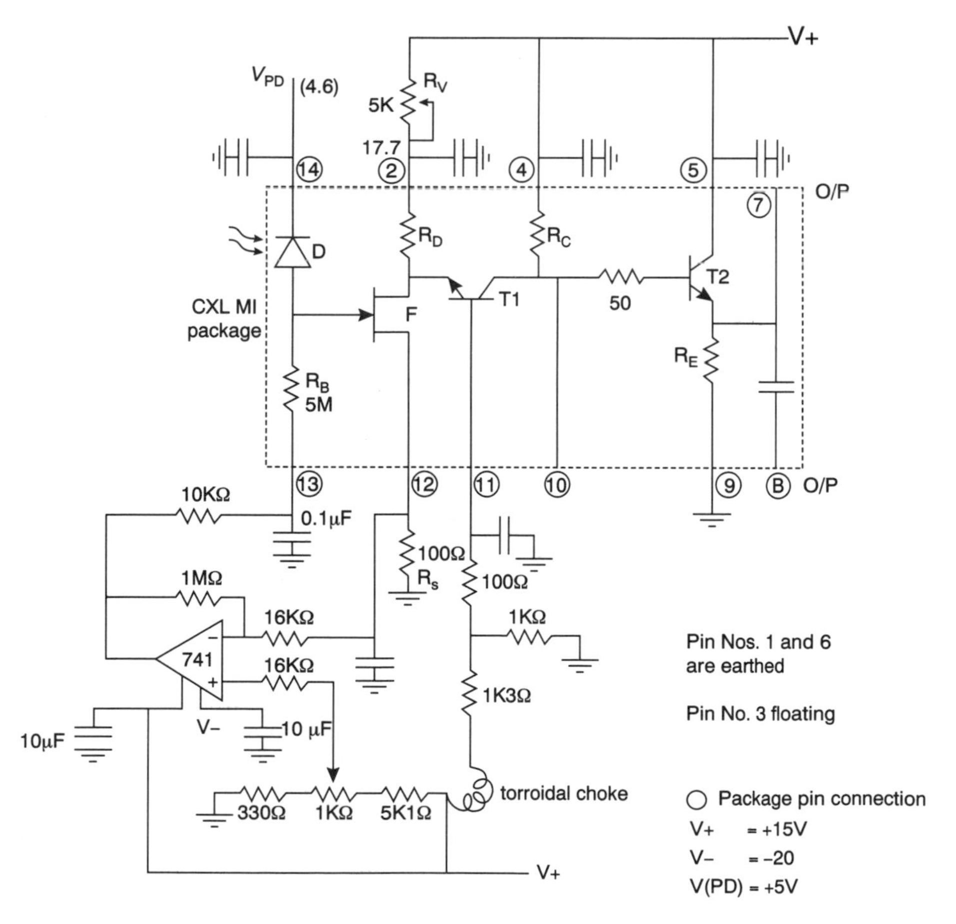

Figure 9.2a Layout of CXL-M1 pinfet module and associated bias circuitry (Plessey information) (all capacitors 10 nF unless otherwise stated)

9.2.1 Performance characterisation of pinfet optical receivers

A pinfet receiver front end is commonly taken to be a hybrid package consisting of a photodiode closely associated with a GaAs mesfet, biasing resistance and possibly a few other passive components, forming a receiver module. There are two popular circuit configurations from which to choose.

(a) The integrating or high impedance front end, in which the frequency roll-off caused by the parallel *RC* combination across the photodiode terminals is compensated by an external equaliser.

(b) The transimpedance amplifier or shunt feedback type in which the frequency response is relatively flat over the wanted band.

In order to characterise such modules there are a number of parameters which must be specified and which will now be considered.

Line signal

This should be specified, because some of the signal parameters are critical and may be adversely effected by certain module characteristics (e.g. unbounded line codes and dynamic range). Typically, the following have been used to define a line signal:

(1) binary or other waveform
(2) return-to-zero or non-return-to-zero, rz or nrz
(3) line code, e.g. 5B6B or 7B8B as in 140 Mbit/s terrestrial systems, or 24B1P as in submarine systems TAT-8 and PTAT-1
(4) line rate, MBd
(5) maximum allowed pulse duration (number of consecutive ones with nrz codes)
(6) predicted extinction ratio.

In practice, the received bit stream will not consist of rectangular pulses as implied in the above list due to various imperfections of the transmitter and system fibres, but at 140 Mbit/s any system impairments due to these causes were considered to be negligible. As remarked earlier, this is not true in Gbit/s systems.

Environmental considerations

Under this heading three physical variables have been taken into account: temperature, pressure and humidity. Take each in turn very briefly.

(1) Temperature: this is usually taken to be the ambient temperature.

(2) Pressure: typically, modules are expected to operate within specification with air pressures in the range from atmospheric to 1620 millibars. In addition, they may be subject to sudden changes of pressure within this range.

(3) Humidity: 20% maximum relative humidity has been assumed.

Reliability and ageing

These terms are usually applied to modules for systems with a mtbf of 4 years per 100 km. This represents a minimum mtbf for the module of 3×10^6 hours, assuming equal reliability in the optical transmitter and in the electronic circuitry. In general, some deterioration in performance occurs with time, as indicated in the section on reliability of pin photodiodes, and a suitable definition of end-of-life must be agreed.

Operating wavelength range

This should be specified to ensure that the sensitivity and other performance criteria are maintained over the full wavelength range expected to be emitted by the optical source throughout its working life, and under the full range of environmental conditions in the specification.

Sensitivity

One manufacturer's definition for the sensitivity of his product was the average light power required at the input to the fibre tail to achieve an error rate better than 1 in 10^9, when the module is incorporated in an ideal regenerator. This ideal regenerator produces a 100% roll-off pulse at the threshold point from an ideal (100% modulated, rectangular) nrz optical input, and adds no further impairments. The noise bandwidth of this ideal regenerator is therefore given by the function:

$$B_{\mathrm{N}} = \frac{fT[1+\cos(fT)]}{2\sin(fT)} \tag{9.1}$$

for $fT \leq 1$. This definition allows an ideal sensitivity to be quoted, and this can be obtained by various methods from actual measurements which require some practical adjustments. By this means a universal reference can be derived regardless of which method of measurement is employed.

Relative sensitivities of high- and trans-impedance pinfet modules

Detailed calculations based on Personick's theory gave the results listed in Table 9.1 for line rates of 160 and 320 MBd, corresponding to 7B8B coding of 140 and 280 Mbit/s, and they indicate realistic values for 1300 nm system calculations.

The differences in sensitivity are not large, 0.9 dB at the lower rate and only 0.7 dB at 320 MBd. These figures might change slightly when non-ideal rectangular drive pulses and intersymbol interference are included in the calculations. The conclusion to be drawn is that a penalty of, say, 0.7 dB is more than outweighed by the much higher overload point of the transimpedance design. In addition, the values for the integrating design sensitivity would need to be reduced slightly to provide marginally adequate overload capability. The values of the resistors can be compared with those used in other receivers, and those computed from the formula given earlier, 500 MΩ or 2 MΩ/(symbol rate) for R_{B} and R_{F}, respectively. The relative merits of integrating or transimpedance configurations concerning parasitic oscillations at GHz frequencies may also require consideration; see below.

Definitions of electrical characteristics of the module are as follows.

Transfer characteristics: two possible definitions were considered, namely output voltage/input photocurrent, or output voltage/input light power level. The second of these calls for a measurement of input light level in the right conditions. In both cases an ac coupled load of 50 Ω is used across the module output.

Table 9.1 Sensitivity comparison between types of pinfet receivers

High impedance (integrating)		
Line rate (MBd)	Sensitivity (dBm)	R_B (MΩ)
160	−41.5	1
320	−37.1	1
Transimpedance (shunt feedback connection at input)		
Line rate (MBd)	Sensitivity (dBm)	R_F (kΩ)
160	−39.6	12
320	−36.4	12

Notes: R_B bias resistance of photodiode. R_F feedback resistance

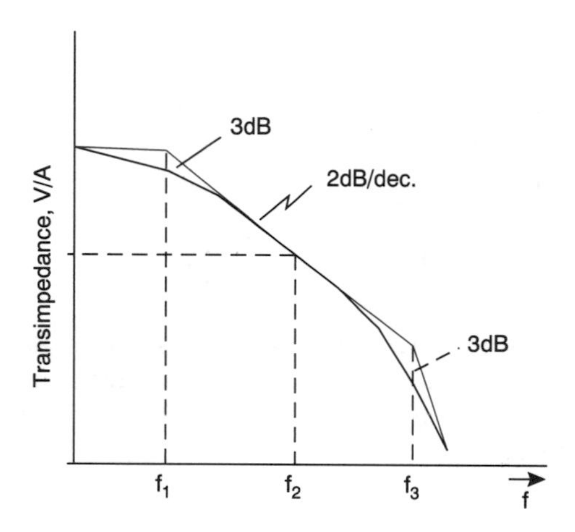

Figure 9.2b Schematic diagram showing the transimpedance (V/A) versus frequency characteristics of a pinfet module: f_1 is the low frequency cut-off, f_2 a reference frequency for midband gain, f_3 the high frequency cut-off)

Bandwidth: broadly speaking, the transfer function had two major cut-offs or break points as frequency increases, denoted by f_1 and f_3, and a midband frequency f_2 is often specified — as illustrated in the sketch of Figure 9.2*b*.

The value of f_1 must be higher than the highest supervisory frequency and f_3 higher than the signalling frequency, despite any variations in these cut-off frequencies as a result of production spreads, ageing, etc. For instance, variations in f_1 may not appear to matter, but may well do so with some optical supervisory schemes. The commonly accepted definition of lowpass cut-off is the frequency at which the response is 3 dB below the dc (or very low frequency) value.

Leakage current: the quantity of interest is the combined leakage current due to the pin diode and the fet gate leakage, both of which depend strongly on temperature.

Dynamic range: this is taken to be the operating range for correct performance of the module, i.e. no limiting, saturation, etc. It is dependent on the supply voltage and load impedance, both of which need to be stated in order to measure dynamic range against specification.

Stability: the module should operate with no instabilities in parameters or parasitic oscillations over wide variations in supply voltages, load impedance and environmental conditions within the specification limits. Module parameters should thus exhibit long-term stability with only changes due to environmental variations allowed. In particular, bias currents should not show any short-term fluctuations.

Overload characteristic and limits: these should be specified in order to ensure, as far as possible, the survival of the module under adverse conditions of use, and to ensure that there are no undesirable characteristics.

Module power supply limits: these allow a wide operating range to be obtained, and are set with the preceding two characteristics particularly in mind.

Signal-to-noise ratio: is defined as the ratio of peak-to-peak signal to r.m.s. noise at the output of the module.

Assessment and characterisation: the characteristics mentioned above should be assessed, but in addition the following values should be measured: bias resistor, module input capacitance, g_m of the mesfet; the first two of which affect both noise performance and the bandwidth characteristic, as shown in the theoretical analysis given earlier.

9.2.2 Directly measurable quantities of pinfet modules

Of the main parameters and characteristics only the following can be measured directly:

(1) gain versus modulating frequency
(2) sensitivity
(3) total leakage current
(4) transconductance, g_m.

Let us now consider (1) and (2) in a little more detail.

Gain-frequency characteristic: values at three spot frequencies should be noted, namely:

f_1 = low frequency cut-off
f_2 = midband gain which provides a reference level, and which should therefore have limits set
f_3 = the 3 dB down frequency as defined.

Sensitivity: there are a number of possible methods for measuring this parameter, of which the following three are commonly used.

(a) Measure the module noise output as a function of frequency, hence calculate the sensitivity performance of the module with an ideal 100% roll-off cosine equaliser.

(b) Use a physical reference filter and make the necessary allowance for the difference between this and the ideal equaliser above. There are two possible ways to proceed from this point:
   (i) measure the snr at the output of the filter; this may be easier said than done, since the peak-to-peak level may prove difficult to measure
   (ii) regenerate the detected output signal and measure the optical power needed to give a prescribed error rate.

It is important to note that there is an inherent difference between the error rate approach and that based on measuring snr, because the former includes any factors contributing to eye closure (e.g., module bandwidth, etc.), whereas the latter will not necessarily do so. If the snr approach is adopted, the bandwidth should also be measured or specified separately. The three methods above should give close agreement if the measurements are carried out with care.

Inferred quantities

Once the module has been sealed, values for the following items can only be inferred: bias resistance and input capacitance.

9.2.3 Design information on photodiodes and optical receivers

Preparatory to the installation of standard optical systems at 1300 nm in the UK in the early 1980s, one of the topics investigated (by means of a questionnaire to potential suppliers and follow-up meetings) was information on optical detectors and receivers judged to be necessary or at least desirable for engineering design purposes. The items listed give some indication of the amount of data needed to design with confidence. Data on devices from two different manufacturers are provided to give readers a feel for the values. Device A is a transimpedance design and B is a high-impedance type. The questionnaire is likely to ask more questions than ordinary data sheets from suppliers will answer, which of course means that a designer must seek out any further information he considers necessary, as was done during the original enquiries.

The questions in Section 9.2.3 are included to enable designers to assess how the overall performance of a module is achieved, and to estimate how sensitive the performance might be to changes of parameters with ageing. The question on the possible use of a cooler is worth noting, since on the basis of information supplied it appears that a 2 dB advantage at 34 Mbit/s can be obtained by cooling the module from 55°C to 20°C and 1.5 dB at 140 Mbit/s. The interpretation of the sensitivity figures needs to be treated with care, and is best discussed with the manufacturer, since the value used in system power budget calculations must be realistic.

		Unit	A	B
1	*Absolute maximum ratings*			
1.1	Storage temperature (min./max.),	°C	−40/+80	−40/+80
1.2	Case temp. in use (min./max.),	°C	−40/+70	−20/+80
1.3	Max. optical powers (peak, cw burnout, sat.),	mW	100, 10, —	
1.4	Max voltage rating,	V	15, −5	−20, 5.5
2	*Electrical/optical characteristics under stated conditions, at 25°C, min./typ./max. where appropriate*			
2.1	Responsivity versus wavelength	A/W	0.7 at 1.3	0.75 at 1.3 0.9 at 1.55
2.2	Sensitivities at bit rates of interest	dBm/rate	−52/8	−53/8(a)

	(a) min. values		−49/34	−50/34
			−43/140	−43/140
			—	−36/565
2.3	Max. temp. sensitivity	dB	n.a.	8/−2.5/−4.0
	derating, rate/55°/70°		n.a.	34/−2.0/−3.5
			n.a.	140/−1.5/−3.0
			n.a.	565/−0.7/−1.4
2.4	Detector cap. at volt.	pF/V	0.5/−5	—
	Device dimensions (see data sheets for device of interest)			
	Detector leakage current at temp., bias volt	nA	< 20/20°/−5	40(typ.)
	Quantum efficiency versus λ (μm)		>0.5 from 1 to 1.6	—
	PIN load resistance at bit rates		1M/8	—
			0.5M/34	—
			80k/140	—
	FET gate-source cap.	pF	≃ 0.5	—
	FET transconductance	mA/V	25/40	—
	FET leakage current at	nA	10/8	—
	bit rates at 20°C,		200/140	—
	APD gain versus voltage		n.a.	n.a.
	K_{eff} at given M		n.a.	n.a.
	case capacitance,	pF	0.1	n.a
3	*Module parameters*			
3.1	Total cap across pin at leakage current, rate,	pF/nA/Mbit/s	< 1/20/8	—
3.2	Total leakage current at bit rate,	nA	< 50/8	40
3.3	Responsivity at	A/W	0.5/8	0.75
	bit rates		0.7/34	—
			0.8/140	—
3.4	Frequency response at	MHz	8/1M	90/8/140
	load resistance or bit rates,		100/50k	700/565
3.5	Dynamic range,	dB	25/30 min	23
3.6	A.G.C. facilities		none	none
3.7	Output load impedance,	Ω	1k min	—
3.8	Output impedance,	Ω	50	50
3.9	Transimpedance,	V/mA	—	n.a.
4	*Life and reliability*			
4.1	Life prediction, burn-in, conditions		n.av.	n.av.
4.2	End-of-life definition		n.av.	n.av.
4.3	Life tests in hand		n.av.	n.av.

5 *Mechanical and other details*			
5.1 Package type		14 pin d.i.l.	14 pin d.i.l.
5.2 Provision for external cooler		yes	n.a.
5.3 Pull-out strength of fibre tail		1 kg	5N
5.4 Type of fibre tail		any std.	900 μm
5.5 SM/MM tails available		yes/yes 50/125	yes/yes 50/125
5.6 Min. bend radius of fibre,	mm	n.a.	30

Notes:

Type A, values based on 1982 information.
Type B comprises a GaInAs/InP pin diode and GaAs fet front end and the entries are based solely on the manufacturer's data published in 1986. Performance data at 25°C unless stated otherwise.
n.a. = not answered.
n.av. = not available at time of survey.

A number of other topics, such as handling precautions, thermal and mechanical, optical safety and a few more, must be considered in practice, but will be omitted here for brevity.

9.2.4 Choice of optical receiver configuration

There have been a great many papers published on configurations for optical receiver amplifiers since the mid-1970s. Some of these are listed in the supplementary references to this chapter. Here we will look at the main features of receiver types which have entered service in operational systems, since these have been chosen as most suitable for the applications on criteria such as reliability, cost-effectiveness and others which seldom feature in papers on design.

A very demanding field of applications is underseas systems, and a very useful source of information on optical receivers for first generation transatlantic optical systems is the proceedings of the conference known as *Suboptic 86* (see Chapter 15). The design of the receiver adopted for the UK portion of TAT-8 is treated therein and also in [2]. Sensitivity is of paramount importance, since the better this is, the longer the section lengths can be, and the fewer the regenerators required.

The amplifier ic was developed as a custom chip to give valuable freedom of choice of design and component values. A number of designs were considered and the choice finally fell on a development of the common-emitter cascode design described in [3]. The performance of this 1984 circuit is described in Chapter 15.

Some of the main factors which influence the choice between two or more competitive modules are listed below:

(1) high guaranteed minimum sensitivity
(2) rate of change of sensitivity with temperature
(3) large-signal handling capacity

(4) availability of a second source
(5) high coupling efficiency between fibre and detector
(6) choice of fibre tail
(7) low reflection
(8) hermetic package
(9) high reliability.

Items (1) and (2) must be taken together to ensure acceptable performance over the full temperature range. Large signal handling capacity, or power to overload, is a more meaningful parameter than dynamic range. It is a function of the line code used, especially for high impedance designs, and it is important because it impacts on the number of attenuator types required. Looking at item (5), a high coupling efficiency maximises sensitivity, and is also needed to minimise modal noise when a multimode fibre tail is fitted. If fusion splicing is required, then the designer must make sure that the tail fibre supplied is consistent with the system fibre to which it will be spliced. A tail specified by the customer might be required, and the willingness of the manufacturer to fit such a tail should be established. There are no questions on reflections, but as shown in earlier chapters lasers are very sensitive to reflections, so item (7) is particularly important for short single-mode systems where the effects are greatest.

Module reliability depends on three major factors:

(a) the reliability of individual components
(b) the reliability of the assembled module, i.e. taking account of the bonding methods and mounting techniques used
(c) the reliability of the package and its impact on the reliability of the contents.

Surface contamination of a pin photodiode (or an apd) can increase the surface leakage current and consequently seriously degrade the noise performance, particularly at the lower bit rates. It is therefore very important to ensure that the package does not contain any contaminants; it is for this reason that no epoxy resins or other volatile organic materials should be used within the package. Another key factor which could affect reliabilty is the pull-out strength of the fibre tail and the sealing method employed. For instance, if a packaged module is fully qualified with a standard tail fitted, it will need to be re-qualified if, instead, a tail specified by the customer is fitted. Potentially the most vulnerable part of the fibre tail is where it emerges from the metal ferrule of the package; to reduce the risk it is considered essential that a tapered compliant ferrule is fitted to restrict the bend radius to safe values as recommended by the manufacturer, e.g. at least 30 mm as given in the table above.

Before a choice is made between a number of possible modules the whole area of reliability must be thoroughly investigated for each candidate, and this may well involve detailed discussions with manufacturers, as well as independent life and reliability tests by equipment manufacturers.

9.2.5 Dynamic range considerations

When the optical power level into a receiver is increased there will come a point at which peak clipping, or overload, occurs and the performance degrades. Similarly, if the signal is reduced sufficiently an input level is reached at which the performance is again degraded. Broadly speaking, the dynamic range of a receiver can be considered as the difference between these two input levels for a specified error rate. An adequate dynamic range is required to allow for system margins, and for variations in length of repeater spans.

In designs such as the integrating or transimpedance front ends where the shunt admittance between the terminals of the diode and the amplifier input can be represented by a parallel R and C, each comprising several physical elements, the voltage swing across R due to the charging and discharging of C dominates. In addition to this peak-to-peak swing, there will be a mean voltage dependent on the signal level because optical signals are unipolar. Careful attention must be paid to this aspect during circuit design to ensure that the mean signal level is not the limiting factor, especially when low disparity codes are used. This topic is treated in detail by Brooks and Jessop [4].

The dynamic range of an optical receiver can be defined in several different ways, one of which is given below:

$$D = 10 \log \frac{P_{\mathrm{MAX}}}{P_{\mathrm{MIN}}} \tag{9.2}$$

where P_{MAX} is the maximum allowable optical input at a given ber, and P_{MIN} is the receiver sensitivity at the same ber. This expression can be rewritten in terms of electrical quantities as:

$$D = 10 \log \frac{|V_{\mathrm{F}}|}{R_{\mathrm{B}} I_{\gamma,\mathrm{MIN}}} \tag{9.3}$$

where V_{F} is the negative voltage available to operate a feedback loop which maintains the correct bias for the fet; R_{B} is the bias resistance; and $I_{\gamma,\mathrm{MIN}}$ is the photocurrent corresponding to P_{MIN}.

The sensitivity P_{MIN} of the module can be defined as the minimum incident power needed to achieve the specified error rate when the module is incorporated in an ideal regenerator. This gives a raised cosine pulse with 100% roll-off from an ideal rectangular on–off pulse at the input. The noise bandwidth is given by:

$$\frac{0.5(1 + \cos f T)}{\operatorname{sinc} f T} \tag{9.4}$$

where $1/T$ is the signalling speed on the line, and $\operatorname{sinc} f T$ is $(\sin f T)/f T$ as usual. This expression is the ratio of the spectrum of a raised cosine pulse to that of a rectangular pulse. This definition in terms of an ideal regenerator enables an ideal sensitivity to be derived from measurements which entail practical compromises, and thus provides a universal reference independent of the actual measurement technique used.

The power supply limits of the receiver module will include values for photodiode bias, e.g. 5 V, gate bias for the fet, e.g. 0 to −2 V, together with bias current and voltage for any following stages of amplification. The manufacturer may also include a setting-up procedure prior to operation of the receiver.

Dynamic range is an important constraint in practice, especially for integrating receivers, and one technique for extending the range was described in 1982. The problem is caused by the current which flows in the bias or feedback resistor at maximum permitted input power, and the consequent voltage drop which may saturate the amplifier. The method is simple, needs no additional components connected to the input node, and it incurs no sensitivity penalty. A 12.6 Mbit/s receiver to which the technique was applied had a sensitivity of −52.6 dBm, and a dynamic range increased from 26.6 dB to greater than 44.8 dB (89.6 dB electrical) [5]. A current shunt to limit the current into the amplifier to a level below saturation is provided by forward biasing the gate-source Schottky diode of the first stage fet, so that it comes into operation at the correct level.

9.3 RECEIVERS FOR INTENSITY-MODULATED DIRECT-DETECTION SYSTEMS

Until the beginning of the 1990s these were the only types of systems in service-carrying traffic. Several million km of fibres had been installed worldwide, so that even with repeater spans of 50 km there were tens of thousands of optical receivers in service. Many of these had been designed on the basis of S.D. Personick's work published between 1977 and 1983, or on simplifications of his work. There have been many advances in technology since 1977, and also in simulation packages for the types of circuits commonly used. Nonetheless, his approach and SPICE were still in use in the early 1990s.

To illustrate the trend in receiver front-end technology and corresponding sensitivities over the period 1988–93 we present some results for submerged plesiochronous systems in Table 9.2 [6].

Because the operating waveband changes, the standard 3 dB reduction in receiver sensitivity for a doubling of bit rate is not seen in the table.

Sensitivities for other bit rates and technologies are quoted in appropriate parts of the chapters on receivers and systems.

In order to impart a realistic flavour to this major section of the chapter, we begin with an account of the optical receiver used in the first privately owned transatlantic system (PTAT-1, 1989), as submerged systems make the most stringent demands on receiver design and technology.

Table 9.2 Trend in receiver front-end technology and sensitivities at ber of 10^{-9} (STC undersea systems, 1988–93)

Bit rate (Mbit/s)	Technology	Sensitivity (dBm)
280	pinfet	−33
280	pinbip	−34
420	pinbip	−34
560	pinfet-bip	−39
2500	apdfet-bip	−32

Key:

pinfet — pin photodiode followed by GaAs field effect transistor

pinbip — pin photodiode followed by silicon bipolar transistor

pinfet-bip — pin photodiode followed by GaAs field effect transistor then bipolar transistor

apdfet-bip — avalanche photodiode followed by GaAs field effect transistor then bipolar transistor

Notes: receivers in first three rows operated in 1300 nm band. Receivers in last two rows operated in 1550 nm band

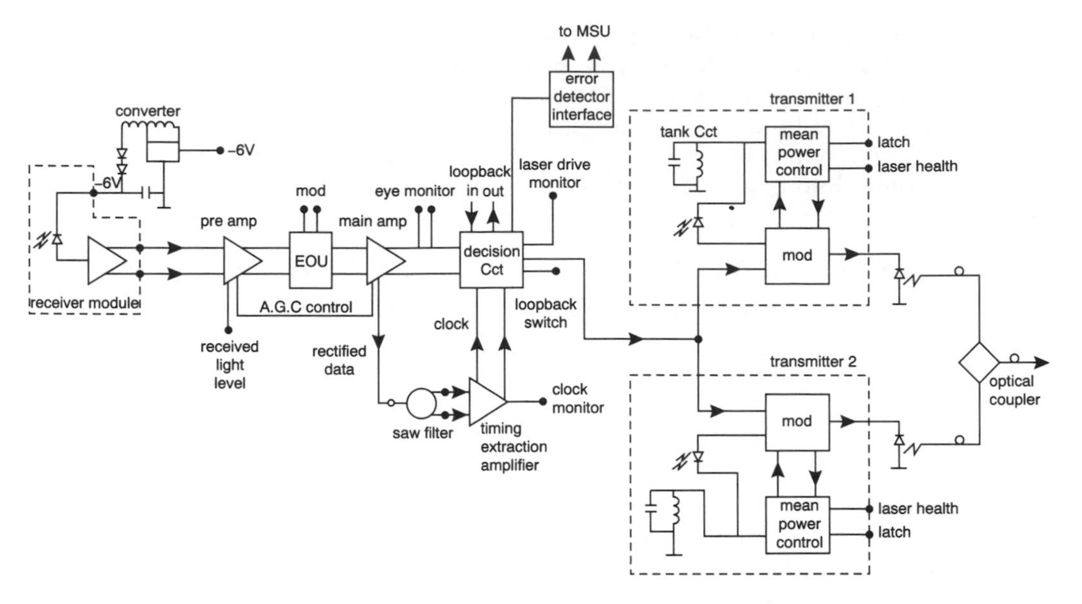

Figure 9.3 Diagram of TAT-8 regenerator showing parts of the main transmission path ([7], Figure 2). Reproduced by permission of Société des Electriciens et des Electroniciens

Optical receivers form only the front end of typical regenerators such as that illustrated in Figure 9.3 [7]. Likewise for receiver terminal equipment, as can be seen from Figure 9.4. [8].

Optical receiver sensitivity is important, since it is a key parameter in determining repeater spacing and hence the number of repeaters required. The receiver consisted of a photodiode connected to a low-noise bipolar silicon integrated circuit configured as a transimpedance amplifier. Both the detector and the chip were mounted in a single package virtually identical to that for the laser. The signal was coupled from the system lightguide on to the active area of the photodiode through a single-mode tail which entered the package through a hermetic metallised gland, as shown in Figure 9.5.

The specially prepared fibre tail was held in position close to the face of the photodiode with a laser-welded fixing. By adopting the same package the many aspects of package reliability and stability in operation already proven for the laser also apply to the receiver. The principal requirements on the photodiode were low capacitance, low leakage current and high responsivity, all of which influenced receiver sensitivity to some extent. Changes in these parameters with temperature and ageing must also be very small. A number of types were appraised for the system and a planar InGaAs pin photodiode manufactured by Fujitsu was selected. This top entry device had an active area of 50 μm diameter; capacitance, leakage current and responsivity are typically 0.35 pF, below 0.5 nA and 0.9 A/W, respectively. This component has been subjected to extensive life-testing at Fujitsu and at BTRL.

9.3.1 Design and performance of PTAT-1 optical receiver

This followed on the successful development and manufacture of the receiver for the first optical transatlantic TAT-8. BTRL was commissioned by STC Submarine

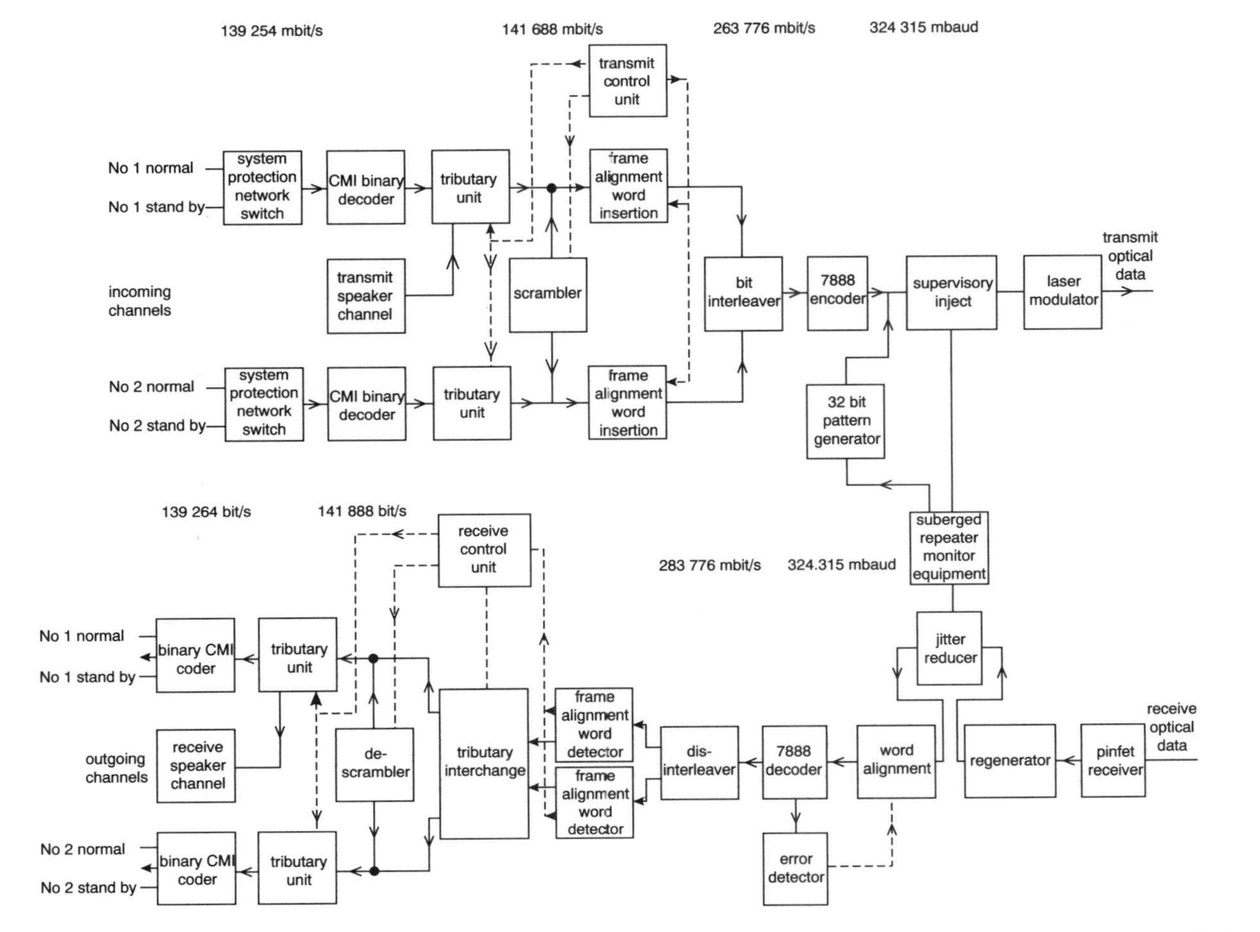

Figure 9.4 Block diagram of terminal transmission equipment ([8], Figure 11). Reproduced by permission of BT

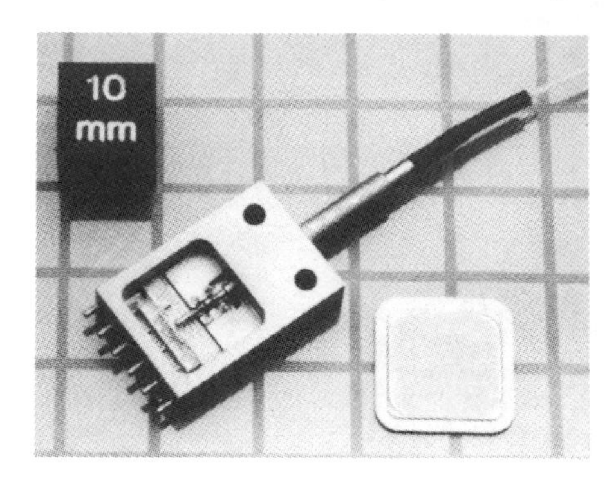

Figure 9.5 PTAT-1 receiver with lid removed ([9], Figure 2). Reproduced by permission of BT

Systems to develop and manufacture the optical receivers for the first privately owned transatlantic system PTAT-1. This system increased capacity from 280 Mbit/s per fibre pair to 420 Mbit/s, had a length of 7000 km, some 140 repeaters each containing 8 digital regenerators, each of which had one optical receiver. The use of the 24B1P code gave a line rate of 443 MBd. The design, performance and testing of these was described in [9].

Design: much of this was derived from the earlier TAT-8 receiver to take maximum advantage of established reliability and qualification pedigrees, e.g. both types contained only two active components: a pin photodiode and an ic low-noise amplifier. The former was a 50 μm diameter Fujitsu device of proven high reliability, low capacitance and high quantum efficiency. The ic amplifier was manufactured in the highly reliable ECL-40 silicon process and incorporated a number of innovations and improvements over the design for TAT-8.

Owing to the larger bandwidth, the circuit details and component values of the PTAT-1 receiver differed from its predecessor. Since sensitivity and the other major performance parameters were largely set by the properties of the amplifier, it was developed as a dedicated custom chip. A transimpedance configuration was chosen to give low noise, large bandwidth and a wide dynamic range as required by the 24B1P line code adopted for the system. The amplifier circuit diagram is shown in Figure 9.6.

The diode-connected transistors forming a bias chain for the input regulates the first stage current to compensate for temperature and supply voltage variations. The active collector load for the input stage also broadens the bandwidth. Another improvement was the removal of the emitter followers at each of the two output ports, which achieved a substantial reduction in the direct current needed by the ic.

Some of the features will now be mentioned. First, the characteristic shunt–shunt feedback connection is clearly shown. The input stage is a cascode which uses a common-base connection to minimise the Miller effect, a necessity because the sensitivity and bandwidth both degrade as capacitance C_T between the input node and ground increases. The amplifier bandwidth is mainly determined by the time constant $R_F C_T$, where R_F is the value of the feedback resistor (8 kΩ). The diode chain forming a bias network at the input regulates the current in the first stage to compensate for variations in temperature (operating range 0–40°C) and supply voltage (6.0 ± 0.2 V). The active collector load of the input stage served to broaden bandwidth, permitting a higher value of R_F than without this load.

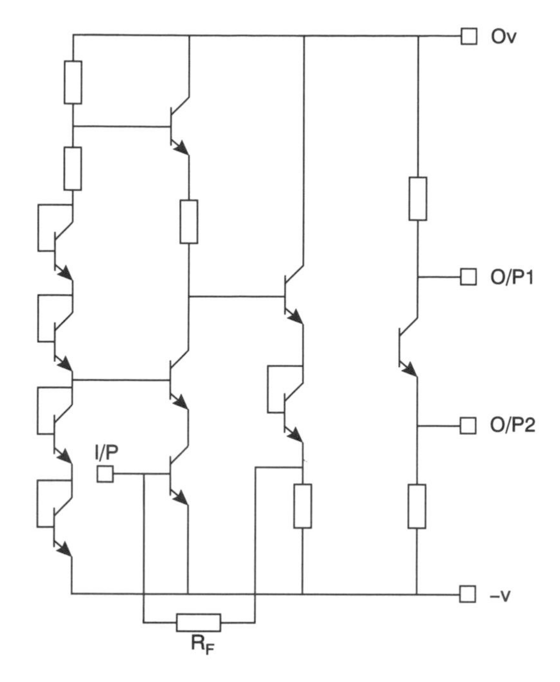

Figure 9.6 Circuit schematic of PTAT-1 receiver ([9], Figure 4). Reproduced by permission of BT

A modified SPICE program was used to optimise component values and circuit topology before fabrication, and to ensure acceptable performance despite normal production tolerances. During this CAD stage, care was taken to account for track and bonding pad strays of the ic, parasitic effects of the photodiode, receiver package and the regenerator, all of which influence circuit behaviour. The final design incorporated bond wire and metallisation options to give some fine-tuning of performance after fabrication.

Performance: the frequency response of a typical receiver is illustrated in Figure 9.7, from which it can be seen that the 3 dB bandwidth is 340 MHz, compared with 200 MHz for TAT-8. This, combined with that of the regenerator, was designed to produce an overall 100% raised cosine spectral response at 443 Mbaud.

Turning to noise performance, this was dominated by contributions due to R_F and the input transistor. The value of R_F was set by bandwidth requirements, but the transistor noise was optimised by suitable values of emitter length, bias current and small-signal current gain using the Personick equations.

Noise consideration: as in most receivers of this type, the dominant sources were thermal noise of the feedback resistor and shot noise from the first transistor. The equivalent shunt current generator representing the former is inversely proportional to R_F, the value of which was fixed by the bandwidth required. The transistor noise could, however, be optimised by suitable choices of small-signal current gain (h_{FE}), emitter length and collector current, based on Personick's noise equations. Some deviation from these optimum values were required in order to meet other performance criteria; for example, a higher collector current was needed to improve bandwidth and ageing characteristics. A comparison between the noise spectral densities of typical production TAT-8 and PTAT-1 receivers can be made from Figure 9.8.

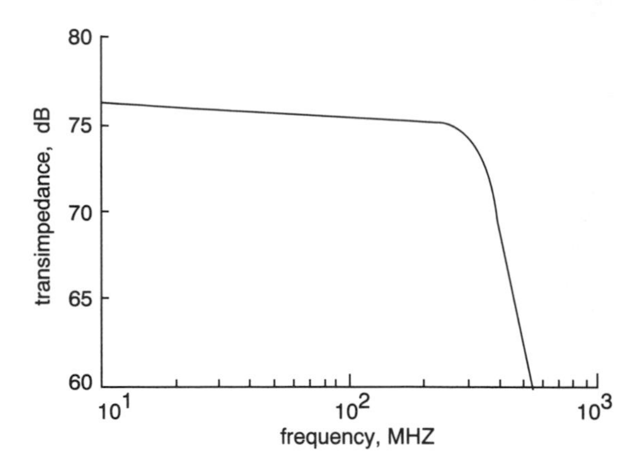

Figure 9.7 Transfer impedance (V/A, dB) versus frequency of PTAT-1 receiver ([9], Figure 5). Reproduced by permission of BT

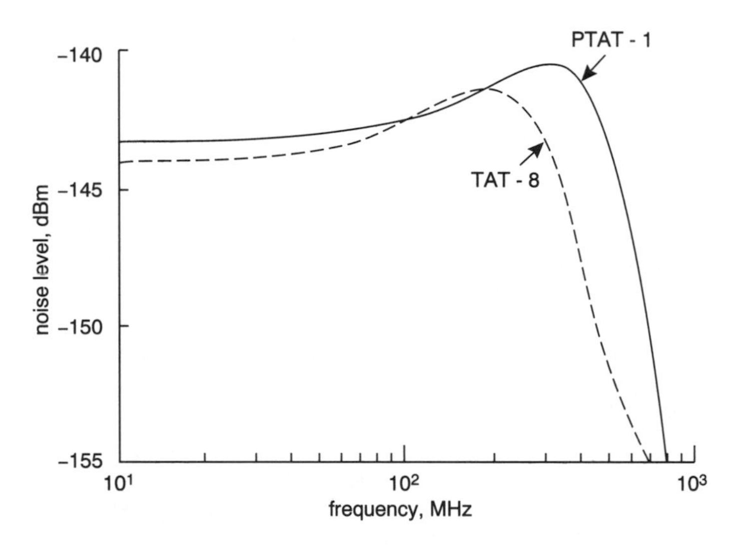

Figure 9.8 Noise spectral density versus frequency, PTAT-1 and TAT-8 receivers ([9], Figure 6). Reproduced by permission of BT

The difference between the two types was almost entirely due to the smaller value of R_F in the PTAT-1 receiver.

The sensitivity and overload performances of a typical receiver at 25°C and 6.0 V supply are −33.6 dBm (ber = 10^{-9}) and −13.1 dBm, respectively, both of which are well within the target values of −31 and −14.5 dBm. ber at other levels are shown in Figure 9.9.

The variations of the sensitivity versus supply voltage at three different temperatures, presented in Figure 9.10, shows a maximum reduction of only 0.1 dB. The equalised receiver eye diagram at this ber was shown in [9].

Life tests and qualification studies have determined that PTAT-1 receivers have a reliability compatible with the 25 year service life of the system.

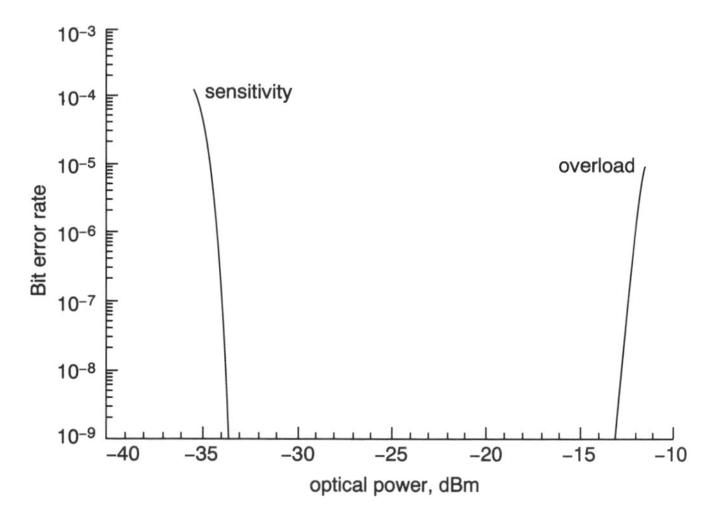

Figure 9.9 Profiles of sensitivity and overload versus frequency for a typical PTAT-1 receiver ([9], Figure 9). Reproduced by permission of BT

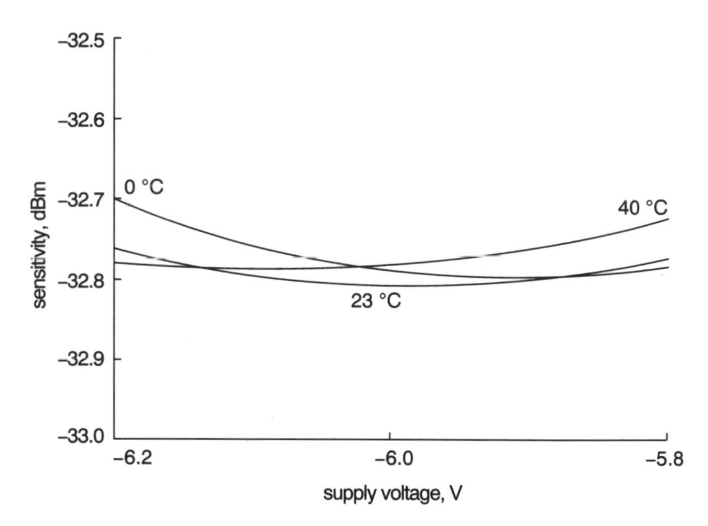

Figure 9.10 Variation in sensitivity versus supply voltage at three values of temperature ([9], Figure 11). Reproduced by permission of BT

9.3.2 Receivers for use in TPON termination equipment

Penetration of optical fibres into the local loop will demand the development of low-cost high performance components and subsystems. The topology, performance and application of a receiver designed for use within the network termination equipment (nte) of British Telecom's telephony over passive optical fibres (TPON) system was described in [10]. The potential success of the TPON system will be determined to a significant extent by the overall cost of the equipment and maintenance, hence the great emphasis on low-cost solutions. Adoption of a highly integrated single package would allow for low piece-part cost, easier assembly and reduced testing overheads. Additional desirable features include low power consumption to enable battery backup, an indication of received optical power to allow network management systems to detect and locate faults, and the adoption of a CMOS level output to simplify interfacing between receiver and associated nte circuits. A highly functional application-specific receiver embobying GaAs ic technology was described. The receiver gave excellent performance and was successfully used as part of a TPON demonstrator. Further evolution of the design will result in a single chip realisation possessing an extended optical dynamic range and low-cost packaging.

Performance targets: the TPON system protocol can support up to 128 optical network terminations. If the corresponding passive optical split is implemented as is required for fibre-to-the-home applications, receiver sensitivities around −49 dBm are required; consequently it was decided to implement the design in GaAs rather than a silicon ic because a semi-insulating GaAs substrate allowed the use of a higher value of on-chip feedback resistor, hence higher sensitivity.

Receiver configuration, technology and performance: the receiver schematic is shown in Figure 9.11 and consists of two chips, hybrid integrated with an InGaAs/InP pin photodiode.

As can be seen, a transimpedance front end was employed, with a 300 Ω on-chip feedback resistor, realised using a 625 Ω/sq. implant. This demanded close attention to both positioning and isolation to ensure that backgating was minimised. Two gain controllable stages each of nominal gain 14.5 dB further amplify the detected signal. The simple *RC* combination shown provided single-ended to differential conversion. The low-frequency response of the receiver was set by the value of the off-chip capacitor.

Noise performance: analysis of the front-end predicted a total equivalent input noise current of 17 $pA/(Hz)^{1/2}$ due mainly to the feedback resistor (81%), flicker (13%) and input fet channel noise (5%). The corresponding calculated sensitivity of −52.0 dBm was in excellent agreement with the measured value of −51.5 dBm at the output of the receiver chip. This value was degraded by non-ideal filtering in the interface chip to −50.0 dBm overall, a penalty of only 1.5 dB.

Interface chip: two more gain controllable stages were followed by two stages of limiting amplification and finally the CMOS output stage. Combined gain of the package containing both chips was linear and greater than 94 dB, and was achieved only through the use of differential circuit structures together with bias point and offset stabilisation techniques to improve tolerances to temperature, ic process and power supply variations. In addition to the main signal path, the interface chip provided a peak detect function which could be used to drive low-signal alarm circuitry. The peak detector outputs represent the peak and trough of the input

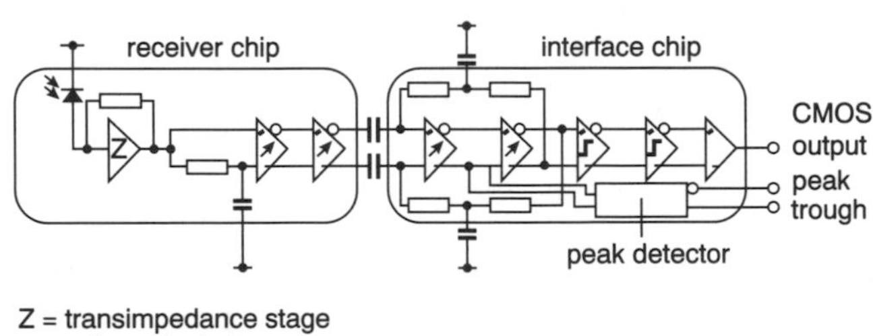

Figure 9.11 Schematic of TPON receiver showing transimpedance front end, four gain control stages and two limiting amplifier stages ([10], Figure 1). Reproduced by permission of IEE

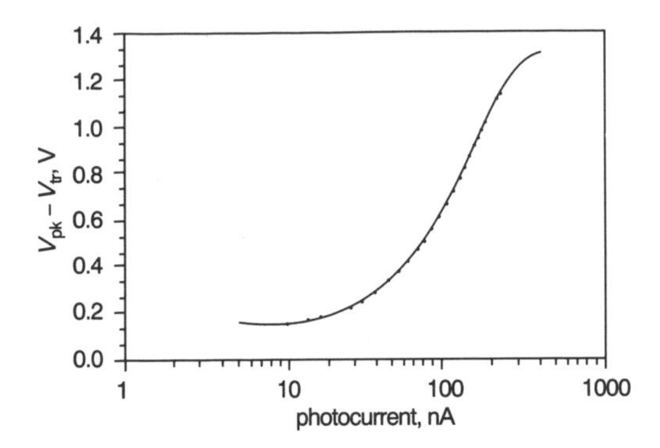

Figure 9.12 Measured voltage versus photocurrent of peak detector ([10], Figure 3). Reproduced by permission of IEE

signal, which when subtracted externally give the peak-to-peak value of the signal. The characteristics of the detector are illustrated in Figure 9.12. Measured power consumption for the two-chip package was below 190 mW.

Packaging: the ics and associated decoupling components were mounted on a thick-film alumina substrate and the complete assembly packaged in a 15-pin carrier, with the photodiode subassembly mounted on a quartz block and bonded to the GaAs ics. For commercial use the quartz block would be replaced by a cheaper and more rugged mount. The Letter [10] went on to describe the complete customer network termination equipment pcb, which comprised five subsystems:

(1) the circuit engine responsibile for assembling and disassembling transmitted/received data packets and for driving up to two customer lines, or one basic rate isdn channel per pcb

(2) the timer chip which controls the timing of transmitted data to the exchange so that the correct interleave with data transmitted from the other customer is achieved

(3) line circuits

(4) transmitter

(5) receiver.

The receiver was mounted adjacent to the transmitter module, although the former must operate at −50 dBm (a photocurrent of 10 nA), whereas the transmitter, mounted less than 10 mm away, was driven with current pulses of 100 mA having rise times of a few ns. Considerable care was taken to ensure that the layout minimised crosstalk between the various subsystems on the pcb. Measures taken to reduce crosstalk included the use of buried power and ground planes, substantial decoupling and inductive filtering of power supply lines. Total power consumption for the customer card was 1.5 W.

9.3.3 Aspects of receiver design and performance for 10 Gbit/s system applications

The development of 10 Gbit/s systems will be well advanced by the time that the 5 Gbit/s TAT-12 system enters service in 1995. In the early 1990s it was not clear which ic technology would be optimum at 10 Gbit/s and above. Prior to 1990, silicon ics were preferred, but after that GaAs technology became an attractive competitor. The main technology options for hybrid wideband optical receivers available in 1993 were discussed in [11]. State-of-the-art high-speed Si and GaAs hbt processes were used to implement the ics, and the results of circuit simulation and preliminary measurements were presented.

IC technology options: transistors with an f_T in excess of 30 GHz and digital ics operating at speeds in excess of 10 Gbit/s were deemed necessary to provide sufficient performance margins. This eliminated the Si mosfet process and greatly reduced the choice of Si process, so that a Si process with adequate performance was unlikely to be commercially available for some years. The majority of commercially available GaAs mmic processes were suitable for bit rates of 2.5 Gbit/s and under, whereas GaAs hbt devices had demonstrated their promise in 10 Gbit/s applications.

Simulated performance of front ends: a simple circuit consisting of an InGaAs pin detector, 5 (Si or GaAs) transistors and a few passive components was modelled by assuming the parameter values listed in Table 9.3.

The corresponding predicted performance of the two front-ends are given in Table 9.4, and they indicate that the sensitivities are equal as one would expect from the nearly equal values of equivalent noise current at input; both dynamic ranges exceed 20 dB. However, the transfer impedance (V/A) and group delay values differ significantly.

Measured results: although examples of both circuits were constructed, measurements were only possible on the GaAs version in the time available.

Table 9.3 Key process and circuit model parameters ([11], Table 1). Reproduced by permission of IEE

	HP25 Si BJT	GaAs HBT
Emitter size	0.46 μm	2 μm
Alignment	Self-aligned	Self-aligned
Interconnection	3-level Al	2-level Au
Complexity	16 mask layers	12 mask layers
TF	4.5 ps	3 ps
BF	125	20–60
	(0.46×1.66 μm^2)	(1.8×1.8 μm^2)
C_{JE}	3.3 fF	6 fF
C_{JC}	5.0 fF	15 fF
RB	393 Ω	100 Ω

Table 9.4 Predicted values for circuit performance ([11], Table 2). Reproduced by permission of IEE

	HP25 Si BJT	GaAs HBT
Bandwidth (GHz)	8.8	10
Transimpedance (Ω)	210	310
Group delay (ps)	15	45
Average noise current (pA/$\sqrt{\text{Hz}}$)	15	16
Optical sensitivity (dBm)	−20.4	−20.2
Dynamic range (dB)	>20	>20

These covered the band 130 MHz to 20 GHz and indicated 3 dB bandwidths in the range 9.5–10.3 GHz and midband transfer impedances in the range 310–340 V/A. Group delay variations less than 30 ps were also measured. All of these are in agreement with the predicted values. A baseline sensitivity of −22 dBm and noise current spectral density of 10 pA(Hz)$^{1/2}$ were calculated from the measured quantities. The improvement over the predicted values was attributed to the transistors being better than modelled.

12 GHz pin-hemt tuned, integrating front end

In general, integrating designs offer slightly better noise performance than transimpedance, because the latter show a tendency to self-oscillation unless the feedback resistor value is sufficiently low. The use of inductor coupling between the photodiode and its bias resistor results in a four-terminal rather than a two-terminal interstage between photodiode and input stage, thus increasing the bandwidth for given values of bias resistor and shunt capacitance. Combined with a simple *CR* differentiator as used to equalise the response of integrating designs, or with an active equaliser which can itself widen the bandwidth by a factor of 5 or 6, the integrator bandwidth can be extended very significantly.

For example, [12] considered a 10 GHz circuit. A two-terminal (untuned) design with an active equaliser extending bandwidth fivefold requires the input *CR* product to correspond to $10/5 = 2$ GHz, and assuming that $C = 0.4$ pF, R must be 200 Ω. Inductor coupling can readily give a threefold increase, thus reducing 2 GHz to 667 MHz, and for the same C increasing R to 600 Ω. Equations for equivalent input noise current spectral density for simple models of both untuned and tuned input were presented. Inserting the numerical values used earlier and others listed in [12] gave the results plotted in Figure 9.13.

The improvement achieved is substantial. The Nyquist bandwidth corresponding to 10 GHz is 5.6 GHz, and the total noise from 0 to 5.6 GHz was found by

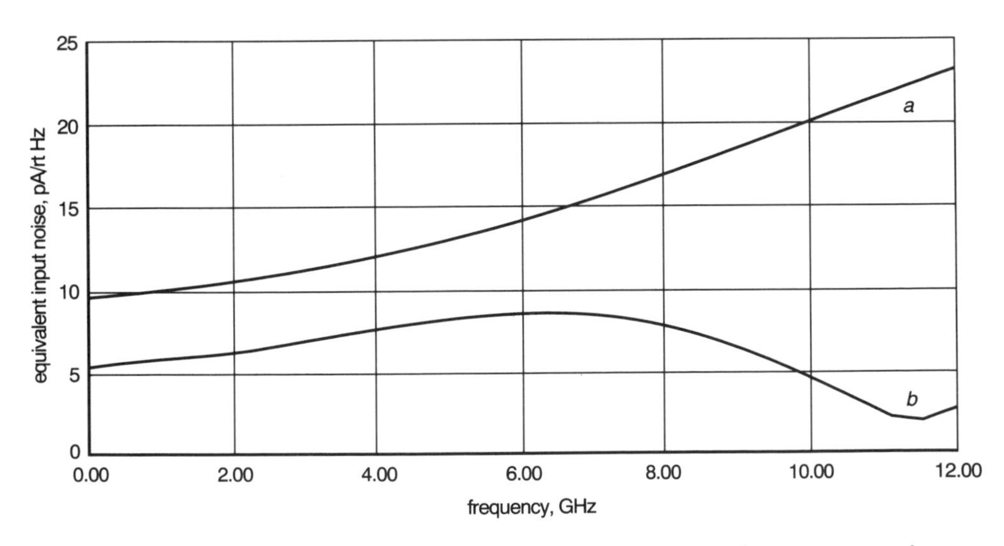

Figure 9.13 Calculated equivalent input noise current spectral density versus frequency for 2- and 4-terminal interstage between photodiode and input transistor (*a*) two-terminal, (*b*) four-terminal interstage ([12], Figure 2). Reproduced by permission of IEE

integrating the curves in Figure 9.13. This resulted in values of 847 and 519 nA corresponding to sensitivities of −22.9 and −25.1 dBm for the untuned and tuned cases, respectively. Mention was also made of the fact that the noise is seldom entirely due to the bias resistor and input transistor (as assumed in deriving the equations), but this is evident from any good noise analysis program. Analysis by the author of simple front-end circuits with thermal noise associated with every resistive component and uncorrelated input and output shot noise with every transistor has indicated that up to 4 or 5 such noise sources may contribute significantly to the total noise.

Experimental results: a three-stage amplifier with inductive coupling interstages and *CR* equaliser was designed, constructed and tested. The photodiode had a 20 μm diameter active area, 70 fF capacitance at a reverse bias of 10 V, responsivity of 1 A/W at 1550 nm and bandwidth exceeding 20 GHz. The transistors were NEC NE32400 (chip) hemts with 0.25 μm gate length, and data sheet s-parameters and noise performance were used to model them; more accuracy would have been achieved by characterising the individual chips. The inductor coupling was realised using bond wires modelled as transmission lines in the optimisation, and the resistor models also included transmission lines for improved accuracy. The input time constant corresponded to about 750 MHz and this was equalised to give an overall bandwidth of about 3 GHz by the differentiator and extended to 12.5 GHz by the bond wires. The front-end transfer impedance (output voltage/photocurrent) was predicted and measured, with the results plotted in Figure 9.14.

The 5 dB difference was attributed mainly to biasing the first stage at a lower drain current than used in the simulation. Corresponding results for the noise performance are presented in Figure 9.15.

The discrepancy between the two was attributed mainly to imperfect modelling of the photodiode shunt resistance and differences between modelled and actual transistor noise behaviour.

The group delay response was also measured and predicted; both showed ±20 ps variations up to 9 or 10.5 GHz for the measured and simulated values, respectively, corresponding to a very low pattern penalty.

Sensitivity: a value of −23.5 dBm (predicted value −24.4 dBm) at a ber of 10^{-9} and 10 Gbit/s was measured using an alternating mark and space signal to minimise

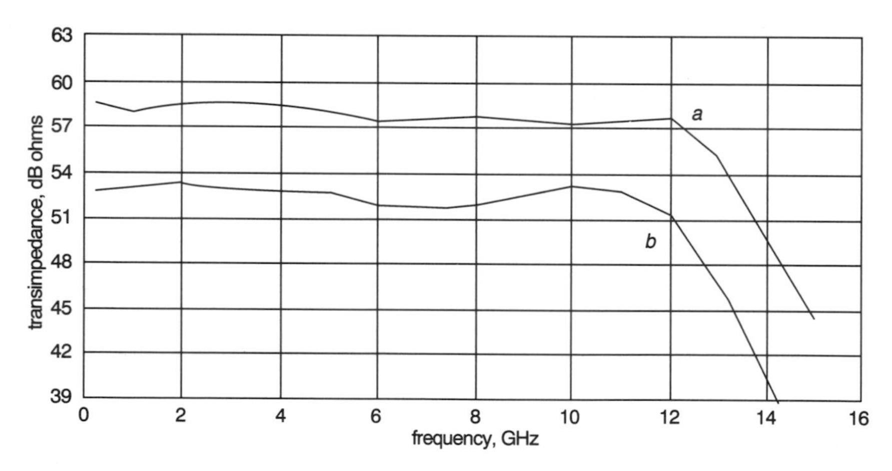

Figure 9.14 Transfer impedance versus frequency: (*a*) simulated, (*b*) measured ([12], Figure 6). Reproduced by permission of IEE

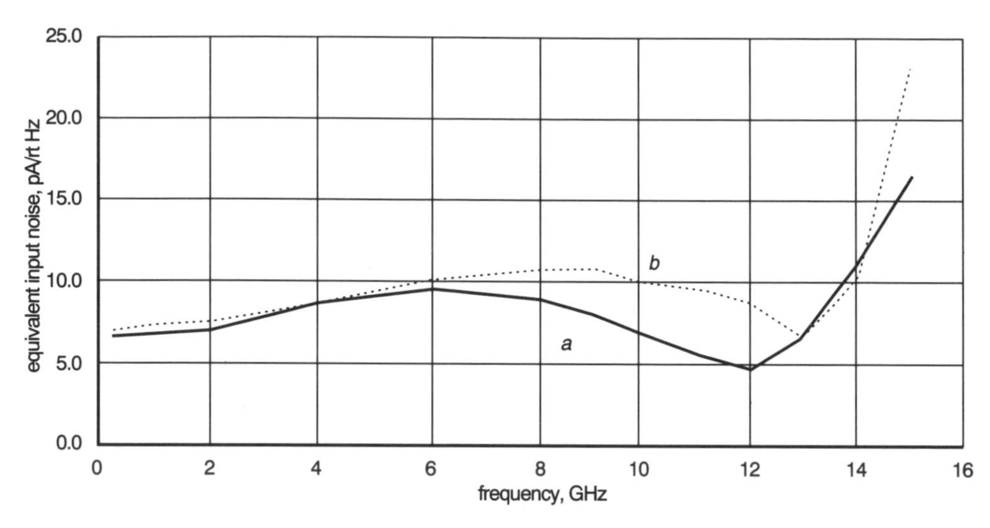

Figure 9.15 (*a*) Simulated, and (*b*) measured, values of equivalent input noise current spectral density versus frequency ([12], Figure 7). Reproduced by permission of IEE

patterning effect and thus give a good indication of intrinsic noise performance. Changing to a $2^7 - 1$ prbs test signal resulted in a sensitivity of -20.7 dBm (2.8 dB penalty relative to ideal case), with no evidence of a noise floor at a ber of 10^{-11}. The predicted and measured eye diagrams were also given, and they showed very good openings and agreement one with the other.

The authors mentioned that hybrid receivers become increasingly impractical at bit rates above 10 Gbit/s and gave a number of reasons for this; some of these limitations could be overcome by the use of an optical preamplifier.

Optical receiver using discrete commercial devices

A simple low-cost design using commercial packaged transistors and a hybrid assembly on a soft substrate has been realised reproducibly, as described in [13]. Many applications require receivers with bandwidths of many GHz; in particular, a bandwidth of about 10 GHz is appropriate for direct detection systems operating up to about 15 Gbit/s, for coherent systems operating at 4 Gbit/s [14] and for subcarrier systems. The receiver proved to be suitable for direct detection systems and as a design basis for balanced and phase diversity receivers.

Introduction: various possible designs had been described in the literature by 1990. A common feature was the use of discrete devices, individually characterised, and assembled on a ceramic substrate. In general each had one or more unsatisfactory features which prevented a simple, low cost, reproducible implementation. With these desiderata in mind a receiver was developed using discrete packaged commercial components with no individual characterisation, and a hybrid assembly on a standard soft substrate. The design ensured that the overall performance was largely insensititive to normal spreads in photodiode and transistor parameters, and to packaging and bonding parasitics.

Design considerations: optical receivers are generally designed around high impedance or transmimpedance configurations as already described. Like microwave amplifiers the upper edge of the passband may exceed 10 GHz; but unlike such amplifiers, the lower edge may be as low as 30 kHz. Techniques normally applied to microwave amplifiers are not suitable at such low frequencies, where the behaviour of open- and short-circuit stubs is very different from that at GHz frequencies. Neglect of such considerations will result in an unacceptable low-frequency response.

In either high impedance or transimpedance types, cascaded or distributed circuit topologies can be used. CAD studies showed that the present objectives could best be realised using the high-impedance configuration with cascaded amplifier. Simulation demonstrated that broader bandwidths could be achieved with this configuration because of the transit time constraints that packaged transistors and a hybrid assembly imposed upon the feedback path of transimpedance designs. Broader bandwidths could also be realised with cascaded topologies than with distributed circuits, for the same packaged devices and substrate. The design of such circuits require that CAD simulators incorporate accurate models which take account of all the important parasitics for the devices to be used. In the present case it was found that the three hemt devices were adequately modelled by the s-parameters given in the manufacturer's data sheets (in other engineering applications this might not be the case). Certainly, by 1990 much progress had been made into the accurate characterisation and modelling of transistors, passive components and various common types of assembly. Similarly, typical values given in data sheets for the thick film resistors and chip capacitors also proved to be satisfactory. Considerably more care was needed in modelling the photodiode, especially as its stray inductance determined the upper band edge frequency. The optimisation procedure in TOUCHSTONE enabled a satisfactory circuit model to be derived. The circuit models for the photodiode and the input of the first hemt are shown in Figures 9.16*a* and *c*, respectively, and Figure 9.16*b* shows segments approximating the modelled and measured s_{11} of the photodiode.

Receivers of the type discussed were assembled as hybrid microwave circuits on a soft substrate; hence there was a choice between microstrip or coplanar

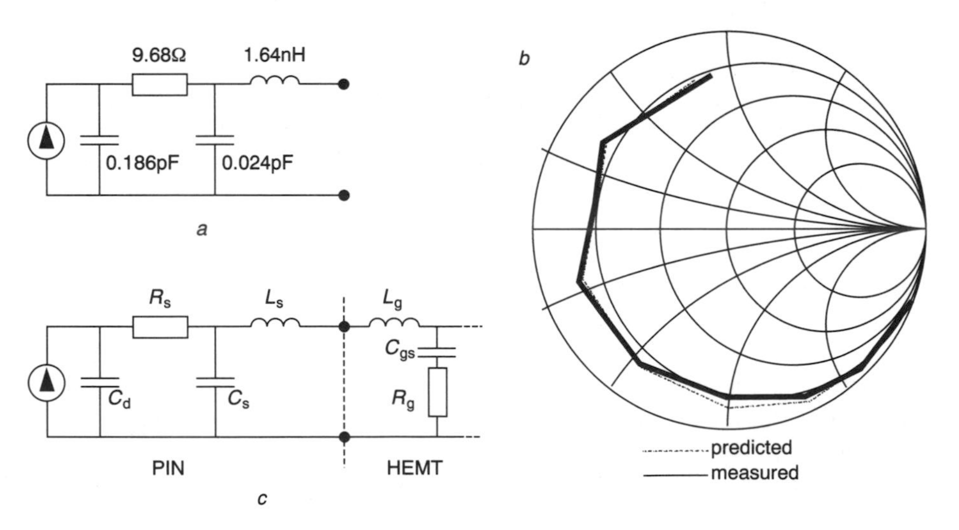

Figure 9.16 (*a*) Optimised circuit model of pin photodiode: (*b*) Smith Chart showing predicted and measured reflection coefficient s_{11}; (*c*) circuit of front end ([15], Figure 2). Reproduced by permission of IEE

waveguide (cpw) technologies. Each have their advantages and drawbacks; for instance, microstrip designs can draw on a wide range of well-established and accurate commercially available tools. A main disadvantage lies in the structural complexity required to ensure good terminations to ground at the appropriate nodes in a circuit. Radiated fields must be sufficiently attenuated to prevent intra-circuit couplings and proximity effects due to the presence of a package for example. cpw is an alternative which eliminates most of the negative features of microstrip; however, according to a late-1990 colloquium contribution [15] there was a 'virtual absence of good, commercially available cad models, despite occasional claims from manufacturers to the contrary. In addition, the literature appears to offer few treatments of cad models for cpw, and of those none appear to address the specific requirements of the broadband receiver work covered in this paper'. Work was being pursued to develop both analytically and experimentally the models necessary, and to include all observed features of cpw transmission lines and discontinuities.

The circuit diagram for the chosen configuration is shown in Figure 9.17 and comprises a single pin diode followed by a three-stage high input impedance amplifier using hemt (high electron mobility transistor) devices and distributed components, as in many microwave amplifiers.

The parallel *RC* combination at the output equalised the effect of the input capacitances and R_1 which corresponded to a frequency of about 1.5 GHz. That portion of the circuit consisting of the pin, input to the first stage and their interconnection produced a resonance at high frequency which was used to extend the passband. This was given by the equation:

$$(2\pi f_{\mathrm{P}})^2 = \frac{(1 + C_{\mathrm{D}}/C_{\mathrm{GS}})}{C_{\mathrm{D}}(L_{\mathrm{S}} + L_{\mathrm{G}})} \tag{9.5}$$

in terms of the circuit elements shown in Figure 9.16*c*, predicting a peaking frequency about 10 GHz. This was confirmed by simulation of the full circuit shown in Figure 9.17. The circuit board was produced on a soft substrate using microstrip technology. The dimensions of the transmission lines and matching networks were obtained by numerical optimisation. Commercial chip capacitors and thick film resistors were used throughout.

Measured and predicted gain-frequency responses: these are presented in Figure 9.18, which shows a 3 dB bandwidth of 10 GHz and gain ripple less than ±1 dB below 8 GHz. Good agreement was obtained between measured and simulated results. Good consistency was also reported between the results from several receivers of the same type.

Measured and predicted noise performance: conventional noise models were used with TOUCHSTONE to produce the simulated results plotted in Figure 9.19 for the r.m.s. noise current spectral densities referred to the input.

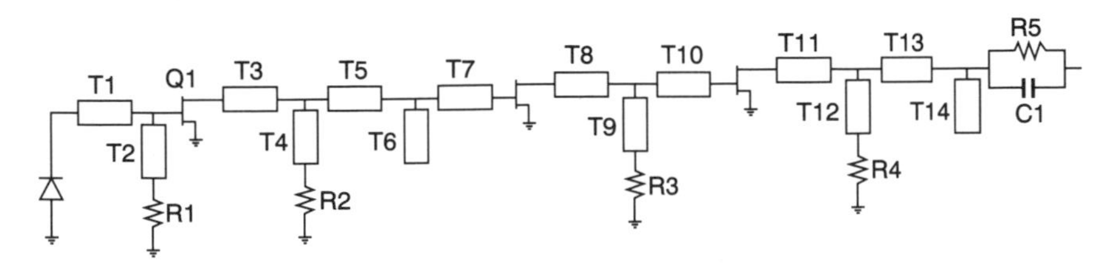

Figure 9.17 Schematic circuit diagram of pin-hemt broadband receiver ([15], Figure 1). Reproduced by permission of IEE

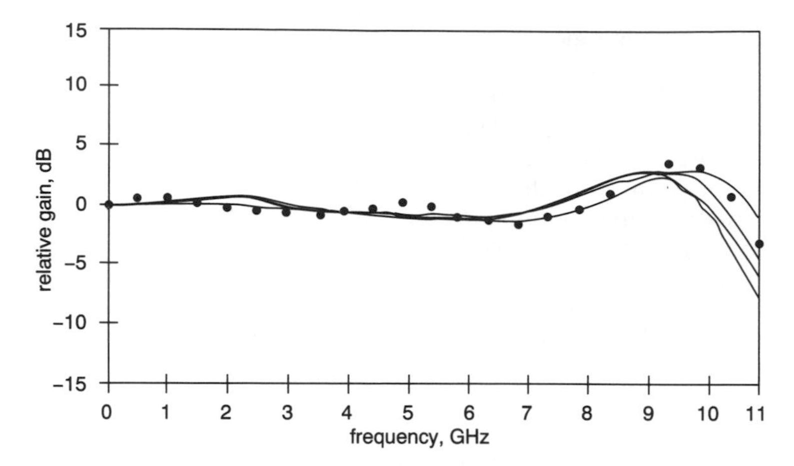

Figure 9.18 Relative gain versus frequency, curves measured, dots predicted values ([15], Figure 4). Reproduced by permission of IEE

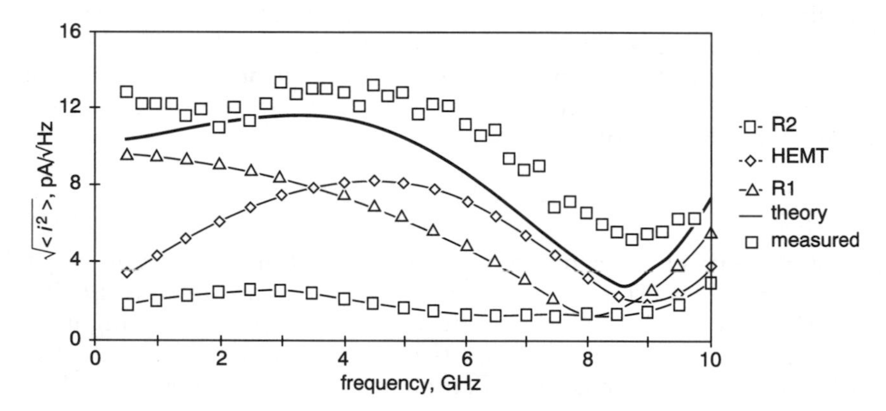

Figure 9.19 Equivalent input noise current spectral density showing predicted contributions due to R_1, R_2 and Q_1. Bold line shows total predicted values, and open squares represent measured values ([15], Figure 3). Reproduced by permission of IEE

These results demonstrate the dominant role played by the thermal noise of the photodiode bias resistor (R_1) and channel noise from Q1. The input hemt Q1 had sufficiently high available power gain that noise contributions from later stages were negligible over most of the band. The predicted worst-case r.m.s. noise current density was 12 pA/(Hz)$^{1/2}$ and the measured value was 13.5 pA/(Hz)$^{1/2}$. Previous work had demonstrated that the results quoted were highly reproducible between individual receivers of the type, particularly below 8 GHz, confirming the robustness of the design.

Sensitivity: this was measured at a selection of bit rates, and the results are shown in Figure 9.20. Also plotted are the values with the penalties attributed to the measuring equipment removed in order to give a more accurate picture. These adjusted values indicate a discrepancy between measurements and simulation of about 1 dB at these rates. In practice, achievable sensitivities were stated to be −24.5 dBm at 1.2 Gbit/s, −22.7 at 2.4 and −20.5 at 5 Gbit/s. Extrapolation

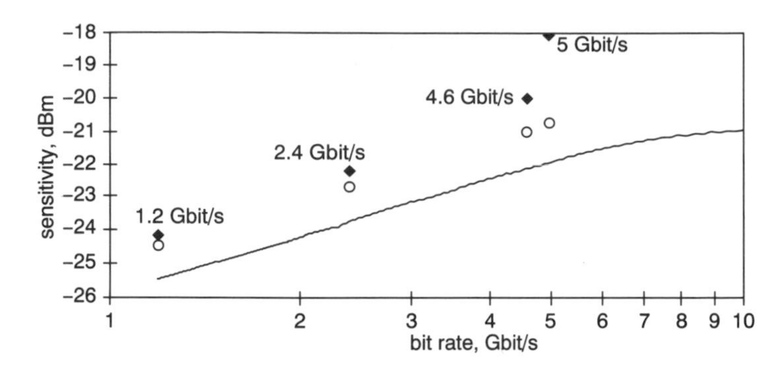

Figure 9.20 Measured and predicted values of sensitivity versus bit rate ([15], Figure 5) (——— theory, ♦ measured, ○ normalised with measurement penalties removed). Reproduced by permission of IEE

suggested an achievable value at 10 Gbit/s comparable with −19 dBm reported in 1988. For the 4 Gbit/s coherent detection application, the sensitivity was determined in a heterodyne system with split contact dfb laser source producing a continuous phase fsk signal, which was then transmitted over 100 km of conventional sm fibre with negligible dispersion penalty to the receiver. The local oscillator was a packaged external cavity device tuned to give an if (intermediate frequency) of 4.6 GHz. This choice, and the flat gain and delay responses below 8 GHz, ensured that the if signal and its 3.2 GHz sidebands suffered negligible distortion. A delay-line frequency discriminator detected the if. At 4 Gbit/s and a ber of 10^{-9} the measured sensitivity was −32.5 dBm for a local oscillator power level of −2.7 dBm; this value of sensitivity included a 3.2 dB penalty due to receiver noise. No error rate floor was seen down to a ber of 10^{-12}.

9.3.4 Broadband receiver modules

Among possible applications for such components will be broadband isdn or bisdn. A key technology will be ofdls operating at 10 Gbit/s and above. Developments of such systems have concentrated on broadening the receiver bandwidth; this requires very low capacitance photodiodes and techniques for parasitic effects. Characteristics of a new pin photodiode and preamplifier circuit were reported in [16] and will be discussed here as illustrative of the state-of-the-art in mid-1991.

Low-capacitance back-illuminated pin photodiode: reduced capacitance was achieved by adopting back-illumination which reduced the active area and hence the capacitance. The structure of the device is shown in Figure 9.21.

An active area diameter of 35 μm corresponded to a capacitance of 0.16 pF; other parameters were a dark current of 30 nA, a quantum efficiency of 70% and a bandwidth of 19 GHz.

Broadband circuit design: face down bonding was adopted to minimise the effect of inductance of the bond wire between photodiode and preamplifier on the frequency characteristic. The preamplifier contained discrete hemt devices in transimpedance configuration, and the salient parameters were a bandwidth of 11 GHz, equivalent average noise input of 10 $\text{pA}/(\text{Hz})^{1/2}$ and transfer impedance of 400 V/A. A simple decoupling circuit was used between the diode and its power supply to reduce resonance.

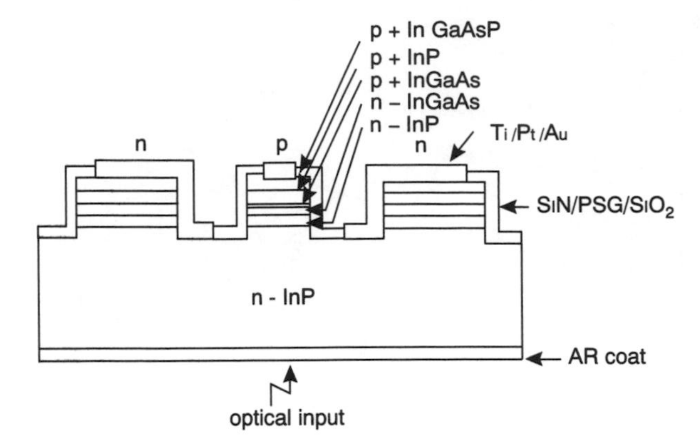

Figure 9.21 Diagram of photodiode structure ([16], Figure 1). Reproduced by permission of IEE

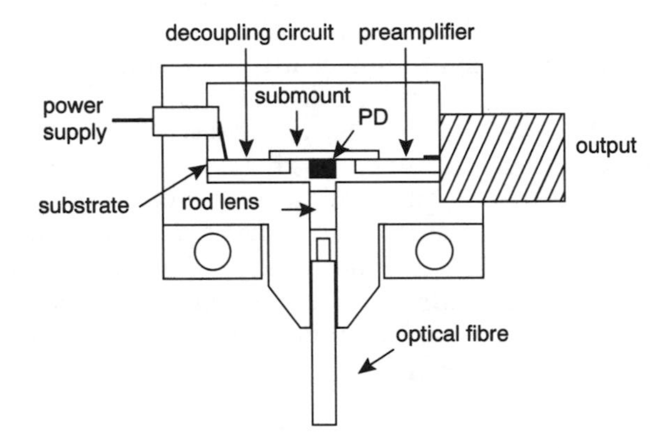

Figure 9.22 Structure of receiver module ([16], Figure 3). Reproduced by permission of IEE

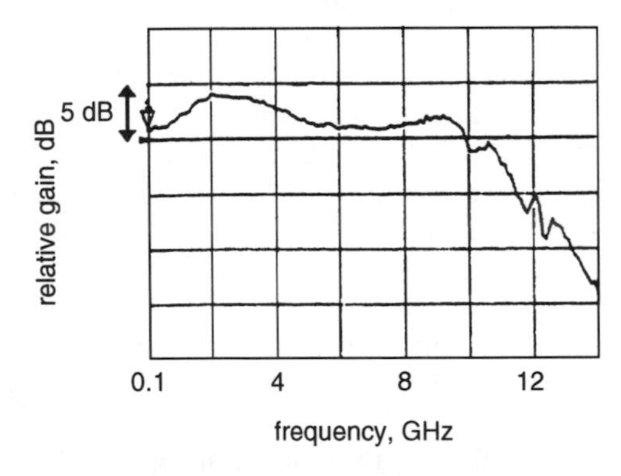

Figure 9.23 Relative gain versus frequency characteristic of receiver module, 3 dB bandwidth 11.6 GHz ([16], Figure 5). Reproduced by permission of IEE

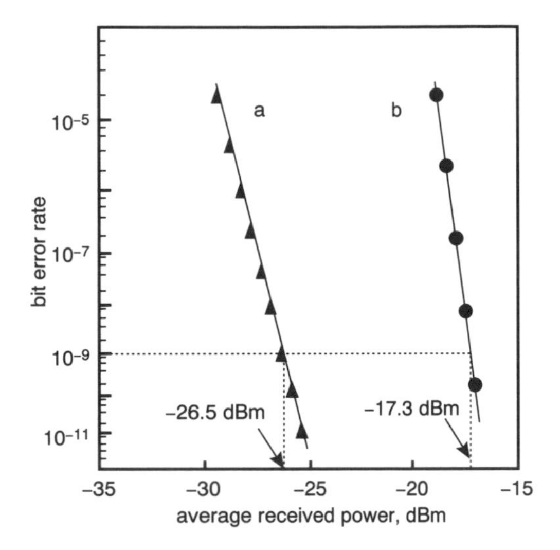

Figure 9.24 ber characteristic of module: (*a*) with edfa preamplifier, (*b*) without preamplifier ([16], Figure 6). Reproduced by permission of IEE

Receiver module: the constructional details are shown in Figure 9.22.

A submount carrying the photodiode chip was bonded face-down on the substrate which contained the preamplifier circuit, the 50 Ω output of which is accessed via a K-connector. The pigtail consisted of a sm fibre and rod lens, which coupled to the photodiode with an efficiency of 97%. The frequency characteristic of this module is shown in Figure 9.23.

The ber performances with and without an erbium-doped optical fibre amplifier (edfa) were also measured, and the results are presented in Figure 9.24.

Sensitivities at 10^{-9} ber and 10 Gbit/s were −26.5 and −17.3 dBm with and without the edfa. Further, more recent information is given later.

A competitive technology involving the use of a three-stage AlGaAs/GaAs hetero-junction bipolar transistor ic front end was described in [17]. This transistor technology also has excellent potential for applications at 10 Gbit/s and above because it combines very high speed (f_T of 41 and f_{MAX} of 44 GHz) with large transconductance. A 300 Ω resistor was connected between the base of Q3 and the base of Q1. On-wafer measurements gave a transfer impedance of 49.8 dB (309 Ω) (49 dB from simulation) and 12.7 GHz bandwidth (13 GHz predicted). Standard deviations for these two quantities over the chips on a 2″ wafer were as low as 0.42 dB and 0.47 GHz. A 50 Ω microstrip line and series 20 Ω resistor connected the ic output to the output pin of the packaged module to improve the match to the following 50 Ω amplifier. The module transfer impedance and 3 dB bandwidth were 45 dB and 9.4 GHz, respectively. The minimum sensitivity at 1552 nm was −15.7 dBm and this was increased to −30.3 dBm when an edfa 30 dB gain preamplifier pumped with 20 mW at 1480 nm was used. 1993 information on this technology is given below [11].

9.3.5 Front-end amplifiers for high-capacity systems

Research and development on systems operating at 10 Gbit/s began in the late 1980s, and since then the receiver sensitivities have improved slowly. Values between −15 and −19.8 dBm were reported in 1989, whereas by 1993 the best reported measured value for a pin-fet receiver was −20.4 dBm. This can be compared with the best measured value of −20.7 dBm reported using a pin–hemt front end [12]. A high-impedance three-stage hemt amplifier was designed using NE32400 active devices in chip form. Bond wires were used between the photodiode and first stage and between successive stages to broaden the passband. The amplifier 750 MHz 3 dB down frequency due to the bias-resistor input-capacitance combination was equalised to about 3 GHz by a parallel *RC* combination, and this was extended to 12.5 GHz by using the CAD package TOUCHSTONE to optimise the lengths of the bond wires.

Results: simulated and measured transfer impedance (i.e. output voltage over input current) characteristics were of similar shape, but the average measured value of some 447 V/A (53 dBΩ) in the passband was considerably lower than the simulated value of 794 V/A. The discrepancy was attributed mainly to biasing the first stage hemt at a lower drain current than used in the simulation. These levels of gain were sufficiently high to make the noise contribution of the main amplifier to the average equivalent input noise current spectral density of about 10 $pA/Hz^{1/2}$ negligible. This noise measure was reduced to about 8 $pA/Hz^{1/2}$ average if the passband was restricted to the Nyquist value for 10 Gbit/s of $0.56 \times$ bit rate, i.e. 5.6 GHz, corresponding to a predicted sensitivity of −24.4 dBm.

Simulations were based on information given in the manufacturer's data sheets rather than on values measured for the devices used in the circuit, so that discrepancies were to be expected. Measurements of sensitivity were made in a 10 Gbit/s system testbed using a 101010... pattern and gave a value of −23.5 dBm at a ber of 10^{-9}, in good agreement with the predicted value quoted above. Close agreement was also achieved between simulated and measured eye diagrams. Finally the authors concluded that the design of hybrid receivers for higher bit rate systems would become increasingly difficult, hence the need for integration. The reduction of sensitivity with increasing bit rate could best be overcome by using optical preamplification, and a 1992 figure of −38.8 dBm at 10 Gbit/s was quoted as the best reported sensitivity to date (early 1993).

However, in 1993 there was still considerable uncertainty surrounding the optimum choice of ic technology needed for such systems. The principal options for hybrid broadband optical receiver implementations using ics available in 1993 were summarised in [11]. State-of-the-art high speed (f_T around 23 GHz) silicon and GaAs hbt processes (f_T exceeding 40 GHz) had been used in a design of 10 Gbit/s receiver front end and the results of preliminary measurements of performance and of simulations using SPICE-type transistor models were reported. The key process features and device parameters were summarised in tabular form. The typical measured performance parameters of transimpedance front ends based on the GaAs hbt ics were 3 dB bandwidths in the range 9.5–10.3 GHz, midband transfer impedances between 310 and 340 V/A and group delay variations of less than 30 ps within the passband. From the measured results an average equivalent input noise current spectral density of 10 $pA/(Hz)^{1/2}$ was calculated and a sensitivity of about −22 dBm was inferred.

The status of available ic technologies in late 1993 suggested to the authors that in the short to medium term the best choice appeared to be GaAs mmics based on hbt technology, and that this technology appeared to possess considerable performance margins at 10 Gbit/s.

9.4 RECEIVERS USING OPTICAL PREAMPLIFIERS

The sensitivity performance limit of optical front ends is set by the noise generated within the photodiode, transistors and any dissipative elements present in the circuit. Coherent detection can provide better sensitivities than either pin or apd devices and direct detection, but it requires narrow linewidth transmitter and local oscillator lasers, polarisation control and increased complexity. The use of an erbium-doped fibre amplifier (edfa) before the photodiode provides better performance than direct detection using an apd, and is much simpler than coherent detection. However, an edfa produces amplified spontaneous emission (ase) and a standing output in the absence of an input signal.

Analysis of receiver performance with an optical preamplifier differs qualitatively from conventional cases where all the noise is electrical, i.e. it originates in the detection process and from shot or thermal noise sources. The presence of ase with the optical signal gives rise to three components of noise in the photodiode output. One is the additional shot noise from the increased mean received optical power, and the other two result from the square-law detection. Beating between signal and ase generates signal-spontaneous noise, and beating between different spectral components within the very wideband ase generates spontaneous-spontaneous beat noise proportional to the optical bandwidth of the edfa. Suitable filtering of the edfa output does not reduce signal-spontaneous noise but it does reduce the other two [18].

A narrowband optical filter can be inserted within the length of erbium-doped fibre, and this novel configuration was described in [19]. The effect of the filter was to stop the build-up of both forward and backward travelling ase power lying outside the signal bandwidth. The component count was increased by the two wavelength division multiplexers needed to provide separate paths for the filtered signal power and for the pump power between the two lengths of the erbium-doped fibre. These multiplexers introduced a loss of between 1 and 1.5 dB in pump power but up to 6 dB of small-signal gain improvement could still be obtained. The noise figure was improved by some 0.8 dB. Simulation results predicted that the optimum split of the overall edfa length was around 48:52%, reckoned from the input end, and was independent of signal wavelength, but that the optimum position for the filter did depend on pump wavelength, being 47% when pumped at 1480 nm but only 42% at 980 nm. The filter insertion loss was taken as 3 dB, and 3 dB down bandwidth was 1 nm.

9.4.1 Receivers using erbium-doped fibre preamplifiers

For use as receiver preamplifiers a good low-noise performance is necessary, but preferably without sacrifice in gain. These desiderata are difficult to satisfy simultaneously in a single edfa. Experience with electronic amplifiers over many years has demonstrated that some two-stage configurations can offer better performance, and this is also the case for optical amplifiers, whether implemented in fibre technology or in semiconductor laser technology. Two examples of these two-stage preamplifiers are treated in this section.

A practical implementation of a transmitter and a receiver operating at 2 × 10 Gbit/s and using a two-stage edfa preamplifier was described in [20]. Only the receiver aspects will be mentioned here. By 1994 commercial mmics based on GaAs hemt or hbt technologies and broadband external modulators with sufficiently low voltage drive requirements were becoming available for use in operational systems. Their advent made the upgrading of 10 Gbit/s imdd systems by tdm an attractive possibility for higher bit rates still. A two-stage preamplifier pumped at 980 nm by two laser diodes provided 41 dB of small-signal gain and a noise figure of 3.8 dB. The bandwidth of the front-end was limited to 14 GHz by the commercially available amplifier which had been designed to give optimum pulse response. This design criterion has been used in amplifiers for applications to radar and television since the 1940s.

Experimental two-stage receiver preamplifier at 10 Gbit/s

This description of the performance of two cascaded erbium amplifiers in a 10 Gbit/s system compared performance with a single-stage preamplifier or none. The use of two devices rather than one has some advantages. The configuration was described in [21] and combined a low noise figure of 4.1 dB with a high net gain of 27.0 dB at a signal wavelength of 1553 nm. The experiment achieved the highest reported (early 1991) 10 Gbit/s receiver sensitivity of −31.1 dBm (606 photons/bit), an improvement of 18.8 dB relative to the non-preamplified receiver sensitivity. Previous work had reported a value of −30.8 dBm but only by operating at the peak gain wavelength (1536 nm) of the erbium amplifier. For practical dispersion-shifted fibre systems the second gain peak at 1550 nm is preferable for two reasons. Firstly, the dispersion is substantially smaller and secondly the gain peak around 1550 nm is broader (>5 nm) thus requiring less stringent tolerances on transmitter wavelength control. The peak gain at 1536 nm is, however, usually higher and so there is a gain penalty in choosing to operate around 1550 nm. This penalty can be avoided by using a two-stage amplifier as a receiver preamplifier.

Experimental arrangement: this is illustrated in Figure 9.25.

The polarisation-independent >30 dB isolators were included to minimise noise arising from Fresnel reflections at the connectors, and the 3 nm wide bandpass filters were used to suppress amplified spontaneous noise. In particular, the first filter reduced gain saturation in the second stage. The major properties of the fibre amplifiers are listed in Table 9.2.

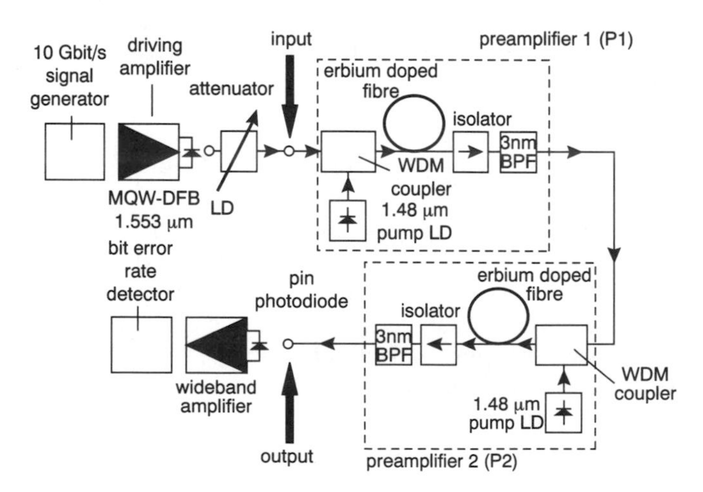

Figure 9.25 Experimental arrangement of receiver with two-stage edfa preamplifier ([21], Figure 1). Reproduced by permission of IEE

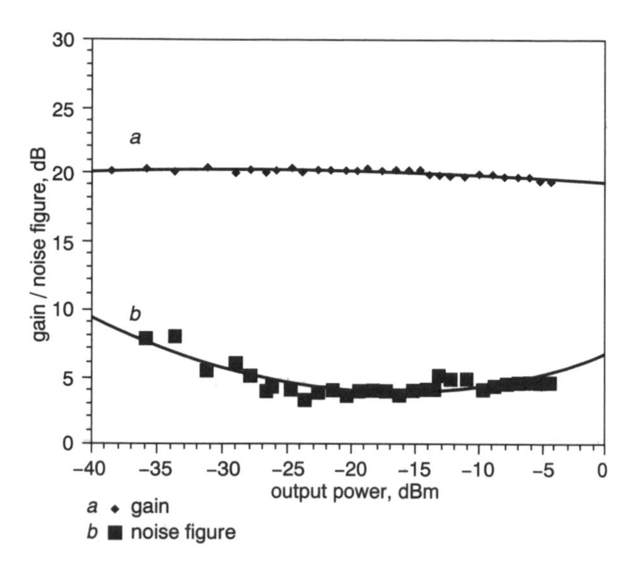

Figure 9.26 First stage of preamplifier: (*a*) gain versus output power, (*b*) noise figure versus output power ([21], Figure 2). Reproduced by permission of IEE

Both stages were forward-pumped at 1480 nm by single laser diodes which delivered 34.9 and 34.3 mW to the input and output edfa, respectively. Polarisation-independent isolators were inserted at the output of each stage to provide more than 30 dB of isolation. The bandwidth of the filters shown in Figure 9.22 was 3 nm.

Results: the gain and noise performances of the input edfa are illustrated in Figure 9.26 (the second stage had similar characteristics) and they indicate that it is suitable as a low-noise preamplifier for input powers exceeding about −47 dBm. The corresponding characteristics of amplifier 2 are very similar but with 1 dB more gain and 0.5 dB larger minimum noise figure. Amplifier 1 was therefore employed as the first stage.

Results: the measured error rates with and without the preamplifier are given in Figure 9.27. These clearly show the improvements in ber obtainable with one or two erbium preamplifiers. The back-to-back sensitivity at 10^{-9} was −12.3 dBm (*c*) and this was improved by 18.8 dB to −31.1 dBm with both edfa inserted (*a*); this value corresponds to 606 photons/bit. Under these experimental conditions the gain of the first stage was 16.8 dB and for both stages it was 27.0 dB. For (*b*) the improvement in sensitivity was 15 dB, giving −27.3 dBm for a net gain of 16.5 dB. The increase in gain of 10.5 dB in going from one to two stages corresponds to an increase of 3.8 dB in sensitivity, some 3 dB less than predicted. This was attributed to the increased relative importance of the noise generated by Fresnel reflections at the connectors under large gain (>20 dB) conditions. The authors stated that the use of a two-stage preamplifier plus a booster at the transmitter suggested the possibility of unrepeatered transmission spans exceeding 200 km at 10 Gbit/s.

Output power and gain characteristics of two-stage edfa preamplifier

By 1990 unsaturated gains of up to 48 dB had been achieved in an edfa by optimising the dopant concentration, the core diameter and the pump wavelength.

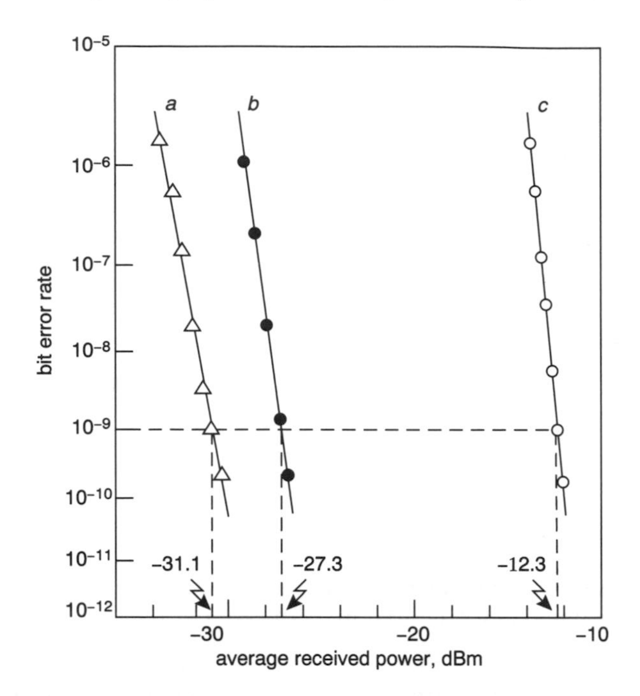

Figure 9.27 ber performance at 10 Gbit/s and 1553 nm: (*a*) with two-stage preamplifier; (*b*) with single stage; (*c*) without preamplifier ([21], Figure 3). Reproduced by permission of IEE

To achieve low noise figure pumping in the forward direction only is the best choice out of the three possibilities. However, the use of a single-stage forward-pumped preamplifier limits the available gain for two reasons. The first is due to saturation caused by amplified spontaneous emission and the second results from the increasing loss of pump power towards the remote end of the fibre. A two-stage configuration which eliminated ase and maintained high pump power intensity in the second stage was described in [22]. Both stages comprised fibre with dopant concentration 25 ppm of erbium, mode field diameter of 10.2 μm at 1500 nm and optimised values of 67 m in length for operation at a signal wavelength of 1535 nm and 100 m when the signal wavelength was 1551 nm. Both ends of the fibre were jointed to isolators and 70 mW of pump power was supplied to the transmit end from two InGaAsP laser diodes, operating at 1480 nm, via polarisation beam combiners and a dichroic mirror. ASE in the output of the first stage was removed by a filter of bandwidth 1 nm and insertion loss 1 dB (note the difference between this value and 5 dB in the previous experiment) and the remaining power was fed to a second erbium fibre via a dichroic mirror and isolator. This fibre was pumped in both directions, again by combining the outputs of two lasers at each end. Powers launched into fibre 2 were 65 mW from the forward pump and 80 mW for backward pumping. The optimum length of fibre 2 at both signal wavelengths was 100 m. This configuration realised a net gain of 51 dB at 1551 nm.

Experimental arrangement: this is shown in Figure 9.28.

The signal wavelength was either 1535 or 1551 nm; at the shorter wavelength the optimum lengths of the first and second stage edfas were 67 and 100 m, respectively. At the longer signal wavelength both were 100 m long. The optical bandpass filter

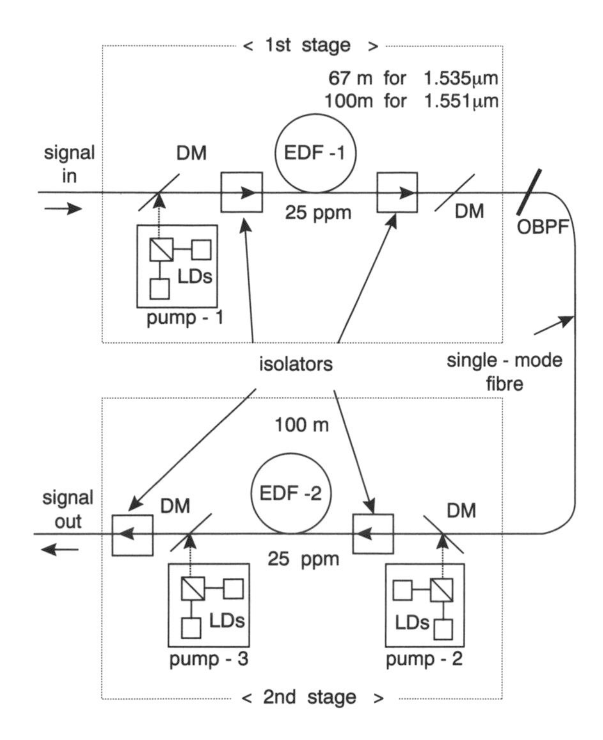

Figure 9.28 Experimental arrangement of two-stage preamplifier configuration ([22], Figure 1). Reproduced by permission of IEE

(obpf) of bandwidth 1 nm and midband insertion loss 1 dB centred on the signal wavelength. For all tests the powers supplied by pumps 1, 2 and 3 were 70, 65 and 80 mW, respectively.

Results: the output power and net gain performances at 1551 nm were evaluated first and the results are given in Figure 9.29.

The first stage unsaturated gain was 26 dB and the 3 dB compression point was +1.5 dBm. With both stages, the unsaturated gain rose from 46 to 51 dB when the filter was inserted, and the corresponding 3 dB compression point improved to +13 dBm.

Results: the output power and net gain performances at 1535 nm were then evaluated, and the results are given in Figure 9.30.

The first stage unsaturated gain was 36 dB and the 3 dB compression point was +3.5 dBm. With both stages the unsaturated gain rose by only 12 dB, but the increase in gain with the 1535 nm filter inserted was 5 dB as before. The ase was concentrated around 1535 nm with a fwhm value of 2 nm, so the 1 nm bandwidth of the filter centred on 1535 nm was not nearly so effective in discriminating against ase as the filter centred on 1551 nm.

9.4.2 Receivers using semiconductor optical amplifiers as preamplifiers

Travelling wave semiconductor laser amplifiers (twsla), or more simply semiconductor optical amplifiers (soa), devices were investigated intensively during the

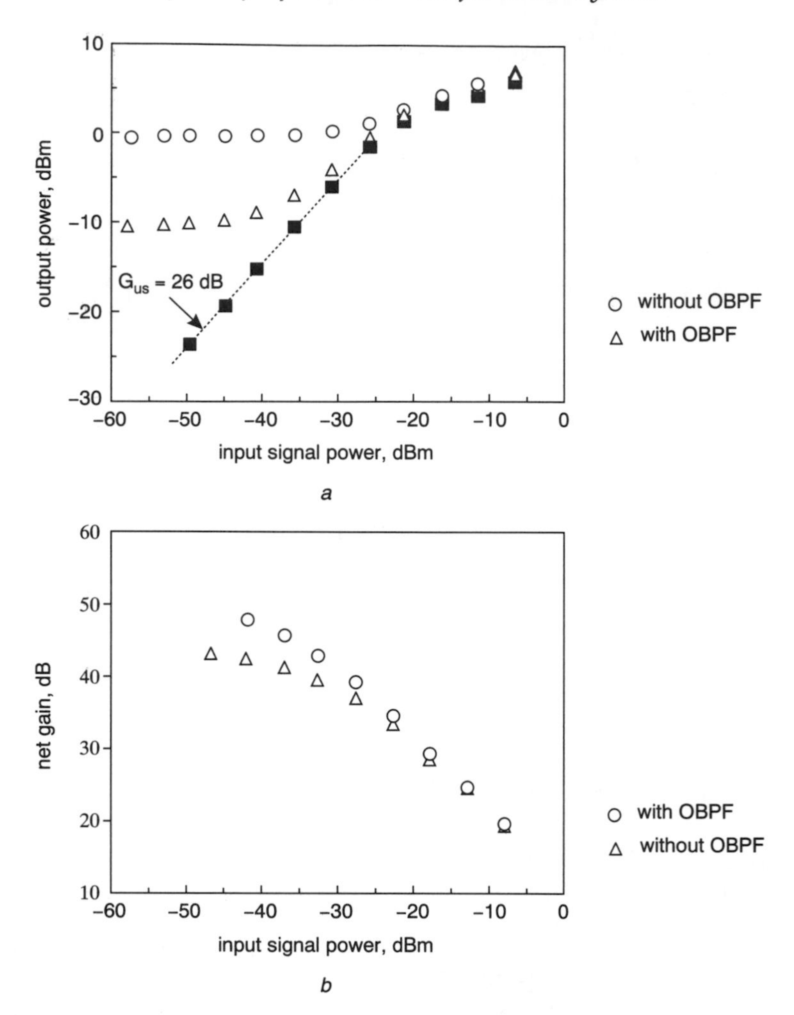

Figure 9.29 Performance with signal at 1551 nm: (*a*) output power of first stage only; (*b*) net gain of two-stage amplifier, both versus input power with and without the filter ([22], Figure 2). Reproduced by permission of IEE

1980s, especially before the development of edfa technology. Broadly speaking, these devices have two main areas of application in systems: as repeaters and as preamplifiers. In both they are operated in their linear region. When employed as preamplifiers they can improve the sensitivity of the detection process, especially for line rates above 1 Gbit/s.

The inherent sensitivity to the sop of the input signal accounts for some 3 dB variation in mean gain, but this can be reduced to acceptable levels by careful design and agc. As with all optical amplifiers, soa devices generate a relatively large amount of spontaneous emission which results in a standing output in the absence of an input signal. The spectrum of this emission typically occupies about 30 nm so filtering can be used to advantage. Practical twsla devices have non-zero facet reflectivities which can lead to instabilities in systems due to the presence of backward travelling waves creating laser action, but again this can be reduced to negligible proportions by careful design.

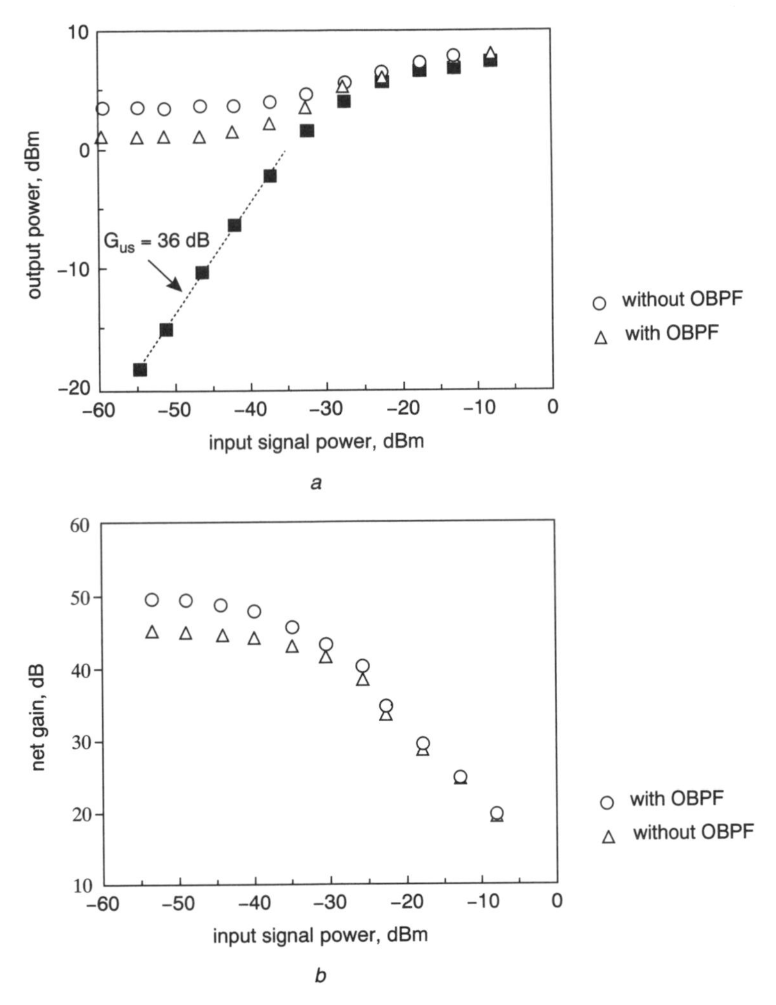

Figure 9.30 Performance with signal at 1535 nm: (*a*) output power of first stage only; (*b*) net gain of two-stage amplifier, both versus input power with and without the filter ([22], Figure 3). Reproduced by permission of IEE

In practical system use, the gain is expected to vary by about ±2 dB with changes in temperature, devices and component ageing, source wavelength and input polarisation. Gain control is therefore required and a number of agc systems have been described. Another limitation which applies particularly in intensity-modulated systems is due to saturation-induced crosstalk. Advantages of the soa are the inherent simplicity, low components count and the key feature of link transparency. Their small size makes them well suited for photonic integrated circuits, whereas fibre amplifiers are not. By 1991 simple, easily manufactured, packaged, single-stage amplifiers with fibre-to-fibre peak gains of 21 dB and gain ripple of 1 to 2 dB had been developed [23]. Similar soa were demonstrated in [24]. We will look first at a 1990 state-of-the-art integrated receiver, then at some other practical aspects.

An example of what was achievable by 1990 was described in [25]. Monolithic integration of components for use in systems possesses many well-known important advantages and the combination of a semiconductor preamplifier and an edge-coupled pin photodiode was described for the first time in [25]. In particular, the technique provides the benefits of a self-aligned and low-loss optical connection between the optical preamplifier and photodiode, as well as a rugged and compact component. Results of a demonstration were also given. A maximum gain (excluding coupling loss) of 20 dB and a 3 dB bandwidth of 35 GHz were achieved. The advantages of such monolithic integrated receiver front ends may make them very attractive in future high-capacity systems.

Structure and fabrication: the structure is based on a ridge waveguide laser, and the essential features are shown in Figure 9.31.

The photodiode absorbing layer is an unpumped extension of the preamplifier laser gain layer giving good optical coupling and low optical feedback between preamplifier and photodiode. Both of these advantages were enhanced by the incorporation of a 0.4 μm thick waveguide layer which forced the peak of the optical field deeper into the structure. This layer also acted to reduce the polarisation sensitivity by minimising the difference between TE and TM confinement factors for both transverse and lateral directions. The essential good electrical contacts between preamplifier and photodiode were also achieved in the chip design.

Device parameters: pin leakage currents were typically below 1 μA at −3 V, and measurements of capacitance at 1 MHz gave a value of 5 pF/mm. A chip incorporating a 600 μm long preamplifier and 35 μm section length photodiode was mounted in a high-speed package which included a Wiltron K-connector microstrip launcher to enable the device to be characterised up to 46 GHz. The reflection parameter s_{11} was measured with a HP 8510 network analyser up to 30 GHz and indicated values of pin parasitic capacitance and inductance of about 0.2 pF and 0.3 nH, respectively.

Measured results: details of the gain and frequency response measuring arrangements can be found in the reference. In particular, the tuned and untuned gains as functions of the bias current for −30 dBm of coupled power around the gain peak wavelength of 1530 nm are illustrated in Figure 9.32 for both polarisations.

Fabry–Perot effects are evident, and they arise from the uncoated input facet and from optical feedback through the short length of the photodiode from the uncoated pin facet. The apparent sensitivity at higher values of bias was caused by higher modal reflectivities for TM polarisations, the single-pass gain displayed no appreciable polarisation sensitivity. Turning to the frequency response, the amplitude–frequency characteristic for the component is shown in Figure 9.33.

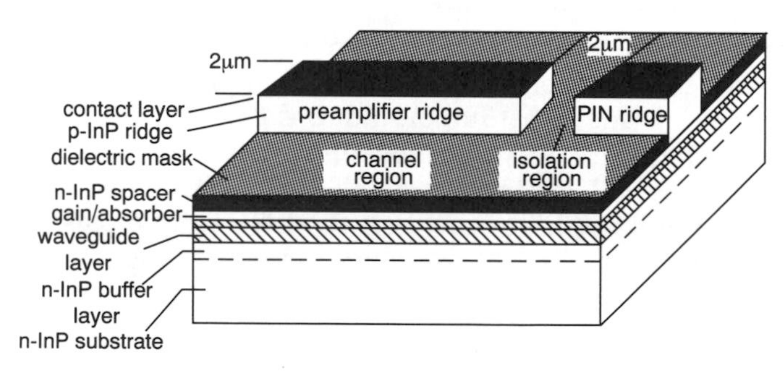

Figure 9.31 Schematic features of selective ridge overgrowth design ([25], Figure 1). Reproduced by permission of IEE

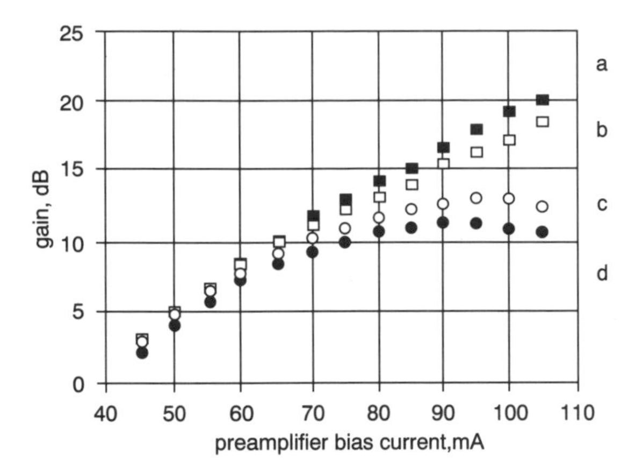

Figure 9.32 Gain versus soa preamplifier bias current: (*a*) TM tuned, (*b*) TE tuned, (*c*) TE detuned, and (*d*) TM detuned ([25], Figure 3). Reproduced by permission of IEE

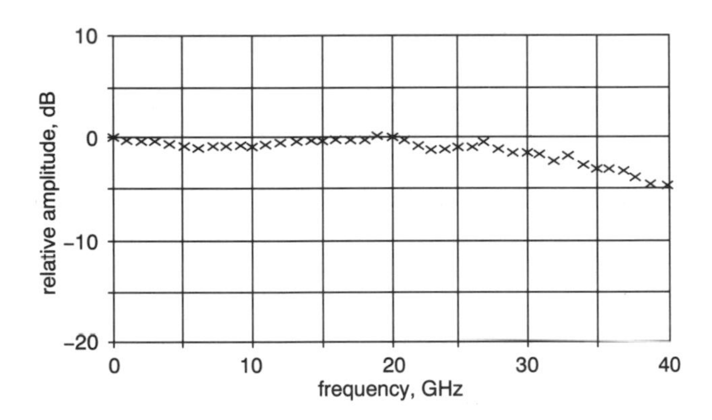

Figure 9.33 Relative amplitude versus frequency with soa biased at 70 mA and pin photodiode at −4 V ([25], Figure 4). Reproduced by permission of IEE

The bandwidth of 35 GHz corresponds to a maximum gain of 20 dB and make the component ideally suited to applications requiring sensitive and wideband detectors.

Some practical aspects were discussed in [26]. The sensitivity and gain-bandwidth performance of pre-amplified pin receivers were compared with apd receivers at rates up to 8 Gbit/s. The technical interest of the former arises from the need for gain to offset the insertion loss of future optical networking elements such as switches, splitters and multiplexers being developed for deployment in ofdls. The loss of these elements must be accommodated within the system power budget. This requirement, together with the need for ever higher transmission capacities, calls for receivers with improved sensitivities and bandwidth. APD receivers are more sensitive than pin types at higher bit rates up several Gbit/s, but the gain-bandwidth product of the apd sets an upper limit on sensitivity. pin receivers offer greater potential bandwidth but poorer sensitivity; however, the sensitivity can be enhanced, without affecting bandwidth, by employing an optical preamplifier.

SOA-PIN receiver: this consisted of a fibre tail coupled to the soa input, the soa itself, an optical filter and a pin receiver in tandem. The amplifier noise output was predominantly spontaneous emission with a bandwidth (typically 30–40 nm) determined by the gain profile. This was substantially reduced by an optical bandpass filter, with an optimum bandwidth primarily determined by

(a) pin receiver noise
(b) bit rate
(c) transmitter chirp characteristics.

As is usual in these types of filters, the narrower the passband the greater the insertion loss, so that a compromise must be made, with resulting passbands typically lying in the range 0.5–3 nm. The sensitivity improvement due to the soa is a function of

(1) internal (cavity) gain
(2) coupling losses between the various optical components
(3) bandwidth of the optical filter.

Signal power dependent noise products ensured that the sensitivity improvement was less than the net signal gain between fibre tail and pin photodiode. Typically, very worthwhile improvements of 10–15 dB were achieved.

The receiver that incorporates the soa, optical bandpass filter and front end is clearly wavelength-selective, and may therefore be employed as a wavelength demultiplexer. Mechanically tuned filters such as solid etalons and Fabry–Perot resonators are particularly suited to perform 1 out of N channel selection because of their relatively slow tuning or switching speeds. Conventional demultiplexing can also be accomplished by means of waveguide filters, which offer significantly higher speeds and hence the prospect of novel networking functions.

A selection of such receivers has been developed at BTRL, each designed for a specific bit rate by choice of passband and the pin receiver electronics. The optical preamplifier in each case was a GaInAsP/InP bh device facet-coated to give a low residual reflectivity. Coupling to the input facet was done by means of a lens-ended sm fibre, whereas the output coupling from the facet to the pin was implemented by high na micro-sphere lenses which provide collimated coupling between the output facet and the photodiode. At optimum bias of the soa the gains between the fibre tail and the pin input were all about 17 dB. Etalons with passbands of 1–2 nm formed the optical filter. The InGaAs photodiode was followed by a low noise ic amplifier fabricated in GaAs technology.

Results: the measured results for bit rates from 600 Mbit/s to 2.4 Gbit/s are plotted in Figure 9.34, together with results for apd-based receivers from AT&T Laboratories published in 1988 and 1989.

Examination of the figure shows that the soa–pin receiver sensitivity offers a worthwhile improvement over an apd receiver at 2.4 Gbit/s and above. This comparison relates to apd receivers which do not have optical filtering; if such filters were inserted to give the same selective properties as the soa–pin receiver, the relative sensitivity advantage of the latter would increase by some 2–4 dB.

The performance of receivers with more than a single-stage soa has been analysed theoretically [26] using a lightwave simulation package which models laser chirp, fibre dispersion, soa and receiver characteristics. The results showed that synchronously tuned stages gave better performance than stagger tuned stages,

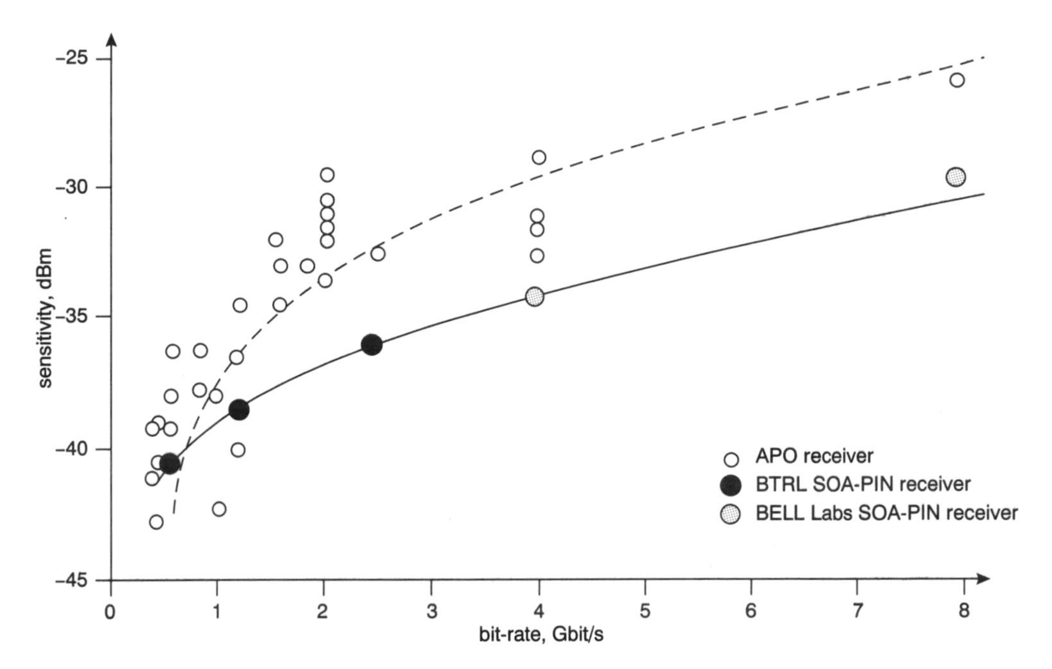

Figure 9.34 1989 Comparison of soa-pin and apd receiver measured sensitivities at bit rates of 600 Mbit/s and above ([26], Figure 3). Reproduced by permission of IEE

despite having a larger gain ripple. The paper did not discuss practical embodiments, effects of temperature variations or production spreads, nor the impact on reliability.

9.5 REFERENCES

1 DAWSON, P.A., and FITCHEW, K.D.: 'Repeater design requirements'. *Br. Telecomm. Engineering*, **5**, part 2, July 1986, pp. 97-103.

2 BROCKBANK, R.G., CALTON, R.L., MOTTRAM, S.W., HUNTER, C.A., and HEATLEY, D.J.T.: 'The design and performance of the TAT-8 optical receiver'. *Br. Telecomm. Technol. J.*, **5**, no. 4, Oct. 1987, pp. 26-33.

3 MITCHELL, A.F., and O'MAHONEY, M.J.: 'The performance analysis of pin bipolar receivers'. *Br. Telecomm. Technol. J.*, **2**, no. 2, Apr. 1984, pp. 74-85.

4 BROOKS, R.M., and JESSOP, A.: 'Line coding for optical fibre systems'. *Int. J. Electronics*, **55**, no. 1, 1983, pp. 81-120.

5 OWEN, B.: 'P-I-N GaAs f.e.t. optical receiver with a wide dynamic range'. *Electronics Lett.*, **18**, no. 14, 8 July 1982, pp. 626-627.

6 JEAL, A.J., and LILLY, C.J.: 'Long-haul systems — technology trends'. *Suboptic '93*, Versailles, France, 1993.

7 WILLIAMSON, R.L., SWANSON, G., and MITCHELL, A.F.: 'The NL2 system'. *Suboptic 86*, Versailles, Feb. 1986.

8 CHANNON, R.D.: 'UK Belgium No. 5, Part 2 — Technical and installation aspects'. *Br. Telecomm. Engineering*, **5**, part 2, July 1986, pp. 138-147.

9 CALTON, R.L., HEATLEY, D.J.T., MOTTRAM, S.W., HUNTER, C.A., and BROCKBANK, R.G.: 'The design and performance of the PTAT-1 optical receiver'. *Br. Telecomm. Technol. J.*, no. 1, Jan. 1989, pp. 78-82.

10 BACH, N.R., WHITE, B.R., and THORP, S.C.: 'Highly integrated 20 Mbit/s GaAs receiver for use in TPON termination equipment'. *Electronics Lett.*, **27**, no. 7, 28 Mar. 1991, pp. 566-568.
11 LEE, W.S., DAWE, P.J.G., SITCH, J.E., and HADJIFOTIOU, A.: 'Hybrid optical receivers with integrated electronics'. *IEE Coll. 'Optical Detectors and Receivers'*, London, Oct. 1993, Coll. Digest 1993/173.
12 KIMBER, E.M., PATEL, B.L., and HADJIFOTIOU, A.P.: '12 GHz pin-hemt optical receiver front-end'. *IEE Coll. 'Optical Detectors and Receivers'*, London, Oct. 1993, Coll. Digest 1993/173.
13 VIOLAS, M.A.R., DUARTE, A.M.O., HEATLEY, D.J.T., and BADDOW, D.M.: 'The design and performance of a 10 GHz bandwidth low noise optical receiver using discrete commercial devices'. *2nd Bangor Symposium on Communications*, University of Wales, 23-24, May, 1990, pp. 77-81.
14 VIOLAS, M.A.R., HEATLEY, D.J.T., XUAN YE GU, CLELAND, D.A., and STALLARD, W.A.: 'Heterodyne detection at 4 Gbit/s using a simple p-i-n h.e.m.t. receiver'. *Electronics Lett.*, **27**, no. 1, 3 Jan. 1991, pp. 58-61.
15 VIOLAS, M.A.R., and HEATLEY, D.J.T.: 'Microstrip and coplanar design considerations for microwave bandwidth pin-hemt optical receivers'. *IEE Coll. on 'Microwave opto-electronics'*, 26 Oct. 1990.
16 KITAMURA, K., ITO, K., MATSUDA, H., KANEKO, T., and HANEDA, M.: 'Broadband receiver module for 10 Gbit/s optical transmission'. *Electronics Lett.*, **27**, no. 16, 1 Aug. 1991, pp. 1435-1436.
17 SODA, M., TAKEOUCHI, T., SAITO, T., SUZUKI, T., FUJITA, S., NAGANO, N., and HONJO, K.: '10 Gbit/s optical receiver module using AlGaAs/GaAs hbt preamplifier ic'. *Electronics Lett.*, **28**, no. 3, 30 Jan. 1992, pp. 336-338.
18 CHAPMAN, D.A.: 'Erbium-doped fibre amplifiers: the latest revolution in optical communications'. *IEE Electronics and Commun. Engineering J.*, **6**, no. 2, April 1994, pp. 59-67.
19 SIMEONIDOU, D., YU, A, O'MAHONEY, M.J., YAO, J., and SIDDIQUI, A.S.: 'Erbium-doped fibre pre-amplifier with an integral band-pass filter'. *IEE Coll. 'Optical Detectors and Receivers'*, London, Oct. 1993, Coll. Digest 1993/173.
20 LEE, W.S., PETTITT, G.A., WAKEFIELD, J., PATEL, B.L., and HADJIFOTIOU, A.: 'Practical implementation of high performance optical transmitter/receiver subsystems for 20 Gbit/s tdm operation'. *IEE Coll. on 'High capacity optical communications'*, London, May 1994, Digest 1994/120.
21 BLAIR, L.T., and NAKANO, H.: 'High sensitivity 10 Gbit/s optical i.m.d.d. receiver using two cascaded edfa preamplifiers'. *Electronics Lett.*, **27**, no. 10, 9 May 1991, pp. 835-836.
22 MASUDA, H., and TAKADA, A.: 'High gain two-stage amplification with erbium doped fibre amplifier'. *Electronics Lett.*, **26**, no. 10, 10 May 1990, pp. 661-662.
23 BOUDREAU, R., MORRISON, R., SARGENT, R., HOLMSTROM, R., POWAZINIK, W., MELAND, E., WILMOT, E., and LACOURSE, J.: 'High gain (21 dB) packaged semiconductor optical amplifiers'. *Electronics Lett.*, **27**, no. 20, 26 Sept. 1991, pp. 1845-1856.
24 MELLIS, J., CULLEN, R., SHERLOCK, G, WEBSTER, S.M., and CAMERON, K.H.: 'Packaged high-gain and polarisation insensitive semiconductor optical amplifiers'. *Br. Telecomm. Technol. J.*, **9**, no. 4, Oct. 1991, pp. 110-113.
25 WAKE, D., JUDGE, S.N., SPOONER, T.P., HARLOW, M.J., DUNCAN, W.J., HENNING, I.D., and O'MAHONEY, M.J.: 'Monolithic integration of 1.5 μm optical preamplifier and pin photodetector with a gain of 20 dB and a bandwidth of 35 GHz'. *Electronics Lett.*, **26**, no. 15, 19 July 1990, pp. 1166-1168.
26 HUNTER, C.A., BARKER, L.N., HEATLEY, D.J.T., CAMERON, K.H., and CALTON, R.L.: 'The design and performance of semiconductor laser pre-amplified optical receivers'. *IEE Coll. 'Optical Amplifiers for Communications'*, Coll. Digest 1989/119, 27 Oct. 1989.
27 FYATH, R.S., and O'REILLY, J.J.: 'Performance of optical receivers incorporating cascaded semiconductor laser preamplifiers'. *IEE Proc.*, **137**, Part J, no. 3, June 1990, pp. 186-192.

10

PERFORMANCE CHARACTERISTICS OF OPTICAL RECEIVERS

From this hour I ordain myself loos'd of limits and imaginary lines,
Going where I list, my own master total and absolute,
Listening to others, considering well what they say,
Pausing, searching, receiving, contemplating, . . .

(*Song of the Open Road*, Walt Whitman, 1819–1892)

10.1 INTRODUCTION

The preceding chapter described in detail aspects of specifications for optical receivers for use in ofdls under the following subsection headings: performance characterisation of pinfet optical receivers, directly measurable quantities of pinfet modules, design information on photodiodes and optical receivers, choice of receiver. The treatment of receivers for imdd systems covered the design and performance of the PTAT-1 receiver, those for use in TPON terminating equipment, aspects of receiver design and performance at 10 Gbit/s, broadband modules and front-end amplifiers for high-capacity. systems. The chapter closed with consideration of receivers using optical preamplifiers. In this second of three chapters devoted to the broad topic of receivers, the treatment is more narrowly focused on photodiodes themselves, and especially those properties of direct interest in evaluating system performance. PIN and avalanche devices each occupy a full section, with a similar pattern of topics: types in service in the late 1980s (i.e. those currently carrying live traffic rather than under development in laboratories), responses in both frequency and time domains, noise considerations and reliability. There are two additional subsections on advanced apd devices and apd components with an integrated dielectric waveguide filter.

Throughout the whole development period of optoelectonic components for optical fibre line systems the receiver has tended to be the poor relative in terms of resources devoted to solving the technical problems ([2] of Chapter 6). This happened despite the demands of system designers for the last fraction of a dB in sensitivity performance from a receiver, while dBs lost elsewhere, such as in the coupling between laser and system fibre were apparently of lesser account. Despite their relative obscurity, the designers of photodetectors and of receivers have worked closely together, since the two functions are inseparable if the best performance is to be achieved. Device designers have risen to the challenges, and significant improvements have been made over the years, contributing to the realisation of cost-effective ofdls and their widespread penetration in national and international networks.

Until 1994 all optical receivers in operational systems had used incoherent or direct detection; that is, the transducer converts mean intensity values of incident radiation into mean values of photoelectric current, assuming a one-to-one correspondence between detected photons and photoelectrons. Fluctuations about the mean value of the radiation are converted into r.m.s. values of photocurrent [1].

The photodetector component must be considered in the context of the receiver front end, and as such its noise performance is of prime concern, since this will determine the repeater spans achievable for a specified bit error rate. Although the noise characteristics are of primary importance, there are others which also need to be considered; in the case of the photodiode itself these are the quantum efficiency and of course reliability, while the preamplifier must be of very low noise as well as meeting reliability criteria. The optoelectronic transducers which are of interest in systems are semiconductor pn junction devices utilising the photovoltaic effect. These devices were made from silicon or germanium for first-generation systems and for 1300 and 1550 nm systems the same III-V compounds as used in lasers. This can most readily be seen from a diagram which displays the responsivity, which is the ratio of photocurrent to mean received power as a function of wavelength, together with the associated quantum efficiency, as shown in Figure 10.1 [2] below.

For the wavelengths of interest, detectors made from InGaAsP have higher quantum efficiencies than those made from germanium. The InGaAsP mixture ratios used differs from those employed in lasers. The photodiodes are of two broad classes pin or avalanche, the pin with unity gain, whereas in the latter each photoelectron is multiplied by a factor which can be varied to optimise performance, but is of the order of 10. A table that compared the relative merits of receivers incorporating

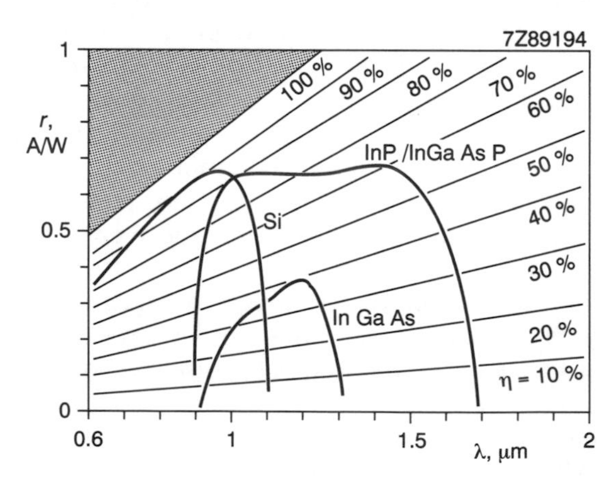

Figure 10.1 Responsivity and internal quantum efficiency versus wavelength for semiconductor photodiodes fabricated from Si, InGaAs and InP/InGaAsP ([2], Figure 19). Reproduced by permission of Philips

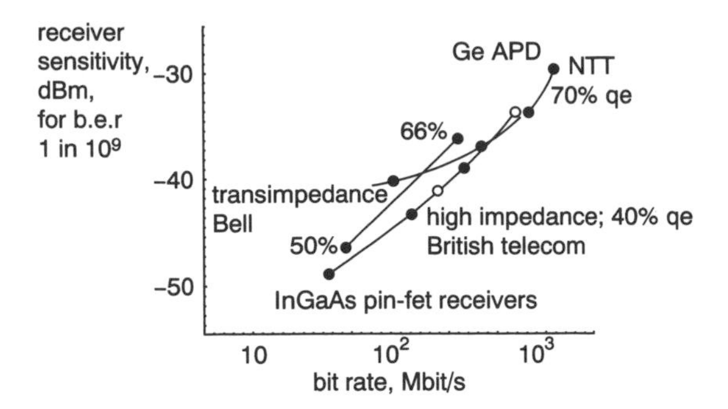

Figure 10.2 Reported performance (early 1980s) of receivers based on pinfet hybrid and Ge apd front ends operating at 1300 nm and at 20°C ([3], Figure 10). Reproduced by permission of IEEE, © 1982

devices available in the mid-1980s was given in [1]. An early and very useful paper which gave a comparison between two Ge avalanche photodetectors for the 1200–1550 nm range [3]. This paper also addressed the question of the features needed in an apd to compete with the very successful pin mesfet hybrid, where the photodiode was fabricated from InGaAs. Receiver sensitivities were calculated for line rates of 8, 40, 160, 294, 600 and 1200 MBd, taking realistic values of the photodiode and transistor parameters; these results will be given later. In the same paper, practical results were displayed comparing performance, and these are shown in Figure 10.2.

10.2 BASIC PRINCIPLES OF SEMICONDUCTOR PHOTODIODE OPERATION

It may help the reader if the basic principles of the photovoltaic effect employed in photodiodes are sketched briefly.

Photoelectric effects are most readily explained in terms of the particle aspect of radiation, i.e. in terms of photons. These quanta of the electromagnetic field were described in Chapter 3; for the present the property of interest is that their energy is proportional to frequency, the constant of proportionality being Planck's constant h, so that $U = hf$. The energy is expressed in units of electron-volts (eV) or joules (J).

If the energy U is greater than the bandgap energy U_G of the material from which the photodiode is made, the photon energy is completely absorbed by an electron in the semiconductor lattice, and one valence electron receives sufficient energy to cross the forbidden band into the conduction band where it contributes to the photocurrent.

The loss of an electron from the valence band leaves a positively charged hole; thus the absorption of one photon gives rise to an electron–hole pair of charge carriers. The number of such pairs generated per incident photon is called the quantum efficiency.

The exponential decay of the intensity of the radiation (i.e. the number of photons) as it travels through the semiconductor is quantified by the absorption coefficient, the reciprocal of which gives the penetration depth at which the intensity has fallen to $1/e$ of its value at the surface. For the wavelengths and materials of interest this depth is of the order of 10 μm.

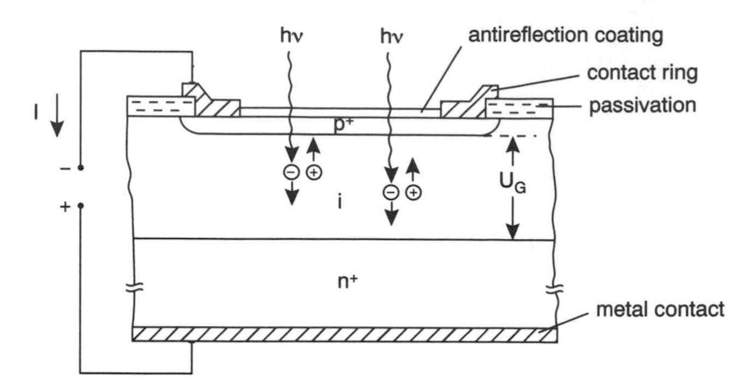

Figure 10.3 Diagram illustrating the principle of operation of a pin photodiode ([4], Figure 1). Reprinted by permission of Schiele & Schön

In the very thin 'active' layer in which the electron–hole pairs are generated the transit time of these carriers is about 10 ps per μm, the value of the saturated drift velocity.

The operation of a photodiode can be most simply explained in terms of the so-called pin diode, shown in schematic form in Figure 10.3 [4].

A lightly doped layer is sandwiched between heavily doped p^+ and n^+ layers, to form the pin junction part of the structure shown in the figure. The diode is reverse-biased so that an electric field and a depletion region extends throughout the intrinsic layer, and in the absence of illumination only the dark current flows in the external circuit. The amount of dark current, and its behaviour with temperature and with time throughout its operating life, are important properties from a system viewpoint.

Theory

Let P_1 be the mean value of the optical power incident on the active surface of the photodiode when a mark is received, and let the fraction η contribute to pair generation; then the relation between incident radiation and corresponding average photocurrent can be written:

$$\frac{I_1}{q} = \frac{\eta P_1}{hf} \tag{10.1}$$

where I_1 is the displacement current [A] produced by a 1 in the symbol interval T[s], q is the electron charge, 1.602×10^{-19}, [As] or coulombs, η is the quantum efficiency, and h is the Planck's constant, or quantum of action, 6.626×10^{-36} [Js].

This process of photon-to-electron conversion is in fact statistical due to the discrete nature of radiation and current. The production of photoelectrons is a random process which obeys Poisson statistics over any interval which is short compared with any time variations of the incident mean power. If we consider an ideal photon detector, electrons will be generated at an average rate Λ which is proportional to the incident mean power, and the constant of proportionality can be written as η_I/hf, and this can in turn be expressed as ϕ_I/q, where ϕ_I is the responsivity of the ideal photodiode. If the incident mean power P is constant, e.g. corresponding to the ideal nrz rectangular 1 or a 0, the average generation rate is constant, and one has:

$$q\Lambda = \phi_{\mathrm{I}}P \tag{10.2}$$

with

$$\phi_{\mathrm{I}} = \frac{q}{hf} = 0.805\lambda(\mu\mathrm{m}), \ \text{units} \ \frac{\mathrm{A}}{\mathrm{W}} \ \text{or} \ \frac{\mu\mathrm{A}}{\mu\mathrm{W}}$$

For a given P, the number of primary photoelectrons n produced in any interval is a discontinuous random variable which is statistically independent of the number produced in any other disjoint interval. This follows from the basic properties of light. The probability that n are generated in the symbol interval T depends only on the mean value N, i.e. the Poisson distribution:

$$\text{prob } (n = N, T) = \frac{N^n}{n!} \times \exp(-N) \tag{10.3}$$

where

$$N = \frac{\phi_{\mathrm{I}}PT}{q} = \frac{\eta PT}{hf}$$

and we note that PT is the energy U of the radiation in the time slot $T = 1/f_S$, where f_S is the signalling speed on the line. The probability distribution given by equation (10.3) possesses the properties that the mean N and the variance σ^2 of the random variable n are equal. The mean corresponds to the signal amplitude, whereas the variance (mean square or dispersion) has the significance of a noise power, since it measures the fluctuations around the mean value.

The Poisson distribution also applies to the generation of dark current, i.e. the generation of electrons in the absence of incident light.

It is important to understand the consequences of this statistical relation for a digital system. If a pulse is received at the photodiode during an interval of duration T, it is possible that no electron will be generated, and the ideal detector will indicate a space, while even one electron will indicate a mark. In the absence of any incident radiation, i.e. a space, the ideal detector will always correctly indicate a space. Returning to the mark received condition, equation (10.3) enables us to evaluate any given probability that no photoelectrons are produced simply by putting $N = 0$. For system calculations it is customary to choose a probability of 10^{-9} or less, and by recalling that $X^N = 1$ and $N! = 1$ when $N = 0$, leads to a minimum received energy equal to 21 photons. For equally probable marks and spaces the average detected power for a ber of 10^{-9} is given by $21hf/2T$, a value often called the quantum limit. In practical direct detection systems thousands of photons are required to meet this error rate as we will see later, so that N becomes very large compared to 21 and the probability distribution becomes Gaussian. For coherent detection systems the numbers of photons required are much closer to the limit. A quantum limit can also be derived for analogue signals [5].

Practical photodiodes produce dark current in the absence of illumination, which is an important constraint on performance, because it increases rapidly with temperature and increases with ageing. The emission of individual dark current electrons also obeys Poisson statistics, and can therefore be represented by an equation very similar to equation (10.3).

Since photodiodes operate on a photon-by-photon basis, the envelope of the signal is recovered by the bandlimited nature of the practical component and transit time effects. This causes noise-in-signal generated by a mark to spill over into the next interval. The noise-in-signal is non-stationary, unlike electronic system noise, and this can be appreciated by the following consideration. Let a single electron be injected into a space-charge region at $t = 0$, and travel with a saturated velocity V_{SAT}. This can be modelled by the impulse response $h(t) = 1/\tau$ for $t \leq \tau$ and 0 elsewhere, with τ the transit time. The corresponding frequency-domain representation is a sin y/y function, where $y = \pi\tau f$, and the 3 dB bandwidth is given by:

$$B_3 = \frac{0.445}{\tau} \tag{10.4}$$

In practice, many electrons are emitted and the overall waveform is obtained by convolving the individual responses; therefore the fluctuations in the numbers of electrons in successive intervals varies, i.e. the noise is non-stationary and bandlimited.

It can be shown that the fluctuations of both photoelectrons and dark current electrons have time functions with spectral densities given by the well-known shot noise formula. The decision circuit is therefore influenced by the noise from adjacent time slots to an extent determined by the characteristics of the photodiode and associated circuitry. Fortunately for mathematical simplicity, the numbers of electrons involved in imdd systems are large and the Poisson process becomes Gaussian.

10.3 PIN PHOTODIODES

These high-performance detectors are relatively inexpensive, need only low operating voltages and are relatively insensitive to variations in temperature. Other properties of major interest in systems are:

(a) good external quantum efficiency
(b) low dark current
(c) small capacitance
(d) fast response time
(e) high reliability

They have thus found widespread application in operational systems at bit rates into the Gbit/s range, and are likely to remain very attractive in many applications. Details of materials, structures and fabrication can be found in the literature. For the purposes of this book we consider a snapshot in time at the end of the 1980s when single-mode fibre systems had become widespread in many developed countries, thanks to the availability of commercial pin detectors in production quantities.

10.3.1 Types in service in late 1980s

InGaAs/InP photodiodes then in common use are illustrated in Figure 10.4. The layer structures have been made using all the conventional growth techniques applicable to III-V semiconductors, i.e. lpe, vpe, movpe and mbe. Of these, lpe is generally favoured because of the relative ease with which the low doping levels ($\simeq 10^{15}$ cm^{-3}) needed to obtain sufficiently low values of capacitance (<0.2 pF). movpe has recently been used to fabricate high-quality devices, and is a technique much better suited to volume production.

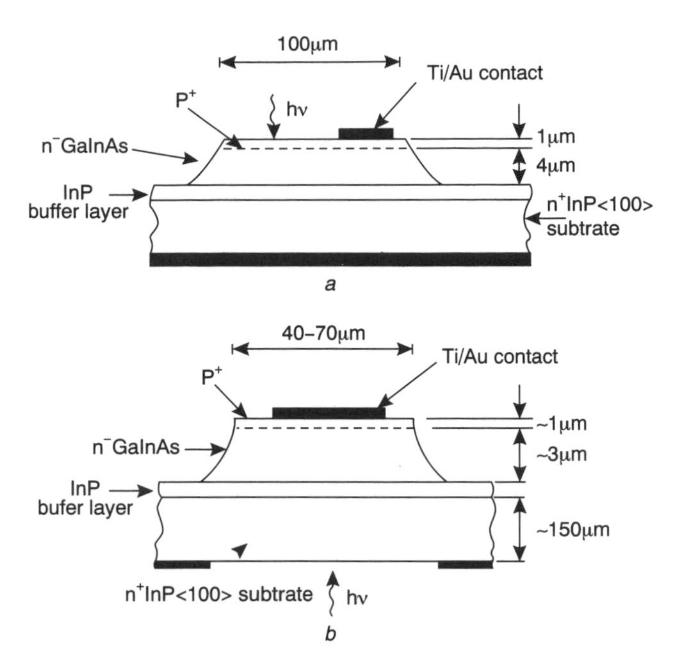

Figure 10.4 Diagram of mesa InGaAs/InP pin photodiodes: (*a*) top entry structure, (*b*) substrate entry structure ([2] of Chapter 6, Figure 14). Reproduced by permission of IEE

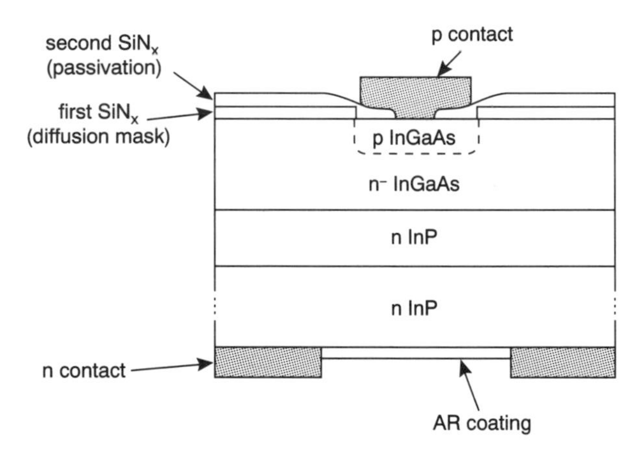

Figure 10.5 Diagram of planar InGaAs/InP pin photodiodes ([2] of Chapter 6, Figure 15). Reproduced by permission of IEE

With the substrate entry configuration, extremely small devices have been made possessing less than 0.1 pF capacitance, quantum efficiency approaching 100% and dark currents below 1 nA. The system implication is that the performance of the optical receiver is no longer limited by the photodiode, but by the associated electronics, particularly the front-end amplifier, which is often a mesfet. As the reader can see from the figure both types are mesa devices, and a drawback of this structure is that it has an exposed pn junction. Devices with protected junctions are preferred for system use, and planar structures are being used increasingly. One such type is shown in Figure 10.5.

A major step towards the commercial production of inexpensive detectors for use in local areas networks and in the local area was described in 1988 [6]. The planar structure is shown in Figure 10.6, and it consists of a GaInAs absorbing layer with an InP capping layer.

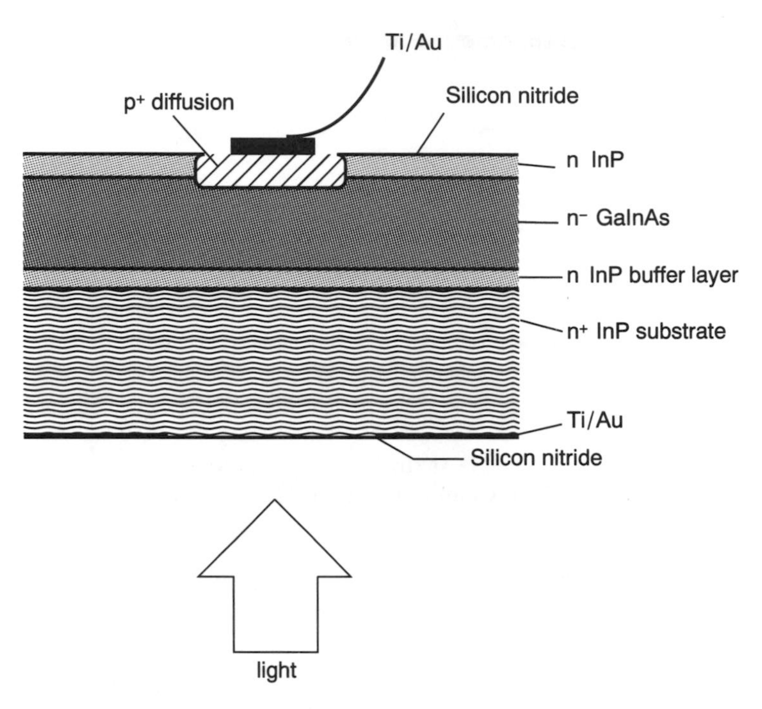

Figure 10.6 Planar InGaAs/InP pin photodiode with substrate entry ([6], Figure 1). Reproduced by permission of IEE

Table 10.1 Typical device parameters [6], Table 1, p. 252 © IEE

	Receiver device Substrate entry	Monitor device Top entry
Active area (diameter)	55 μm	250 μm
Dark current	300 pA (best 8 pA)	10 nA (best 140 pA)
Capacitance	<0.2 pF at 5 V	<5 pF at 5 V
Yield	>75%	>75%
Quantum efficiency	$95 \pm 5\%$ (with AR coating)	$60 \pm 5\%$ (no AR coating)
Speed (rise time)	<90 ps	

Substrate entry devices with active areas of diameter 55 μm for use in low-capacitance (<0.2 pF at −5 V) diodes in receiver packages, and top entry devices with a diameter of 250 μm for use as monitor diodes in transmitter packages. Typical device parameters are listed in Table 10.1.

The detection and characterisation of microwave modulated optical signals or picosecond optical pulses require ultrahigh-speed photodetectors. Devices with light entry perpendicular to the plane of the structure suffer from the need imposed by the thickness of the absorber region to trade-off speed of response and quantum efficiency. Edge-coupled (or waveguide) pin photodiodes can overcome this limitation, because with this geometry thickness has little effect on internal quantum efficiency. An example of such a device was described in 1991 [7] which had a measured bandwidth of 50 GHz at 1530 nm and, combined with a standard 12 μm radius lens-ended sm fibre, gave an external efficiency as high as 40%. In addition, it was stated that a bias-free photodiode with a bandwidth 20 GHz should prove to be a useful component enabling novel receiver design.

The device design was described in detail and the structure was illustrated. Calculated curves of internal, external and coupling efficiencies versus absorber layer thickness were given. The wideband device mount consisted of a brass support block with a Wiltron K connector; it and was shown in cross-section and the precautions taken to minimise parasitics were stated. The measurement method was explained, measured points plotted and fitted using smoothing splines to produce graphs of the relative response versus frequency of a typical device over the band 0–50 GHz with and without bias. The smoothing was done to remove the ripples caused by electrical reflections between photodiode and bias tee.

10.3.2 Responses in frequency and time domains

As a rule-of thumb, the receiver 3 dB bandwidth needed in an imdd system is half the symbol rate on the line, i.e. given roughly by $f_S/2$. By 1995 optically amplified submerged systems operating at 5 Gbit/s will have been laid on a number of transoceanic routes, and experimental systems operating at 10 Gbit/s will have been developed for some years.

The availability of fast photodiodes dates back to the early 1980s, at least. The performance of one type of InGaAsP pin photodiode in a millimetre wavelength coaxial mount was reported in 1986. The values obtained seem almost incredible, and are a dramatic example of the capabilities of the technology [8]. The measured response of the device biased at 5 V is shown in Figure 10.7, by the dash lines, and the computed impulse response of the diode itself by the full lines. The corresponding frequency characteristic obtained by Fourier transformation is given in Figure 10.8.

Using measured values of parasitic capacitances and inductances, and assuming these to be constant throughout the operating frequency range, gave a predicted 3 dB frequency of 69 GHz, compared with 67 Ghz derived from measurements. This bandwidth was, however, only achieved at a low efficiency of about 27%.

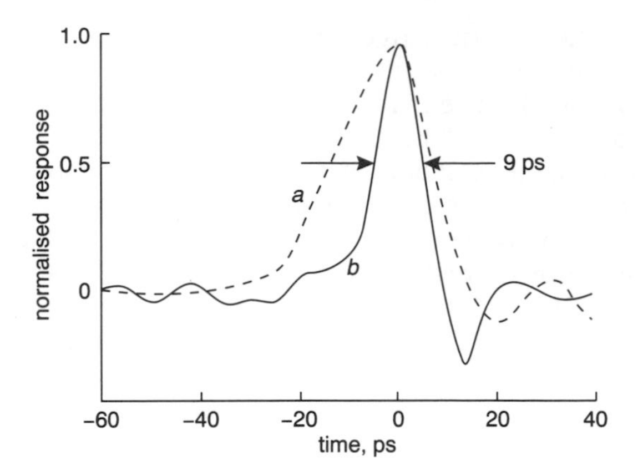

Figure 10.7 Measured impulse response of photodiode biased at 5 V: (*a*) sampled signal, (*b*) deconvolved response ([8], Figure 1). Reproduced by permission of IEE

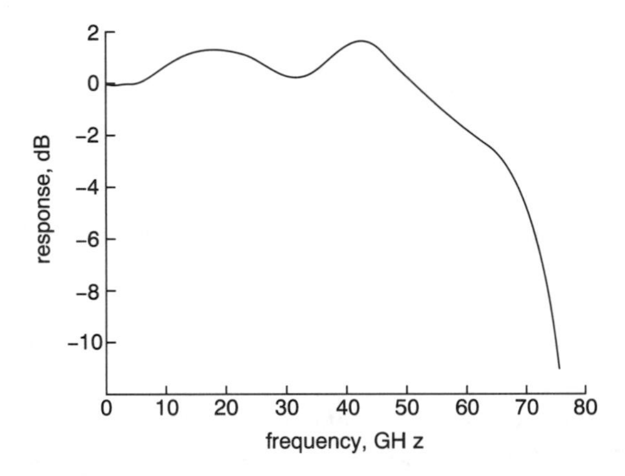

Figure 10.8 Computed frequency response of photodiode obtained by Fourier transformation of the impulse response ([8], Figure 2). Reproduced by permission of IEE

Planar InGaAsP/InP photodiodes with a flat frequency response up to at least 20 GHz were described by authors from Bell Laboratories [9] in 1986. Very low chip capacitance values down to 20 fF were achieved by reducing the pn junction diameter to 25 μm. At 1300 nm a best external quantum efficiency of 37% was obtained. The authors described their high-frequency studies on passivated planar photodiodes, which have assured and predictable reliability, but which do not achieve quite the frequency performance demonstrated by etched mesa devices (mesa structures can be susceptible to problems of dark current drift and short lifetime). These have reached 36 GHz in a coaxial mount and 60 GHz in waveguide, due to a number of factors: material availability, device structures largely free of parasitics, improved understanding of bandwidth limitations imposed by transit-time effects, and the development of optical microwave techniques.

A schematic cross-section of the diode is given in Figure 10.9.

The reverse current–voltage characteristics and the capacitance–voltage characteristics for 10 randomly selected devices from one wafer are also shown in the figure. Because of the very high purity of the absorbing layer the leakage currents are

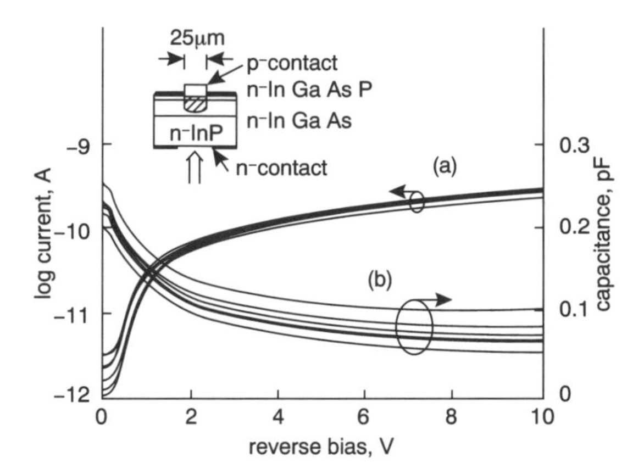

Figure 10.9 Curves of (*a*) current versus reverse bias, and (*b*) capacitance versus reverse bias for substrate entry planar InGaAs/InP pin photodiode ([9], Figure 1). Reproduced by permission of IEE

very low, as expected, typical values being 0.2 nA at −10 V. This can be compared to the corresponding value for 'standard' 75 μm diodes, which is some four times as great. A histogram of chip capacitances measured on seven device wafers peaks at 0.04 pF at −10 V bias, with a large fraction of the sample below half this value. From a system designer's viewpoint, these values are conveniently associated with a load resistance, usually taken as 50 Ω, to give the high frequency cut-off. For the median value this results in a 3 dB frequency of 80 GHz. Under such conditions, the minority carrier transit time through the absorbing layer of thickness 0.4 μm dominates the performance. Another noteworthy feature of these diodes is that they are fully depleted at between 4 and 6 V negative bias.

A planar-junction top-illuminated device with the largest bandwidth (25 GHz) and highest quantum efficiency (80% at 1550 nm) was reported in mid-1989 by staff at British Telecom Research Laboratories and BT&D Technologies [10] for potential application in multi-Gbit/s systems in the late 1990s. The device combined wideband operation and high responsitivity with stability, reliability and ease of use. The chosen structure greatly simplified packaging and also made the device suitable for monolithic integration. Optimum bandwidth was achieved by closely matching the current response due to carrier transit time effects to the network response of the photodiode and its associated packaging parasitics. When housed in a well-designed package the photodiode junction capacitance is the major influence on the component's frequency response.

Coupling of the input light signal was simplified by choosing the active area diameter as 30 μm, and the corresponding variations of bandwidth and efficiency as functions of depletion layer thickness are shown in Figure 10.10.

Calculations were based on a circuit model which included some given values, and the current response from carrier transit time effects in a 1964 paper. The corresponding values of quantum efficiency at 1300 nm exceeded 90%. Process yield and uniformity were excellent and preliminary lifetest results showed no change in dark current.

Photodiode frequency response was measured using two different sources and was also computed; the results are presented in Figure 10.11.

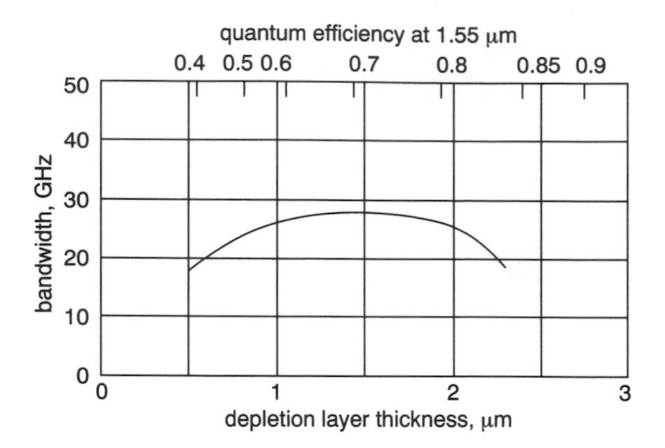

Figure 10.10 Calculated 3 dB bandwidth versus depletion layer thickness of packaged photodiode and associated measured values of quantum efficiency at 1550 nm ([10], Figure 1). Reproduced by permission of IEE

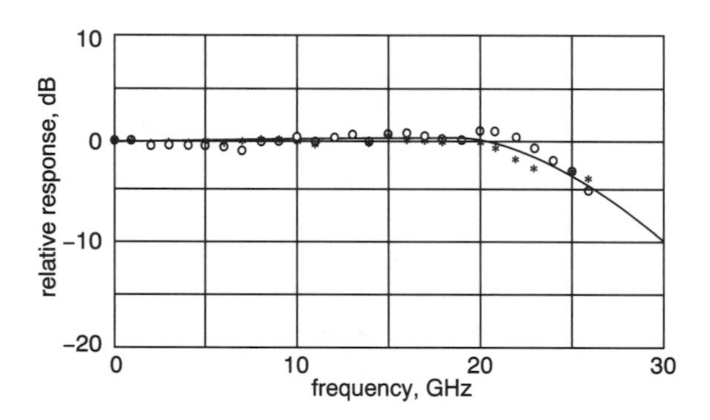

Figure 10.11 Measured and calculated (solid curve) relative response of photodiode packaged on modified Wiltron K-connector versus frequency ([10], Figure 4). Reproduced by permission of IEE

Measurement considerations

The measurement of such high-speed devices poses many problems. Very briefly the technique adopted by Temkin *et al.* was optical heterodyning. The light from two external cavity 1500 nm lasers were combined in an optical fibre; these lasers were chosen in preference to dfb lasers because of their narrower linewidth, in the 10 KHz range as opposed to tens of MHz. The measured beat power was −34 dBm, compared with a calculated value of −29.5 dBm. Earlier, in 1983, picosecond InGaAsP film lasers had been used by workers at the Crawford Hill Laboratory to measure the performance of diodes for the 1300 and 1550 nm regions [11]. Film lasers have three major advantages over sources with longer pulses:

(a) the pulse width is known accurately from measurements which are independent of those of the detector devices under test (dut)

(b) pulse durations of 5–15 ps are much shorter than any dut and oscilloscope response times, so that the source contribution to the measured values of pulse broadening was very small

(c) film lasers can be operated over a very wide waveband of interest, namely 770–1590 nm.

The detectors under test were substrate-illuminated InGaAsP/InP punch-through pin photodiodes. For devices of 25 μm diameter mounted in 50 Ω stripline, the fwhm pulse duration was 30 ± 1 ps, obtained by deconvolving the contributions from the oscilloscope and the source. This duration was the same throughout the band 1150–1540 nm, and did not vary for bias voltages between 5 and 15 V; the agrees with the estimated transit-time limitation of the devices.

10.3.3 Noise in pin photodiodes

The sensitivity of a photodiode is the minimum optical power required for a specified error rate, and it is limited by noise. This originates from a variety of sources, and their individual contributions depend on wavelength, diode material and design, and on the associated electronic circuit. At 1300 and 1550 nm detection is limited by either thermal or shot noise, but the only significant photodiode noise mechanisms give rise to shot noise. One representation for the noise currents in diodes is shown in Figure 10.12 [12].

The optical input is taken to consist of two parts, the wanted signal and a background illumination. The fluctuations in these quantities are converted into noise currents of r.m.s values I_{NS} and I_{NB}. In addition to these input-dependent quantities there is a contribution from fluctuations in the dark current given by I_{ND}. In accordance with the standard formula for shot noise in terms of the mean value I of the current whose fluctuations it characterises, we can write:

$$I_{NS}^2 = 2qI_S df = 2q\left(\frac{q\eta}{hf}\right)P_S df \tag{10.5}$$

$$I_{NB}^2 = 2qI_B df = 2q\left(\frac{q\eta}{hf}\right)P_B df \tag{10.6}$$

$$I_{ND}^2 = 2qI_D df \tag{10.7}$$

where df is the noise equivalent bandwidth in which these values are measured, and the spectrum I^2/df of each noise current is therefore constant or white, assuming that P_S and P_B are constant.

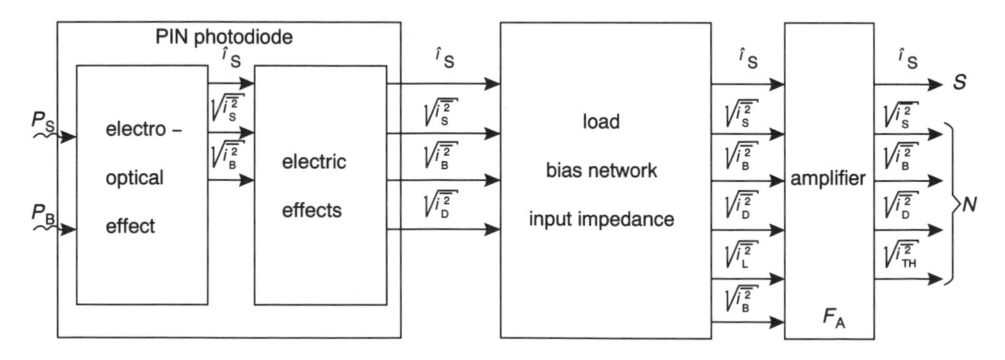

Figure 10.12 Diagram illustrating the shot noise rms currents accompanying signal and dark photocurrents, reverse dark current, and thermal noise currents arising in bias resistor and amplifier input resistance ([12], Figure 18*b*). Reprinted by permission of Academic Press

Calculations of receiver sensitivity are usually done with commercial circuit analysis packages which include noise analysis, such as SPICE © or TOUCHSTONE ©, most of which are based on nodal analysis and have a library of active and passive component models (often inadequate for real devices). The most useful noise representation of a pin photodiode is by means of a number of current generators in parallel, one for the mean value of the signal photocurrent, and one for each of the noise sources, with m.s. values given by equations (10.5)–(10.7). The generators representing the noise sources are uncorrelated, i.e. the total noise power dissipated in unit resistance is simply the sum of the mean square values of the generators. Formulae will be derived and used later in this chapter.

10.3.4 Reliability of pin photodiodes

The most recent reliability figures and calculations available were quoted in an authoritative paper entitled 'Reliability aspects of optical fibre systems and networks' in the *BT Technology Journal* for April 1994 [13]. Table 3 quoted results for a single one-way 3R repeater which included sm fibre systems with pin detectors (90/140 Mbit/s) or with apd detectors in 0.56/1.8 Gbit/s plesiochronous, and 2.4/10 Gbit/s synchronous digital hierarchy. Both had been allocated the same failure rate in terms of operating time normalised to 10^9 hours (fits) of 10 fits and contributed less than 0.1% to the total fits.

Atmospheric pressure movpe is a technique which possesses the capability of providing large areas of uniform, defect-free material with yields of around 80%. The authors of [6] supply a wafer map of device dark currents at 5 V bias. Lifetests using thermal overstress at 125, 175, 200 and 240°C, with 5 V bias applied, have demonstrated the reliability of the photodiodes. A burn-in of 168 hours at 125°C has been found to remove cases of infant mortality and is applied to all devices. For the small area diodes the dark currents were measured and a typical value of 0.3 nA at −5 V obtained. The variation of the room temperature dark currents with time for a batch of devices on lifetest at 200°C is given in Figure 10.13.

For these photodiodes the failure limit was defined as 10 nA. Turning now to the larger monitor photodiodes, the dark currents measured at room temperature and 5 V bias had typical values of 6 nA. The variation of dark currents at room temperature with time for a batch of devices at 240°C is shown in Figure 10.14.

The failure limit for the monitor diodes was taken as 50 nA, but precise values of failure rates could not be calculated, because no devices had failed at the time that the letter was written. However, data to data for the devices logged in Figures 10.13 and 10.14 are equivalent to about 100 years and 300 years, respectively, of service life at 70°C. From the combined results for both types of photodiodes, the random failure rate prior to the onset of any wear-out failure mechanism has been evaluated as less than 0.4 fits to 90% confidence level at an operating temperature of 20°C, whereas at 70°C this has increased to only 13 fits. These extremely low values for predicted failure rates are expected to improve as more data become available.

The long-term stability of planar InGaAs pin photodiodes with InP cap layer and SiN_x, passivation was reported by authors from Siemens at the end of 1988 [14]. Elevated temperatures up to 200°C had been used in the study. An increase of the dark current with saturation at a level of a few nA was observed as the only chip-related degradation mode. Additional experiments enabled this increase to be identified as surface leakage current flowing due to charge transfer in the SiN_x, and

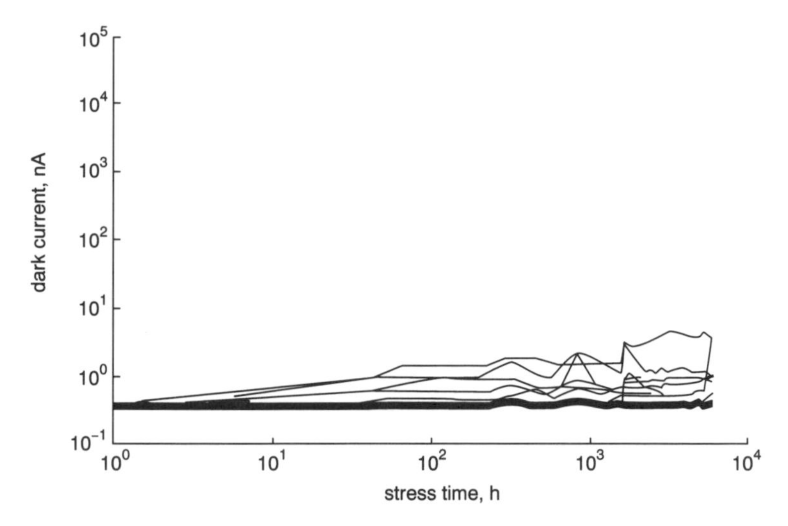

Figure 10.13 Dark current versus stress time of devices on lifetest at 200°C and 5 V bias (measurements at 25°C) ([6], Figure 3). Reproduced by permission of IEE

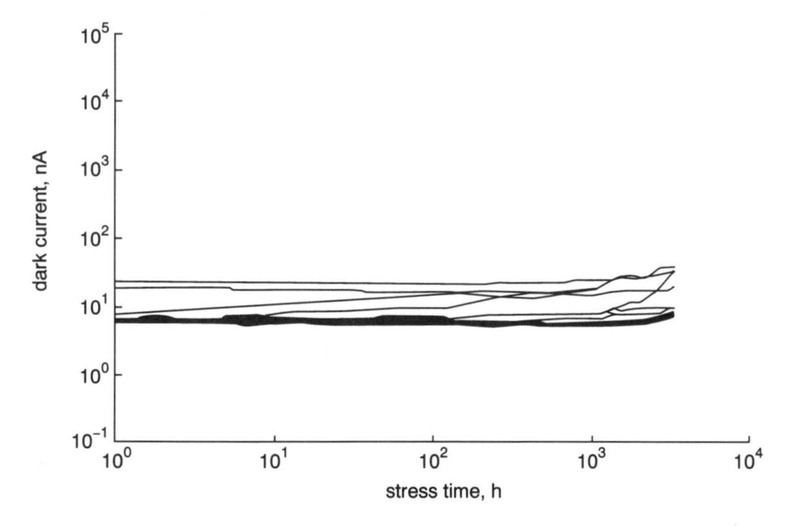

Figure 10.14 Dark current versus stress time of devices on lifetest at 240°C and 5 V bias (measurements at 25°C) ([6], Figure 4). Reproduced by permission of IEE

to change in the InP–SiN_x interface state density. Defining a dark current increase to 1 nA at 15 V reverse bias at room temperature as the criterion of failure, a median lifetime of 10^9 hours was predicted by extrapolation. A total of 112 devices were used in the study. The increase in the dark current is activated both thermally, with an activation energy of 1.3 eV, and by the operating bias voltage.

Reliability methodology

A description of the methodology used by British Telecom Research Laboratories to predict the reliability of InGaAs/InP pin photodiodes was given in early 1989 [15].

Results of accelerated life tests at elevated temperatures were used to predict the mean-time-to-failure (MTTF), the wear-out failure rate and the random failure rate, all at the service temperature. Comparison of the Weibull and the lognormal distribution functions for the analysis of failure data showed that the Weibull had several advantages. The methodology was demonstrated for some very high reliability planar InGaAs/InP pin photodiodes developed at BTRL for use in 1300 and in 1550 nm wavelength systems and manufactured by BT&D Technologies. The life tests predicted that the MTTF will exceed 10^{11} hours at 20°C, and that the total failure rate in service will be less than 0.3 FITs at 20°C.

The paper was divided into five main sections covering:

(1) methods used to predict the reliability
(2) fabrication and performance of the photodiodes
(3) life tests and measurements
(4) reliability results
(5) conclusions.

(1) contained subsections which described: accelerated life-tests, the variation of failure rate during component life, reliability parameters, Weibull and lognormal distribution functions, calculation of MTTF, calculation of wear-out failure rate during service life, and calculation of random failure rate.

(2) was headed 'Photodiode fabrication and performance'.

(3) was titled life testing and measurements.

(4) The section heading was 'Life test results and reliability calculations', and contained subsections: 'life test results', 'recovery under unbiased baking, and 'reliability calculations'.

(5) stated the conclusions.

10.4 AVALANCHE PHOTODIODES

As stated earlier, this type of photodiode is, or will be, used in 3*R* repeaters in plesiochronous and in synchronous digital imdd systems at rates of 565 Mbit/s and above, where their sensitivity advantage over pin detectors outweighs their disadvantages. They are also deployed on the receive side of the main transmission path in terminals, not only in such systems but also in optically amplified systems. In the future they may well find application in wdm systems.

A solid-state analogue of the photomultiplier tube was realised in the first half of the 1960s by means of the silicon avalanche photodiode (apd), which had a structure very similar to that of the simple pn junction diode. Absorbed photons generate electron–hole pairs, which are separated by an electric field, and flow in an external circuit. By operating the apd very close to the avalanche breakdown point the primary generated carriers may produce more carriers by impact ionisation of atoms in the lattice, and thus gain is achieved. Receivers employing these silicon apd devices had some 15 dB better sensitivity than unity gain detectors, so that in first-generation systems at 850 nm with fibre losses of say 2 dB/km, the use of these devices allowed an extra 7 km in length, making all the difference between an economic and an uneconomic system [6.2 ([2] Chapter 6)]. Some of the important features of receivers based on pin and apd structures using various materials are highlighted in Table 10.2.

Table 10.2 Relative merits of receivers incorporating III-V PIN diodes, Ge APDs, and III-V APDs. Reproduced by permission of IEE

	III-V PIN diode	Ge APD	III-V APD
Device structure	simple	difficult especially for 1.55 μm operation	very difficult
Fabrication	new technology but relatively simple	established technology	still in research stage, yields poor
Receiver sensitivity	very good at low bit rates progressively worse above 565 Mbit/s, trade-off with dynamic range	better at high bit rates, especially if device cooled	potentially gives best sensitivity, but by only a few dB. Still considerably inferior to Si APD
Associated circuitry	low voltages, high performance FETs needed, great care with circuit layout to minimise stray capacitances	high voltage operation simpler circuitry, but more of it, extra circuitry to stabilise gain, and compensate for temperature variations	
Applications	chosen for submarine systems because of simplicity, will be preferred for coherent systems	preferred for high bit rate land systems at present problems with saturation in coherent systems	still at research stage

As the authors note, perhaps the most important point to mention is that the ultimate in receiver performance should not be taken as the *sole criterion* (author's italics) in assessing a photodetector. Other factors such as dynamic range, performance over an operating temperature range, associated circuit complexity, cost and reliability all need to be taken into account.

10.4.1 Types of apd in service in the 1980s

Because silicon becomes transparent at wavelengths longer than 1100 nm or so, other semiconductor materials are needed for second-and third-generation applications, and experience with transistors pointed to germanium. As can be seen from Figure 10.15 this material, which has a direct bandgap, is suitable for use to around 1500 nm.

The behaviour of the absorption coefficient with wavelength is of prime importance in the design of photodiodes. This coefficient describes the exponential decay of the intensity of light along its path in the semiconductor, and the reciprocal is the penetration depth at which the intensity has fallen to $1/e$ of the initial value. With indirect semiconductors (Si, Ge) the coefficient increases relatively slowly as λ decreases from the values corresponding to the bandgap energy for each material (see along bottom edge). Only after the bandgap transition becomes direct does it rise sharply, as can be seen from the figure. For germanium only the direct transition is utilised in photodiode applications. Direct bandgap materials like InP, GaAs and most of the ternary and quaternary alloys of these compounds reach very large

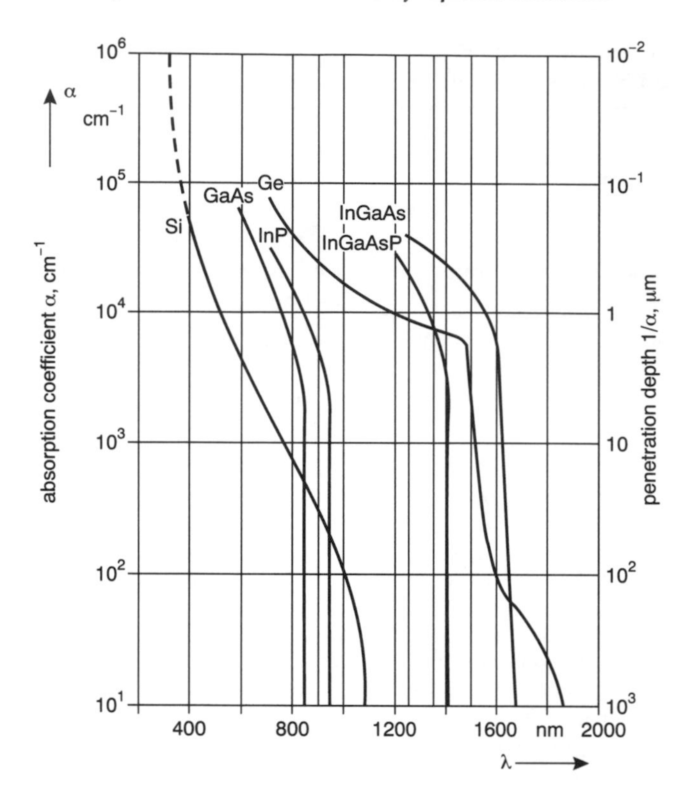

Figure 10.15 Absorption coefficient α versus wavelength and penetration depth $1/\alpha$ versus wavelength of semiconductor materials of interest in photodetectors ([2] of Chapter 3, Figure 12.14). Reproduced by permission of Publicis MCD Werbeagentur

values immediately below the bandgap wavelength. Again, this behaviour is clear from the curves.

Ge apds have been developed over a 20 year period, and they are used in operational systems, especially by Japanese system manufacturers. Germanium does, however, have a small bandgap energy and a high density of states in the conduction band, both of which contribute to a large reverse bias dark current. Since 1978 there has been a gradual evolution and improvement in these devices, mainly in Japanese laboratories. A few of the stages in the evolution, covering the years from 1978 to 1984, are illustrated in Figure 10.16.

The latest of the structures, called the Hi-Lo type (Figure 10.16*f*), was among the highest-performing Ge apds for use in the 1550 nm window. The estimated penalty relative to zero dark current was only 1.5 dB at 25°C and 3 dB at 45°C. State-of-the-art Ge apds were very competitive with pinfet receivers and with the recent InGaAsP avalanche photodiodes, but it seemed that their performance had reached a limit set by the inherent noise properties of the impact ionisation process. More recently, spurred on by the developments in material systems for lasers, apds have been fabricated in the quaternary compound InGaAsP, but with mixtures which differ from those used in sources. Devices made from the III-V compounds have reverse bias dark currents one or two orders smaller than germanium, and are therefore attractive from a systems viewpoint, so they will also be considered in the next few sections.

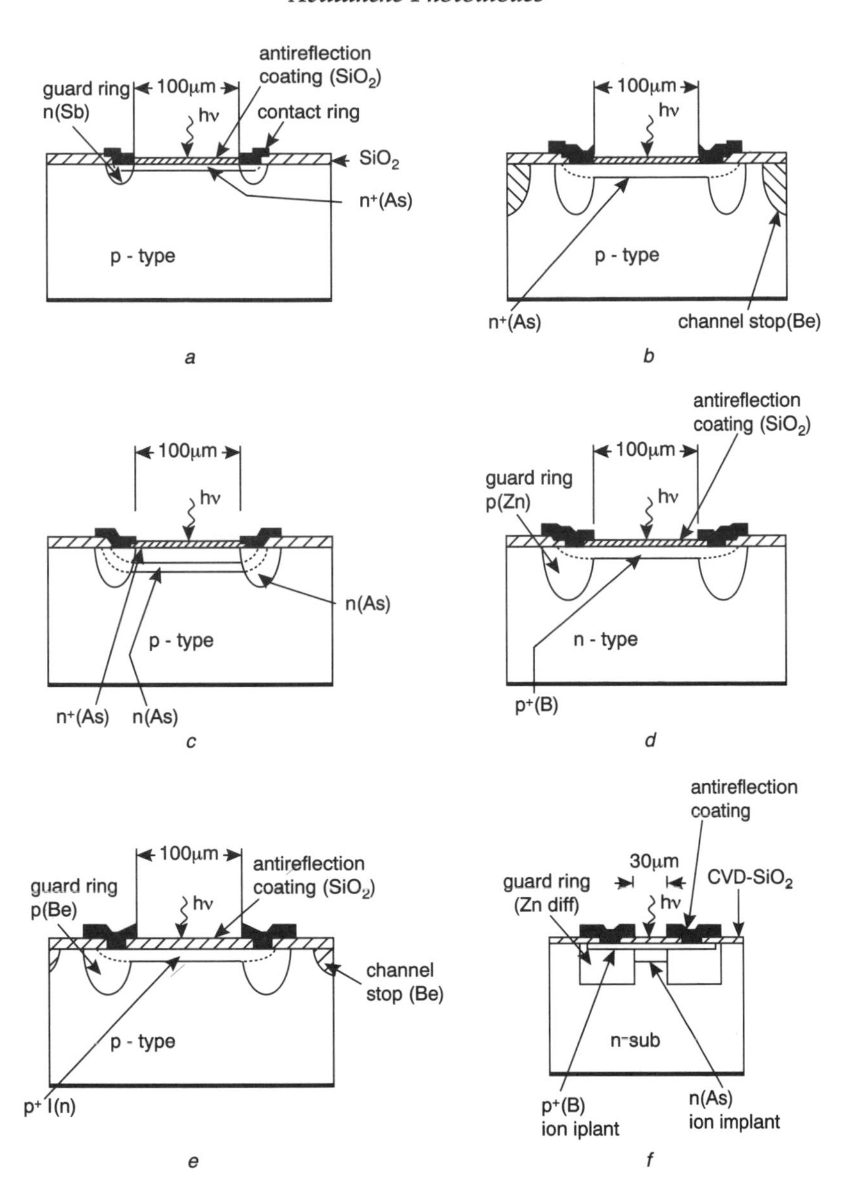

Figure 10.16 Figure showing six stages in the evolution of germanium apds ([2], Figure 12). Reproduced by permission of IEE. *a* Basic n^+p structure (from reference 64, IEEE Journal of Quantum Electronics, Copyright © 1978 IEEE), *b* Improved n^+p structure (from reference 65, Fujitsu Sci. & Tech. Journal), *c* n^+np structure reference 66, IEEE Journal of Quantum Electronics Copyright, © 1980 IEEE), *d* p^+n structure (from reference 67, Applied Physics Letters), *e* all-implanted p^+n structure (from reference 68, Applied Physics Letters), *f* p^+nn^- structure (Hi-Lo) (from reference 70)

The classic paper on the properties of avalanche photodiodes is that by P.P. Webb, R.J. McIntyre and J. Conradi of R.C.A. published in 1974 [16]. Under the heading 'basic properties', the authors considered ionisation rate, gain or multiplication, excess noise factor, gain probability distribution, linearity and frequency response. They then went on to compare apd structures, followed by the properties

Table 10.3 Comparison of optoelectrical characteristics at 25°C

Characteristic	symbol	min	typ	max	unit	test conditions
Reverse breakdown volt.	V_{BK}	60	90/40	120	V	$I_D = 100$ μA
dark current	I_D	—	30/250	100	nA	$V_R = 0.9\ V_{BK}$
terminal cap.	C_T	—	0.5/1.7	1.0	pF	$V_R = 0.9\ V_{BK}$
multiplication factor (gain)	M	20	40/100	—		$\lambda = 1550$ nm /$\lambda = 1300$ nm
quantum eff.	η	70	80/85 70	—	%	$\lambda = 1300$ nm $\lambda = 1550$ nm
cut-off freq.	f_{co}	1	/1.6		GHz	M = 10
effective radius of active area		30	40/100		μm	M = 10, 80% of peak
excess noise factor			/8.5			/$\lambda = 1300$ nm, M = 10

Notes:
InGaAs device/Ge device values
All Ge device values listed in Table 1 of [10.17] have been taken as typical test conditions refer to both types unless distinguished by /

of reach-through apds for which quantum efficiency, speed of response, noise equivalent power were treated. Temperature effects were described, then gain uniformity. In an appendix two further topics were treated: the derivation of an equivalent circuit for the apd and transit time effects in the device. All these topics are still matters of concern to the designer of optical receivers. In high rate systems very good receiver sensitivity is an essential requirement, and this needs low values of dark current and high values of quantum efficiency, as well as minimum noise.

Optical receiver designers need details of commercially available apds. As an early but detailed example, Table 10.3 compares the data given by workers from NEC in [17] with that given in a NEC preliminary specification for the InGaAs NDL 5500P device, dated April 1987.

10.4.2 Advanced avalanche photodiodes

As with semiconductor lasers, the move from bulk structures to structures embodying quantum wells has enabled enhanced performance to be achieved in certain respects. An illustration of the characteristics of one such device will now be given, before advanced bulk devices are considered.

Multiquantum well avalanche photodiodes

Development of advanced apd structures with superior performance characteristics to conventional bulk devices took place throughout the late 1980s. A graded avalanche photodiode with separate undoped GaInAs absorption layer and an AlInAs/GaInAs quantum well avalanche region in which pure electron injection was achieved under top or rear illumination was developed at CNET, Bagneux. The potential of quantum wells as a multiplication layer was evaluated, and the results demonstrated the potential of the structure for high bit rate systems [18].

In apd devices a limitation on noise performance and hence sensitivity is set by the ratio of the ionisation rates for electrons (α) and holes (β), which when close to unity gives a large value of excess noise factor and limits the gain × bandwidth product. These limitations also partly apply to GaInAs/InP separate absorption, grading and multiplication devices (sagm-apd). Research into other materials and structures which will provide a high value for the ratio has resulted in devices using AlInAs/GaInAs quantum wells as a multiplication layer. A larger electron ionisation rate was obtained than in bulk GaInAs material, whereas the rate for holes was close to that for the bulk material, resulting in α/β approaching 10. The structure also ensured that the absorption layer was fully depleted under operating conditions, which reduced the carrier transit time to a very small value, hence a very large bandwidth.

Results: the use of suitable doping levels produced a minimum capacitance of 0.5 pF for a 40 μm diameter diffused junction when reverse biased with more than 19 V. The dark current (upper curve) and the photocurrent measured at 1300 nm with a 10 MHz square wave is shown in Figure 10.17.

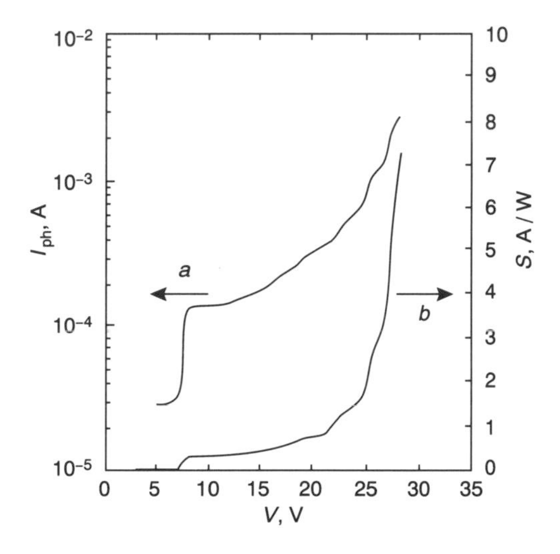

Figure 10.17 (*a*) Photocurrent and (*b*) responsivity versus reverse voltage; $M = 1$ at 20 V ([18], Figure 3). Reproduced by permission of IEE

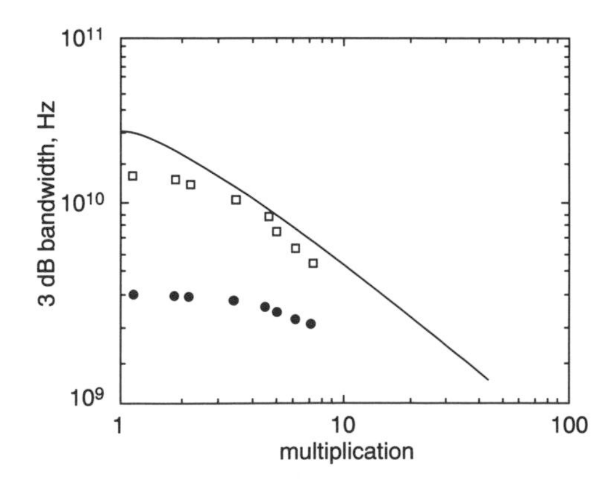

Figure 10.18 3 dB bandwidth of sagm-apd versus multiplication: • as measured, □ deconvolved ([18], Figure 4). Reproduced by permission of IEE

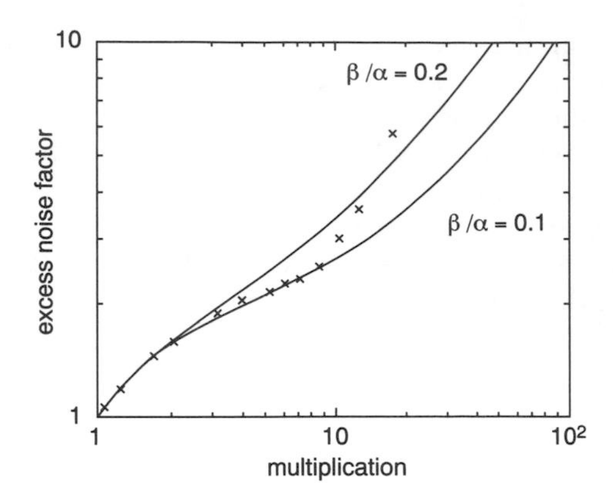

Figure 10.19 Excess noise factor versus multiplication for sagm-apd; solid curves are from theory due to McIntyre, crosses represent values measured in a 1 GHz bandwidth ([18], Figure 5). Reproduced by permission of IEE

$M = 1$ corresponded to 20 V reverse bias. At this voltage the responsivity was 0.6 A/W, which rose to about 5 A/W at 24 V as shown in the lower curve. The low transit time corresponded to an estimated intrinsic gain-bandwidth product of 50 GHz; the behaviour of the 3 dB bandwidth versus M is illustrated in Figure 10.18, estimated by measuring the increase in the impulse response time as M varied. The contribution of the avalanche build-up time to bandwidth reduction became apparent for $M > 3$ in agreement with the (Emmons) curve; for smaller values the 3 dB bandwidth was estimated as 9 GHz. Finally, the excess noise factor versus M is given by the plotted points in Figure 10.19; note the good agreement with the McIntyre curve for $\beta/\alpha = 0.1$ when $M \leq 9$. The departure above this value was explained.

Advanced avalanche photodiodes using bulk III-V materials

Despite appearing as a natural development of the pin diodes made from the same material and the work devoted to making devices suitable for optical systems over the past 10 years or so, results have been meagre. Problems have been encountered and their extent and difficulty seem out of proportion to the relatively small sensitivity improvements achieved. However, they do show promise for systems use at bit rates above 565 Mbit/s, and by 1986 the best sensitivity results had been achieved with such devices.

Among the most recent types of device are those which separate the absorption and multiplication regions, the so-called samapd, as with the much earlier reach-through silicon apd. A selection of recent structures of this type are illustrated in Figure 10.20.

Figure 10.20*a* shows a structure which can be grown by lpe, the best developed of all the growth techniques. This device is optimised for high speed and a gain × bandwidth product of 80 GHz has been reported. The second of the three structures is also grown by lpe and, in addition, has the very attractive feature of planar construction, thus having no exposed pn junction. Data given for this type include a multiplied unity-gain dark current I_{D1} of only 10 nA, nearly low enough to have insignificant effect on receiver sensitivity, an excess noise factor of 5–6 at a gain of

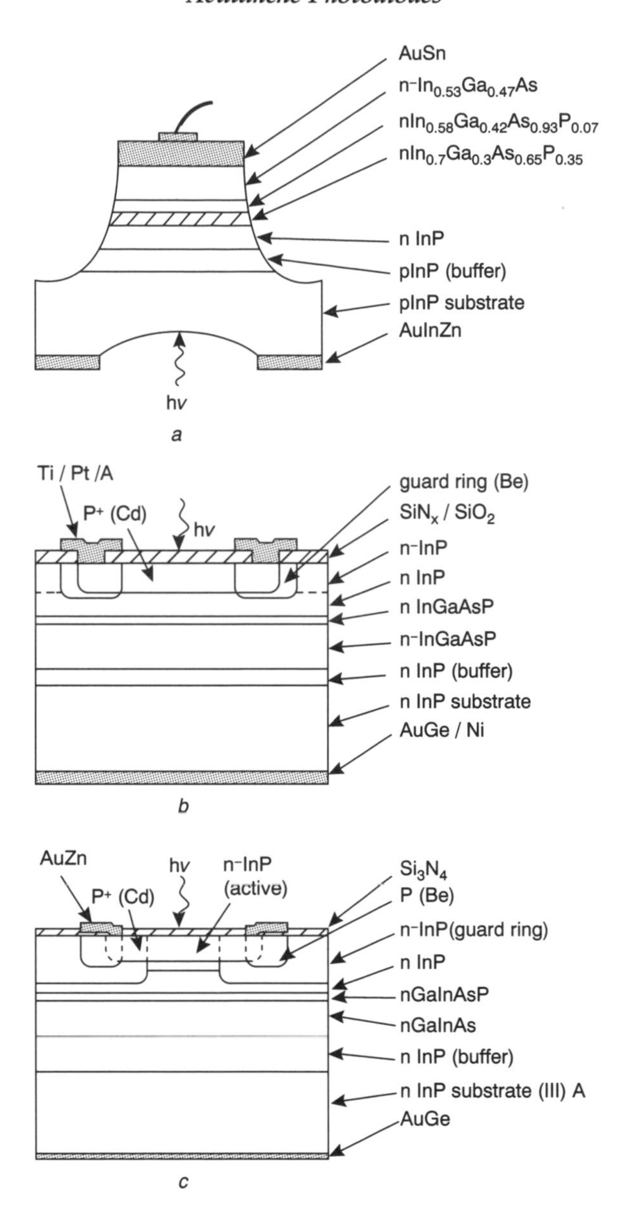

Figure 10.20 sam-apd structures as at mid-1980s ([2] of Chapter 6, Figure 18). *a* LPE-grown with p^+ substrate (from reference 86, Copyright © Istituto Internazionale Delle Comunicazioni, 1985), *b* VPE-grown planar structure (from reference 87), *c* LPE-grown buried-structure APD (from reference 88, Applied Physics Letters). Reproduced by permission of IEE

10 which is slightly better than a Ge apd, and 35 GHz gain × bandwidth product. The third structure is evidently the most complicated of the three and it gives an improved guard ring and gain uniformity, with a marginal lower dark current. It perhaps represents the complicated technology which will be needed to demonstrate the highest performance in future very-high-capacity systems. It remains to be seen how much performance benefit will be realised, on what time scale and, perhaps most importantly, at what high-yield production cost and with what reliability.

For applications above 5 GHz suitable devices require large gain × bandwidth products, and this feature is also needed to minimise the degradation of frequency response induced by the avalanche build-up time at modest gains. InP-based devices with separate absorption, grading and multiplication layers (sagm) have received much attention. In 1990 work was published which demonstrated that charge and multiplication layers could themselves be separated, giving a sequence of separate absorption, grading, charge and multiplication layers, which resulted in better performance. Gain–bandwidth products in excess of 100 GHz were discussed in [19] for such a device. In addition to this result the devices were easier to fabricate than sagm structures, and many of their key characteristics such as gain and intrinsic bandwidth were less affected by the thickness of the multiplication layer.

10.4.3 Responses in time and frequency domains

The response of an apd at a fixed value of avalanche gain depends on a number of factors:

(1) the time constant of the junction-capacitance shunt-resistance combination
(2) the transit time in the depletion layer
(3) the avalanche build-up time
(4) the RLC parasitics
(5) the hole trapping at heterojunction surfaces.

In well-designed devices the transit time effect will dominate.

Time-domain responses

Many long-haul high bit rate systems operating in the 1300 and 1550 nm bands have deployed apds fabricated from the material system whose frequency response was described in [20]. The evolution of rates to 10 Gbit/s and above requires very wideband frequency response to about $f_S/2$, where f_S is the signalling speed on the line as usual. Equations were developed for the frequency response function which took account of all the significant effects, from which results were calculated using tabulated device parameter values. The computed values agreed well with the measured frequency response. The same parameters were used in a numerical inverse Fourier transform of the frequency response of a pin photodiode to yield relative current impulse responses for the pin diode and apd for various values of gain. The curves are presented in Figure 10.21.

For $M_o = 1$ only the primary electron and hole currents are present; for all larger values additional current components are also present. The family of curves shows that the pulse response is dominated by transit time effects at low values of M_o and avalanche build-up at higher values, with $M_o = 10$ as a 'boundary' between them.

Frequency-domain responses

For the same back-illuminated mesa structure photodiode as before the frequency response was calculated from the mathematical model and was also measured at spot frequencies between 100 MHz and 18 GHz. The bandwidth of the modulated

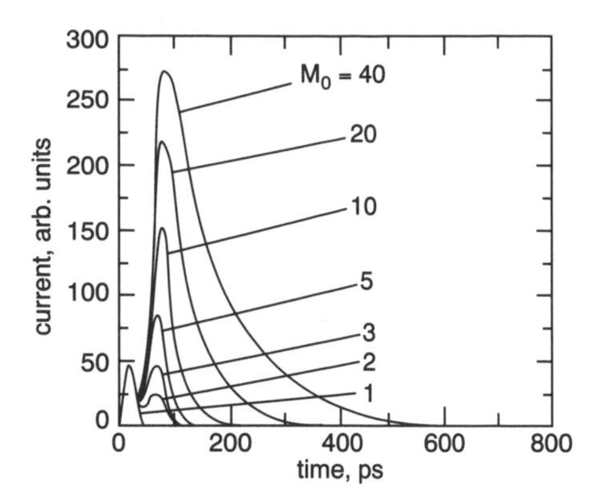

Figure 10.21 Calculated impulse response of an InP/InGaAsP/InGaAs apd for $M_o = 1$, 2, 3, 5, 10, 20 and 40 ([20], Figure 5). Reproduced by permission of IEEE, © 1989

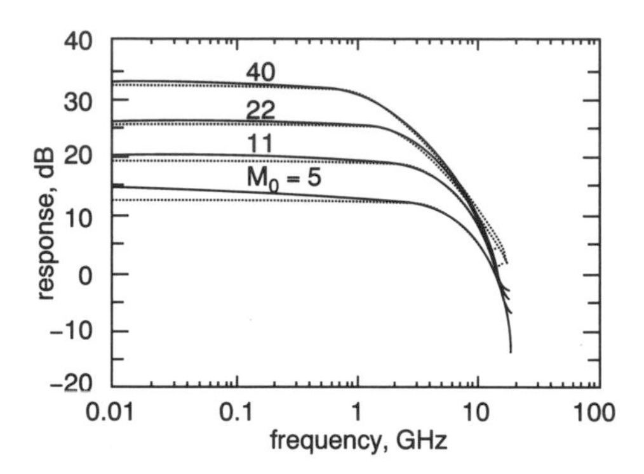

Figure 10.22 Measured (solid lines) and calculated (dotted lines) responses versus modulating frequency of a back illuminated InP/InGaAsP/InGaAs apd for $M_o = 5$, 11, 22 and 40, carrier wavelength 1300 nm ([20], Figure 7). Reproduced by permission of IEEE, © 1989

laser itself was about 14 GHz, and both apd and a fast pin photodiode used to normalise the data were mounted in Wiltron K-connector packages. Considerable care was taken to ensure that the measurements were as realistic and free from errors as possible. The results are presented in Figure 10.22.

The agreement between measured and predicted results is excellent over much of this frequency range.

A new (1988) approach to modelling apd frequency response was put forward in [21], based on solving the transport equations (for the first time) to obtain the relations between carrier concentration in the multiplication layer, the device parameters and the response. The calculations involved knowledge of the profile of the electric field and were applied to a conventional separated absorption layer and multiplication layers (sam) InGaAs apd. The predicted upper limit to the gain × bandwidth product was 140 GHz. The type of structure mentioned three years later in [19] was much more complicated, and achieved a gain × bandwidth product of 107 GHz.

We now look at the frequency response of a commercially available apd rather than that of a laboratory device, because the former is a much more realistic guide to the sort of values a system designer must use. Again, the curves are taken from manufacturer's data for the NDL5500 type of device, and are shown in Figure 10.23.

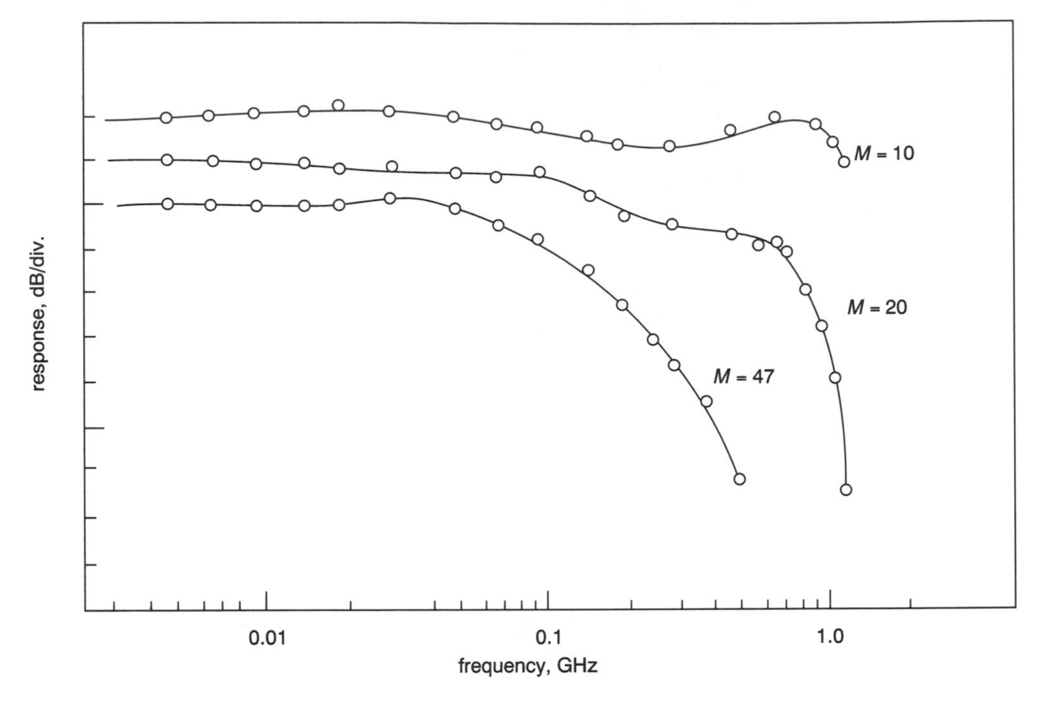

Figure 10.23 Measured amplitude versus frequency response of NDL5500 apd at $M = 10$, 20 and 47 (NEC data, frequency response)

Note that the shapes of the curves as well as their low-frequency gains all differ from one another. The cut-off frequency of the $M = 10$ curve is about 1 GHz, in agreement with the typical values listed in Table 10.2 above. The data sheet notes that the values of the curves are in good agreement with those expected from calculations and are dominated by the carrier transit time.

A closely related quantity is the gain × bandwidth product of the InGaAs device. This is conveniently displayed as 3 dB bandwidth versus avalanche gain, and the characteristics for the NDL5500 at 1540 nm are illustrated in Figure 10.24.

Dots represent measured points, the full line curve corresponds to a product of 20 GHz, and the asymptotic slope shown by the dashed lines is derived from the approximate expression which assumes one dominant time constant, namely:

$$M(f) = \frac{M(0)}{[1 + (2\pi f)^2 \times (M(0) \times \tau_{AV}\tau_I]^{1/2}} \tag{10.8}$$

where: $M(0)$ is the gain at very low frequency, τ_{AV} is the average value of the time constant, and τ_I carrier transit time.

10.4.4 Noise performance

The signal-to-noise ratio of an optical receiver using an apd detector can be approximated by:

$$\text{snr} = \frac{M^2(\phi P)^2}{(2q[\phi P + I_D]M^2F(M)B_N + 2qI_D''B_N + I_E^2)} \tag{10.9}$$

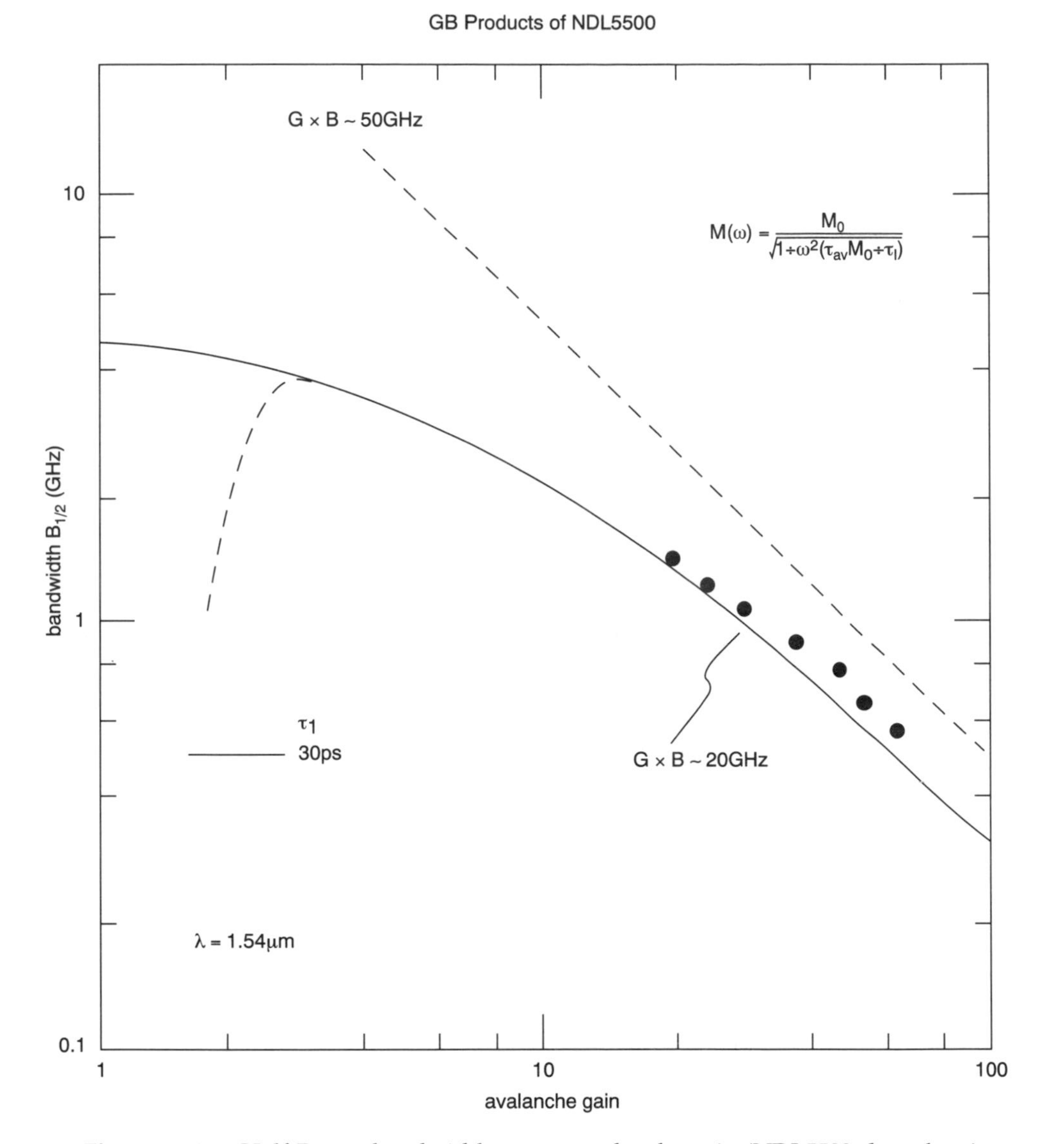

Figure 10.24 Half Power bandwidth versus avalanche gain (NDL5500 data sheet)

where:

M = mean value of multiplication factor
ϕ = responsivity, equal to $\eta q/hf = 0.805\eta\lambda(\mu\text{m})$, [A/W]
P = received optical mean power
I_D = component of dark current which is multiplied
$F(M)$ = effective excess noise factor, a function of M
B_N = receiver noise bandwidth
I''_D = component of dark current which is not multiplied
I_E = r.m.s. value of external noise across diode terminals.

Theoretical expressions for the excess noise factor when both electrons and holes initiate the avalanche process, as in all practical devices, tend to be much too complex to be of direct use, and tend to require knowledge of parameters which is seldom available, e.g. [16], so that a number of simpler, but approximate, equations have been derived for use by system designers. One such has already been given, and is repeated below for convenience:

$$F(M) = M\left\{M - (1 - k_{EFF})\left(1 - \frac{1}{M}\right)^2\right\} \tag{10.10}$$

Another assumption inherent in equation (10.10) is that F is the same for both the photocurrent and the bulk dark current, which is not strictly true, but again for simplicity the distinction is ignored and a 'worst case' is taken. Further consideration of noise performance will be found below in the sections dealing with receiver noise performance. A 1989 paper on the multiplication noise in sagm–apd devices [22] indicated that measured values were lower than predicted and that k_{EFF} depended weakly on M, rising from less than 0.4 to approaching 0.6 between $M = 10$ and 30. Corresponding values of F were about 5 (14 dB) and 15 (24 dB), respectively.

The excess noise factor as a function of the gain can be displayed graphically rather than be found from a formula. As an example, for a commercially available apd using InGaAs material, the data sheets for the NDL5500 include Figure 10.25. The dots and circles refer to 1300 and 1550 nm operation, respectively, and are measured at 35 MHz in a bandwidth of 1 MHz for a range of values of M from about 5 to 35, and the dashed lines show the computed results using two different values of k_{EFF}. Similar behaviour has been reported for Ge devices [17] for 1300 nm operation with the same noise bandwidth but a measuring frequency of 30 MHz. If anything, the characteristic of the Ge apd was closer to a straight line through the origin.

Non-Gaussian approach to apd receiver design

The sensitivity of such receivers depends upon the thermal noise (assumed Gaussian) and on the signal-dependent and signal-independent noise mechanisms in the detector modified by the avalanche multiplication processes. Both noise and multiplication processes are known to be non-Gaussian, but for simplicity are often assumed to be Gaussian. The resultant error in predicted sensitivity is usually less than 1 dB, but the predicted optimum gain and threshold level tend to be inaccurate [23]. Earlier attempts to account for the non-Gaussian nature of the pdf of the avalanche gain had been unsatisfactory; the paper presented a more complete version of a theory due to McIntyre in which non-Gaussian avalanche noise from both dark current and photocurrent was convolved with the Gaussian thermal noise in an approximate manner. This theory of receiver design takes account of the salient differences between the Gaussian and non-Gaussian approximations to the pdf associated with the avalanche process. It provides more accurate predictions of optimum avalanche gain and of decision thresholds. In addition, it gives simple results for the influence of background photocurrent or dark current on optimum gain, receiver sensitivity and decision threshold. The author stated that it should serve as a convenient starting point for avoiding numerically intensive calculations of sensitivity in situations where these were previously unavoidable.

An additional complication concerns the theoretical noise behaviour of diodes; the two physical sources of noise are the shot noise associated with the average

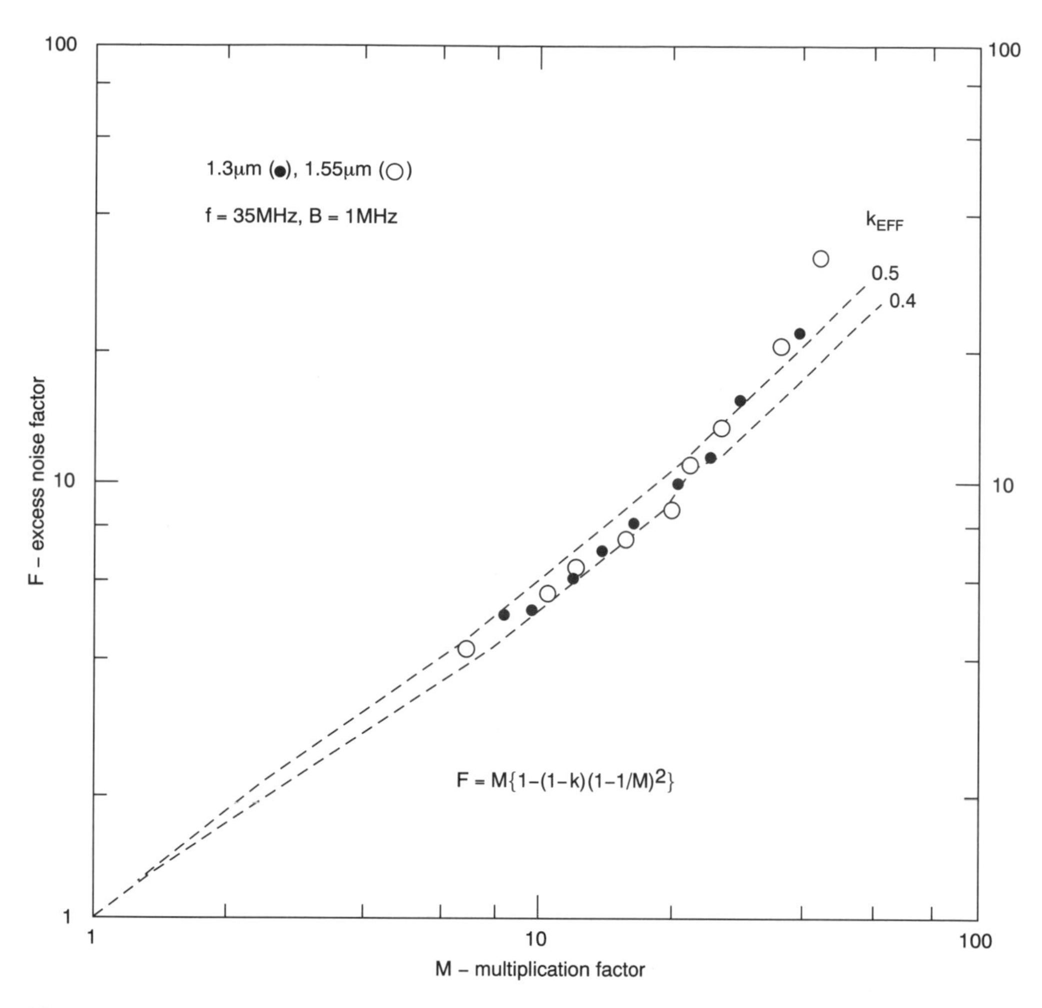

Figure 10.25 Excess noise factor F versus multiplication factor in second and third window (NDL5500 data sheet)

current through the diode, and the thermal noise of the resistance (usually negligible). The most accurate representation of a noisy diode is by means of a pair of generators, an emf in series with the signal current generator, and a noise current generator in parallel with it [24]. These noise generators are uncorrelated, but when transformed to the accessible terminals through the package equivalent circuit for comparison with measurements or for calculations of the noise performance of the circuit in which the diode is embedded, then the transformed noise generators at the accessible terminals become correlated.

10.4.5 Reliability predictions

In the 1994 paper [13] 10 fits were allocated to both pin and apd devices, a value 20 times lower than that allocated to lasers. Many papers on both types of components quote values around 1 fit. For instance, a typical rogue failure rate below 1 fit could

Table 10.4 Bias-temperature test conditions for InP/InGaAs apds

Test conditions		Sample size
temp. °C	reverse current μA	
200	100	9
225	100	6
250	100	5

be expected for optoelectronic receivers for use in submerged repeaters [25], but of course the reliability requirements for inland networks are much less stringent.

Some information published in late 1988 by authors from NTT relates to the surface degradation mechanism [26]. The authors stated that the failure mechanism of these types of photodiode remained quite unknown, and must be clarified before their reliability can be assessed. From experience, apd failure modes can mainly be classified into surface failures (e.g. dielectric breakdown, surface leakage, electrode failure). The paper first reported the results of bias temperature testing of InP/InGaAs apds at temperatures above 200°C. Next, they described the results of a failure analysis for the devices which demonstrated a wear-out failure mode during these tests. Thirdly, based on the results, a two-stage surface degradation mechanism was proposed.

The devices used in the investigation were 50 μm diameter planar structure InP/InGaAsP/InGaAs apds with guard rings. The epitaxial layers were grown on an n-type InP substrate by the vpe method, the junction surface is coated with a SiN_x-based passsivation film deposited by the pcvd process. The p^+ metallisation scheme was Ti/Pt/Au with Au bonding wire, and the chips were hermetically sealed in dry nitrogen. The test conditions are summarised in Table 10.4.

During the tests, the dark current and the breakdown voltage were both monitored at room temperature, because they are the most sensitive characteristics as far as degradation is concerned.

Test results covered the ageing characteristics of the reverse voltage corresponding to a reverse current of 100 μA, and of the dark current at 90% of the breakdown voltage, and those for the nine devices at 200°C are given in the paper. At all test temperatures all devices exhibited a gradual increase in dark current with time, followed by a rapid increase, without any appreciable change in breakdown voltage. The reverse current versus reverse voltage characteristics of a degraded apd after 1000 hours at 200°C showed two to three orders of magnitude larger currents than the initial characteristic for the same device. Behaviour characteristic of the formation of microplasma was not observed. Failure was defined as the time taken for the dark current to exceed 1 μA, chosen because there is a rapid rise above this value. With this definition, the time-to-failure data for the small sample versus the percent of cumulative failures plotted on lognormal paper obeyed the lognormal distribution reasonably well, and indicated that the failure mode was of the wear-out type. An Arrhenius plot of the median lifetimes of 50, 300 and 2150 hours at 250, 225 and 200°C, respectively, versus the reciprocal of the ageing temperatures yielded a thermal activation energy of 1.7 eV. Using this value, the extrapolated median life under normal working conditions was predicted to exceed 10^{10} hours. The authors went on to clarify that the surface degradation mode was responsible for the value of thermal activation energy.

10.4.6 Effects of temperature

The values listed in Table 10.1 all referred to an ambient temperature of 25°C, but equipment must operate within specification over a range of temperatures, and be capable of being stored over a wider range. It is therefore necessary to provide an example, and for this we return to the NDL5500 apd, for which the ranges are −40 to +70°C and −55 to +100°C respectively. Referring to Figure 10.26, the value $M = 3$ corresponds to a reverse bias voltage of 60 V approximately, and at this bias the temperature range −20 to +70°C corresponds to a change in dark current of some two orders of magnitude, according to Figure 10.27.

Another quantity strongly affected at lower temperature is the quantum efficiency, for which typical characteristics at 1550 nm are shown in Figure 10.28 at four different values of M.

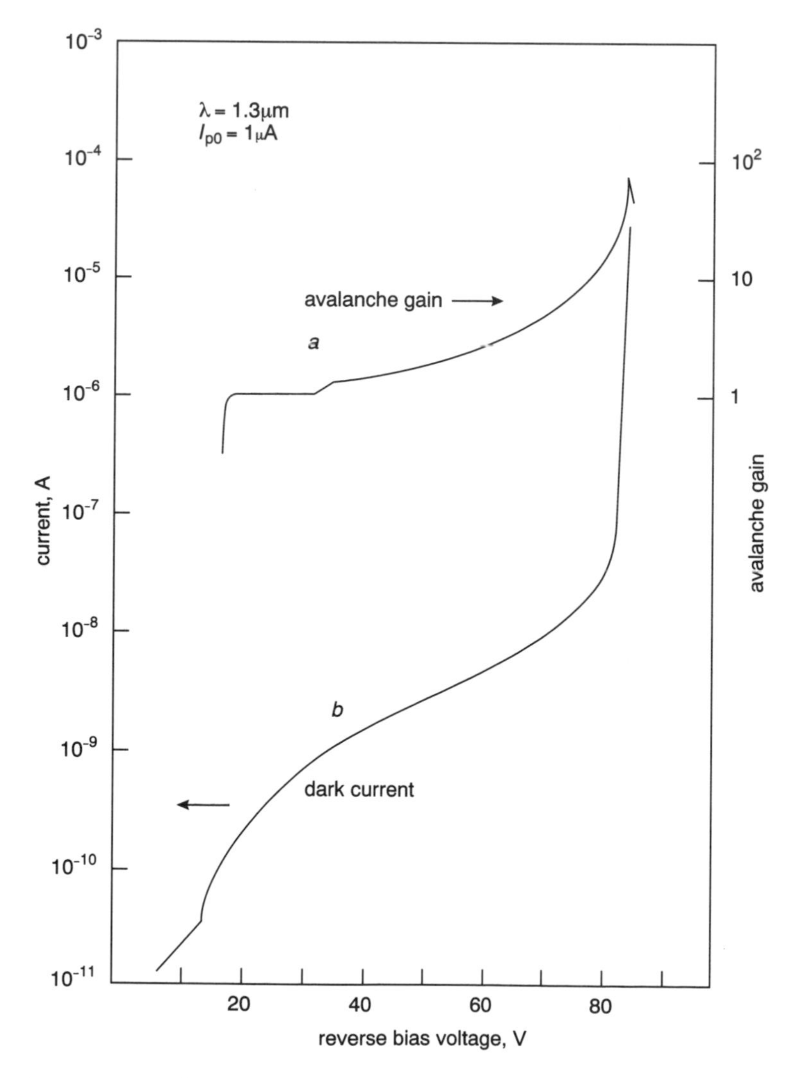

Figure 10.26 (a) Avalanche gain and (b) dark current, both versus reverse bias voltage at 1300 nm and a photocurrent of 1 μA

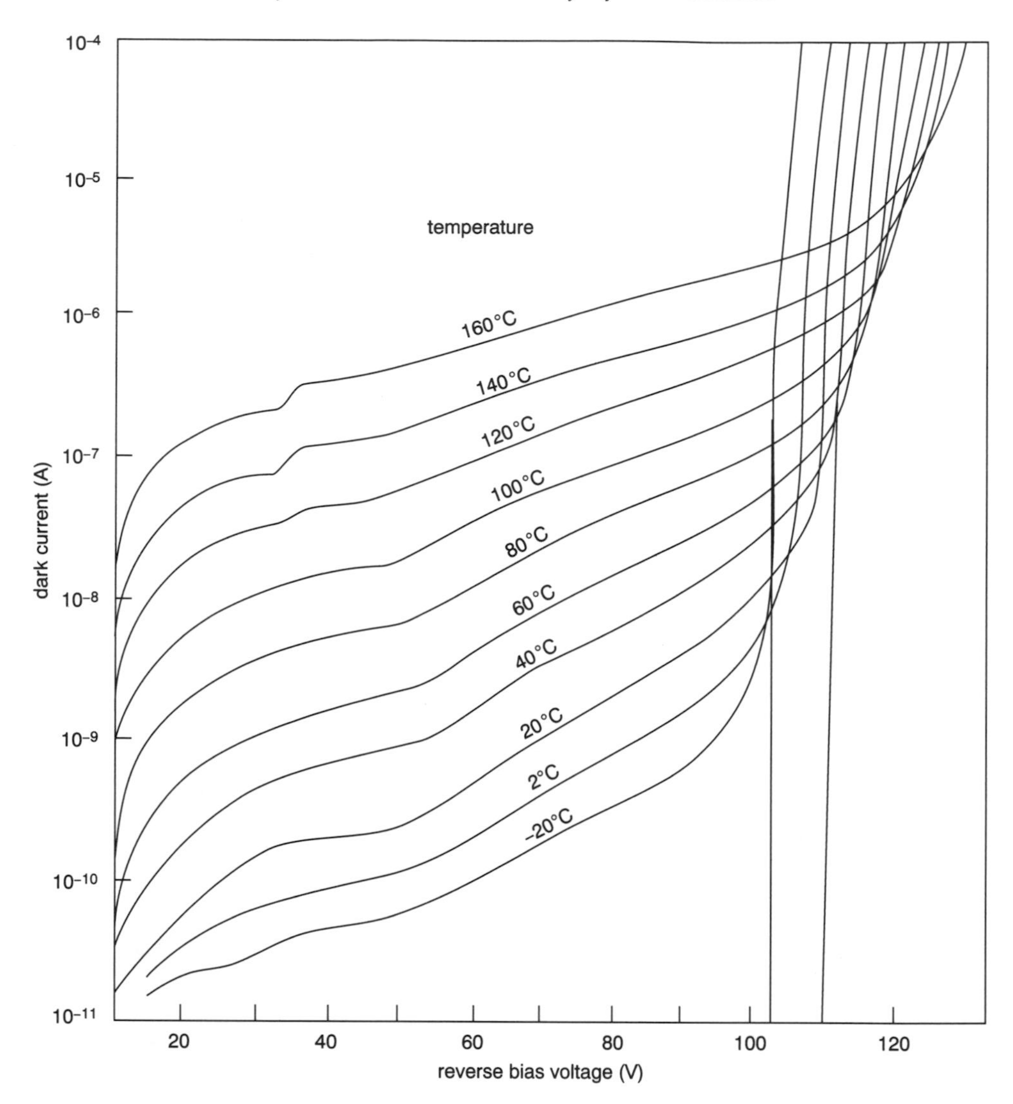

Figure 10.27 Dark current versus reverse bias voltage characteristic of an NDL5500 apd with temperature as parameter

This behaviour might cause concern in deep-sea submerged systems where the ocean bed temperature is about 3°C.

10.5 PHOTODETECTORS WITH INTEGRATED DIELECTRIC WAVELENGTH FILTER

WDM techniques are likely to find application in telecommunication transmission systems in the second half of the 1990s. Transmitters in such systems usually consist of several narrow linewidth lasers with their operating wavelengths suitably spaced. An example of a source producing eight channels separated by 2 nm was described in Chapter 8. WDM demultiplexers or receivers employ discrete transmission filters

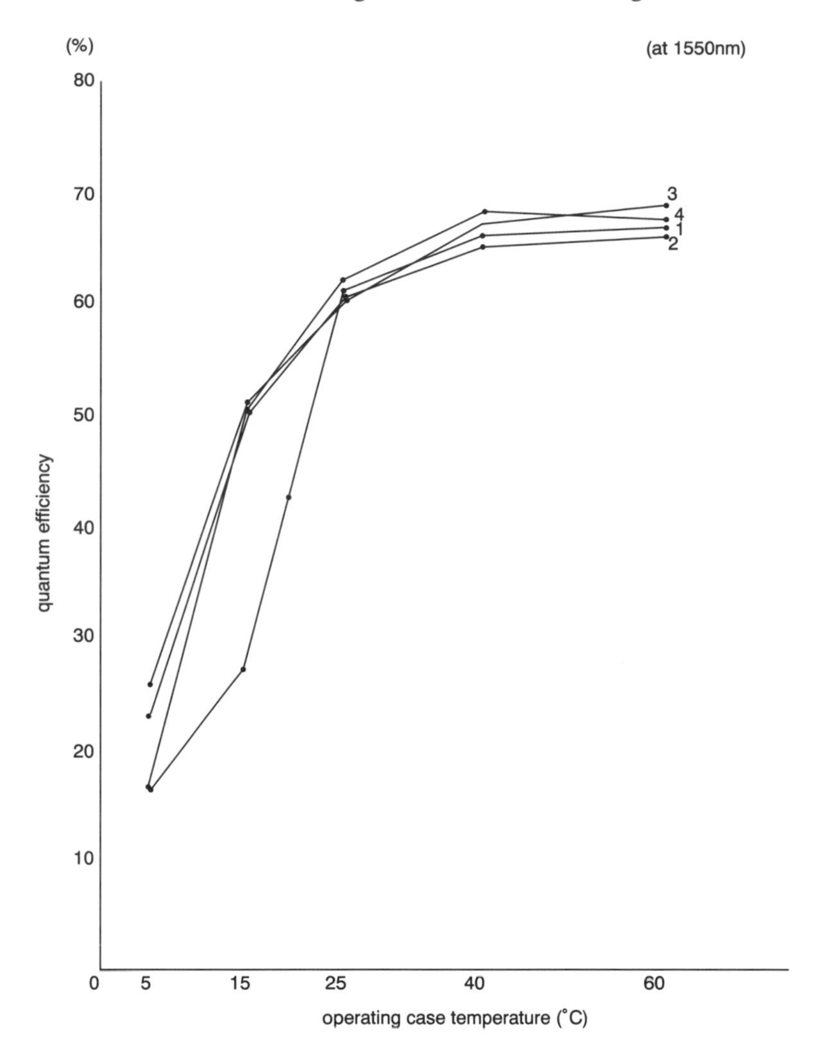

Figure 10.28 Quantum efficiency at 1550 nm versus operating case temperature for $M = 1, 2, 3$ and 4

or gratings to separate wavelengths, but these suffer from disadvantages such as the need for precise mechanical alignment of many components and high cost make such demultiplexers unsuitable for mass production. Methods of producing a dielectric wavelength filter integrated on to photodiodes were described in 1990. This alternative approach reported dielectric wavelength filters being directly deposited on to substrate entry InGaAs pin photodiodes. The extension of the technique to apds was reported later in 1990.

10.5.1 Narrowband pin photodiodes

A cross-section of the detector [27] is shown in Figure 10.29 and it has an active area of 55 μm diameter. The movpe technique was used to grow the device because of its advantages in this application. The filter consisted of multiple layers of different dielectrics, each $\lambda/4$ in optical thickness, with spacer layers of $\lambda/2$. It was designed to provide a bandpass filter with high transmission, narrow linewidth and high peak rejection. Current–voltage characteristics of the devices were similar to photodiodes without the filter showing that the presence of the filter did not affect the pin diode pn junction on the far side of the InP substrate (see Figure 10.29). The performance of the filtered pin detector is illustrated by the quantum efficiency versus wavelength characteristic given in Figure 10.30.

The peak quantum efficiency was 73% at 1234 nm and the fwhm width was 23 nm. Peak rejection was greater than 32 dB, which was the limit of the measuring equipment.

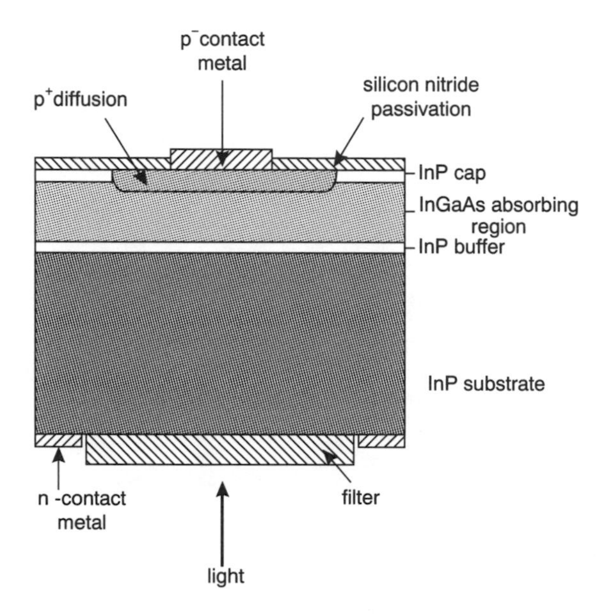

Figure 10.29 Substrate entry pin photodiode with integral dielectric filter ([27], Figure 1). Reproduced by permission of IEE

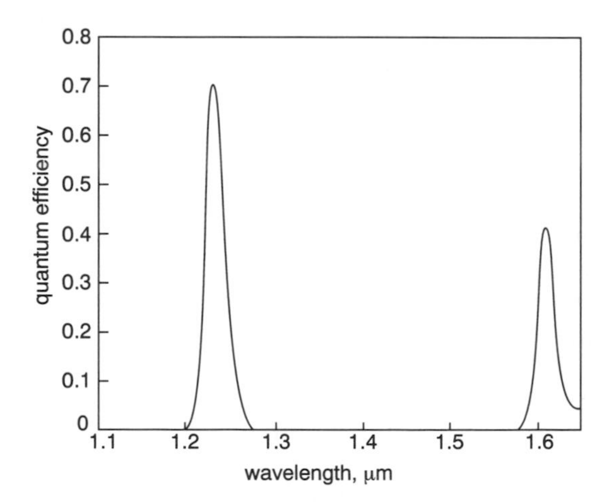

Figure 10.30 Quantum efficiency versus wavelength for substrate entry pin photodiode with integral filter ([27], Figure 2). Reproduced by permission of IEE

10.5.2 Narrowband avalanche photodiodes

Patterned bandpass dielectric interference filters with midband wavelength around 1300 nm were applied successfully [27] to substrate entry photodiodes to produce high-speed, high responsivity detectors for wdm applications such as very-large-capacity systems or passive optical distribution networks. The extension to apd devices was described in [28] to complete the demonstration that wavelength selective, as opposed to broadband, detectors with very high responsivity can be fabricated for wdm applications that require high receiver sensitivities.

The design and fabrication of the apd had been reported earlier, and a sketch of a cross-section of the device is shown in Figure 10.31.

The very controlled high yield process resulted in devices that consistently possessed high, spatially uniform gains without a separately formed guard ring, low noise and 3 dB bandwidths in excess of 1 GHz for gains above 10.

Device characteristics: the dark current and multiplication factor were both very similar to those of unfiltered devices at < 100 nA ($0.9V_B$) and V_B in the range 50–60 V and are illustrated in Figure 10.32.

Spatially uniform high gain was achieved and maximum gains above 40. The responsivity of the filter apd was measured for three bias points corresponding approximately to gain of 1.6, 5 and 37, and for all of these the filter peak was at 1258 ± 1 nm, as shown in Figure 10.33.

The fwhm was 19 nm and the peak rejection exceeded 25 dB. At unity gain, the quantum efficiency was approximately 65% (filter transmission 73% and quantum efficiency of standard ar coated device $\simeq 90\%$). Other attractive features of the device were low noise and a 3 dB bandwidth greater than 1 GHz when $M > 10$.

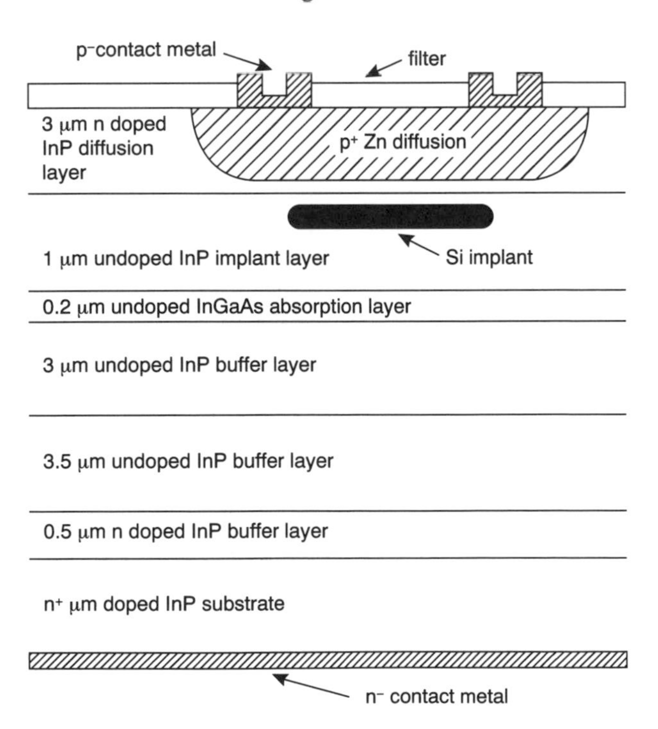

Figure 10.31 Cross-section through top entry avalanche photodiode integrated with dielectric filter ([28], Figure 1). Reproduced by permission of IEE

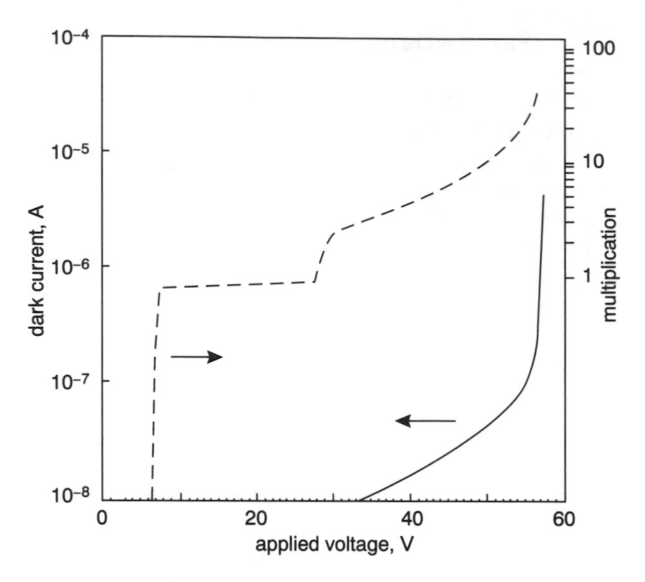

Figure 10.32 Dark current and multiplication factor versus applied voltage at a wavelength of 1260 nm ([28], Figure 2). Reproduced by permission of IEE

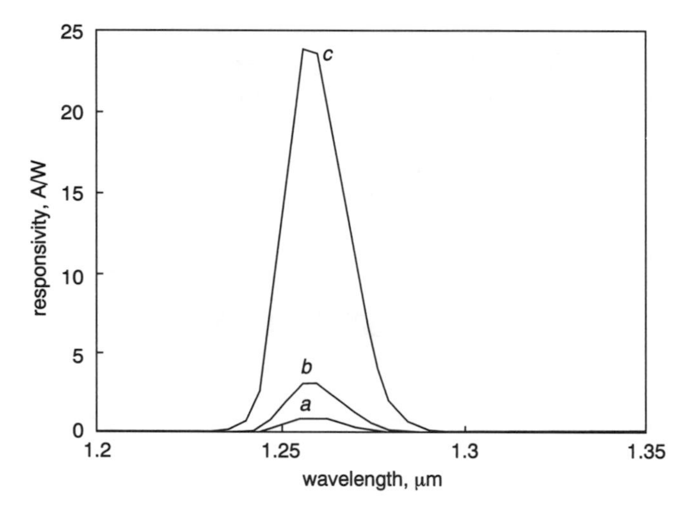

Figure 10.33 Measured responsivity of filtered apd versus wavelength ([28], Figure 3). *a* DC gain = 1.6, *b* DC gain = 5, *c* DC gain = 37. Reproduced by permission of IEE

10.6 REFERENCES

1 HADJIFOTIOU, A.: 'Receivers for direct detection'. *IEE Third Vacation School on Optical Fibre Telecommunications*, 31 Aug. to 5 Sept. 1986.

2 OTTEN, H.J.M.: 'Fibre-optic communications'. *Electronic Components and Applications*, **3**, no. 2, Feb. 1981, pp. 87–100.

3 BRAIN, M.C.: 'Comparison of available detectors for digital optical systems for the 1.2–1.55 μm wavelength range'. *IEEE J. Quantum Electronics*, **QE-18**, no. 2, Feb. 1982, pp. 219–224.

4 SCHLACHETZSKI, A., and MÜLLER, J.: 'Photodiodes for optical communication'. *Frequenz*, **33**, no. 10, Oct. 1979, pp. 283–290.

5 PERSONICK, S.D.: 'Receiver design for optical fiber systems'. *Proc. IEEE*, **65**, no. 12, Dec. 1977, pp. 1670–1678.

6 ROBERTSON, M.J., RITCHIE, S., SARGOOD, S.K., NELSON, A.W., DAVIS, L., WALLING, R.H., SKRIMSHIRE, C.P., and SUTHERLAND, R.R.: 'Highly reliable planar GaInAs/InP photodiodes with high yield made by atmospheric pressure m.o.v.p.e.'. *Electronics Lett.*, **24**, no. 5, 3 Mar. 1988, pp. 252–254.

7 WAKE, D, SPOONER, T.P., PERRIN, S.D., and HENNING, I.D.: '50 GHz InGaAs edge-coupled pin photodetector'. *Electronics Lett.*, **27**, no. 12, 6 June. 1991, pp. 1073–1075.

8 TUCKER, R.S., TAYLOR, A.J., BURRUS, C.A., EISENSTEIN, G., and WIESENFELD, J.M.: 'Coaxially mounted 67 GHz bandwidth pin photodiode'. *Electronics Lett.*, **22**, no. 17, 14 Aug. 1986, pp. 917–918.

9 TEMKIN, H., FRAHM, R.E., OLSSON, N.A., BURRUS C.A., and MCCOY, R.J.: 'Very high speed operation of planar InGaAsP/InP photodiode detectors'. *Electronics Lett.*, **22**, no. 23, 8 Nov. 1986, pp. 1267–1269.

10 WAKE, D., WALLING, R.H., HENNING, I.D., and PARKER, D.G.: 'Planar-junction, top-illuminated GaInAs/InP pin photodiode with bandwidth of 25 GHz'. *Electronics Lett.*, **25**, no. 15, 20 July 1989, pp. 967–969.

11 WIESENFELD, J.M., CHRAPLYVY, A.R., STONE, J., and BURRUS, C.A.: 'Measurement of very high speed photodetectors with picosecond InGaAsP film lasers'. *Electronics Lett.*, **19**, no. 7, 6 Jan. 1983, pp. 22–24.

12 MÜLLER, J.: *'Photodiodes for optical communication'*. Advances in Electronics and Electron Physics, **55**, 1981, Academic Press.

13 COCHRANE, P., and HEATLEY, D.J.T.: 'Reliability aspects of optical fibre systems and networks'. *BT Technol. J.*, **12**, no. 2, Apr. 1994, pp. 77–92.

14 BAUER, J.G., and TROMMER, R.: 'Long-term operation of planar InGaAs/InP pin photodiodes'. *IEEE Trans. Electron Devices*, **35**, no. 12, Dec. 1988, pp. 2349–2353.

15 SUTHERLAND, R.R., SKRIMSHIRE, C.P., and ROBERTSON, M.J.: 'A reliability methodology applied to very high reliability planar InGaAsP/InP pin photodiodes'. *Br. Telecom Technol. J.*, **7**, no. 1, Jan. 1989, pp. 69–77.

16 WEBB, P.P., McINTYRE, R.J., and CONRADI, J.: 'Properties of avalanche photodiodes'. *RCA Rev.*, **35**, June 1974, pp. 234–278.

17 HINO, I., TORIKAI, T., IASAKI, H., MINEMURA, K., and NISHIDA, K.: 'Ge apd characteristics for optical communication'. *NEC Res. & Dev.*, no. 67, 1982, pp. 67–72.

18 LE BELLÉGO, Y., PRASEUTH, J.P., and SCAVENNEC, A.: 'Double junction AlInAs/GaInAs multiquantum well avalanche photodiode'. *Electronic Lett.*, **27**, no. 24, 21 Nov. 1991, pp. 2228–2230.

19 TAROF, L.E.: 'Planar InP/InGaAs avalanche photodetector with gain-bandwidth product in excess of 100 GHz'. *Electronics Lett.*, **27**, no. 1, 3 Jan. 1991, pp. 34–36.

20 CAMPBELL, J.C., JOHNSON, B.C., QUA, G.J., and TSANG, W.T.: 'Frequency response of InP/InGaAsP/InGaAs avalanche photodiodes'. *J. Lightwave Technol.*, **7**, no. 5, May 1989, pp. 778–783.

21 SHIBA, T., ISHIMURA, E., TAKAHASHI, K., NAMIZAKI, H., and SUSAKI, W.: 'New approach to the frequency response analysis of an InGaAs avalanche photodiode'. *J. Lightwave Technol.*, **6**, no. 10, Oct. 1988, pp. 1502–1505.

22 CAMPBELL, J.C., CHANDRASEKHAR, S., TSANG, W.T., QUA, G.J., and JOHNSON, B.C.: 'Multiplication noise of InP/InGaAsP/InGaAs avalanche photodiodes'. *J. Lightwave Technol.*, **7**, no. 3, Mar. 1989, pp. 473–477.

23 CONRADI, J.: 'A simplified non-Gaussian approach to digital optical receiver design with avalanche photodiodes: theory'. *IEEE/OSA J. Lightwave Technol.*, **9**, no. 8, Aug. 1990, pp. 1019–1030.

24 VAN NIE, A.G.: 'Representation of linear passive noisy 1-ports by two correlated noise sources'. *Proc. IEEE,* **60**, no. 6, June 1972, pp. 751–753.
25 BOUSSOIS, J.-L., GOUDARD, J-L., GUEGUEN, M., KRAMER, B., and SAUVAGE, D.: 'Optoelectronic device selection for submarine repeaters'. *Electron. Commun.*, **63**, no. 3, 1989, pp. 240–248.
26 SUDO, H., and SUZUKI, M.: 'Surface degradation mechanism of InP/InGaAs a.p.ds.'. *IEEE J. Lightwave Technol.*, **6**, no. 10, Oct. 1988, pp. 1496–1505.
27 LEARMOUTH, M.D., HEWETT, N.P., REID, I., and ROBERTSON, M.J.: 'Integrated dielectric wavelength filters on InGaAs pin photodiodes'. *Electronics Lett.*, **26**, no. 9, 26 Apr. 1990, pp. 576–577.
28 MACBEAN, M.D.A., HEWETT, N.P., LEARMOUTH, M.D., and REID, I.: 'Planar InP/InGaAs avalanche photodetector with integrated dielectric wavelength filter'. *Electronics Lett.*, **26**, no. 12, 7 June 1990, pp. 804–806.

11

OPTICAL RECEIVER DESIGN CONSIDERATIONS

Design is the heart of the engineering process — its most characteristic activity

William H. Roadstrum, *Excellence in Engineering*, 1967

11.1 INTRODUCTION

The two preceding chapters devoted to the very broad area of optical receivers have dealt firstly with information about receiver properties and performance which are of direct concern to system designers. The background to the selection of material for Chapter 9 was pioneering work in the UK leading to the design, development and deployment of standard optical systems in the British Telecom inland network. Each system was required to be capable of operating with equipment supplied by different manufactures. The remainder of this chapter was devoted to intensity-modulated direct-detection receivers using examples from operating systems. Bit rates of 10 Gbit/s and above may be used in the latter half of the 1990s; therefore some material was included on receivers for 10 Gbit/s and above. Finally, receivers with optical preamplifiers were considered. Chapter 10 was focused on the properties and performance of pin and avalanche photodiodes in sufficient detail for use in system design. Again, practical considerations governed the choice and emphasis given to the various topics. The present chapter turns to receiver design in more detail and puts forward a design approach more in keeping with the capabilities of current hardware and software for the testing, characterisation, and simulation of both passive and active components and circuits used in receivers up to the highest bit rates likely to be of interest. The topics of noise and its effects on performance are given due weight.

At the input of an optical receiver in the receive side of the main transmission path of a regenerative repeater or terminal in trunk and junction networks of an ofdls, signal levels may be within a few dB of the design sensitivity. Moreover, the transmitted pulses will have been distorted by any pulse broadening and intersymbol interference effects present, so that the received waveform will differ from that sent. Before making a decision on the presence or absence of a mark or 1, a strong electrical signal which embodies a good estimate of the signal transmitted in the individual symbol intervals is required. Intensity-modulated direct-detection systems rely solely on the detection of amplitude variations in optical mean power, there being no information in the phase or frequency variations in the received signal.

Receivers for imdd systems must perform three functions:

(1) conversion from the pulse-modulated optical carrier to an electrical multiplex signal
(2) amplification of the very weak electrical signal
(3) estimation of the transmitted signal.

The first is performed by an optoelectronic transducer in the form of a pin or avalanche photodiode. The second item is electrical amplification, which for high bit rate systems calls for very wideband electronic amplifiers, with passbands covering frequencies from, say, 30 kHz to half the line rate, involving feedback, unintentional or deliberate, and the consequent potential for self-oscillations. The electronic circuits used to perform this function and the third one tend to impose the major constraints on speed of operation, as stated earlier, and as one might have gathered from the response times quoted earlier for current sources and detectors. The third function is that of signal estimation which involves dealing with noise and various other signal impairments.

Choice of front-end amplifier technology

We consider the amplification requirements of front ends for bit rates of 10 Gbit/s and above, therefore bandwidths exceeding 5 GHz or so. Accurate characterisation of components and circuits, simulation and optimisation of small signal and noise performance at microwave frequencies are needed [1].

In the early 1980s the GaAs mesfet and the silicon bipolar transistor were the two types of devices available, and since then the bandwidths achievable with either technology have kept roughly in step and were fully adequate for bit rates up to 565 Mbit/s. For 10 Gbit/s and above other technologies are preferable; in particular high electron mobility transistors (hemts) based on modulation-doped GaAs/AlGaAs structures are well suited to preamplifier requirements. Extrinsic parameters of interest are gate leakage current (60 nA), transconductance (40 mA/V at 10 mA drain current) and noise figure (1.8 dB at 20 GHz). The noise figure of hemts is lower and the available power gain is higher than for mesfets. This paper from British Telecom included a section on broadband receiver design followed by another on practical assessment of performance of an apdhemt two-stage front-end receiver optimised for use at 4.8 Gbit/s.

Later in 1989 a contribution from Bellcore discussed state-of-the-art receiver technology, design trade-offs and implementation techniques for ultra-wideband receivers [2]. A low-noise pinhemt with a 16 GHz bandwidth was described in detail and results of measurements and of simulations were given. Future directions were considered in the final section under the headings of 'optical preamplification' (fibre or semiconductor amplifiers) and 'improved device technology and increased integration'. Further and more recent information is given in [11].

11.2 SENSITIVITY OF OPTICAL RECEIVERS FOR IMDD SYSTEMS

The system design approach starts with the performance required of the receiver as a subsystem and finishes with the receiver component performance necessary to meet the specification. The design procedure is similar in concepts and approach to the design of any other baseband receiver.

11.2.1 Basic requirements

The most important property of any receiver is its sensitivity for a given bit error rate, that is, the minimum received optical power which will ensure correct decisions with the specified probability, often taken to be 10^{-9}. Several factors prevent arbitrarily small signals from being detected; these include the quantum nature of the transmitted signal, the waveform distortion in the fibre discussed in earlier chapters, the statistical nature of photodetection itself, and the electronic noise present in the circuitry.

Optical systems differ from radio and microwave systems in the relative magnitudes of three fundamental properties of the EM field in their respective parts of the spectrum which profoundly influence receiver design, hardware and performance.

Table 11.1 clearly shows the enormous range of relative magnitudes. The most significant comparison is that between the energy of the quantum, hf, and the thermal energy associated with an electron at temperature θ, namely $k\theta$. At room temperature, $k\theta/h$ is about 6000 GHz, well above frequencies used in millimetric bands and well below optical frequencies, so that the quantum nature of the field does not show up in conventional systems; whereas in optical systems the quantum nature dominates.

Another way in which the importance of the relative energies can be seen is from Planck's classical expression for the average energy per mode, or the lowest noise spectral density attainable at the input to the optical receiver, namely:

$$\frac{hf}{\exp(hf/k\theta) - 1} + hf, \quad \frac{\text{J}}{\text{Hz}} \tag{11.1a}$$

The first term is the thermal noise associated with the non-zero temperature of the input, the second is purely quantum mechanical in nature and sets the ultimate lower limit. An equivalent form which shows the relation between the noise energy

Table 11.1 Some characteristics of electromagnetic transmission systems

Item	Radio	Microwave	Optical	Units
frequency	3×10^7	3×10^9	3×10^{14}	Hz
wavelength	10	10^{-1}	10^{-6}	m
energy of quantum	2×10^{-26}	2×10^{-24}	2×10^{-19}	J
of radiation field	1.2×10^{-7}	1.2×10^{-5}	1.2	eV

spectral density per mode U_N/df is:

$$\frac{U_N}{df} = \frac{hf}{1 - \exp(-hf/k\theta)} \tag{11.1b}$$

Here, and throughout the chapter, the subscript N denotes a noise quantity.

Although conventional error rate performance is of immediate concern in digital systems as can be judged by the many examples in this book, the signal-to-noise ratio defined as $20 \log (V_P/\sigma)$, where V_P is the peak voltage and σ is the r.m.s. noise voltage at the input to the decision circuit, is closer to the physics of detection in noise, and is of course the criterion used in the design of low-noise front-end amplifiers. Because photodiodes are reverse-biased they have a very high impedance; consequently the most convenient circuit representation for noise is by means of noise current generators across the photodiode terminals, and that for the thermal noise in the photodiode is very small. Conventional noise figure definitions are ratios of total noise at an output to contributions at this output from the thermal noise of the signal source. In optical receivers it is thus inappropriate to use noise figure; instead ratios of mean square currents are used, and can be referred to the input in terms of r.m.s. values, the commonly accepted spectral densities being expressed in units of $\text{pA}/(\text{Hz})^{1/2}$, giving values of about 3 in 140 Mbit/s system receivers.

The general problem of calculating the performance of an optical receiver with all the factors taken into account is very difficult, and is best avoided. In practice, designers tend to use some simple equations which describe the system performance in terms of the critical parameters. This is particularly appropriate for digital systems which, by their very nature, use nonlinear signal processing and are designed for worst-case conditions, i.e. the required signal for a given ber is computed on the assumption that all the signal impairments are correlated, and the noise component of this impairment is a normal, stationary uncorrelated process. The general approach adopted here and by many other authors is to treat the noise as a quasi-stationary process in which the noise statistics for a mark and a space are different, but within the respective symbol intervals the statistics can be considered as stationary. This is not strictly correct, but it allows conventional shot noise expressions to be used, while retaining reasonable accuracy.

11.2.2 General assumptions in receiver design

The following assumptions are commonly made and are restated here for completeness. They will be employed in this chapter to derive the sensitivity equation for an optical receiver.

(1) The noise process at the receiver is Gaussian, wide sense stationary, and independent when sampled at the symbol rate.
(2) The receiver performance is noise-limited, and other impairments such as intersymbol interference are negligible.
(3) The decision detector operates on a sample-by-sample basis; that is, it has no memory.
(4) The criterion of performance is the bit error rate, and this is assumed to be limited by noise rather than any other factor.

(5) A binary system is considered in which the symbols 1 and 0 are equally probable.
(6) The mean received optical power corresponding to a space or 0 is itself not zero, so that the extinction ratio defined as $P(0)/P(1)$ is also greater than zero.

The first assumption implies that if the circuitry between the photodiode and the decision circuit is linear the noise powers can be referred to the input because they remain invariant to the linear transfer functions involved. This is a very useful technique, and has been applied for many years in receiver design.

The second assumption implies that the receiver has been equalised so that the effect of isi on the ber is negligible, and that other impairments such as clock recovery, timing, etc., have only second-order effects on the ber.

The third assumption indicates that decisions are made on the basis of the present signal-plus-noise statistics, and taken together with the assumed independence of the noise in disjoint intervals, makes the use of simple error probability calculations possible. The correctness of the assumed independence of noise samples depends on the system bandwidth; for a bandwidth of half the sampling rate the first zero of the noise autocovariance is at the sampling interval and the samples are then independent. For other bandwidths, the samples can be correlated, but it is assumed in such cases that the correlation is negligible.

Assumption (5) means that the information source is of maximum entropy.

11.2.3 Calculation of receiver sensitivity

The seminal paper on optical receiver design is undoubtedly Personick's 1973 paper published in the BSTJ [3] some four years before the first operational multimode fibre 850 nm system was installed in the UK. This was followed by other papers by him as sole or as co-author. Part I presented a systematic approach to the design of a regenerative repeater with particular attention to a proper choice of front-end amplifier, of biasing circuit for the photodiode and how to achieve a specified ber for give bit rate, optical pulse shape at photodiode input and desired electrical pulse shape at input to decision circuit. Lower received power and/or lower photodiode gain are obtained with a high impedance (integrating) amplifier followed by an appropriate equaliser, compared with a non-integrating amplifier. When isi is present in the received bit stream and the channel is linear in power the baseband equaliser can be designed to separate pulses at the cost of increased input power compared with no isi present. This increase in power (penalty) is calculated as a function of the two pulse shapes mentioned earlier. Part II applied the results to specific receivers and presented the numerical results. The receiver parameters were then varied and their effects on sensitivity were plotted.

A simplified approach by D.R. Smith and I. Garrett of BTRL followed in 1978 [4] and an extension of this to cases of non-zero extinction ratio by R.C. Hooper and B.R. White [5], also of BTRL. It was shown that reduced-width transmit pulses permit the noise bandwidth of the receiver to be smaller than for full-width pulses, thus increasing the sensitivity. Other benefits are reduced source power consumption and improved lifetime, and in timing recovery. Their results are used in the following paragraphs.

Specific assumptions

In calculations of sensitivity of integrating front-end receivers followed by an equaliser and decision circuit the following specific assumptions are made.

Although the shot noise has a Poisson distribution we use the Gaussian approximation for convenience, because the inaccuracy is acceptable.

Mean square values of the shot noise voltage at the input to the decision circuit at the sampling instants are related to corresponding values of the unity gain photocurrent in the photodiode averaged over one time slot by the standard shot noise relation between voltages and currents. The relation also involves the modulus squared of the transfer function relating the current excitation at the photodiode output port to the response voltage at the decision circuit input port in the receiver.

Voltages at the decision circuit at the sampling instants are Gaussian random variable with mean values V_1 and V_0 for a mark and a space, respectively, and the corresponding r.m.s. values are σ_1 and σ_0. The r.m.s. values take account of shot noise contributions from any isi present, and therefore depend on the shape of the received optical pulse and on the bit stream. The approximation made in [4] was to equate the r.m.s. shot noise voltage at a specific decision instant to the shot noise corresponding to a specific value of current by the standard expression $\sigma^2 = 2qIB$. The value of current to be used is the mean photocurrent in the time slot when the photodiode gain is one, transformed to the decision circuit input.

We consider a specific sampling instant in which either a mark or a space are present. Worst-case shot noise is taken to occur when all pulses outside this specific time slot are ones; thus there are two bit streams 1111111 and 1110111 to be considered. Denote the corresponding noise variables with double subscripts, thus V_{11}, V_{01}, σ_{11} and σ_{01}, where the first subscript refers to the pulse in the specific timeslot and the second to all the remaining pulses.

The decision threshold level V_T is set to give equal error probabilities pr for marks and spaces; then for the worst case:

$$\frac{V_{11} - V_T}{\sigma_{11}} = \frac{V_T - V_{01}}{\sigma_{01}} = Q \tag{11.2}$$

Since Gaussian statistics have been assumed the probability of error pr is related to Q by the equation:

$$2\text{pr} = \text{erfc}\left(\frac{Q}{\sqrt{2}}\right) \tag{11.3}$$

Sensitivity equations

For the calculation of sensitivity it is more convenient to work in terms of energies in bit intervals, since power and energy are very simply related. The mean photocurrent I_{11} in the time slot when the photodiode gain is one is very simply related to the corresponding energy U_{11} as follows:

$$I_{11} = \frac{\eta \times q f_S \times U_{11}}{hf} \tag{11.4}$$

The discrepancies between values of U_{11} derived in this way and values obtained from Personick's more exact analysis in the case of Gaussian and rectangular received pulse envelopes and raised cosine shape at the decision circuit were plotted in Figure 6 of [4]. The curves showed the error (dB) versus the shape parameter α of pulses with unit area. A full-width rectangular (nrz) pulse has $\alpha = 1$ and error <0.1 dB, a half-width (rz) pulse has $\alpha = 0.5$ and error <0.5 dB. For Gaussian pulses just confined within one timeslot of duration T, $\alpha \simeq 0.25$ and the error <0.3 dB.

Case 1: $U_{01} = 0$

Assuming that the shot noise is negligible compared with thermal noise when the photodiode gain is unity, and that the energy in the time slot is zero for a space, i.e. $U_{01} = 0$, the number of photons needed for a prescribed ber (hence Q) is given by:

$$\text{Unity gain photodiode: } \frac{U_{11}}{hf} = \frac{2Q}{\eta} \times Z^{1/2} \tag{11.5}$$

where Z is a dimensionless parameter representing the signal-independent noise terms in the equivalent circuit and defined ([3], equation (30)) as the ratio of the amplifier input noise current to the photocurrent for a single photon. Personick also calls it the thermal noise parameter. Equation (11.5) can be used to give an indication of the numbers involved. Take a ber of 10^{-9}, then $Q = 6$, and let $\eta = 1$ for simplicity; Z could be, say, 10^5 then the number of photons needed to achieve this ber is about 3800.

The parameter Z can be found from the circuit constants and noise generators shown in Figure 11.1, using the expression:

$$\frac{q^2}{2k\theta} \times Z = \left(G_N + G_B + \frac{R_N}{R^2}\right) B_2 + (2\pi f_s C)^2 R_N B_3 \tag{11.6}$$

or for $\theta = 291$ K

$$Z = 3.13 \times 10^{17} \times \left[\left(G_N + G_B + \frac{R_N}{R^2}\right) B_2 + (2\pi f_s C)^2 R_N B_3\right]$$

where

G_N = equivalent noise conductance = $S_J/2k\theta$

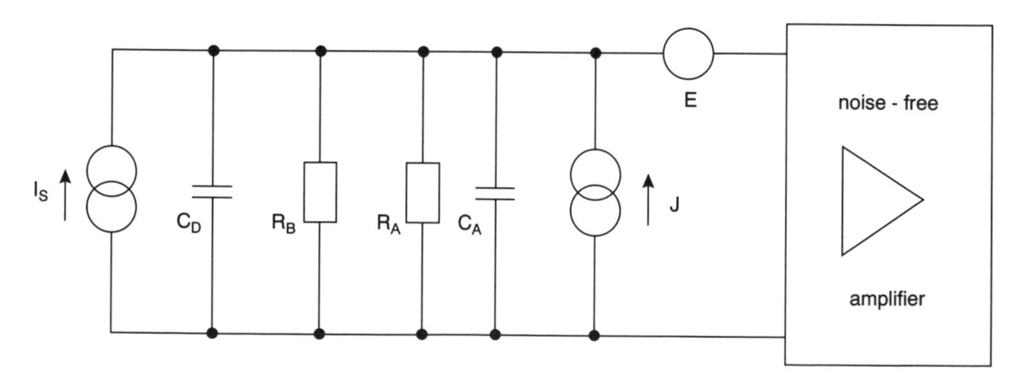

Figure 11.1 Circuit model of photodiode (C_D), bias circuit (R_B) and front-end amplifier (R_A and C_A) showing circuit constants used in the definition of the parameter Z. Note: R_A is assumed to be noise-free since the noise of the amplifier is represented by E and J ([3], upper part of Figure 5). Reproduced by permission of AT&T

S_J = two-sided spectral density of noise generator J
$G_B = 1/R_B$, bias resistance (see caption to Figure 11.1)
R_N = equivalent noise resistance = $S_E/2k\theta$
S_E = two-sided spectral density of noise generator E
C = total capacitance across photodiode terminals = $C_D + C_A$
$B_2 = I_2 T$
I_2 = Personick integral I_2
$B_3 = I_3 T$
I_3 = Personick integral I_3

The Personick integrals are discussed in Appendix B.

The energy U_{11} depends on the width of the received optical pulses and is a minimum when these are impulses. For all other shapes there will be a penalty which can be evaluated from an equation given in [4]. This reference included a graph of the penalty (Figure 3) when Gaussian pulses were received under stated simplifying assumptions; the penalty exceeds 2 dB if more than 6% of the energy in a pulse spills into adjacent timeslots.

Case 2: $U_{01} \neq 0$

We define an extinction ratio in terms of average powers in the specific time slot as:

$$\varepsilon = \frac{\text{average power all zeros}}{\text{average power all ones}} = \frac{U_{00}}{U_{11}} \tag{11.7}$$

where U_{00} is the energy in the pedestal in one time slot of duration T and U_{11} the total energy in a mark (including adjacent time slots if isi present).

The corresponding mean pin photocurrent in one time slot for a bit stream consisting of all marks is:

$$I_{11} = \frac{(\eta q/hf) \times U_{11}}{T} = \frac{\eta q}{hf} \times U_{11} \times f_S \tag{11.8}$$

Now U_{11}/hf is simply a ratio of energies, or normalised energy, and qf_s clearly has dimensions of current; thus we can write:

$$I_{11} = \frac{\eta \times q f_S \times U_{11}}{hf} \tag{11.9}$$

For a space when all the other pulses in the bit stream are marks, the unity gain mean photocurrent expression is:

$$I_{01} = \frac{\eta(1-\gamma') \times q f_S \times U_{11}}{hf} \tag{11.10}$$

where

$$\gamma' = \gamma(1-\varepsilon) \tag{11.11}$$

It was also shown that the worst-case total mean-square noise expression for a mark at the sampling instant can be written as:

$$\left(\frac{\eta\sigma_{11}}{hf}\right)^2 = \left[M^2 \frac{I_{11}}{q f_S} I_2 + \frac{Z}{M^2}\right] \tag{11.12}$$

and for a space:

$$\left(\frac{\eta\sigma_{01}}{hf}\right)^2 = \left[M^2\frac{I_{11}(1-\gamma')}{qf_S}I_2 + \frac{Z}{M^2}\right] \tag{11.13}$$

These four energy terms are related to the quantity Q by the equation:

$$U_{11} - U_{01} = Q \times [\sigma_{11} - \sigma_{01}] \tag{11.14}$$

This can be used to obtain an equation for $U_{11}(1-\varepsilon)$ from which the partial derivative with respect to M gives U_{11} (min) for a specified value of Q, and the corresponding optimum value of M. The result is:

$$\frac{\eta U_{11}(\varepsilon, \text{min})}{hf} = \frac{Q^a \times Z^b \times I_2{}^c \times L'}{(1-\varepsilon)^a} \tag{11.15}$$

with

$a = (2+x)/(1+x)$
$b = x/2(1+x)$
$c = 1/(1+x)$

where x is the excess noise coefficient of the apd in the empirical relation between the mean value M and the mean square value = $M^{(2+x)}$ of the avalanche gain (a random variable).

$$L'K' = \left[\frac{2(1-\gamma')}{(2-\gamma')}\right] \times \left\{\left[\frac{(2-\gamma')K'}{2(1-\gamma')} + 1\right]^{1/2} + \left[\frac{(2-\gamma')K'}{2} + 1\right]^{1/2}\right\} E(2+x) \tag{11.16}$$

$$K' = -1 + \left[\frac{1 + 16((1+x)/x^2)(1-\gamma')}{(2-\gamma')^2}\right]^{1/2} \tag{11.17}$$

The relation between the parameter L′ and γ' for three values of the excess noise coefficient x is shown in Figure 11.2.

The corresponding sensitivity P (ε, min.) is given by:

$$2P(\text{min.}) = (1+\varepsilon) \times f_S \times U_{11}(\varepsilon, \text{min.}) \tag{11.18}$$

The extinction ratio power penalty is given by:

$$\frac{P(\varepsilon, \text{min})}{P(0, \text{min})} = (1+\varepsilon)(1-\varepsilon)E\frac{1+x}{2+x} \times \frac{L'}{L} \tag{11.19}$$

Appropriate values of L and L' can be found from Figure 11.2 using values of γ and γ'. The corresponding expressions for the case when $U_{01} = 0$, i.e. a zero extinction ratio, are found by putting $\varepsilon = 0$.

11.2.4 The influence of photodiode design on receiver sensitivity

Conventional InP/InGaAs apds are inherently noisy and this fact has stimulated the search for material systems or device designs which will exhibit much less noise.

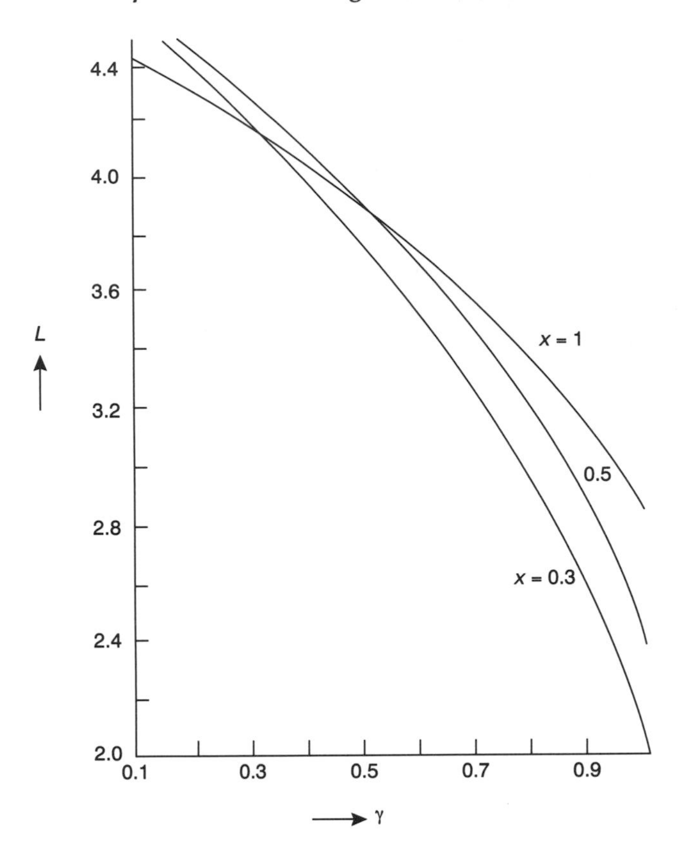

Figure 11.2 Curves of L versus γ (and L' versus γ') for $x = 0.3$, 0.5 and 1 ([5], Figure 1). Reproduced by permission of Chapman & Hall

A 1989 theoretical study to systematically compare the performance of several pin and apd receivers and predict the probable performance of future apd receivers was described in [6]. Two types of front end were used: a low-noise high-cost transimpedance design with fet input stage; and a higher-noise, low-cost high-impedance amplifier with silicon bipolar input stage. Three devices were used: an idealised InGaAs pin photodiode with no significant dark current; a conventional InP/InGaAs sagm apd with characteristics typical of devices available in 1990 from world manufacturers (gain × bandwidth in the range 10–50 GHz); and an advanced multilayer apd of a type proposed in the literature (superlattice apd) which is potentially capable of dramatically lower excess noise for a given gain.

The results of the simulations predicted that conventional apd receivers were superior to pin receivers for bit rates in the range 0.2–10 GBd, and even when a fully optimised pinfet was compared with a conventional apd and low cost receiver a 5 dB advantage at 3 GBd was predicted for the latter. Four figures graphing calculated receiver sensitivity versus bit rates between 0.1 and 10 GBd were presented in which the optimised pinfet receiver values lay on a straight line and ranged from −53 dBm to −23 dBm, respectively. The remaining curves showed: a conventional apdfet and conventional apdbjt both with 50 GHz gain-bandwidth product; a conventional apdbjt with various values of gain-bandwidth product; an advanced apdbjt with dark current component of 100 nA and various values of ionisation ratio; and advanced apdbjt with various values of dark current component and ionisation ratio of 0.1. At 10 GBd the sensitivity was independent of dark current at about −31 dBm (−23 dBm for optimised pinfet), whereas with 100 nA and a ratio of 0.01 the sensitivity improved to −36 dBm. The value read from Figure 2 for a product of 80 GHz at 5 Gbit/s was −34 dBm, which compared with −31.8 dBm realised in a laboratory model utilising a planar GaInAs apd with that product and a Si preamplifier ic with 6 GHz bandwidth. The corresponding optimum apd gain was 15 [6].

Advanced apds were predicted to offer greater benefits than conventional devices, provided that the ratio of carrier ionisation probabilities was kept below 0.1 and the multiplied dark current was below 100 nA, together with the possibility of extending the upper usable symbol rate to well in excess of 10 GBd.

11.3 AN IMPROVED DESIGN APPROACH FOR GBIT/S RECEIVERS

The earliest papers on the design of receivers for optical fibre systems appeared in the early 1970s, when first-generation systems based on multimode fibres had not even reached the stage of field trials. Moreover, line rates of 2, 8, 34 and 140 Mbit/s (and the corresponding North American and Japanese rates) were very modest compared with rates up to 5 Gbit/s on systems installed by the mid 1990s, and even more so for 10 Gbit/s systems which were then in the advanced planning stages. The corresponding receiver bandwidths needed have increased by a factor of approximately 35 times, e.g. from about 70 MHz for 140 Mbit/s to around 2.4 GHz for 4.8 Gbit/s. Receiver design and hardware for use in the Gbit/s range require microwave technologies, CAD tools and measuring and test equipment.

Perhaps the best-known author on optical receiver design is S.D. Personick of Bell Laboratories, e.g. [3], but there are also an number of other well-known names who have made valuable contributions. The present treatment will incorporate the proven simplification given in [4] and [5].

11.3.1 Novel features of the present treatment

The present treatment departs from all published approaches in that the model for the optical receiver is slightly more complex than normal, but a lot more realistic, especially in view of the extension of operating frequencies into the microwave region, and the tremendous advances in measuring equipment and computer-aided design tools which have taken place since the early 1970s. The treatment can exploit the unprecedented accuracy and ease-of-use now available from both test equipment and circuit and system simulation packages. However, the reader can see from some of the following examples that relatively old and crude programs were still in use at the start of the 1990s.

There are four novel features, namely:

(1) the introduction of a general one-port or a two-port interstage between the photodiode and the front-end amplifier.

(2) the use of correlated frequency-dependent noise generators at the input of the amplifier two-port to represent all the noise sources attributable to the components within the two-port.

(3) the third innovation is to assume neither that the amplifier transmits signals only in the forward direction, nor that it has an infinite input impedance
(4) another important aspect seldom touched upon in the literature is that of stability or freedom from self-oscillation; a true assessment can be found by means of the embedding network method.

Let us consider these in turn.

Firstly, the introduction of a general one-port or a two-port interstage between the photodiode and the front-end amplifier, rather than simple combinations of a shunt resistor and capacitor. The use of four-terminal (or two-port) interstages goes back many years, to H.A. Wheeler and others who pioneered the design of amplifiers for early television receivers about 50 years ago; their use is treated fully in H.W. Bode's classic book *Network Analysis and Feedback Amplifier Design*. In the present context such an interstage separates the shunt capacitance of the diode from the capacitance across the input of the amplifier, leading to higher gain and/or increased bandwidth. This is done so that the use of a tuning inductor at the input to the amplifier, for example, or bond wire inductances can be accommodated, but the interstage can readily be reduced to a simple shunt admittance, or even to a combination of shunt R and C elements should that be appropriate in a given design situation. Advantages of tuned two-port interstages between the photodiode and the front-end amplifier of receivers in coherent systems have been reported in many papers.

Consider the use of correlated frequency-dependent noise generators at the input of the amplifier two-port to represent all the noise sources attributable to the components within the two-port, so that there is no need to identify these noisy components in a circuit model containing active and passive elements. The r.m.s. values of the two-port noise generators can be determined from four real numbers measured at each frequency from a suitable set of four noise quantities. Previous treatments generally assume no correlation, but this is simply not true for mesfets, nor is it true for bipolar transistors at the frequencies of interest in high-speed systems.

The third innovation is to assume neither that the amplifier transmits signals only in the forward direction, nor that it has an infinite input impedance. The resultant treatment is at one and the same time more accurate, more general and scarcely more complex. It is also necessary to point out that the construction of circuit diagrams which are accurate representations of physical embodiments above about 100 MHz or so is a formidable task, sometimes involving man-years of skilled engineering effort, and is best avoided by using the more appropriate means now available.

Another important aspect seldom touched upon in the literature is that of stability in the sense of freedom from self-oscillation. A true assessment can be found by means of the embedding network method [7], for frequencies up to 10 or 20 GHz, or even higher, using commercially available software and precision test equipment which can measure and characterise transistor chips and passive components and functional blocks on integrated circuits by means of on-wafer dc and RF probes.

11.3.2 Preliminaries

Before plunging into the sometimes complicated algebra and expressions which are necessary for design purposes, a summary of what each part is about will be given, and in general only results and not full derivations will be stated. Let us begin with some basics of optical receivers; these will cover the following topics:

(a) a block schematic of the optical receiver which will be used for analysis
(b) mathematical representation of the received optical bit stream
(c) output current from the photodiode
(d) voltage expression at input to decision circuit
(e) relation between received pulse shape and that desired at the decision circuit.

The next major topic is the noise in the receiver, divided into shot and thermal noise components. This is followed by calculation of the receiver sensitivity; then the calculation of optimum avalanche gain if apds are considered in the design. Results are then given for a number of pulse shapes, and the question of optimum transmitted pulse discussed. Finally, comparisons with Personick's more exact results are provided.

Block schematic of optical receivers

A typical environment in which a receiver is required to operate is in a regenerative repeater. As an example, consider the block schematic of the main transmission path in a regenerator of the UK segment of the first transatlantic optical fibre link which entered service in December 1988, as shown in Figure 11.3 (repeat of Figure 9.3), which provides the reader with a realistic setting for the remainder of the design approach. Within the regenerator, the optical receiver module comprises the photodiode and its bias circuitry, together with the ic amplifier, all mounted within a single package fitted with a single-mode fibre tail for the optical input and connections for the electrical input and signal output. The module amplifier converts the low-level photocurrent input to an output voltage at a suitable level to serve as

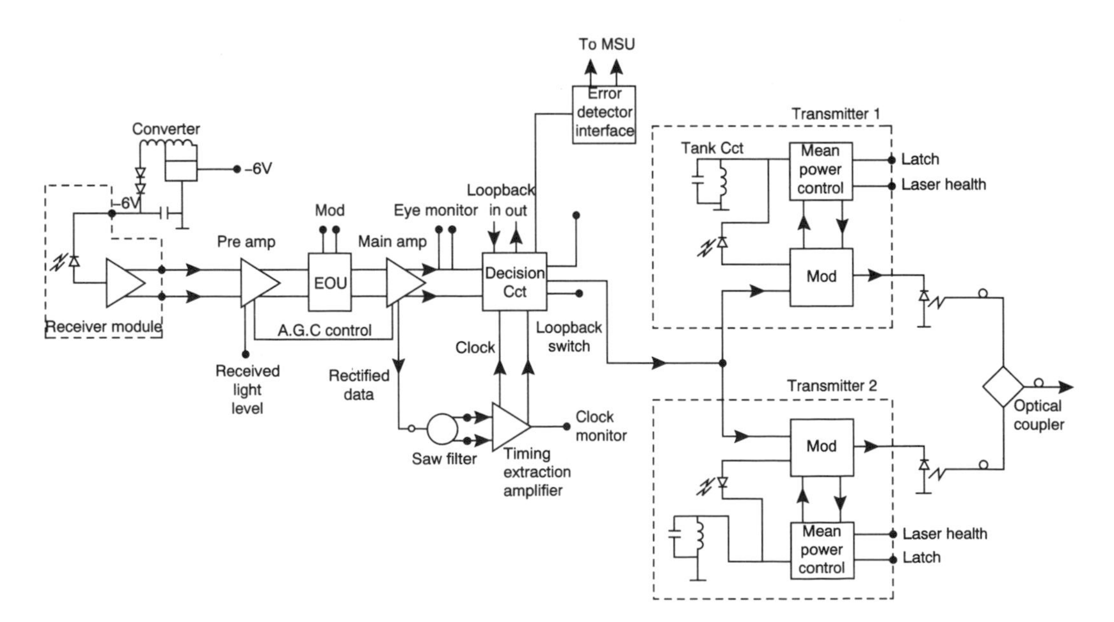

Figure 11.3 Block diagram of a regenerator pair ([9.8], Figure 5)

input to the agc amplifier. The performance of the optical receiver is set mainly by three factors: sensitivity, bandwidth and overload. Attention will be focused on the properties of signals at:

(a) the optical input
(b) the input to the receiver module amplifier
(c) the input to the decision block (the eye monitor).

A simplified circuit diagram of the pin-bipolar optical receiver showing only components, but no parasitics or strays is given in Figure 11.4.

This transimpedance configuration is basically a feedback amplifier with current shunt feedback via the resistor R_F, the value of which sets the transimpedance (output voltage over photocurrent). The cascode gain stage is followed by a phase splitter to provide a balanced output. The design ensured the correct amount of feedback needed to produce a flat frequency response over the required range for the line rate of 295.6 MBd, and also provided an adequate overload power level at 20 dB above the sensitivity. For the purposes of signal and noise analyses these will be simplified to the block schematic of Figure 11.5. The module amplifier and the preamplifier in Figure 11.3 are represented by the amplifier block, which realistically is neither unilateral nor is the input admittance zero; the equaliser and main amplifier by the block labelled equaliser, and finally the decision circuit itself represented by its input impedance. Between the photodetector and the amplifier there is a shunt admittance $Y_B = G_B + jB_B$ representing the bias network, where in general G_B and B_B may both be frequency-dependent and rational functions of frequency

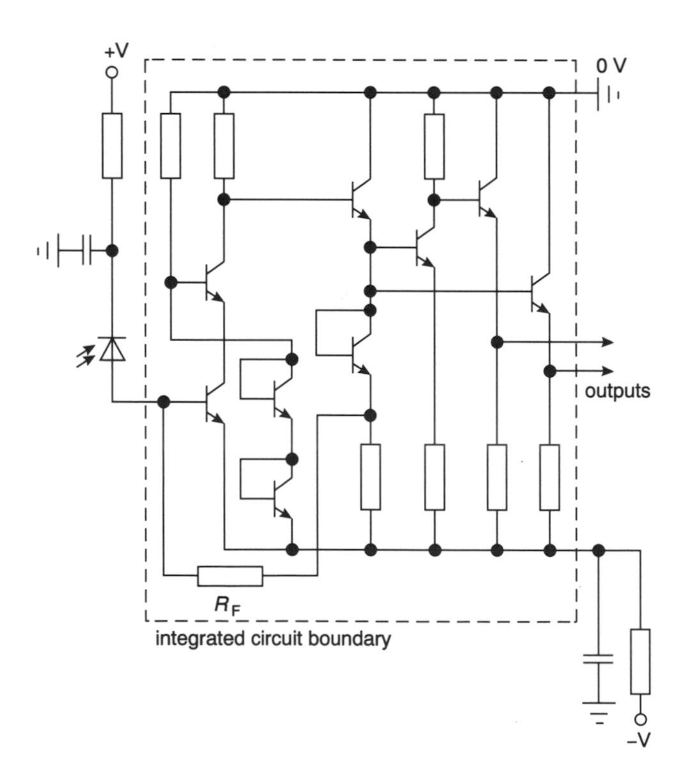

Figure 11.4 pin bipolar optical receiver circuit for UK segment of TAT-8 ([9.7], Figure 3). Reproduced by permission of Société des Electriciens et des Electroniciens

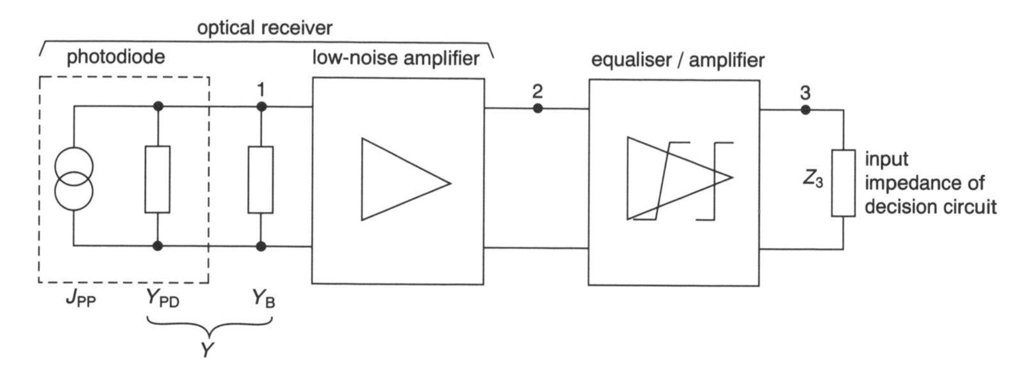

Figure 11.5 Block schematic representing photodiode, amplication up to the equaliser, and the equaliser plus main amplifier

as a result of strays and layout effects. The admittance may even correspond to distributed behaviour. This admittance includes any stray capacitance across the diode, amplifier input or bias resistor. The only significant element in G_B is usually the bias resistor itself. If necessary, a two-port can be substituted for this one-port.

11.3.3 Representation of received optical signal

The manner in which the received signal is represented mathematically must take account of the physics of the detection process. We recall that all photodetectors respond only to the modulation $m(t)$ of the lightwave carrier and not to the instantaneous value of the carrier itself, so that the shortest time intervals of interest lie in the picosecond range. The modulation of interest is a digital bit stream, which in operational systems is usually encoded and scrambled and therefore pseudo-random in nature. Following common usage but with new notation, the input signal will be a stream of optical pulses, represented by:

$$m(t) = \sum P_{(r)} \times h(t - rT) \geq 0 \tag{11.20}$$

where:

$m(t)$ = received optical power in envelope of carrier, [W],
$P_{(r)}$ = power in time slot r, P(1) for a mark, P(0) for a space, [W],
$h(t - rT)$ = dimensionless quantity defining the shape of the received optical pulse, normalised such that the integral between ± infinity is:

$$\frac{1}{T} \int h(t)\, \mathrm{d}t = 1. \tag{11.21}$$

The variable under the summation sign is r, where r is an integer ranging from $-\infty$ to $+\infty$, and including zero. Note that two-sided integrals will be used throughout, e.g. for noise spectral densities for mathematical convenience. In general, $m(t)$ will have a non-zero mean value $\langle m(t) \rangle$, and this will result in a mean primary photocurrent $\langle i(t) \rangle = \phi \langle m(t) \rangle$; for a pin, but will be multiplied by the mean value of the avalanche gain (a random variable) for an apd. The output current from the photodiode is the excitation and the voltage at the decision circuit is the response in the electrical parts of an optical receiver. The photodiode converts the optical pulse stream of equation (11.20) into the primary photocurrent, which can thus be represented by the expression:

$$\sum \phi P_{(r)} \times h(t - rT) \tag{11.22}$$

and for an apd this will be multiplied by the avalanche average gain to give the internal signal current. This forms the internal signal current generator of the photodiode, but it is not the current which flows through a short-circuit across the terminals giving the Norton equivalent current generator, except at very low frequencies where all the diode parasitics are negligible. It is the time- and frequency-domain properties of the Norton equivalent generator at the terminals which are of interest in signal-to-noise and error-rate calculations. One would expect these to be similar to those of the theoretical photocurrent generator, suggesting that we represent the Norton equivalent current generator by i_{PD}, where with the same range of values of r as usual:

$$i_{\mathrm{PD}}(t) = \sum \phi P_{(r)} \times h_{\mathrm{PD}}(t - rT) \tag{11.23}$$

The corresponding Fourier transforms are $I_{\mathrm{PD}}(\mathrm{j}f)$, and for the shape $H_{\mathrm{PD}}(\mathrm{j}f)$, and the latter can be taken to be a mathematically simple shape, e.g. Gaussian.

The average optical power P can be linked to the probabilities introduced in Appendix A, the relation being:

$$P = P_{(r)}(0) \times \mathrm{pr}(0) + P_{(r)}(1) \times pr(1) \tag{11.24}$$

and for equiprobable symbols both probabilities equal one half; omitting the subscript for brevity, this is simplified to:

$$P = 0.5\,(P(1) + P(0)) \tag{11.25}$$

Finally, if the extinction ratio is zero so that the power in a space is zero, one obtains the obvious result that the long-term mean power P is one-half of the power in a mark.

Sometimes it is more convenient to work in the frequency domain so that the Fourier transforms of the time functions are required. As elsewhere in this book, these will be denoted by capital letters, e.g. $i(t)$ and $I(\mathrm{j}f)$ are transform pairs, with the argument of I taken as $\mathrm{j}f$ as a reminder that, in general, real waveforms have complex Fourier transform mates.

11.3.4 Voltages at the input to the decision circuit

The voltage response at the input to the decision circuit to the signal at the photodiode terminals can also be represented in the time domain by a stream of pulses described by the sum of individual marks or spaces. The voltage at the decision circuit will be characterised by an average, or expected, value as well as by fluctuations from this value resulting from the randomness of photon arrival times and the random nature of the detection process itself. The response is the product of the transfer function from the photodiode to the decision circuit, and the photodiode signal current, and is therefore proportional to the product of the power in the optical pulse and a shape factor normalised according to equation (11.21). As before, the integer variable r in all the following summations take values ranging from minus to plus infinity. It is customary to omit the transfer function (voltage/current) in the following derivation for brevity, but it can be reinserted wherever necessary to obtain the correct dimensions. Alternatively, the expressions can be read as voltages per unit transimpedance. With this understanding we then have the following values.

The long-term average voltage of the bit stream

$$\langle v(t)\rangle = \sum V_{(r)} \times h_{\mathrm{D}}(t - rT) \tag{11.26}$$

where $V_{(r)}$ is the average voltage response to a single mark or space, with only two possible values $V(1)$ and $V(0)$, and $h_{\mathrm{D}}(\cdot)$ is the shape factor chosen so that $h_{\mathrm{D}}(0) = 1$, i.e. the maximum values occur at the sampling instants. The relation between this voltage bit stream and the optical input is given mathematically by:

$$\langle v(t)\rangle = \phi \int P_{(r)} \times h(t - rT) \times z(t)\mathrm{d}t \tag{11.27}$$

with $z(t)$ the impulse response of the receiver between photodiode output and decision circuit input, and the limits of integration are from minus to plus infinity.

The instantaneous voltage

$$v(t) = \langle v(t)\rangle + v_{\mathrm{N}}(t) \tag{11.28}$$

where $v_{\mathrm{N}}(t)$ stands for the fluctuations of the voltage about the average value $\langle v(t)\rangle$; clearly this is given by:

$$v_{\mathrm{N}}(t) = v(t) - \langle v(t)\rangle \tag{11.29}$$

The variance or mean square value of the fluctuations

$\langle v_{\mathrm{N}}{}^2(t)\rangle$ is found by squaring both sides then taking the mean. Since the terms on the right of equation (11.29) are uncorrelated, there is no product term, and by omitting the explicit dependence on time for brevity, we have:

$$\langle v_{\mathrm{N}}{}^2\rangle = \langle v^2\rangle - \langle v\rangle^2 \tag{11.30}$$

which is a function of r and of t from equation (11.27). Without loss of generality, we now restrict the values of t to correspond to the sampling instants when the decisions are made, i.e. $t = rT$, with r a positive or negative integer.

Noise voltages at sampling instants

Since we have assumed that $h_{\mathrm{D}}(0) = 1$, we also assume that $h_{\mathrm{D}}(rT) = 0$ for r nonzero (since no isi), resulting in:

$$v_{\mathrm{N},(rT)} = v_{(rT)} - V_{(r)} \tag{11.31}$$

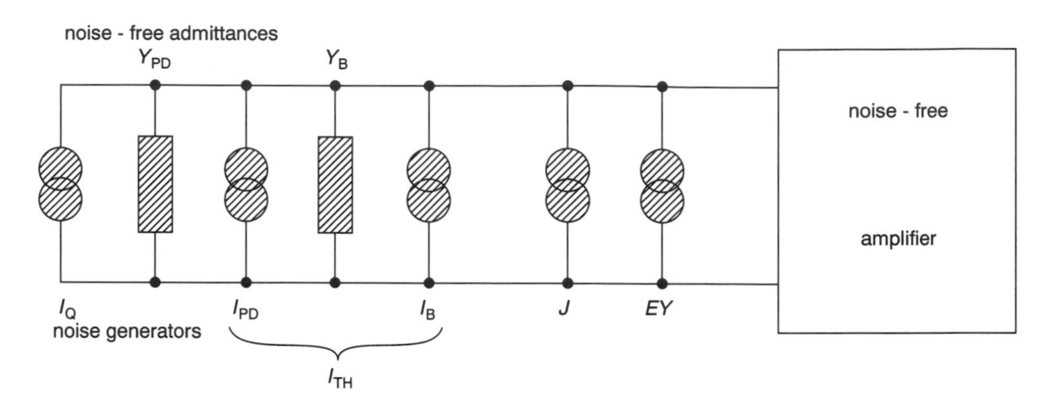

Figure 11.6 Representation of optical receiver noise sources by current generators in parallel with noise-free admittances and amplifier

Note that the noise voltage v_N at the sampling instants depends on all possible pulse sequences as well as on t; this property distinguishes optical systems from all others in which noise is independent of the signal.

The mean-square noise voltage (or noise power in unit resistance)

The quantity given by equation (11.30) can be regarded as having a number of components arising from the various noise sources in Figure 11.6.

Taking the terms in the same order as the current generators in Figure 11.6 and assuming no correlation between the equivalent noise generators E and J of the amplifier two-port, the mean square noise voltage defined by equation (11.30) is the sum of four components and can be written as:

$$\langle v_N{}^2 \rangle = \langle v_Q{}^2 \rangle + \langle v_T{}^2 \rangle + \langle v_E{}^2 \rangle + \langle v_J{}^2 \rangle \tag{11.32}$$

where the components are: $\langle v_Q{}^2 \rangle$ caused by quantum noise associated with the statistical nature of the photodetection (and for an apd the gain) process; $\langle v_T{}^2 \rangle$ caused by the thermal noise arising in all the dissipative components in the receiver, e.g. bias resistors, lossy microstrip; $\langle v_E{}^2 \rangle$ caused by the noise equivalent emf E in series with the input to the noise-free preamplifier; $\langle v_J{}^2 \rangle$ caused by the noise equivalent current generator J across the input to the preamplifier.

Each of these mean square voltages is obtained from the product of the m.s. noise current at the point of origin and the modulus squared of the transfer function from that point to the input of the decision circuit. Later, the last three terms will be grouped together, as they are independent of the signal, leaving the first term as the only one which is signal-dependent, and which will now be discussed.

11.4 TRANSFER FUNCTIONS BETWEEN PHOTODIODE AND DECISION CIRCUIT

In order to relate the signal and noise spectra and waveforms at the photodiode terminals to the corresponding quantities at the input to the decision circuit given above, the signal and noise properties of the amplifier and equaliser blocks must be described. This is possible in the time domain using convolution, but for design purposes a frequency-domain approach based on transfer functions is almost mandatory, because design in the time domain is far less advanced than our knowledge of amplifier, equaliser and filter theory which is expressed in terms of frequency characteristics.

11.4.1 Model of receiver

The model which we will use for the receiver is a cascade of one-port or two-port networks, each of which we assume to be linear, as shown schematically in Figure 11.5.

The photodiode and the necessary bias circuitry are fairly simple, but in practice the low-noise amplifier is usually an integrated circuit comprising a number of transistors, such as shown for the PTAT-1 receiver in Figure 9.6. At frequencies beyond, say, 100 MHz the representation of the amplifier block (e.g. hybrid integrated circuit or monolithically integrated circuit) by a simple circuit diagram which neglects circuit board or substrate strays and parasitics is almost useless for predictive purposes. In any case, the actual amplifier will be a lot more complex than an idealised single bipolar or fet model assumed in many treatments. The equaliser itself may, however, be relatively simple, e.g. a simple differentiator, but would still need careful modelling at frequencies above, say, 100 MHz.

Representation of photodiode: realistic electrical circuit models of real photodiodes contain quite a number of elements to represent the chip, bond wires and package parasitics. In view of this fact, and the availability of microwave test equipment able to measure the reflection coefficient and noise of laser diodes and photodiodes in small-signal conditions, the photodiode will be represented by the Norton equivalent at its electrical output. This consists of a signal current generator of r.m.s. value $J_S(jf)$ in parallel with an output admittance Y_{PD}. The corresponding noise sources are represented by a single generator $I_Q(jf)$, which is dominated by the quantum noise associated with the mean value of the photocurrent.

Representation of bias circuitry: these generators and passive elements are followed by a shunt admittance $Y_B = G_B + jB_B$, the admittance of the bias network (possible including high-impedance microstrip lines as well as discrete components) as seen from the photodiode terminals. The conductance G_B represents the dissipative elements, e.g. bias resistance R_B, and B_B the total shunt susceptance, mainly composed of capacitance C (photodiode, strays, parasitics associated with resistors). Note that the amplifier input admittance is not included in this definition, because for Gbit/s systems this admittance will, in general, be complex and frequency-dependent since frequencies into the microwave region are present in the signal. For convenience in the following analysis, the small-signal admittance of the photodiode and the bias network will be combined into a single admittance $Y = G + jB$. In parallel with Y_B there is a thermal noise current generator $I_{NT}(jf)$ which accounts for the dissipative elements in the bias network.

Representation of receiver low-noise amplifier input: early treatments and many later ones model the amplifier input admittance by a resistor and capacitor in parallel with R_B and C, but this is a quite inaccurate and unnecessary in view of the availability of good microwave circuit analysis packages. Instead, the small-signal properties of the amplifier block are most accurately characterised by its reflection and transmission properties as determined from measurements, e.g. of two-port s-parameters, failing which a circuit diagram valid up to microwave frequencies and a suitable simulation package for analysis and design.

Representation of noise in amplifiers: for system design and analysis we are only interested in the noise and signal behaviour of a black box, generally a two-port. The noise behaviour of a linear two-port is completely characterised by four real quantities at each frequency, compared with eight required for the small-signal properties. Four real noise quantities can be represented by pairs of real-valued generators with a complex-valued correlation between them, and are most accurately characterised by values derived from measurements since sophisticated equipment and software are now available to do this.

For optical receivers the most convenient pair consists of two correlated noise generators $E(jf)$ and $J(jf)$ which fully describe the noise characteristics, and which are connected in series and in parallel, respectively, at the input of the noise-free linear amplifier. The r.m.s. values of these noise current and voltage generators depend on several active and passive components within the amplifier and not just on the first transistor. The values also depend on the configuration adopted for the first few stages, and on whether a high-impedance or transimpedance design is being analysed. These generators can be defined in terms of any one of a number of sets of four real and measurable quantities which are themselves frequency-dependent. The noise-free amplifier is then followed by an equaliser, which is taken to be noise-free and which shapes the pulses to have some desired properties; a common choice of shape is one of the raised cosine family. For brevity, the explicit dependence on frequency denoted by (jf) will normally be omitted, where no confusion can arise with time-domain constant quantities. For convenience, the noise generator E connected in series with the low noise amplifier (lna) input is replaced by its Norton equivalent current generator EY connected across the input.

11.4.2 Transfer function between photodiode terminals and input of the decision circuit

This transfer function is taken to be the ratio of voltage response across the input impedance of the decision circuit to the independent current generators across the photodiode terminals. It is thus a transfer impedance in nodal analysis terms, which is a very common basis for microwave and lower frequency circuit analysis packages. The signal analysis will be based on the simple, three-node, two two-port diagram shown in Figure 11.5.

The independent current generator J_1 represents any one of the signal or noise current generators in Figure 11.6. The nodal admittance matrix corresponding to Figure 11.7 can be written down by inspection, and it is:

$$\begin{pmatrix} y_{11}+Y_1 & y_{12} & 0 \\ y_{21} & y_{22}+Y_{22} & Y_{23} \\ 0 & Y_{32} & Y_{33}+Y_3 \end{pmatrix} = [Y] \tag{11.33}$$

where the y refer to the optical receiver two-port and the Y with double subscripts to the amplifier–equaliser two-port; note that a subscript 1 has been added to the composite admittance Y introduced earlier to denote its location in the circuit. From standard nodal analysis using determinant Δ and cofactors Δ_{JK}, each of which is a polynomial in jf for lumped element networks, the voltage responses at nodes 2

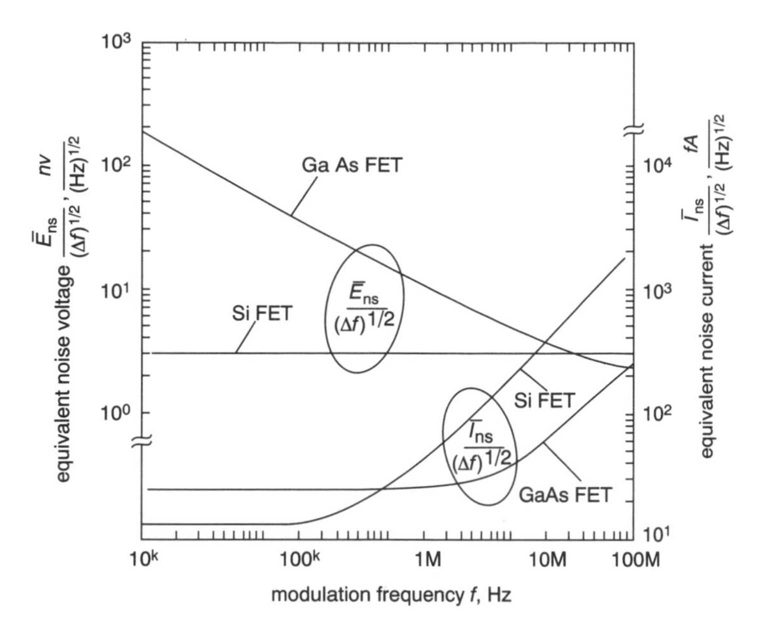

Figure 11.7 Equivalent input noise voltage and current versus modulation frequency of a typical Si fet SN4416 and a typical GaAs fet CFY 13 ([10], Figure 2). Reproduced by permission of IEE

and 3 are:

$$V_2 = \frac{\Delta_{12}}{\Delta} J_1 \text{ and } V_3 = \frac{\Delta_{13}}{\Delta} J_1 \tag{11.34}$$

where the coefficients of J_1 are sometimes called the transimpedances from port 1 to ports 2 or 3. In particular, the ratio V_3/J_1 relates the desired pulse spectrum at the decision circuit input to the spectrum of the current generator J_1. For instance, the spectrum of V_3 is often chosen to correspond to one of the family of raised-cosine pulse shapes. The spectrum of J_1 is often assumed to correspond to a Gaussian pulse. The transimpedance V_3/J_1 can be written as the product of a current ratio I_3/J_1, often written in the form exp $(\alpha + j\beta)$, where α and β are frequency-dependent, and the load impedance Z_3, since $V_3 = I_3 Z_3$. The transfer impedance (transimpedance) can be found by standard nodal analysis of the circuit as it stands, and it depends on every element, whether the amplifier front end is an integrating or a transimpedance configuration. The quantity α is a gain or loss in nepers, and β a phase shift in radians: in practice dBs and degrees would be used in most countries. The transfer impedance between nodes 1 and 3 can usefully be written as a product of the transimpedance of the amplifier two-port times the voltage gain of the equaliser two-port from the determinant identity:

$$\frac{\Delta_{13}}{\Delta} = \frac{\Delta_{12}}{\Delta} \times \frac{\Delta_{13}}{\Delta_{12}} \tag{11.35}$$

and the first term can, in turn, be expanded, yielding:

$$\frac{\Delta_{13}}{\Delta} = \frac{\Delta_{11}}{\Delta} \times \frac{\Delta_{12}}{\Delta_{11}} \times \frac{\Delta_{13}}{\Delta_{12}} \tag{11.36}$$

Examining the terms on the right-hand side in the order given, each has a simple physical meaning, namely:

$$\frac{\Delta_{11}}{\Delta} = \text{the reciprocal of the admittance seen from the generator } J_1,$$

$$\frac{\Delta_{12}}{\Delta_{11}} = \frac{V_2}{V_1}, \text{ the voltage gain of the amplifier two-port} \tag{11.37}$$

$$\frac{\Delta_{13}}{\Delta_{12}} = \frac{V_3}{V_2}, \text{ the voltage ratio of the equaliser two-port} \tag{11.38}$$

The determinant and cofactors are readily evaluated from the simple 3 × 3 admittance matrix above, and the reader can easily prove that:

$$\frac{V_2}{V_1} = -\frac{y_{21}}{y_{22} + Y_{22} - y_{23}{}^2/(Y_{33} + Y_3)} \tag{11.39}$$

for the voltage gain of the optical receiver two-port and

$$\frac{V_3}{V_2} = -\frac{Y_{23}}{Y_{33} + Y_3} \tag{11.40}$$

for the voltage ratio of the amplifier–equaliser two-port.

The two voltage ratios are conveniently written as exp θ_A and exp θ_E for the amplifier and equaliser, respectively, where in both cases $\theta = \alpha + j\beta$ as before. Thus, given the frequency characteristic of the transfer impedance between nodes 1 and 3, and that of the receiver two-port, the frequency characteristic of the amplifier–equaliser voltage ratio can be determined.

The transfer impedance of the optical receiver, i.e. between nodes 1 and 2, is given by Δ_{12}/Δ and is a function of frequency. It will be illustrated by an actual example, a typical production optical receiver of the PTAT-1 system for which a simplified circuit diagram was given in Figure 9.6, and the modulus of the transimpedance in dB relative to 1 Ω was shown in Figure 9.7. The value of about 76 dBΩ corresponds to a transimpedance of 6310 Ω or 6.31 mV/μA. These 443 MBd receivers had a typical sensitivity of −33.6 dBm (corresponding to a pin photocurrent of about 0.2 μA for a responsivity of 0.5 A/W) and a dynamic range of almost 20 dB (giving a maximum photocurrent of about 20 μA). Sensitivity measurements were made using a $2^{23} - 1$ prbs data sequence at 443 MBd to simulate the spectral characteristics of the 24B1P line code. Both TAT-8 and PTAT-1 lna receiver amplifiers used silicon transistors; by 1990, high-speed (20 GHz f_T) silicon processes had enabled optical receivers to be realised for use at 2.5 and 5 Gbit/s [8]. In one such example of an apd-bip design, a bandwidth of 4.2 GHz was achieved at an apd optimum gain of 20 (measured sensitivity −35.5 dBm at 2.5 Gbit/s), and with an optimum gain of 15 the sensitivity was −31.8 dBm at 5 Gbit/s.

11.5 NOISE CONSIDERATIONS

The reason why the above notation was chosen in the present context is that noise voltages due to the different noise current generators must be added in terms of mean square values across the same load, here Z_3. It follows that mean square (m.s.) values of the current generators are involved, and the modulus squared of the current ratio, rather than the current ratio itself. In the present notation this is simply $\exp(2\alpha)$, multiplied by the modulus squared of the input impedance of the decision circuit. Since this is, however, the same for all terms, this impedance can be taken as 1 Ω without loss of generality.

We again take the PTAT-1 optical receiver as our example. The noise spectral densities of both TAT-8 and PTAT-1 typical receivers were shown in Figure 9.8, and they extend well beyond 100 MHz. In both the dominant contributions to amplifier noise comes from the feedback resistor and the first-stage transistor. The value of the resistor was fixed by the bandwidth required, but the noise contribution from the first stage was optimised by the choice of transistor and its operating conditions, e.g. bias current. The noise equations due to Personick were used in the optimisation.

11.5.1 Analysis when noise generators E and J are uncorrelated

With the above explanation and representation, it is clear from inspection that the frequency-domain description of the total m.s. noise current across the photodiode is given by:

$$\overline{(I_Q + I_T + J + EY)^2} = \overline{I_Q I_Q^*} + \overline{I_T I_T^*} + \overline{JJ^*} + \overline{EE^*YY^*} \tag{11.41}$$

if the two noise generators are uncorrelated, as most treatments assume. The mean square values of E can be written in terms of the thermal noise expressions $2k\theta R_N\,df$, and those of J can be written $2k\theta G_N\,df$ for double-sided spectra, where G_N and R_N are so-called noise quantities, and of course the thermal, or Johnson noise m.s. current associated with the real part of the composite admittance Y is $2k\theta G$; thus we can write:

$$\overline{I_Q I_Q^*} + 2k\theta(G + G_N + (G^2 + B^2)R_N)df \tag{11.42}$$

Note that G includes a photodiode bias resistor. In the customary representation, where Y consists of R_B and C in parallel, $G^2 + B^2$ becomes $R_B^{-2} + (2\pi f C)^2$, which have the simplest possible frequency dependence.

Since the representation by two noise generators is less well known, Figure 11.7 has been included to help the reader to appreciate the frequency dependence of these generators between 10 kHz and 100 MHz and the magnitudes involved. Two pairs of curves are shown: one for a typical Si and one pair for a typical GaAs fet device [9]. The reason why this particular frequency range was chosen for illustration is that it covers most of the band occupied by a 7B8B (the spectrum for the 5B6B code is very similar) encoded 140 Mbit/s main traffic signal and by the supervisory channels, as can be seen from Figure 11.8 [10]. This spectrum corresponds to a stream of impulses with equiprobable ones and zeros. The spectrum should be borne in mind when, later, bandwidths and limits of integration are discussed in connection with the noise performance of receivers.

11.5.2 Analysis when noise generators E and J are correlated

In the general and more realistic cases where E and J are correlated, two more terms must be added to the expression, namely:

$$\overline{EYJ^* + E^*Y^*J} = G\overline{(EJ^* + E^*J)} + jB\overline{(EJ^* - E^*J)} \tag{11.43}$$

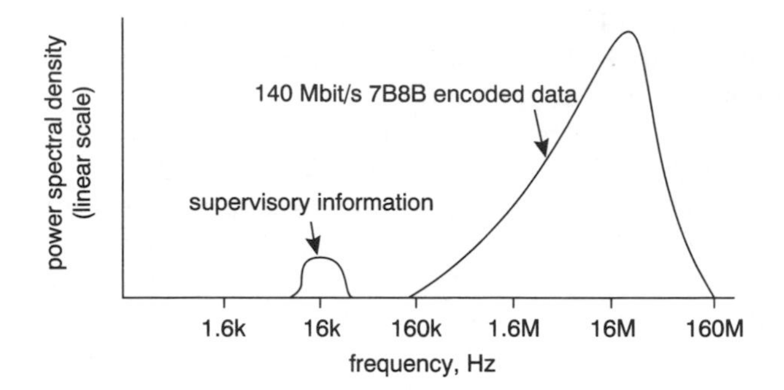

Figure 11.8 Power spectral densities of main traffic (140 Mbit/s 7B8B encoded data) and supervisory information which are frequency division multiplexed. Reproduced by permission of BT

The expressions in brackets can, in turn, be written in terms of the real ζ (zeta) and imaginary $\mathrm{j}\kappa$ (kappa) parts of the complex correlation coefficient determined from measurements (or theory) by the equations:

$$\overline{(EJ^* + E^*J)} = 2k\theta \times \zeta \times df \tag{11.44}$$

$$\overline{(EJ^* - E^*J)} = 2k\theta \times \mathrm{j}\kappa \times df \tag{11.45}$$

Following the approach above, these correspond to two further generators of m.s. values proportional to $G\zeta$ and $B\kappa$, respectively. Thus, in the general case the noise properties of the receiver can be represented by one shot noise generator plus five fictitious independent thermal noise current generators across the photodiode, and the m.s. output voltage from each of these is found by multiplying by the same function exp (2α). In this mode of representation the snr of the receiver can be expressed by the ratio of m.s. signal current to the sum of the m.s. noise currents at the front-end. This is necessarily the same as the corresponding m.s. voltage ratio at the input to the decision circuit since the common multiplier exp (2α) cancels out when the ratio is taken.

It is important to note that the above expressions are all in terms of frequency intervals df, supposed relatively small. Actual m.s. values of current, voltage or power must be found by integration between appropriate limits, usually determined by the noise bandwidth of the circuit under discussion, but in theory between minus and plus infinity, as we will see later. Another point of note is that the four noise quantities G_N, R_N, ζ and κ are in general functions of frequency. Since all the integrands are even functions of frequency, the limits can be taken as 0 and infinity, so-called single-sided spectra, in which case $2k\theta$ becomes $4k\theta$, which the reader may recognise more readily.

11.5.3 Summary of expressions for noise contribution

Let us now collect all the noise current generator terms together, using the appropriate shot or thermal noise forms given earlier, obtaining,

$$2q \times I_{AV} \times df + 2k\theta(G + G_N + (G^2 + B^2)R_N + G\zeta - B\kappa) \times df \tag{11.46}$$

which has dimensions of A^2. For Gbit/s systems where receiver bandwidths extend into the GHz range, the behaviour of the noise quantities with frequency determined by measurements on the optical receiver lna may be available. As an illustration, the frequency dependence of one set of noise quantities derived from measurements on a 0.25 μm gate length GaAs mesfet is shown in Figure 11.9 [11].

The quantities G_N and R_N appear in equation (11.46), but the correlation conductance and susceptance is given in this figure are clearly not the same as the correlation coefficients used in that equation. The noise data on a transistor or amplifier can also be listed, as given for the same device in Table 11.2. Each of the terms in equation (11.46) can be identified term by term with the corresponding m.s. noise voltages at the decision input. Each of the letters denotes a positive quantity; therefore worst-case noise corresponds to each term being at its upper limit, and κ at its lower limit. These limits depend upon the components and circuits used and on the production tolerances on the noise quantities of the amplifier two-port. Bearing in mind the simplifying assumption that the input impedance of the decision circuit is taken as 1 Ω, the m.s. noise voltages can be written down immediately in terms of definite integrals with the usual limits, as follows.

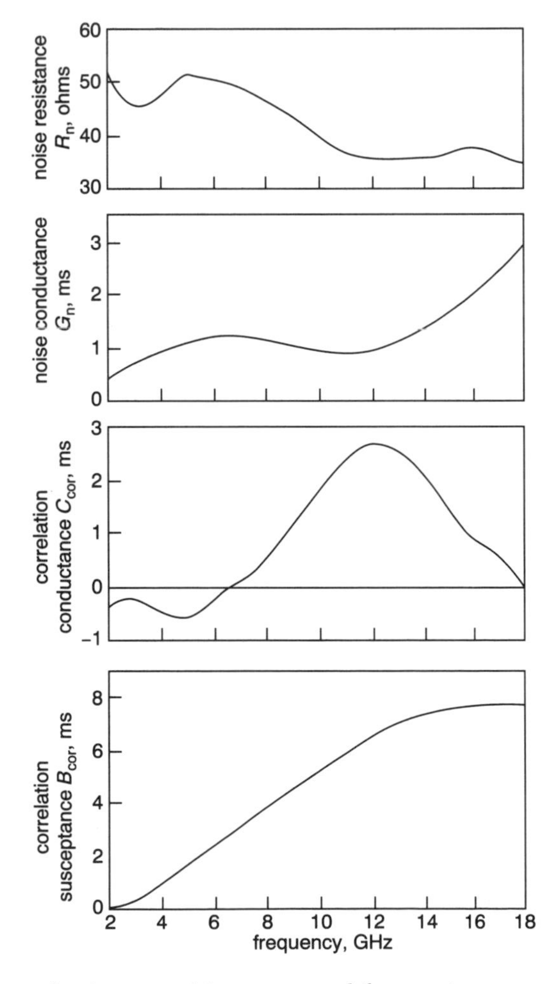

Figure 11.9 One set of noise quantities computed from noise measurements on a 0.25 μm gate length GaAs mesfet ([11], Figure 2). Reproduced by permission of IEEE, © 1989

Table 11.2 Noise quantities of 0.25 μm gate length mesfet derived from measurements ([11], Table 1). Reproduced by permission of IEEE, © 1989

Freq, GHz	F_{min}, dB	Γ_{opt}, Mag.	Ang.	F50, dB	R_n, Ohm
2.0	1.00	0.75	0.5	3.05	51.7
3.0	1.20	0.69	1.5	2.85	45.7
4.0	1.35	0.65	6.0	2.90	47.0
5.0	1.50	0.63	10.0	3.10	52.3
6.0	1.65	0.62	14.5	3.10	48.7
7.0	1.75	0.61	18.5	3.15	48.4
8.0	1.80	0.61	22.5	3.15	46.4.
9.0	1.85	0.61	27.0	3.15	44.2
10.0	1.95	0.60	31.0	3.10	39.4
11.0	2.00	0.60	36.0	3.10	37.0
12.0	2.05	0.59	39.0	3.10	35.7
13.0	2.05	0.58	42.0	3.10	35.8
14.0	2.10	0.56	44.0	3.10	35.5
15.0	2.10	0.54	46.0	3.10	36.7
16.0	2.15	0.51	47.0	3.10	37.7
17.0	2.15	0.48	48.0	3.00	36.1
18.0	2.15	0.44	50.0	2.90	35.2

Signal-dependent term

$$\frac{|V_{NQ}(jf)|^2}{\Omega} = 2q \times I_{AV} \times \int \exp(2\alpha)\, df \quad \text{[units of power]} \tag{11.47}$$

for the quantum noise component. The integrand is the ratio of the pulse shape at the decision circuit to the pulse shape at the photodiode which Personick defines as his integral I_2. Values of this integral have been computed for several different shapes, and are shown. By substituting the expressions for I_{AV} from equation (11.46) one has two expressions for this component, namely:

$$2q \times \phi P(1) \times \int \exp(2\alpha)\, df \qquad \text{for a mark} \tag{11.48}$$

$$2q \times \phi P(1) \times (1-\gamma) \times \int \exp(2\alpha)\, df \quad \text{for a space} \tag{11.49}$$

In both, the only term in the integrand is $\exp(2\alpha)$, that is, the Personick integral referred to above. Typical results are presented later.

Terms which are independent of the signal

The remaining components are given by:

$$2k\theta \times \int (G + G_N + (G^2 + B^2)R_N + G\zeta - B\kappa) \times \exp(2\alpha)\, df \tag{11.50}$$

That part of the integrand in brackets will, in general, not be known in terms of polynomials or rational functions of frequency, so the designer must resort to numerical techniques. Expression (11.50) is a generalisation of the signal-independent terms in the simple noise analysis, and we can therefore equate it to q^2Z, introduced by Personick in his 1973 paper. The relation between this approach and that commonly given can be seen by disregarding the correlation completely, and by replacing G by $1/R_B$ and B by $2\pi fC$. The resulting expression for the m.s. thermal noise contains three terms independent of frequency, and proportional to R_B, G_N and G^2R_N, and one term proportional to frequency squared, namely B^2R_N. All four of these terms are multiplied by the transfer function and the product denoted by q^2Z, where Z is the mean square number of thermal noise electrons per symbol interval.

Many authors state that the quantities E and J for GaAs mesfets are such that $\zeta = 0$ and $\kappa = 1$, i.e. the correlation coefficient is purely imaginary. The result of this assumption is that the mean square noise from these types of transistor is less than the value with no correlation; whereas if the coefficient is real, the net amount of noise power may be larger or smaller than without correlation. Measurements indicate that the imaginary part tends to dominate.

11.5.4 Relation to the parameter Z

The importance of the quantity Z arises from the simple relation between it and pin receiver sensitivity, which is the performance parameter of direct concern to a system designer.

$$\frac{q^2Z}{2k\theta} = \int (G + G_N + (G^2 + B^2)R_N + G\zeta - B\kappa) \times \exp(2\alpha)\, df \tag{11.51}$$

The terms in this expression are easily related to more familiar forms, e.g. that given in [4]:

$$q^2Z = \left[T\left(\frac{2k\theta}{R_B} + S_I + \frac{S_E}{R^2}\right) \times I_2 + (2\pi C)^2 \times S_E \times \frac{I_3}{T}\right] \tag{11.52}$$

written in the same order as the preceding equation. S_I corresponds to $\overline{JJ^*}$ and hence G_N, S_E to $\overline{EE^*}$, and hence R_N, $1/R$ to G and $2\pi fC$ to B.

Values of Z between 10^5 and 10^7 have been quoted for packaged components and apd receivers, and around 10^4 for pin receivers. The noise quantities R_N and G_N can be found from measurements on the low-noise amplifier in the receiver module or, with reduced accuracy, from a circuit diagram of the amplifier. If the amplifier is taken as a single transistor, bipolar or fet, the noise quantities are expressible in terms of the elements of hybrid-π like circuit models for the active device.

Returning to the expression for the total noise at the input to the decision circuit, and using the frequency-domain expressions introduced above, we may write the Fourier transform of $v_N(t)$ in equation (11.26) as:

$$|V_N(jf)|^2 = \text{shot noise term} + \text{thermal noise terms}$$

or, using Z for brevity

$$|V_N(jf)|^2 = (\phi q)^2\left(\frac{M^{\times}u(1)\phi B_2}{q} + \frac{Z}{M^2}\right) \tag{11.53}$$

$$\text{sensitivity (dBm)} = 10\,\log\left(\frac{Nf_S}{\lambda\eta}\right) - 97 \tag{11.53a}$$

11.6 WORST-CASE PINFET MODULE NOISE PARAMETERS

These parameters are included to provide readers with examples of values found for practical optical receiver modules when worst-case values are taken for the important parameters in the noise expressions; e.g. the ambient temperature is taken as 50°C, so that the operating case temperature is taken as 62°C. The value of g_M is taken as the minimum, the break or corner frequency as the highest, and so on. The values of the parameters should also be taken as those corresponding to end-of-life, not the values expected on day 1. Results will be pessimistic as intended. Values quoted refer to a receiver for use in 1300 nm, 140 Mbit/s systems.

pin photodiode noise

Shot noise current density is $2q(I_G + I_D)/df$, where the sum of the two currents is 160 nA at the case temperature; this gives the value 0.512 $(\text{pA}/(\text{Hz})^{1/2})^2$.

Flicker noise with corner frequency at 100 kHz is given by the formula $10^4/f$ $(\text{pA}/(\text{Hz})^{1/2})^2$.

Thermal noise power density

That associated with 1 MΩ bias resistor given by the relation $4k\theta/R$ is 0.185 $(\text{pA}/(\text{Hz})^{1/2})^2$.

Both extinction ratio current and minimum mean signal current are negligible compared with the worst-case pin leakage, and the fet gate leakage currents.

Mean-square noise current densities of fet noise model

(a) drain-source conductance, $4k\theta g_{DS}$
(b) shot noise of gate current, $4k\theta\omega^2C^2\tilde{p}/g_M$
(c) shot noise of drain current, $4k\theta g_M\tilde{q}$.

Both (a) and (b) appear at the output of the fet model and must therefore be referred to the input; this is done simply by dividing both by the transconductance squared, giving m.s. noise voltage generators in series with the gate terminal (assuming a common source configuration). These can be transformed into current generators across the fet input by the relation $\overline{EE^* \times YY^*}$, where $Y = G + jB$ so that YY^* is simply $G^2 + B^2$, Combining (a) and (b) because they are uncorrelated yields:

$$\frac{(G^2 + \omega^2C^2)4k\theta(g_{DS} + g_M\tilde{q})}{g_M{}^2}$$

Taking $G = 10^{-6}$, $C = 1$ pF, $g_{DS} = 1/274$, $g_M = 12$ mA/V and $\tilde{q} = 1.2$ gives the frequency-independent contribution as 2.32×10^{-6} $(\text{pA}/(\text{Hz})^{1/2})^2$.
The f^2 contribution will be negligible at flicker noise frequencies.

Flicker noise in mesfet

When the gate is ac shorted to the drain (earth) a noise power appears at the output which increases as $(f_C/f)^{3/2}$ below some corner frequency f_C. This can be referred to the input as a shunt current generator, as described in the preceding paragraph. Taking worst-case values of 1.5 for the exponent and 60 MHz for the corner frequency results in the expression:

$$(G^2 + \omega^2 C^2)4k\theta\tilde{q}\left(\frac{f_C}{f}\right)^{1.5}$$

Clearly this does not have a frequency-independent term, and so no values are given.

The above information is sufficient to enable the reader to complete the calculations at frequencies of interest with the above values, or to substitute values appropriate to other devices.

11.7 EFFECT OF FLICKER NOISE ON RECEIVER PERFORMANCE

The frequency band covered by the spectrum of an encoded line signal extends downwards from f_S and may have significant components down to frequencies of the order of $f_S/100$, being shaped to avoid energy at low frequencies. Within the band covered by the spectrum the noise spectrum of the optical receiver may well be frequency-dependent, in particular at the low end. For example, the noise spectrum at the output of a pinfet module with the gate of the fet ac shorted to ground behaves as shown in Figure 11.10 [13].

The noise power spectral density begins to increase with decreasing frequency somewhere between 100 and 20 MHz, depending on type and individual transistor, with a corresponding range of values of the 3 dB frequencies above the noise floor. Apart from the spread among different types it appears that there is considerable variation between devices of the same type [13]. The slope is around 15 dB/decade at 1 MHz, and the break frequency between the slope and the noise floor locates the $1/f$ characteristic. The slope continues down to frequencies of interest for supervisory channels (perhaps up to 20 or 30 kHz) where the noise performance of the photodiode may not be as simple as considered so far.

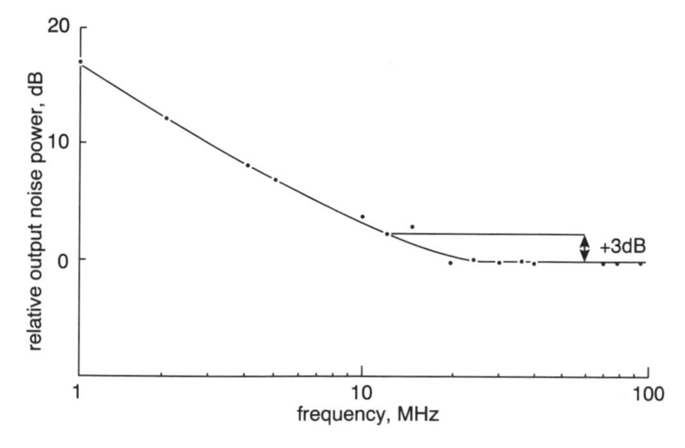

Figure 11.10 Relative output noise power versus frequency of a pinfet module with the fet gate ac shorted to ground (Monham *et al.*, Figure 3). Reproduced by permission of IEEE

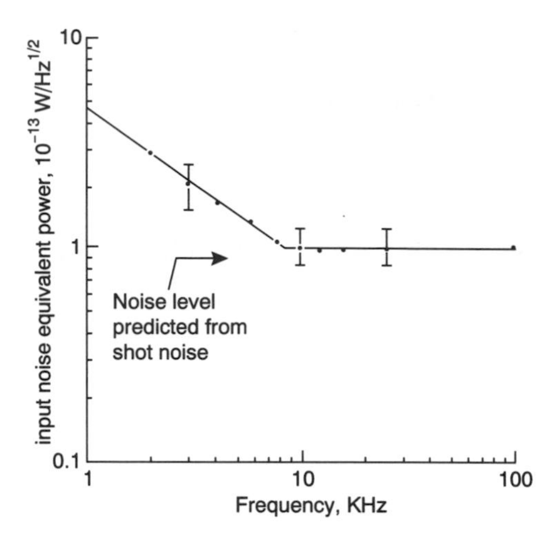

Figure 11.11 Equivalent noise power density at input versus frequency for a GaInAs pin photodiode (Monham *et al.*, Figure 2). Reproduced by permission of IEEE

11.7.1 Expressions for flicker noise

The noise of pin photodiodes has been described by shot noise expressions earlier in the chapter, but below about 100 kHz or so, some devices also exhibit noise equivalent powers which rise with decreasing frequency, as illustrated in Figure 11.11. This contribution seemed to be independent of bias voltage between 5 and 15 V. The m.s. current was proportional to a constant, the experimentally determined maximum value of which was 10^4 (pA)^2, divided by f; hence the m.s. noise current density is:

$$\frac{10^4}{f}, \text{ in units of } \left(\frac{\text{pA}}{(\text{Hz})^{1/2}}\right)^2$$

Relatively little has been published on this type of noise in GaAs mesfets because they are normally manufactured for use at microwave frequencies; consequently data sheets do not supply information about this aspect of the transistor. What little data available in the early 1980s suggested break frequencies in the range 20–60 MHz. However, in a sample of commercially available receiver modules of a given type the $1/f$ noise contributions at a given frequency, e.g. 10 kHz, had a spread of 100 to 1 in values of spectral density. More information and closer control of the factors affecting this type of noise are very desirable from a receiver designer's point-of-view.

11.7.2 Effect of flicker noise on optical receiver sensitivity performance

The importance of this topic was discussed in two papers published in 1987 [14] and 1988 [15]. In the second of these two contributions, the authors calculated the sensitivity and the total equivalent noise current at the input of an optical receiver, including a model for flicker noise. The power penalties resulting from flicker noise were computed from the equivalent noise generator. Flicker noise was found to degrade the sensitivity over the whole range of bit rates examined. The power penalty, which was serious in the case of high-impedance pinfet type receivers, was strongly affected by the total input capacitance. The authors suggested optimum values to be used in the design of practical pinfet receivers.

Theory: this took account of the following noise mechanisms in optical receivers incorporating GaAs mesfets:

(1) shot noise due to photodiode dark current
(2) thermal noise of bias and feedback (if present) resistors
(3) shot noise due to gate leakage current
(4) thermal noise in channel
(5) induced noise in the gate
(6) correlated noise in the mesfet
(7) flicker noise.

The authors followed the common practice of basing their noise analysis of the fet on Baechtold's 1972 paper [16]. In this model, the sources in the intrinsic fet are (4), (5) and (6) above, and all are modelled as thermal noise as follows:

(4) channel noise $4k\theta \times g_M \times \tilde{p} \times df$
(5) induced gate noise $4k\theta[\omega^{2C2}{}_{GS}/g_M] \times \tilde{r} \times df$
(6) correlation between gate, drain sources $j4k\theta \times \omega C_{GS} \times \tilde{q} \times df$

where $\tilde{p}$, $\tilde{q}$ and $\tilde{r}$ are used in place of P, Q and R in [16] to avoid confusion with their meanings in this book, namely power, the factor related to ber and resistance. These three quantities have been evaluated for various fet parameters and gate bias conditions and their values are given in graphs. Unfortunately, the designer may find it very difficult to assign values to these quantities because the necessary information is unlikely to be generally available. In most practical cases, it will be much more satisfactory to rely on measurements of a set of noise quantities and derive values for the two external correlated noise generators E and J. However, these expressions can be transformed into an equivalent noise current generator across the photodiode of mean square value:

$$G^2\tilde{p} + (\omega C)^2 \left[\tilde{p} + \left(\frac{C_{gs}}{C}\right)^{2\tilde{r}} - 2\left(\frac{C_{gs}}{C}\right)\tilde{q}\right] \tag{11.54}$$

all to be multiplied by $4k\theta \times df/g_m$. The quantity inside the square brackets is often denoted by Γ.

At frequencies below about 50 MHz, mesfets exhibit flicker noise, which can be represented by an equivalent mean-square noise voltage per Hz of value:

$$\frac{S_N(f_L) \times G^2}{f} \tag{11.55}$$

where the multiplier is the mean-square noise current spectral density at some suitable frequency f_L well below the break point between the $1/f$ asymptote and the flat portion of the noise characteristic ('bathtub' shape). The authors give the range of values as 10^{-11}–10^{-9} V^2.

Effect of the flicker noise on pinfet receiver sensitivity

This is expressed in terms of a flicker noise power penalty in dB, defined as the difference between the sensitivities with and without the flicker noise contribution. Computer simulations covering bit rates between 1 Mbit/s and 10 Gbit/s, various 1300 nm photodiodes, high impedance and transimpedance types of receivers, 50% rz and nrz formats and a transconductance of 15 mA/V and gate leakage current of 2 nA for the mesfet have been performed. Details are given in the references, but are too numerous to be reproduced here. Instead, a few of the figures will be given so that the reader can obtain some quantitative information of the penalty.

We begin with the pinfet transimpedance (TZ) receiver treated in [14] and use the total shunt capacitance C as a parameter for the penalty versus bit rate curves and sensitivity versus bit rate family. Two values of S_N have been assumed; 5 ×

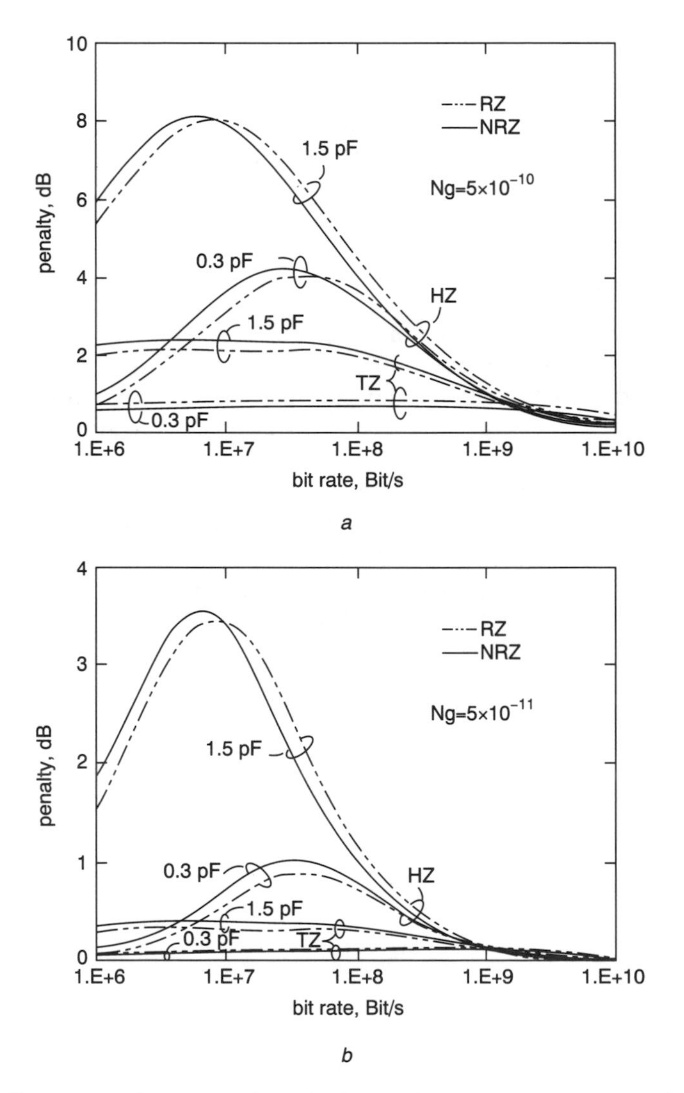

Figure 11.12 Power penalty versus bit rate for pinfet receivers: (*a*) noisy devices, (*b*) quiet devices ([15], Figure 3). Reproduced by permission of IEEE, © 1988

10^{-10} for noisy receivers, and 5×10^{-11} for quieter modules. Other numerical values were Γ (equation (11.54) *et seq.*) $= 1.78$, $g_M = 40$ mA/V, sum of gate leakage and dark currents 50 nA, and feedback resistance 2 MΩ/f_S. The results are given in Figure 11.12, respectively.

Considering the effect on sensitivity first, it can be seen that increasing capacitance degrades the sensitivity, particularly at high bit rates; for example, at 10 Gbit/s by about 5 dB between 0.3 and 1.5 pF. The penalty is greatest over a mid-range of bit rates, peaking around 34 Mbit/s at 2.5 dB for the noisy device but only 0.4 dB for the quiet receiver. At 10 Gbit/s the penalty is about 0.3 dB for the noisy case, and negligible for the other.

In the figure given in [15] each pair of curves was labelled HZ or TZ for high impedance or transimpedance and referred to either rz or nrz formats, as indicated. The values of bias and feedback resistances used in the calculations have been derived from the relations 500 MΩ/f_S and 2 MΩ/f_S, with f_S in Mbit/s for HZ and TZ designs, respectively. A transconductance of 15 mA/V and gate leakage current of 2 nA were taken for the GaAs mesfet; photodiode dark current of 1 nA for InGaAs pin and apd and 50 nA for the Ge apd. These two values were also taken for the multiplied dark currents, and $k = 0.3$ ($x = 0.7$) for the InGaAs photodiode and $k = 1$ ($x = 1$) for the Ge device.

The authors provide a useful comparison between pinfet sensitivities obtained by simulation based on their equation (12) using typical values of receiver parameters, and the sensitivities of commercially available (1987) HZ and TZ receivers. The simulated results are shown in the full line curves and the sensitivities by individual points in Figure 11.13.

Equation (12) gave an expression for the total m.s. noise current generator across the photodiode terminals for the usual lumped constant R and C elements representing the photodiode and amplifier input impedance. In addition to the customary shot and thermal noise contributions there were two terms associated with flicker noise.

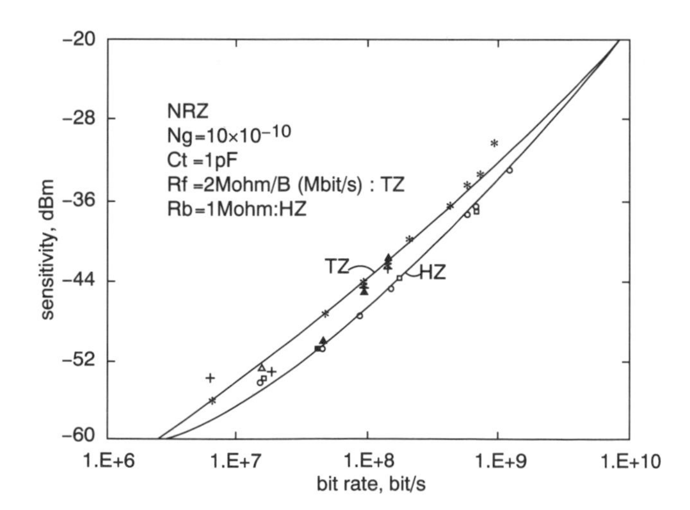

Figure 11.13 Sensitivities versus bit rate, curves simulated; points plot typical results from measurements on modules from several manufacturers listed in key ([15], Figure 7). —: Simulated results, □: HZ (Lasertron), ○: HZ (STC), △: TZ (STC), +: TZ (Lasertron), ∗: TZ(PCO). Reproduced by permission of IEEE, © 1988

As a result of their work on flicker noise penalty, the authors' conclusions for 0.3 pF total capacitance (an achievable target) can be amplified and restated as follows.

For HZ pinfet receivers the peak value of the penalty versus bit rate occurred at about 34 Mbit/s, and was 4 dB for noisy components and about 1 dB for quiet ones. Values at 2.5 Gbit/s were about 0.7 dB for noisy receivers and negligible for quiet ones.

For TZ pinfet receivers the corresponding values were less than 0.7 dB for noisy devices and were negligible for quiet receivers, but the behaviour at 2.5 Gbit/s and above was indistinguishable from that of HZ types.

There is a broad minimum in the penalty versus feedback resistance of less than 1 dB, giving an optimum value of feedback resistance for each bit rate for both nrz and rz pulses. At 565 Mbit/s the penalty was 0.5 dB and the optimum resistance was 10 kΩ. Note: a value of 8 kΩ was used in the optical receiver of the 420 Mbit/s PTAT-1 system and 10 kΩ in the TAT-8 280 Mbit/s system.

For InGaAs apdfet receivers of high-impedance design the maximum penalty was about 1.2 dB at about 34 Mbit/s, whereas for TZ designs the value was about 0.3 dB. For both types the penalty above 1.2 Gbit/s was less than 0.4 dB.

Taking the 565 Mbit/s rate and 0.3 pF total capacitance, Figure 5*d* in [15] showed that the sensitivity improved by about 7 dB when the feedback resistance increased from 1 to 10 kΩ, with the improvement some 1 dB better for rz as compared with nrz pulses. At 8 kΩ, the sensitivities were about −37.5 and −38 dBm, whereas the value for the typical PTAT-1 receiver was almost −34 dBm.

APPENDICES

APPENDIX A: BER IN DIGITAL TRANSMISSION SYSTEMS

The current practice is to work with worst-case designs, thereby ensuring that the performance of an operational system in the presence of isolated random errors can only be better (the effect of bursts of errors is treated elsewhere). Operational systems can, in fact, show no measured errors over considerable time intervals, as mentioned elsewhere. In this approach, the received signal for a prescribed error rate performance must exceed a value derived by assuming that all the signal impairments are correlated, and that the noise component is a normal, stationary, uncorrelated process. The assumption that a decision circuit without memory is employed simplifies the derivation of this value to some extent.

Starting from the assumption that normal hypothesis testing is applicable, the error probability of a two-level digital signal can be expressed in terms of *a priori* probabilities of 1, pr(1) and 0, pr(0) decisions and the conditional probabilities of error, pr(1|0) and pr(0|1) (see, for instance, [17]); hence the definition of the probability of error, pr, that Gaussian noise will cause the signal plus noise at the decision instant to cross the threshold level to the opposite side from the signal alone is:

$$\mathrm{pr} = \mathrm{pr}(1|0)\,\mathrm{pr}(0) + \mathrm{pr}(0|1)\,\mathrm{pr}(1) \qquad (11.\mathrm{A1})$$

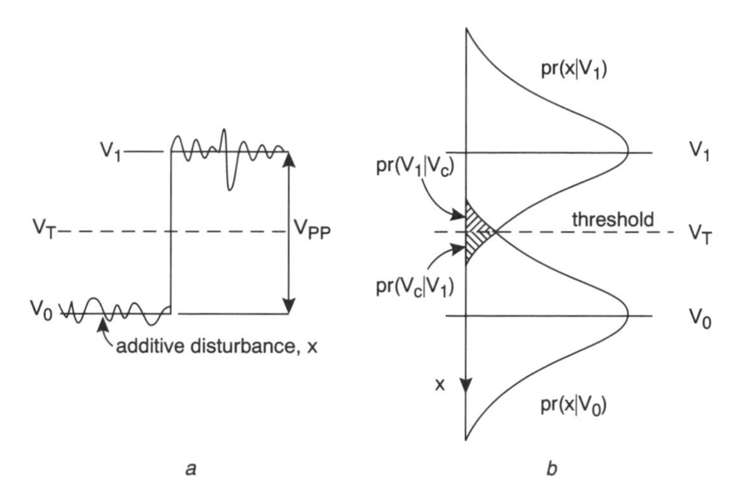

Figure 11.14 Diagrams illustrating error mechanism (*a*) waveform of two levels with additive Gaussian noise n superimposed (*b*) corresponding probability density functions centred on a mark (or 1) and on a space (or 0) and threshold midway between them. Reproduced by permission of IEE

By symmetry, the conditional probabilities are equal, and if ones and zeros are equally probable, the probabilities pr(1) and pr(0) are both 0.5. These quantities are illustrated in Figure 11.14.

It can be shown that since pr = pr(1|0) because of these assumptions, this can be written as:

$$\mathrm{pr} = \mathrm{pr}(1|0) = \int \mathrm{p}(x)\,dx \qquad (11.A2)$$

where the integrand is a Gaussian distribution, and the definite integral has a lower limit equal to half the peak-to-peak value $V_{PP}/2$, say, between a one and a zero, and an upper limit of infinity. Because the distribution is Gaussian, the right-hand side can be expressed by:

$$\mathrm{pr} = \frac{1}{(2\pi)^{1/2}} \int \exp\left(\frac{-x^2}{2}\right) dx \qquad (11.A3)$$

where the lower limit of integration is now $V_{PP}/2\sigma = V_P/\sigma$. This ratio is customarily denoted by Q, hence we obtain

$$\mathrm{pr} = 0.5\,\mathrm{erfc}\left(\frac{V_{PP}}{2\sigma\sqrt{2}}\right) = 0.5\,\mathrm{erfc}\left(\frac{Q}{\sqrt{2}}\right) \qquad (11.A4)$$

For the case of binary signalling the probability of error, pr, is more commonly called the binary error rate or ber.

The relation between ber and Q is often given graphically, but in some respects a table of values is preferable, and one is shown as Table 11.3. The reader may wonder why two columns of values of Q are presented; the answer lies in the fact that the equation relating Q and ber is nonlinear and is thus best solved numerically. The differences are negligible in practice since only two or three significant figure values of Q are normally used. This probability and the quantity Q can be related to voltages at the input to the decision circuit, which is also the output from

Table 11.3 Values of Q for given error probabilities. Reproduced by permission of Nortel

P_e	Q (from eq. (A.3))	Q (from Ref. [A-2])	Error %
10^{-4}	3.72003	3.71902	0.0200
10^{-5}	4.26530	4.26489	0.0096
10^{-6}	4.75362	4.75342	0.0042
10^{-7}	5.19944	5.19934	0.0019
10^{-8}	5.61206	5.61200	0.0010
10^{-9}	5.99784	5.99781	0.0005
10^{-10}	6.36136	6.36134	0.0003
10^{-11}	6.70603	6.70602	0.000
10^{-12}	7.03450	7.03448	—
10^{-13}	7.34880	7.34880	—
10^{-14}	7.65063	7.65063	—

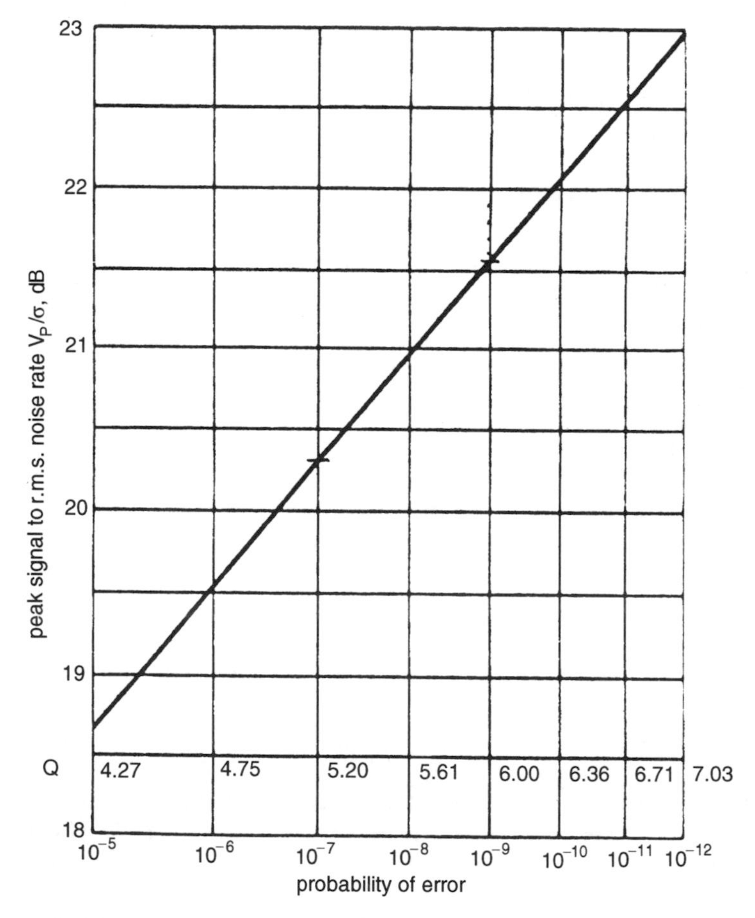

Figure 11.15 Peak signal to rms noise ratio in dB versus ber and Q for binary baseband transmission. Reproduced by permission of IEE

the equaliser. We assume that these voltages at the decision instants are Gaussian random variables with mean values V_0 and V_1 for a zero and a one, respectively, and with variances $\sigma_0{}^2$ equal to the worst-case noise for zeros and $\sigma_1{}^2$ equal to the corresponding value for ones, as shown graphically in Figure 11.14. It can be shown that the optimum value of the threshold V_T, say, is a linear function of V_1, the interference v_I, together with a third term proportional to the mean square noise voltage, the actual expression being:

$$V_T = v_I + \frac{V_1}{2} + \frac{\sigma^2}{V_1} \ln \left(\frac{\text{pr}(0)}{1 - \text{pr}(1)} \right) \tag{11.A5}$$

but for simplicity the interference term will be neglected.

The optimum decision threshold must be set between V_0 and V_1, so that:

$$\frac{V_1 - V_T}{\sigma_1} = Q = \frac{V_T - V_0}{\sigma_0} \tag{11.A6}$$

i.e. V_T must be $Q\sigma_1$, volts below V_1 or, equivalently, Q standard derivations σ_0 above V_0 to obtain the desired ber. The corresponding value of Q can be taken from Table 11.3 and the signal peak-to-peak voltage from equation (11.A6):

$$V_{PP} = V_1 - V_0 = 2Q\sigma \tag{11.A7}$$

assuming that $\sigma_1 = \sigma_0$. These calculations apply directly to baseband systems, but in a carrier system the effect of the channel on the carrier enters the equation through the mean and r.m.s. values. Further details can be found in a number of papers or books, e.g. [18].

A useful graph of peak signal to r.m.s. noise ratio versus ber for baseband transmission was given in [17]. Corresponding values of Q have been added to give the results shown in Figure 11.15.

APPENDIX B: THE PERSONICK INTEGRALS

Let us now look at two of the Personick integrals in more details, as they form an important part of his seminal contributions to optical receiver theory. Personick defined four integrals in his 1973 paper [3], which he denoted by I_1, I_2, I_3 and Σ_1 in terms of the Fourier transforms of the input pulse shapes and the equalised output pulse shape. Three classes of input pulse shapes were considered: rectangular, Gaussian and exponential; and one class of output shapes, namely raised cosine. Graphs were given for each of these families as functions of two shape parameters, and he indicated that other shapes could easily be used and programmed on hand calculators. This has been done repeatedly since, and one set of results is given Figures 11.16–11.19.

The waveform at the decision circuit input is the convolution of four terms:

(1) the impulse response of the interstage between photodiode and front-end amplifier
(2) the impulse response of the amplifier itself
(3) the impulse response of the equaliser itself
(4) the shape of the received optical pulse, hence that of the photodiode signal current.

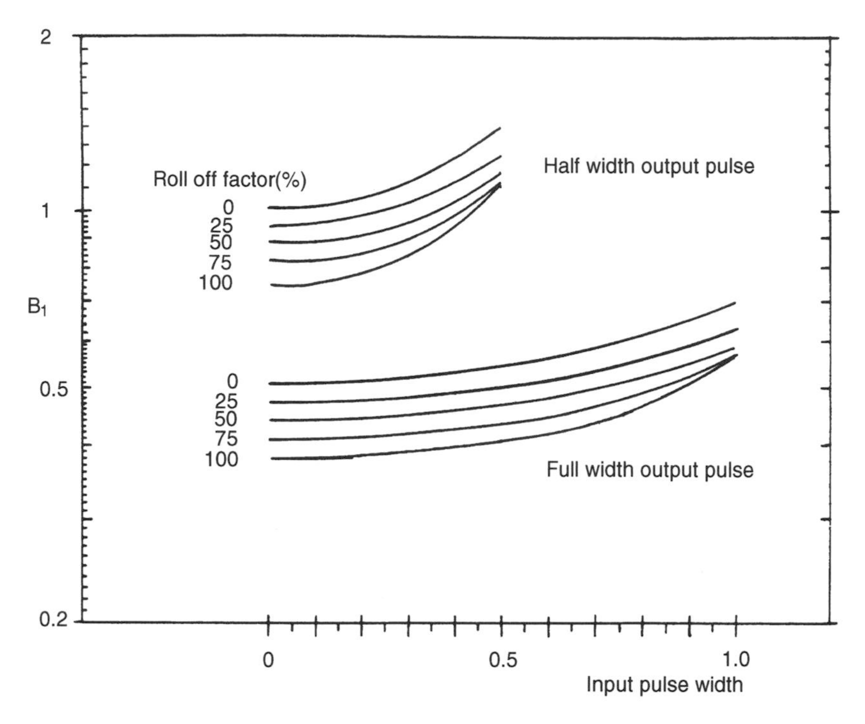

Figure 11.16 $B_1 = I_2/f_S$ versus input pulse width for half and full width rectangular pulses with roll-off factor as parameter). Reproduced by permission of Nortel

Convolutions of the first three correspond to the transfer function of the receiver. By taking Fourier transforms of all four functions and adapting his notation to suit, let $H(\mathrm{j}\omega)$ refer to the decision circuit input and $H'(\mathrm{j}\omega)$ to the optical pulse; then the transfer function is simply $H(\mathrm{j}\omega)/H'(\mathrm{j}\omega)$. For brevity, denote this by TF(jω); then m.s. noise quantities at the photodiode output are multiplied by the definite integral between appropriate ω limits of $|TF|^2$ to give the corresponding quantities at the decision circuit. Personick then defined normalised transforms which depended only upon the shapes of H and H' and not on the symbol interval; thus $H_\mathrm{n} = H(2\pi\omega/T)$ and $H'_\mathrm{n} = (1/T)H'(2\pi\omega/T)$, where n denotes normalised. He then changed the variable from ω to f. The integrals I_2 and I_3 were then defined as:

$$I_2 = \int \left| \frac{H_\mathrm{n}(\mathrm{j}f)}{H'_\mathrm{n}(\mathrm{j}f)} \right|^2 \times df \tag{11.B1}$$

$$I_3 = \int \left| \frac{H_\mathrm{n}(\mathrm{j}f)}{H'_\mathrm{n}(\mathrm{j}f)} \right|^2 \times f^2 \times df \tag{11.B2}$$

with lower and upper limits of minus and plus infinity in both cases. Later in his 1973 paper, a numerical example gave values for both integrals as pure numbers, but examination of the associated expression showed that TI_2 must have dimensions of Hz because of the given dimensions of the amplifier series S_E and shunt S_I noise spectral densities of V^2/Hz and A^2/Hz, respectively.

Later workers sometimes use a different normalisation in which the variable f is replaced by (fT) and the corresponding integrals become

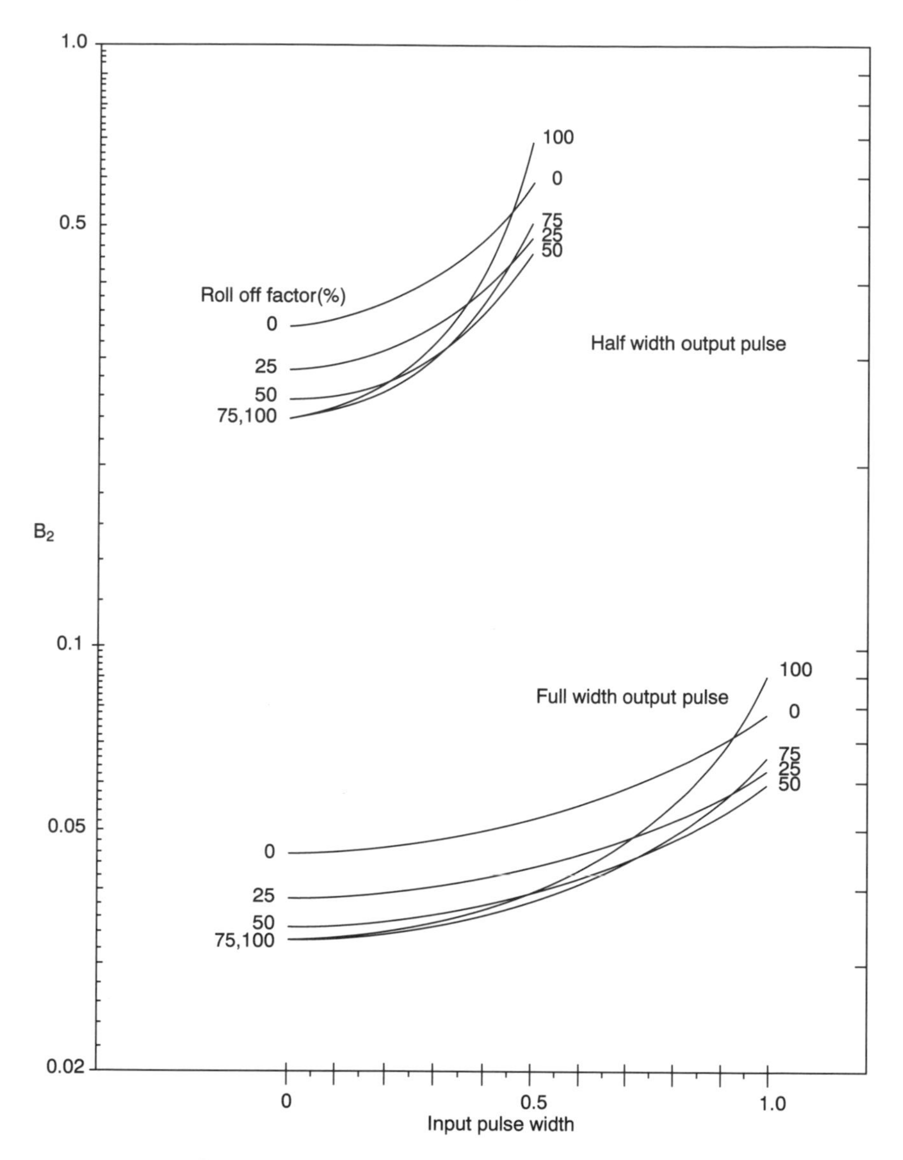

Figure 11.17 $B_2 = I_3/f_S{}^3$ versus input pulse width for rz and nrz rectangular pulses with roll-off factor as parameter. Reproduced by permission of Nortel

$$I_2 = \frac{1}{T}\int \left|\frac{H(\mathrm{j}f\,T)}{H'(\mathrm{j}f\,T)}\right|^2 \times d(f\,T) = \frac{B_1}{T} \tag{11.B3}$$

$$I_3 = \left(\frac{1}{T}\right)^3 \int (f\,T)^2 \left|\frac{H(\mathrm{j}f\,T)}{H'(\mathrm{j}f\,T)}\right|^2 \times d(f\,T) = \frac{B_2}{T^3} \tag{11.B4}$$

B_1 is a noise equivalent bandwidth normalised to the bit rate f_s. The values of B_1 and B_2 have been calculated and are given in Figures 11.16–11.19.

As mentioned above, the voltage pulse shape at the decision circuit is usually chosen to be one of the raised-cosine family. In turn, the integrals can be related to

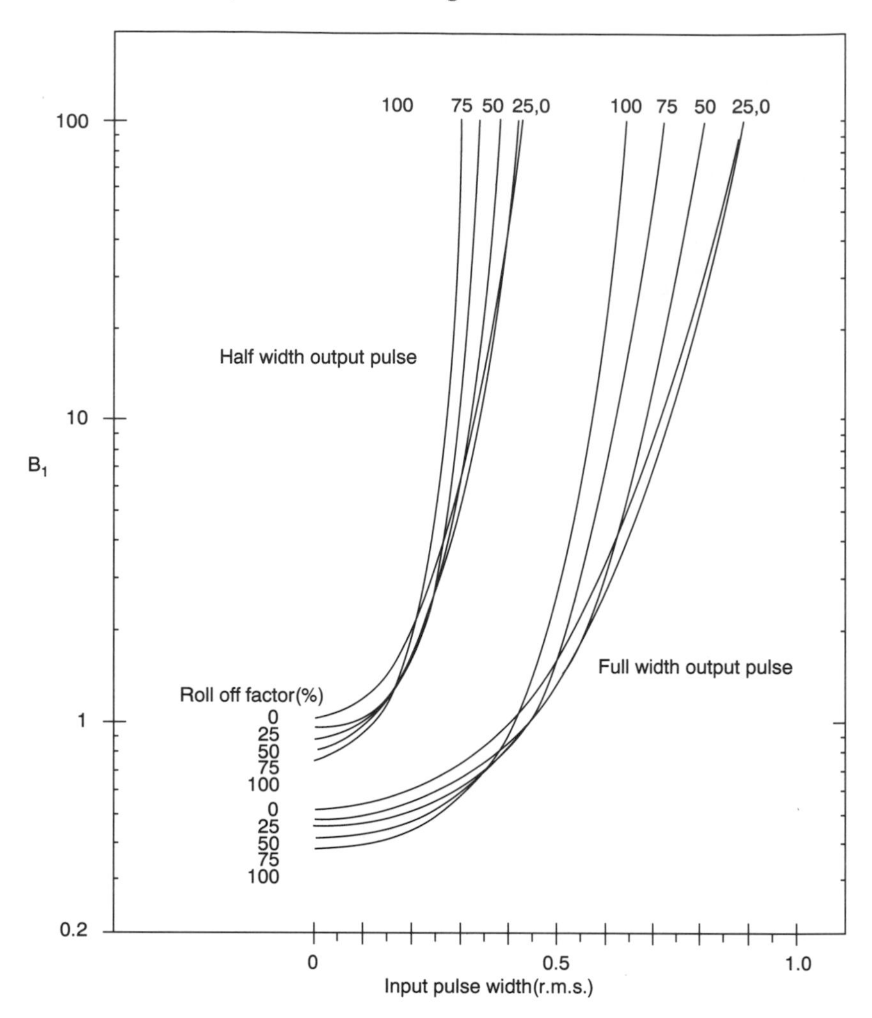

Figure 11.18 $B_1 = I_2/f_S$ versus input pulse width for half and full width Gaussian pulses with roll-off factor as parameter. Reproduced by permission of Nortel

corresponding bandwidths B_1 and B_2 as follows:

$$B_1 = I_2 \times T = \frac{I_2}{f_s} \tag{11.B5}$$

$$B_2 = I_3 \times T^3 = \frac{I_3}{f_s^3} \tag{11.B6}$$

APPENDIX C: ASPECTS OF THE DESIGN OF APD OPTICAL RECEIVERS

Avalanche photodiodes were described in Chapter 10 and the mean gain and related quantities were introduced in equations (10.8) and those that followed. We now provide further information to assist system designers.

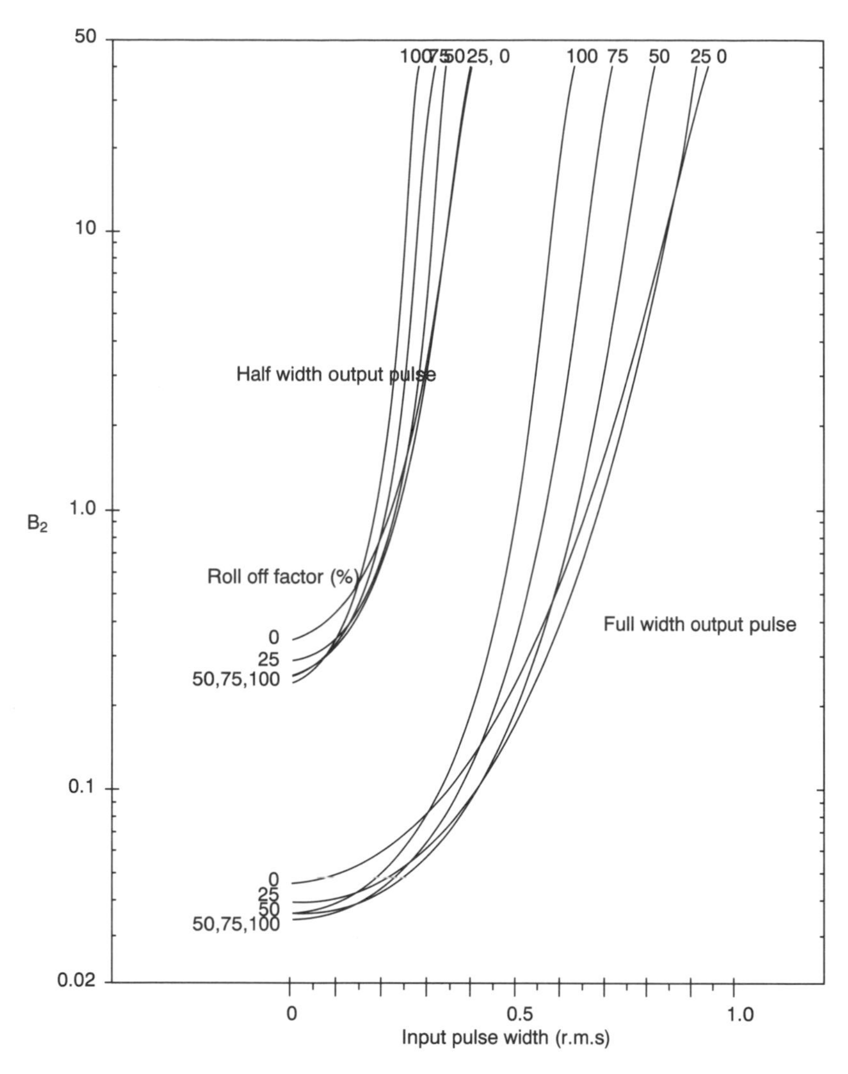

Figure 11.19 $B_2 = B_2 = I_3/f_S{}^3$ versus input pulse width for half and full width Gaussian pulses with roll-off factor as parameter. Reproduced by permission of Nortel

The avalanche process is statistical and therefore noisy. The number of charge carriers (hence current) produced at a given nominal gain has a mean value (which determines the associated shot noise), and fluctuations in gain which are equivalent to an excess noise factor. The statistics of the avalanche gain and the resulting noise spectral density are of central importance to the theory of apd receivers, particularly at very low signal levels. The best-known author of papers on multiplication noise in apd devices is probably R.J. McIntyre, e.g. [19]. The treatment in these papers tends to be more detailed than necessary for system design, where it is sufficient to assume that the effect of avalanche noise can be accounted for by increasing the mean square shot noise current (which is proportional to M^2) by a factor $F = M^x$ known as the excess noise factor, where M is the mean value of the gain at a given frequency and x depends upon the device design, structure, materials, etc. In addition, it is assumed that the resulting noise is white with Gaussian statistics.

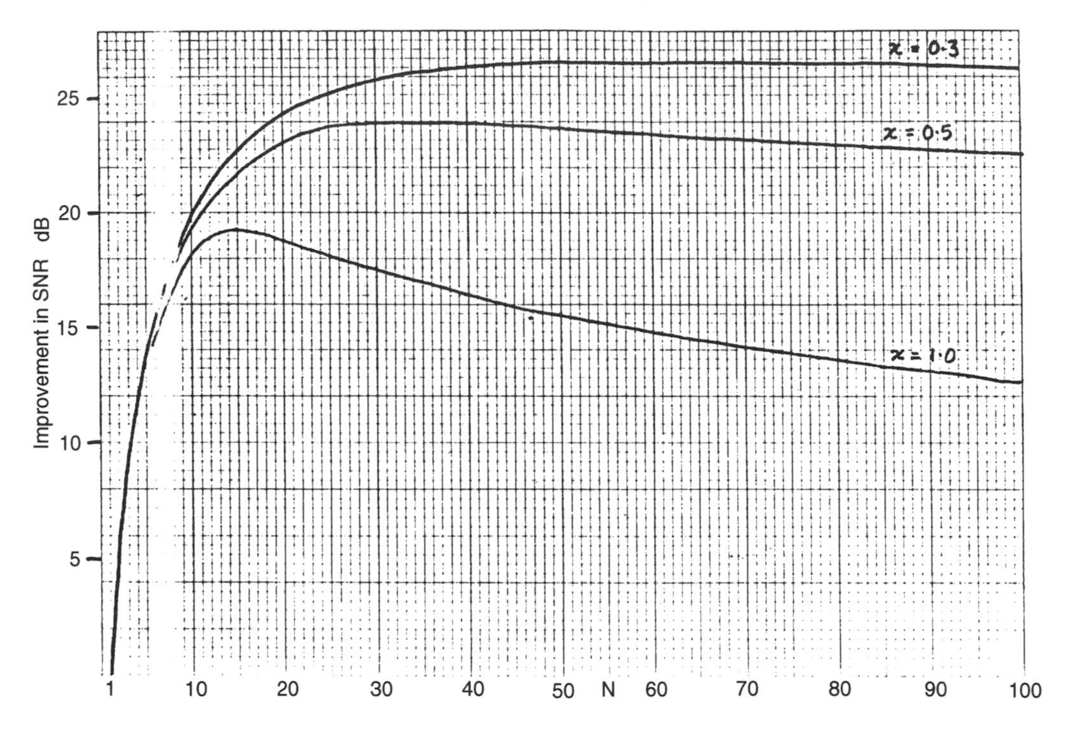

Figure 11.20 Improvement in snr versus avalanche gain M for $x = 0.5$ and $x = 1$

McIntyre gave two more accurate expressions for F for the extreme cases of all injected carriers being electrons or all holes, which were functions of the ratio of hole to electron ionisation rates, denoted by k. For apds fabricated from Ge or from 3 - 5 compounds, $0.3 < k < 1$ and the same values of F can be obtained by taking $0.7 < x < 1$ in the simple relation stated above.

The snr of apd receivers is a function of M and x; for $M = 1$ thermal noise dominates the total noise power, thus snr increases with M; for large values of M shot noise dominates and the snr decreases as M^x. There is an optimum value of M giving the maximum improvement in gain, and which results in the ratio of shot noise power to thermal noise power being equal to $2/x$. The improvement in snr as a function of M is shown in Figure 11.20. Other relations of interest are the variations of sensitivity and of optimum gain versus the thermal noise parameter Z with x as parameter.

Receiver analysis can be broadened to include several impairments, such as isi resulting from dispersion in the fibre and laser chrip, non-zero extinction ratio and photodiode leakage current, both of which result in noise on the 0 level.

The output pulse from practical lasers modulated by rectangular current pulses is approximately rectangular in shape, and if the device operation is peak power limited, maximum power will be launched into the system fibre if the pulse width is T, the symbol interval. The family of rectangular pulses of unit area symmetrical about $t/T = 0$ can be parametrised by a real constant a denoting the pulse width, where $0 < a \leq 1$; $a = 1$ corresponds to a full-width (nrz) pulse ($t/T = \mp 0.5$) and $a = 0.5$ to a half-width (50% rz) pulse. The corresponding pulse height is $1/a$. The use of rz pulses (i.e. $a < 1$) produces a stronger spectral component at the

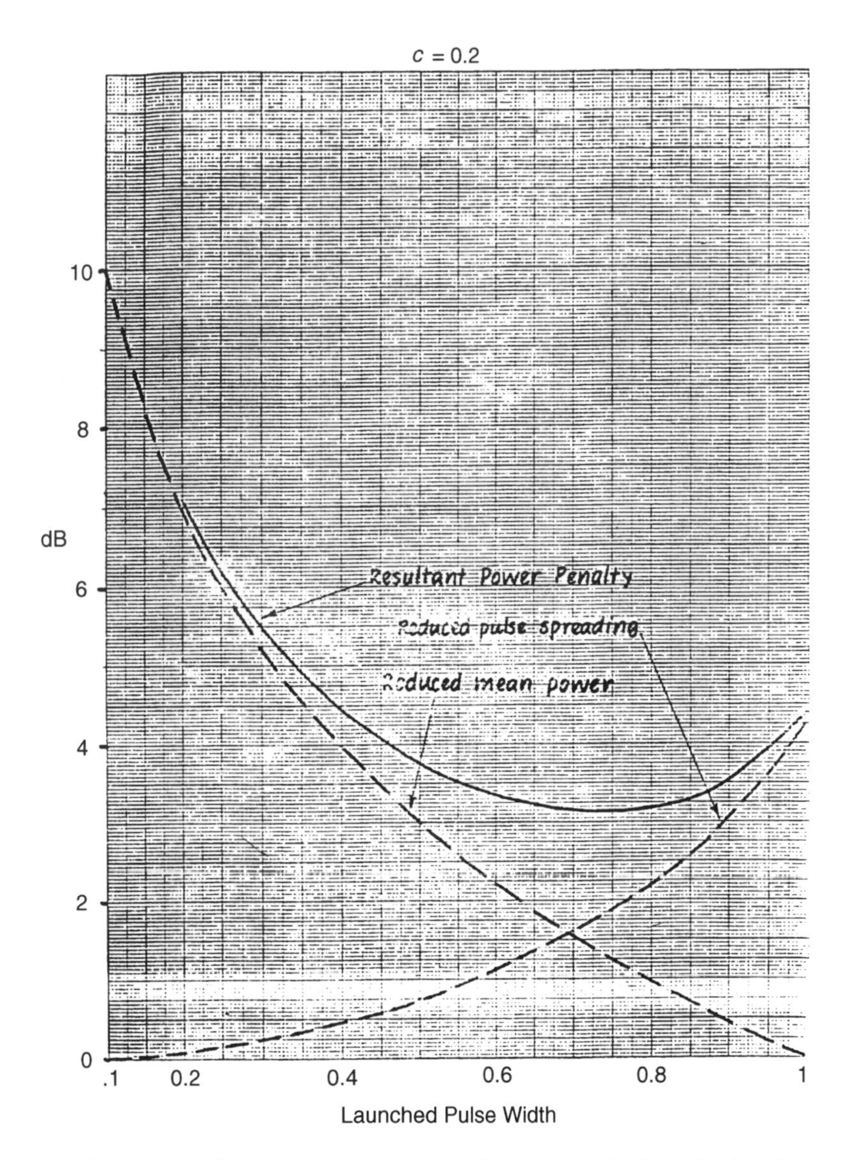

Figure 11.21 Power penalty versus parameter a for rectangular launched pulses and fibre impulse response parameter $c = 0.2$, zero extinction ratio

symbol rate, hence a stronger timing signal and consequently less jitter at the decision gate.

The impulse response of realistic spans of fibre (e.g. 40–200 km) can be approximated by one member of a family of Gaussian shapes of unit area represented as before. The parameter is denoted by c, where $0 < c \leq 1$ with values of c in the lower half of this range of interest.

The convolution of an nrz launched pulse shape and the fibre impulse response will exhibit considerable pulse spreading unless c is sufficiently small. Single-mode fibres have enormous bandwidths which correspond to very narrow impulse responses, but if bandwidth is a limitation there is an optimum value of the parameter a which depends upon the fibre bandwidth. The penality for reducing the mean power in the launched pulse and the benefit of reducing isi can be shown for a given value of c. An example is shown in Figure 11.21. The curves shown that the optimum value of a is about 0.75 under these simplified conditions. In practice lasers are usually biased to just below threshold, in which case the extinction ratio is no longer zero and patterning can arise and was expected to be serious if $(1 - a)T$ is a small fraction of the spontaneous carrier lifetime (about 3 ns).

Concerning performance with impairments, again, an optimum value of M can be found analytically be solving the equation for the sensitivity and that for $dN/dM = 0$ simultaneously, where N is the mean number of photoelectrons per bit. The optimum gain required for a specified sensitivity P (min.) is a function of x, the number N_D of dark current electrons per bit Z, the pulse-spreading parameter (γ) and the extinction ratio (ε) defined in Chapter 11, equation (11.11) and those that follow. Sensitivity is related to N as follows:

$$P(\text{min.}) = 10 \log \left(\frac{N f_S}{\eta \lambda} \right) - 97 \quad [\text{dBm}] \tag{11.C1}$$

Dark current present, but $\gamma = 0$ and $\varepsilon = 0$: N versus x with N_D as parameter and optimum gain versus x with N_D as parameter are displayed in Figure 11.22. Note that only regions with $x > 0.7$ are of interest.

Extinction ratio nonzero and pulse spreading: the variations of N and of optimum M were shown for $N_D = 0$, $x = 0.3$ (Si apd) and a half-width rectangular pulse ($a = 0.5$), and the results are reproduced in Figure 11.23. The curves show that N, and hence the received power needed for a specified ber, increase quite rapidly as ε and as c increase from zero. Conversely, the optimum gain decreases as ε and as

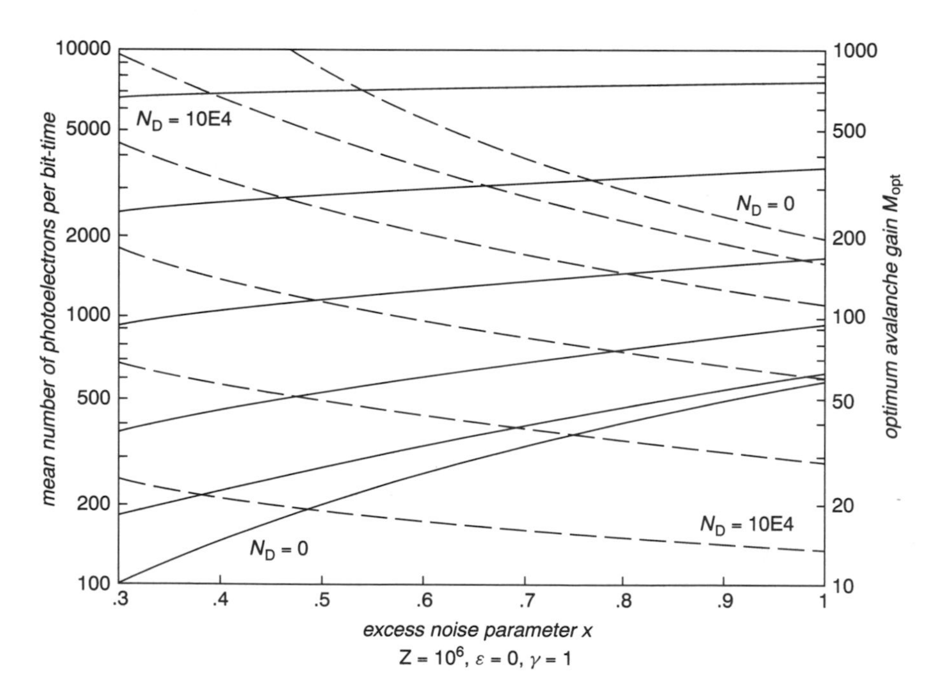

Figure 11.22 Full lines: N versus x with N_D as parameter. Dash lines: optimum gain versus x with N_D as parameter for $Z = 10E6$, $\varepsilon = 0$, $\gamma = 1$

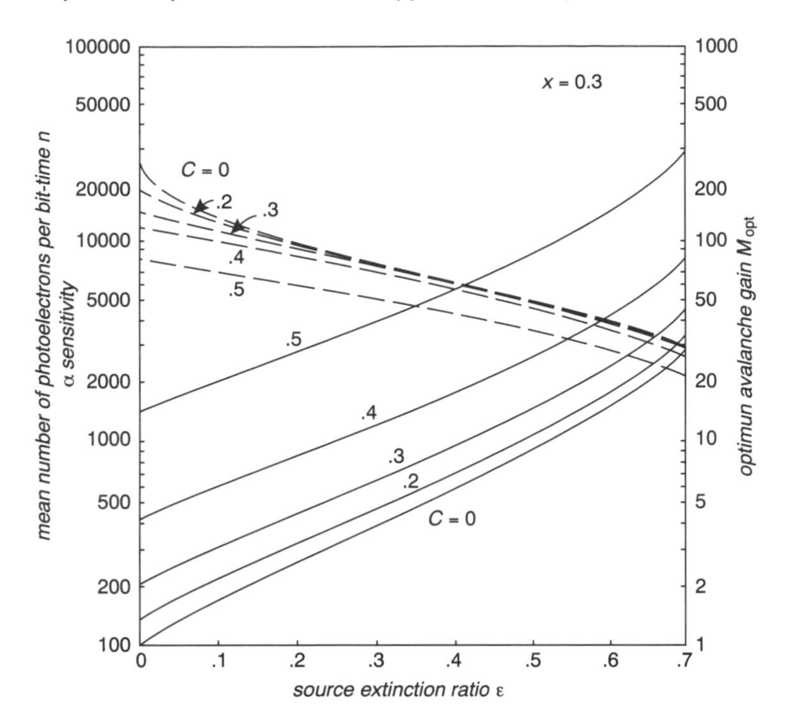

Figure 11.23 Full lines: N versus ε with c as parameter. Dash lines: optimum gain versus ε with c as parameter for $N_D = 0$, $a = 0.5$

c increase from zero. Note that since Gaussian noise statistics were assumed, the values of optimum gain tend to be overestimated; however, relative magnitudes are predicted more accurately.

To give some indication of realistic values of N we can use equation (11.5) which refers to a unity gain photodiode, and which gives $N = U_{11}/hf = (2Q/\eta) \times \sqrt{Z}$. For $Z = 10^6$, a ber of 10^{-9} hence $Q = 6$ and 100% efficiency, $N = 1.2 \times 10^4$ photogenerated electrons per mark.

APPENDIX D: OPTIMUM EQUALISERS AND THEIR APPLICATION TO OPTICAL RECEIVERS

A matched filter allows the optimum detection of a known signal in noise; the first part is a network which 'whitens' the r.m.s. spectral noise density, the second network has an impulse response which is a time-translated and reversed replica of the input signal. In digital transmission systems in which the noise is not white, a matched filter often exhibits intersymbol interference (isi). The design of an equaliser for such systems involves;

(a) maximising the snr at the output
(b) constraining the time-domain waveform to minimise isi
(c) constraining the time domain waveform to minimise pattern-dependent jitter.

A raised cosine frequency-domain response is often taken as the target output response, as in Chapter 11, but does not take account of specific signal or noise characteristics. An optimum equaliser structure for binary on–off digital transmission systems was described in [21]. This structure was a synthesis of the classical matched filter and transversal equaliser, and results were presented for optical receivers comparing optimum equalisers with raised cosine designs. These latter were shown to be near-optimum for noise spectral density characteristics of the form $S_N = S(1 + (\alpha f)^2)$, with S the noise density coefficient and α the noise factor of the coloured noise generally found in optical receivers.

The time-domain waveforms and frequency-domain transfer functions of an optimum equaliser for nrz pulses and parameter $T/\alpha = 2/\pi$, and the raised cosine, were displayed in Figures 4 and 5 of the publication [21], respectively. Despite the closeness of the waveforms to one another over the range of $t/T = \mp 2$, the transfer functions were not nearly so close below the normalised frequency $f/f_S = 1$; and above this value the raised cosine function is zero (and is not physically realisable), whereas the optimum equaliser has a sidelobe about 19 dB down on the dc frequency response.

A function F (defined as a sum of two integrals with the second divided by $(T/\alpha)^2$) that represents the impact of noise on receiver sensitivity was used for the equaliser and for the raised cosine network, and their ratio in dB was plotted versus $0.1 < T/\alpha < 5$ for nrz and for 50% rz rectangular pulses (corresponding parameter values giving closure of the eye at the equaliser output were ≤ 0.26 and ≤ 0.2, respectively). The two curves were of similar shape and each had a minimum at around $T/\alpha = 0.7$. For white noise $\alpha = 0$, hence T/α is infinite and the ratio takes the values 0.52 and 0.23 dB for the nrz and rz formats, respectively. The corresponding differences in sensitivity for raised cosine and optimum equaliser based receivers were half these values due to the square law photodetection process.

APPENDIX E: DYNAMIC RANGE IN OPTICAL RECEIVERS

Such receivers must be capable of accepting a range of input light levels to accommodate difference in section lengths and tolerances in the photodiode and regenerator electronics. The dynamic range requirements on these electronics is unaffected by whether an agc amplifier chain or a limiting amplifier chain is used. A convenient definition of dynamic range is the range of optical received power over which the error rate does not exceed 10^{-9}.

The 'working margin' should not be confused with dynamic range. This margin is the difference between the maximum and minimum error rate working margins observed on a large number of optical systems (pairs of fibres) subject to the complete range of operating temperature and pressure conditions and to initial tolerances, ageing and repair. The worst case corresponds to a 0 dB working margin, and in this case the received power level equals the effective sensitivity for a 10^{-9} error rate. The received power level depends on a combination of the launched power, the path loss and repair margin allocated, and the difference between the best and worst cases could amount to perhaps 8 dB. The two received power levels must then be combined with corresponding numbers for the overall worst and best case effective receiver sensitivities to give the working margins.

Let D be the working optical dynamic range; then for no variations in the receiver the corresponding electrical range would be $2D$. Owing to production spreads there will be a variation between the most sensitive and least pinfet receiver, and the effect that this has on the overall transfer function must be investigated.

We first consider the snr at the low-noise amplifier (lna) output. Using the symbols introduced elsewhere in this book, the electrical signal power at the lna output, port 2, is:

$$P_2' \propto \left(\frac{\phi P \times g_M}{C}\right)^2 \tag{11.E1}$$

provided that the shunt capacitance at the lna fet input is a fixed proportion of the total effective capacitance at that point. If fet noise is the dominant contribution the equivalent electrical noise power at the lna input port 1 is:

$$N_1' \propto \frac{C^2}{g_M} \tag{11.E2}$$

Combining the signal current ϕP and noise terms, the snr at the output is

$$\text{snr} \propto \frac{(\phi P)^2}{C^2/g_M} \tag{11.E3}$$

Let the received power for a reference ber of 10^{-9} be P_R; then

$$P_R \propto \frac{C^2}{g_M \phi^2} \tag{11.E4}$$

From equations (11.E1) and (11.E4) changes in P_2' due to changes in say P_R can be found. If a change in P_R is caused by a change in only one circuit constant at a time, i.e. capacitance, transconductance or responsivity, there will be a corresponding change in P_2'. For instance, a 1 dB degradation in P_R would result in net changes of −2, −4 and −2 dB in P_2' if due to changes in C, g_M or ϕ, respectively, and *pro rata* for other changes in P_R. If instead all three of the circuit constants are allowed to change, the results can be found from combining the two equations to obtain:

$$P_2' = g_M \times \left(\frac{P}{P_R}\right)^2 \tag{11.E5}$$

11.8 REFERENCES

1 WALKER, S.D., BLANK, L.C., GARNHAM, R.A., and BOGGIS, J.M.: 'High electron mobility transistor lightwave receiver for broad-band optical transmission system applications'. *J. Lightwave Technol.*, **7**, no. 3, Mar. 1989, pp. 454–458.

2 GIMLETT, J.L.: 'Ultrawide bandwidth optical receivers'. *J. Lightwave Technol.*, **7**, no. 10, Oct. 1989, pp. 1432–1437.

3 PERSONICK, S.D.: 'Receiver design for digital fiber optic communication systems'. *Bell Sys. Tech. J.*, **52**, no. 6, July/Aug. 1973, Part I pp. 843–874, Part II, pp. 875–886.

4 SMITH, D.R., and GARRETT, I.: 'A simplified approach to digital optical receiver design'. *Optical & Quantum Electronics*, **10**, 1978, pp. 211–221.

5 HOOPER, R.C., and WHITE, B.R.: 'Digital optical receiver design for non-zero extinction ratio using a simplified approach'. *Optical & Quantum Electronics*, **10**, 1978, pp. 279–282.

6 MACBEAN, M.D.A.: 'The influence of photodiode design on receiver sensitivity'. *IEE Coll. 'Optical Detectors'*, London, 19 Jan. 1990, Coll. Digest 1990/014.

7 MACLEAN, D.J.H.: *'Broadband Feedback Amplifiers'*. Research Studies press, a Division of John Wiley & Sons, Chichester, UK, 1982, and subsequent papers.

8 FUJITA, S., SUZAKI, T., MATSUOKA, A., MIYAZAKI, S., TORIKAI, T., NAKATA, T., and SHIKADA, M.: 'High sensitivity 5 Gbit/s optical receiver module using Si ic and GaInAs apd'. *Electronics Lett.* **26**, no. 3, 1 Feb. 1990, pp. 175–176.
9 SCHMIDT, W. auf ALTENSTADT: 'Optimum choice of Si f.e.t. or GaAs f.e.t. as first stage active device for very low noise optical receivers with medium bandwidths'. *Electronics Lett.*, **19**, no. 22, 27 Oct. 1983, pp. 937–938.
10 SHARLAND, A.J., BROOKS, R.M., COOMBS, J.P., and WHITE, S.: 'An optical fibre supervisory system'. *Br. Telecom Technol. J.*, **1**, no. 1, July. 1983.
11 NICLAS, K.B., PEREIRA, R.R., and CHANG, A.P.: 'A 2–18 GHz low noise/high gain amplifier module'. *IEEE Trans. Microwave Theory and Techniques*, **37**, no. 1, Jan. 1989.
12 OGAWA, K.: 'Noise caused by GaAs m.e.s.f.e.ts. in optical receivers'. *Bell Sys. Tech. J.*, **60**, no. 6, July/Aug. 1981, pp. 923–928.
13 MONHAM, K.L., BURGESS, J.W., AND MABBITT, A.W.: 'Pin-fet receivers for long wavelengtth fibre optic systems'. *'Communications 82'*, pp. 280–284.
14 PARK, M.-S., HWANG, J.-A., and CHUN, Y.-Y.: 'Flicker noise effect on sensitivity of p-i-n GaAs f.e.t. fibre optic receivers'.,*Electronics Lett.*, **23**, no. 12, 4 June. 1987, pp. 631–633.
15 PARK, M.-S., SHIM C.-S., and KANG, M.-H.: 'Analysis of sensititvity degradation caused by the flicker noise of GaAs m.e.s.f.e.ts. in fiber-optic receivers'. *IEEE J. Lightwave Technol.*, **6**, no. 5, May 1988, pp. 660–667.
16 BAECHTOLD, W.: 'Noise behaviour of GaAs field-effect transistors with short gate lengths'. *IEEE Trans. Electron Devices*, **ED-19**, no. 5, May 1972, pp. 674–680.
17 BYLANSKI, P., and INGRAM, D.W.: *Digital transmission systems'*. Revised second edition, Peter Peregrinus Ltd., London, 1990.
18 SMITH, R.G., and PERSONICK, S.D.: 'Receiver design for optical fiber communication systems'. *Semiconductor Devices for Optical Communication*, Ed.H. Kressel, Springer Verlag, 1980, pp. 89–160.
19 McINTYRE, R.J.: 'Multiplication noise in uniform avalanche diodes'. *IEEE Trans. Electron Devices*, **ED-13**, no. 1, Jan. 1966, pp. 164–168.
20 *'Optical receivers'* IEE Vacation School, University of Essex, Mar. 1982.
21 HOOPER, R.C.: 'Optimum equalisers for binary transmission systems and their application to optical receivers'. *IEE Proc.*, **136**, Pt J, no. 2, Apr. 1989, pp. 137–140.

12

SINGLE-MODE CABLES, SPLICES AND CONNECTORS

Oh tell me, when along the line
From my full heart the message flows,
What currents are induced in thine?
One click from thee will end my woes

Electric Valentine, verse three of four, which appeared in the journal *Nature*, in 1875, occasioned by reports of a wedding between the bride in the telegraph office in Boston and the man she loved in the telegraph office in New York, reproduced in [1]

12.1 INTRODUCTION

Many aspects of electro-optic sources used in ofdls were considered in Chapters 6–8 and the following three chapters gave extensive coverage to aspects of optical receiver design and application. Both lasers and their associated electronics and photodiodes and their associated electronics form part of terminal equipment and of regenerative repeaters. We now turn to the cables containing the single-mode fibre pairs which form the line sections of systems between repeaters or between repeaters and terminals.

The importance of cables and jointing in system power budgets and reliability is not as well known as it should be, perhaps because cables have been used in line communications for about 150 years, and also perhaps because of the more glamorous nature of lasers and, to a lesser extent, receivers. The material in this chapter will, it is hoped, help to redress the balance somewhat, and provide system designers with information of use. As with lasers and detectors, advances in cable design, manufacture and performance have come from laboratories and plants in a number of countries throughout the world. There are numerous types of cables to suit different application areas, but in order to restrict the amount of text, attention will be concentrated on cables for use in ducts or directly buried in the ground and on cables for submerged systems.

The development of cables goes back to the early days of the electric telegraph, and the most striking achievements were probably the great submarine lines which united the Old World with the New World. Their installation was often a story of high drama and great tenacity. In about 1843 both Wheatstone and

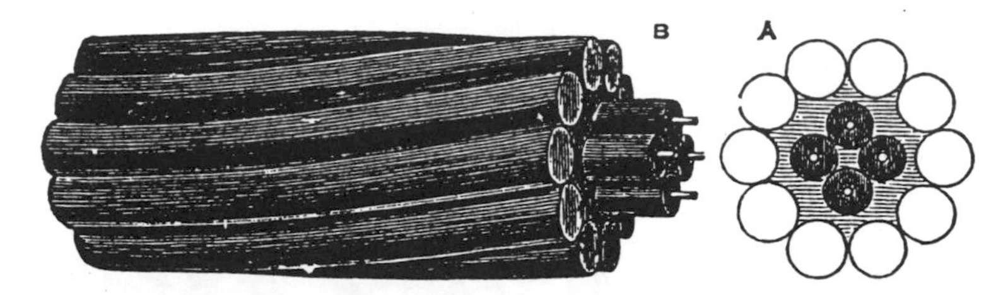

Figure 12.1 Drawings showing the construction of the first practically successful submerged cable laid between England and France in 1851 ([1], Figure 297)

Morse independently experimented with underwater insulated wires, and their success gave rise to numerous projects for submerged systems. Some years later the insulation properties of gutta-percha were recognised, and after one short-lived attempt the first practically successful cable was laid between Dover and Cap Gris Nez in 1851. A view and cross-section of this cable is shown full-size in Figure 12.1.

This cable had four separate copper wires each insulated with a covering of gutta-percha, and the whole was spun with tarred hemp into the form of a rope. This in turn was protected by an outer covering of ten of the thickest iron wires wound spirally upon it, the first example of an armoured cable. Some of the features of this cable will be familiar to cable engineers today. The cable itself was 43.2 km (27 miles) long and weighed about 4.4 tonnes per km (7 tons/mile) [1].

The first transatlantic cable entered service on 5 August 1858, but failed after a month due to overvoltage on the line. Succeeding years saw advances in cable design (and disasters, e.g. of the 5500 miles of cable manufactured for the attempt to link the continents in 1865, 4000 miles lay at the bottom of the Atlantic, and an enormous sum of money had been swallowed up in the endeavours (the cable was subsequently recovered in better electrical condition than before the lay!)). However, the following year a new cable was laid successfully and the two continents have remained linked by submerged cables ever since. Drawings of the 1866 Atlantic cable are given in Figure 12.2.

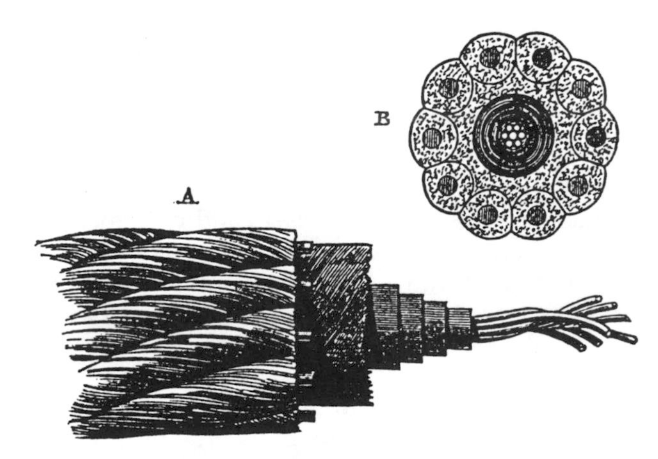

Figure 12.2 Drawings showing the construction of the first practically successful submerged cable laid between England and the United States of America in 1866 ([1], Figure 301)

This cable consisted of seven twisted copper wires as before, forming the conductor around which were four coatings of gutta-percha, which were in turn surrounded by a layer of jute, and the whole protected by ten iron wires.

Leaping over the intervening years and the careers of many cable engineers who contributed to the development and engineering of copper cables deployed from the deepest oceans to every telephone customer throughout the world, this chapter is devoted to those aspects of optical fibre cables which are pertinent to optical fibre digital transmission systems.

To provide a realistic introduction to the effects of cabling and jointing on transmission characteristics, we summarise the results of an extensive series of measurements to establish the transmission characteristics of a cabled and jointed 31.6 km fibre link. These were reported in 1982 [12.2] at a time when there were technical difficulties implementing single-mode (technology, including problems associated with cabling and jointing.

Measurements of loss

Cut-back loss measurements were made over the band 1100 to 1750 nm on the fibres and on the 31.6 km link at various stages of manufacture, taking care to ensure the maximum reproducibility. The link consisted of 14 fibres, each 2.25 km long, and the sum of the losses of all 14 lengths is shown by the curves labelled fibre in Figure 12.3.

The fibre diameters were 100 μm with an average core radius of 4.77 μm, and an average Δ of $3.18 \times 10E - 3$. Individual fibres were then loose-tube secondary coated and the total loss for all fibres again measured, and results are shown by the curve labelled packaged. The next stage was to strand the 14 loose-tube fibres, and remeasure the loss of each fibre in each of the three cables which were made. The summed results of this set of measurements is displayed in the curve labelled taped. Finally, the 14 fibres were fusion spliced to form a continuous system fibre, the loss of which was measured and the result graphed in Figure 12.3 by the curve labelled jointed, the loss is about 18 db (actually 17.3 dB) at 1300 nm. The incremental losses due to packaging and to taping both increased monotonically with wavelength, as shown in Figure 12.4 along with the incremental jointing loss.

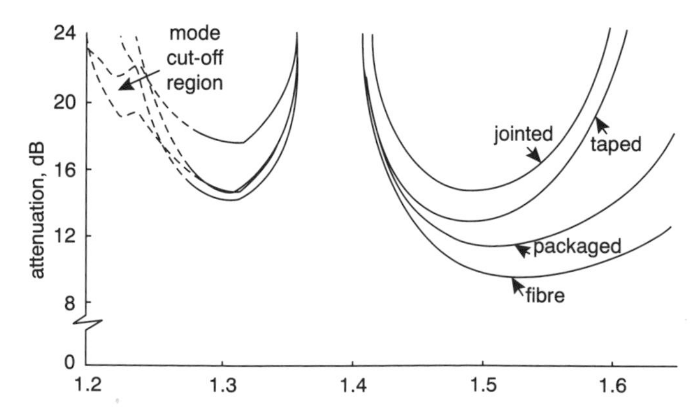

Figure 12.3 Attenuation versus wavelength showing spectral loss curves for 14 fibre, 31.6 km link at all stages of assembly. Reproduced by permission of IEE

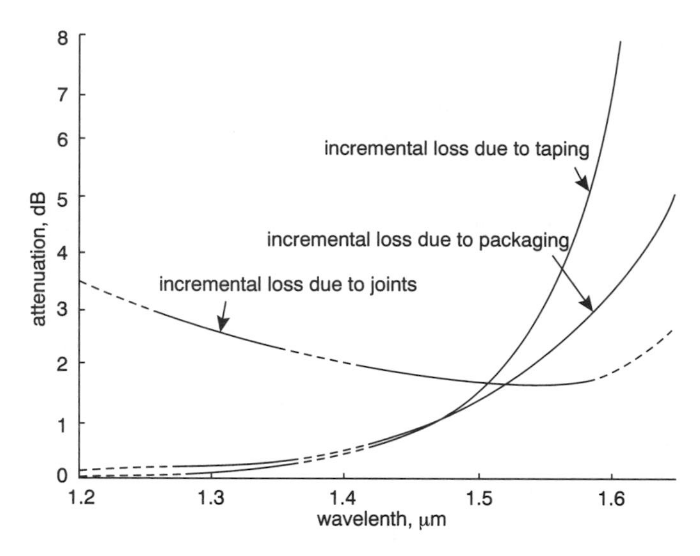

Figure 12.4 Attenuation versus wavelength showing incremental loss curves after each of the three stages for 14 fibre, 31.6 km link at all stages of assembly. Reproduced by permission of IEE

Measurements of total dispersion

The output of a mode-locked Q-switched Nd:YAG laser operating at 1060 nm was launched into 200 m of sm fibre to produce an output spectrum covering the band 1060 to 1600 nm due to the stimulated Raman effect. A monochromator selected the wavelength of the signal launched into the fibre and the delay differences between pulses of different wavelength recorded. A linearised Sellmeier equation was fitted to the results, and the total dispersion obtained by differentiation. All 14 fibres were measured and found to have similar characteristics, this is best expressed in tabular form as listed in Table 12.1.

The similarity of the results was attributed partly to the similarity in fibre dimensions and partly in the relatively small waveguide dispersion associated with the relatively large core radius (average value 4.77 μm). The spread in radius of 30% made little difference to the dispersion from which the authors stated that results were inconclusive in determining whether ESI measurements were suitable for predicting dispersion.

Measurements of pulse broadening showed that birefringence in this design of fibre did not appear to limit bandwidth × distance product. For this link and a laser with 4 nm linewidth calculated values were 2.6 GHz at 1300 nm and 250 MHz at 1550 nm, thus both length and bit rate could be increased.

Table 12.1 Maxima and minima of dispersion measurements on 14 fibres ([2], Table 1). Reproduced by permission of IEE

	max.	min.
Dispersion at 1.55 μm	18.5 ps/km/nm	14 ps/km/nm
Dispersion at 1.30 μm	−1.3 ps/km/nm	−2.5 ps/km/nm
Wavelength of zero dispersion	1.33 μm	1.31 μm

An AT&T survey of sm fibre cable for long-haul, trunk and loop networks appeared in 1987 [3]. A brief overview of this article which summarised the optical and mechanical performance of twelve-fibre ribbon unit and Lightpack cables will be given as an introduction. Ribbon fibre cable technology was extended to single-mode fibres in late 1984 and was used for links between central offices up to 20 km in length, with factory-fitted connectors to provide fast array splicing (less than 3 hours for 144 fibres) and installation in the field. For long-haul applications individual fibre splicing was employed to minimise splice loss. The Lightpack cable was introduced in 1985.

The basic construction of both types of cable was described and illustrated, a short mechanical specification tabulated and the mechanical performance and relevant EIA test requirements were stated. The main geometrical and optical parameters of AT&T standard depressed-cladding sm fibre are given in Table 12.2.

Histograms showing the per-unit loss in both windows for both types of cables are reproduced in Figures 12.5 and 12.6.

Environmental stability: three temperature ranges were quoted: for cables in ducts or directly buried in the ground; for aerial cables in most parts of America; and aerial

Table 12.2 Main parameters of AT&T single-mode fibres (1987) ([3], Table III). Reproduced by permission of AT&T

Characteristic	Value
Core diameter	8.3 μm
Cladding diameter	125 μm
Coating diameter	245 μm
Total index difference	0.37%
Cladding index depression	0.12%
Zero dispersion wavelength	1300–1325 nm
Coating type	Ultraviolet-cured acrylate
Proof test level	50 000 lb/in^2

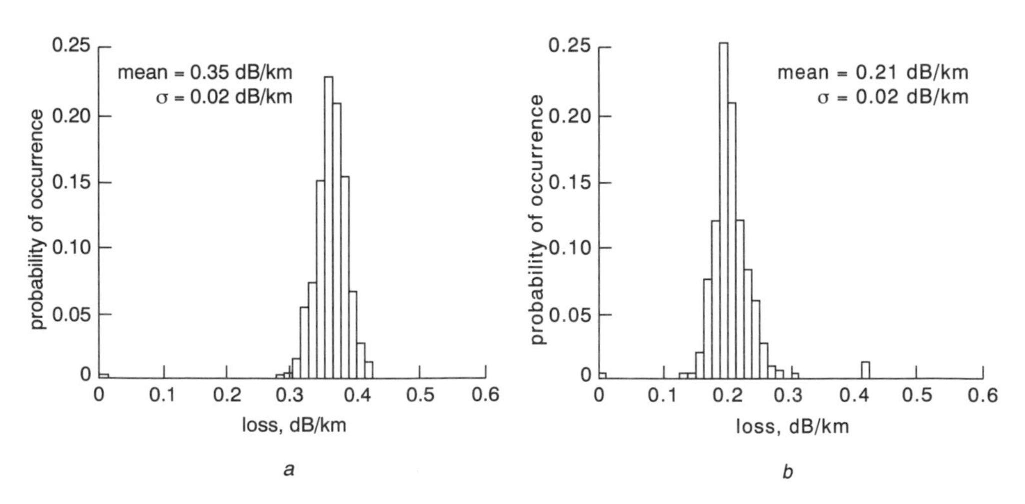

Figure 12.5 Per-unit loss of production ribbon cabled fibres: (*a*) at 1310 nm; (*b*) 1550 nm, and corresponding means and standard deviations ([3], Figure 6). Reproduced by permission of AT&T

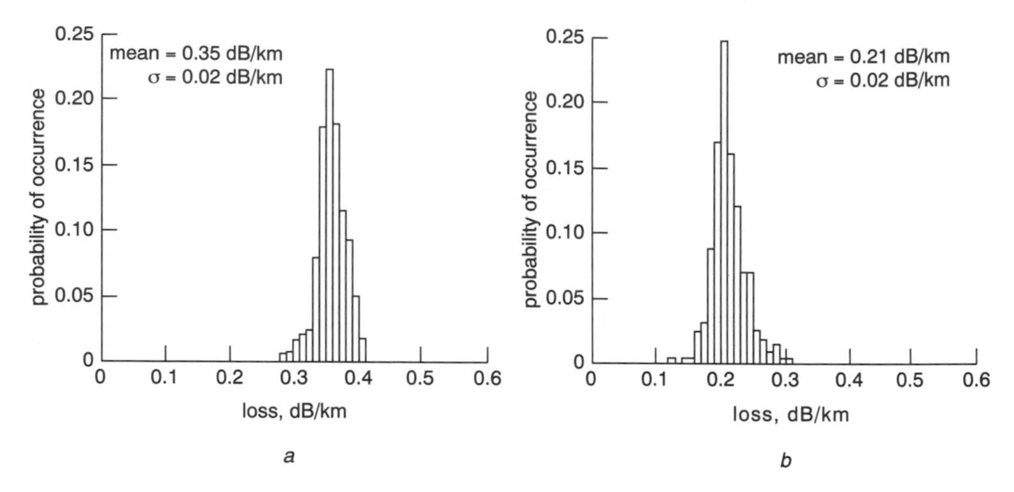

Figure 12.6 Per-unit loss of production Lightpack cabled fibres: (*a*) at 1310 nm; (*b*) 1550 nm, and corresponding means and standard deviations ([3], Figure 7). Reproduced by permission of AT&T

cables for use in extreme climates. Optical performance was evaluated in cyclic environmental tests on entire cables, chosen to be much more severe than would occur in the field. Both ribbon and Lightpack cables exhibited zero change in loss between −20 and +80°C; mean losses increased with decreasing temperature below −20°C by less than 0.02 dB/km for ribbon cables and less than 0.03 dB/km for Lightpack cables at −40°C. The increase in mean loss was slightly greater at 1550 nm than at 1300 nm due to the greater microbending loss at the longer wavelength. The generation of hydrogen was eliminated by the selection of suitable materials for the cable components.

12.2 OPTICAL FIBRE CABLES

Consider an idealised cable design in order to introduce the components which make up real cables without the need to refer to specific designs or types. This idealised structure is shown in Figure 12.7; it shows the nine (labelled) components, and the small circles inside the buffers represent coated fibres [4].

12.2.1 Cable core

This consists of a number of components which will be introduced shortly. The transmission paths are the fibres themselves in their buffers, but these do not provide the mechanical strength required. This is provided by stranding the buffers around a central member which serves both as support against buckling and kinking and as a strain relief member. Stranding contributes largely to providing the fibres with a well-defined free space within which strain buckling, pressure and bending stresses will have no influence on the transmission characteristics. Buffers themselves can be of the single or multifibre loose tube types, tight buffered types, composite buffered types or ribbon designs. Fillers can also be stranded with them; these could be empty buffers or solid polyethelene fillers, but also copper wires

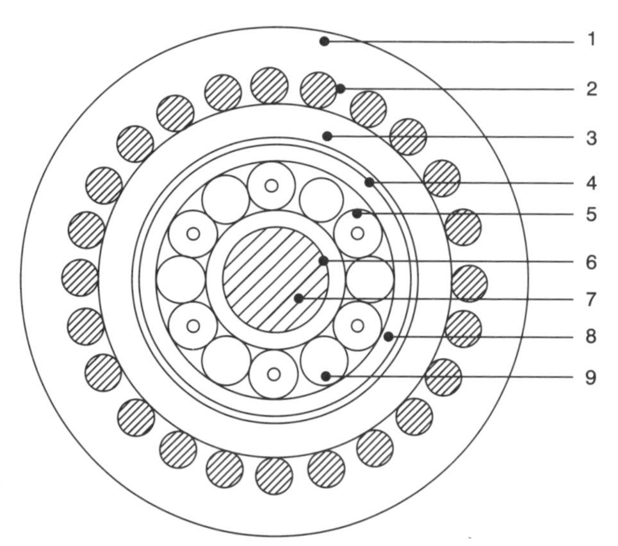

Figure 12.7 Idealised optical cable design ([4], Figure 2). Reproduced by permission of IEE

as pairs or quads. The entire assembly of the stranding components in a cable, together with the anti-buckling and strength members and the wrapping around them, if present, constitute the cable core.

Elongation and contraction: the fibres can stretch or contract within the buffer tubes, and as with bending these changes must be limited as that within a specified temperature range and for a specified tensile load neither the transmission properties nor the integrity of the fibre are adversely affected. Within loose buffer tubes fibres can move freely and in the neutral state, i.e. no stress, they are located on or around the tube axis, with some nominal clearance between them and the inner wall. The relation between elongation and contraction of a cable and the clearance of a fibre in a loose tube buffer is shown in Figure 12.8. The effects of contraction are shown on the left side of the diagram. The fibre touches the top of the inner surface at 12 O'clock for a relative change of 0.6% in cable length with temperature. Greater contraction, corresponding to lower temperature, would cause the fibre to buckle. On the right of Figure 12.8 the reverse situation is portrayed, in which elongation occurs due to tensile forces, and the fibre touches the inner surface at about 6 O'clock for about 0.6% elongation. A further increase in tensile force causes the fibre itself to stretch.

Cable core filling: all outdoor cables need a barrier against longitudinal penetration of water; therefore the interstices in the core are filled with a suitable compound at very high pressure, about 15 bar. For example, a barrier layer of petroleum-resistant and relaxing thermoplastic adhesive is extruded around the filled core which serves as an additional barrier for the filling compound, and as a seamless connection between the tension resistant wrapping and the cable sheath without detriment to flexibility. Cables for indoor use and others not requiring protection from longitudinal water penetration usually have the core wrapped with one or several layers of thin plastic foil to keep components outside the core from entering the interstitices.

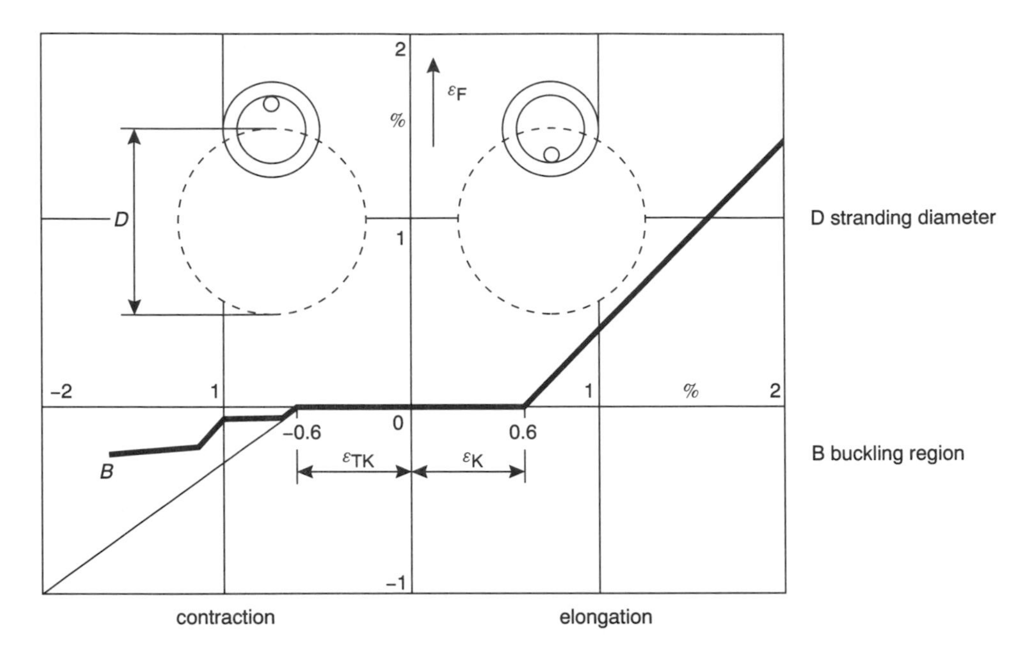

Figure 12.8 Optical fibre elongation ε_F, cable elongation ε_C and cable contraction ε_{CC} for stranded single-fibre loose buffers ([2] of Chapter 3, Figure 9.11). Reproduced by permission of Siemens

12.2.2 Cable sheaths and protective coverings

The functions of a sheath are to protect the fibres in the core from mechanical, thermal and chemical effects and against moisture from the outside. Polyethelene with or without a barrier layer of, for example, aluminium is used frequently, but some other materials such as polyvinyl chloride, fluorine plastic, or halogen-free materials are in common use. Metal-free cable construction is sometimes required and in such cases a barrier layer of polyamide thermoplastic adhesive is added between the sheath and the yarns and filling compound. This is done to ensure that the filling compound does not migrate from the core into the sheath. It also ensures a very tight connection between the core with the aramide yarns and the sheath itself, which is very desirable when cables are pulled into ducts.

Outdoor and special cables such as shallow water cables in submerged systems sometimes have armour wires, as shown in Figure 12.7 as one component in the cable construction. Outer sheaths or protective coverings are used to protect the wires from corrosion or external damage which may occur during ploughing in on land or on the seabed. Armouring is found not only in submarine cables, but in applications such as mines, self-supporting aerial cables and in some parts of the world for protection against rodents. Aramide yarns (Kevlar) and steel are used in various forms, with Kevlar having a much higher Young's modulus to weight ratio than steel. Further details on armouring and other topics can be found in the references cited in [4] or in more recent works.

In summary, it can be said that for all applications of copper cable technology developed since 1851 there are adequate optical fibre cable replacements, and that these have the following major advantages (among others): smaller cable diameter, lighter weight, longer delivery lengths, and smaller reel dimensions and weights.

Table 12.3 Comparisons of diameter, weight and delivery lengths for optical and copper cables. ([2] of Chapter 3, Table 9.4). Reproduced by permission of Siemens

Number of fibres		Outside diameter (mm)		Weight (kg/km)		Standard delivery lengths (m)	
Fibre	Cu twisted pair 0.6 mm	Fibre	Cu	Fibre	Cu	Fibre	Cu
2,4,6	6	11.5	12	100	140	2000	1000
10	10	12	13.5	115	190	2000	1000
20	20	14.5	16.5	160	315	2000	1000
40	40	14.5	21.5	160	355	2000	1000
60	60	16	25.5	195	800	2000	1000
100	100	20	31.5	295	1245	2000	1000

A comparison of optical and copper cables which provides quantitative evidence of these advantages is given in Table 12.3, reproduced from an excellent book by authors from Siemens ([2] of Chapter 3).

A review of important issues involved in advancing mid-1980s state-of-the-art fibre cable technology from a telecom operator's point-of-view was given in an invited paper by authors from Bellcore [5] in a special issue of the *Journal of Lightwave Technology* in August 1986. Advances required to extend fibre cables into the subscriber (local) loop to provide cost-effective implementation of B-ISDN in future (mid- to late 1990s) were discussed. Cable structures optimised for applications in different parts of public networks, e.g. trunk, junction (interoffice) and distribution plant, were identified. The paper covered the technologies of fibre and coating designs, cable structures, installation, splicing and connectors, and it included reliability issues.

Following on the Introduction the second section dealt with fibre designs and characteristics, mainly for 1300 nm optimised sm fibre (depressed cladding or matched cladding) designed in accordance with CCITT Rec. G.652, but including dispersion-shifted fibre and mentioning dispersion-flattened fibre. Aspects of coating materials and design were mentioned. Section 2 ended by comparing transmission characteristics of cabled versus uncabled fibres. Geometrical parameters, mode field diameter and chromatic dispersion were unaffected by cabling, microbending loss and cut-off wavelength could differ. The effective cut-off wavelength of a 2 m length of cabled fibre may be higher or lower than that of the uncabled fibre, depending on the cable design and deployment conditions. An example illustrating the difference in behaviour for depressed- and matched-cladding fibres is given in Figure 12.9.

In practice, therefore, it is very difficult to relate the effective cut-off wavelength of a cabled fibre to that before cabling; however values for long lengths (as in system fibres) are always well below corresponding values at 2 m, thus giving an adequate performance in 1300 nm systems. The condition that the effective cut-off wavelength be less than the lowest operating wavelength also applies to short sections such as interconnection cables and pigtails on active or passive components. Fibre loops of small radius are commonly used as mode filters to ensure single-mode operation

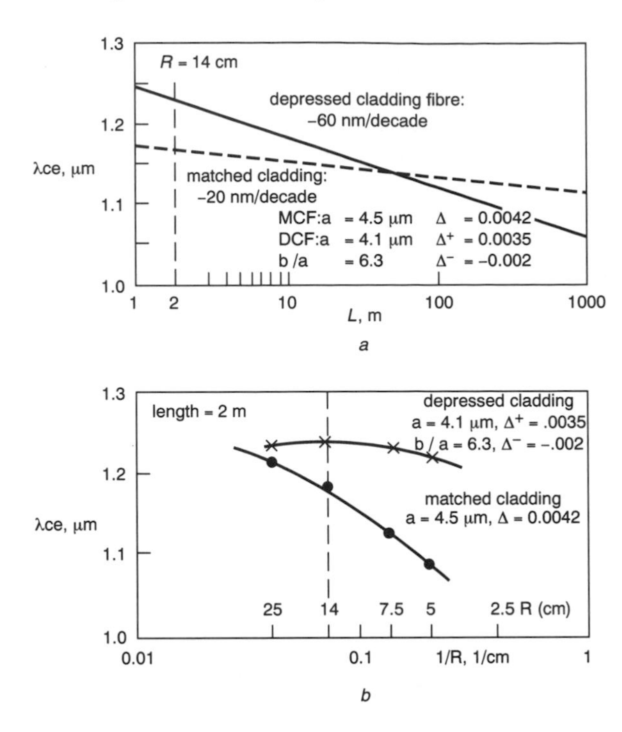

Figure 12.9 Comparison between matched cladding and depressed cladding standard sm fibres: (*a*) effective cut-off wavelength versus fibre length (log scale); and (*b*) effective cut-off wavelength versus reciprocal of bend radius (log scale). Note: a = core radius, b = cladding inner radius, R = radius of curvature ([5], Figure 1). Reproduced by permission of IEEE, © 1986

in such short lengths, but care must be taken to avoid too small a radius which would cause excess loss at 1550 nm.

Section 3 was devoted to cable design, and it contained subsections on design objectives and on designs covering loose-tube, slotted core, fibre bundle, tight buffer and ribbon types. Section 4 concerned the (then) less advanced characterisation of mechanical performance, which was still in the formative stage. A table listed mechanical characteristics and corresponding EIA test procedures, and noted that a test for corrosion/hydrogen was under development.

Installation and applications formed the subject of Section 5, where the four main methods of outside plant installation were discussed and the corresponding requirements on cables. These four were duct, directly buried, aerial and submerged types of installation. Section 5 described briefly certain aspects of the interrelation between cable design and jointing (splices or connectors). Future trends were also discussed.

12.2.3 The blown fibre cable

A radically new approach to the problems caused by the large end stresses associated with traditional cable drawing, particularly on tortuous routes in metropolitan areas, was introduced in 1984 with the first trial installation in Leeds, England [6]. By taking advantage of the small size and weight of packaged optical fibres it was found possible to install them in cable bores using compressed air. Units consisting of several fibres are drawn into the bores by the viscous flow of air. These units can be routed through various sizes of pre-installed empty cable without the need for fibre joints as and when they are needed, and the distributed nature of the drag force along the cable route avoids the high stresses formerly encountered. Empty cable can be made cheaply and installed without undue care, and the fibre unit could become a standard item to be used in any size of cable. British Telecom awarded its prestigious Martlesham Medal for 1989 to M.H. Reeve as the inventor of the blown-fibre technique.

The important features of this installation method are as follows.

Empty polyethelene bores are pulled into the ground using traditional techniques and with no special care. Each bore has an inside diameter of 6 mm and an outside diameter of 8.3 mm. Seven such bores are assembled in a 'six round one' configuration to form a cable with an outer diameter of 29 mm which includes an outer sheath of polyethelene.

Individual bores are connected together using small pressure couplings to form a chosen continuous route.

The 2 mm outside diameter fibre package is blown into the route using compressed air.

Packages containing several fibres could easily be blown over distances up to 500 m, making the technique particularly suitable for local area distribution. Longer installation lengths could readily be installed by fleeting operations at one or more intermediate points.

'Blown fibre' cable offers a number of advantages over conventional cables.

(a) Packages can have different numbers of fibres, with seven being popular, but smaller and larger numbers can be used. This results in a low-cost cable with high flexibility.

(b) Packages can be installed gradually as demand increases, thus spreading costs. Packages could even be recovered and reused.

(c) At branching points packages can divert from high fibre count to low count cables without using joints, and this greatly reduces the total number of fibre joints needed in a large local area network (LAN).

(d) Fibre can be installed quickly into complicated routes with very little fibre strain.

12.2.4 Installation of conventional cables in conduits using viscous air flow

Many countries use pre-installed ducts (conduits) into which cables are drawn. Manually pulling cables into ducts is labour-intensive and time-consuming, and therefore costly, but has been widely used in the UK by cable television companies in setting up their networks. Using a viscous flow of air in a scaled-up version of the 'blown fibre' technique, cables can be installed in cheap high-density polyethelene (HDPE) ducts with an inner diameter of 26 mm, even when there are many bends. A simple compressor (75 l/s, 8 bar) is used for the air supply and a special 'snaking gun' or cablejet was developed which could handle the large air flows required, without significant pressure drops. The technique was described in [7].

Results from field trials showed excellent agreement with predictions from the theory, and showed that it was possible to install maximum cable lengths of 800 m in 26 mm ducts using one cablejet and paraffin oil lubricant, with speeds up to 60 m/min reached with standard cables. 530 m of cable were installed in 40 mm conduit using a 130 l/s compressor. The results demonstrated that right-angled bends and windings (zig-zag curves) had very little effect on cable installation. Cablejets can easily be used in cascade for longer routes at distances depending only on properties of the cable, duct and lubricant used, and there are no synchronisation problems when so used. An example was quoted where a 2000 m circuit was installed in under one hour using several cablejets in tandem.

12.3 LONG-TERM RELIABILITY OF OPTICAL FIBRE CABLES

'... it is fundamentally impossible to improve upon the reliability dictated by cable damage or failure on an individual link. Alternative routing (i.e. redundancy) within the network is often used to improve end-to-end reliability but once again this improvement is fundamentally constrained. The bottom line on transmission system reliability is the cable.' ([13] of Chapter 10). Cable and line plant was considered in this excellent publication, which included the typical MTBFs and MTTRs on which the study was based in Tables 1 and 2, giving cable failure and repair data for a long-distance link of 1000 km and for a 20 km local-loop link. Hydrogen was not mentioned as a failure mechanism in either table.

([13] of Chapter 10) was published in 1994, and among the information given was the end-to-end reliability of long lines using 100 km, 1000 km and 10 000 km systems for the numerical examples. It was shown that cable reliability always dominated in undersea systems. In 1000 km long repeatered landlines, the reliability of line, terminal mux and repeaters are broadly similar if duplicated power supplies are not used; with duplicated supplies cable reliability dominates. The reliability of 10 000 km long systems is heavily cable-dominated. The analysis of reliability in the local loop found that the overall balance between the various MTBFs and unavailabilities is shifted dramatically. Comparing the overall reliabilities of the local loop and 100 km systems, yielded similar values for today and for the future, with the optimum situation that in which reliability is evenly distributed between the local loop and long line. A pessimistic example was given of the mttr values and unavailability figures for a cable containing a number of fibres in parallel, each of which carries the same share of the total traffic on the cable. The time taken to detect and locate a break, despatch repair staff and prepare the fibres for splicing was assumed to be 24 hours, and the resulting values are listed in Table 12.4.

A reliability concern in cable manufacture in the 1980s was static fatigue in the fibres, for which design criteria were established for fibre strength based upon proof testing.

12.3.1 Effect of hydrogen on attenuation in optical cables

A second concern in the early 1980s was the system implications of hydrogen-related degradation in optical fibres. Problems arising from hydrogen were reported in 1982, and this was followed by numerous papers demonstrating that hydrogen can cause significant increases in attenuation of silica-based fibres. There were also

Table 12.4 Cable MTTR and unavailability figures for a given amount of traffic concentrated on N fibres. Reproduced by permission of BT

	Splice time per fibre = 30 minutes				Splice time per fibre = 15 minutes			
Number of fibres	Repair time (hours)	Repair time (days)	U_v (%)	A_v (%)	Repair time (hours)	Repair time (days)	U_v (%)	A_v (%)
1	24.5	1.02	0.028	99.97	24.25	1.01	0.028	99.97
2	25	1.04	0.029	99.97	24.5	1.02	0.028	99.97
10	29	1.21	0.031	99.97	26.5	1.1	0.030	99.97
100	74	3.08	0.085	99.92	49	2.04	0.056	99.94
200	124	5.17	0.142	99.86	74	3.08	0.085	99.92
500	274	11.42	0.313	99.69	149	6.21	0.170	99.83
1000	524	21.83	0.598	99.40	274	11.42	0.313	99.69

Notes: Availability $A_V = (MTBF)/(MTTR + MTBF)$ Unavailability $U_V = (MTTR)/(MTTR + MTBF) = 1 - A_V$

several reports of hydrogen generation in optical fibre cables either by electrolytic action, or by evolution from cable components.

The magnitude of the various hydrogen-related optical ageing mechanisms was considered in [8]. Data were presented from tests of fibres in constant hydrogen environments as a function of fibre composition, hydrogen pressure, temperature and time. Methods of extrapolating these short-term data to system lifetimes were discussed and estimates were made of the optical ageing. The impact of these long-term predictions on system reliability were considered for both single- and multimode fibres. Various cable and fibre design options which reduce these hydrogen-related effects were discussed, including the application of hermetic coatings to the fibre to act as a diffusion barrier against the ingress of hydrogen.

Two distinct types of loss effects were observed.

(1) Absorption due to molecular hydrogen which diffuses into interstitial sites in the silica network.

(2) Absorption due to chemical reaction of the diffused hydrogen with the glass constituents to form, for example, OH groups (not observed in sm fibres at ambient temperatures).

The interstitial loss was the same for all (1983) sm and multimode production fibres tested but larger values were found in experimental fibres of different composition. The amounts of loss at 20°C are illustrated in Table 12.5 for 5°C the values should be increased by about 10%.

The effect of the application of a ceramic hermetic coating to the fibre, which would act as a diffusion barrier against the ingress of hydrogen, was investigated and predictions of long-term implications based on a diffusion model. A 20 nm thick silicon oxynitride coating (such coatings had been investigated previously to improve the mechanical performance of fibres by preventing static fatigue due to attack by water) was applied on-line and the resulting difference in loss increase between uncoated and coated fibres is shown in Figure 12.10

The cables manufactured by STC in the early 1990s for use in submerged systems all included a hermetically sealed hydrogen barrier consisting of a copper inner sheath (see 3 in Figure 12.7). The manufacturer's data sheet did not state anything about a silicon oxynitride coating on the fibres themselves.

Table 12.5 Loss at 20°C in system operating window due to interstitial hydrogen ([8], Table 1). Reproduced by permission of BT

Wavelength, μm	Loss, dB/km/atm H_2
1.275	0.6
1.30	0.3
1.325	0.15
1.525	0.25
1.55	0.6
1.575	1.5

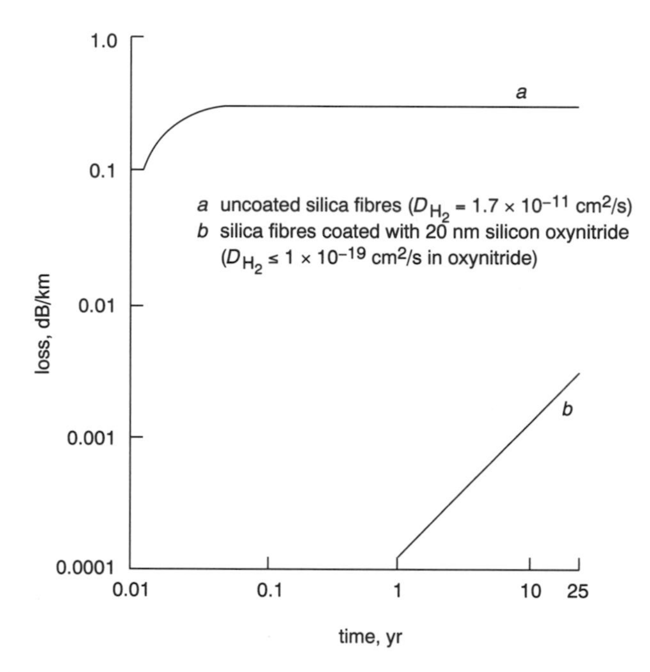

Figure 12.10 Increase in loss at 1300 nm due to interstitial hydrogen. Note: $D(H_2)$ is the effective diffusion coefficient ([8], Figure 16). Reproduced by permission of BT

Interest in the subject continued, and in 1988 two authors from Sumitomo Electric Industries described a model for the evolution of hydrogen in optical cables from the organic components as a function of time and temperature, and for the outward permeation of hydrogen through the cable sheath. Only in the case of laminated aluminium polyethelene sheaths was the characterisation of hydrogen diffusion complicated. They also discussed the design criterion for cable materials necessary to ensure their long-term reliability [9]. Another possible source of hydrogen evolution, metal in the cable e.g. a sheath, but this source was excluded on the ground that in well-designed cables all metal surfaces are protected from water and humidity. The validity of the proposed model was confirmed by long-term tests at room and high temperatures.

Long-term test of stored cables: the hydrogen concentration in two drum wound cables left in the open air for five years was measured and the results were compared with the predicted values using the model. The experiments confirmed the validity of the present model for estimating hydrogen concentrations in optical cables.

On the basis of their work some guidance is given for the selection of materials for use in optical cables. Present materials such as low hydrogen silicone resin and most of the uv cured materials satisfy the criteria for low hydrogen concentrations, and suitability for a 25 year service life.

The diffusion of hydrogen molecules in fluorine-doped silica sm fibres was studied by authors from a cable manufacturer and KDD [10] and the results were published in a letter in 1989. Four types of sm fibres were used in the investigation, all of them fabricated by the vad process.

Evaluation of irreversible increase in loss in silica-based fibres by means of high-temperature tests was discussed by workers from KDD in [11]. The test results demonstrated that fibres with germanium-doped cores behaved differently from those with pure silica cores; and in pure silica core fibres fabricated under optimum conditions the formation of OH was negligible, and so the attenuation was very stable.

12.4 OPTICAL FIBRE SPLICING

The function of a splice in a lightguide is to permanently connect two fibres with as little loss as possible, although some splice designs offer the possibility of separation and rejoining [12]. The most important systems property of splices and of connectors is their optical loss. This depends on a number of factors, including alignment of the two fibres, their end conditions, and parameters of the fibre cores (diameter and peak index difference in mm fibres, and spot size in sm fibres). These parameters are not under control of the splice or connector design, and will not be discussed here, but have been covered in Chapters 4 and 5. Since multimode fibres are no longer of so much technical interest and space is limited, attention will be concentrated on the splicing of single-mode fibres.

Splice loss is due to several factors which can be grouped into extrinsic and intrinsic factors. Extrinsic factors are peculiar to the splicing method used, and thus they include many factors and procedures e.g. stripping and cleaving procedures for a given type of fibre, type of splicer equipment, operator skill, environmental conditions and method of measuring loss. Intrinsic factors include mismatches between optical, geometrical and material characteristics of the fibres being spliced. In the results which follow actual production probability densities of the geometrical parameters of the fibres used were employed to determine the appropriate intrinsic loss distribution, which was then used as simulation input data. The other set of input data was the extrinsic loss distribution appropriate to the splicing method being simulated.

12.4.1 Optimisation of fusion splicing in the field

A paper [13] submitted in mid-1988 analysed the criteria for choosing parameters which influenced the distribution of splice loss. The proposed method adopted two criteria, average loss and repetition ratio, and allowed a designer to determine the maximum permissible loss and the maximum number of repeat splices to secure the best trade-off between average splice loss and number of repetitions in a very large number of splices. It was suggested that the method could also be used to compare different splicing equipments.

Monte Carlo simulations were performed to determine the average loss versus maximum permissible loss and the repetition ratio (number of splices/number of combinations) versus maximum permissible loss for splicing in the trunk network, where in general it is possible to minimise the average loss. Having chosen a particular splicing method, the initial condition for the simulation was the extrinsic loss distribution as determined in the laboratory, with the procedure following that used in the field. Over 5000 combinations were simulated and field results for more than 600 combinations are shown in Figure 12.11.

Based on these curves a maximum splice loss of 0.12 dB and a maximum of four attempts were chosen. Results obtained in the field for five sm links confirmed the simulation results, as can be seen by comparing the solid dot (representing the average splice loss) below the number 6 with the corresponding curves in Figure 12.11*a*. It could happen that an unacceptably high loss result from the fourth splice attempt, so four attempts was not taken as an absolute limit; in fact this situation arose in only six out of 624 combinations. When fibres with their primary coating were cleaved and spliced it was found necessary to allow seven attempts for the same average loss as before.

12.4.2 Optimisation of performance of mechanical splices

The optimisation of the return loss and insertion loss of mechanical splices was the subject of [14], in which a combination of oblique end faces and index matching fluid was used to avoid the need for complicated alignment, selection of optimum refractive index matching material or special polishing techniques. A later paper [15] by the same authors from Bellcore stated that in some applications it may be preferable to replace the index matching substance by an air gap. Multiple beam interference can occur in such joints, and if present it causes variations in transmitted power (i.e. insertion loss variations) and return loss variations. The authors demonstrated that when the oblique end faces are parallel, the worst-case value of return loss could still be sufficiently high that the effects of variations in return loss could be neglected.

The axes of the two fibres were taken as the z-axis as usual and the x-axis was taken vertically up, so that the oblique angle α was the angle made by the normal to the end faces and the z-axis. The end faces could be separated by an amount Δz along the z-axis only to form an air gap (in which the incident beam is refracted), the axis of each fibre displaced vertically upwards by an amount Δx (to minimise the insertion loss), or a combination of both displacements used. Lowest insertion loss and highest return loss were both achieved when the end faces are parallel, therefore only this orientation was considered. The incident signal was taken as a plane wave propagating from left (transmit fibre) to right (receive fibre) so that Snell's law could be applied to reflection and refraction at the oblique end faces. Both fibres were assumed to be identical and the fundamental mode power distribution to be Gaussian.

Measurements were made using a 1306 nm laser and transmitted power versus Δz plotted for $\alpha = 0, 7$ and 10° and for $\Delta x = 0$, showing that the fringes (i.e. variations in transmitted power) disappear much more rapidly when α is not zero. Graphs were also presented to show transmitted power versus Δz for $\alpha = 7^\circ$ and

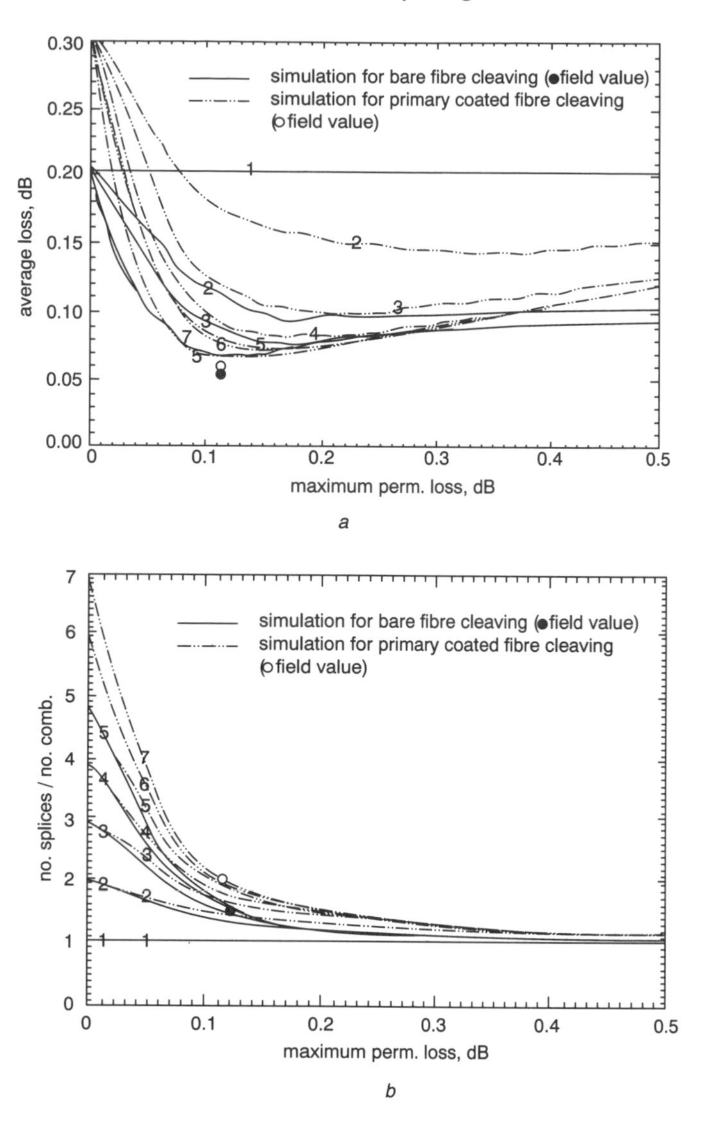

Figure 12.11 Simulation results for (*a*) average loss and (*b*) repetition ratio versus maximum permissible loss, using the distribution of extrinsic loss determined experimentally in the laboratory. Heavy lines indicate the simulation results with conventional splicing procedures, and the solid dots the corresponding field values. Dashed lines give the simulated results for modified splicing procedure in which the fibres were cut before removal of the primary coating, and the open circles are the corresponding field values ([13], Figure 3). Reproduced by permission of IEEE, © 1989

$\Delta x = 4$ μm, and for $\alpha = 10°$ and $\Delta x = \Delta y = 2$ μm. In both cases there was a broad maximum centred on 37 and on 25 μm, respectively, with fringes present below the maximum and absent above. For 7° and $\Delta z > 25$ μm the (peak-to-peak) variations in transmitted power were less than 0.1 dB compared with $\Delta z > 100$ μm for joints with end faces perpendicular to the fibre axis, and excellent return loss was achieved. For various combinations of Δz and Δx the lowest value of return loss (corresponding to the worst case) was greater than 57 dB.

12.4.3 Array splicing of single-mode fibre ribbons

Cables containing multiple ribbons each with many fibres (e.g. 144 fibres in a 12×12 arrangement, as illustrated in Figure 8.11 of ([2] of Chapter 3) showing the structure of such a cable) are expected to find a number of applications in subscriber loops, LANs and building distribution systems.

Array splicing was introduced in the US in 1979 for multimode fibre applications and by the mid-1980s over half a million field splices had been made and no splice failures reported. An average loss of 0.15 dB with a standard deviation of 0.15 dB had been achieved. Productivity and installed costs had been lower than for individual splicing and an array of 12 fibres could be jointed and tested in the field in five to ten minutes. The success of the method stimulated interest in sm fibre array splicing, but this had to wait until the technologies needed were sufficiently advanced.

Array splicing was the subject of a number of papers as early as 1984, and the technique used by AT&T was described in a late 1986 paper [12]. A new (1988) stuffing–pulling fusion technique was described in [16] in which a new array fusion splice machine had been manufactured and is used to produce an average splice loss of 0.024 dB in agreement with the calculated value. The authors from NTT examined the end face gap variance, and splice loss and failure ratio problems, reducing the splice loss by optimising the two methods (called constant stuffing and pulling (CSP) and variable stuffing and pulling (VSP)), and the calculation of average loss. The penultimate section dealt with the performance of the new splicing machine. A comparison was given between three methods in Table 12.6 reproduced below. Note the improvement in performance of the VSP over the CSP method.

In the discussion of performance of the new splicing machine two histograms were given and are reproduced as Figure 12.12.

The four-fibre ribbons used in the experiment were identical and the time taken was about 60 seconds per splice. The cumulative probability distribution for the three methods was given and it again showed the greatly improved performance obtainable with the VSP technique.

A companion paper which appeared in the same journal was devoted to the estimation of loss in array splicing of sm fibre ribbons [17]. The method presented was based on measuring the fibre axis discrepancy and the stuffing and pulling strokes during splicing. Errors in loss estimation were due mainly to errors in measurement of the axis discrepancy, and the distribution of these measurement errors was found to agree with that from calculations. The predicted distribution of loss estimation

Table 12.6 Comparison of results from conventional, CSP and VSP methods of array splicing ([16], Table 2). Reproduced by permission of Fachverlag Schiele und Schön GmbH

method	α_S	R_F	S_0	L_0
conventional	0.06 dB	5%	50 μm	0 μm
CSP	0.05 dB	5%	50 μm	46 μm
VSP	0.03 dB	0%	variable	variable

Key: α_S is average splice loss, R_F is probability of failure of splice, S_0 is predetermined stuffing stroke, and L_0 is optimum pulling stroke.

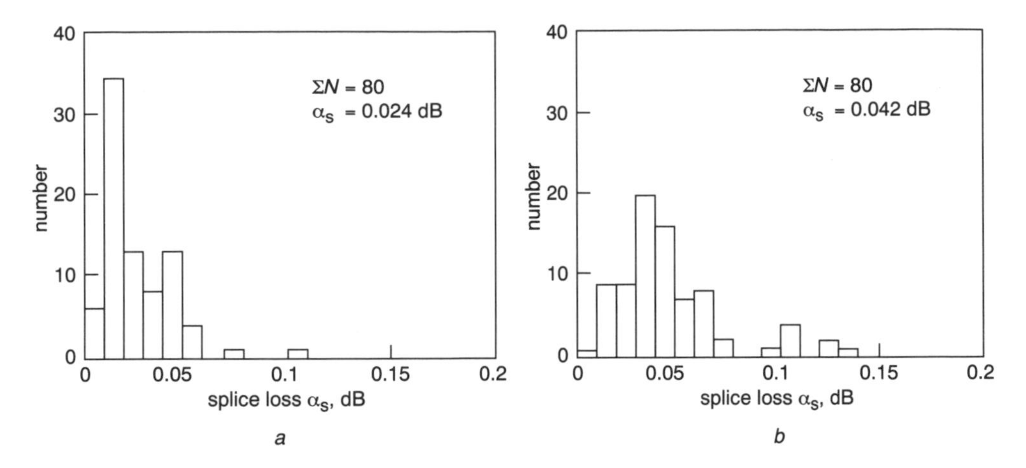

Figure 12.12 Histograms showing: (*a*) splice loss obtained with VSP method, (*b*) values obtained with the conventional technique ([16], Figures 13 and 14). Reproduced by permission of Fachverlag Schiele und Schön GmbH

errors also agreed with that found experimentally. An average splice loss estimation error of 0.039 dB with a standard deviation of 0.036 dB were obtained. Actual splice loss could be estimated with an accuracy shown in Figure 12.13.

Several of the same authors also wrote a paper devoted to the analysis of the loss factors present in splicing sm fibre without the precise alignment of core axes used for trunk and junction systems which was published in an American journal [18]. Under these conditions the dominant sources of loss are core deformation caused by discrepancies between core axes and axial misalignment due to core eccentricity. To reduce the average splice loss to below 0.1 dB the average core eccentricity must be less than 0.35 μm. The authors described a new method and splicing machine for quick and accurate setting of gaps between the fibre end faces which enabled

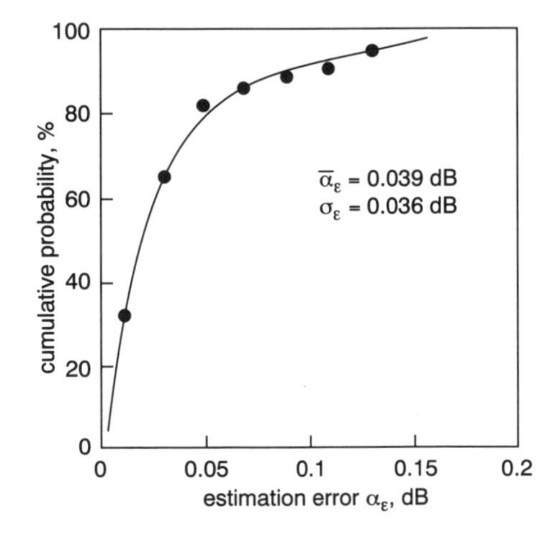

Figure 12.13 Cumulative probability of error (cumulative distribution function) versus error for given values of mean and standard deviation ([17], Figure 7). Reproduced by permission of Fachverlag Schiele und Schön GmbH

an average loss of 0.012 dB and standard deviation of 0.006 dB to be achieved. The loss factors were first analysed, then the conditions needed for making splices with losses below 0.05 dB, followed by a simple and accurate gap setting method and the results achieved with the machine.

12.4.4 Jointing of submarine optical fibre cables

The jointing of submarine optical fibre cables introduces a number of other complications, and is a more specialised subject. A detailed description of a joint developed for the installation and repair of undersea optical cables was given in 1985 [19]. Beginning with a description of the cable design, the authors then considered the constraints on joint design, followed by the joint construction. A complete joint in an armoured cable took about 15 hours. The cable preparation was treated, then joining cable ends followed by fibre preparation and by the fusion splice. The inspection of the encapsulated splice and evaluation of the loss followed, then fibre storage, polyethelene moulding, radiographic examination and armour reinstatement. Sixteen joints were to be manufactured for a stringent type approval programme to check the reliability of the joints required for the first international optical link (UK–Belgium No. 5, 3 × 280 Mbit/s, entered service 1986) and other later systems. The tests were to include:

(a) tension tests to check the optical transmission and mechanical performance
(b) tension cycling to simulate a joint hanging in suspension over the bows of a cableship
(c) a simulated laying and recovery from the bows of a ship around a sheave
(d) simulation of deep water pressure
(e) temperature cycling tests
(f) high-voltage breakdown tests on the moulded joint
(g) accelerated high-voltage life tests
(h) flexing of the PE moulding to check the amalgamation to the cable insulation
(i) humidity ingress.

The last section described the development of tooling for the tests. In conclusion, the authors state that the development of this optical fibre submarine cable joint involved a wide range of engineering disciplines to demonstrate that a stable and reproducable jointing system has been developed, which meets all the engineering requirements for the first North Sea optical cable system. Similar work on jointing of submarine cable has taken place in America, France and Japan, the other countries which design and manufacture undersea systems.

Jointing techniques in the future

One of the key advantages of an extensive optical fibre network, or optical aether, is the ability to 'weld it down' on day 1, and then not have to 'break in' at a later date to provide new routings and changes [20]. This advantage would, however, be lost if under fault conditions fibres had to be broken to facilitate testing or fault location. To obviate this, 'clip on' devices are needed that provide an optical 'crocodile clip'. By 1988 several devices have been developed that allow both a passive and active access [21]. Operations such as engineering speaker, otdr, individual fibre and user identification were feasible on an 'in-service' basis. In any future optical aether network it will be vital for staff to be able to positively identify the correct fibre before starting any maintenance work. Demonstration systems and feasibility studies had already been reported.

12.4.5 Summary of optical fibre splicing

Splicing requirements: the following list is not meant to be exhaustive, but it does cover the important points:

(1) low splice loss

(2) adequate splice strength, with care splices can be produced which can easily be handled before being re-protected locally

(3) easy assessment of splice loss, so that the operator can quickly reject high-loss splices; this can now be done by means of the local injection and detection (lid) technique described later

(4) suitable techniques for the removal of fibre coatings, which are compatible with the environment in which the splice will take place, and with the skill of the operator

(5) suitable ancillary equipment to prepare the fibre ends and re-protect the splices.

Principal causes of loss in single-mode splices are:

(1) spot size mismatch
(2) misalignment of fibre cores, which can arise from core-cladding non-concentricity or incorrect alignment of the fibres before splicing
(3) angular misalignment
(4) core deformation through the splice
(5) quality of the cleaved end faces
(6) microbending in the splice.

As an example of the type of formula which can be used to predict the loss associated with any one of the above, that for the loss due to mismatch between the profiles of the two fibres is:

$$-10\lg\frac{(2w_1w_2)^2}{(w_1{}^2+w_2{}^2)} \qquad (12.1)$$

where w denotes the spot size and the subscripts 1 and 2 refer to the transmit and receive fibres, respectively. One can show that a 10% difference or mismatch leads to a loss of less than 0.05 dB.

A dramatic example of the sizes of the cores to be jointed is given in the reference, where an 8 μm core is compared to the thickness of ordinary notebook paper of about 50 μm. The joint loss with such fibre dimensions is directly related to the accuracy of alignment, and for a given method the cost increases with the accuracy required. Joint loss requirements are not the same for all application; for instance, losses in the trunk network must be minimised to obtain maximum repeater spacings, whereas in the junction and local networks loss can be traded off against cost or shorter installation times in some optimum manner. The reference gives a review of the design and performance of AT&T splices and connectors which will now be outlined as the latest available information from the largest optical fibre system operator.

Splicing techniques: two basic approaches to splicing lightguides in the trunk and junction networks are used, fusion welding and mechanical jointing, both of which require three basic steps:

(1) preparation of the fibre ends
(2) alignment of the fibres to be jointed
(3) retention of the aligned fibres in position.

In the local area, cheaper and quicker methods are desirable if optical fibres are to compete successfully with copper wires for telephony.

12.4.6 Local injection and detection (LID)

Equipment based on this system makes it possible to align both mm and sm fibres quickly and easily, without the need for remote light launching or detection, and without the need to rely on the dimensions of the fibre for minimising the loss. The Siemens/Corning equipment works on the following local transmission principle. Bends in fibre destroy the guiding properties so that for a sufficiently small bend radius the majority of the light in the core is radiated at the bend and can be detected by a photocell. This effect is reciprocal, i.e. light can be injected through the coloured coating and the cladding glass of a bent fibre into the core. A schematic of the Siecor system is illustrated in Figure 12.14. Any change of the relative positions of the two ends which are to be spliced changes the transmission.

Active alignment techniques have been mentioned above and are often used in both fusion and in mechanical splices in preference to far end transmission monitoring. Local detection can be done by sensing the light transmitted through the splice or sensing the light lost at the splice, and the latter gives a sensitivity 200 times greater than the former. Thus, it is the loss technique which is used in the rotary splice. A further advantage is that local loss detection equipment can easily be calibrated, thus permitting accurate splice loss measurements to be made. The two fundamental processes which are available to inject or couple light into a sm fibre

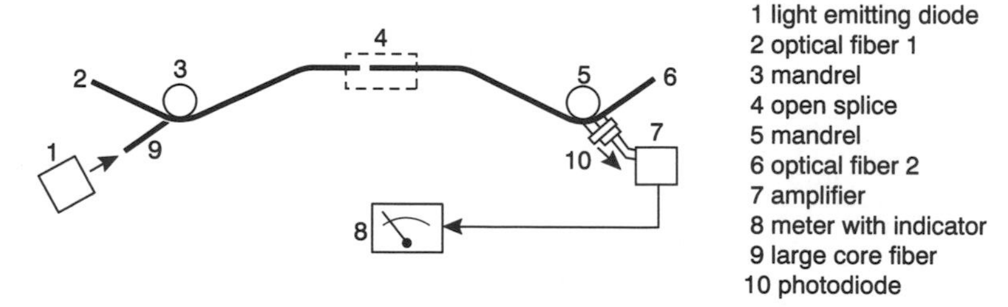

Figure 12.14 Diagram of the local injection and detection LID system ([2] of Chapter 3, Figure 10.6). Reproduced by permission of Siemens

core are microbending and macrobending. These bends couple light from the fundamental LP_{01} mode into the LP_{11} leaky mode, or vice versa. The method preferred by AT&T is local light injection using short-period, small-amplitude microbends, which have several advantages over other possibilities; these advantages are as follows:

(1) the modulated 1300 nm laser source excites the sm fibre properly
(2) more than −35 dBm can be coupled into the fibre, almost totally into the core
(3) bending stresses are kept to below 344.7 N/mm^2
(4) the variation from fibre to fibre is low
(5) the method is compatible with local loss detection techniques.

Among the disadvantages of local injection and detection techniques are the risk of permanent damage to the fibre coating, the risk of breakage and the need for plenty of spare fibre.

Other telecommunication operators have adapted the basic techniques to their own requirements, and some are described in the literature.

12.5 OPTICAL FIBRE CONNECTORS

The jointing of optical fibres covers both splicing and the use of connectors. By the late 1980s the Japanese were perhaps 3 to 5 years ahead of the West, owing to the interesting and impressive advances in jointing technologies they had designed, developed and produced commercially. A contributory factor might well have been the attention devoted to ensuring a supply of production engineers of superior quantity and quality who were highly valued and well managed.

Connectors can be divided into two broad classes, depending upon their principle of operation: butt-coupled (ferrule connectors) and lens-coupled (expanded beam connectors). As the names suggest, the butt coupling requires the two fibre ends to be placed very close together with the axes accurately aligned. The lens coupling on the other hand allows for larger tolerances in separation but introduces more loss. The principles involved are illustrated in Figure 12.15.

The reader should consult a manufacturer's data sheets for information on all aspects of current optical connectors and their performance.

12.5.1 Loss in connectors for single-mode fibres

In studies for UK-standard ofdls in the early 1980s the losses in single-mode connectors were evaluated from standard formulae, with the results listed in Table 12.7.

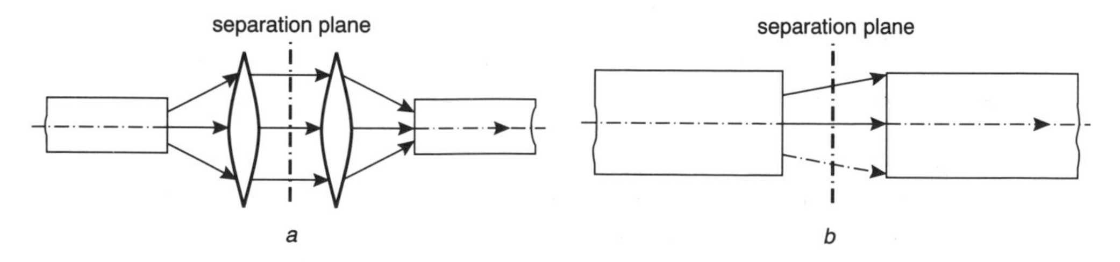

Figure 12.15 Diagrams illustrating the principles of: (*a*) expanded beam connectors, and (*b*) butt-coupled ferrule connectors ([2] of Chapter 3, Figures 10.12 and 10.11). Reproduced by permission of Siemens

Table 12.7 Loss mechanisms in connectors for single-mode optical fibres

Item	Loss (dB)
(1) Fresnel loss ($n_1 = 1.47$) say	0.32
(2) Loss due to tolerances in sm fibres:	
(a) concentricity error (1 μm)	1.1
(b) tol. on core diameter (10%)	0.04
(c) tol. on reference surface diameter (av.2.4%) for 125 μm od fibre (worst case offset of core diameters assumed)	3.0
(3) Loss due to connector tolerances	
(a) lateral misalignment (1 μm)	1.1
(b) axial misalignment	na
(c) tilt loss	0.20

In the years since these values were given, connector technology has advanced considerably with resulting improvements in performance. By the late 1980s, for instance, the optical performance of the widely used ST II sm fibre connectors was quoted as follows:

Insertion loss: mean	0.35 dB
std. dev.	0.2 dB
Loss repeatability	<0.2 dB (500 reconnections)
Temperature range	−40 to 80°C
Temperature stability	±0.1 dB

The lowest insertion loss was achieved by using radiused polished ends (rather than cleaved ends perpendicular to the axis), which reduced the loss by 0.35 dB.

12.5.2 Connectors for 1300 and 1550 nm systems (late 1980s)

By far the most popular configuration was the precision-aligned butt joint with ground and polished end faces. Most of these connectors used secondary and tertiary alignment, i.e. the fibres were positioned and held in supporting ferrules as the ends are ground and polished. The ferrules, which were usually cylindrical or conical, were then aligned in sleeve couplings. To achieve good alignment of the fibre cores required precise diameter and concentricity control of the cylindrical ferrule, or in the case of the conical type, precise taper length and concentricity control of the conical parts.

Connectors have larger losses than splices because they are not tuned and because matching gels were not normally used. Connector performance is strongly affected by wear, damage to end faces of the fibres and the ingress of dirt. Low losses are obtained when the end faces are flat, perpendicular to the alignment surface axis, and in contact in a dry butt joint connection. Under these conditions high performance mm and sm connectors have typical losses in the range 0.3–0.5 dB. If an air gap is present between the end faces, the loss increases by 0.3 dB due to Fresnel reflections at the surfaces (see Table 12.7 earlier). Table 12.7 lists a number of other sources of loss in connectors. The following contributions can be discussed without specifying the shape of the alignment surfaces:

(1) eccentricity between bore and alignment surface axes
(2) diametral clearance between the largest bore and smallest fibre
(3) angular misalignment between bore and alignment surface axes.

Typical limits for these three contributions in high-performance connectors were given as:

(a) sm (1) 1 μm, (2) 2 μm, (3) 0.5°
(b) mm (1) 5 μm, (2) 5 μm, (3) 1°

12.5.3 Transmitted power variations in sm fibre joints with obliquely polished fibre end faces

Optical feedback from reflections caused by index discontinuities at connectors and splices adversely affect laser performance, and should therefore be minimised. Placing an isolator at the laser output reduces optical feedback, but does not solve problems caused by interferometric conversion of phase noise into intensity noise, nor crosstalk in bidirectional systems.

Improved performance requires higher return losses at discontinuities, and various techniques have been tried to achieve this end, for example: close contact between adjacent fibre end faces; index matching fluid between adjacent faces; radiused polished ends referred to earlier; and the use of oblique end faces. The use of the first technique demands very close control of mechanical tolerances; in the second, small departures from the optimum index value causes large variations in return loss, while not showing appreciable insertion loss variations. Use of oblique faces leads to more complex alignment procedures, and can give rise to variations in transmitted power due to small changes in separation between faces causing multiple beam interference.

This problem was the subject of a paper by authors from Bellcore [15]. Multiple beam interference effects at oblique but parallel end faces were studied experimentally and theoretically, and they demonstrated the large variations in power transfer through the joint, despite the excellent return loss. These variations and the average transmitted power both decrease rapidly with increasing separation of two axis-aligned fibres. Theory predicted and experiment confirmed that the amplitude of these variations was independent of the oblique angle, and of the transverse offset between axes between butt-jointed (zero separation) end faces. An appropriate combination of oblique angle and transverse offset can cause the variations to become larger with increasing separation until they exceeded the variations found in normal butt-jointed fibres with a small air gap.

The performance of a joint with obliquely polished end faces at 7° off perpendicular to the fibre axis and aligned axes was investigated for the separation corresponding to the first minimum in transmitted power (hence lowest return loss). Under these conditions the lowest return loss was larger than the 62 dB limit of the otdr measuring equipment. Taking into account other possibilities and imperfections, the authors concluded that the worst-case return loss exceeded 57 dB. We note that a value of 7° was also widely used to reduce facet reflectivity in travelling wave semiconductor laser amplifiers developed in the late 1980s.

12.5.4 Power penalty due to beat noise in connectors

In high-speed systems using single-mode lasers, optical beat noise can be induced by interference between the incident beam and reflections from discontinuities within a connector. This degrades the system sensitivity. One approach adopted to understand and quantify the effect was to treat it as a phase-to-intensity noise conversion. The theoretical and experimental approach adopted in [22] was directed towards deriving an effective rin value. This approach provides a good understanding of the relation between rin value and power penalty, and eases the derivation of permissible and acceptable connector reflectivity.

In the experimental work a dfb laser was followed by an isolator, then through a coupler to the two connectors separated by a 4 m length of fibre. The output of the downstream connector fed a receiver. The maximum reflectivity of the connectors was about 10%, so that the maximum interfering signal was about 1% of the transmitted signal at the receiver input. Photographs clearly showed the beat noise commencing 40 ns into the rectangular pulse waveform at the receiver (the 40 ns delay corresponds to the light transmit time through the 4 m of fibre between the two connectors).

The theoretical approach started from the maximum receiver output which occurred when both inputs are in phase and have the same polarisation, giving:

$$P_{\mathrm{BM}} = 2I_{\mathrm{S}}I_{\mathrm{R}} \tag{12.2}$$

where P_{BM} is the maximum beat intensity, I_{S} is the photocurrent corresponding to signal, and I_{R} is the photocurrent corresponding to reflected input.

The transmitted and reflected beam photocurrents are related by:

$$I_{\mathrm{R}} = R_1R_2 \times I_{\mathrm{S}} \tag{12.3}$$

The effective relative intensity noise density in dB/Hz within the receiver bandwidth df is given by:

$$-20\log\frac{P_{\mathrm{BM}}}{(I_{\mathrm{S}}{}^2/\delta f)} \tag{12.4}$$

where δf is the fwhm bandwidth of the laser. Owing to chirping, $\delta f \gg df$ and it was assumed that the beat noise spectrum in the receiver bandwidth df was flat. The theoretical rin as defined above was plotted against the product of the reflectivities, R_1R_2, as shown in Figure 12.16. The measured values follow the theoretical curve for 0.1 nm laser spectral width well.

The beat noise spectra were measured for 90–100% sinusoidal modulation at 1 GHz and for 1010 pulse modulation at about 2.4 Gbit/s. The dependence of the rin on bit rate and number of connectors is illustrated in Figure 12.17.

The peaks in the two spectra correspond to the transmitted and reflected beams interfering in phase; thus the spectra show minimum spread and maximum noise power density at each modulation frequency corresponding to a peak. As one would expect, the beat noise increases when the number of connectors increases, because the number of beat combinations increases. With two connectors there is one combination, with three there are three (5 dB noise increase), and with four connectors the theoretical increase in beat noise is 8.5 dB higher than with two connectors.

To determine the power penalties due to beat noise, ber measurements were performed at 2.412 Gbit/s and at 595 Mbit/s with nrz and rz format $2^{15} - 1$ prbs. The higher of these two rates was chosen because it gave the maximum values of

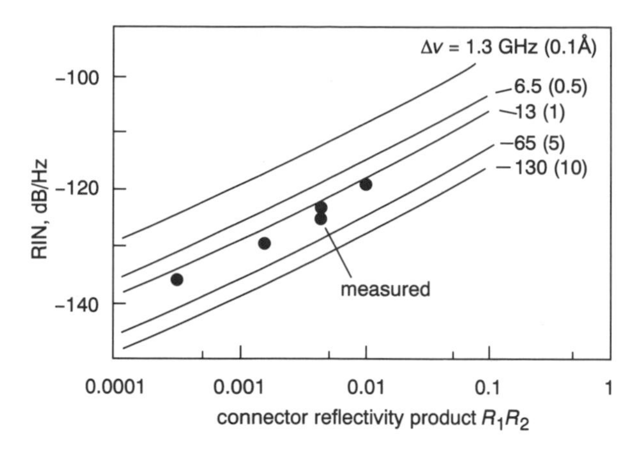

Figure 12.16 RIN versus value of product R_1R_2 with source linewidth as parameter; curves are theoretical values, dots are measured values. Laser spectral linewidths in GHz(nm) ([22], Figure 2). Reproduced by permission of IEE

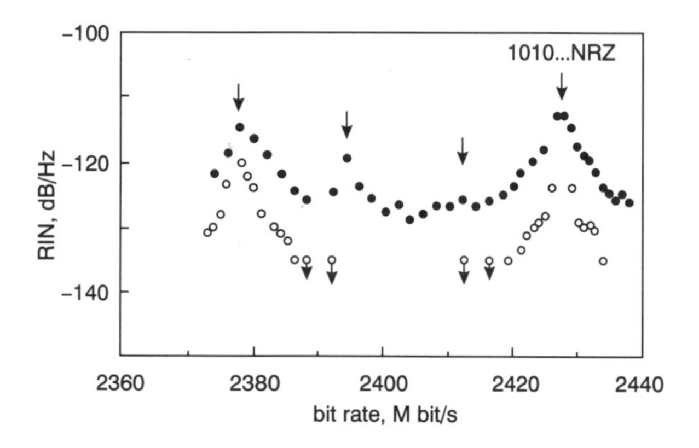

Figure 12.17 RIN versus bit rate for 1010 nrz pulse modulation for two and three connectors in tandem ([22], Figure 3). Reproduced by permission of IEE

rin, thus a worst-case condition. Experimental results gave rin levels of −120 and −113 dB/Hz with 2.4 Gbit/s nrz sequences; these rin values correspond to penalties of 0.4 and 2.0 dB, respectively. The agreement between measured and theoretical values can be read from Figure 12.18.

Discrepancies between theoretical curves and experimental points were attributed to the simplicity of the assumptions underlying the calculation. To ensure a penalty of less than 0.1 dB at 2.4 Gbit/s from beat noise required that the rin must remain below −135 dB/Hz.

From their results the authors suggested that the reflectivity of individual connectors should not exceed 0.01 and the reflectivity product for two connectors should not exceed 0.0001, i.e. return losses of better than 20 and 40 dB, respectively. These performance targets are achieved in commercial connectors in which the two fibre ends are in contact. As the penalty rises steeply for more than two connectors, the authors recommended that not more than two should be used if the above conditions were to be satisfied.

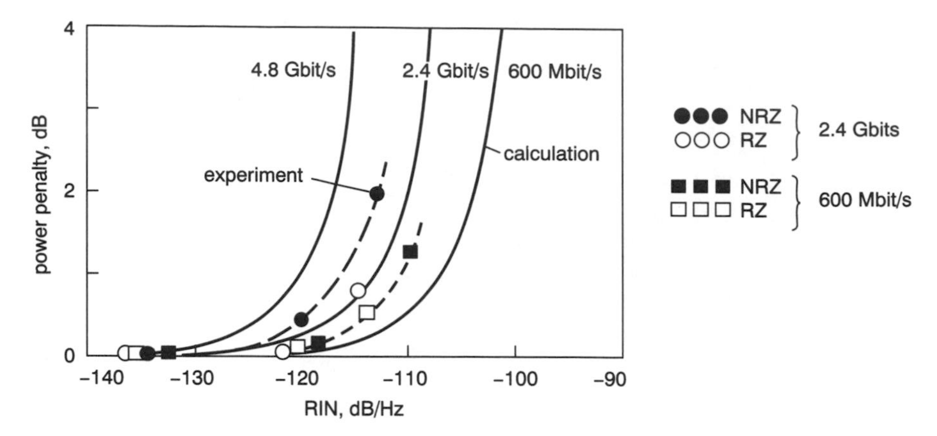

Figure 12.18 Power penalty versus rin with bit rate as parameter: full curves theoretical, points and dash curves measured values ([22], Figure 4). Reproduced by permission of IEE

12.5.5 Plug-in connectors

Low insertion and high return loss connectors of small size are required to connect printed circuit boards to back panels in equipment such as the transmitter/receiver in synchronous digital hierarchy (SDH) systems. Such equipment also needs many electrical interconnections for wideband digital signals and to distribute clocks through a back panel to a pcb. Clearly maintenance and testing are facilitated if electrical and optical connections are made simultaneously. These operational requirements were realised by the design and development of two types of sm plug-in connectors: one with two plugs, two jacks and 64 electrical terminals; the other with four plugs and four jacks. The design concept and structural features of both connectors were described in a paper from NTT [23]. Another connector, known as SC-type, had been described in a 1989 issue of the same journal [24].

Design: the concept was summarised in three statements of requirements which the new connector was designed to satisfy; an insertion loss below 0.5 dB and return loss exceeding 25 dB formed the third requirement. A description of the structural features illustrated by figures and photographs of the new connector followed. The next section dealt with the alignment mechanism. Butt-coupled sm fibres must be aligned axially to better than 1.6 μm to achieve less than 0.5 dB insertion loss, and this must be done in the context of printed board mounting jack tolerances of about 1 mm in both radial and axial directions in a plug-in unit. The primary requirement to achieve high performance is to ensure accurate alignment between plugs and jacks in which the tolerances and misalignments are automatically absorbed. This was achieved by mechanisms which ensured accurate radial and axial alignment, both of which were discussed. The third part of the first section dealt with adjustment.

Characteristics: this section was divided into four parts: insertion loss and return loss, repeatability, mechanical and environmental test, and transmission tests.

Insertion and return losses: both were measured using a 1300 nm laser source and the resulting values obtained on a 10/125 μm sm fibre are summarised in Table 12.8 and in Figure 12.19. Histograms of both losses are given in Figure 12.19 for a set of randomly concatenated sm fibres.

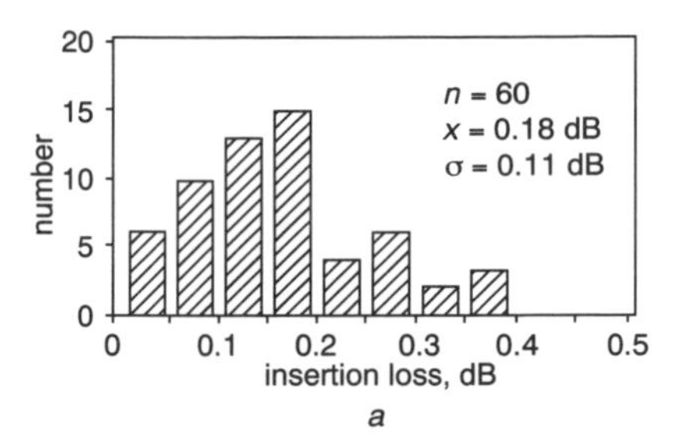

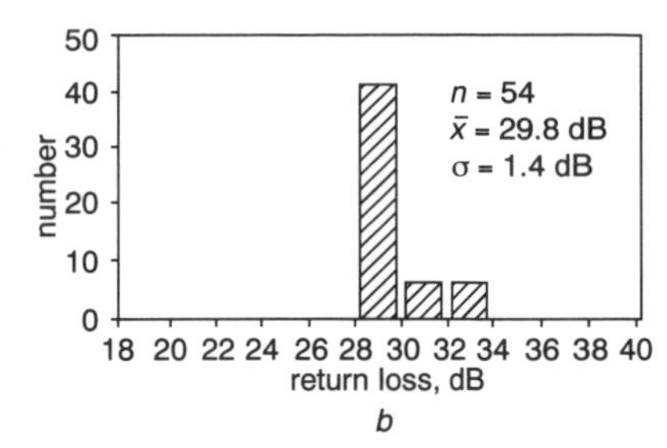

Figure 12.19 Histograms showing (*a*) insertion loss and (*b*) return loss of assembled plug-in connectors and 10/125 μm sm fibre with randomly concatenated sm fibres ([22], Figure 11). Reproduced by permission of IEEE, © 1988

Insertion and return losses were also measured for the worst-case condition in which the printed board housing was 1.2 mm from the front and back ends of the housings of the back panel. No change in average value of either type of loss between the random and the worst-case condition was seen.

Repeatability: 200 insertions of a printed board into a back panel and 500 matings of a plug into a back end housing were performed. The results are summarised in Table 12.7 and in Figures 12.20 and 12.21. During these tests the return loss never fell below 25 dB and insertion loss changes were smaller than 0.2 dB.

Mechanical and environmental tests: temperature and humidity cycles were applied and the results are given in Figures 12.22 and 12.23, respectively, and are also listed in Table 12.8, together with vibration and shock test results. All the results revealed that the connector characteristics were extremely stable under the conditions of the tests.

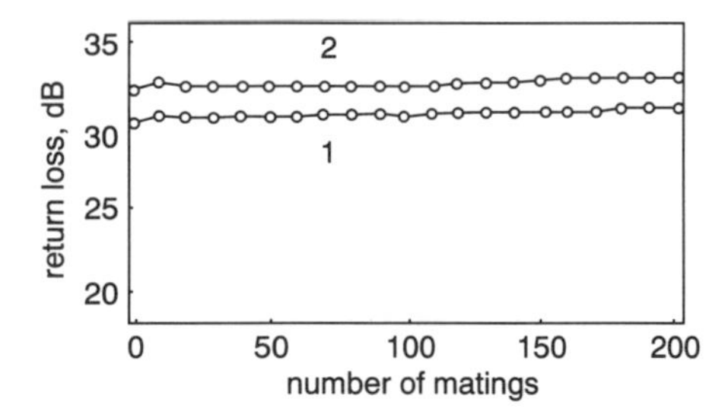

Figure 12.20 Return loss versus number of matings between pcb and back panel showing repeatability ([23], Figure 12). Reproduced by permission of IEEE, © 1990

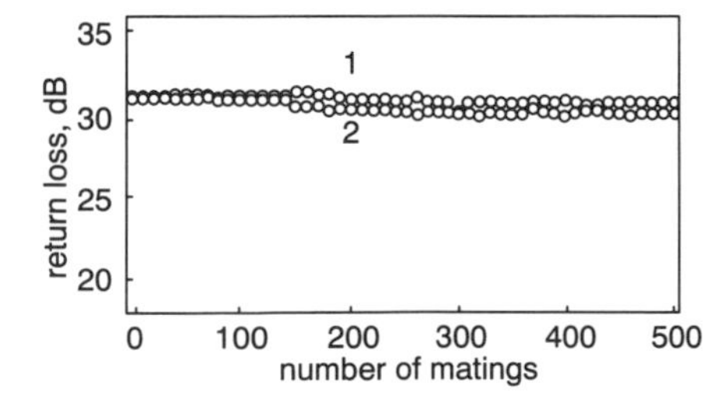

Figure 12.21 Return loss versus number of matings between plug and back-end housing showing repeatability ([23], Figure 13). Reproduced by permission of IEEE, © 1990

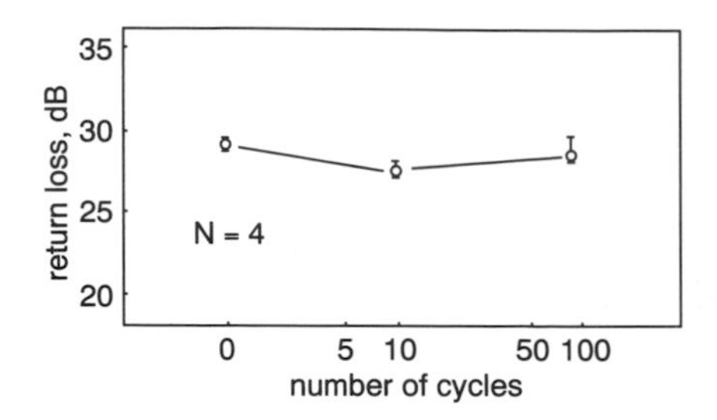

Figure 12.22 Return loss versus number of temperature cycles ([23], Figure 14). Reproduced by permission of IEEE, © 1990

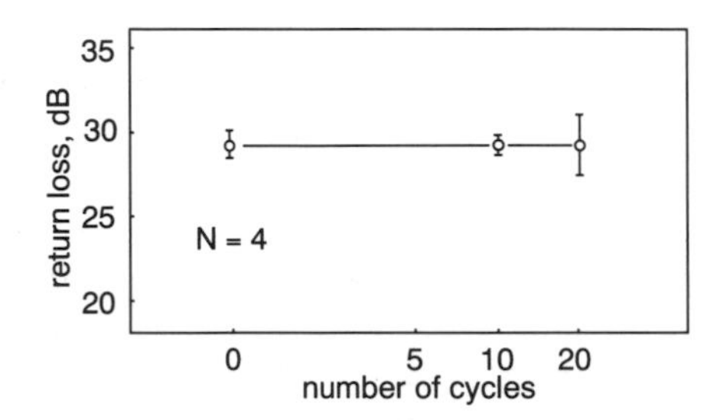

Figure 12.23 Return loss versus number of high-humidity temperature cycles, with error bars shown ([23], Figure 15). Reproduced by permission of IEEE, © 1990

Table 12.8 Characteristics of sm plug-in connectors ([23], Table 1). Reproduced by permission of IEEE © 1990

Items		Conditions	Results
Optical characteristics	Insertion loss	$\lambda = 1.3$ μm n = 60	Ave. 0.18 dB σ 0.11 dB
	Return loss	$\lambda = 1.3$ μm n = 54	Ave. 29.8 dB σ 1.4 dB
Mechanical tests	Mating	500 (plug element) 200 (printed board)	$\Delta L_i < 0.2$ dB* $L_r > 25$ dB
	Vibration	10 ~ 55 Hz, Amplitude 1.5 mm 2 hours × 3 directions	〃
	Shock	100 G, 6 ms duration 10 shocks × 3 directions	〃
Environmental tests	Temperature cycling	−25 ~ 70°C, 4 hours/cycle 100 cycles	〃
	Thermal ageing	85 ± 2°C 960 hours	〃
	Temperature & humidity cycling	−10 ~ 25 ~ 65°C, 93 ± 3% R.H 1 cycle/day, 20 cycles	〃

*ΔL_i; Insertion loss increase
L_r; Return loss

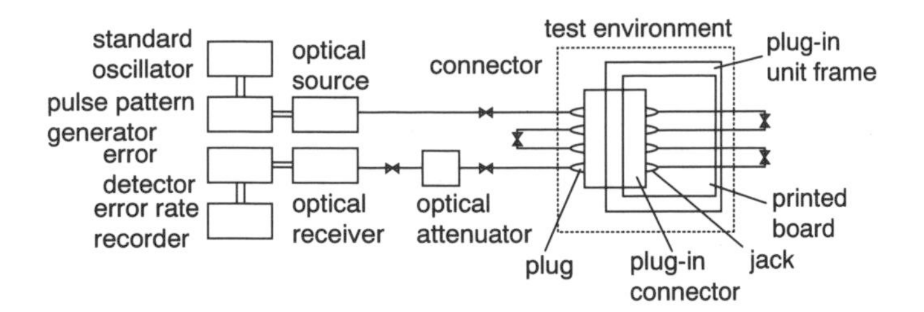

Figure 12.24 Arrangement of equipment for transmission tests measurements ([23], Figure 16). Reproduced by permission of IEEE, © 1990

Table 12.9 Transmission test conditions and results ([23], Table 2). Reproduced by permission of IEEE © 1990

Items	Conditions	Results
Temperature cycling	$-25 \sim 70$°C, 4 hours/cycle 100 cycles	
Temperature & humidity cycling	$-10 \sim 25 \sim 65$°C, 93 ± 3% R.H. 1 cycle/day, 20 cycles	
Thermal aging	85 ± 2°C 96 hours -25 ± 2°C 96 hours	no error (error rate less than 10^{-11})
Vibration	$10 \sim 55$ Hz, amplitude 1.5 mm 2 hours × 3 directions	
Shock	100 G, 6 ms duration 10 shocks × 3 directions	

Transmission tests: the arrangement shown in Figure 12.24 was adopted to make measurements of ber under the conditions listed in Table 12.9. Six pairs of plugs and jacks were connected in tandem and a 1.8 Gbit/s, $2^{11} - 1$ prbs rz format signal was applied. A 3 dB power margin was set from the minimum system level. Table 12.8 indicates that error rates were below 10^{-11}.

12.6 REFERENCES

1 ROUTLEDGE, R.: '*Discoveries and Inventions of the Nineteenth Century*, George Routledge & Sons, London, 1876.

2 NELSON, B.P., HORNUNG, S.: 'Transmission characteristics of a cabled and jointed 30 km monomode optical fibre link'. Electronics Lett., **18**, no. 6, 18 Mar. 1982, pp. 270–2.

3 GARTSIDE, C.H., PANUSKA, A.J., and PATEL, P.D.: 'Single-mode cable for long-haul, trunk, and loop networks'. *AT&T Tech. J.*, **66**, no. 1, Jan./Feb. 1987, pp. 84–94.

4 FOORD, S.G., and LEES, J.: 'Principles of fibre-optical cable design'. *Proc. IEE*, **123**, no. 6, June 1976, pp. 597–602.

5 KAISER, P., and ANDERSON, W.T.: 'Fiber cables for public communications: state-of-the-art technologies and the future'. *J. Lightwave Technol.*, **LT-4**, no. 8, Aug. 1986, pp. 1157–1165.

6 CASSIDY, S.A., and REEVE, M.H.: 'A radically new approach to the installation of optical fibre using the viscous flow of air'. *Br. Telecom Technol. J.*, **2**, no. 1, Jan. 1984.
7 GRIFFIOEN, W.: 'The installation of conventional fiber-optic cables in conduits using the viscous flow of air'. *J. Lightwave Technol*, **7**, no. 2, Feb. 1989, pp. 297-302.
8 RUSH, J.D., BEALES, K.J., COOPER, D.M., DUNCAN, W.J., and RABONE, N.H.: 'Hydrogen related degradation in optical fibres — system implications and practical solutions'. *Br. Telecom Technol. J.*, **2**, no. 4, Sept. 1984, pp. 102-111.
9 TANAKA, S., and HONJO, M.: 'Long-term reliability of transmission loss in optical fibre cables'. *IEEE J. Lightwave Technol.*, **6**, no. 2, Feb. 1988, pp. 210-217.
10 IINO, A., MATSUBARA, K., OGAI, M., HORIUCHI, Y., and NAMIHIRA, H.: 'Diffusion of hydrogen molecules in fluorine-doped single-mode fibres'. *Electronics Lett.*, **25**, no. 1, 5 Jan. 1989, pp. 78-79.
11 KUWAZURU, M., NAMIHIRA, M., MOCHIZUKI, K., and IWAMOTO, Y.: 'Estimation of long-term transmission loss increase in silical-based optical fibers under hydrogen atmosphere'. *J. Lightwave Technol.*, **6**, no. 2, Feb. 1988, pp. 218-225.
12 ANDERSON, J.M., FREY, D.R., and MILLER, M.: 'Lightwave splicing and connector technology'. *AT&T Tech. J.*, **66**, no. 1, Jan/Feb 1987, pp. 45-64.
13 SERAFINI, E.: 'Statistical approach to the optimisation of optical fibre fusion splicing in the field'. *J. Lightwave Technol.*, **7**, no. 2, Feb. 1989, pp. 431-435
14 YOUNG, W.C., SHAH, V.A., and CURTIS, L.: 'Optimisation of return loss and insertion loss performance of single-mode fiber mechanical splices'. *Proc. 37th IWCS*, Reno, NV, 1988.
15 SHAH, V.A., CURTIS, L., and YOUNG, W.C.: 'Transmitted power variations in single-mode fiber joints with obliquely [polished endfaces'. *J. Lightwave Technol.*, **7**, no. 10, Oct. 1989, pp. 1478-1483.
16 ISHIKURA, A., KATO, YASIYUKI., and OOYANAGI, T.: 'Mass splice method for single-mode fiber ribbons'. *J. Opt. Commun.*, **10**, no. 2, 1989, pp. 61-66.
17 ISHIKURA, A., KATO, Y., and OOYANAGI, T.: 'Loss estimation in mass splicing of single-mode fiber ribbons'. *J. Opt. Commun.*, **10**, no. 2, 1989, pp. 56-60.
18 ISHIKURA, A., AKTO, Y., OOYANAGI, T., MIYAUCHI, M.: 'Loss factors analysis for single-mode fiber splicing without core axis alignment'. *J. Lightwave Technol.*, **7**, no. 4, Apr. 1989, pp. 577-583.
19 THOMAS, C.D., MARR, D., FREEMAN, R.A., and JENKINS, P.D.: 'The jointing of submarine optical fibre cable'. *Br. Telecom Technol. J.*, **3**, no. 1, Jan. 1985.
20 COCHRANE, P., and O'MAHONEY, M.: 'Future demands in network management'. *Coll. Digest 1989/5*, The Changing Face of Telecommunication Transmission, IEE, London.
21 JAMES, S.M.: 'Non-intrusive optical fibre identification using a high-efficiency macro-bending 'clip-on' optical component'. *Electronics Lett.*, **24**, no. 19, 15 Sept. 1988, pp. 1221-1222.
22 SHIKADA, M., and HENMI, N.: 'Evaluation of power penalty due to beat noise induced by connector reflection'. *Electronic Lett.*, **24**, no. 18. 1 Sept. 1988, pp. 1126-1128.
23 IWANO, S., SUGITA, E., KANAYAMA, K., NAGASE, R., and NAKANO, K.: 'Design and performance of single-mode plug-in type optical-fiber connectors'. *IEEE/OSA J. Lightwave Technol.*, **8**, no. 11, Nov. 1990, pp. 1750-1755.
24 SUGITA, E., NAGASE, R., KANAYAMA, K., and SHINTAKU, K.: 'SC-type single-mode fiber connector'. *IEEE/OSA J. Lightwave Technol.*, **7**, no. 11, Nov. 1989, pp. 1689-1696.

FURTHER READING

25 REINAUDO, CH.: 'Undersea cables; state-of-the-art technology'. Electrical Communication, 1st quarter 1994, pp. 5-10.

26 BONICEL, J,-P.: 'Optical fiber ribbon cable technology'. Electrical Communication, 1st quarter 1994, pp. 45–51.
27 HUNWICKS, A.R., RUSHER, P.A., BICKERS, L. STANLEY, D.: 'Installation of dispersion-shifted fibre in the British Telecom trunk network'. Electronics Lett., **24**, no. 9, 28 Apr. 1988, pp. 536–537.

13

GENERAL SYSTEM CONSIDERATIONS

In the order of the universe
Time and space are the two-faced god
Inseparable
Like the two sides of the same coin
The twin routes
Through which all things must pass
Through which knowledge travels and traverses
Is received and recorded and answered
To pave the way for things
We've neither dreamt about
Nor yet know of
By the shaping and reshaping of matter
And the magnification of the human mind.

(Author unknown)

13.1 INTRODUCTION

Earlier chapters have described in detail many systems-related aspects of the transmitters, receivers and cables which form the three main subdivisions of a transmission system. The performance and parameters of individual types of components, their minimum, typical and maximum values in some cases, mean and standard deviations in others, are all part of the data which are needed in order to design systems and predict their performance. Some examples of such data have been supplied which give values in specific cases, but in practice data from manufactures on current commercially available components, supplemented by further characterisation in the laboratory or in the field if necessary, would be used.

This chapter deals with some system aspects of optical fibre digital line systems (ofdls), and will introduce a number of general topics in appropriate depth. Attention will be focused on 1300 nm systems, but third-generation systems at 1550 nm will also be mentioned, as they entered service in the late 1980s. First applications operating at 1550 nm were in submerged links such as the Ireland–UK system mentioned later, and will become more widespread when optical fibre amplifiers operating in the 1550 nm band became commercially available and made their advantages overwhelming.

By the mid-1980s more than 400 ofdls had been installed in the British Telecom trunk and junction networks, and in the following ten years the number grew enormously; this growth pattern was repeated in many countries. Up until the early 1990s they were based on the European or the North American CCITT plesiochronous digital hierarchies, since then on the world standard SDH. In the UK, and in particular in the BT network, systems operating at 140 Mbit/s provided the optimum solution for the trunk network during the 1980s.

In 1988 the longest unrepeatered 140 Mbit/s system anywhere in the world at that time was installed between Ireland and the UK, and it could not have been implemented using coaxial cable technology. This system was accepted from the French manufacturer Submarcom, on 31 May 1988, and it has carried live traffic since September that year. An excellent description appeared in [1]. This article outlined the planning, surveying, construction, installation, overall testing and performance of the system. BT-TE No. 1 Submarine Cable System provided the major interconnection between the telephone networks of the two countries.

Systems are installed to provide the following network requirements:

(a) additional capacity in national networks or between national networks
(b) system diversity between networks
(c) a wider range of services by means of a higher grade digital bearer
(d) route diversification
(e) improved security.

Application of a given system operating at a particular bit rate is a function of many parameters, including economics of provision, growth rates, nature of traffic to be carried and any operational factors. The AT&T approach serves as an example. AT&T determined, in the mid-1980s, that the best long-term strategy for ofdls was the development of a single-mode technology, with a carefully orchestrated upgrade and evolution plan [2]. From this plan the FT Series G 417 Mbit/s and 1.7 Gbit/s products were born, and a substantial portion of the development proceeded in parallel. The first 417 Mbit/s system entered service in January 1986, and the upgrade to 1.7 Gbit/s some two years later. The system was of modular construction and the 417 Mbit/s version could provide up to nine two-way 44.736 Mbit/s (DS3) streams per fibre pair. Field experience with this technology was excellent, with virtually error-free performance, except for bursts caused by protection switching operation. Data from circuit-pack field-tracking reliability studies has shown that failure rates are well within the predicted limits. The receiver for the Gbit/s systems used a Ge apd of proven reliability from Japan, at least initially, but an intensive development effort in AT&T was directed towards an InGaAsP apd replacement (see Chapter 10). The technology for the 417 Mbit/s system enabled AT&T to make a giant stride in its quest for the ubiquitous high-performance digital transmission capability.

13.2 OPERATIONAL REQUIREMENTS

Before looking at systems in broad outline we consider briefly an earlier stage in the complete process of providing any telecommunication service to business and domestic customers. This example from the early 1980s refers to types of advanced service which are or will be of interest in the 1990s.

The first step in the design of any transmission system is the definition of the operational requirements; that is, a statement of the users' needs in terms of services they require (and are willing to pay for) and of the constraints of significance to the providers of the service. For telephony, these have been established for many years. However, the unlimited bandwidth of single-mode optical fibres has made the provision of a much wider range of services possible and probable; for instance, the broadband integrated services digital network (BISDN). The operational requirements for advanced services are still the subject of international investigation and discussion. The definition of the operational requirements (ORs) is an essential prelude to the remaining steps in the design of a transmission system, namely:

(a) the derivation of an outline system specification in terms of technical, rather than operational, parameters such as bandwidth, reach and so on
(b) the establishment of cost and reliability targets
(c) the identification of candidate systems, or system principles, which appear to be capable of meeting these requirements
(d) comparison of their technical and economic aspects
(e) selection of a system or family of systems for development, and evaluation of an equipment specification.

As an illustration, we list some headings of the operational requirements for broadband optical subscriber systems as seen in the early 1980s, which are becoming part of telecommunication networks in the 1990s. The traffic to be carried by such systems includes audio, visual, and digital data which will have been supplied via, or bound for, the PSTN. Other information, e.g. entertainment may be from other sources. The operational requirements cater principally for the domestic customer but would also meet the needs of businesses.

Modes of use

The system at its simplest will provide a direct substitute for the metallic link to the local exchange. It could also provide a broadband link to the centre of a community cluster. Two-way usage of the fibre is highly desirable.

Flexibility

The wide variety of services such a system could supply requires great flexibility, combined with provision that is as simple and economical as possible.

Environment and reliability

The total environment in which the system must operate needs separate and detailed study, and will tend to vary from one country to another due to different organisational, legal, social as well as climatic conditions. At the customer's premises it will have to be user-friendly, yet robust and very reliable.

Electrical power and emergency operation

The subscriber terminal (optical transmitter/receiver, electrical line signal generator/receiver and electrical multiplexer/demultiplexer) must demand the minimum of electrical power, as must the primary communication terminal, the telephone handset. It is virtually essential that the telephone service should remain usable during periods of mains power failure.

Communication facilities

Apart from at least one speech circuit, a visual telephone service may become attractive, both of these must be two-way services. Text transmission may be provided, including facsmile. A wide variety of transactions may be implemented.

Entertainment

Up to four simultaneous television channels and perhaps two high-quality audio channels may be sufficiently attractive to domestic subscribers to become commercially viable.

Education and games

The provision of educational services via the broadband link, e.g. the Open University in the UK, either non-interactively as at present or interactively. TV games could perhaps develop into two basic types: against another subscriber, or group of subscribers; and against a central machine.

These last paragraphs have indicated just some of the many considerations which form the operational requirements in telecommunications. Whatever the eventual decisions about such systems, it was thought likely that the following technical aspects, among others, would be important:

(a) simple and cheap forms of optical multiplexers
(b) cheap and efficient optical hybrids/directional couplers, allowing two-way transmission on a single fibre
(c) similar couplers allowing channels to be dropped or inserted *en route*
(d) cheap, flexible and versatile electrical multiplexers in modular form
(e) operation in the event of mains power failure
(f) reliability.

Let us now turn from operational requirements to an outline of current 140 Mbit/s optical fibre PDH systems in the plain old telephone service (POTS) which still forms the vast bulk of the traffic.

13.3 OVERVIEW OF AN OPTICAL FIBRE DIGITAL LINE SYSTEM

The basic design, which must be agreed between operators and manufacturers, includes:

(1) choice of type of system, e.g. optical landline, optical submerged, satellite, radio relay
(2) for optical systems the number of fibre pairs (ofdls)
(3) unrepeatered or repeatered; if repeatered, regenerative or optical amplifier
(4) initial circuit capacity (64 kbit/s channels) and possibility of upgrades
(5) number of ofdls (i.e. fibre pairs) to be equipped (including standby) in the transmission system.

In the basic arrangement an ofdls closely resembles its technology predecessor, coaxial cable systems, but there are a number of differences in the implementation of the two technologies.

When the basic system design is finalized the parameters involved in cable routeing can be established. This is relatively simple in the case of landline systems, but much more complicated in the case of submerged systems. A number of papers have been published which describe this system aspect in some detail. Overall route planning will not be considered further.

Four categories of design parameters which must be considered and evaluated can be identified, and are associated with the following:

(1) overall route
(2) type of optical fibre
(3) cable electrical and mechanical data
(4) optical fibre digital line system data.

Items (1) and (2) cover attenuation and dispersion, and thus the choice of operating wavelength and type of source. (3) depends on the type of system (inland, submerged, aerial).

Item (4) is of direct interest in this chapter, and the main parameters of one such system are listed in Table 13.1. A more detailed list of parameters for the optically amplified 5 Gbit/s TAT-12 system is supplied in Chapter 15.

Table 13.1 Main parameters of 140 Mbit/s optical fibre digital line system ([1], Table IV). Reproduced by permission of John Wiley

Number of systems	2 + 1 hot stand-by
System bit rate	140 Mb/s (139.264 Mb/s) — CCITT G703
System capacity	1920 circuits
System MTBF — years	13.7
Traffic interface code	CMI
Optical line code	5B/6B
Optical line bit rate	167 MBaud
Optical source	Laser diode O/P level −2.5 dBm
Optical bandwidth, nm	2.5
Optical detector	APD III V (InGaAsP) Sensitivity ≤ -44.5 dBm Overload −12 dBm
Bit error rate (BER)	10^{-9} (CCITT G821)
Jitter	Complies with CCITT G823 and 921

13.3.1 Terminal transmission equipment (TTE)

A simplified block diagram of the traffic path terminal equipment for a 2 × 140 Mbit/s short-haul system, using UK-Belgium No. 5 as an example, is shown in Figure 13.1. The 140 Mbit/s line interface is in accordance with CCITT Recommendation G.703. ([8] of Chapter 9) The design requirements of this system were discussed and the implementation of the design is detailed in [3].

Transmit terminal equipment

A principal function of this equipment is to translate the electrical input signal into a format suitable for transmission over the system fibres. The principal functions are outlined, starting with the incoming channels through to the transmit optical data.

CMI binary decoder: this converts the incoming channel cmi bit stream to binary, and extracts the clock pulses.

Scrambler: this spreads the energy of the signal spectrum evenly across the frequency band, irrespective of the data sequence being transmitted. It ensures that there are ample transitions in the received signal from which timing signals can be recovered by preventing the occurrence of repetitive sequences of marks (ones) or spaces (zeros), to make the transmitted data sufficiently random for the receiver to function correctly [4].

7B8B Line encoder: this introduces the following properties regardless of the structure of the input signal: limited power at low frequencies; limited power at high frequencies; the sequence must be suitable to maintain circuits with agc at the correct level; ease of translation between input signal and line code; and some means of providing in-service monitoring. A balanced-disparity binary block code such as 5B6B or 7B8B translates each n-bit traffic block into a $(n+1)$-bit line block using a code translation table, thus giving an encoded line signal of slightly higher rate. Other codes are sometimes used, e.g. 24B1T on TAT-8 and later submerged systems.

Laser modulator: this comprises a laser drive circuit, control circuits which provide bias and modulating current for the laser source, and appropriate control circuits to minimise the effects of the more variable environmental conditions found on inland systems. This could include the use of a Peltier cooler to actively regulate the laser to a stable temperature of $27 \pm 1^{\circ}$C to minimise mode hopping. The encoded line signal modulates the bias current for the laser directly, which in turn produces the corresponding intensity-modulated transmit optical data as on–off modulation of light. Matching the optical source to the system fibre with a minimum coupling loss is necessary. The off state usually corresponds to a bias level just below threshold, and hence to a small fraction of the light output when in the on state.

Receive terminal equipment

The functions required in the receive path are largely complementary to those in the transmit path, so only the principal differences will be described.

Optical detector: this demodulates the received modulated optical carrier to provide an output current proportional to the modulation. The photodiode (apd or pin) and front-end amplifier are often realised as an ic. A high-gain agc amplifier is employed to raise the signal to the correct level, and together with internal control

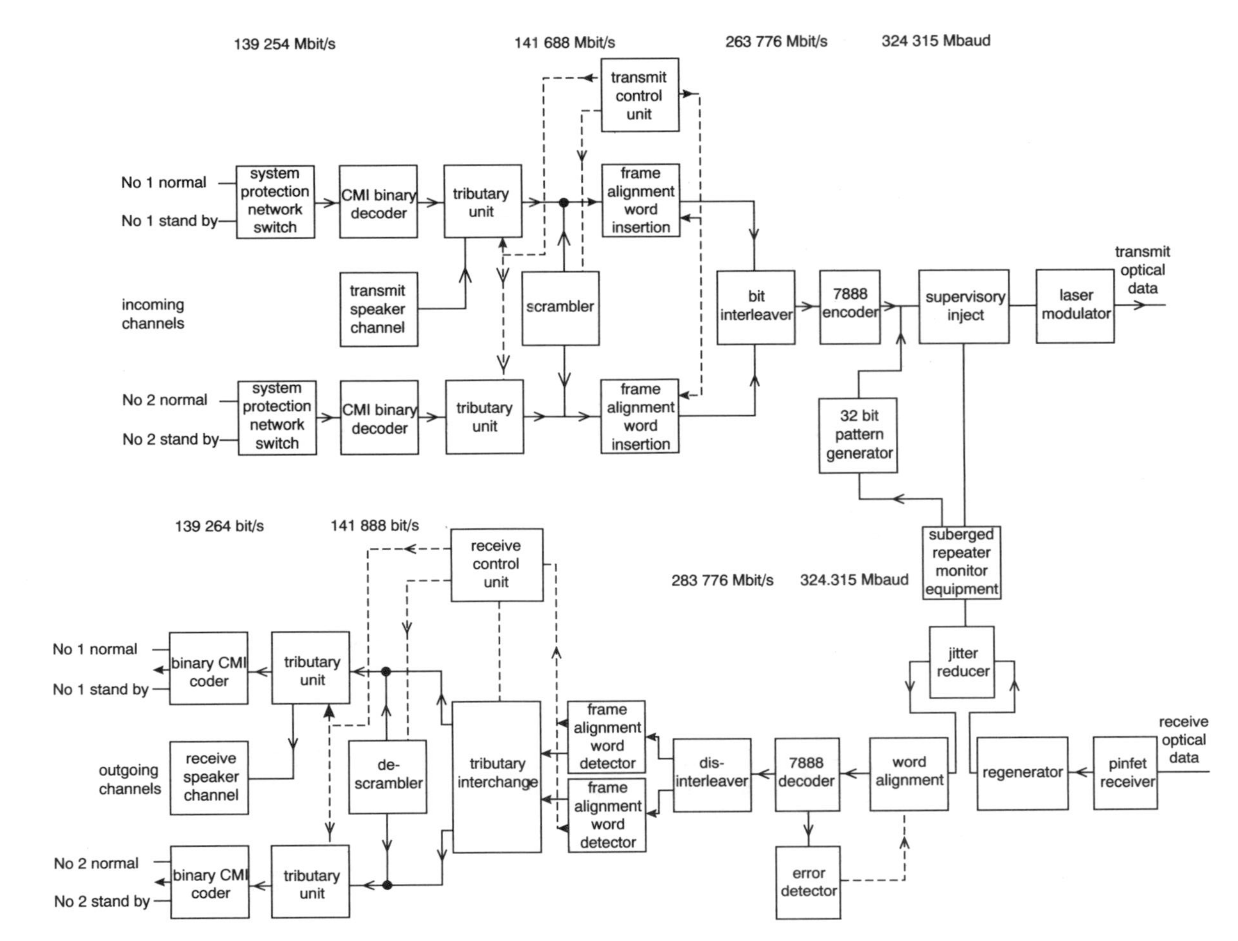

Figure 13.1 Block diagram of terminal transmission equipment (Channon, Figure 11)

of the photodiode current it provides the required input sensitivity and dynamic range over the life of the system.

Regenerator: this retimes, reshapes and regenerates the line signal by means of a master–slave bistable driven by the clock signal extracted from the incoming signal. The serial data are reassembled into parallel 8-bit words before passing to the decoder.

7B8B decoder: this translates $(n + 1)$-bit blocks back to the original n-bit traffic blocks in accordance with complementary alphabet (conversion table). An alignment circuit provides an error output which is connected to the error monitoring circuits.

Disinterleaver: the decoded data are serialised and divided into two tributary paths. Frame alignment errors on the system are detected, and frame timing and control signals are generated for the ongoing decoding process.

Descrambler: this is reset to obtain synchronism with the scrambler at the transmit terminal. Dejustification of the data stream is carried out and the frame alignment words and justification bits are removed. The resultant data is a complete replica of the incoming binary data and at the same rate.

Binary cmi coder: this encodes the 140 Mbit/s stream into cmi format for onward transmission over the repeater station distribution to the next line system or multiplex.

In addition to the terminal traffic-path equipment, provision is also made for terminal alarms, error-rate monitoring, power feed and supervisory and speaker subsystems. Some of these functions are also found in repeater equipment.

13.3.2 3R regenerative repeater equipment

This consists essentially of a receive and a transmit terminal (without line coder and decoder) back-to-back. Intermediate regenerative repeater stations are required to reshape, retime and regenerate a noise-free replica of the transmitted signal to send down the line. The process has to be performed such that the overall system error rate over a 2500 km route would be better than 2×10^{-7}, which corresponds to less than one error in 10^{10} bits per km. The error rate achievable in a repeater is essentially a function of the received power level. On the other hand, the sensitivity is determined by the noise performance of the optical receiver front end and the following repeater transfer function, as explained in Chapter 11.

Another example from a real system will be used, in this case the regenerator from the STC segment of the TAT-8 system ([7] of Chapter 9). A diagram showing the main transmission path is given in Figure 13.2.

Optical receiver: this comprises a pin photodiode and low-noise silicon bipolar ic front-end transimpedance amplifier. Both chips are mounted in a single package fitted with a sm fibre tail to form the receiver module. The output voltage level is appropriate for the agc amplifier.

AGC amplifier: this contains a variable gain preamplifier ic followed by a main amplifier ic, giving a combined gain of 52 dB and an unequalised 3 dB bandwidth of 320 MHz. A peak detector provides a dc control voltage for agc and timing impulses are supplied by a rectifier to a saw filter.

Timing extraction: the high Q (800) transversal saw filter produces the 295.6 MHz sinusoid (the line rate is 295.6 MBaud) which drives a single chip limiting amplifier. The output is a uniform clock signal and is supplied to the decision circuit.

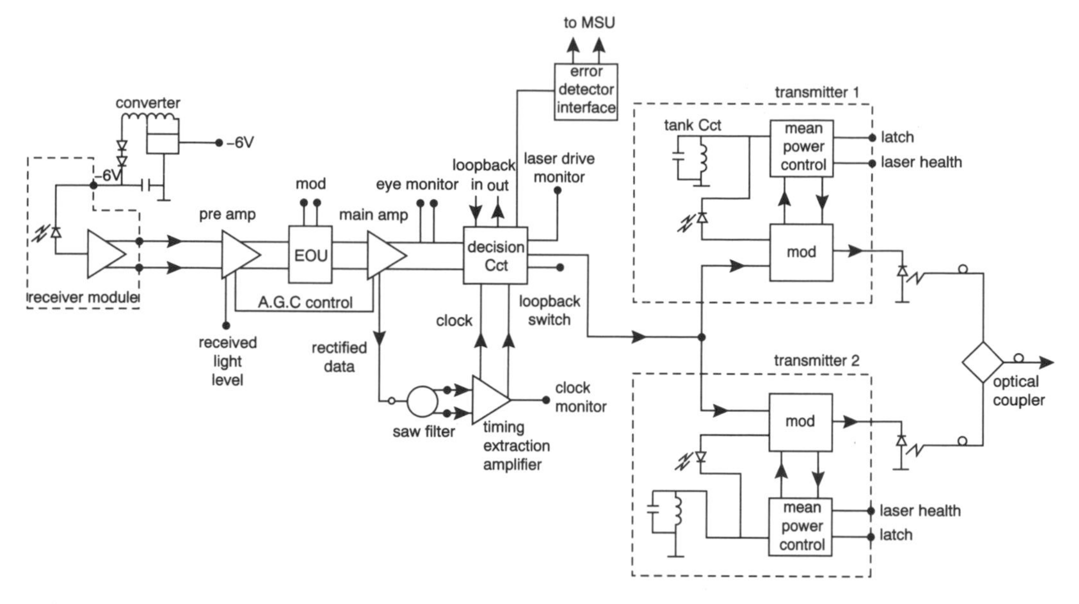

Figure 13.2 Diagram of TAT-8 regenerator showing parts of the main transmission path (Williamson *et al.*, Figure 2)

Decision circuit: this includes a D-type flip-flop to regenerate the data and to drive the loopback gating circuit. The error monitoring circuitry consists of a clock, gate and modulo-2 divider, and it exploits the even-mark parity line code (24B1P).

Transmitter: two modules are provided, one of which is on cold standby and draws no current. Each transmitter contains an IRW laser and a monitor photodiode which drives a mean power control loop and a loop which corrects for any change in laser slope efficiency. Feedback ensures that the modulation depth is adjusted until the data zeros are at the knee of the laser L/I characteristics.

13.3.3 Evaluation of maximum repeater spacing

By 1990 several publications had stated that 1550 nm systems operating at bit rates above 2 Gbit/s were dispersion-limited when standard sm fibres were used. However, experiments reported in [5] demonstrated that with dfb lasers then available and standard technologies it was possible to achieve up to 153 km at 2.488 Gbit/s. With an extinction ratio as low as 11 dB the measured value of the dispersion penalty was 0.5 dB and the extinction ratio penalty was less than 0.7 dB. The terminal regenerator was described in some detail.

The evaluation of maximum repeater spacing requires knowledge of the dispersion characteristics of the system fibres and the frequency chirping of the directly or externally modulated sources. Values of the chirp parameter α to be used in design calculations on 1550 nm systems are listed in Table 13.2 [5]. A broad guide to the type of system was provided in Figure 13.3.

The repeater spacing of 1300 nm systems was predicted to be half that of a 1550 nm unrepeatered system because of larger loss at the shorter wavelength. This parameter occurs in the relation between transmitted pulse width and other parameters given as equation (1) in [5]. For a dispersion of 2 ps $nm^{-1}km^{-1}$ at 1550 nm

Table 13.2 Values of chirp parameter for combinations of modulation scheme class of laser and type of external modulator ([5], Table 1). Reproduced by permission of IEEE, © 1992

Type of modulation	α parameter/class of laser/type of modulator	
Direct	4 to 6	(bulk lasers)
	1 to 2	(strained layer mqw lasers)
External	0	($LiNbO_3$ Mach-Zehnder)
	0.5 to 2	(Electro-absorption)

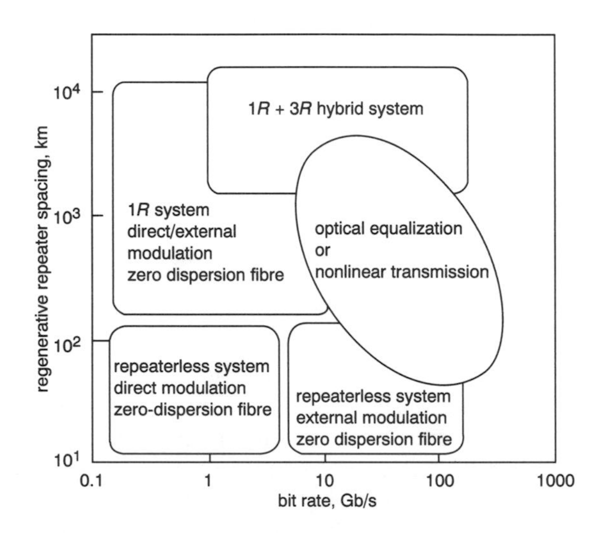

Figure 13.3 Regenerative repeater spacings versus bit rate showing areas of application for various types of 1550 nm systems ([5], Figure 5). Reproduced by permission of IEEE, © 1992

the dispersion-limited section length L_S between regenerative repeaters is approximately given by:

$$L_S = \frac{10^5}{f_S{}^2} \tag{13.1}$$

where f_S is the signalling rate on the line (GBd).

Where loss limits the unrepeatered system length, the use of many cascaded edfa repeaters removes this limitation, but the snr of these amplifiers limits system length. The attainable section lengths with 1*R* regenerative repeaters and with optical amplifiers are shown in Figure 13.4. The difference between section lengths for direct and external modulation is very significant.

For dispersion-limited systems the attainable section lengths of dsf for a fixed value of dispersion and range of values of α are shown in Figure 13.5.

13.3.4 Multiplexing in ofdls

The utilisation of line bandwidth by stacking channels side-by-side in time or in frequency by time division multiplexing or frequency division multiplexing has also played a very important role in electrical communication since the early days, and

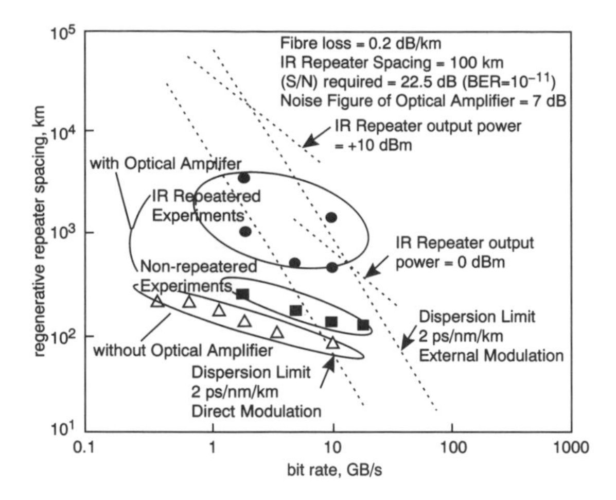

Figure 13.4 1*R* regenerative repeater spacings with dsf versus bit rate, under various conditions compared with experimental results with and without edfa repeaters ([5], Figure 3). Reproduced by permission of IEEE, © 1992

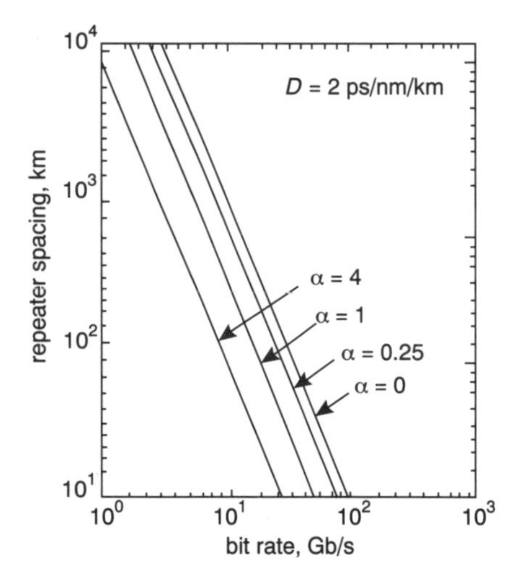

Figure 13.5 Attainable regenerative repeater spacings versus bit rate for 2 ps nm^{-1}km^{-1} of dispersion and $\alpha = 0, 1/4, 1$ and 4 ([5], Figure 4). Reproduced by permission of IEEE, © 1992

continues to do so in optical systems. For completeness we list in Table 13.3 the rates in the three PDH systems in current use.

In the North Atlantic, where the densest transoceanic traffic routes are found, there must be interworking between the European and North American PDHs, and this is done as shown in Figure 13.6.

The use of 12×140 Mbit/s tributaries to form the 1.7 Gbit/s level instead of 36×45 Mbit/s tributaries simplifies the high-speed multiplexing circuitry. The 1.7 Gbit/s system contains the same intensive maintenance capability as the lower-speed one, but a major new maintenance milestone will be introduced with directed line monitoring. This novel feature allows in-service selection and monitoring, via parity error detection, of any of the system's 45 Mbit/s tributaries.

Table 13.3 The three PDH recommended by the CCITT

	1 k/bits	2 Mbit/s	3 Mbit/s	4 Mbit/s	5 Mbit/s
Europe	64	2	8	34	140
N. America	64	1.5	6.3	45	274
Japan	64	1.5	6.3	32	100

Note: accurate values are 1.544, 2.048, 6.312, 8.488, 34.368, 44.736 and 139.264 Mbit/s

Recommendation G.707 SDH level	Synchronous digital hierarchy bit rates 1 Mbit/s	 4 Mbit/s
	155.520	622.080

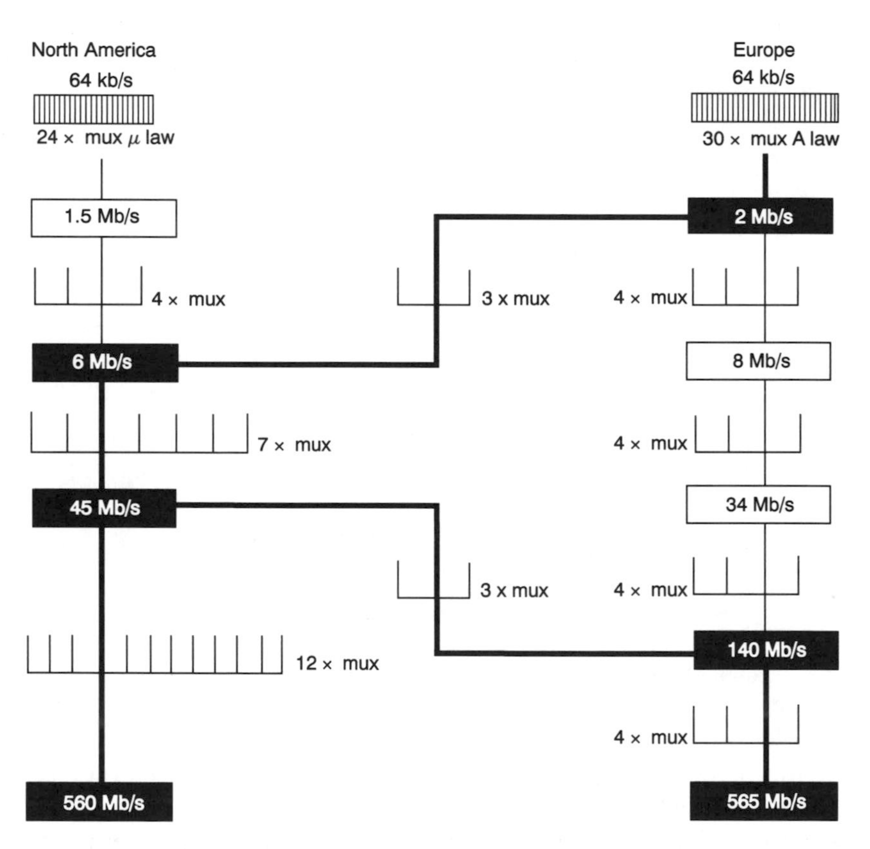

Figure 13.6 Hybrid digital hierarchy used on submerged routes between Europe and North America. (Reprinted from *Zodiac*, the house magazine of Cable & Wireless)

This feature is considered to be sufficiently significant that retro-fitting into existing equipment may be desirable. Examples of other multiplex arrangements will appear later.

The well-documented difficulties and deficiencies of the earlier plesiochronous system and dramatic advances in technology led to the development of the

Table 13.4 Bit rates (*a*) recommended and (*b*) suggested for the SDH (G.707)

Level	kbit/s
STM-1	155 520 (a)
STM-4	622 080 (a)
STM-8	1244 160 (b)
STM-12	1866 240 (b)
STM-16	2488 320 (b)

synchronous digital hierarchy and to its adoption as a world standard in the early 1990s. The bit rates for the SDH are as shown in Table 13.4.

Synchronous transmission systems address the shortcomings of the PDH, and using essentially the same fibre are able to increase significantly the available bandwidth and reduce the amount of equipment in the network. Significantly more flexibility is also achieved by means of sophisticated network management which becomes possible ([7] of Chapter 2).

13.4 CONSIDERATIONS IN SYSTEM DESIGN

Various design approaches have been used in the many papers on optical fibre systems over the past 15 or so years, e.g. design based on average or typical values, design based on standard deviations of individual distributions, design based on measured values, and a mixed design approach. At the end of the day, the system must perform to specification, within a wide range of environmental conditions, throughout its life. Regenerative repeater systems are intended unless otherwise stated.

13.4.1 Worst-case design

It is sound practice in digital systems to design for worst-case conditions, despite the negative image these create compared with typical figures, and some manufacturers quote worst-case performance figures. Typical performance figures are quite useless for a realistic assessment, unless accompanied by some indication of the system margins required to accommodate changes caused by varying operating conditions and ageing of components. Thus, for example, the output of a transmitter is quoted as that value to be expected at the end of life, with low limit input voltages at the worst extreme of the operating temperature range, etc. Similarly, a receiver figure would be the lowest input power that would achieve a given snr with everything against it. In short, Murphy's Law is assumed to be valid from the start, and things can only get better [6]. This approach may be unduly pessimistic, leading to over-engineered systems, and as experience is gained of the operational performance of ofdls in many countries, a more realistic and cost-effective basis may emerge. In particular, [6] was written in 1980, long before software became such an integral and important part of telecommunications.

There are two aspects of system design which must always be borne in mind: the dispersion or bandwidth limitations; and the optical loss budget. Fibre manufacturers' data sheets give values for dispersion, e.g. a fibre to the British Telecom specification CW1505E has a dispersion of less than or equal to 3.8 ps $km^{-1}nm^{-1}$ over the waveband 1285–1330 nm.

The second aspect is much more significant and, as explained above, typical values of the data necessary to compile the budget are not very useful. The data must therefore represent the worst-case losses, and with this information to hand the available margin between worst-case transmitter outputs and receiver inputs can be evaluated, and a minimum system performance guaranteed. For convenience, the individual losses are usually expressed in optical powers relative to 1 mW, i.e. dBm. The actual items which are included in the budget tend to vary, and detailed examples will be given. One suggested list is derived from the following considerations.

(1) Look up the worst-case output power from possible transmitters, e.g. laser fitted with a fibre pigtail.

(2) Calculate the coupling loss between the fibre tail and the worst-case system fibre.

(3) Subtract the coupling loss from the output power to yield the power launched into the system fibre.

(4) Select a possible optical receiver and note the worst-case sensitivity at the chosen operating wavelength.

(5) Check that the transmission window (passband) of the fibre matches that of the receiver.

(6) If necessary, allocate a value of loss to this mismatch.

(7) Subtract the worst-case sensitivity of the receiver from the launched power to give a value for optical path loss.

(8) Multiply the desired system length and the worst-case per-unit attenuation of the chosen fibre to obtain the fibre loss of the section.

(9) Account for all connectors (e.g. one at each end), and the number of splices required for the chosen cable lengths (manufacturers quote a range of standard lengths in steps, for example from 4.4 to 25.2 km) which make up the section length. Obtain worst-case loss values for each of the joints.

(10) Add the total fibre loss to the total joint loss to give the section loss.

(11) Subtract the section loss from the optical path loss to obtain a margin.

(12) If the margin is positive, is it sufficient to allow for additional splices for repairs during the design life, for any penalties due to pulse formats, etc? If so, the system will work under all circumstances for which its components are specified, and over the specified life of the equipment.

(13) Repeat steps (1)–(12) using best-case values wherever appropriate to establish that the receiver will not be overloaded in these circumstances. The resulting margin should be negative for there to be no danger of overloading the receiver input.

(14) If desired, repeat steps (1)–(12) using typical values to obtain an indication of the average performance of the system.

These steps lead to a choice of repeater spacing which ensures that the operating margin remains positive under the worst-case combination. By the late 1980s several authors had pointed out that this approach involved significant cost penalties.

13.4.2 Statistical design of long-haul systems

Substantial savings in equipment costs can be made if repeater section lengths are increased above the worst-case value. One design strategy is based on minimising the sum of the installation cost and repair cost over the lifetime of the system, and leads to a nonzero probability of system failure and necessary repair. The probability is not arbitrary but is that which achieves an optimum balance between these two costs. Possible mechanisms of system failure were identified, and the appropriate repair strategies (and the corresponding costs) were identified for each type of failure in a paper by authors from Telecom Australia [7], an administration whose network includes some very long routes. The treatment was deliberately kept general and a number of simplifying assumptions were made, in order to emphasis general principles and point out significant trends. The strategy involves repeater spans which are longer than those calculated using the worst-case basis of design, and also accepting the possibility of the system margin becoming negative. When this span is chosen correctly, the most likely cost of repairing these failures, added to the reduced installation costs, is significantly lower than the worst-case design.

This statistical strategy was applied to a 565 Mbit/s single-mode fibre system operating at 1300 nm, and it was shown that for typical values of device parameters a saving of about 30% in equipment installation and repair costs over the lifetime of the system is predicted. Unlike previous methods of statistical design, in the present one the probability of the system failing and requiring repair is not arbitrary as in the other methods, but is that probability which achieves an optimum balance between the costs of installation and of repair. Possible mechanisms of system failure are identified, and these failure modes are treated separately, because different repair strategies and therefore costs may be required for each mode. The paper illustrates general principles and points out significant trends, rather than develop a design strategy in detail.

As mentioned in the reference, a truly optimum design procedure for an ofdls could be defined as that which minimises the cost per bit of information transported over the given route over the lifetime of the system. This would require knowledge of many factors, including:

Capacity and lifetime of the route, type(s) of fibre used, number of fibres installed, type of repeaters, their spacing and number, tests carried out before, during and after installation, repair strategy adopted, random nature of device characteristics, effects of ageing, and cost of money over the lifetime.

Some of these, and of the many other relevant factors, are simply not known, so a less general approach must be adopted. The authors assumed that the lifetime of the ofdls, including all the transmission equipment as well as the cable, had been chosen, so the only remaining parameter left to vary was the repeater spacing. The strategy assumes that the route length is very much longer than the repeater span, as may be the case in Australia, the US and Russia for instance.

The authors investigated a 565 Mbit/s, 1300 nm inland system. Their results indicated that the maximum repeater span could be increased from about 35 km based on a worst-case design to around 55 km (note that at *Suboptic'93* repeater

Table 13.5 Parameter values used in the calculations ([7], Table 1). Reproduced by permission of IEEE, © 1989

Parameter	Value
Ratio of Repeater Housing Cost to Repeater Cost	$k_H = 25$
Number of Repeaters per Housing	$N = 10$
Line Rate	$B = 0.678$ (Gbit/s)
Mean Laser RMS Spectral Width	$\mu_W = 1.5$ (nm)
Standard Deviation LD Spectral Width	$\sigma_W = 0.5$ (nm)
Parameter Characterizing Rayleigh Distribution of LD RMS Spectral Width	$w_0 = 1.0$ (nm)
Maximum Possible LD RMS Spectral Width	$w_{max} = 2.0$ (nm)
Mean Available Power	$\mu_{AV} = 32.3$ (dBm)
Standard Deviation of Available Power	$\sigma_{AV} = 1.67$ (dB)
Truncation Coefficient of Available Power	$C_{AV} = 4.4$
Mean LD Centre Wavelength	$E[\lambda_C] = 1308$ (nm)
Standard Deviation of LD Centre Wavelength	$\sigma_C = 10.0$ (nm)
Maximum LD Centre Wavelength	$\lambda_C^{max} = 1323$ (nm)
Minimum LD Centre Wavelength	$\lambda_C^{min} = 1293$ (nm)
Ratio of Average Cost of Repairing a failure due to Negative System Margin to Cost of a Repeater	$K_m = 4.5$
Ratio of Average Cost of Repairing a failure due to Excessive Dispersion to Cost of a Repeater	$K_d = 0.75$
Ratio of Average Cost of Repairing a failure due to both Insufficient System Margin and Excessive Dispersion to Cost of a Repeater	$K_{md} = 0.75$
Dispersion Slope	$s = 0.1$ (ps/(km.nm^2))
Mean Fibre Loss	$\mu_\alpha = 0.545$ (dB/km)
Standard Deviation of Fibre Loss	$\sigma_\alpha = 0.0396$ (dB/km)
Fibre Loss Truncation Coefficient	$C_\alpha = 2.65$
Proportion of Repeaters which fail for $L-$independent Reasons over System Lifetime	$n_i = 0.33$
Ratio of Average Repair Cost of an $L-$independent Failure to Repeater Cost	$k_i = 4.0$
Standard Deviation of Zero-Dispersion Wavelength of a 5 km length of Fibre	$\sigma_F = 0.696$ (nm)
Maximum Fibre Zero-Dispersion Wavelength	$\lambda_F^{max} = 1320$ (nm)
Minimum Fibre Zero-Dispersion Wavelength	$\lambda_F^{min} = 1300$ (nm)

spacings of 110 km in early systems and 130 km in later ones were quoted for 565 Mbit/s submerged links operating at 1550 nm). The parameter values used in the calculations are listed in Table 13.5. The entries in this table can, in some cases, be compared with those given elsewhere and provide the reader with data.

The problem formulation and related factors were discussed in Section II, and failure modes and cost in Section III under the headings: 'failure of a repeater section'; 'effect of mode partition noise'; and 'calculation of most likely cost of system'. Section IV treated numerical considerations in the applications of the theory under the headings: 'truncation of the probability density function'; 'worst case calculations'; and 'evaluating three functions for cost optimised design'. The conclusions included the identification of the fibre attenuation and the available power output from the transmitter and at the receiver as the major limiting factors, and not the dispersion penalty. Sensitivity tests showed that the cost of a single repair affected the overall system cost significantly but did not strongly influence the optimum repeater spacing.

The statistical approach had been described earlier in the regeneratively repeatered submerged system context in a special issue [8] published in 1984. Then the final CCITT recommendation for Equipment Design Objectives for transmission systems had not been published, so existing network requirements (G.821) were taken and suitable values were apportioned. Under the heading of design philosophy, statistical design approach, an outline was given, followed by a comparison based on examples.

The starting point was the fact that all components in a repeater section will vary in their value and performance during the system lifetime according to some form of distribution, generally Gaussian. Therefore, if many regenerator sections are constructed their performance will form a distribution which results from convolution of all the relevant individual component distributions which make up a section. Provided that care is taken to ensure that the tails of distributions that describe received power and receiver sensitivity and receiver overload point do not overlap adversely, a system can be constructed that has an end-to-end long-term mean ber, ⟨ber⟩, at a certain confidence level. The value of ⟨ber⟩ can be obtained from the value of errored seconds for the system, which in turn is derived from the current CCITT Recommendation G.821 for the Hypothetical Reference Connection (HRX), assuming errored seconds to be the most demanding of the criteria. The approximate relation quoted was that the equivalent ⟨ber⟩ is equal to the errored seconds divided by the number of bits per second (64 000). The system ⟨ber⟩ value must then be transformed to a value appropriate to the chosen regenerator section length on either a worst-case or statistical basis. Two more values of ⟨ber⟩ are derived by conversion from the degraded seconds budget and from the degraded minutes budget, and the most onerous of the three values of ⟨ber⟩ taken for design purposes. This design approach was then compared with the worst-case approach in examples.

Example of worst-case design approach

The end-to-end ber performance requirement was taken as 1×10^{-8} and 200 repeater sections were assumed. The corresponding repeater performance is then $10^{-8}/200$ or 5×10^{-11} at the 100% confidence level.

Statistical design approach

Regenerator section loss: this calculation requires the pdf of all relevant system parameters to be known. These may be derived from measurements on components and manufactures sub-systems, and they are expressed in terms of mean μ and standard deviation σ of the parameters.

Example: assume that the transmitter output power P_T (dBm) is a random variable with mean μ_T and variance $\sigma_T{}^2$ derived from a sufficiently large number of measurements. Likewise, the receiver sensitivity P_R (dBm) has mean μ_R and variance $\sigma_R{}^2$ and the cabled fibre a per-unit attenuation α (dB/km) of mean μ_C and variance $\sigma_C{}^2$. If follows that the system margin is simply:

$$\text{system margin, } M = P_T - P_R - \alpha L \qquad (13.2)$$

where L is the section length. The mean and variance of the margin is the convolution of the distributions for P_T, P_R and for α, giving:

$$\mu_M = \mu_T - \mu_R - \mu_C L \tag{13.3}$$

$$\sigma_M{}^2 = \sigma_T{}^2 + \sigma_R{}^2 + \sigma_C{}^2 \tag{13.4}$$

assuming that the variance of α is proportional to the length. The essence of the statistical approach is that the span margin should not be less than the design value at the 99.97% confidence level (this value being an engineering decision), which corresponds to 3.4 standard deviations from the mean, hence $\mu_M = 3.4\sigma_M$. The optimum value of section length L is found from

$$\mu_T - \mu_R - \mu_C L - 3.4L(\sigma_T{}^2 + \sigma_R{}^2 + \sigma_C{}^2)^{1/2} = 0 \tag{13.5}$$

Receiver design: values of ⟨ber⟩ versus mean received power are plotted and extrapolated, with a noise floor limit if applicable. The shape of such a curve is well known.

pdf of ⟨ber⟩: with a fixed system margin distribution and the curve of ⟨ber⟩ found in the preceding paragraph, a composite curve of ⟨ber⟩ pdf can be constructed, and it takes the form shown in Figure 13.7.

Multiple convolutions of this curve with itself according to the number of repeater spans, together with a suitable technique such as Monte Carlo analysis, result in a histogram giving the number of systems with a given number of repeaters with a chosen value of ⟨ber⟩. The examples given were for systems with 200 repeaters, for which

1000 systems would result in ⟨ber⟩ $= 10^{-12}$
180 systems would result in ⟨ber⟩ $= 10^{-10}$
80 systems would result in ⟨ber⟩ $= 10^{-9}$

Such information enables the designer to plot confidence level versus ⟨ber⟩ which will give the level corresponding to, say, ⟨ber⟩ $= 10^{-9}$. During manufacture and testing any repeater spans giving a ber below this value with the required margins would be rejected, but this would be a rare occurrence because the confidence level is chosen to be very high.

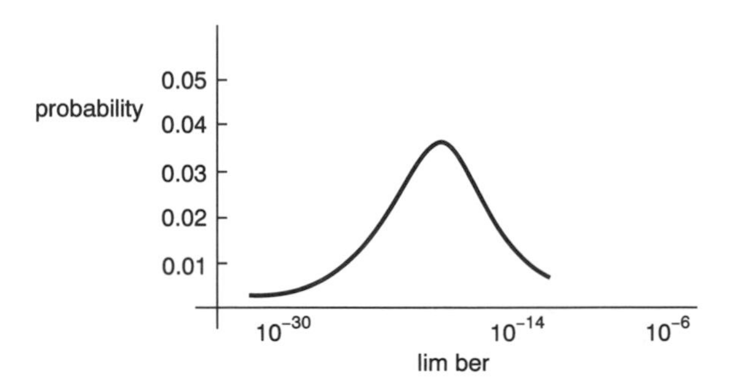

Figure 13.7 Probability density function versus long-term mean value of ber ([8], Figure 2). Reproduced by permission of IEE

13.5 POWER BUDGETS AND MARGINS IN OFDLS

In designing a practical system it is necessary to combine the characteristics of the terminal and repeater (regenerative or optical amplifier) equipment with those of the cable so that adequate margins are built-in. Effects caused by temperature variations, ageing and many other factors experienced throughout the lifetime of the system must be incorporated from the beginning. The above characteristics are expressed in terms of loss penalties which are combined in a system design power budget. The following treatment is based largely on [9] and on earlier work on regenerative repeatered systems.

Performance requirements and measures

Optically amplified systems will form part of the SDH/SONET network for which performance measures were given in CCITT Specification G82x. These measures are listed (see Section 13.4) but need to be interpreted to derive a single calculable measure such as ber which can be used in budgets. However, ber is a very coarse indicator of performance, and a much more suitable measure is the quantity Q introduced in Chapter 11 which is a measure of the electrical signal to noise ratio at the receiver (decision circuit input). It is given by equation (11.A6) in general and by equation (11.14) for the worst case; for optically amplified systems we use equation (11.A6) in terms of the mean and r.m.s. values of voltages at the sampling instant, hence:

$$Q = \frac{(V_1 - V_0)}{(\sigma_1 + \sigma_0)} \tag{13.6}$$

where the subscripts 1 and 0 denote a mark (1) and space (0), as usual. The fluctuations on the marks and spaces must be due to random noise and not to their positions in the bit stream. Q can be measured directly from the data eye or deduced from measurements of ber, and because it is a measure of electrical snr it can be used directly in a power budget. It is sometimes expressed in dB.

Power budgets

Power budgets are used in the planning of all terrestrial and submerged systems, and they enable owners and suppliers to identify the margin that exists in the system design and to compare this with the specified performance objectives.

Key requirements of a power budget are that:

(a) it should allow the effect of changing a single parameter to be seen

(b) it should provide start-of-life and end-of-life margin assessments whose basis can be agreed between suppliers and purchasers

(c) it should allow individual specifications to be written for components and subsystems

All entries in the budget should be verified experimentally.

In regenerative repeatered systems the signal is reformed at every regenerator, and this has the effect of limiting the build-up of all impairments, except for jitter, to a single section. The overall performance is thus dominated by the worst regenerator section, and this is reflected in power budgets for such systems which are effectively for the worst section.

In systems with optical amplifier repeaters the signal is not reformed at every repeater, and impairments such as noise, nonlinearity and so on accumulate over the entire length of the system, a feature of all analogue systems. Such systems are relatively immune to individual degradations on a single span; instead, the overall performance is dominated by average values of amplifier parameters such as noise figures. Power budgets must therefore be produced for the entire system, rather than for the worst section.

13.5.1 Power budgets for optically amplified systems

To draw up a power budget requires an initial evaluation of system performance without taking impairments into account. Two possible starting points for this step are system simulations, and in the case of optically amplified systems a model of amplifier noise concatenation. Impairments must then be identified and their effects must be calculated individually; the results (in dB) are then subtracted from the initial budget to give a design budget from which actual performance can be predicted.

The general format of optically amplified system power budgets is illustrated in Table 13.6 and 13.7.

Table 13.6 Start of life power budget, taking optical snr as starting point ([9], Table 1). Reproduced by permission of IEE

Line	Optical Signal to Noise Ratio Budget	SNR(dB)
1	SNR from calculations or simulations	a
2	Optical impairments not included in line 1	b
3	Start of life optical SNR	a-b
	Electrical Signal to Noise Ratio Budget	Q(dB)
4	Equivalent electrical Q	d(use equation 3)
5	Electrical impairments not included in line 1	e
6	Start of life electrical Q	d-e
7	End of life Q requirement	f
8	Start of life Q margin	d-e-f
9	Start of life BER	Use equation 2

Table 13.7 Start of life power budget taking simulated value of Q as starting point ([9], Table 2). Reproduced by permission of IEE

Line	Electrical Signal to Noise Ratio Budget	Q(dB)
1	Q from calculations or simulations	a
2	Impairments not included in line 1	b
3	Start of life Q	a-b
4	End of life Q requirement	f
5	Start of life Q margin	a-b-f
6	Start of life BER	use equation 2

The relation between Q and the optical signal to noise ratio (snr′) in optically amplified systems is

$$Q = \left[\frac{2\text{snr}'}{1 + (1 + 4\text{snr}')^{1/2}}\right] \times \left(\frac{B'}{B}\right)^{1/2} \tag{13.7}$$

where B' is the optical 3 dB bandwidth and B is the electrical 3 dB bandwidth.

This equation can be used to change from electrical to optical domains, and vice versa, at will.

Simulation-based power budgets for optically amplified systems

The complexity of the model determines what properties of the model elements are included, and the following list of properties are typically included:

(1) amplifier noise
(2) amplifier saturation effects
(3) nonlinear effects (e.g. four-wave mixing) and dispersion in fibre
(4) extinction ratio of transmitter
(5) rise and fall times of transmitted pulses
(6) optical filtering at receiver
(7) electrical filtering at receiver

Simulators modelling these properties are generally time-consuming (e.g. up to 12 hours on an HP 9000/730 workstation for a 6000 km long system).

Alternatively, impairments may be estimated for each of these for some sacrifice in accuracy, but a better insight is gained. This approach also enables 'what if' calculations to be done rapidly.

The effect of amplifier noise on signal depends on the amplifier control scheme used (three were outlined). Only the first maintained a fixed signal level at the receiver; in the other two the signal level drops with distance and so introduces another variable. The third scheme involves operating the amplifiers in saturation, and in this regime the 'steady-state' corresponds to the output power remaining approximately constant, as shown in Figure 13.8.

An estimate of the noise at the receiver can be based on measurements of edfa noise figure under typical operating conditions (wavelength, gain, output level,

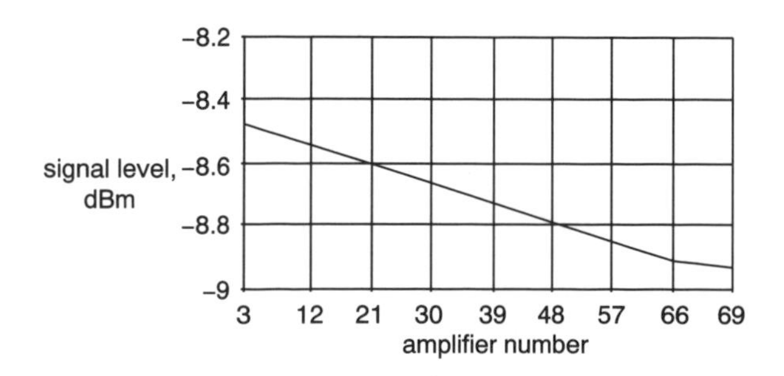

Figure 13.8 Measured signal levels at input to each amplifier on a 3100 km laboratory test system ([9], Figure 1). Reproduced by permission of IEE

pump power configuration). The definition of noise figure was chosen in [9], so that the noise at an edfa output in a bandwidth B′ was given by:

$$N_2 = NF \times G \times hf \times B' \quad [\mathrm{W}] \tag{13.8}$$

If A identical amplifiers are concatenated and the gains are chosen to match exactly the preceding section loss, the noise level after A amplifiers will be AN_2. Let the input signal power at the first amplifier be $S_1(1)$ in mW; then snr′(A) is given by:

$$\mathrm{snr}'(A) = 10\log(S_1(1)) - 10\log(NF) - 10\log(A) - 10\log(10^3 \times hf \times B') \quad [\mathrm{dB}] \tag{13.9}$$

An upper limit to $S_1(1)$ in long-haul systems is set by fibre nonlinearity, so snr′ can only be reduced by making NF smaller.

Any nonlinearity in the fibre will broaden the signal spectrum. A very brief explanation based on [10] is as follows. We restrict our attention to four wave processes associated with the third-order polarisation which can be conveniently modelled by a refractive index which is time- and intensity-dependent, and can be modelled by the equation:

$$n(|E|^2, t) = n_0 + n_2 \times |E|^2 \tag{13.10}$$

where n_0 is the linear value of refractive index, n_2 is the Kerr coefficient and $|E|^2$ is the optical intensity, and is related to the power $P(z,t) = |E|^2 \times A_\mathrm{E}$, the effective area of fibre. The sign of n_2 is positive, so that increasing $|E|^2$ increases n, an effect called self-focusing (a form of positive feedback). The simplest effect which occurs with this nonlinearity is self-phase modulation (SPM).

SPM alone results in a broadening of the spectrum, but the waveform (temporal) shape is unchanged. Let $E(z,t)$ be the pulse envelope propagating in the z direction; then for the leading edge of the pulse $|E|$ is increasing, hence n increases, the wave velocity decreases, and hence the wavelength increases (red shift). This increases the rate of change of phase shift, which corresponds to new frequency components being generated. At the trailing edge $|E|$ is decreasing, thus n, hence velocity, increases; the wavelength shortens (blue shift) and again frequency components are generated.

FWM is the intermodulation (beating) between components of amplifier noise and the signal, or between the noise components themselves, and it attenuates the signal by absorbing signal power.

The effect of nonlinearity on a spectrum is illustrated by simulation results reproduced in Figure 13.9.

The authors stated that increasing A_E will become a key fibre parameter in future systems, as it will permit more power to be launched for a given amount of nonlinearity. It is important to ensure that the offset between these two wavelengths is never zero under any conditions throughout the system lifetime.

This can be compared with the simulated results shown in Figure 13.10 [9], and in combination with fibre dispersion it can give rise to a significant penalty if not well controlled. Predicting this penalty without measurements or simulation is very difficult, as it depends on power levels and on the exact dispersion map that is present. To overcome the effect of dispersion in long-haul systems, the spectral width of the transmitter should be close to that of the modulating signal (e.g. about 0.04 nm for a 5 Gbit/s signal). All such system demonstrators reported by 1993 had deployed external modulators for this reason, and with these there is no significant penalty from source spectral width.

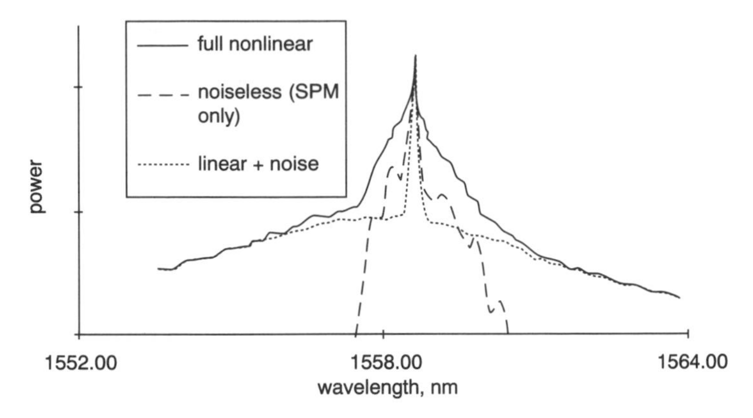

Figure 13.9 Simulation of spectral broadening due to noise only, spm only and full nonlinear in a 6000 km long system with 45 km repeater spans and average 2 nm offset between zero dispersion and signal wavelengths ([10], Figure 2). Reproduced by permission of IWCS

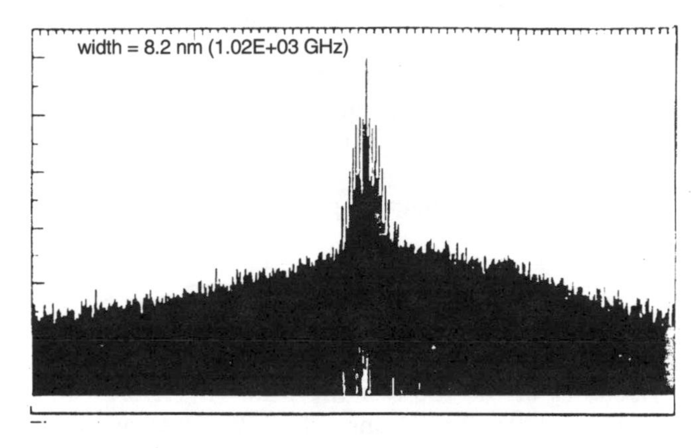

Figure 13.10 Simulated optical spectrum after 6300 km ([9], Figure 2). Reproduced by permission of IEEE, © 1993

Dispersion management in long-haul optically amplified systems

It is evident from the above discussion that this is an important topic in such applications [10]. The effect of pulse broadening can be reduced if the average dispersion over the route is zero. This can be achieved by cancelling the slightly negative dispersion per unit length of the dispersion-shifted fibres forming most of the route with the large positive dispersion per unit length from short lengths of standard fibre. The operating wavlength is chosen (within the band 1530–1560 nm) to produce the negative dispersion. The technique and practical considerations were described in [10] but for the present it is sufficient to illustrate the idea in Figure 13.11.

The dispersion equaliser fibre cables are several km long but would typically form less than 1% of the route length.

Simulations can also be used to account for a non-zero extinction ratio, but an alternative approach is to modify the equation (13.7) for Q to account for signal-spontaneous beat noise present in the spaces.

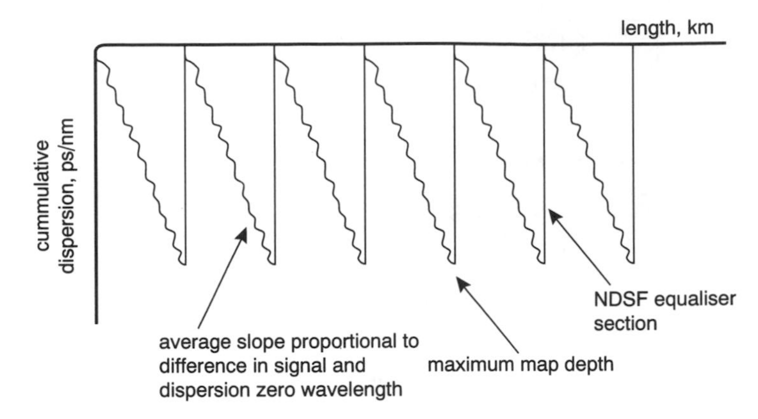

Figure 13.11 Cumulative dispersion of a 'dispersion managed' long-haul system ([10], Figure 3). Reproduced by permission of IWCS

Other impairments not included so far

Bandwidth imperfections in the transmitter: these result in the heights of individual marks and spaces being data-pattern-dependent, an effect illustrated in Figure 13.12. In such cases it is the values of the smallest mark and largest space which dominate the ber, and if we represent the corresponding voltages at the decision circuit input by V_1 (min.) and V_0 (max.), the penalty is given by:

$$P = 10 \log \frac{V_1 - V_0}{V_1(\text{min.}) - V_0(\text{max.})} \tag{13.11}$$

where the voltages on the numerator are nominal values.

Receiver imperfections: at the receiver the following effects result in penalties: patterning due to bandwidth limitations; jitter; baseline wander; decision level uncertainty; and receiver noise. The assessment and reduction of these penalties are very important in drawing up power budgets for long-haul systems. Note that if an optical preamplifier is used, receiver sensitivity is no longer important and the design need only operate with high incoming light levels of about 0 dBm.

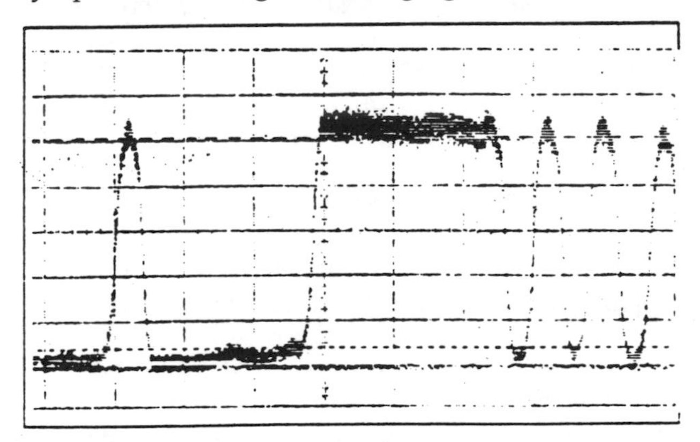

Figure 13.12 Typical 5 Gbit/s optical data pattern measured by means of a high-speed pin receiver ([9], Figure 3). Reproduced by permission of IEE

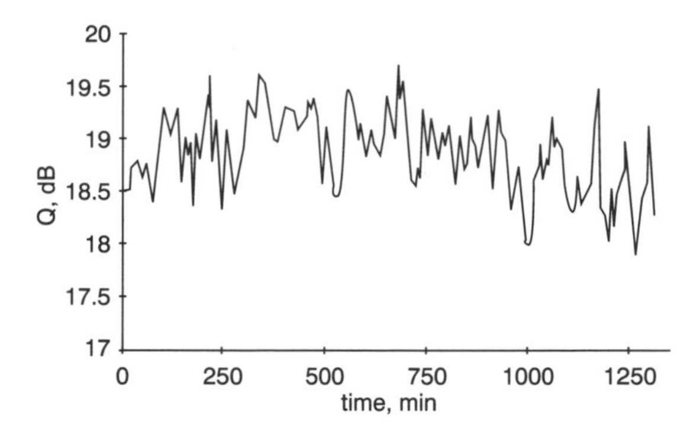

Figure 13.13 Measured Q versus time for a 6300 km long system ([9], Figure 4). Reproduced by permission of IEE

Polarisation fluctuations: all the previously considered impairments are assumed to be constant with time, whereas long-haul systems show a temporal variation of Q and of ber due to polarisation effects. An example of the first of these is shown in Figure 13.13.

In 1993 this was thought to arise from the statistical nature of the polarisation mode dispersion and polarisation-dependent loss along the system. The corresponding penalty can only be characterised if the pdf of the fading is known. In practice it may not be possible to guarantee that the performance limits for ber are always met. An allowance must be made in the budget for a penalty which will only be exceeded very rarely (say 1 minute in 25 years).

Comparison between measured and predicted performance

Experience of long-haul laboratory systems has shown agreement between typical predicted and measured Q values to be within 1 dB. Comparisons of the spectra at the output of the system have shown that the effects of nonlinearity and dispersion can be predicted accurately by computer simulations. Improvements in the accuracy of some items in the power budget were expected to come from better modelling of transmitter and receiver effects, and from more detailed knowledge of the Q fading phenomenon.

13.5.2 Power budgets for regenerative systems

The approach taken here is to present a systems analysis which, as will be seen, includes numerous items. In the following groupings which together form the Systems Analysis for regenerative systems, all the numerical values would be in dB unless stated otherwise. Where $\pm$ appears it denotes the same numerical value but with either sign; where $+, -$ appears it denotes a positive value and a different negative value for the item. Where two or more numbers appear addition or subtraction are meant, e.g. $1 - 2 - 3$ means subtract the value of entry 2 and entry 3 from the value of entry 1.

Additional paragraphs extending the discussion to optically amplified systems are inserted where appropriate.

Power budget

Items which can be included under this heading for regenerative systems are:

(01) source level		nom. mean optical power
(02) receiver sensitivity		nom. mean optical power
(03) system design penalty		nominal
(04) path loss capability	1 – 2 – 3	nominal
(05) path loss		nominal
(06) system margin	4 – 5	nominal

System margins

Items which can be included under this heading for regenerative systems are:

(07) design margin	±	worst cases
(08) degradation margin	+, –	dimmer, brighter
(09) repair margin		extra cable, splices
(10) subtotal (darkest)	7+, 8+, 9+	sum
(11) subtotal (lightest)	7–, 8–	sum
Minimum dynamic range	10 – 11	

Design margin

Items which can be included under this heading for regenerative systems are:

(12) source level	±
(13) receiver sensitivity	±
(14) system design penalty	±
(15) path loss	±
(16) subtotal (linear addition)	all +
(17) subtotal (r.m.s addition)	

Degradation margin

Items which can be included under this heading for regenerative systems are:

(18) source level	+, –	worse, better
(19) receiver sensitivity	+, –	worse, better
(20) system design penalty	+, –	worse, better
(21) path loss	+, –	worse, better
(22) subtotal (darkest)	+	all plus values
(23) subtotal (lightest)	–	all minus values

Production tolerances of passive and active components such as fibres and edfa will be reflected in their performance, and the resulting variability must be allowed for. In addition, the ageing of components of a given type, e.g. line amplifiers will have a spread. Simulation results with component variations included have shown that it is the mean, rather than the individual values which dictate the penalty. An allowance must be made which reflects variations in mean values. This approach is radically different from the pass or fail criterion used for components in regenerative systems.

Repair margin

Items which can be included under this heading for regenerative systems are:

(24) length L_1 of extra fibre plus two splices

(25) length L_2 of extra fibre plus two splices

(26) one large or two small repairs

Optically amplified systems are very resilient to single faults or deviations from nominal values. The example given in [9] was for increased loss in some of the sections, where the $a < A$ section had an increased loss given by $10\log(x)$ dB. Equation (13.7) was modified appropriately to give the optical snr in dB:

$$\mathrm{snr}'(A) = 10\log(S_1(1)) - 10\log(NF) - 10\log(A - a + ax) - 10\log(10^3 \times hf \times B') \tag{13.12}$$

Combining this equation and equation (13.3) gives the repair penalty P_{REP} caused by these additional losses as:

$$P_{\mathrm{REP}} = 10\log\left[1 + \frac{a(x-1)}{A}\right] \text{ [dB]} \tag{13.13}$$

Predictions based on this formula have been found to agree well with measurements of simulated repairs on a 3000 km laboratory system, despite the assumptions of constant NF and output power. For example, $P_{\mathrm{REP}} = 0.5$ dB would result from $a/A = 0.12$ and $x = 2$.

Design penalty

Items which can be included under this heading for regenerative systems are:

(27) jitter	± angle	± time
(28) mistiming	± angle	± time
(29) frequency modulation	± angle	± time
(30) set-up, hold times		time
(31) chromatic dispersion		time
(32) variation in turn-on/turn-off delay		time
(33) subtotal (r.m.s. addition)	dB	time
(34) subtotal (linear addition)	dB	time
(35) reflections	dB	

Table 13.8 Power budgets for initial specification, factory test and overall commissioning on site ([1], Table IX). Reproduced by permission of John Wiley

System	Initial spec.	Factory test	On site
Transmitter			
Mean Power at +45°C			
Point S of CCITT, dBm	−2.50	−1.4	−2.0
Receiver			
Mean receiver sensitivity, dBm	−44.50	−46.7	−46.1
Gross budget, dB	42.00	45.3	44.1
Fibre			
Loss at 1560 nm and +10°C, dB	31.70	30.2	30.2
Mean loss, dB/km	0.25	0.24	0.24
Component penalties, dB			
(including installation and laying splices,			
distribution panels, pigtails, connectors, etc.	2.30	2.30	1.9
Chromatic dispersion penalty, dB	1.00	0.5	0.8
Ageing margin, dB			
Cable	0.10	0.10	0.10
Electronics	0.60	0.60	0.60
Electro-optics			
transmitter	1.00	1.00	1.00
receiver	1.30	1.30	1.30
Repair margin, dB	4.0	9.3	8.2

(36) amplitude modulation dB
(37) misequalisation dB
(38) polarisation dispersion dB
(39) total (r.m.s. addition) dB
(40) total (linear addition) dB

Power budgets can be drawn up at various stages in the progress of a system from design to commissioning. An example of such a sequence was given in [1] in tabular form, reproduced as Table 13.8, which enables the reader to compare the values of certain items at initial specification, factory test and overall commissioning.

13.6 ASPECTS OF ERROR DETECTION IN OFDLS

Digital networks are susceptible to error bursts from various sources ranging from thermal noise producing random errors or short bursts, through to external influences such as environmental effects which can result in long bursts of error activity. The propagation of such bursts through the European PDH has been studied and the error-extension mechanisms involved had been identified by 1990. Following on from this work, the emerging standard for the SDH was analysed to determine its performance during error activity. The conclusions and recommendations from this investigation were stated in [11].

The error activity in a digital system has been used as a performance parameter from the beginning. A first attempt to establish a means to use it as a diagnostic tool was described in 1990. Such a tool would allow ailing systems to be identified and removed from service prior to their performance becoming unacceptable. It would also provide an indicator to the root cause, which would also expedite repair to the system and the early restoration of service. The test bed for burst error acquisition and analysis was described and some typical results were presented in [12]. A bit error correlation technique was devised which related variations in recorded error activity to variations in the parameters of an interference waveform injected into the system under test.

Reasonable agreement between computer simulation and laboratory measurements was demonstrated for the mean event case.

The general inaccessibility of repeater circuitry in systems often makes a remote signal level or eye monitor desirable. If sufficiently sensitive, this could provide another diagnostic tool for system fault location, and could allow performance certification during installation and repair. It could also be used to give a continuous measure of the snr margin available at each repeater. A technique applicable to 2*R* and 3*R* systems consisting of a number of cascaded dependent regenerators, which offered simple realisation and which allowed both in-service and out-of-service operation, was described in [13]. An outline was given of the system, then of the theory, followed by a description of the test bed and results which demonstrated operation on a single repeater using a simulated agc circuit. The performance quoted was expected to be satisfactory for most system supervision applications.

13.6.1 Bit-error-rate floors resulting from transient partitioning

Systems operating in the 1550 nm band over standard fibre use dfb lasers to minimise degradations due to the combination of chirp and fibre dispersion. Tests performed on the laser include measurement of the penalty arising in a short length (1–10 m) and in the maximum section length (hence maximum dispersion), perhaps 90 km for a 2.5 Gbit/s system. The penalty arising from chirp shifts the ber curve to higher received powers, thus reducing the sensitivity. Although it is true that the dispersion penalty due to laser chirp is proportional to chirped linewidth times fibre dispersion, degradation can also result if there are modes that partition with the main dfb mode. The penalty associated with partitioning can result in an abrupt ber floor at a level determined by the statistics of the partitioning events, the decision threshold and the length of fibre involved. Some dfb lasers modulated at 1.7 Gbit/s exhibit ber floors for certain lengths, e.g. not at 43 km but at 27 km; hence such floors are difficult to detect [14].

13.6.2 Burst noise

The presence of bursts of errors has great significance for the design of error control codes, so let us look briefly at the topic of error bursts. The most fundamental statistic to study is the error rate in a system, i.e. the expected number of errors in unit time, or the probability that a given bit will be in error. Unfortunately, this does not completely describe the error process, for it is known that in many real systems there can be long periods with no, or very few, errors. This fact of experience has already been mentioned in connection with the AT&T FT Series G cable on a section of the Philadelphia to Chicago route. However, it is convenient to repeat the information at this point. The ber of a 45 Mbit/s channel was measured and compared with the 900 errors per day corresponding to a ber of 10^{-9}. During a 28 day stretch of the test, no errors were measured at all! [15].

The tendency for errors to cluster or come in bursts can greatly affect the design of error control codes and transmission protocols. The causes of clustering are not entirely known, but the operation of the switch protection network switches is known to be one cause of error bursts. A mathematical description of noise in digital communication channels was published in 1987 [16]. Two parameters are described: the burstiness which describes the tendency for errors to cluster, and the noise rate which describes the rate at which errors occur, and is proportional to the familiar error rate. The usefulness of a proper statistical description of digital noise was discussed.

13.6.3 Error rate in end-to-end system performance predictions

Undersea systems may span distances up to 10 000 km and contain more than 100 regenerators; both figures are much larger than in most inland systems, but more typical values would be those for a transatlantic route 6000 km in length with 60 regenerative repeaters. In addition, undersea systems tend to use multiplexing, coding and supervisory subsystems which differ from landlines. Multiple landings rather than the point-to-point cables of the coaxial era have been introduced; for instance, TAT-9 was planned to terminate in five different national networks and to incorporate branching units containing active multiplexers. These will need to be considered in predicting the overall performance of the system [17]. This paper discussed some of the problems of predicting final error performance and jitter of a long optical submarine system. It also examined some of the measurement and modelling techniques used to make predictions and validate system performance. The error performance of regenerative repeater SDH-compatible systems must comply with CCITT Rec. G.821.

Accumulation and planning

The end-to-end ber of a system with several regenerated sections can be expressed in terms of the individual section rates ber_J by

$$\text{ber} = 1 - (1 - \text{ber}_1) \times (1 - \text{ber}_2) \times (1 - \text{ber}_3) \times \dots \qquad (13.14)$$

which for small values of ber becomes

$$\text{ber} = \Sigma(\text{ber}_J) \qquad (13.15)$$

Undersea systems are usually planned on the basis of ber $< 10^{-11}$/km, which is a little better than required by the CCITT. For section lengths of 100 km and identical regenerators, $\text{ber}_J < 10^{-9}$.

Likely ber performance

Submerged systems are well screened from usual sources of interference over most of their length; therefore one might assume Gaussian noise statistics to dominate and the snr to increase directly with signal level. This is roughly true for pin receivers but not apd detectors, which tend to exhibit significant shot noise dependent on both received signal level and multiplication factor. An accurate determination of ber requires observing several events, and for low values of ber requires long observation times. For ±50% accuracy at the 90% confidence level one should wait for about 12 errors; the corresponding ber versus measurement times at 590 Mbaud are listed below:

1 minute	3.4×10^{-10}
1 hour	5.6×10^{-12}
1 day	2.4×10^{-13}

Such measurements are valid only if the ber is stable during the period; however, as mentioned earlier, measurements on systems have revealed error-free performance over several days.

Measurement and description of burst errors is extremely difficult, but these are covered by describing the short-term ber. The CCITT specifies the following measures:

errored seconds	> 0 errors
severely errored second	ber $> 10^{-3}$
degraded minute	ber $> 10^{-6}$

A severely errored second or a degraded minute can only be produced by a burst of errors, and for a 5000 km link likely allocations were given as:

errored seconds	< 292 per day
severely errored seconds	< 2.7 per day
degraded minute	< 5.8 per day

Long-haul optically amplified systems will form part of the SDH network for which performance measures are given in CCITT Specification G82x using quality of service (QOS) measures as seen by the end user, namely: errored second ratio (ESR), severely errored seconds ratio (SESR), background block errors (BBER), and outage.

To derive a single error measure (e.g. ber) requires that these QOS measures be interpreted, which in turn requires knowledge of the way in which errors occur. Random errors are assumed to have a Poisson distribution. Burst errors can be modelled by assuming that the time of arrival and the number of errors in a burst are not dependent on the time or number of errors in the previous bursts; these assumptions give a Neymann Type A distribution. An early paper [8], which included a discussion of relations between QOS criteria (from Rec. G.821) and long-term mean ber, gave numerical examples for a 7000 km submerged system design.

13.6.4 Error propagation through plesiochronous demultiplexers

Until the late 1980s there appeared to have been few attempts to quantify the type of errors that exist in digital systems and how they propagate through the multiplex hierarchy. Theoretical studies and equipment deployed in the network were initiated by British Telecom to capture error events occurring across a wide range of technologies. Some of the key results of these studies were reported in [18]. This important paper covered a large number of relevant topics in the introduction, demultiplexer basics and effect of errors, software models, model results, single run example, single burst measurements, hardware measurements and synchronous operation.

In conclusion, the authors stated that the use of 'activity time' illustrated the nonlinear way in which errors can propagate through a plesiochronous hierarchy. Frame structures at each level were originally specified to limit the propagation of error bursts, by ensuring that loss and recovery times of the higher-level demultiplexers were generally shorter than the loss time of the following level. The occurrence of uncontrolled slips effectively defeated the original design concepts and appeared to be the critical mechanism by which errors propagate through the demultiplexer stacks. This was manifest in the sensitivity to both the input burst length and to the propagation of a loss of alignment condition. Despite the fact that the model was based upon CCITT Recommendations only and made no attempt to model specific hardware implementations or phase-lock loop recovery times, comparison of model results and hardware measurements showed good agreement down to 2 Mbit/s.

Slightly earlier, the same authors had given an analysis of the effects of error bursts on a 140 Mbit/s line system terminated by a PDH demultiplexer stack, which detailed the extended error activity at 8 Mbit/s and demonstrated good agreement between software models and theory [19]. A suite of computer programs were developed to model the European PDH employing CCITT defined multiplexers spanning 140/34/8/2 Mbit/s. An extensive study had been made on how errors propagate from 140 Mbit/s down to 64 kbit/s.

The application of error bursts to a 140/34 Mbit/s demultiplexer was considered with every bit in error and in the context of the linear error propagation assumption of CCITT Recommendation G.821 in [20]. Results generated by theory, software and hardware showed good agreement, and they demonstrated that the above assumption is only valid for error burst lengths shorter than one-third of the frame length.

13.7 ASPECTS OF JITTER IN OFDLS

The ability to regenerate signals periodically as they propagate down a line has been essential to the growth of digital network. Few engineers currently working in the telecommunications industry will be aware that a system of regeneration was perfected before they were born, and that the foundations were laid for digital systems. In 1923 the Eastern Telegraph Company used the system at its relay stations to signal a new element of theoretically perfect shape and amplitude into the next cable section. The new technique allowed as many as 14 cable sections to be joined in series and was introduced at 120 stations working over some 232 000 km within four years. Accurate timing was the key to the process of sampling and regeneration. Problems associated with these processes were overcome electromagnetically, as were problems with network synchronisation, jitter, bit interleaving, combining of channels operating at unequal speeds, etc., without a valve or transistor in sight! Even store and forward techniques were available ([8] of Chapter 1).

By 1995 transoceanic systems with complex network topologies were in service, carrying traffic at 4.8 Gbit/s, and systems operating at 10 Gbit/s will be nearing the end of the development phase. Such systems place increased emphasis on verification prior to installation. The accumulation of timing jitter is of particular concern on such long routes and those with submerged multiplexing facilities [21]. An equivalent-time jitter synthesis experiment was described which allowed laboratory evaluation of transoceanic regenerated ofdls. Results were reported for the systematic jitter accumulation obtained over a 591.2 Mbit/s 6000 km route derived from the simulated transatlantic span containing 60 regenerated sections. These results provided insight into regenerated link performance at the laboratory prototype stage.

Unrepeatered systems jitter is derived from the terminal equipment, and it can be divided [1] into intrinsic jitter, jitter transfer function and maximum input jitter. These components of the jitter performance of the terminals can, taking a 140 Mbit/s systems as an example, be measured as follows:

(a) intrinsic jitter measurements were made using a cmi code with $2^{23} - 1$ prbs injected at a nominal rate of 139.264 Mbit/s at the optical line terminating equipment input.

(b) the transfer function was measured by injecting 1 UI jitter at the transmit terminal using a 1000 word long sequence at 139.264 Mbit/s and measuring the received jitter.

(c) maximum input jitter was also measured using a cmi code with $2^{23} - 1$ prbs injected at a nominal rate of 139.264 Mbit/s. The input jitter was increased until bit errors occurred.

13.7.1 Power penalty due to jitter

The degradation due to imperfect timing becomes more serious as the bit rate increases and the resulting power penalty was discussed in [22]. It was shown that some simplifications made in earlier work gave misleading results and that a more rigorous approach was necessary. The authors concluded quite generally that the performance assessment of an ofdls in the presence of jitter requires a more rigorous approach than a simple Gaussian approximation, and it requires the use of numerical techniques to compute power penalties. Two pairs of graphs were presented that showed that power needed to maintain ber at 10^{-9} (power penalty) versus bit rate multiplied by twice the r.m.s. jitter: the first pair referred to a uniform distribution and a Gaussian approximation; the second pair to a Gaussian distribution and a Gaussian approximation. The difference between members of a pair was particularly marked in the case of jitter with a Gaussian pdf. It was inferred from reported results that realistic jitter distributions were far from Gaussian.

13.7.2 Jitter

This falls into two distinct and almost independent categories: line system jitter, and tributary jitter, and we will look at these in turn.

Line system jitter

Passage through many regenerators in tandem can result in a pronounced bunching of bits due to imperfect timing recovery; this effect can be viewed as phase modulation (jitter) of an ideally extracted clock signal. Jitter may be expressed in either the frequency or the time domain, and in many cases only the peak-to-peak value is of interest as being the most likely to cause decision errors. An estimate of the operating margin against jitter can be obtained at an early stage in a system design from modelling techniques based on simplifying assumptions.

The small amount of jitter caused by a single regenerator causes a serious measurements problem, but this can be alleviated by the use of suitable measuring equipment such as the jitter measurement system developed by British Telecom. This equipment has enabled the effects of production variations and various data patterns on the jitter level in a single regenerator to be seen. The choice of test patterns is important and must therefore be considered carefully. Although de-jitterisers can be used at terminals, they cannot correct any errors that jitter may have caused in submerged regenerators; thus the repeaters must operate error-free with a suitable margin at the highest foreseeable jitter level.

Tributary jitter

Justification processes used in plesiochronous systems are characterised by the input tributary bit rate varying within certain limits. The corresponding output rate at the remote station will follow these variations in accordance with the presence or absence of justification digits in the line signal, i.e. the tributary path through the system is largely transparent to jitter. Therefore it is important to be certain that the inland multiplex hierarchy specifications are compatible. If they are not, the tributaries will need to be de-jitterised either before or after transmission over the line, and this process is quite independent of any jitter on the line. Since only an integer number of time slots can be added to the incoming tributary by the justification process, when these are deleted as appropriate the signal arrives in bursts, separated by gaps of one time slot. This signal is read into a buffer store then read out regularly at a rate controlled by the average relative positions of the read and write pointers. Because of circuit imperfections, the amount and spectral distribution of jitter in the resulting bit stream depends on the characteristics of the phase-lock loop.

An additional component of jitter comes from the fact that a justification time slot is not always immediately available when a decision to insert a justification digit is made; hence a delay can be introduced. Theory shows that a change in the nominal justification ratio will change the jitter spectrum after de-justification and buffering. A reduction in this ratio might necessitate redesign of the (de-justified) clock recovery circuit if the original design was barely adequate.

In conclusion, the fact that tributary paths in the undersea link are largely transparent to jitter exposes the inland networks connected to the link terminal stations to one another. The link jitter specification is bounded solely by the need to operate error-free with suitable margins.

13.7.3 Accumulation of jitter

A novel technique for the simulation of jitter accumulation in regenerated submarine optical transmission systems was also described in [23]. Laboratory jitter measurements systems have been described in the literature based on either equivalent-time or recirculating loops containing one or more regenerated optical spans. An innovative equivalent-time jitter synthesis technique based on

Manley's tape recorder experiment was presented which allowed straightforward extension to high-speed optical system evaluation. Illustrative system simulation results were described which showed the effects on timing jitter accumulation of increasing saw timing recovery filter delay from 0.89 μs to 10 μs on a synthesised 591.2 Mbit/s, 3000 km optical transmission system comprising 30 concatenated 100 km regenerated sections.

13.8 REFERENCES

1 DOYLE, T., and STYLES, J.: 'A 126 km unrepeatered optical fibre submarine cable between Ireland and the U.K.'. *Int. J. Digital and Analog Cabled Systems*, **2**, no. 1, Jan./Mar. 1989, pp. 35-48.

2 ROGALSKI, J.E.: 'Evolution of gigabit lightwave transmission systems'. *AT&T Tech. J.*, **66**, no. 3, May/June 1987, pp. 32-40.

3 MYALL, J.V.W.: 'The design of production terminal equipment for digital optical-fibre submarine cable systems'. *British Telecommunications Engineering*, **5**, part 2, July 1986, pp. 113-123.

4 RIDOUT, I.B., and HARVIE, I.: 'The principles of scramblers and descramblers designed for data transmission systems'. *British Telecommunications Engineering*, **1**, July 1982, pp. 111-114.

5 HAGIMOTO, K., MIYAMOTO, Y., ONO, T., KATOKA, T., and NAKAGAWA, K.: 'Ultra-high-speed optical modulation and detection schemes for long-haul fiber transmission'. *IEEE Int. Conf. on Communications*, June 1992.

6 JONES, D.L.: 'An engineering approach to fibre optic link design'. *Electronic Engineering*, mid-Apr. 1980.

7 BATTEN, T.J., GIBBS, A.J., and NICHOLSON, G.: 'Statistical design of long optical fiber systems'. *IEEE J. Lightwave Technol.*, **7**, no. 1, Jan. 1989, pp. 209-217.

8 BORLEY, D.R., WALKER, S.D., CARPENTER, R.B.P., and KITCHEN, J.A.: 'Some design aspects of long-haul digital submarine optical fibre systems'. *The Radio and Electronic Engineer*, **54**, no. 4, Apr. 1984, pp. 163-172.

9 TAYLOR, N.H.: 'Power budgets for long-haul optically amplified systems'. *4th IEE Conf. Telecommunications*, Manchester, 18-21 Apr. 1993.

10 RUSSELL, J.N., TAYLOR, N.H., BYRON, K., and BLEWETT, I.J.: 'Dispersion management in long-haul optically amplified submarine systems'. *Int. Wire and Cable Symp.*, St. Louis, USA, 15-18 Nov. 1993.

11 DESLANDES, J.E., JONES, E.V., and COCHRANE, P.: 'Analysis of SDH pointer corruption due to error activity'. *Electronics Lett.*, **27**, no. 5, 28 Feb. 1991, pp. 453-455.

12 BUTLER, R.A., and COCHRANE, P.: 'Correlation of interference and bit error activity in a digital transmission system'. *Electronics Lett.*, **26**, no. 6, 15 Mar. 1990, pp. 363-364.

13 COCHRANE, P., and POWELL, W.: 'Simple received signal/eye amplitude monitor for digital repeaters'. *Electronics Lett.*, **18**, no. 9, 29 Apr. 1982, pp. 388-390.

14 FISHMAN, D.A.: 'Elusive bit-error-rate floors resulting from transient partitioning in 1.5 μm dfb lasers'. *J. Lightwave Technol.*, **8**, no. 5, May 1990, pp. 634-641.

15 SANFERRARE, R.J.: 'Terrestrial lightwave systems'. *AT&T Tech. J.*, **66**, iss. 1, Jan./Feb. 1987, pp. 95-107.

16 BOND, D.J.: 'A theoretical study of burst noise'. *Br. Telecom Technol. J.*, **5**, no. 4, Oct. 1987, pp. 51-60.

17 FRISCH, A., and FLETCHER, I.: 'End-to-end system performance predictions and problems'. *IEE Coll. 'Submarine optical transmission systems'*, London, Mar. 1988, Coll. Digest, 1988.

18 DESLANDES, J.E., COCHRANE, P., and JONES, E.V.: 'Error propagation through plesichronous demultiplexers'. *IEEE Globecom '90*, San Diego, USA, 2-5 Dec. 1990, pp. 1111-1115.

19 DESLANDES, J.E., COCHRANE, P., and JONES, E.V.: 'Error propagation through the European demultiplexer hierarchy — an analysis'. *Electronics Lett.*, **26**, no. 18, 30 Aug. 1990, pp. 1469-1470.

20 DESLANDES, J.E., COCHRANE, P., and JONES, E.V.: 'Effect of 100% density error bursts on a 140-34 Mbit/s plesiochronous demultiplexers'. *Electronics Lett.*, **25**, no. 7, Mar. 1989, pp. 473-474.

21 McNALLY, B.I., CARTER, S.F., and WALKER, S.D.: 'Jitter accumulation in a simulated 591.2 Mbit/s 6000 km optical transmission system'. *Electronics Lett.*, **24**, no. 11, 26 May, 1988, pp. 676-678.

22 SCHUMACHER, K., and O'REILLY, J.J.: 'Penalty due to jitter on optical communication systems'. *Electronics Lett.*, **23**, no. 14, 2 July 1987, pp. 718-719.

23 CARTER, S.F., McNALLY, B.I., and WALKER, S.D.: 'A novel technique for simulation of jitter accumulation in regenerated submarine optical transmission systems'. *IEE Coll. 'Submarine optical transmission systems'*, London, Mar. 1988.

14 SINGLE-MODE SYSTEMS

Specifications are a complete and precise written statement of what is to be accomplished on an engineering project or task.

W.H. Roadstrum, 'Excellence in Engineering'

14.1 INTRODUCTION

In the preceding chapter some general systems considerations were discussed, which were intended to provide background to the topics chosen. Operational requirements are familiar to most system engineers, but the example chosen was very forward-looking, dealing as it did at the beginning of the 1980s, with the delivery of broadband services which are only entering service in any number in the mid-1990s. After the section devoted to an overview of optical fibre digital line systems two design philosophies were described: worst-case design, and the use of a statistical design approach, which can yield some worthwhile benefits. A major part of the chapter was taken up with detailed examples of power budgets appropriate for regenerative ofdls and for optically amplified ofdls; from initial budgets involving only a few parameters, to others which take into account many sources of impairments. Recent material on errors and their effects on performance formed the penultimate section, and the chapter concluded by considering some aspects of jitter.

In the present chapter there are only three main sections, which deal with the contents of a specification for an ofdls, with details of a 565 Mbit/s system specification as representative of core trunk systems of the early 1990s, and lastly with some changes to specifications necessitated by the introduction of optical amplifiers on a network-wide basis in the mid- to late 1990s.

Commercial exploitation of single-mode technology in the UK and USA began during 1983, and large-scale production of cables and equipments started that year. The decision to change from multimode to sm, and to go all the way from subscriber to subscriber on single-mode fibres was made during the 1983–4 period by both administrations. During the next two years the first commercial undersea systems were installed, Optican 1 by AT&T and UK–Belgium No.5 by STC. The years 1985–6 also witnessed a number of 565 Mbit/s and 1.7 Gbit/s trial systems, which were followed by commercial installation in the next two years. By 1987 the pace of installation of trunk (long-haul) and junction (short-haul) systems had slowed down, and the deployment of local line access systems became significant in both countries. In the UK the conversion of some routes to duplex (same wavelength, bi-directional) and diplex (different wavelengths, unidirectional) single fibre working was done to achieve greater utilisation of existing cable routes. In addition, this provided a low cost add-drop facility on these routes [1]. Some five years after the decision was taken to go to sm fibre, an estimated 2 500 000 km of optical fibre, the vast majority of it single-mode, had been installed world-wide in the long and short-haul networks, with some 480 000 km in the UK alone.

14.2 OFDLS SPECIFICATIONS

Operational systems must conform to national specifications, which in turn take account of the recommendations of international bodies such as CCITT and CCIR. These specifications are necessarily very detailed to ensure the service reliability and performance.

Headings in system specifications

General, construction, cables, electrical interface, optical interface scramblers, regenerative repeaters, receiver, transmission system performance, maintenance and alarms, supervisory subsystem, protection from lightning and induced 50 Hz interference, error ratio detection strategy, service protection network (SPN), alarm indication signal (AIS).

To indicate the comprehensive nature of specifications for operational systems, the following list of main headings and topics from relevant British Telecom specifications of the 1980s may be of interest.

(1) Overall system configuration and description
(2) Overall system performance
(3) Interfaces: 3.1 Electrical (system) interfaces; 3.2 Optical (fibre/equipment) interfaces
(4) Main transmission fibre and cable
(5) Miscellaneous design aspects: 5.1 Line code; 5.2 Scrambler and descrambler; 5.3 Timing extraction circuitry; 5.4 Alarm indication signal; 5.5 System budget; 5.6 System margin; 5.7 Optical connectors
(6) Maintenance philosophy and alarms
(7) Supervisory subsystem
(8) Engineering speaker subsystem
(9) Power feed subsystem
(10) Equipment configuration
(11) Optical safety
(12) Optoelectronic devices
(13) Reliability
(14) Test equipment
(15) Measurement techniques.

Such a document could consist of some 30 single-spaced pages.

The Specification for 140 Mbit/s Single-Mode Systems included the following topics.

Specific design features: 1 SPN facilities; 2 electrical interface (traffic signal through ofdls); 3 optical interface; 4 transmit apparatus; 5 receive apparatus.

Cable section parameters: internal coaxial cable, external cable section length.

System margins and power budgets: transmission system performance limits, bit sequence independence (error ratio, jitter, word realignment).

Supervisory and speaker facilities.

14.3 565 MBIT/S SINGLE-MODE SYSTEMS

By mid-1988 the British Telecom long lines network was capable of providing a wholly digital service on a network dominated by 140 Mbit/s transmission over single-mode fibre. The further expansion and development of this network will take place by the progressive introduction of 565 Mbit/s and higher rate systems to exploit the latent capacity of installed and new fibre cables. Among the key objectives in the provision of this network are the achievement of high reliability and performance through long repeater spans and integrated equipment. In the United States and Japan similar trends took place; the upgrading of the AT&T 417 Mbit/s system to 1.7 Gbit/s has already been mentioned.

The use of a high level of integration confers many benefits, among which improved reliability, reduced component count, lower overall power dissipation and reduced cost are the most important [2]. Unfortunately, a key problem with mid-1980s high-speed logic families, such as bipolar ECL or GaAs, was the high power dissipation per unit volume. This was a major constraint on the level of circuit integration which can be achieved in practice; typically, such ics contained only a few thousand gates, but the dissipation could be as high as 8 W. The concentration of such high powers in such small volumes demanded careful attention to packaging and thermal management, particularly because natural convection is often specified in telecommunication systems. These requirements often conflict with the need for good electromagnetic screening and protection against electrostatic discharge. A new thermal management system which overcame these problems was specifically designed for the 565 Mbit/s system to be described.

14.3.1 System outline

Basically, the system comprises the equipment for multiplexing and coding four plesiochromous tributary bit streams at 139.264 Mbit/s and for the transmission of the composite signal over a single-mode fibre at 1300 nm. Each tributary interface is provided with service protection switching options to interface to the automatically switched digital SPN. Ancillary equipment provides network management facilities for supervisory and fault-location purposes. To meet the needs of future broadband services as they arise, optional tributary interfaces at 278.528 and 564.992 Mbit/s can be provided. Regenerative repeaters with high optical gain of more than 33 dB are used, thus allowing long section lengths, and a direct upgrading of existing 140 Mbit/s sm systems without the need for additional repeater sites. As all the distances between surface buildings in the BT network are less than 35 km, this high gain also eliminates the need for buried repeaters and the necessary power feed equipment. The optical receiver has an exceptionally large dynamic range of

Table 14.1 Main parameters of 565 Mbit/s system ([2], Table 1). Reproduced by permission of BT

Interface Options:	4×140 $2 \times 140 + 1 \times 280$ 2×280 1×565 } Mbit/s
Line code:	7B8B
Fibre:	Single mode, 1.3 μm
Repeater Gain:	33 dB
Repeater Spacing:	>40 km
Length:	96 intermediate repeaters
Supervisory:	Comprehensive, in-service sub-band signalling over fibre
Integration:	BTRL ECL semi-custom arrays microcomputer controlled supervisory
Equipment Practice:	BT type TEP-1E

greater than 30 dB, which does away with the requirement for optical attenuators on short low-loss sections. All the optical components are hermetically sealed and temperature-stabilised to ensure the required performance, and the reliability is achieved over a wide range of operating conditions.

The high-speed circuits of the system employ two new types of uncommitted cell array developed by British Telecom Research Laboratories, which are based on the ultra-reliable ic developed for submerged applications enhanced in speed and extended in size. These, together with a number of commercially available ics used for the slower speed functions, combine reduced equipment volume and power consumption with improved reliability.

The microprocessor-based supervisory subsystem provides comprehensive in-service monitoring and fault-location facilities. Transmission of supervisory and fault information between repeaters and terminals is on the same fibres as the traffic signals. A sequential polling strategy has been developed which simultaneously and automatically polls the system from each line terminating equipment, giving an improved fault-finding capability, and eliminating the need for a master–slave strategy. The main parameters of the system are listed in Table 14.1.

Multiplexing and coding concepts

For systems based on the 2 Mbit/s hierarchy the highest level interface defined by the CCITT is 140 Mbit/s, so that systems of higher capacity are simply defined as $n \times 140$ Mbit/s. The lack of interest in higher rate systems in the mid-1980s stemmed from the fact that in most countries, 140 Mbit/s was more than adequate to meet customer needs for all foreseen services, with the possible exception of high-definition television. Higher-level interfaces also present technical problems for in-station cabling and service protection. For example, in 1986 BT was installing a national digital service protection network, fully automatic in operation, and at the 140 Mbit/s level. Clearly, it was economically attractive in the short-term to exploit this facility for higher-capacity trunk systems, therefore all 565 Mbit/s systems designed for the 2 Mbit/s hierarchy are configured as 4×140 Mbit/s systems.

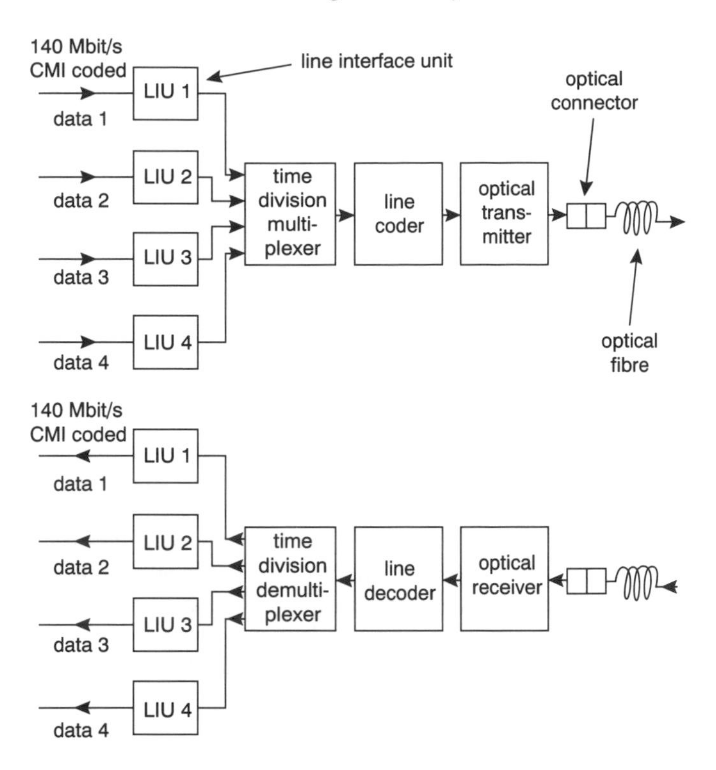

Figure 14.1 General schematic of the signal transmission path of a 4 × 140 Mbit/s tdm ofdls terminal ([2], Figure 1). Reproduced by permission of BT

A general schematic of the signal transmission path of such a tdm optical line system is given in Figure 14.1.

In accordance with CCITT recommendations each of the system input signals operates at a nominal information rate of 139.264 Mbit/s ±15 ppm, but has been encoded using coded mark inversion (cmi), a 1B2B code. Before decoding, it is necessary to regenerate and retime the incoming signal, because it is usually attenuated and distorted by the in-station coaxial cabling. Although the tolerance on the clock frequency may at first sight seem small it is equivalent to more than 4000 bits per second difference between the fastest and the slowest permitted tributary rates. Clearly it is not possible to undertake a simple time division multiplex of the four tributaries; instead, it is necessary to synchronise the bit streams. This is achieved by the positive justification, or pulse stuffing, technique, whereby additional bits are inserted into the bit stream in a controlled manner which does not corrupt the information, and which can be extracted at the receiving terminal. Normally information is partitioned into frames with a fixed number of time slots per frame, known as justifiable digits, being used either to transmit information from the original signal or stuffed digits. A number of additional digits (justifiable service digits) are added to each frame to transmit information concerning the status so that stuffed digits can be removed at the far terminal. The penalty for making a single mistake in the operation can be catastrophic, in that it will cause all subsequent demultiplexers to lose alignment. To guard against this possibility, and to provide some protection against random and burst errors, five justifiable service digits for each justifiable digit spread throughout the frame are used per justifiable digit with majority decision. The service digits provide a facility for the signalling of remote alarms, supervisory and end-to-end engineering services/speaker. Referring to Figure 14.1, it can also be seen that line coding and time division multiplexing functions are shown separately, but for high-speed systems there are many advantages in physically combining the two processes because this reduces the amount of circuitry which is required to operate at high speeds. This approach has been adopted in the present system.

Optical technology

This section begins with the optical power budget for the system, given in Table 14.2.

The mean launched power is −3 dBm and the pulse format is nrz, which is readily obtained from modern lasers, and under normal operating conditions does not exceed the safe working limits of a Class 1 laser as derived from British Standards. Turning now to the receiver, a sensitivity of −36.0 dBm (−39 dBm in 1993 for 565 Mbit/s systems) at the line rate of 647 MBd is achieved using a front-end stage based on a hermetically sealed 100 μm Ge apd. Since these photodiodes suffer from a large bulk leakage current at high temperatures (see Section 10.4), a Peltier cooler is incorporated within the receiver to keep the temperature constant at 25°C over the full range of operating conditions. The importance of this can be judged from the fact that without the cooler a sensitivity penalty of approximately 2.5 dB would be incurred at the maximum rack temperature of 60°C. The sensitivity value quoted in the table is the expected measured performance at an ambient temperature of 60°C using an optical attenuator instead of fibre.

In practice there are additional impairments which are due to fibre dispersion, reflection noise, jitter and component tolerances, collectively accounted for under the heading of design penalty. At 1300 nm chromatic dispersion and the related pulse broadening, mode partition noise and reflection noise are potentially serious problems for rates of 565 Mbit/s and higher. These effects can be contained within acceptable limits by tightly specifying the centre wavelength and the spectral width of the laser, and by providing temperature stabilisation. This latter precaution normally has the added bonus of improving the laser reliability and preventing thermal runaway. Because dispersion-related penalties increase extremely rapidly

Table 14.2 Optical power budget for 565 Mbit/s system ([2], Table 2). Reproduced by permission of BT

Mean launch power (NRZ)	−3.0 dBm
Receiver sensitivity (10^{-10} EBER)	−36.0 dBm
Repeater gain	33.0 dB
Fibre loss	23.5 dB
Cable repair margin	2.0 dB
Cable allocation	25.5 dB
Design penalty	3.0 dB
Ageing and temperature margin	2.5 dB
Connector loss (20 connectors)	2.0 dB
	7.5 dB
Optical receiver dynamic range	>30.0 dB

with the consequent increase in the error floor, it is good design practice to restrict this penalty to less than 2 dB. We now consider this in more detail. Although the dispersion penalty depends on the particular receiver configuration and on the laser characteristics in a given system, for 565 Mbit/s transmission using a 7B8B code the dispersion should be constrained to below about 500 ps. Single-mode fibres installed by BT in 1986 had dispersion zeros around 1310 nm, and in the band 1300–1325 nm the dispersion was less than 3.5 ps $nm^{-1}km^{-1}$. The laser wavelength was specified to lie within this window, and for a 2σ spectral width of less than 4 nm, repeater spans exceeding 40 km are realisable. Experience of installed fibres showed that they often have lower dispersion so that even longer spans are possible. In general, installed fibres had per-unit loss below 0.6 dB/km, which for the values quoted in Table 14.2 allow spans of $\geq$40 km. Combining the dispersion and attenuation values leads to the result that, for this particular system, repeater section length could be either dispersion- or loss-limited. Which of the two applied depended on the precise characteristics of the installed fibres on any given route, but in general a span of 40 km could be achieved with an adequate system margin.

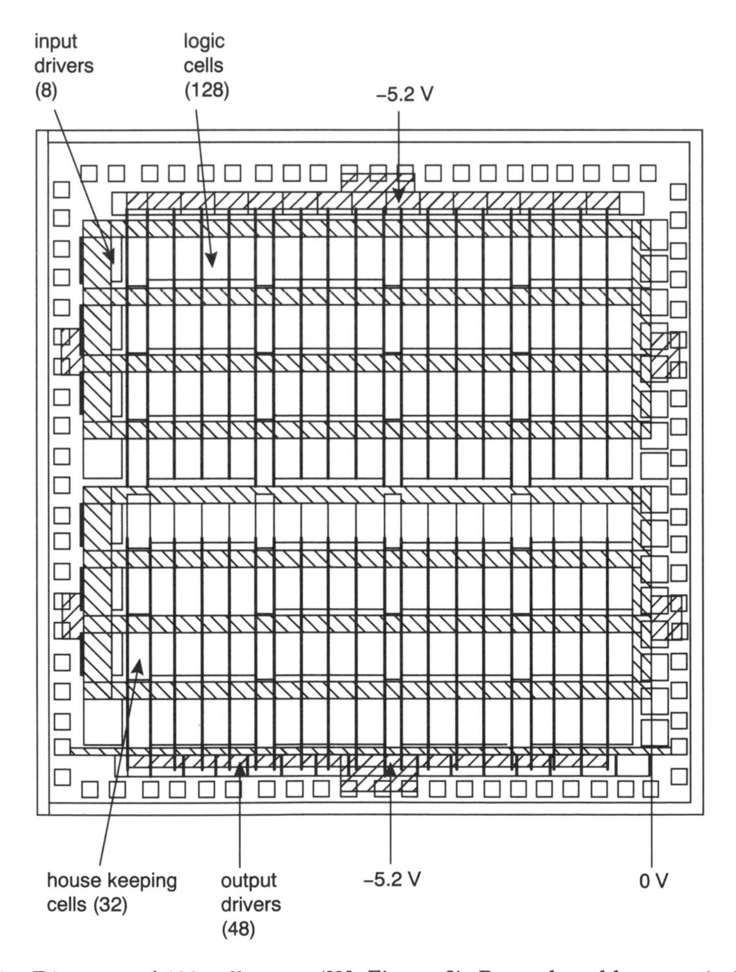

Figure 14.2 Diagram of 128 cell array ([2], Figure 3). Reproduced by permission of BT

Electronic technology

A number of bipolar cell arrays based on the well-established BT ECL40 technology with an f_T of 4 GHz are used in the system. This process uses an implanted buried layer for collector contacts and isolation between components, and has been in use in submarine systems for at least a decade. Two types of cell are employed, and they will now be described briefly.

128 cell array

These are based on the standard ECL process which uses a 5 μm emitter width, with the addition of a double-level metallisation system, copper–aluminium–silicon to obtain a higher level of integration and interconnection. The overall dimensions of the chip are 4.5 × 4.5 mm, providing 1624 transistors in 8 rows of 16 macrocells, together with input/output driver and housekeeping cells, as illustrated schematically in Figure 14.2. The purpose of the housekeeping cells is to generate temperature and voltage compensated bias voltages for current sources, reference voltages and clock distribution buffers.

24 cell array

This embodies enhancements to the ECL40 technology with reduced geometry transistors giving a significant improvement in switching speed. A 3 μm emitter size

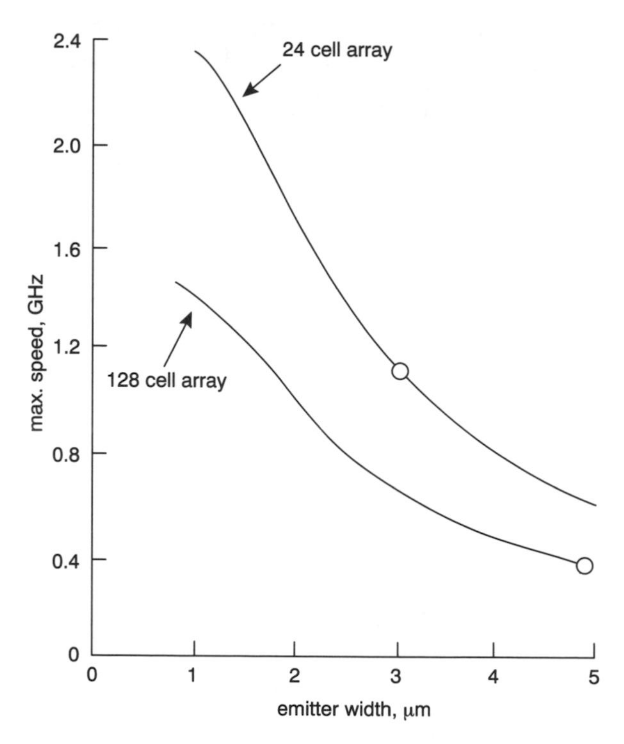

Figure 14.3 ECL40 technology speed predictions and array operating points ([2], Figure 4). Reproduced by permission of BT

Table 14.3 BT bipolar chip array performance ([2], Table 3). Reproduced by permission of BT

	128 Cell	24 Cell
Output risetime	1 ns	600 ps
Output falltime	1 ns	600 ps
Maximum clock rate	>450 MHz	>850 MHz
Power dissipation (typical)	<4 W	<1 W

is used, but the minimum dimension is only 1.2 μm for both contact masks and metallisation. The chip dimensions are 2.5 × 4.5 mm and there are 524 transistors arranged in 3 rows of 8 macro-cells, together with input/output driver cells.

Typical performance results for these chips are listed in Table 14.3 and in Figure 14.3.

14.3.2 565 Mbit/s system design details

Multiplex and coding

A key feature of the LTE is that multiplexing and coding functions are combined, as the reader can see from Figure 14.4. This figure shows the ic architecture of the transmit half of the line terminating equipment, and the use of parallel processing in the manner illustrated provides a robust and reliable design, possessing the following advantages:

(a) a 7B8B alphabetic block code which can be realised using proms
(b) minimum circuitry is needed to operate at rates above 140 MBd
(c) the only circuits working at the line rate are the 2–1 interleaver and disinterleaver
(d) there are few critical timing areas.

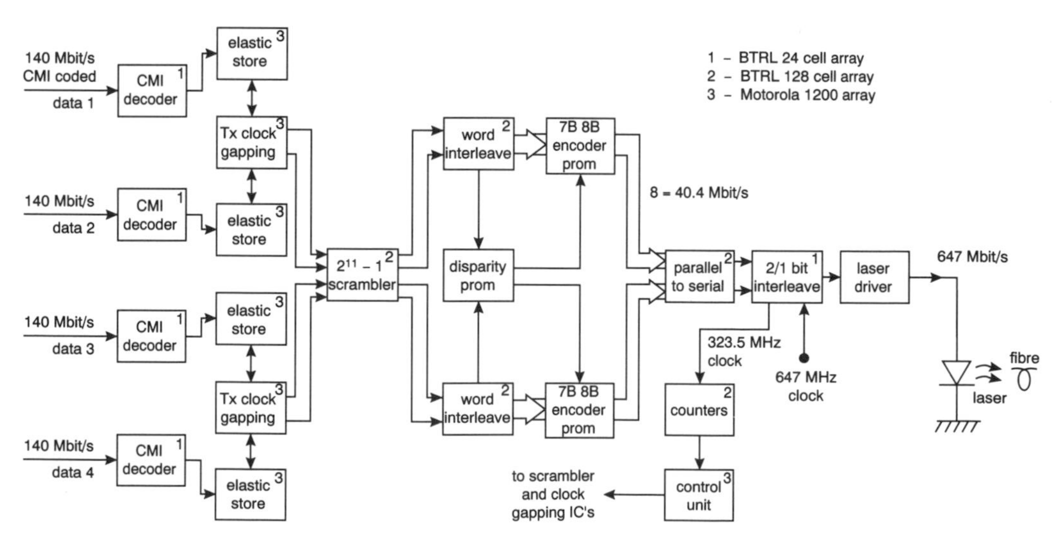

Figure 14.4 565 Mbit/s transmit terminal architecture ([2], Figure 5). Reproduced by permission of BT

Each interface unit decodes the cmi in-station interface code, synchronises the signal via a 16-bit elastic store and inserts gaps into the bit stream for the frame alignment signal (FAS) and end-to-end communication channel. With the exception of the FAS each synchronous 140 Mbit/s tributary is scrambled using a $2^{11} - 1$ set–reset scrambler. A single scrambler is used with quadrature phases being accessed for the four parallel signals. Scrambling, although not essential, is desirable since randomising the data improves the performance of the demultiplexer and decoder, error detection, repeater control circuits and ac coupling.

There are two main paths in parallel through the equipment, which are in turn subdivided into a number of parallel paths for the operation of coding. This approach results in a clock period of 24.5 ns, which has the advantage of allowing standard ECL prom to be employed. Although two separate encoders are used, the disparity control is common and results in a genuine 7B8B line code with a digital sum variation of 5. The two 8-bit parallel outputs from the encoder are interleaved, passed through a parallel-to-series converter, giving a 650 MBd symbol stream. The reason for the choice of this code for the present system is that the word rate is sufficiently low to require only two parallel paths through the coder, and the size of the coding table is compatible with commercially available proms. An interesting feature of the frame structure adopted is that it is composed of seven sets, each divisible by seven; hence the FAS always occupies a known position in the 7B8B coding table, thus allowing detection at the relatively slow speed decoder. The resolution required of the framing circuit is therefore 24.5 ns, compared with 1.8 ns which would be needed if detection was performed at the line, or multiplexed rate. In addition, the number of possible positions of the FAS is reduced by a factor of seven, leading to a fast realignment time when using a conventional slip technique.

At the receiver the inverse operations are carried out, and this embodies a similar number of integrated circuits

Repeaters

Intermediate repeaters are of the 3*R* type, i.e. they reshape, retime and regenerate the main channel data as well as providing comprehensive remote supervisory and local alarm facilities. A schematic of the traffic path and the architecture of the ics is given in Figure 14.5. As shown in the figure, there are two types of ic employed, both of which have been implemented in the 24 cell array described earlier. ic 1 performs analogue functions such as amplification and peak signal detection, as well as containing a nonlinear element for the timing extraction circuit. The other, ic 2, performs digital functions associated with retiming the data, the autophase clock alignment circuit, and an invalid data inhibit which prevents the repeater 'free-running' in the absence of data, as this could degrade the supervisory telemetry path.

Turning to the front end, this uses a commercial 100 μm Ge apd, temperature stabilised to 25°C, feeding bipolar transimpedance preamplifier of bandwidth approximately equal to 0.7 times the symbol rate. The input stage together with a variable gain linear amplifier, supervisory interface and high voltage bias generator for the apd, are enclosed within a well-screened module. Inside the module, the electro-optics and the high-frequency electronics are separately screened from the power supply circuitry and the supervisory interface. A very wide dynamic range exceeding 30 dB is achieved by the use of a two-stage agc circuit, operating on both the apd and the second stage of the linear amplifier.

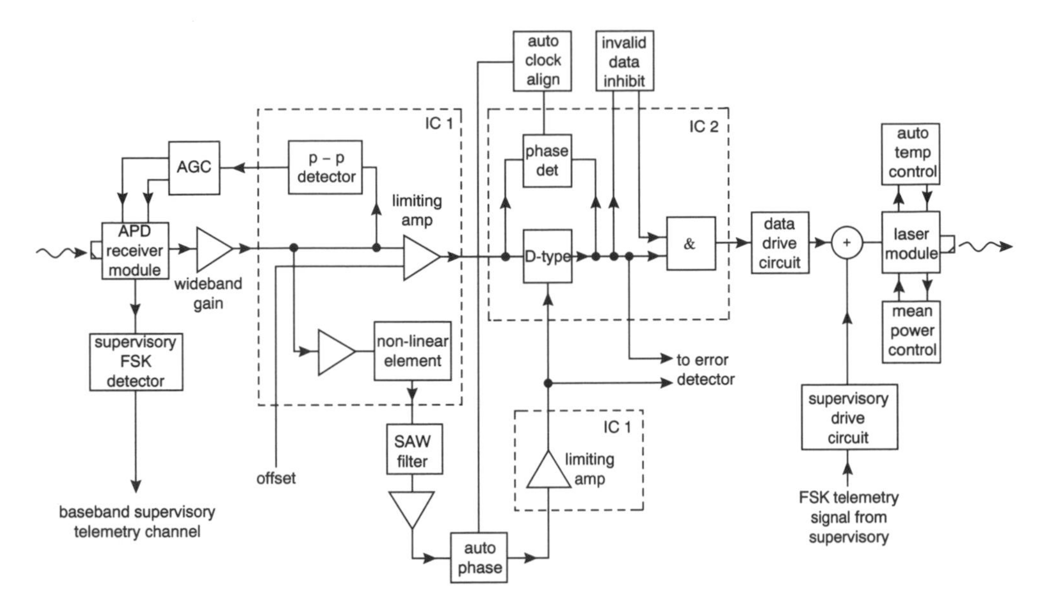

Figure 14.5 Block schematic of regenerative repeater ([2], Figure 6). Reproduced by permission of BT

As far as the receiver is concerned the 7B8B code provides an abundance of timing information, and at the same time it does not unduly restrict the Q or dynamic range of the timing extraction circuit. A quartz transversal surface acoustic wave (saw) filter has been developed having the advantages of high reproducibility, no manual adjustments and excellent long-term stability, e.g. $< \pm 0.1$ rad over 15 years. The saw filter is centred on the fundamental frequency, has a Q of 150, and an out-of-band rejection of greater than 30 dB. A novel and interesting feature of the timing extraction loop is that an automatic phase alignment network is used which gives dynamic decision point adjustment throughout the system life. Compensation is obtained for initial circuit misalignment and differential delay variations between the data and clock path caused by temperature variations and ageing. The circuit is shown in Figure 14.6, and it comprises two phase correlators (one to sense the clock, the other to serve as a reference with a fixed delay), a feedback loop and phase shift circuit.

The optical tramsmitter is a bh InGaAsP semiconductor laser with integral Peltier cooler and back face monitor. Mean power control and temperature control are employed to maintain a constant light output, and to stabilise the operating centre wavelength.

Supervisory subsystem

A comprehensive in-service system has been developed which operates automatically over the same fibres as the traffic, and uses sub-band signalling. This possibility has been treated in Chapter 9, and it relies on the fact that the power spectral density of 7B8B encoded data is very small at low frequencies, as indicated schematically in Figure 14.7.

The supervisory channel operates at 1200 Bd with 2 fsk, with a modulation index of 0.4 on an unmodulated carrier frequency of 15 kHz. Only a relatively small

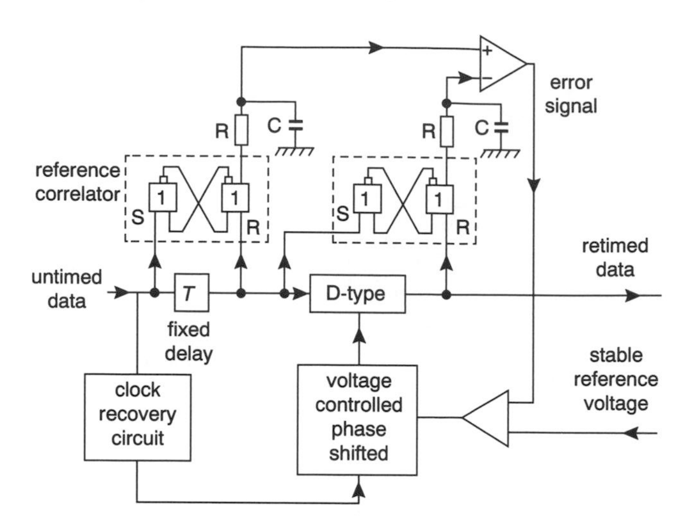

Figure 14.6 Automatic timing alignment circuit ([2], Figure 7). Reproduced by permission of BT

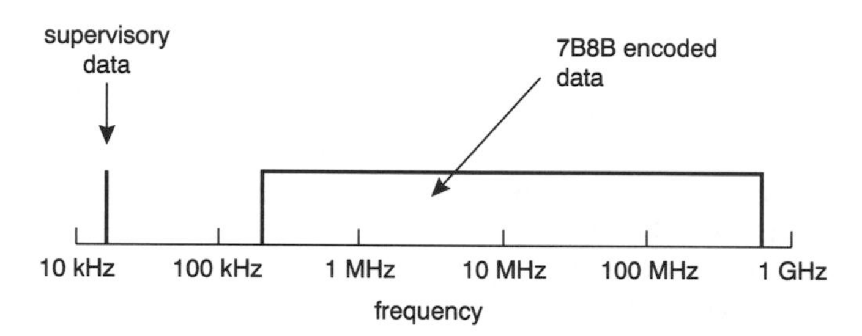

Figure 14.7 Traffic and supervisory spectrum allocations ([2], Figure 9). Reproduced by permission of BT

supervisory signal is required, and this causes a negligible main system penalty of less than 0.1 dB. Typically, the ratio of supervisory to traffic signal r.m.s. currents in the laser drive circuit is less than 2% for satisfactory operation of the supervisory channel over the range −6 dBm to −46 dBm. A property of the supervisory channel which is especially important under fault conditions is that it continues to operate satisfactorily in the absence of the traffic signal.

Messages which are either interrogation commands or status reports are sent over the channel in the form of asynchronous packets, in which redundancy is inbuilt to guard against mimicking of the interrogation commands, and to provide error checking. Both system terminals independently and simultaneously interrogate repeaters according to an automatic sequential polling strategy. On receipt of an interrogation command a repeater sends its status report, relating to both directions of transmission, towards both terminals before forwarding the interrogation command to the next repeater. Status reports are regenerated and checked at each repeater before retransmission. Under normal operating conditions two complete sets of status information are available at both terminals, and the sets are distinguished by their direction of polling. One advantage of this approach is that the terminals function independently, thereby avoiding master–slave working arrangements. In the event of a line break both terminals will be able to function and locate the fault to within one repeater section; whereas if a break occurs in only one direction of transmission, i.e. in one fibre, then one of the terminals will still receive status information for the complete system, but the other will only receive status information from repeaters upstream of the break.

Each terminal is equipped with two display systems. One is an alphanumeric LCD which is menu-driven through a set of software definable keys, the other is an LED display. An advantage of this dual display arrangement is that the LEDs can be used to continuously monitor a specific repeater, while the LCD can be employed to examine adjacent repeaters or the rest of the system. In addition to providing information in real time each terminal retains a historic log of the system status. Input/output ports have been provided to allow computer access, facilitating remote surveillance and network management.

Thermal management

Many circuit and system advantages flow from the use of a high level of circuit integration, but there is one major disadvantage, high power density, which leads to thermal dissipation limitations. The problem is compounded by the need to provide for good electromagnetic screening, which unfortunately tends to restrict the flow of air. Another important consideration is the need to eliminate mechanical stress caused by thermal mismatch between ic, package, heat sink and printed circuit board. The heat sinking arrangements which have been adopted are shown in Figure 14.8.

Ceramic packages have been used for their high reliability and low thermal resistance of less than 3°C/W. Heat is extracted through the base of the package by means of the aluminium pillar which protrudes through the pcb and is attached to the common heat sink. Measurements made on this arrangement have demonstrated that the thermal resistance is below 1°C/W between ic and the pillar. A small air gap between the pcb and the heat sink improves convection and enables both sides of the pcb to be utilised. Good electrical screening is ensured by a metal cover fitted over the component side of the pcb and fixed to the heat sink. The

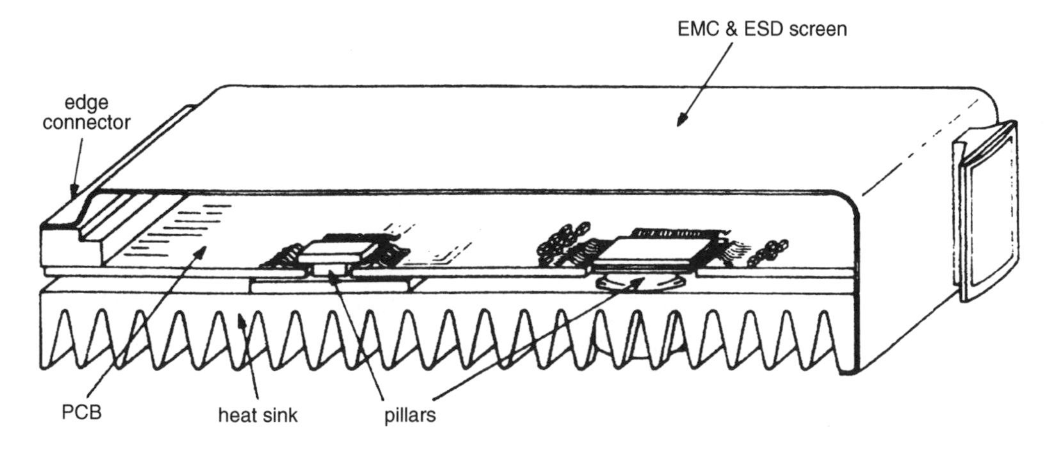

Figure 14.8 Heat sinking arrangements ([2], Figure 10). Reproduced by permission of BT

system thermal design relies on natural convection and ensures that ic junction temperatures remain below 100°C, and internal rack temperatures below 60°C for room temperatures up to 40°C.

Equipment

The line terminal equipment occupies only $2\frac{1}{2}$ shelves of TEP-1E type construction. The bottom shelf contains the optical transmission and receiver cards, error monitor card, alarm card, two supervisory cards (system processor card and local monitor/supervisory repeater card) and power units. The top shelf accommodates all the multiplexer/demultiplexer and coder/decoder cards. The half-shelf positioned between these two shelves has three important functions:

(1) it provides access to the optical connectors
(2) it allows mixing of the heated air and evens out the temperature
(3) it permits access to backplane sockets for interface coaxial cables, and extension of supervisory and alarms for remote management.

A twoway repeater is also constructed in TEP-1E practice and it occupies one shelf, but an additional half-shelf directly above the repeater is needed to provide access to the optical connectors.

14.3.3 Field trial system

Part of the system verification programme included a field trial, which began in December 1985 on the 77 km route illustrated in Figure 14.9. The values of loss recorded in the figure are the results of measurements carried out when the 565 Mbit/s system was installed to upgrade the existing 140 Mbit/s system. The average over the route is 0.52 dB/km, and it includes the splices. Measurement of dispersion in the window 1300–1325 nm was found to be less than 1.5 ps $nm^{-1}km^{-1}$, and the minimum occurred at 1315 nm. Based on these results, no significant dispersion-related penalties were expected, and this was found to be the case on each fibre span.

Measurements of repeater performance over the installed sections with a variable attenuator inserted gave sensitivities between −36.0 and −37.5 dBm for the main

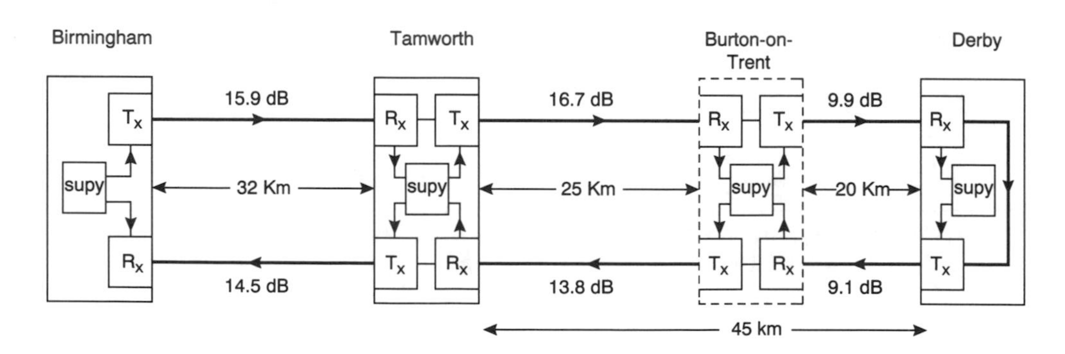

Figure 14.9 Field trial route ([2], Figure 14). Reproduced by permission of BT

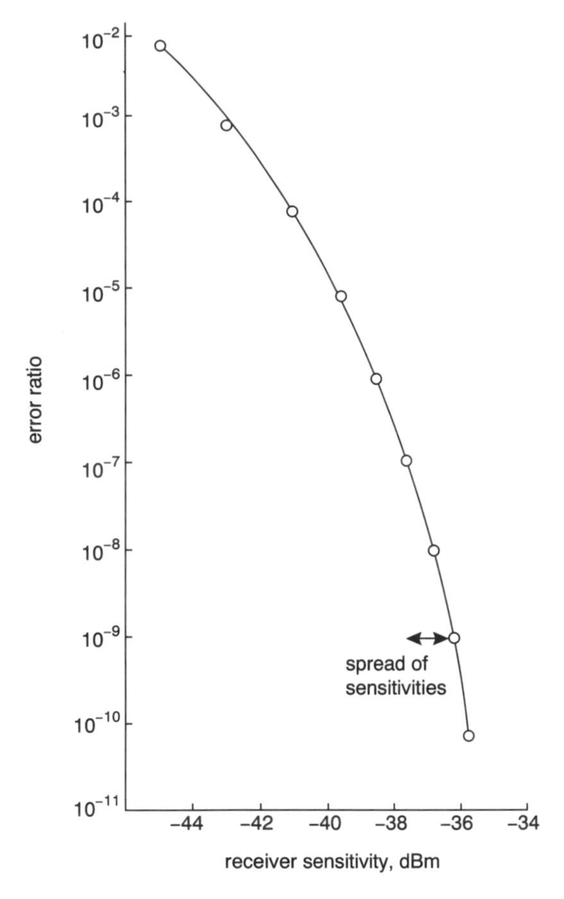

Figure 14.10 Typical measured repeater sensitivity ([2], Figure 15). Reproduced by permission of BT

channel for a ber of 10^{-9}, and between −46 and −51 dBm for the supervisory channel at 10^{-6} ber. A typical error ratio measurement of a repeater is given in Figure 14.10.

Referring to Figure 14.9, the low loss of the Burton to Derby section resulted, as expected, in received power levels of −14 and −15 dBm for the two fibres, but this did not present an overload problem for the receivers. As a demonstration of the potential of the system to achieve a long repeater span, the bothway repeater

Table 14.4 Line terminal performance ([2], Table 4). Reproduced by permission of BT

99% probability alignment time:	
Word	9.5 μs
Word + frame	13.0 μs
Receive tributary output jitter:	
200 Hz–3.5 MHz	0.061 UI p-p (typical)
	0.14 UI p-p (worst case)
10 kH–3.5 MHz	0.043 UI p-p

at Burton was bypassed, giving a double hop of 45 km with fibre losses of 22.9 and 26.6 dB. The system continued to operate error-free with margins of 7.1 and 5.7 dB on the 45 km section, and this established a world record for an installed 565 Mbit/s system at that time.

A summary of the realignment and output jitter measurement results is presented in Table 14.4. The word and frame realignment time is very much faster than the 63 μs specified for the system, and the receive tributary output jitter is very much less than the CCITT recommendations of 1.5 UI p-p for the band 200 Hz to 3.5 MHz and 0.075 UI p-p for the band 10 kHz to 3.5 MHz. The maximum tolerable

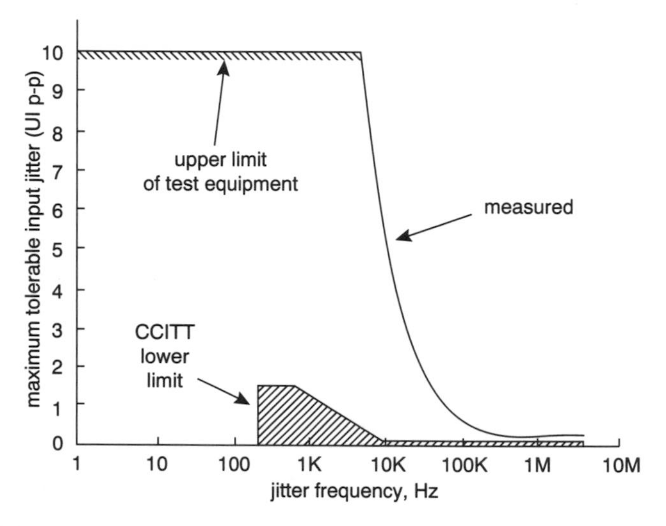

Figure 14.11 Maximum tolerable input jitter versus jitter frequency showing measured results and CCITT lower limit ([2], Figure 16). Reproduced by permission of BT

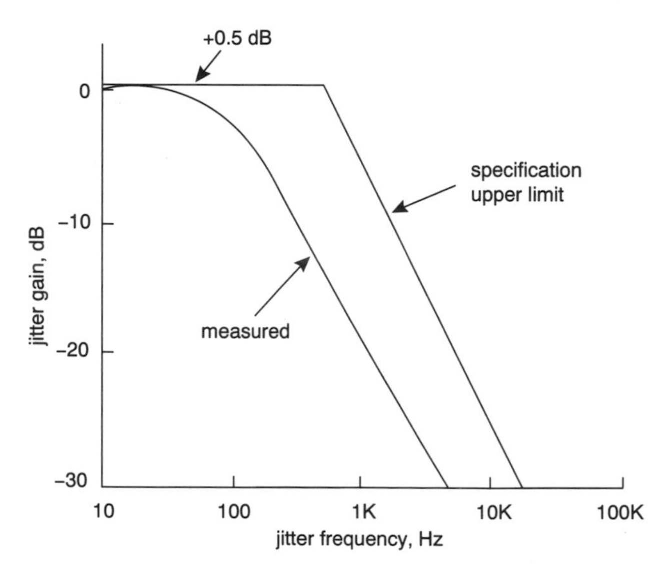

Figure 14.12 Measured jitter gain versus jitter frequency and upper limit of specification ([2], Figure 17). Reproduced by permission of BT

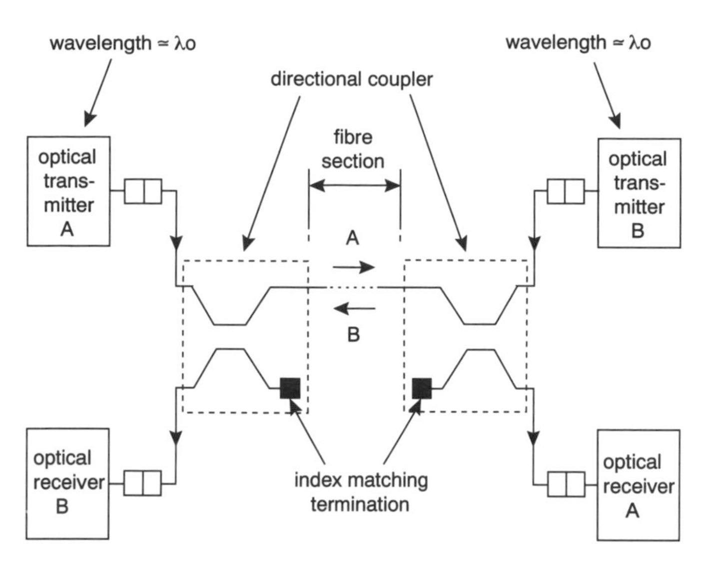

Figure 14.13 Arrangement of equipment for duplex transmission ([2], Figure 18). Reproduced by permission of BT

jitter at the input cmi interface is shown in Figure 14.11, and the jitter gain in Figure 14.12.

The reader can see from these figures that the system performance conforms to both the CCITT recommendations and the system specifications.

14.3.4 Use of route as a test bed for duplex operation

Around this time the feasibility of single-fibre duplex transmission of 565 Mbit/s signals over 50 km had been established in the laboratory, and BT decided to use the modified route as a test bed. The route was successfully converted to duplex operation in March 1986, by splicing directional couplers into the fibre at each end of the repeater sections. This made it possible to transmit signals in both directions simultaneously over the same fibre at the same operating wavelength. The principle is illustrated in Figure 14.13. This was the first significant field demonstration of duplex transmission anywhere in the world, and it showed that the capacity of optical fibre cables could be doubled at low cost.

14.4 APPLICATION OF OPTICAL AMPLIFIERS TO SINGLE-MODE SYSTEMS

By the early 1990s it had been shown that an amplified link can always outperform the equivalent regenerated link, and that optically amplified links are transparent to a range of data rates and wdm upgrades. The first such links will enter service in the mid-1990s on submerged routes such as TAT-12 and TAT-13 and on transpacific routes. A description of one such 5 Gbit/s system operating in the 1550 nm band is presented in Chapter 15; amplified systems operating in the 1300 nm band using praseodymium-doped fluoride fibre amplifiers may enter service about the same time. Here we will concentrate on the use of a comprehensive model of an optical amplifier in a detailed theoretical analysis of amplified two-section 622 Mbit/s optical transmission links.

Regenerators rely on the use of optoelectronic interfaces (photodiodes and lasers) and integrated electronics to detect, regenerate and retransmit optical signals. A link using regenerators consists of a number of independent optical sections, isolated by electronics, with each section comprising a laser, lossy and dispersive cable and a photodetector. A network-wide system specification can therefore be used to define the operational requirements (maximum loss, minimum launched power and so on) of the various elements of a section, so that arbitrary lengths can be installed using components from different suppliers. The major limitations of this type of link are due to the complex and costly embedded electronics and optoelectronics, which reduce reliability and which are dependent on the format of the optical signals (modulation, coding, etc.).

The transparency and absence of in-line optoelectronic interfaces of fibre amplifiers contrast dramatically with those listed in the preceding paragraph. These features confer operational advantages over regenerators; in particular, the ability to handle a range of transmission formats and data rates, and including wdm. Fibre amplifiers are, however, complex analogue devices in which the computed gain depends on the average power of the input signal, as shown in Figure 14.14, based on the improved average power analysis technique. Such devices generate spontaneous noise which affects the gain (as shown in Figure 14.14) and degrades the photodetection process in the receiver.

The deployment of optically amplified links could thus involve complex planning and design procedures because there are a wide range of options and associated input/output transmission loss combinations, even in links with only two sections.

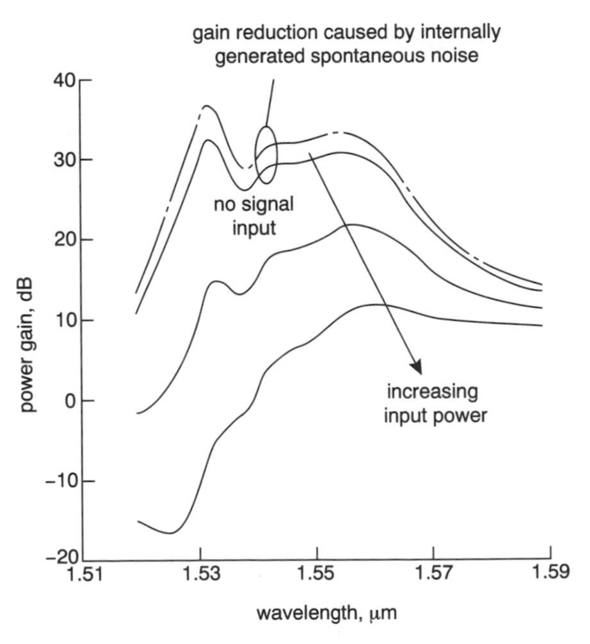

Figure 14.14 Computed curves of power gain versus wavelength, showing the effects of input power level and spontaneous noise on the gain profile of an erbium-doped fibre amplifier ([3], Figure 1). Reproduced by permission of BT

Satisfying this wide range of operating conditions could entail the availability of different amplifier designs (power, in-line repeater, and receiver preamplifier) and/or amplifier control techniques. Consequently, optically amplified links would not be considered practical on a network-wide basis unless it could be demonstrated that a simple amplified link specification exists which is compatible, in term of performance, with its regenerated equivalent. Even if this could be shown, amplifed links would need to offer some unique operational advantages to justify the costs incurred with the introduction of new technology into the network.

The existence of a network-wide specification for an amplified two-section link and the assessment of its potential performance advantages were investigated in [3]. A qualitative discussion of regenerative and optically amplified links was followed by a network-wide specification for the latter derived from that of the equivalent regenerated link. The two specifications were investigated by comparing the calculated performance of a regenerated and an amplified two-section 622 Mbit/s link for a full range of possible input and output section loss combinations. Since one advantage of amplified systems is their upgradability, this was investigated for three specific cases: 2.5 Gbit/s tdm, four-channel 622 Mbit/s wdm, and lastly four-channel 2.5 Gbit/s wdm. Results of the study showed that a network-wide specification can be used, that the two-section amplified link always outperforms the regenerative equivalent, and that the amplifed link provides increased positional flexibility and an inherent capability for upgrading. In order to cover the full range of combinations of sections with gain saturation, amplifier noise and suboptimum decision thresholds present, it was necessary to develop a comprehensive model of the fibre amplifier and the receiver.

14.4.1 Amplified two-section link analysis

The procedure adopted for deriving the specifications began from that for the equivalent regenerated system, followed by assuming that the regenerator was replaced by the amplifier, and that values of transmitter launch power and section loss equalled the maximum specified values. Finally, using the laser, fibre amplifier and receiver models discussed in the appendix, the pump power was adjusted iteratively until the amplifier power gain was just sufficient to produce the specified system margin. The resulting value of gain and associated value of input power were included in the regenerated link specification, effectively giving a specification for the amplified link based on using a constant value of pump power.

Assumptions and procedure: in addition to constant pump power maintained by a local stabilisation loop, it was assumed that the transmitter performance was such that, when combined with fibre dispersion, it resulted in only a small penalty. Isolators were also taken to be included within the amplifier. Worst-case analysis conditions were satisfied by allocating the system margin totally to transmitter output power degradation. Since it is impractical to optimise the decision threshold on a link-by-link basis, the threshold was fixed at the normalised value of 0.5 for the study. This value is of course the standard setting for regenerative systems and it also gives the best immunity from the possible range of amplifier noise levels.

The pump power was set up so that the power gain satisfied the specification given in Table 14.5, and then the launched power was reduced to −8 dBm (i.e. maximum power minus the system margin). With these values fixed, the worst-case operating window was derived by repeatedly analysing the link for various

Table 14.5 Specification for an amplified two-section link ([3], Table 2). Reproduced by permission of BT

	622 Mbit/s TDM	2.5 Gbit/s TDM	4 × 622 Mbit/s WDM	4 × 2.5 Gbit/s WDM
Maximum launch power	−2 dBm	−2 dBm	−2 dBm/channel*	−2 dBm/channel*
Maximum section loss	20 dB	20 dB	20 dB	20 dB
Minimum receiver power @ 10^{-10} BER	−28 dBm	−25 dBm†	−28 dBm	−25 dBm†
System margin	6 dB	6 dB	6 dB	6 dB
Demux loss	0 dB	0 dB	6 dB	6 dB
Amplifier power gain	20 dB (−22 dBm)•	23 dB (−22 dBm)•	26 dB (−13 dBm)•	29 dB (−13 dBm)•

* the power per channel after having assembled the multiplex
† 1 dB dispersion penalty assumed
• total average power input to the amplifier at the specified power gain

combinations of input and output section losses, while the total power at the receiver was monitored to ensure that saturation was avoided. The operating window was taken to enclose those combination of losses for which the link was just operational (i.e. tolerant of the 6 dB drop in launch power. Typical erbium-doped fibre cross-section data were used and also typical regenerated link specifications, but this was not expected to affect the generality of the analysis results. The trends and overall conclusions were thus expected to be representative of most if not all such links.

14.4.2 622 Mbit/s link analysis

The operating window in plotted in Figure 14.15.

The reduction of launch power by 6 dB causes the curved part of the window to miss the (20, 20) dB point as the associated increase in amplifier gain allows an increase in the loss of the output section. Note also the closure at the two extremes of the window (on the abscissa and ordinate) compared with the behaviour of an ideal (no saturation, no spontaneous noise) 20 dB gain amplifier, illustrating the inherent nonlinear performance of the amplified link. The resulting loss in operating window could be recovered either by overpumping the erbium fibre to make it operate as a power amplifier or by inserting an optical filter before the receiver (preamplifier operation). Both of these possibilities bring their own disadvantages which would probably outweigh the gain in window area, and would not justify the complexities of moving from a network-wide specification.

The extended range of the amplified compared with the regenerated link is shown in Figure 14.16, which was derived from the preceding figure.

Upgrade options: the 622 Mbit/s specification only requires the amplifier to provide a given power gain. If the chosen device is capable of providing sufficient extra gain the simplest upgrade option would be to increase the pump power and upgrade the terminal equipment. Three upgrades (2.5 Gbit/s, 4 × 622 Mbit/s wdm and 4 × 2.5 Gbit/s wdm channels at 1547, 1550, 1553 and 1556 nm) using this option

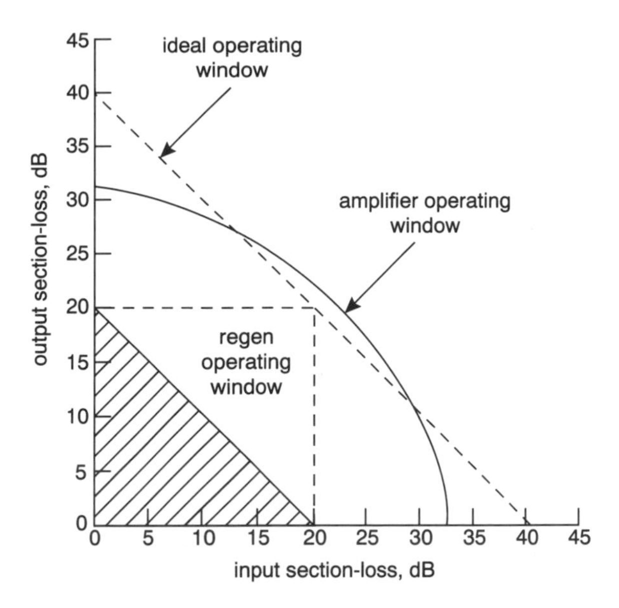

Figure 14.15 Operating window for an amplified two-section 622 Mbit/s link ([3], Figure 4). Reproduced by permission of BT

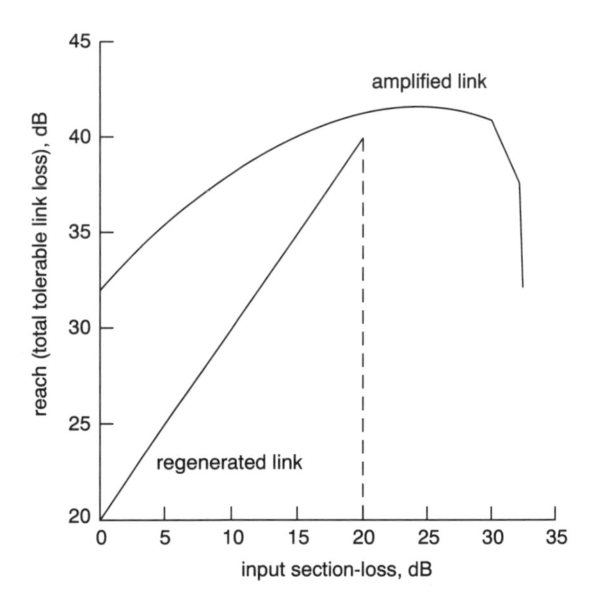

Figure 14.16 Total tolerable link loss (reach) versus input section loss for a regenerated and for an amplified two-section link operating at 622 Mbit/s ([3], Figure 5). Reproduced by permission of BT

were analysed for the specification given in Table 14.5. The specified gains for the wdm options were derived for the worst-case (lowest gain) wavelength with all launched powers at their maximum specified values.

The performance of each channel was analysed with the launch power reduced to −8 dBm as before, and the corresponding operating window was obtained for each

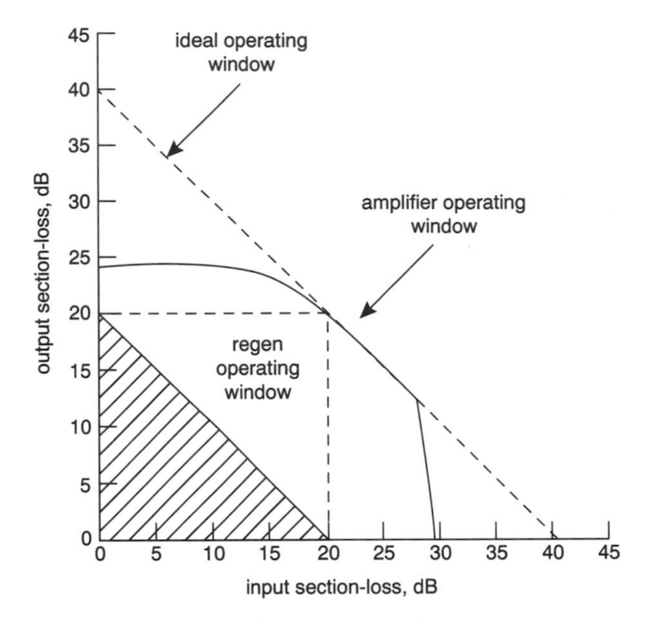

Figure 14.17 Overall operating window for the various options listed in Table 14.5 for an amplified two-section link ([3], Figure 7). Reproduced by permission of BT

case. Figure 6 in [3] showed two windows, one for each of the extreme multiplex channels in the 4 × 622 Mbit/s option. The slight differences between them were due to the gain saturation characteristic of the amplifier being wavelength-dependent. The results also showed that the higher total input power to the amplifier increased saturation significantly, and that the optical filtering suppressed the spontaneous noise at the receiver. Upgrading each channel to 2.5 Gbit/s required additional gain to overcome reduced sensitivity and the small dispersion penalty. The increased gain is accompanied by more noise at the receiver, thus reducing the maximum input section loss that can be used with a given output section loss. Finally, superimposing the individual windows for all of the scenarios in Table 14.5 gives the result depicted in Figure 14.17.

14.4.3 Models used in the simulation

Model of fibre amplifier

For the link analysis it was necessary that the amplifier model enabled accurate predictions of performance under all operating conditions to be made, and in reasonably fast run times. The accuracy requirement inhibited the use of existing (1992) empirical and simplified-analytic models because both sacrifice accuracy for simplicity. The speed of most computers was then inadequate for designers to use the accurate full-numerical analysis model. An excellent compromise was to employ the average power analysis technique, which predicts performance using the total average-core-power along the erbium fibre. This is an iterative analysis technique which is much faster (a few seconds on a Macintosh SE/30) than full numerical analysis and which overcomes the limitations of the simpler models mentioned above by accounting for gain profile changes caused by both the level and the wavelength distribution of the input power, and by gain saturation caused by internally generated spontaneous noise (see Figure 14.14).

Model of laser spectral profile

The lasers were modelled as single-mode narrow-linewidth devices. Pump lasers operated at 1490 nm and transmit lasers oscillated in the 1550 nm band. This model gave slightly different results to those obtained from the actual power spectral density when applied to the pump laser, but it was shown that this discrepancy only resulted in a small difference between predicted and actual values of pump power when the amplifier power gain was specified.

Model of receiver

This covered all noise sources by including terms for thermal noise, signal and spontaneous shot noise and for (spontaneous) emission–signal and (spontaneous) emission–emission beat noise. Some loss in accuracy was accepted by assuming these to have Gaussian distributions to facilitate computation. Received pulses were taken to be rectangular and were equalised to 100% raised cosine at the input to the decision circuit. It was necessary to take $V_T/V_{1P} = 0.5$, where V_T denotes the decision threshold voltage and V_{1P} is the peak value of a mark for all scenarios to avoid making the threshold dependent on the link configuration. The notation used differs from that in [3] to agree with the notation used elsewhere in this book; subscripts E denote (spontaneous) emission, S signal, P peak and T total. Beat noise terms have double subscripts which are combinations of these.

The noise power double-sided spectral densities for a mark and for a space were taken as follows:

$$S_1 = S_{TH} + S_S + S_E + S_{ES} + S_{EE} \tag{14.1}$$

$$S_0 = S_{TH} + S_E + S_{EE} \tag{14.2}$$

where

S_1 = double-sided spectral density for a mark
S_0 = double-sided spectral density for a space
S_S = double-sided spectral density of signal shot noise
S_E = double-sided spectral density of emission shot noise at amplifier output
S_{TH} = double-sided spectral density of thermal noise
S_{ES} = double-sided spectral density of emission-signal beat noise
S_{EE} = double-sided spectral density of emission-emission beat noise

all referred to the output of the amplifier.

For analysis S_{TH} was derived from the specified value of minimum receiver power, and the shot noise spectral densities were obtained from signal and noise powers at the amplifier output:

$$S_S = \frac{\eta q^2}{hc} \times \lambda_S \times \alpha L_2 \times P_{SP} \tag{14.3}$$

and

$$S_E = \frac{\eta q^2}{hc} \times \lambda_S \times \alpha L_2 \times P_{ET} \tag{14.4}$$

where P_{SP} is the signal peak power at the amplifier output, P_{ET} is the sum of the spontaneous emission noise power in both polarisation states at the amplifier output, and L_2 is the length of output (second) section; αL_2 is section loss.

The beat noise spectral densities were obtained from the frequency translated emission noise power spectral density $S_E(f \pm f_S)$ at the output of the amplifier:

$$S_{ES}(f) = 2 \times C \times P_{SP}\{S_E(f + f_S) + S_E(f - f_S)\} \tag{14.5}$$

where

$$C = \left[\frac{\eta q}{hc} \times \lambda_S \times \alpha L_2\right]^2$$

assuming a polarised signal; hence only the corresponding emission noise sop is included. For the emission-emission beat noise, we have:

$$S_{EE}(f) = 4 \times C \times \{S_E(f) \otimes S_E(f)\} \tag{14.6}$$

where $\otimes$ denotes convolution, and for this equation both sop must be taken, as the emission noise is randomly polarised.

The improved power analysis technique for the amplifier inherently yields S_E, so that the beat noise contributions can be found without recourse to amplifier noise figure or the profile of the spontaneous emission output noise power, both of which have their limitations.

14.5 REFERENCES

1 COCHRANE, P., and BRAIN, M.C.: 'Future optical fiber transmission technology and networks'. *IEEE Communications Magazine*, **27**, no. 11 , Nov. 1988, pp. 45-60.

2 BROOKS, R.M., COCHRANE, P. BICKERS, L., BLAU, G., COOMBS, G.P., DAVIES, R., DAWES, R., HAWKER, I., HILTON, G., STEVENSON, A., and WHITT, S.: 'A highly integrated 565 optical fibre system'. *Br. Telecom Technol. J.*, **4**, no. 4, Oct. 1986, pp. 28-40.

3 O'NEILL, A.W., and HODGKINSON, T.G.: 'Analysis of an optically amplified two-section link specification incorporating an erbium-doped fibre amplifier'. *BT Technol. J.*, **11**, no. 2, Apr. 1993, pp. 108-113.

FURTHER READING

4 YAMASHITA, K., KINOSHITA, T., MAEDA, M., NAKAZATO, K.: 'Simple common-collector full-monolithic preamplifier (optical receiver) for 560 Mbit/s optical transmission'. *Electronics Lett.*, **22**, no. 3, 30 Jan. 1986, pp. 146-147.

15
SUBMERGED SYSTEMS

The Deep Sea Cables

The wrecks dissolve above us; their dust drops down from afar —
Down to the dark, the utter dark, where the blind white sea-snakes are.
There is no sound, no echo of sound, in the deserts of the deep,
Or the great grey level plains of ooze where the shell-burred cables creep.

Here in the womb of the world — here in the tie-ribs of earth
Words, and the words of men, flicker and flutter and beat —
Warning, sorrow, and gain, salutation and mirth —
For a Power troubles the Still that has neither voice nor feet.

They have wakened the timeless Things; they have killed their father Time;
Jointing hands in the gloom, a league from the last of the sun.
Hush! Men talk today o'er the waste of the ultimate slime,
And a new word runs between: whispering, 'Let us be one!'

Rudyard Kipling (1865–1936)

15.1 INTRODUCTION

The first verse of Kipling's poem was given great poignancy by the discovery of the wreck of the Titanic, and perhaps even more so of the German battleship Bismarck in summer 1989. No one who has seen the extraordinary photos of the latter some 5 km deep and 1000 km west of Brest on the French coast can fail to have been moved at the sight of this war grave where so many brave men lie.

Previous chapters have covered many aspects of landline (or inland) systems which form the bulk of the international telecommunications network. The impact of the change from copper to glass has already affected national networks, especially those in the developed countries, and considerable operating experience has been gained. This chapter deals with submerged (often called submarine) systems where in the mid-1980s (when planning began) very little operational experience was available on optical systems, in contrast with the vast amount of experience with coaxial terrestrial systems before their introduction to undersea systems in 1956.

A major difference between land and most submarine systems stems from the effects of a failure. A failure underwater involves a very special kind of ship, costing perhaps £40 000 a day (at 1987 prices), 'and the total bill for putting things right might come to half a million pounds. If you then add the lost revenues for 8000 customers at $1 a minute, a breakdown becomes very expensive.' *Financial Times,*

20 May 1987. For these simple economic reasons, the components used in such systems need to be about ten times as reliable as on land, and a systems has a design life of 25 years of uninterrupted operation, 'in which it will need minimum maintenance and repair.' This concern with reliability will be evident throughout the chapter as various topics are considered.

Since submerged systems are designed and manufactured in specialist divisions of relatively few telecommunication equipment suppliers, most readers are probably not familiar with such systems. The major part of this chapter will therefore be a description of the first transatlantic system, known as TAT-8, in sufficient detail to highlight the engineering challenges faced by designers. This is followed by a brief description of installed and planned undersea links in the Pacific Basin. The second part deals with development aspects and system details of the first SDH-compatible transoceanic systems to deploy large numbers of erbium amplifiers, and which are due to enter service in 1995/6. Finally, the rapidly growing use of unrepeatered submerged systems as extensions of national networks is treated.

A historical introduction, which links the present and near future to a long and distinguished past of Man's endeavours to span the seas then the oceans for telecommunication is given as an Appendix. This past began within living memory of the American and French revolutions, the age of Napoleon, and wooden sailing ships. Mankind's first steps on the Moon are a part of the memory of many engineers practising today.

By 1990 six transatlantic, three transpacific and two major systems linking Australia to Hawaii and Guam had been planned and announced, all to be commissioned by 1996. This represents a multibillion dollar investment, which was predicted to grow with as many as twenty more systems announced and installed by 1996. The number of countries using submerged optical systems was 27 in early 1993, but was expected to rise to 67 by the end of 1994. Demand for undersea cable systems was predicted to grow at annual rates between 9.5% and 19.5%, substantially higher than rates for landline systems. Projections indicated that the share of international traffic carried by submarine cable systems would grow from 38% in 1986 to 60% a decade later, a substantially larger share of a bigger market. The provision of end-to-end services over high-quality optical line systems will be very profitable. During the coaxial era, the channel capacity increased at an average rate of 27% per annum between 1950 and 1986; one prediction for the corresponding figure for channel capacity on optical systems for the period 1988–96 was 41%.

In submerged systems the 'optical pipe' remains the ultimate solution. (by optical pipe is meant a transparent (i.e. lossless) bidirectional guided medium of infinite bandwidth). The ability to transmit 565 Mbit/s over 300 km without a repeater would satisfy some 30–40% of the total number of undersea systems. Repeatered systems with a range of 300–2000 km would cover a further 50% or so, leaving the remaining 20 to 10% for transoceanic systems.

Optical amplifier repeaters offer many advantages over 3R (reshape, regenerate, retime) regenerative repeaters, leading to: a reduction in physical size which would correspond to repeater housings only one-tenth of their present volumes; a possible reduction in cable size, hence better handling and storage properties and lower costs; a corresponding reduction in shipboard handling and laying costs; substantial reduction in power feed current from around 1600 mA. Traffic capacity increases only need changes in the terminal equipment, thus the costliest part of the system is to some extent future-proofed.

Table 15.1 Comparison of repeater spans achieved in submerged systems 1988–93 ([9.6], Table 1). Reproduced by permission of SEE

Bit rate	wavelength	repeater span	
		early systems	later systems
280	1300	45	55
420	1300	55	65
560	1550	110	130
2500	1550	87	—
Mbit/s	nm	km	km

The growth and technical development of optical fibre regenerative and optically amplified submarine systems are briefly described, giving due prominence to the parts played by STC Ltd., British Telecom (formerly part of the British Post Office), and to Cable and Wireless, the largest international telecommunication organisation in the world.

As an example of the progress made by manufacturers of submerged systems in the five years since 1988 when TAT-8 was laid, consider the STC Ltd regenerative repeater spans listed in Table 15.1 ([6] of Chapter 9). The corresponding values of receiver sensitivities were given in Table 9.2.

15.1.1 The international transmission network (ITN)

This is the worldwide dynamic and ever-developing means by which international communication services are relayed between countries and continents [1]. Its purpose is to provide the means by which customers can communicate quickly, reliably and in a cost-effective manner. Individual components of the ITN are satellite links, submarine cable systems, microwave radio links, and the national networks of countries through which international services are routed or terminate. Important considerations in the provision of these services include the following: the design life of satellites is seven years, the design life of undersea cable systems is 25 years, there is a transmission delay of satellites in geostationary orbit at 250 ms (two such links in tandem are unacceptable for speech traffic), and the transmission delay of submarine cable links is about 30 ms.

The main elements of developing ITN plans include:

(a) the range of services to be provided and the forecast circuit demand
(b) the protection of these services (redundancy, diversity, restoration, repair)
(c) financial considerations.

Reasons for initiating planning for new transmission facilities include:

(a) provision of additional capacity to meet forecast demand
(b) path or route diversity
(c) provision of more economical means of satisfying demand
(d) provision of a direct link where none existed
(e) replacement of a transmission facility due to retirement.

Cable system planning falls broadly into two parts as follows. There are independent studies by individual entities, followed by discussions either bilaterally or within *ad hoc* planning to establish the possible need for a new system. These discussions result in agreement on the broad definition of the project and agreement in principle to proceed. Then there is a series of formal meetings of potential co-owners to define the project in detail and reach agreement on the key factors, such as system size, landing points, timing, sharing of costs, methods of procurement and project management.

15.1.2 Undersea optical fibre cable systems: future trends

By 1989 the demand for undersea cable transmission capacity was progressing at a pace only previously experienced when the first repeatered coaxial systems overtook HF radio telephony [2]. By 1989 one could see that the installed and planned optical systems were set to similarly eclipse their coaxial predecessors in terms of global capacity within the next two years.

Within a five-year period the capacity will be over double that installed during the previous 35 years of the coaxial era. Systems will include second (1300 nm) and third (1550 nm) generation technology offering repeater spacings of 40–100 km. Although this realises a greater than fifteen-fold reduction in the number of repeaters compared with a coaxial system, the complexity of regenerative repeaters in optical systems maintains the component density at a similar level for 1300 nm systems, with approximately a two-fold reduction for a 1550 nm, but with a capacity increase of at least double. Overall, therefore, component densities are reduced by more than five times per channel-km.

Although such advances in capacity and repeater spacing, together with the facility of cable branching, have vastly improved the cost-effectiveness of submarine systems, the physical construction of the submerged plant has changed very little. Repeater housing diameters are substantially the same, although their length has been reduced in some cases, which gives better handling properties; power feed requirements have increased by two-fold for deep-sea applications, but have remained virtually unchanged for near-shore applications, with direct burial now preferred over mechanical armouring. For any dramatic economic and system performance improvements new technological solutions need to be realised, as the further refinement of second generation systems will only realise marginal gains.

The quality of the digital service provided by the first transatlantic optical fibre system TAT-8 was far superior to the quality of all earlier (coaxial) cable systems carrying analogue traffic [3], so that its capacity was fully utilised within 18 months of entering service in 1988. Moreover, the technology supported digital service connectivity and had the capability to carry broadband services. Enhancements in the technology yielded circuit multiplication factors up to five, thus increasing capacity from 8000 to 40 000 simultaneous 64 kbit/s circuits.

15.1.3 Installed or planned transoceanic optical systems (1986–96)

Maps showing the Atlantic and Pacific Oceans, and all the major installed or planned international systems as at 1991, are given in Figure 15.1. These are based on three generations of optical systems:

(1) 280 Mbit/s per fibre pair in the 1300 nm band, e.g. TAT-8 and TPC-3

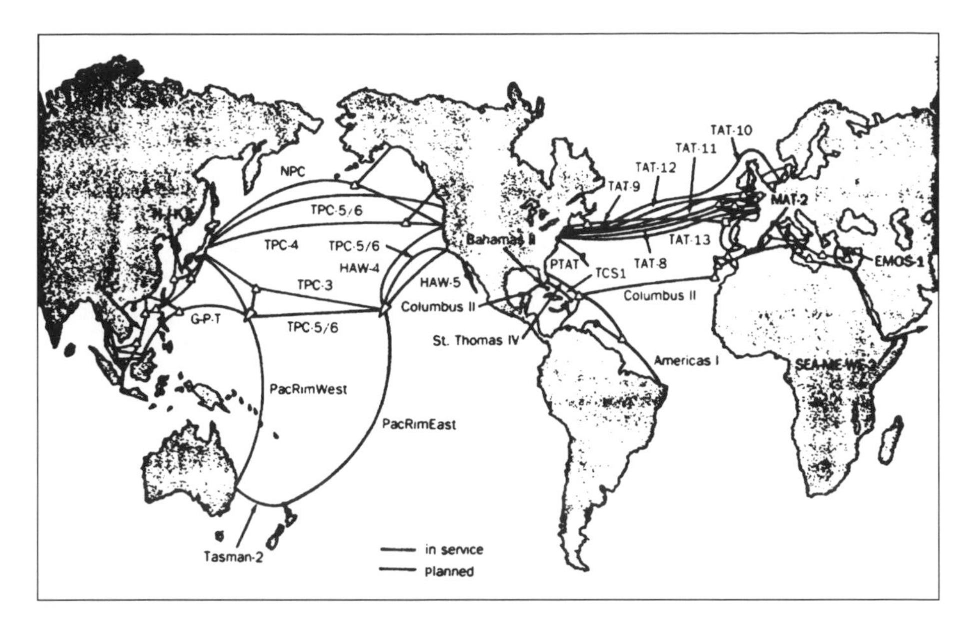

Figure 15.1 Map showing all major installed or planned systems in the international telecommunication network up to 1996 ([3], Figure 4). Reproduced by permission of AT&T

(2) 560 Mbit/s per fibre pair in the 1550 nm band, e.g. TAT-9/10/11 and TPC-4
(3) 5 Gbit/s per pair of dispersion-shifted fibres at 1550 nm, e.g. TAT-12, TAT-13, TPC-5 and TPC-6.

A comparison between the estimated capacities of transoceanic and satellite systems over the decade 1986–1996 is presented in Table 15.2. Information on the North Atlantic region which has the highest traffic density in the world is presented in Table 15.3 [4].

First-generation systems (e.g. TAT-8, TPC-3)

The transmission quality of systems such as TAT-8, TPC-3 and HAW-4 is measured continuously from the terminal stations. Their performance according to measures such as outage, degraded minutes, severely errored seconds and errored seconds (these quality measures were mentioned in Section 13.6.3) far exceeded the system requirements set by the owners. The performance of the digital services as judged by customers is excellent.

Such has been the success of first-generation lightwave undersea systems that since 1988 they have been installed at an average rate of 18 000 km per annum. The capacity of undersea optical systems was growing at about 2×10^8 64 kbit/s circuit-km annually by the end of the 1980s and the beginning of the 1990s. Projections by AT&T based on the deployment of second- and third-generation systems show that these growth rates will continue well beyond 1995.

Second-generation systems (e.g. TAT-9/10/11, TPC-4)

These will operate at 560 Mbit/s per fibre pair in the 1550 nm band. DFB single-frequency laser sources will be employed, which possess very low chirp. APD

Table 15.2 Estimated capacities of transoceanic cable and satellite systems (1986–96)

Year	Cable voice paths	Satellite voice paths
Trans-Atlantic (North America to Europe)		
1986	22 000	78 000
1987	22 000	78 000
1988	60 000	78 000
1989	145 000	93 000
1990	145 000	283 000
1991	221 000	283 000
1992	346 000	496 000
1993	471 000	496 000
1994	471 000	540 000
1995	640 000	720 000
1996	809 000	720 000
Trans-Pacific (North America to Japan via Hawail or Guam)		
1986	2000	39 000
1987	21 000	39 000
1988	21 000	39 000
1989	21 000	39 000
1990	21 000	39 000
1991	106 000	27 000
1992	183 000	27 000
1993	183 000	117 000
1994	183 000	207 000
1995	183 000	207 000
1996	783 000	207 000

Estimates based on year cable/satellite facilities begin service. Estimate of cable voice paths assume that live virtual voice paths can be derived from one 64 bps digital circuit; cable estimates include circuits held in reserve for cable/satellite restoration services. Estimate of satellite voice paths based on intelsat satellite only and intelsat's July 1991 deployment and launch schedule; satellite estimates exclude one satellite in each region held in reserve. Satellite estimates also assume one voice path per channel until 1989 deployment of intelsat VI series with 24 000 channels or 120 000 voice paths using Digital Code Multiplication Equipment (DCME). The Intelsat VII series, to be deployed in 1992, will have a nominal capacity of 18 000 channels or 90 000 voice paths using DCME. A small number of additional satellite voice paths in the Atlantic and Pacific likely will be available to 1996 from PanAmSat (PAS-1; ORBX-2); Columbia Communications (using NASA's TDRSS system); and InterSputnik. Regional capacity estimates do not necessarily imply that full capacity is available to satisfy demand on any given bilateral route.

Source: *Financial Times*, Survey of World Telecommunications, 7 Oct, 1991

Table 15.3 Installed and planned transoceanic systems in the North Atlantic (1956–96)

System	Circuits	Cumulative circuits	Service date
TAT-1	48	48	1956
TAT-2	48	96	1959
CANTAT-1	80	176	1961
TAT-3	138	314	1963
TAT-5	845	1159	1970
TAT-6	4000	5159	1976
TAT-7	4200	9359	1983
TAT-8	3840	13 199	1988
PTAT-1	17 280	30 479	1989
TAT-9	15 360	45 839	1991
TAT-10	11 520	57 359	1992
TAT-11	11 520	68 879	1993
TAT-12	30 720	99 599	1995
TAT-13	30 720	130 319	1996

detectors will be used in the receivers to provide greater sensitivity than pin photodiodes. This combination of source and detector will permit repeater spacings to be extended to about 150 km. The economics of these systems have enabled a new trend in design to be adopted in place of the point-to-point nature of previous systems. TAT-10 and TAT-11 (RFS 1993) are planned as networks with a mutual restoration capability between several systems to increase the availability. This design trend is expected to become standard practice as cable traffic capacity increases so that a fault can affect more circuits.

Third-generation systems (e.g. TAT-12/13, TPC-5/6)

These systems will operate at 5 Gbit/s and be SDH compatible, and all will deploy erbium amplifiers as repeaters, making TAT-12 and TPC-5 the first optically amplified transoceanic links in the Atlantic and Pacific, respectively. TAT-12/13 will be a four-fibre ring network linking the United States, Great Britain and France. The ring will include two transatlantic links, each of two fibre pairs, and in the event of a fault in either, full restoration facilities will be provided by the ring configuration. TAT-12 (US–UK and UK–France) will be 6100 km in length and use repeater spacings of about 45 km (STC and Alcatel) or 33 km (AT&T) and is due to enter service in 1995, and TAT-13 (US–France) will do so in 1996. Several different choices of technology were possible, all of them challenging, one of which was the use of optical amplifiers as repeaters.

If the demand in the mid-1990s is sufficient to justify further advances in technology, 5 Gbit/s and higher rates can be used with soliton pulses in an amplifier based-system. Because the spectrum of solitons does not change with distance, in theory wdm techniques could be utilised to send five simultaneous 5 Gbit/s streams of solitons in one fibre over transoceanic spans, resulting in a capacity of 25 Gbit/s per fibre pair.

Submerged optical systems have had a major impact on the ITN, and they demonstrate vividly the feedback interaction between increasing demand for services driving improvements in technology and capabilities, which in turn provide better service, which further stimulates demand.

15.1.4 Submerged repeaters

The reliability of submerged systems is largely determined by the integrity and reliability of the submerged repeaters over their design life of 25 years.

Submerged electronic regenerative repeaters

In the case of TAT-8 each regenerative repeater can contain up to six one-way optical regenerators, one for each of the three fibre pairs. In addition, the repeaters also include supervisory and power circuitry, as illustrated in Figure 9.3 by the

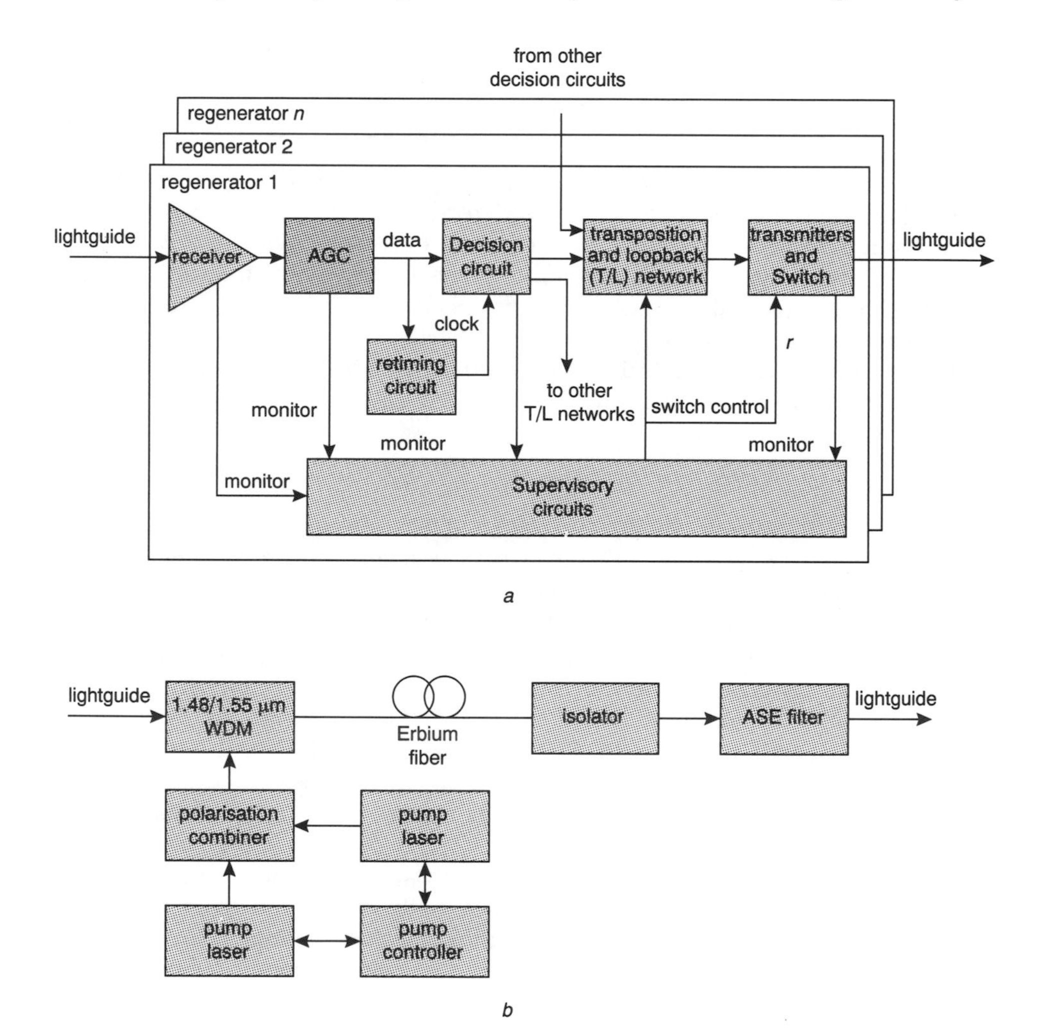

Figure 15.2 *a* Block diagram of regenerative repeater ([3], Figure 3*a*) *b* Block diagram of repeater using erbium amplifier ([3], Figure 3*b*). Reproduced by permission of AT&T

block diagram, showing only the main functions. The transmitter and optical switch combination provides two available transmitters per outgoing fibre. In the first two generations of submerged systems all the electronic circuits were implemented in custom silicon integrated circuits. Details are given below.

Submerged optical amplifier repeaters

The relative complexity of the electronic regenerator repeaters in Figure 15.2*a* can be compared with the relative simplicity of optical amplifier repeaters shown in Figure 15.2*b*, again showing only the major functional blocks. Repeaters of this type operate at 2.5 and 5 Gbit/s and do not require electronic circuits to perform the 3R functions indicated in Figure 15.2*a*. The edfa is forward-pumped using two pump lasers to give enhanced reliability. Details are given later.

15.2 OVERVIEW AND GENERAL DESCRIPTION OF TAT-8

As noted above, the quality of the digital service provided by TAT-8 was far superior to the quality of all earlier (coaxial) cable systems carrying analogue traffic. Various performance parameters of the 85% portion of TAT-8 supplied by AT&T were measured during installation and the tight distributions demonstrated the excellent control of the numerous processes used to realise the system. Comparable results would be expected from the other two suppliers STC (8%) and Submarcom (7%). [3] took as an example the repeater section margin, i.e. the allowable change in received optical power before transmission becomes unacceptable. The histogram of these margins is given in Figure 15.3.

Other statistical results quoted were the mean repeater output power of 0.2 dBm (on ocean bed at 4°C) and standard deviation of 0.6 dB, and the distribution of receiver sensitivities with standard deviation of 0.2 dB about a mean of −37.2 dBm.

In 1986 an overview of TAT-8 was published which dealt with the following aspects: pre-contract planning arrangements, configuration, investment method and unit of investment, technical and procurement arrangements, contract placing, implementation phase, route selection and marine considerations, system description and finally status [4].

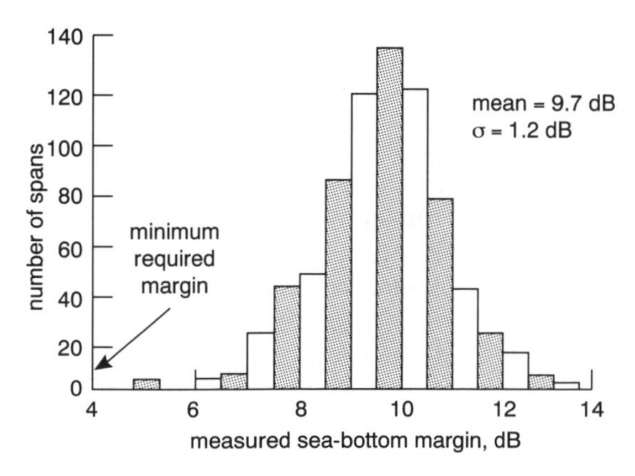

Figure 15.3 Repeater section margins for the AT&T portion of TAT-8; average value 9.7 dB, standard deviation 1.2 dB ([3], Figure 4). Reproduced by permission of AT&T

15.2.1 General system description [5]

The TAT-8 system comprises two pairs of fibres. One pair runs from Tuckerton. USA to Penmarch, France, the other from Tuckerton to Widemouth, UK, and the pairs separate at a branching repeater off the European Continental Shelf. To keep symmetry within the cable and facilitate restoration of service another pair is provided between UK and France. TAT-8 also saw the first commercial application of submerged branching technology within such systems. The simple branching unit enabled the two fibre pairs in the main cable to be split so that one pair each terminates in the UK and France. An auxiliary purpose was to provide static switching for restoration of service. The special three-way branching repeater developed by AT&T provided switching of the fibres between different locations for reasons of operational security, and it ensures that power in the submerged plant can be fed from either the UK or France. As with coaxial links it is necessary to feed power from both ends of the cable. The spur not feeding to the USA feeds to an earth in the branching repeater.

Each fibre pair carries two bothway 139.264 Mbit/s digital line sections, which are combined into a nominal 280 Mbit/s which, along with engineering facilities, is transmitted to line at a rate of 295.6 Mbaud. The 140 Mbit/s line sections are made by multiplexing 3 × 45 Mbit/s tributaries, each obtained from 21 × 2 Mbit/s blocks, thus giving a total system capacity of 3780 × 64 kbit/s digital bearers per fibre pair. Digital circuit multiplication schemes (dcms) can be used to provide the equivalent of 40 000 speech circuits. The interface between the inland network and the TAT-8 terminals is at 140 Mbit/s, and the performance of each 140 Mbit/s system is specified according to CCITT Rec. G703. The line current is 1.6 A and the line code is 24B1P.

The system of each manufacturer differed from the other two in many respects, but they had some of the major parameters in common, as illustrated in Table 15.4 ([7] of Chapter 9), listing the major parameters of the STC-BT segment.

The cable design of each manufacturer is also different, although each design places the fibres along the neutral axis of the cable. Features such as tensile strength, protection from pressure, hydrogen and water ingress are provided by other concentric layers of the cable. Hydrogen, an ever-present hazard in submerged cables, was a major worry during the design stage when the effect on optical fibre became known. As in metallic coaxial systems different designs are used for the deep water and shallow water portions of the route.

One other hazard emerged: AT&T reported in February 1986 that three faults had occurred on the Canary Islands Optican-1 system between depths of 1000 and 1500 m over a three month period [6]. This system had been manufactured and installed by AT&T, with the final splice made on the 30 September 1985, and it was the world's first deep-water repeatered submerged lightguide system. It used the same technology as that employed in the AT&T segment of TAT-8, and before entering service with Telefonica it served as a test bed for Bell Laboratories engineers. Investigation revealed 46 shark's teeth, some of which had penetrated the insulation layer and caused the current supplying power to the submerged repeaters to short-circuit. Coaxial cables laid in the same area had not suffered any damage due to this cause, so an investigation began to find why some species of shark are partial to optical fibre cables. The absence of an outer coaxial sheath and a much thinner dielectric layer may be to blame. One theory is that ripple on the power feed is sensed by the sharks, because the field is no longer confined within the

Table 15.4 Major system parameters of STC-BT portion of TAT-8 ([5], Table 1). Reproduced by permission of BT

Wavelength	:	1310 nm*
Bit rate	:	280 Mbit/s*
Line code	:	24B1P*
Line rate	:	295.6 MBaud*
No. of fibres	:	1 to 8
Repeater span	:	40 to 50 km
Supervision		
in-service	:	Laser status Received light level Error count
Out-of-service (loop-back)	:	Fault location
Standby options		
Submerged plant (remotely controlled	:	1 + 1 laser n + 1 line
Terminal (auto/manual control)	:	1 + 1 set
Power feed	:	single/double end Load sharing Cold standby

(*These parameters are common to the NL2, SL, S280 and OS-280 systems).

cable. Meanwhile the British company STC has bitten back and designed a new cable which should be more than a match for such jaws. AT&T have also produced a cable protected against fish bites and have now replaced the original Optican cable. Other manufacturers have followed suit.

15.2.2 Regeneration

A diagram of the main transmission path is shown in Figure 15.4 [9.7], and the power budget for the corresponding path loss appears in Table 15.5.

These figures and then-current STC cable design led to repeater spans of 40–50 km (21.6–27 nautical miles), depending on repair margins and whether a standby laser was fitted. These spans compared very favourably with values in comparable coaxial cable systems where the spans were about 10 km. Some details of the regenerator are given below; these differ from the AT&T and Submarcom designs and also from the Japanese OS-280M system.

Optical receiver

To some extent this was the major determinant of system performance, and it consisted of a photodiode connected to a low-noise bipolar silicon integrated circuit configured as a transimpedance amplifier. Both the detector and the chip were mounted in a single package virtually identical to that for the laser. The signal was coupled from the system lightguide on to the active area of the photodiode through

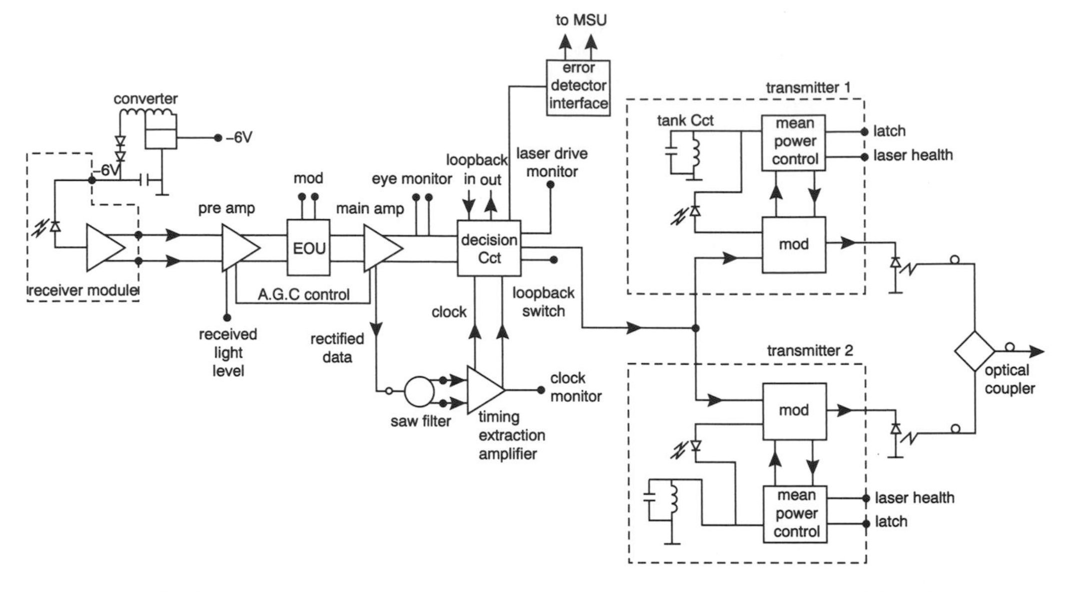

Figure 15.4 TAT-8 regenerator diagram showing main transmission path ([9.7], Figure 2)

Table 15.5 Power budget of STC-BT portion of TAT-8 ([5], Table 3). Reproduced by permission of BT

Tx source level (l + 1 spare)	−5.5 dBm
Rx sensitivity	−35.0 dBm
	29.5 dB
Splicing and gland losses	−0.7 dB
System penalties	−1.1 dB
Manufacturing tolerances	−2.6 dB
Path loss capability	25.1 dB
Ageing allowances	−2.1 dB
Repair allowances	−1.6 dB
End of life margin	−3.0 dB
Nominal initial path loss	18.40 dB+

(+ For configurations with a standby laser; the figures for single laser operation are approximately 3.3 dB higher).

a single-mode tail which entered the package through a hermetic metallised gland. The specially prepared fibre tail was held in position close to the face of the photodiode with a laser welded fixing. By adopting the same package the many aspects of package reliability and stability in operation already proven for the laser also apply to the receiver. The principal requirements on the photodiode were low capacitance, low leakage current and high responsivity, all of which influenced receiver sensitivity to some extent. Changes in these parameters with temperature and ageing must also be very small. A number of types were appraised for the system, and a planar InGaAs pin photodiode manufactured by Fujitsu was selected. This top entry device had an active area of 50 μm diameter; capacitance, leakage current and responsivity are typically 0.35 pF, below 0.5 nA and 0.9 A/W, respectively. This component has been subjected to extensive life testing at Fujitsu and at BTRL.

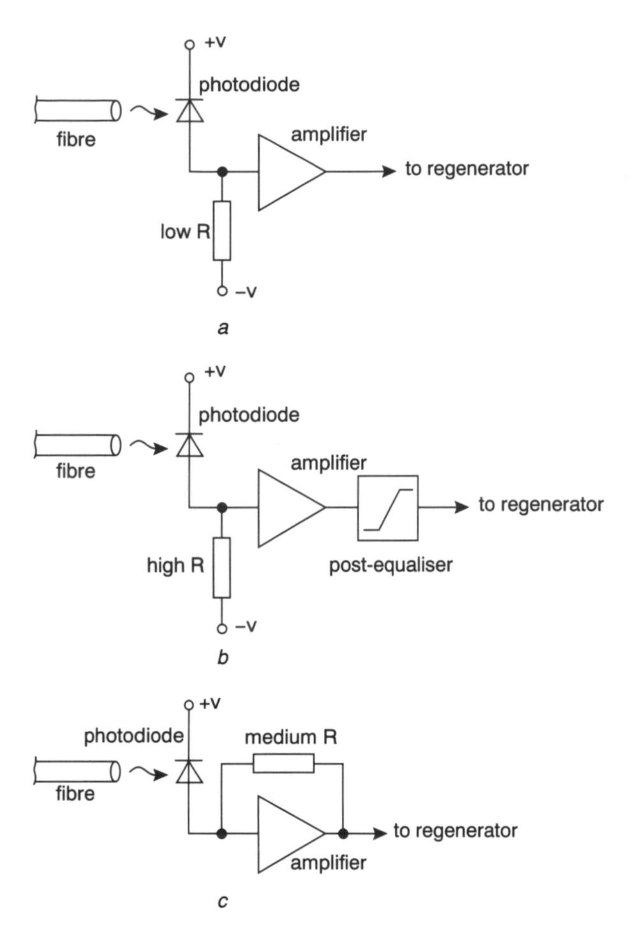

Figure 15.5 Diagrams of the three common types of receiver front ends (*a*) low impedance (*b*) high impedance (*c*) transimpedance ([2] of Chapter 9), Figure 3). Reproduced by permission of BT

Receiver front end: as described in earlier chapters, there are three categories from which to choose: low impedance, high impedance or transimpedance, as shown in Figure 15.5 ([2] of Chapter 9).

The transimpedance configuration is in some senses a compromise between the other two. It possesses good sensitivity, a wide dynamic range, is transparent to all codes and requires no post-equalisation, features which are particularly beneficial in submarine systems. With this as the configuration, there were still a number of choices to be made, one of which was between GaAs mesfets or silicon bipolar transistors. The sensitivity of the former generally exceeded that of the bipolar by some 3–4 dB; however, against this must be set the availability of the proven ECL-40 process and to lesser extent the greater tendency of GaAs mesfets to self-oscillation and low-frequency noise as discussed in the chapters on receivers. The reliability advantages of the bipolar approach were judged to outweigh the sensitivity advantages of the GaAs mesfet pre-amplifier, and the ECL-40 was selected.

A number of circuit designs were examined and the corresponding sensitivities were calculated to see if the target value could be achieved, and to provide

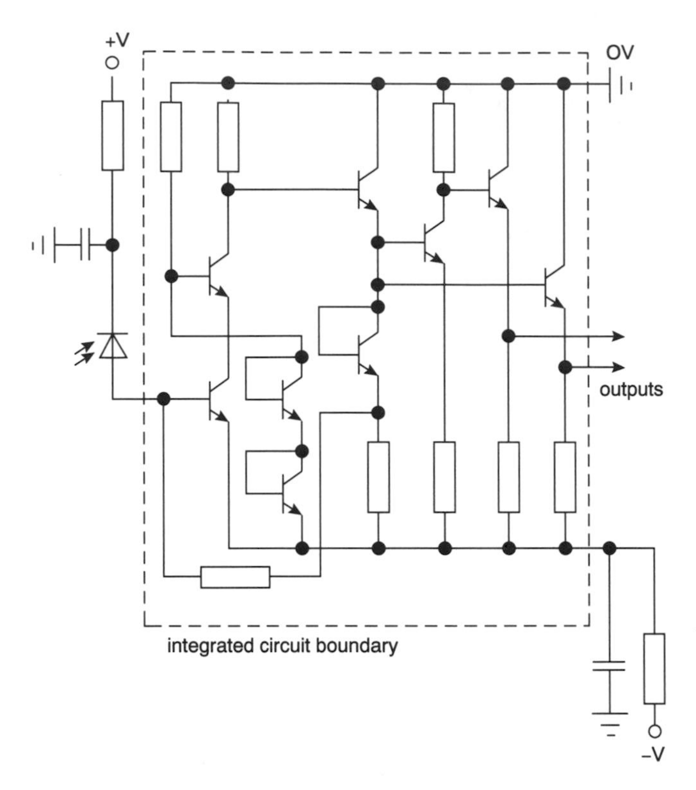

Figure 15.6 TAT-8 circuit diagram of receiver based on ECL-40 technology ([2] of Chapter 9, Figure 4). Reproduced by permission of BT

preliminary values for the optimisation of such key parameters as transistor gains and first stage current. The most promising designs were found to be a common emitter cascode, and a modified version of a common collector design [7–8], but later work demonstrated that the cascode design was more suited to the particular requirements of TAT-8. Figure 15.6 shows the circuit diagram.

The cascode first stage used a common base configuration in order to minimise the overall input capacitance. Figure 15.7 illustrates the transimpedance versus frequency characteristic.

The frequency response of the receiver was designed to combine with that of the regenerator to equalise the 295.6 Mbaud data stream, giving a raised cosine 'eye'. The second stage shifts the dc bias level to improve the overload performance of the receiver, and it also buffers the high output impedance of the cascode. Subsequent stages provided an inverter and emitter-followers which gave a balanced low-impedance output as needed for the regenerator. Figure 15.8 shows the noise spectral density of a typical production receiver.

Measurements on receivers

After assembly each receiver was subjected to an extensive programme of measurements to establish its performance in a system environment, including tests to validate the integrity of each new photodiode and amplifier ic wafer, and to determine receiver performance with temperature. To ensure that measurements

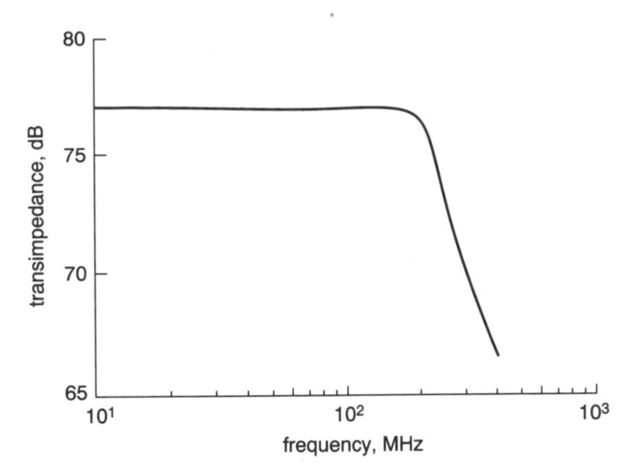

Figure 15.7 Optical receiver transimpedance versus frequency response ([2] of Chapter 9, Figure 5) TAT-8. Reproduced by permission of BT

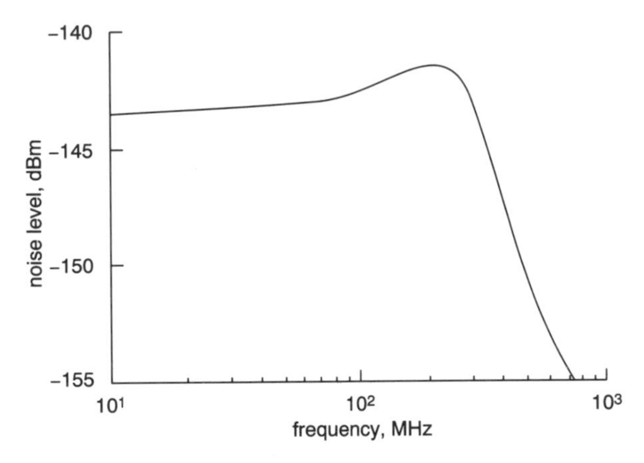

Figure 15.8 TAT-8 optical receiver noise power level versus frequency ([2] of Chapter 9, Figure 6). Reproduced by permission of BT

of sensitivity and overload gave a true indication of receiver performance in a system, each receiver under test was coupled to a test regenerator essentially identical to those to be used in TAT-8 repeaters. These measurements were carried out in a RF-screened room to minimise disturbances from mains transients and from radiated interference. Automated measurement facilities were used to reduce test times and ensure reproducibility during the entire production phase. A $2^{23} - 1$ pseudo-random data sequence ensured that each receiver was subjected to signal conditions representative of the 24B1P line code. To comply with the quality assurance requirements all test equipment was calibrated against national standards at regular intervals, and a reference TAT-8 receiver was used periodically to check that the test facility was functioning correctly.

Sensitivity, overload and dynamic range

Sensitivity was taken as the minimum mean optical power in the receiver fibre tail which produces an error rate of 1 in 10^9 from the regenerator, and overload as the maximum power level for which the same error rate is measured. The dynamic range was the difference between these two values, and it is illustrated in Figure 15.9. This shows ber as a function of optical power for a typical TAT-8 receiver; the sensitivity and the overload are −34.8 dBm and −12.9 dBm, respectively, giving a dynamic range of 21.9 dB. These values compare favourably with the targets of −34.0 dBm, −14 dBm and 20 dB. Figure 15.10 shows the equalised 'eye' from the receiver at −34.9 dBm.

The receiver was designed to operate over the range 0–40°C, to cover deep-water conditions where the mean temperature is 4°C, to shallow water on continental shelves where large temperature variations occur. Allowance had been made for possible fluctuations in the regenerator power supply and the photodiode bias voltage.

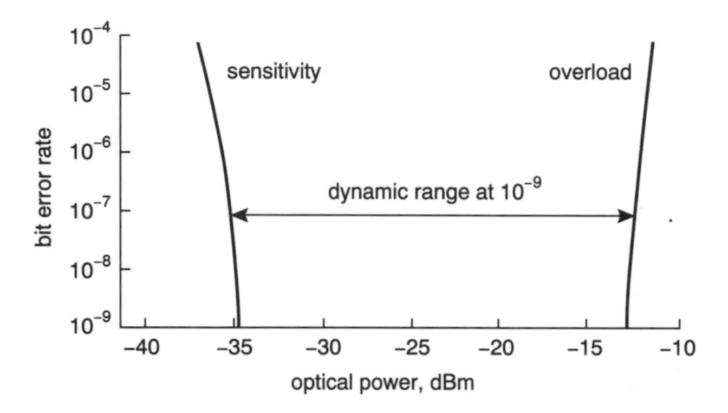

Figure 15.9 TAT-8 sensitivity and overload profiles for a typical receiver and dynamic range ([2] of Chapter 9, Figure 9). Reproduced by permission of BT

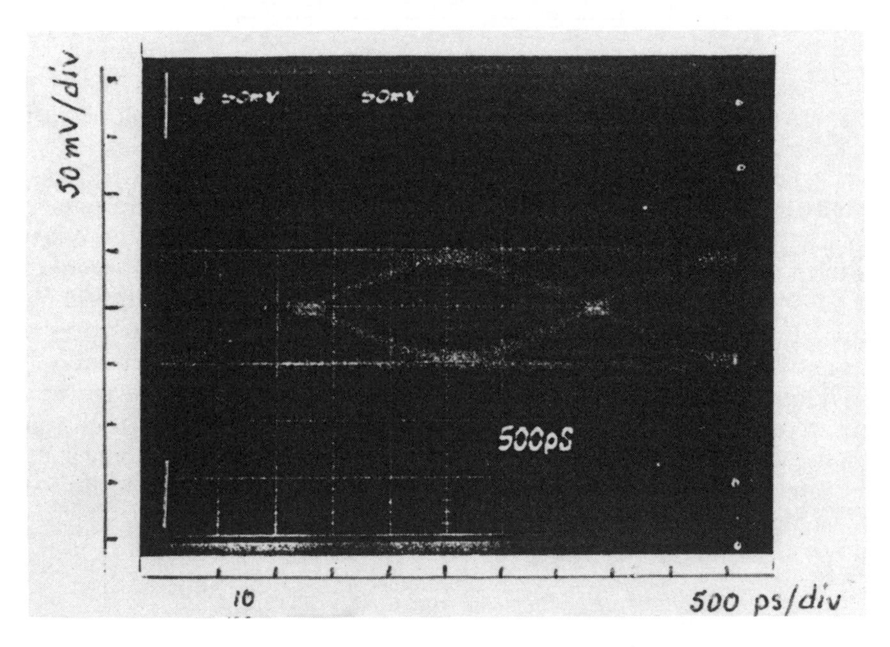

Figure 15.10 TAT-8 eye diagram of a typical receiver ([2] of Chapter 9, Figure 10). Reproduced by permission of BT

AGC amplifier

The variable output of the receiver is amplified to a constant level suitable for the decision circuit by an agc amplifier. This consists of two ics: a variable gain pre-amplifier, and a main amplifier of maximum gain 52 dB and an unequalised 3 dB bandwidth of 320 MHz. A peak detector in the main amplifier ic provides a dc control voltage which is able to maintain the output level within 10% over a dynamic range of 40 dB. The main amplifier ic also has a rectifier circuit which generates timing impulses to drive the saw filter. A dc voltage on the variable gain stage provides an indication of the gain setting of the stage and hence of the received light level. Finally, a simple *CR*-filter located between the pre- and main amplifiers equalises the signal to the target 100% raised cosine response.

Timing extraction

Referring to Figure 15.4 the reader can see that one output from the main amplifier is a stream of rectified data to the transversal surface acoustic wave (saw) filter. The filter Q is about 800 and its insertion loss between 50 Ω terminations is some 17 dB. It has been designed to have a very good phase versus temperature characteristic, which results in a timing loop having very good thermal stability. A typical variation is less than 3° phase shift over the range 0°–40°C. The 295.6 MHz output sinusoid drives a single-chip limiting amplifier which in turn supplies a uniform clock signal to the decision circuit with minimum jitter penalty. The measured clock jitter is less than 1.0° r.m.s.

Decision circuitry

The functions available on the decision chip are shown in Figure 15.11.

The D-type flip-flop regenerates the data stream and also drives the loop-back gating circuit. Typically, 10–90% transition times of 600 ps are achieved, and total set-up plus hold times of 320 ps. The error monitoring circuitry operates on the even mark parity line code (24B1P), and it comprises a clock, gate and modulo-2 divider circuit.

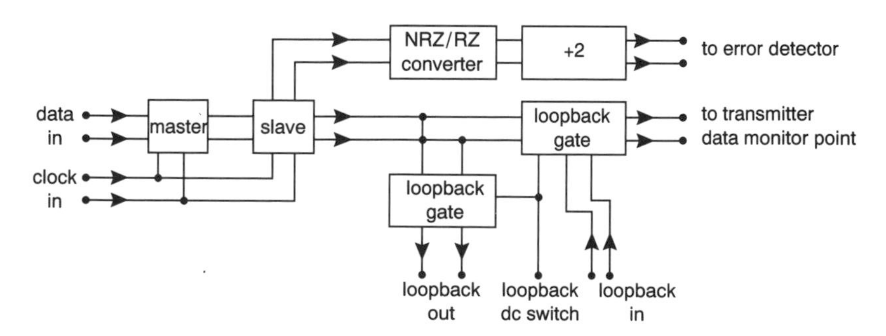

Figure 15.11 TAT-8 diagram of the decision circuit chip ([1] of Chapter 9, Figure 5). Reproduced by permission of Société des Electriciens et des Electroniciens

Transmitter

In TAT-8 two IRW laser modules are provided, one of which is on cold standby and draws no current, and both outputs are commoned by means of a passive coupler, as shown in Figure 15.12 [9].

The 3 dB coupler is fabricated from two single-mode fibres twisted together, fused and tapered. In this structure the two lower-order modes are excited, and phase slippage between them produces the power transfer between fibres. Coupling is also dependent on the wavelength and polarisation of the light input. Results achieved on production components for TAT-8 are given in Table 15.6.

The above limits include measurement uncertainty, all possible states of input polarisation in the input fibre, changes over the operating temperature range (0°C to +40°C), and variations during storage. A diagram of one laser module is given in Figure 15.13.

There are two separate control loops driven by the monitor photodiode; the first controls the mean power by adjusting the laser bias current to correct for any changes in threshold; the second loop corrects for any changes in laser slope efficiency by adjusting the depth of modulation. This depth ($I - I_0$) is controlled by injecting a low-frequency ripple on to the data zeros, as is shown in Figure 15.14.

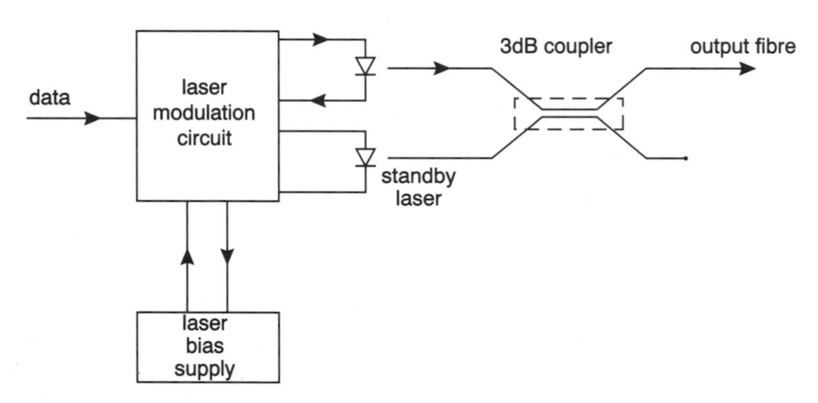

Figure 15.12 TAT-8 redundant laser arrangement using 3 dB optical coupler ([9], Figure 1). Reproduced by permission of IEE

Table 15.6 TAT-8 major parameters of production 3 dB couplers ([9]). Reproduced by permission of IEE

Parameter	Specification
Crossover wavelength	: 1317 nm ± 16 nm
Excess loss at crossover wavelength	: 0.3 dB max
Coupling ratio gradient at crossover wavelength ± 30 nm	: 0.04 dB nm^{-1} max
Back reflections	: Better than −38 dB

The above limits include measurement uncertainty, all possible states of input polarisation in the input fibre. Changes over the operating temperature range (0°C to +40°C) and variations during storage. The optical loss measurements were made using a cut-back technique.

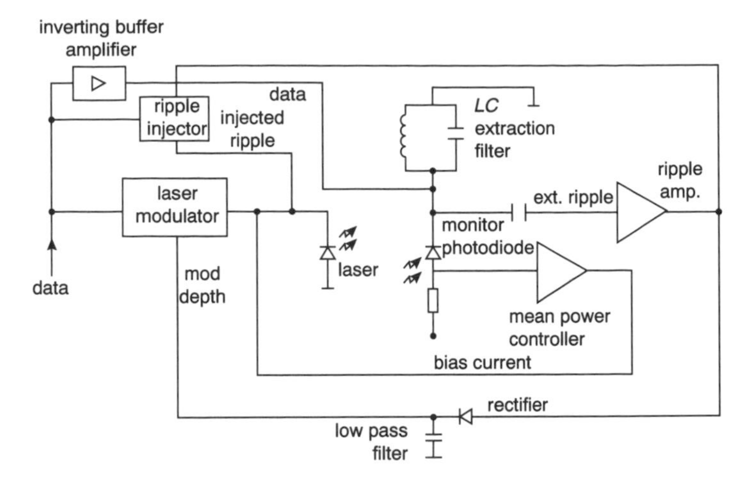

Figure 15.13 TAT-8 schematic of control loops for transmit laser ([7] of Chapter 9, Figure 6*a*). Reproduced by permission of Société des Electriciens et des Electroniciens

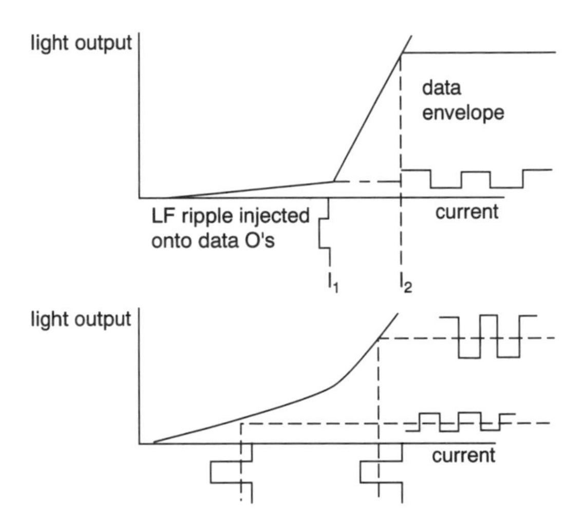

Figure 15.14 TAT-8 transmitter, operation of the ripple control loop ([7] of Chapter 9, Figure 7). Reproduced by permission of Société des Electriciens et des Electroniciens

The feedback ensures that the modulation depth is adjusted until the data zeros are at the knee of the laser output characteristic with the ripple at a constant low amplitude. The use of an unbounded line code with significant energy at the ripple frequency can cause a problem, but this is overcome by injecting some data in anti-phase to cancel the unwanted component.

Test results

The results of measurements on a sample of eight regenerators are presented in Table 15.7 with, for comparison, the design targets which were set in all cases. Figures 15.15 and 15.16 illustrate the jitter transfer functions and the accumulation

Table 15.7 Test results on sample of eight regenerators and design targets ([7] of Chapter 9, Table 4). Reproduced by permission of Société des Electriciens et des Electroniciens

	Measurements	Target
Trans. output	−6.2 to −4.9 dBm	−5.5 nom.
Rec. sensitivity	−34.3 to −35.9 dBm	−34.0 worst
Rec. overload	−13.2 to −12.7 dBm	−16.0 worst
rms jitter	0.9 to 0.6 deg	1.5 max.
ptp jitter	12.0 to 6.9 deg	22.5 max.
Jitter spectral density at 17.59 kHz (in deg^2/MHz):-		
Random	0.26 to 0.16	5.0 max.
Systematic	3.25 to 0.49	15.0 max.

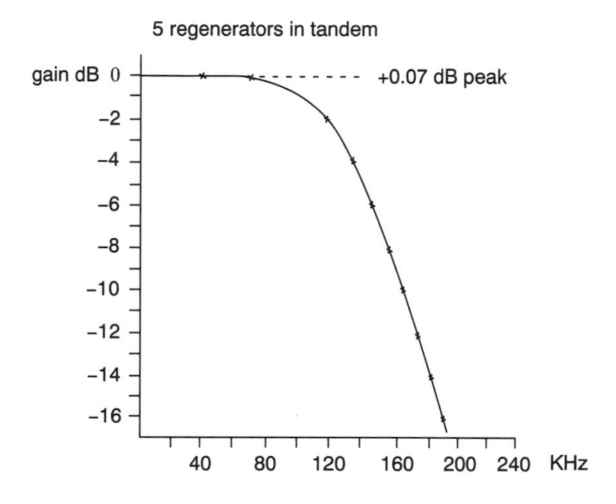

Figure 15.15 TAT-8 receiver, jitter transfer function ([7] of Chapter 9, Figure 9). Reproduced by permission of Société des Electriciens et des Electroniciens

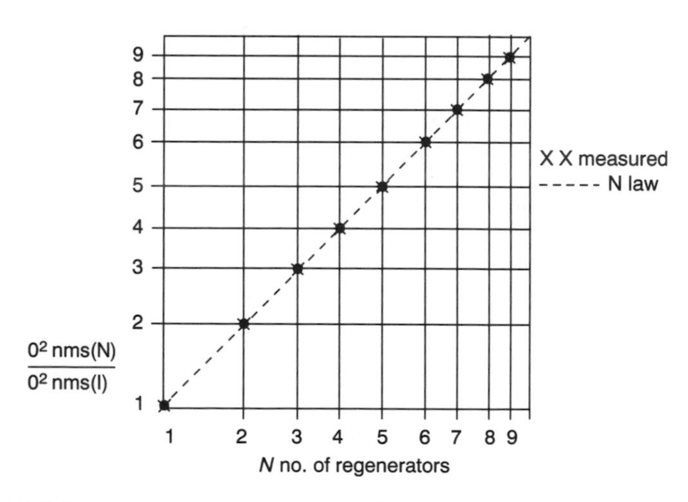

Figure 15.16 TAT-8 receiver, systematic accumulation of jitter ([7] of Chapter 9, Figure 10). Reproduced by permission of Société des Electriciens et des Electroniciens

of systematic jitter for several of the units connected in tandem. Both sets of results conform closely to the theoretical predictions.

15.2.3 Terminal equipment

Transmission path

Two landline 140 Mbit/s tributaries are interfaced, with each pair of the 280 Mbit/s submarine fibres as shown in Figure 15.17 and 15.18 for the transmit and receive directions, respectively.

The main functions to be performed in the transmit direction are:

select the normal or standby tributary inputs

convert from cmi to binary code

retime with system clock and justify (deriving control bit information)

scramble and add frame alignment, control and telemetry bits

interleave and add parity bits (with required violations for supervisory signalling)

convert the electrical to optical signals.

In the receive direction the sequence is essentially the reverse of the above.

Both the transmit and receive functions for one pair of submerged fibres are housed in a single cabinet that conforms to standard UK equipment practice. The cabinet is duplicated in order to provide the near 100% availability over the 25 year system design life, and suitable switching is installed.

Supervisory functions

Laser status, received light level and error count on the line, and terminal alarms (both transmission and power feed) are monitored continually. Appropriate corrective action of switching to standby units occurs automatically when that mode of

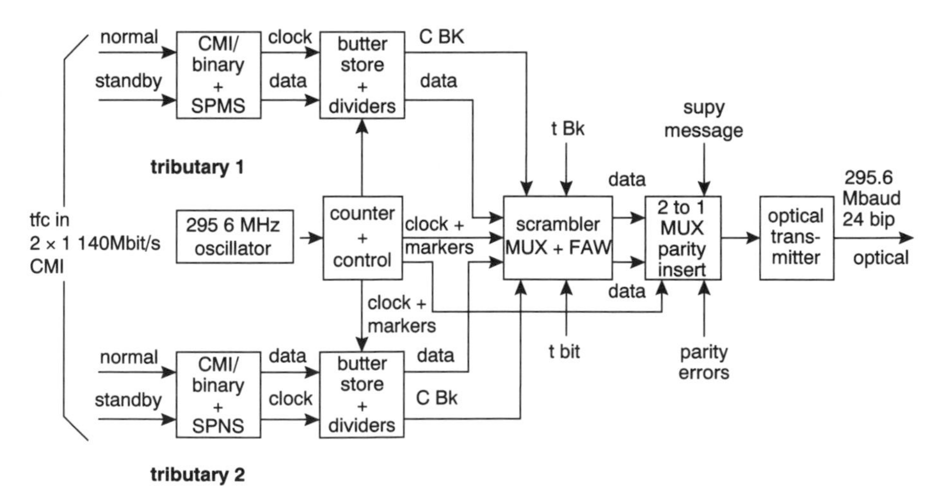

Figure 15.17 TAT-8 terminal, transmit direction ([7] of Chapter 9, Figure 13). Reproduced by permission of Société des Electriciens et des Electroniciens

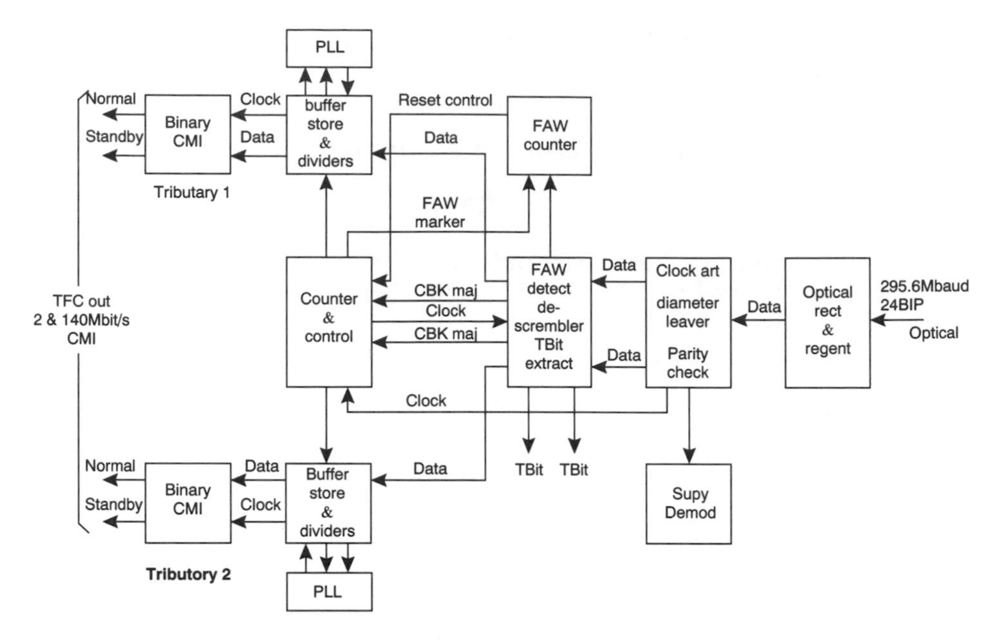

Figure 15.18 TAT-8 terminal, receive direction ([7] of Chapter 9, Figure 13). Reproduced by permission of Société des Electriciens et des Electroniciens

operation is selected, or manual operation can be used. The equipment records all performance readings, alarm signals and actions taken, together with any outages or degraded bit error ratios. Facilities are also available to make alarms remote and for signalling to and from other terminal stations, both via the line and, in the event of a fault, via the switched public network.

As with other equipment, the whole of the supervisory equipment is duplicated, with two sets housed in a single rack; either set can monitor the whole of the system, and the standby set will automatically take over in the event of the working set developing a fault.

15.2.4 TAT-8 supervisory subsystem

This provides comprehensive facilities for system maintenance and the ability to switch in redundant transmitters at each repeater. These facilities can be divided between in-service monitoring and out-of-service control functions. In-service monitoring at each regenerator includes measurement of: error ratio, received light level, and bias current in the spare laser. Out-of-service electrical loopback in both directions is provided at any regenerator and on either pair of fibres. Up to 1024 regenerators can be addressed, and access to any regenerator is available from any system fibre, so that all supervisory functions can be performed from either terminal. In addition supervisory facilities are not lost in the event of a failure affecting one pair of fibres. Loopback facilities, and in some configurations line finding facilities, are also available; both are implemented by electronic switching. In the next subparagraphs we will briefly address the following topics:

(a) terminal-to-repeater telemetry
(b) repeater-to-terminal telemetry

(c) implementation
(d) error monitoring
(e) loop-back and laser switching
(f) laser health and received light level.

Implementation of supervisory subsystem

A complete supervisory subsystem serves each regenerator in the repeater and is capable of performing all the measurement; signalling and control functions. Figure 15.19 [10], is an overall block diagram of the subsystem and it shows how it connects with the main transmission path. To provide the 17.595 kHz tone described above, the decision circuit produces an rz data signal from a clocked gate, and this signal drives a modulo-2 divider. The divider output is fed to the error detector where it is filtered, leaving only the required tone. This passes to the command decoder which reads and recognises address words and function codes then sends instructions for measurements, loop-back or return signalling. Upon instruction the two identical analogue-to-digital converters and latch circuits read voltages in the agc amplifier or 'laser transmitter', and pass timed pulses to the return signalling unit. This responds by forming an eight bit word representing the voltage measured. The error detector can also be instructed to signal into the return signalling unit to produce an eight bit word corresponding to the number of transmission errors over a prescribed period.

Terminal-to-repeater telemetry: Signalling is carried out by modulating a carrier generated in the main transmission path by the introduction of parity violations to

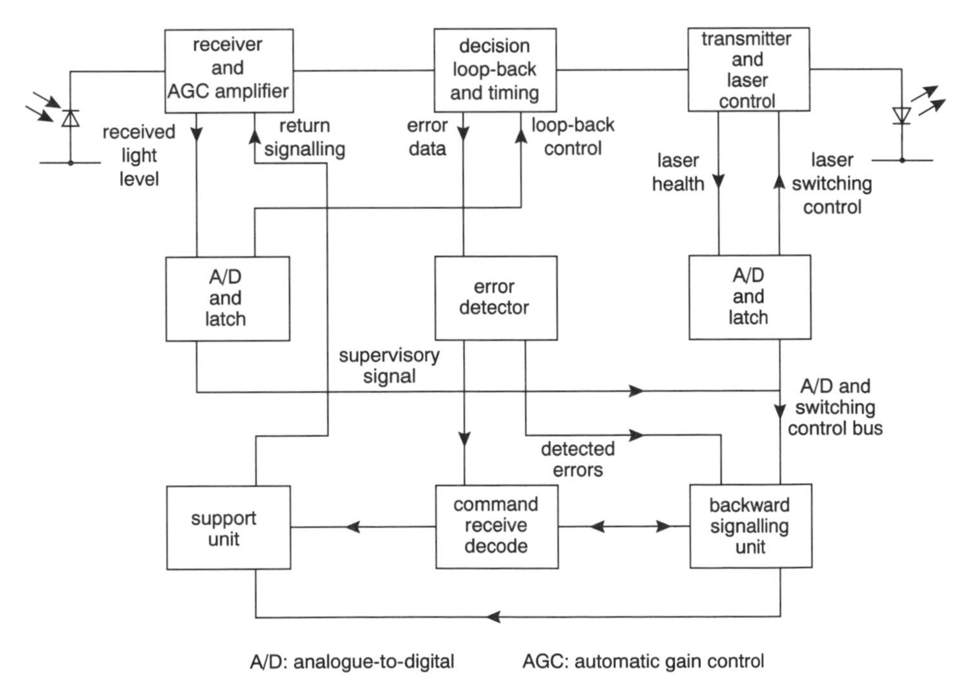

Figure 15.19 TAT-8 block diagram of supervisory subsystem ([10], Figure 1). Reproduced by permission of BT

the 24B1P line code. An odd mark parity bit is introduced for one 25 bit word in every sixth frame, resulting in a detected carrier at the repeater at 17.595 kHz. This is then pulse-width-modulated to provide a binary asynchronous channel at 85 baud, with '1' represented by the carrier on for 7.8 ms then off for 3.9 ms, and '0' by 3.9 ms of carrier followed by 7.8 ms for off. This channel exists on each fibre pair and allows commands in repeaters to be sent in the form of two identical and consecutive 15 bit words. Further details are given in [10]. Following the transmission of the 30 bit command word a continuous burst of 17.595 kHz carrier is sent and it acts as an enabling signal to implement the function requested and to provide a sub-carrier for return signalling.

Repeater-to-terminal telemetry: This is achieved by phase modulation of the system clock with a peak-to-peak amplitude of about 20 degrees and is applied by a subcarrier derived from the enable signal. A pulse-width-modulated binary channel at 30 baud is obtained by on–off modulation of the subcarrier. Digital ones and zeros have the following formats: on for 22 ms, off for 11 ms; and on for 11 ms, off for 22 ms, respectively. In response to a request for information, data are transmitted asynchronously over this channel in an eight bit word, preceded by a silent period of duration 400 ms or so to allow for the transmission and decoding of the command word.Then there is a preamble, followed by a second pause, and only then is the eight bit word transmitted containing the information about received light level, laser bias, or the contents of the error counter.

Error monitoring: Errors are detected in a circuit using that property of the line code which determines the average (dc) value of the output of the divider fed from the rz data signal. This value is constrained to one of two values when even mark parity is preserved, but once parity is lost as the result of a single bit error the level moves to the other allowed level. The change in level is detected and used to trigger an error counter.

Loop back and laser switching: When the loop back function is selected the command decoder sends a signal to the return signalling unit, which in turn operates latches causing the electrical loop back circuits to provide loop back paths in both directions for the fibre pair selected. The latches are held on only long enough for the enable signal to be received by the command decoder, when the supervisory signal ceases loop back is removed and the direction of transmission reverts to normal. Laser switching is initiated in a similar manner, except that the enable signal need only be held on long enough for the command to be recognised and executed. The newly selected laser continues operating unless a second switching signal selects the other device.

Laser health and received light level: The laser health command initiates a measurement by the a/d converter of the spare current available to the operating laser. An eight bit word representing this current is assembled in the return signalling unit, then pulsed out in serial form to the support unit. This unit converts its input to pwm of the 17.595 kHz subcarrier enable signal sent from the terminal via the command decoder. The pwm carrier is then applied to the phase modulator in the agc amplifier to complete the return signalling path.

Information about the received light level is obtained by monitoring the voltage in the feedback loop of the agc amplifier, since this voltage is a measure of the optical input to the receiver. The voltage is measured and then returned in an identical manner to that described above for the spare laser current. Both quantities are measured in a two-part process consisting of a calibrate and a measure phase.

Hardware: Each of the functional blocks in Figure 15.19 corresponds to a hybrid module consisting of an ic and a small number of discrete resistors and capacitors. Each of the six integrated circuits is packaged in a leadless chip carrier and mounted on a ceramic substrate with thick film resistors and interconnections, and the ceramic capacitors are mounted directly on the substrate. The three ic which interface directly with the main transmission path use ECL 40 eight cell arrays of exactly the same construction as those in the main transmission path, thus ensuring that the reliability of this path is not compromised by the supervisory circuitry. The remaining three ic (the support unit, command decoder and the signalling unit) which interface indirectly are more complex and so use an 80 cell array packaged in a 68 way leadless chip carrier. However, the array itself is constructed in the same ECL 40 technology, and is subject to the same test and screening procedure as the eight cell devices.

Power Feed: The essential parts of the arrangements within a repeater are shown in the simplified diagram of Figure 15.20.

Each one-way regenerator dissipates about 10 watts at 6.5 V and 1600 mA + 2%, including the currents in the zener diodes and in the bypass resistors not shown in the diagram. The overall voltage of a four-fibre repeater is 28 V, which includes an allowance for the surge protection units.

Sets of static converters operating from the station battery supply in the terminal building provide the 1600 mA of current. Each converter has a maximum rating of 1.5 kV and 2 A in a constant current mode, and in TAT-8 a stack of five in series is used to supply the 7.5 kV required. The converters operate at an ultrasonic frequency to avoid aural discomfort, and use power fet switching transistors with high-efficiency and low-drive requirements and relatively low levels of rfi. Heavy screening surrounds the converter cubicles to minimise any emc problems.

For long-haul high-voltage systems like TAT-8 nearly 100% availability can be assured by a combination of active redundancy and cold standby arrangements. In TAT-8 the load current is shared by two stacks of five converters, although either stack could deliver the whole load, and will do so automatically in the event of a failure. The complete set of seven cubicles which includes one common unit to control the load sharing is duplicated on what is normally cold standby. When

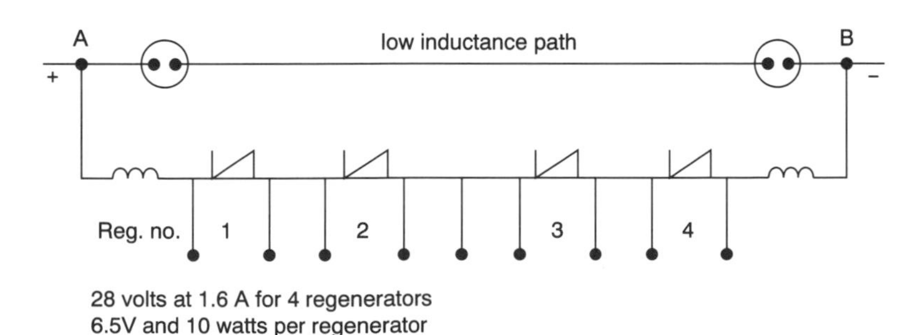

Figure 15.20 TAT-8 simplified diagram of repeater power feed ([7] of Chapter 9, Figure 12). Reproduced by permission of Société des Electriciens et des Electroniciens

Table 15.8 Supervisory facilities ([10], Table 1). Reproduced by permission of BT

In-service	Main path error count Received light level Laser health Laser switching
Out-of-service	Main path loop-back

required for maintenance, etc., this standby set can be powered up using a dummy load and a hot transfer of the system load takes place without interrupting traffic.

Further information on the TAT-8 supervisory subsystem is given in [10]. The choice of 24B1P line code required the use of saw filters with Q of 800 (as used by AT&T), thus precluding the use of the fm supervisory system developed for the UK–Belgium No. 5 system, where a 7B8B line code was used. The new system used digital repeater addressing. Violation of the line code parity bits is used in the outbound direction for signalling, and angle modulation of the data in the shorebound direction. This paper contains paragraphs grouped under the section headings 'facilities', 'techniques' and 'realisation'.

A brief summary of the facilities available are given in Table 15.8, and this was followed by paragraphs on analogue measurements (received light level indication, laser health indication, error counting), loop-back, redundant laser switching and return signalling.

Under 'techniques' the topics covered were: outbound telemetry, shorebound telemetry, analogue measurements (received light level indication, laser health monitor, analogue-to-digital conversion, error counting (even mark parity error detection, error rate measurements 'the upper limit, error rate measurements' the lower limit, long-term error rate measurement range (Table 2, error detector performance 24B1P code, 294.4 Mbaud)).

The section on realisation was: hardware, measured error-detector performance (Figure 8 measured characteristic).

15.2.5 Quality assurance

[11] described the QA methods practised by BTI on systems purchased by BT and when BTI acts as a consultant for other international telecommunication authorities. An all-embracing control activity is necessary throughout every stage of the project, covering all activities and functions concerned with the attainment of quality: BTI quality policy; quality strategy; quality costs; some financial considerations for implementing a quality-management system; past practice, current practice; pre-contract QA (invitation to tender, quality requirement (contract conditions); adjudication of tenders; vendor appraisal; in-contract QA (concept of quality audit, Figure 9, typical audit procedure for the design activity); quality of design; QA in manufacture: Table 1 Quality requirements, progress in the application of BS5750 to submarine systems; QA in installation: submerged plant installation, dry-land-based plant installation; post-contract QA; QA in maintenance (Figure 7 the QA loop).

A related paper entitled 'Considerations for the lifetime of submarine optical cables' also contains material of interest [12], but a more recent and wide ranging publication is 'Reliability aspects of optical fibre systems and networks' [13] of Chapter 10 by P. Cochrane and D.J.T. Heatley of British Telecom.

15.2.6 Digital circuit multiplication systems (DCMS)

Speech interpolation equipment was introduced into the international network in the mid-1960s on TAT-1, and was used on later systems to exploit the silent time which occurs during conversation, and thereby to increase circuit capacity [13]. The following TAT-8 DCMS requirements were discussed: overall objectives, fundamental interworking requirements (interface conversion, time-slot interchange, pcm standards conversion), traffic-carrying requirements. It then went on to consider network operation of the DCMS and dealt with traffic management, selective traffic management, DCMS-International Switch Centre (ISC) control link and management support-cluster operation.

The TAT-8 network configuration for DCMS

Three fundamental requirements which must be satisfied to enable PSTNs and other providers operating with the North American and European PDHs shown in Figure 15.21, to be interconnected are:

(1) interface conversion
(2) time-slot interchange
(3) PCM standards conversion.

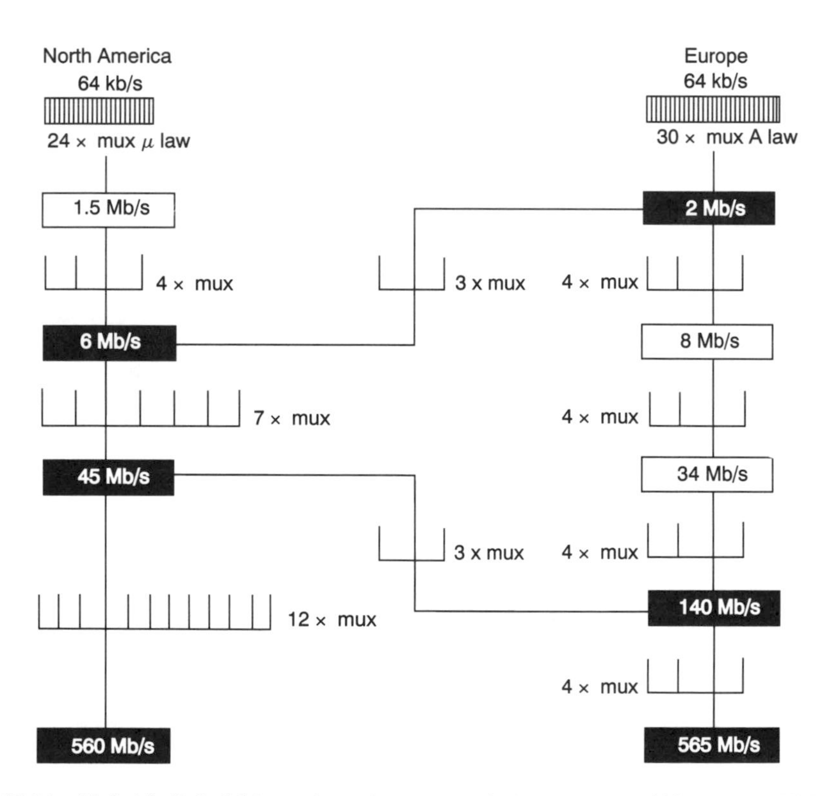

Figure 15.21 Hybrid digital hierarchy relating North American and European PDHs (Cable and Wireless)

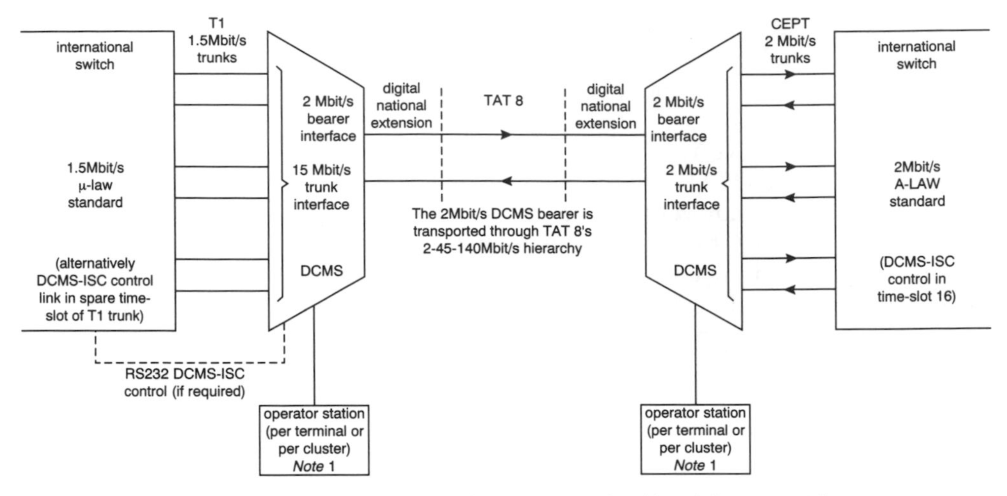

Figure 15.22 TAT-8 typical system configuration ([13], Figure 1). Reproduced by permission of BT

For brevity, only (1) will be mentioned here. The 2 Mbit/s digital path, which is bit sequence independent, forms the bearer between the two DCMS terminals, one with a 1.5 Mbit/s trunkside interface (North America), the other a 2 Mbit/s trunkside interface (Europe), as illustrated in Figure 15.22. Apart from the DCMS equipments, this diagram also shows other major parts of the configuration.

15.3 SUBMERGED LINKS IN THE PACIFIC BASIN

It is always wiser to prophesy after the event, but in any business requiring long-range planning a view must be taken in order to draw up more detailed plans. By

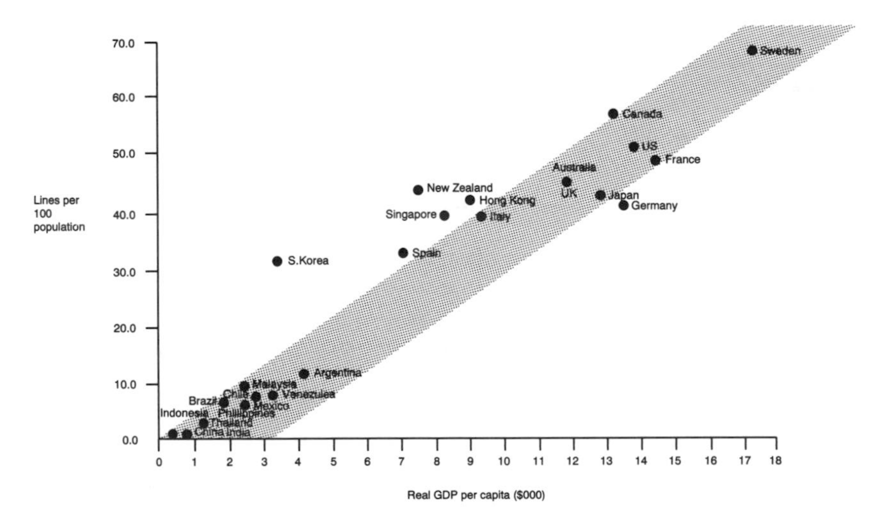

Figure 15.23 Telephone penetration versus real GDP per capita (in thousand $) for a number of countries worldwide ([14], Figure 1). Reproduced by permission of KMI

the early 1990s it appeared that the centre of gravity of world economic activity might move increasingly to the countries bordering the Pacific Basin as the 21st century approaches. With the well-known correlation between economic indicators such as real GDP per head and penetration of telephones it may be that an increasingly large share of growth in telecommunication services and networks will occur there. The remarkably linear relationship between the number of telephone lines per 100 people in the population of a number of countries versus the real GDP per capita (in thousand $) is shown in Figure 15.23 [14].

By 1994 research had extended the above relationship to project how the global telecommunication infrastructure might evolve to the year 2000 or so. The data base included that used in Figure 15.23, and was extrapolated using various economic growth rates appropriate to the various country categories. The general conclusions are summarised in the Figures 15.24 and 15.25.

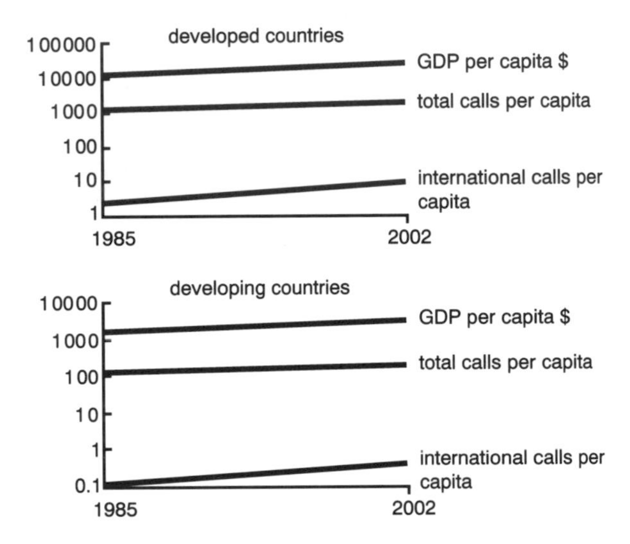

Figure 15.24 Projected growth in numbers of exchange lines and trunk circuits ([14], Figure 2). Reproduced by permission of KMI

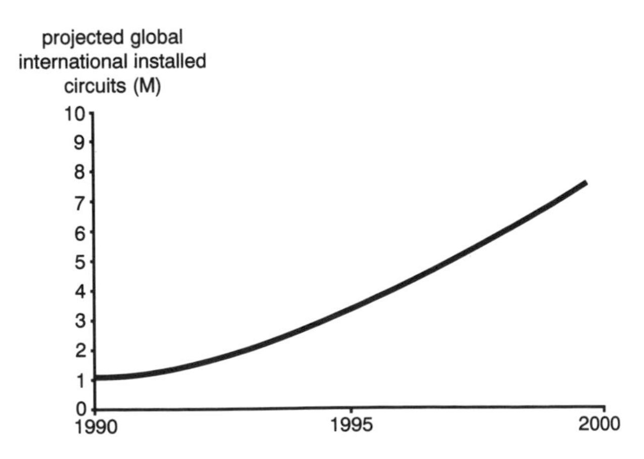

Figure 15.25 Projected growth in numbers of global international circuits ([14], Figure 3). Reproduced by permission of KMI

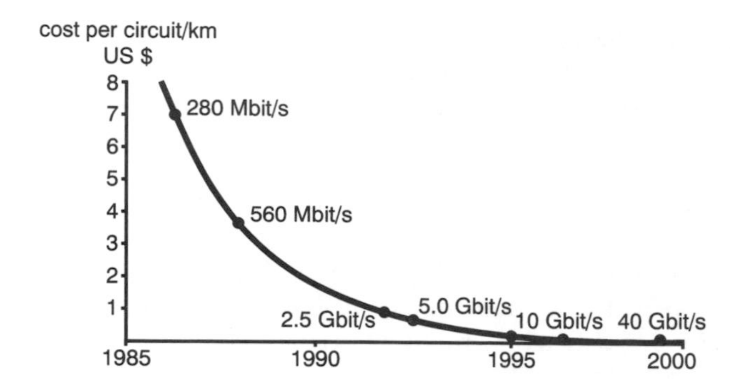

Figure 15.26 Actual and predicted cost per circuit km in US $ of submerged systems of various capacities versus year of (projected) system purchase ([14], Figure 4). Reproduced by permission of KMI

Note that the growth rates assumed were based entirely on economic factors and did not take account of future wideband or other services which might result in a jump in demand for traffic capacity. Figure 15.25 suggests a continued dramatic growth in demand for international circuits over the period. Actual and predicted costs per circuit/km over the same period have fallen as shown in Figure 15.26.

15.3.1 Existing and planned links

The plan to forge a modern optical fibre network in the Pacific Basin developed rapidly in 1987–88, and the network of installed and planned routes to 1996 as at end 1991 was given in Figure 15.1. Traffic is growing at about 20% per year in the region, and the web of cables will provide valuable route diversity in the event of a break, and as an alternative to the satellites which serve the region.

Optical systems link Malaysia, Singapore, Brunei and the Philippines in the West with Hong Kong, South Korea and Japan in the North, Australasia in the South, Guam and Hawaii in the centre and the United States at the Eastern rim. HAW3 (3 × 280 Mbit/s)/TPC3 (3 × 280 Mbit/s) entered service by end 1988, and were followed by about 10 other systems. TPC-5 will be the first optically amplified system in the Pacific, will operate at 5 Gbit/s and is due to enter service in 1995. It will have effectively four times the circuit capacity of its predecessor TPC-4, which entered service in 1992. A 25 000 km extension of TPC-5 known as the Asia-Pacific Cable Network (APCN) was under study in 1993 and was due for completion in 1996–98. A later and longer (26000 km) system known as FLAG (Fibreoptic Link Around the Globe) to link Europe with Japan via the Mediterranean and Indian Ocean was in the early planning stages by 1993, with a target completion date in the mid-1990s [15].

PACRIM(W) (1 + 1 × 560 Mbit/s) was completed in 1993 and PACRIM(E) (1 + 1 × 560 Mbit/s) is due for completion three years later. The link between Australia and New Zealand, TASMAN 2 (3 × 560 Mbit/s), should be completed in the mid-1990s, and the contract had been awarded to Alcatel of France. Europe was linked to the region by the 18 000 km long SEA-ME-WE 2 (2 × 560 Mbit/s) which has 160 submerged regenerative repeaters and which entered service in June 1994. There are submerged branching units giving access to Colombo, Bombay, Jeddah, Palermo, Bizerte and Marseilles. Construction was shared between AT&T SSI, STC and Alcatel Submarcom, and was the latter's single biggest project up to 1994. In

1993 Alcatel bought STC to form Alcatel Cable, which then became the world's largest underwater cable group.

Of these, FLAG and the North Pacific Cable (NPC $(3 + 1 \times 420$ Mbit/s)) are the odd ones out, because the pattern of ownership is different. FLAG is privately financed rather than on the usual intergovernmental basis and can be more commercial and innovative on tariff structure. NPC is owned jointly by Cable and Wireless, International Digital Communications of Japan and Pacific Telecom of America, and it creates a new way in which such a facility can be provided.

15.3.2 North Pacific Cable (NPC) system

The record-breaking North Pacific Cable (NPC) system links Miura, near Tokyo in Japan and Pacific City, Oregon in the USA, a distance of 8300 km, with a 1100 km spur to Seward in Alaska. The Japanese company NEC laid the segments from the branching point in the Pacific to both US landings, plus a short length of the cable from the branching point to Japan. STC supplied and laid the longer segment from there to Japan itself. When completed in 1990, NPC-1 was the longest system at 9400 km, and it contained the longest continuous submerged cabling. STC supplied and laid 5670 km of the total with 95 repeaters, and the cable contained eight fibres. The system operates at 420 Mbit/s in the 1300 nm band. It has three owners, Cable and Wireless plc, Digital Communications Inc. of Japan and Pacific Telecom Cable Inc. of America.

'Cable team battle to lay ocean link' was a headline in the STC Gazette published in March 1991. The article described how courageous cable teams had battled 60 ft seas as STC laid its section of NPC, in appalling weather conditions reminiscent of those encountered by the 'Great Eastern' over a century earlier in the North Atlantic (1856). In one heart-stopping moment a series of massive waves hit the ship while it attempted to cut the cable and run from the storm. At one point the 'Cable Venture' took a terrifying roll of about 48°! Further delay occurred when the cable ship arrived at the linking point of the cable previously laid by the Japanese, where the end was buoyed off. When the buoy was picked up prior to recovering the laid cable, the rope attaching it to the cable 5 km below snapped with the additional strain after all the stresses due to the long spell of bad weather. Several days of grappling for the cable, buoying-off and joining the two ends passed before the job was complete, and the first transmission took place on Christmas Day 1990. However as the Gazette reported, STC Submarine Systems successfully completed one of its toughest-ever assignments. The lay had taken more than 90 days instead of the scheduled 32 days because of the terrible weather conditions.

15.4 SDH-COMPATIBLE SUBMERGED SYSTEMS

STC Submarine Systems was chosen in 1992 as the sole supplier of the first direct optical fibre link (CANTAT-3) between Canada and Europe. This will deploy the NL16 regenerative repeatered system and will be the world's first implementation of 2.5 Gbit/s technology in a submarine environment. It is fully compatible with the new synchronous digital hierarchy standards and will be completed in late 1994. Linking Canada with Denmark, Germany, Iceland and the UK, the 7500 km long CANTAT-3 will enhance international connectivity by linking to the Western European RIOJA (1837 km, 17 edfa repeaters, RFS 1994) and ODIN (985 km, RFS

1995) systems, as well as TAT-12/13 (RFS 1995/6). It will be capable of carrying multiservice traffic, such as data, facsimile and video, as well as speech.

TAT-12/13 and TPC-5 will all be SDH compatible, will operate at 5 Gbit/s and use optical amplifier technology when they enter service in the mid-1990s. They are likely to be the forerunners of many more such systems.

15.4.1 TAT-12/13

This will be a four-fibre ring network linking the US, UK and France and will be supplied by Alcatel, AT&T and STC. As with other multisupplier cases hardware integration development tests were carried out using the line subsegment of concatenated erbium amplifiers and ds fibres, transmitter and receiver from each supplier. These tests took place at the AT&T facility at Freehold, NJ, in the United States. Interim results on this phase, which demonstrated that 5 Gbit/s nrz data could be transmitted successfully through an optically amplified line segment comprising subsegments from two or three suppliers, were reported in [16]. The purpose was also to reconcile test results with each supplier's own system simulator. The tests:

(1) characterised the transmitters and receivers
(2) measured the performance of each supplier's line subsegment
(3) measured the end-to-end performance of the line segment.

Computer simulations of the tests were also performed.

Transmitter (tx) and receiver (rx) testing

Transmitter testing covered measurements of linewidth, rise and fall times, extinction ratio, intensity noise and chirp. Receiver testing included sensitivity, overload, dispersion penalty and optical/electrical bandwidth measurements. In general, similar performance was found for all the tx/rx combinations.

Performance of each supplier's line subsegment

Tests covered wavelength of gain peak, chromatic dispersion, polarisation mode dispersion, ber and Q (a measure of electrical snr used as a figure-of-merit; see equations (11.3), (13.2), (13.3)). Q was evaluated using a variable decision threshold method in which ber was measured as the decision threshold was scanned through the 1 and 0 levels of the received eye, and by extrapolation of the optimum decision threshold level the associated value of Q was found. Close gain wavelength matching was achieved between different line subsegments. Control of chromatic dispersion was demonstrated. Polarisation mode dispersion below 0.22 ps/(km$^{1/2}$) was achieved.

Measured end-to-end performance of the line segment

BER and Q tests on each transmitter paired with the STC receiver were made and error-free operation was observed on all subsegment tests. In general, values of Q derived from measurements were lower than values from simulations; discrepancies were attributed, in part, to isi, thermal and shot noise and other effects not included in the simulations. In addition, Q varied with time as the sop of the system varied (see Figure 13.9), due to polarisation-dependent loss within amplifiers and the polarisation mode dispersion of amplifiers plus transmission fibres. 'Optimum' segment performance was dependent on amplifier output power levels, demonstrating the trade-off between nonlinear degradations (at high levels) and noise degradations (at low levels).

Independent computer simulations were run, which included:

(1) fibre group velocity dispersion
(2) fibre attenuation
(3) ase from amplifiers
(4) self-phase modulation (spm) and signal-ase noise interaction due to the Kerr effect
(5) receiver effects (e.g. bandshape, optical bandwidth).

The edfa gain and noise were simulated using a 1991 model. The tests demonstrated that an optically amplified system based on a common set of integration specifications (which did not require common detailed system and component design) and embodying hardware from different suppliers is feasible. No multi-supplier impairment was observed, and the performance exceeded the requirements of CCITT G.82X.

15.4.2 Development work on the STC portion of TAT-12

STC will supply nearly 4000 km of cable and 85 submerged optical amplifiers for TAT-12, which will link Greenhill, Rhode Island, USA, to Land's End, UK. The product to be used is designated the OAL32 system, which can be deployed for short-, medium- or long-haul applications. It will embody the fifth generation of technology since 1986, and will increase the 64 kbit/s channel capacity per fibre pair from 3800 of TAT-8 to over 60 000 (in round numbers) without a corresponding increase in system cost.

The development programme [17] consisted of the following.

(1) Construction of a 6300 km, 138 edfa repeater system demonstrator to confirm performance and provide valuable statistical data on component and manufacturing tolerances.
(2) Development of a recirculating loop measurement facility.
(3) Construction of a 20 amplifier, 1500 km experimental system and its assessment at 0.6, 2.5, 5 and 10 Gbit/s.
(4) Development of a software model for the simulation of amplifier systems.
(5) Development of a long-haul supervisory subsystem based on gain modulation of repeater amplifiers.
(6) Development of transmitter and receiver optoelectronics and their optimisation for use in optically amplified systems.

6300 km, 138 edfa repeater system demonstrator

A line diagram of the demonstrator is shown in Figure 15.27.

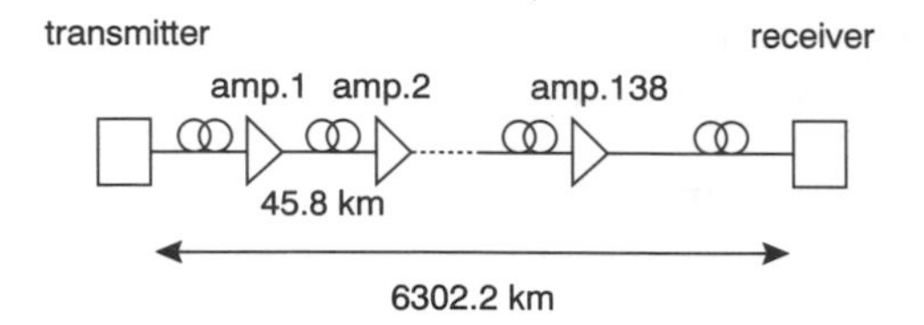

Figure 15.27 Line diagram of system demonstrator ([17], Figure 1). Reproduced by permission of Société des Electriciens et des Electroniciens

Dispersion-shifted fibre with a nominal zero dispersion wavelength of 1558.25 nm was used throughout. The repeater configuration is given in Figure 15.28.

Open-loop control of the amplifiers (in which the mean power of each pump laser was controlled) was used for all experiments. Stability of output power level was achieved by operating the amplifiers in gain saturation. Return supervisory signalling was implemented by gain modulation of the polled line amplifier. DBR pump lasers were employed because of their excellent environmental stability. Measured Q versus system length is plotted in Figure 15.29, with $Q = 10$ (20 dB) at the input to the receiver, corresponding to an implied ber which was calculated to be 10^{-23}.

The arrangement of Figure 15.27 was also used to investigate:

(a) the system behaviour over a range of transmitter wavelengths

(b) the choice of output levels for the repeaters to optimise system performance

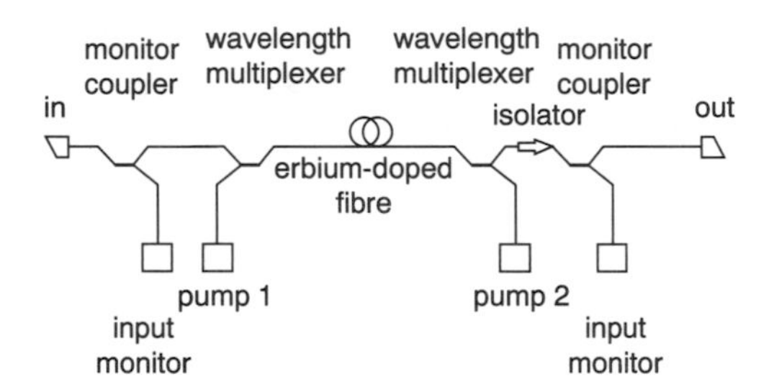

Figure 15.28 Block diagram of optical amplifier ([17], Figure 2). Reproduced by permission of Société des Electriciens et des Electroniciens

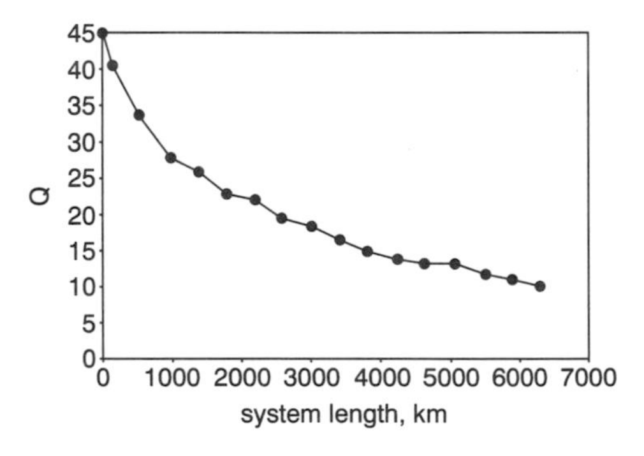

Figure 15.29 Measured Q versus system length ([17], Figure 3). Reproduced by permission of Société des Electriciens et des Electroniciens

(c) the suitability of chosen supervisory subsystem

(d) the system performance under fault conditions.

Recirculating loop measurement facility

The development of this facility provided a means to simulate long-distance transmission without the investment in a whole-system test bed, and enabled rapid results to be obtained for changes in amplifier design. A diagram of the arrangement is shown in Figure 15.30.

An example of measurements made on this facility was the wavelength response of the loop made to verify that the peak gain wavelength of 150 amplifier systems would remain within a specified narrow range. Figure 15.31 shows the measured points. The wavelength of peak gain is 1557.9 nm, confirming that the amplifier design was within specification. The corresponding wavelength of amplifiers manufactured to this design was 1558.25 nm. This is but one example of many where a recirculating loop can be used in the characterisation of optically amplified systems.

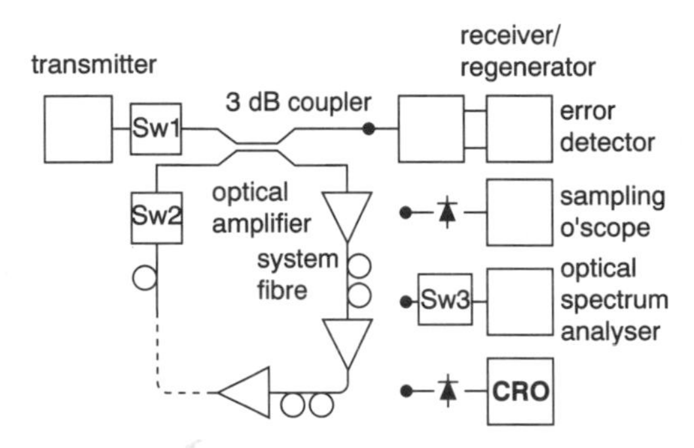

Figure 15.30 Diagram of recirculating loop measurement arrangement ([17], Figure 5). Reproduced by permission of Société des Electriciens et des Electroniciens

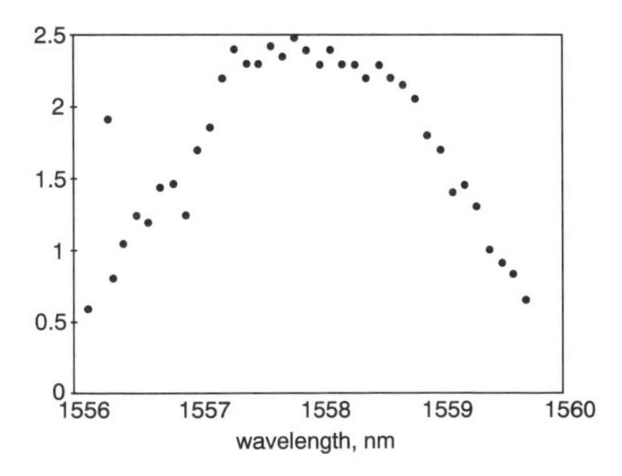

Figure 15.31 Recirculating loop measurements of wavelength response after 50 orbits in a loop containing three amplifiers ([17], Figure 6). Reproduced by permission of Société des Electriciens et des Electroniciens

Development of a software model for simulation of amplifier systems

This resulted in a simulator comprising transmitter, transmission path and receiver. The first part generated a pulse train, then filtered it to obtain a realistic transmission signal. The receiver converted the optical signal to its electrical analogue then displayed the received data pattern, if necessary after shaping, and derived from the pattern any performance measures required. The simulated transmission path required many parameters to model the effects of dispersion, nonlinearity, loss, polarisation and amplification on the transmitted signal for each repeater section in turn. For example, the effect of self-phase modulation was found by solving the nonlinear Schrödinger equation. Simulator results fell into two distinct categories: those which could be compared with measured results to build confidence in the accuracy of the simulator; and simulation of a proposed design to indicate whether a design performance was achievable. Both analytical and experimental results have shown that it is possible to construct an accurate computer simulator for optically amplified systems. When applied to transatlantic systems the primary requirement was to establish optimum operating conditions, in particular, acceptable values for amplifier output power levels, repeater spacing, polarisation and fibre dispersion characteristics. The utmost importance was attached to obtaining agreement between simulator and system test bed results.

Development of a long-haul supervisory subsystem based on gain modulation of repeater amplifiers

This was needed to provide the ability to monitor remotely the performance of any oa repeater and to locate the cause of a system degradation or fault to a specific repeater or cable section. Unlike systems with regenerative repeaters this must be achieved without access to the actual data signal. A detailed account was given of the subsystem adopted.

Further information is given in [18].

15.4.3 Main parameters of the STC OAL32 optically amplified system

SDH compatibility: the system is fully compatible with the Synchronous Digital Hierarchy, and is based on the STM-16 format in accordance with CCITT Rec. G707, G708, G709.

Traffic capacity: two traffic streams are accepted at the STM-16 level. However, standard SDH multiplex equipment can be used to provide the following interfaces: 8×STM-4 (optical), 32×STM-1 (optical), and 32 × 140 Mbit/s (cmi, electrical). Combinations of these tributaries can be accommodated.

Circuit capacity: up to 61 440 × 64 kbit/s both-way circuits per fibre pair.

System bit rate: interface bit rate 2.488 Gbit/s with two ports providing a total equivalent capacity of about 5 Gbit/s.

System configuration: cable and repeater housings can cater for up to eight working fibres, i.e. four systems per cable.

System length: up to 10 000 km.

Repeater spacing: up to 75 km depending on system length.

Repeaters: erbium-doped fibre amplifiers pumped at 1480 nm with dual pump laser redundancy. Maximum of eight edfa and associated supervisory circuitry per housing.

Line terminal equipment: line interface equipment uses an STC-produced mqw-dfb ridge lasers, operating at 1550 nm nominal. Output power is +2 dBm approximately. Receiver uses an STC-produced apd package with minimum sensitivity about −32 dBm with an optical preamplifier.

Fibre: precise type of dispersion-shifted fibre dependent on application and upgradeability requirements. Typical loss is 0.21 dB/km at 1550 nm, CCITT Rec. G.653 applies (insofar as applicable).

Cable: a full range of submerged cables are available, from which types suited to individual application are chosen.

Deep water: lightweight (NL/LW), lightweight screened fish bite protected (NL/LWS).

Shallow water: single wire armour light (NL/SAL), single wire armour medium (NL/SAM), single wire armour heavy (NL/SAH), double wire armoured (NL/DA), rock armoured (NL/RA).

System redundancy: various options are possible depending upon the overall networking arrangements and availability requirements. All terminal equipment can be duplicated and/or provided with spare equipment. Dependent repeaters have redundant pump lasers supplied as standard but stand-by systems can be provided if necessary.

Engineering order wire: terminal-to-terminal connectivity is provided via first STM-1 within STM-16 signal using E1, E2 overhead bytes accessible from associated SDH multiplex equipment.

System control: Repeater monitoring and control using a unique system developed by the manufacturer. Local control at LTE: Q interface as per CCITT Rec. G.773 via local terminals and/or hand-held terminal. Remote monitoring and control: via an Element Manager operating as part of an SDH Management Network (CCITT Rec. G.784 applies).

System performance: Error performance to CCITT Rec. G.821 (and G82X). Jitter performance to CCITT Rec. G.958. Submerged plant design reliability: less than three ship repairs in 25 years for a 7 500 km system. System availability: 99.99% (excluding damage from external sources).

Safety: In accordance with IEC 825.

EMC: The system complies with all EC and IEC standards.

Equipment practice: Line terminal equipment (LTE): Sub racks 482.6 mm (19 inches) series to IEC 297. Rack dimensions: width 600 mm, depth 300 mm, and height 2200 mm.

(Source: data sheet from STC Submarine Systems Ltd., 1994.)

15.4.4 10 Gbit/s wdm transmission with optical add/drop multiplexing

This was demonstrated during trials lasting several days in mid-1995 over the longest segment of the first optically amplified submerged system in Europe (RIOJA). This 930 km link between Santander, Spain and Porthcurno, UK was used to evaluate the transmission of four 2.5 Gbit/s channels at four different wavelengths over 3700 km by concatenating the four fibres in the cable. Error-free performance was obtained in all channels.

Using advanced components under development for submerged system applications, add/drop multiplexing was successful demonstrated by wavelength branching. After 1850 km data was dropped from one of the channels and new data inserted at the same wavelength for propagation over a further 1850 km. A schematic of the trial arrangement is shown Figure 15.32.

This approach permits wavelength routeing (i.e. use of wavelength to define paths through an optical network), thus giving additional flexibility to configuration. A wavelength routeing approach to optical communication networks was described in [19]. Other work has shown that it is possible to switch optical channel signals in a wavelength routed network for either routeing or protection purposes by means of erbium amplifiers, without the need to provide an opto-electronic interface [20].

Error-free transmission over this distance was achieved by using a proprietary repeater supervisory technique which enabled repeater outputs to be optimised

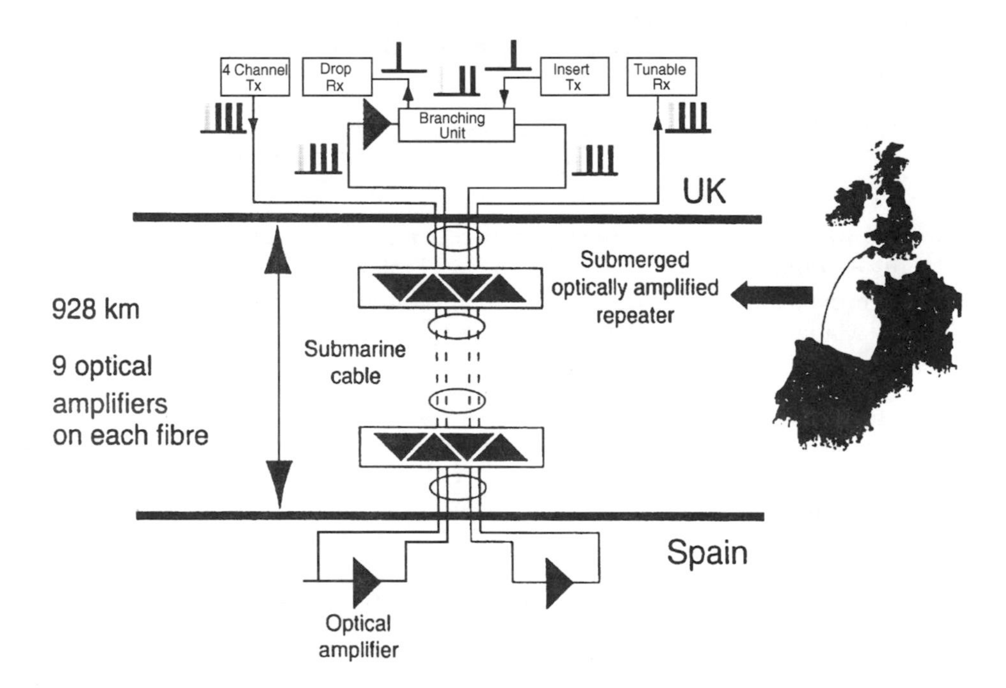

Figure 15.32 10 Gbit/s wdm transmission with add/drop multiplexing over the UK-Spain segment of the 2 × 2.5 Gbit/s RIOJA submerged cable system. Alcatel Submarine Networks Press Release

to produce a minimum ber under actual traffic conditions. The natural bandwidth limitation of the erbium amplifiers was overcome by a channel power pre-emphasis technique applied at the transit terminal.

15.5 UNREPEATERED SUBMARINE SYSTEMS (USS)

These novel systems were installed from the late 1980s in short point-to-point sea routes and they operate in the 1550 nm band. Between 1987–8 and 1992 the value of this market had quadrupled, and new applications were emerging as a result of very significant technology advances embodied in this new generation of systems (so-called unrepeatered submarine systems, USS). These offer operators in all but the most landlocked of nations a significant alternative in their network planning strategies, because such systems interface directly with standard existing SDH or PDH landline terminals and thus become extensions of the national network. The new submerged systems provide a 'wet' optical pipe with economic, operational and strategic properties not possessed by satelite systems. As an illustration, the main parameters of a system trial held during three days in February 1992 are listed in Table 15.9. [21]. The length of 180 km was made possible by the use of a commercially available erbium-doped optical fibre amplifier which presented a transparent interface to the terminals, thus allowing the system to appear as a standard link. The amplifier thus appears as another card or shelf in the SDH terminal equipment independently of the terminal supplier. Indeed, the use of optical amplification can increase the length of links based on USSs to about 300 km.

In mid-1994 the world's longest seabed 2.5 Gbit/s SDH-compatible USS was CELTIC, commissioned on 8 August 1994, a six fibre-pair cable linking Ireland and England. Initially, 1 + 1 pairs of the standard sm fibres were equipped, giving a capacity of more than 30 000 speech circuits. The system has 23 co-owners. This 270 km (including land cable) system operates at 1550 nm and links Ireland directly into the ITN, which will include TAT-12/13 and the intra-European RIOJA system by 1996 (source: STC Press Release, 1994). The RIOJA system will include a 284 km

Table 15.9 UK-France 2.5 Gbit/s STM-16 unrepeatered system trial ([21], Table 1)

Path length	180 km (including 60-km land section with splices every 1.5 km) (equivalent to 220 km of all low-loss cable)
Path loss	42.5 dB
System gain	44 dB
BER	Better than E(-11), measured over one week
SDH terminal	Standard Northern Telecom STM-16 product designed for 1550 nm terrestrial systems
Optical amplifier	30 m length of erbium-doped optical fibre, pumped by two 1480 nm lasers. Fully alarmed and monitored. Output power +13 dBm
Optical receiver	III–V APD
Optical transmitter	Low chirp DFB laser

long unrepeatered segment linking Belgium and The Netherlands which will include a submerged passive device.

15.5.1 Actual and potential applications

Conventional terrestrial wayleaves such as in underground ducts or alongside railway lines are becoming congested in many areas; underwater routes in rivers, lakes, and canals as well as those in coastal waters can offer alternative wayleaves which are economically attractive. For example, a 100 km typical USS link could be installed in about six days, whereas an equivalent landline could take six weeks. The dependence of such systems on a 'wet route' is often not a limitation because the sites of many large human settlements through the ages have been determined by the availability of good river or maritime transport and communication. One need only think of the US, Japan, Italy and Norway, but there are many more countries and indeed continents where this is valid. By mid-1994 STC alone had installed, or was contracted to install, more than 6000 km of unrepeatered systems worldwide, more than any other supplier.

[21] listed the following: coastal festoons: these are loops of cable following routes close to the coastline and landing at or near coastal towns. A number of separate

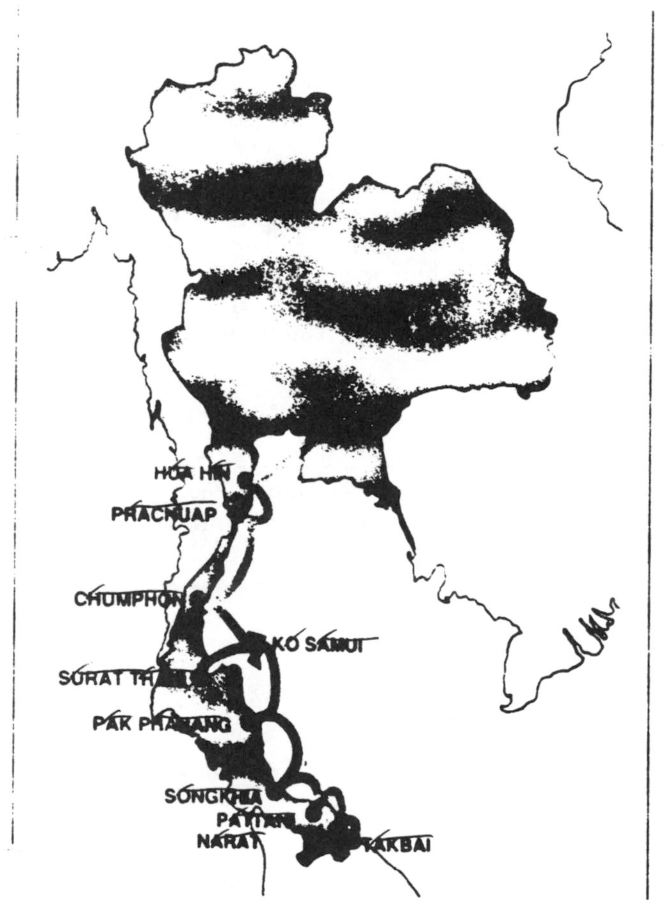

Figure 15.33 Thailand festoon ([21], Figure 1)

loops are combined to form a system (see Figure 15.33 below). Mainland to island links: separations of up to 300 km can be spanned to link islands into a national network. Island-to-island links: either in shallow or deeper water, commonly in archipelagoes (e.g. Micronesian system mentioned later). Direct links between coastal towns: these can avoid the need to deploy circuituous land links (e.g. Swansea to Brean, Severn Estuary below). Rivers, lakes and canals: links between places sharing common waterways.

An illustration of the first application is shown in Figure 15.33. This interesting application of submerged systems is in Thailand, where in 1993 a series of loops of cable was installed around the coastline linking eleven landing sites. The total system is 1300 km in length, with the longest single span 185 km. This system is known as the Thailand Festoon, and it operates at 565 Mbit/s.

Two other routes are the 240 km Micronesian system linking Guam, Rota, Tinian and Saipan operating at the SDH 622 Mbit/s rate, and the Severn Estuary 2.5 Gbit/s, STM-16 link which is 275 km long. Another company in the USS business, BT Marine, had installed systems in Scotland and Iceland among other countries by 1993.

15.5.2 Relative advantages of USS links

Attractive features of USS links *vis-à-vis* landlines include better protection against weather and damage, very high capacities (e.g. up to 180 000 × 64 kbit/s circuits per link), excellent error-rate performance (e.g. as in Table 15.9), transparency, and rapid nondestructive installation. Future-proofing is also a feature, since only the terminals need be changed when bit rates are increased. A further advantage in synchronous systems is the relative ease with which ring structures can be implemented to provide route diversity and self-healing capabilities. The economic attractiveness of USS technology rests on re-engineering conventional submerged systems to reduce costs and make installation cheaper, and the development of optical amplifiers which can be added to existing landline terminals to provide a

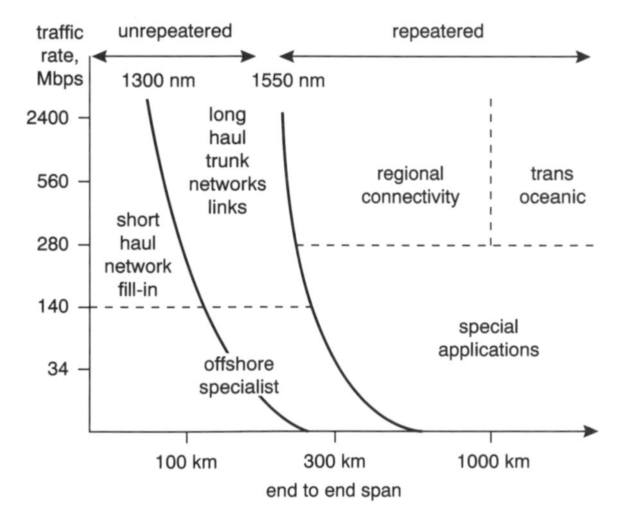

Figure 15.34 Underwater optical cable systems application areas plotted on a bit rate versus end-to-end span basis ([21], Figure 2)

seamless interface between terrestrial and submarine sections of the network. For instance, a much lighter and more flexible cable with the same traffic capacity at half the cost of standard repeatered transoceanic cables, and which can be laid from container-stored equipment mounted on a flat-backed ship rather than a cable ship. A typical 100 km USS link can be installed in about six days, whereas an equivalent landline would take about six weeks.

As a result of all these developments, the range of applications (both actual and potential) of submerged cable systems has expanded enormously, as can be seen from Figure 15.34.

APPENDICES

APPENDIX A: THE GROWTH OF SUBMERGED CABLE SYSTEMS

The seas and oceans that girdle the globe have been a barrier to communication for centuries. It is only our great grandfathers' lifetime that Man has been able to communicate electrically across such large expanses of water. An important political example of this barrier was the lack of a telegraph link between Great Britain and India at the time of the Indian Mutiny in 1857; although the telegraph era began around 1840, the first overland cable between these two countries was only opened in 1865.

A convenient starting date is 1851, when the world's first workable submarine telegraph cable linked England and France across the Channel [22]. It was insulated with gutta-percha, a material which was in universal use for over 100 years on submerged cables, and this cable was still in service 40 years later. In 1851, Guest and Company of England (now Guest, Keen and Nettleford, or GKN) was the largest manufacturing company in the world, with 10 000 employees. The base of its operations was steel making. The achievements of submarine cable makers like Glass Elliot of Greenwich and Gutta Percha of Stratford, in perfecting cables of light weight and flexibility with a high standard of insulation, physical protection and electrical efficiency, did much to advance the technology. Oliver Heaviside was born in 1850, and became one of the outstanding telegraphists. The mid-1800s was a time of great scientific and technological ferment in Great Britain which received great publicity in the First International Exhibition (London, 1851). In the next two years a service between London and Paris and between London and Brussels was established by means of submarine cables. Before the decade ended there was a link between Lowestoft in England and Zandoorst in Holland.

The 1850s also saw the start of heroic attempts to span the North Atlantic. The first, in 1857, met with failure. The following year, on 12 August HMS Agamemnon and USS Niagara succeeded in linking Ireland to Newfoundland. Queen Victoria and President Buchanan exchanged messages, the Queen writing 'The Atlantic Telegraph has half undone the Declaration of 1776 and has gone far to make us once again, in spite of ourselves, one people ... ' The event had the same impact on the British public as the moon landing in 1969, and *The Times* wrote 'Since the discovery of [America by] Columbus [1492], nothing has been done in any degree comparable to the vast enlargement which has thus been given to the field of human activity'. A few far-sighted people saw the enormous potential of such links across seabeds where no one has jurisdiction; thus global bypass was implemented from the very beginning of international telecommunications. The cable worked for a few weeks under a signalling voltage which was 10 times too large for the insulation, then it went dead. The Atlantic Telegraph Company, whose directors included Sir William Thomson, Sir Charles Wheatstone, Cyrus Field and John Pender refused to give up. They saw that one of the obstacles to success had been due to the size of the ships used, since neither could carry enough cable to span the route. The failure of the 1857 attempt had resulted from the great storm which arose as the two warships were splicing the cable, and it was lost in the mountainous seas. Another mistake has been the use of the Morse code for signalling on a line of such capacitance. Sir W. Thomson invented a new form of code, which became known as 'cable code', based on the Morse code but representing a dot by current in one direction and a dash by current in the opposite direction. Cable code was in universal use for over a century. In 1858 the pioneering engineer I.K. Brunel built the 'Leviathan'. Designed to carry 4000 passengers and 6000 tons of cargo it was by far the largest vessel in the world, and 40 years later was still among the largest. Unfortunately for its owners it was not a commercial success, because it was so far in advance of its time. By 1864 it had been renamed the 'Great Eastern', was laid up and was for sale.

In 1860 a new decade began with a telegraph service between England and Germany. The first commercial link between Britain and America was finally established in 1866, the first slender thread of what was to become a web of cables criss-crossing the North Atlantic. J. Pender had arranged a merger between the cable manufacturer and the owners of the 'Great Eastern', the only ship in the world able to carry 4000 km of cable and lay it in one piece. She left the Thames in 1864 with rather more than 25 000 tons of cable in her hold, 'a burden almost as great as the whole fleet with which Nelson fought the battle of Trafalgar (1805) could have carried'. Unfortunately this 1865 attempt failed because the cable broke and sank in 3600 m of water. However, in July 1866 the great ship successfully linked Valentia, Ireland to Hearts of Content, Newfoundland. Furthermore, she returned to the spot where the cable had been lost in 1865, and recovered it from a depth of 1969 fathoms! There was no looking back now, the 'Great Eastern' had become the first cable ship in the world, and since then more than 600 cable ships, many of them built in Glasgow, have been registered. The concept of communication with the far regions of the world began to occupy the minds of a handful of scientists and businessmen. They foresaw the enormous advantages which would accrue, both strategically and commercially, from a network of submarine cables which might finally encircle the globe. It was a decade of much activity in cable laying in and around Europe, mainly by British companies. In 1868 Oliver Heaviside began work as a telegraph operator at Fredericia, Denmark, where the terminal station of the newly-laid Anglo-Danish cable was situated. His starting salary was £150 per annum [23]. In British colonies, the local cable manager was an important member of society. Eastern Telegraph operators were credited with the introduction of soccer to Spain.

1870 saw the start of a great enterprise led by the entrepreneur John Pender, who like A.G. Bell was a Scot. He was born in the year of Napoleon's final overthrow at the battle of Waterlo, 1815, and was educated at the High School, Glasgow [24]. On 7 February 1870 the cableship (CS) 'Chiltern' laid the buoyed out the shore-end at Bombay; a week later the CS 'Great Eastern' spliced on and started paying out towards Aden, thence to Suez, and so began the system of submarine cables to serve the need for communication in the far-flung Empire over which Victoria had now reigned for 32 years. The network had begun with a cable to India, reached in 1870, Australia and the West Indies by 1872, Brazil and Argentina by 1874 and South Africa by 1879. His was a remarkable achievement, even by today's standards. The telegraph companies which he founded became part of Cable and Wireless plc.

Eventually the telegraph network spanned a total of 303 892 km (164 000 nautical miles), an enormous distance equal to about 7.5 times the circumference of the Earth at the Equator. In fact, the total length of underseas cable links for telephony traffic worldwide was only 153 799 km (83 000 nautical miles) in 1970, a hundred years after Pender began his work.

APPENDIX B: ROUTE SELECTION FOR SUBMERGED SYSTEMS

In TAT-8 (and all other transoceanic and undersea links) route selection and marine considerations formed one of the key decision areas relating to the choice of a secure route and especially the location of the branching point. In shallow waters an early decision was taken to bury the cable for protection. British Telecom International (BTI) had developed a cable plough which has a number of advantages over the American and French versions. This was used to bury most of the STC made part of the system from the coast at Widemouth, Devon to a point some 950 m deep at a distance of some 500 km from shore. The route from the UK to the branching unit was surveyed by BTI using the latest navigational aids and sea survey methods. Typically, side-scan sonar techniques and the use of the detrenching grapnel over the route enable the conditions on the sea bed to be found. Information concerning fishing and other sea bed activities is also used in selecting the best route.

By the early 1990s submerged systems formed an increasing proportion of the International Telecommunications Network referred to elsewhere. The selection of routes for such systems is evidently very different and more challenging then for most inland routes, and so may be of special interest to the many communication engineers who are keen small boat sailors.

The first international repeatered optical fibre submerged system was laid across the North Sea early in 1986, and is known as UK–Belgium No. 5 [25]. The North Sea is one of the two regions of highest traffic density; it has other problems which concern the route itself. Most failures of submarine cable systems result from mechanical damage, the main culprits being the fishing industry and in particular those vessels which engage in bottom trawling. There is a continuing battle waged between more lethal fishing equipment and better cable protection, and burial of the cable is used increasingly on these routes.

Once the landing points for the cable have been selected and accepted by all concerned, the initial route planning can begin. Landsmen may not appreciate some of the hazards which have to avoided. Attention must be paid to features which appear on current charts, e.g. British Admiralty charts; these features will include: wrecks, sea marks (navigation buoys, lightships etc.), anchorage areas, traffic separation zones, sand banks and shallow waters, rugged seabed contours, areas of seabed instability or seismic activity, areas of strong tidal activity.

The general objectives of the physical route survey are to confirm;

(a) the depth of water, submarine contours and in particular any steep slopes
(b) the nature of the sea bed and features such as rock outcrops and sand waves
(c) the sub-bottom strata and geology down to 2 m
(d) sediment erosion in the form of scar trails and depressions
(e) the existence of any wrecks or obstructions
(f) previous fishing activity indicated by trawl scars on the sea bed.

All data acquired about the vertical datum (sea level) must obviously be related to the corresponding position of the survey ship. Unlike survey work on shore, position fixing at sea was much more difficult (before the advent of navigational satellite systems) for a number of reasons; thus accuracies of 10–20 m at 50 km from the shore were considered good. Suitable equipment is needed for navigation, echo sounding, sea floor mapping, sub-bottom probing and bottom sampling. There is a world of difference between land surveying, using a stable theodolite and visible fixed marks, and the rolling heaving platform that a marine surveyor has to use. Allowing for the fact that marine surveying equipment was not as precise or infallible as might be desired, the route selected for TAT-8 was that considered to be the best available with regard to all the known physical conditions and man-made hazards.

APPENDIX C: INTEGRATED CIRCUITS IN SUBMERGED SYSTEMS

[26] described the development of ics for first-generation submerged optical systems, showed the design methodology, discussed some limitations and the very demanding reliability assurance required. Following the trend in the early 1980s in terrestrial systems of higher operational speed, power constraints and increased complexity, first-generation submerged systems were designed for 280 Mbit/s. The number of transistors in regenerators for such ofdls was about 100 times the number in analogue systems using discrete devices.

Constructional analysis: the approach adopted used only fully assessed circuit elements in the final chip design. High-reliability discrete silicon planar transistors known as Type 40 had been in submarine system service since 1976, and were adopted as the foundation elements for the new family of ics. Accelerated life tests at 60°C predicted a time to first failure of 1×10^9 years and 1.6×10^7 years at the 99.9% confidence level. Diffused resistors formed the passive elements in the fabrication of ECL arrays used in the construction of a range of ics.

Integrated circuit design: methodology for eight cell arrays covered emitter-coupled logic circuits, hierarchical design elements, circuit types, and a table of categories of circuit for regenerative electronics. Larger cell arrays were then considered, with a table listing uncommitted logic array sizes, and another table giving ECL 40 array power dissipations.

Simulation models: both analogue and digital simulations were required. The integrity of a simulation depended upon the quality of characterisation of all elements in the simulated array. The expected variations in these values over a 25 year life must be also be known. Electrical circuit design was modelled at the component level and circuits evaluated for dc and small-signal ac behaviour. An example was given of a clocked bistable circuit (one cell in an eight cell array) and the corresponding simulated and measured results when operated as a master–slave D-type bistable. The software package SPICE was used, and tabulated SPICE model parameters were given for a single transistor used in the eight cell array and the scaled input transistor for the custom ic.

Integrated workstations were required to have the following capabilities:

(a) interactive schematic entry

(b) logic, analogue and fault simulations

(c) automatic placement and routing programmes that provide mask pattern generation data leading to designs on silicon that are initially correct

(d) Placement and routing checkers, electrical rule check, design rule check, network connectivity check.

Wafer fabrication: ic package hermetic sealing and leak testing. Results of hermeticity testing on 18 pin leadless chip carriers used to house the eight cell array products were presented in tabular form. Other topics covered were:

(a) higher levels of integration
(b) speed limitations
(c) integrated circuits in submarine systems in 1986
(d) test and verification
(e) ic reliability: overstress testing, and a table of potential failure mechanisms
(f) reliability demonstration by means of extended operational life test and accelerated life test.

Finally, the requirements for next-generation systems were discussed. The development of integrated circuits of increasing complexity, functionality and speed continued throughout the 1980s and into the 1990s. For use in submerged routes these must have the necessary reliability pedigree. The increasing role of software in systems demands that this too be subject to very stringent reliability and performance criteria. The development and engineering of test methods and equipments have proceeded throughout the period.

15.6 REFERENCES

1 BALL, D.: 'The international transmission network and the role of submarine cable systems'. *Br. Telecom. Engineering*, **5**, July 1986, pp. 72–79.
2 COCHRANE, P.: 'Future directions in submerged fibre optic system technology'. *Proc. IOOC '89*, Kobe, Japan, Paper 21B1-2 (invited), July, 1989.
3 RUNGE, P.K.: 'Undersea lightwave systems'. *AT&T Tech. J.*, Jan/Feb 92, pp. 5–13.
4 LILLY, C.J.: 'Cables under the sea'.
5 SMITH, R.L. and WHITTINGTON, R.: 'TAT-8: An overview'. *Br. Telecom. Engineering*, **5**, July 1986, pp. 148–152.
[9.1] WILLIAMSON, R.L., SWANSON, G., and MITCHELL, A.F.: 'The 1NL2 system'. *Suboptic 86*, Versailles, France, Feb. 1986.
6 SMITH, R.: 'BT bites back. . .new cable more than a match for 'Jaws''. *Br. Telecom. J.*, **8**, no. 2, 1987, pp. 38–41.

[9.2] BROCKBANK, R.G., CALTON, R.L., MOTTRAM, S.W., HUNTER, C.A., and HEATLEY, D.J.T.: 'The design and performance of the TAT-8 optical receiver'. *Br. Telecom Technol. J.*, **5**, no. 4, Oct. 1987, pp. 26-33.
7 O'MAHONEY, M.J.: 'The analysis and optimisation of a pin photodiode GaAs fet cascode optical receiver'. *Br. Telecom Technol. J.*, Part 1 **1**, no. 1, July 1983, pp. 38-42.
[9.3] MITCHELL, A.F., and O'MAHONEY, M.J.: 'The performance analysis of pin-bipolar receivers'. *Br. Telecom Technol. J.*, **2**, no. 2, Apr. 1984, pp. 74-85.
8 UNWIN, R.T.: 'A high speed optical receiver'. *Optical & Quantum Electronics*, **14**, 1982, pp. 61-66.
9 FIELDING, A., BRICHENO, T., LEACH, J.S., BAKER, V., and JOHNSON, R.: 'High performance couplers for submarine fibre optic transmission systems'. *IEE Coll.* 'Submarine optical transmission systems'. Mar. 1988, Digest 1988/45.
10 DAWSON, P.A., and KITCHEN, J.A.: 'TAT-8 supervisory subsystem'. *Br. Telecom. Engineering*, **5**, July 1986, pp. 153-157.
11 WILLIAMS, A.: 'Quality assurance of submarine cable systems'. *Br. Telecom. Engineering*, **5**, July 1986, pp. 124-129.
12 WRIGHT, J.V., SIKORA, E.S.R., SCOTT, J.M., and PYCOCK, S.J.: 'Considerations for the lifetime of submarine optical cables'. *Fibre Optics '90*, SPIE, **1314**, pp. 24-26, Apr. 1990.
13 COTTERILL, D.A.: 'User requirements and provision of digital multiplication systems for TAT-8'. *Br. Telecom. Engineering*, **5**, July 1986, pp. 158-164.
14 LILLY, C.J.: 'From global village to global connectivity'. *1994 KMI Fiberoptic Submarine System Symposium*, New York, Mar. 1994.
15 McCLELLAND, S., and BRIGHT, J.: 'Twenty thousand leagues under the sea'. *Telecommunications*, Int. edn., **27**, no. 6, June 1993.
16 BALLAND, G., PASKI, R.M., and BAKER, R.A.: 'TAT-12/13 integration development tests — interim results'. *Suboptic '93, Second International Conf. on Optical Fiber Submarine Telecommunication Systems*, Mar./Apr. 1993, Versailles, France.
17 DAVIS, F., JOLLEY, N.E., TAYLOR, M.G., BRANNAN, J., BAKER, N., WRIGHT, M.C., and BAKER, R.A.: 'Optical amplifiers for submerged systems'. *Suboptic '93, Second International Conf. on Optical Fiber Submarine Telecommunication Systems*, Mar./Apr. 1993, Versailles, France.
18 BAKER, R., BUTLER, D., DUFF, I., HARRISON, D., JONES, K., WEBB, BRANNAN, J., PENTICOST, S., ROBINSON, N., and MARTIN, P.: 'A long-haul, high bit rate submarine transmission demonstration using optical amplifiers'. *Suboptic '93, Second International Conf. on Optical Fiber Submarine Telecommunication Systems*, Mar./Apr. 1993, Versailles, France. Printed in *L'Onde Electrique*, **73**, no. 2, Mar./Apr. 1993.
19 HILL, G.R.: 'A wavelength routeing approach to optical communication networks'. *Br. Telecom Technol. J.*, **6**, July 1988, pp. 24-31.
20 CHIDGEY, P.J., and HILL, G.R.: Diverse routing of multiplexed signals in an amplified wavelength routed network' *ECOC '90, Amsterdam*, 16-20 Sept. 1990.
21 MACLEOD, A.: 'Underwater networking solutions'. *Telecommunications*, Int. edn., **26**, no. 6, June, 1992.

FURTHER READING

22 GAUTHERON, O., GRANDPIERRE, G., GABLA, P.M., BLONDEL J.-P., BRANDON, E., BOUSSELET, P., GARADÉDIAN, P., HAVARD, V.: '407-km, 2.5 Gbit/s repeaterless transmission using an electroabsorption modulator and remotely pumped erbium-doped fiber post- and pre-amplifiers'. *IEEE Photonics Lett.*, **7**, no. 3, Mar. 1995, pp. 333-5.
23 SIPRESS, J.M.: 'Undersea Communications Technology'. *AT&T Technical J.*, **74**, no. 1, Jan./Feb. 1995, pp. 4-7.

24 ZSAKANY, J.C., MARSHALL, N.W., ROBERTS, J.M., ROS, D.G.: 'The application of undersea cable systems in global networking'., *AT&T Technical J.*, **74**, no. 1, Jan./Feb. 1995, pp. 8-15.

25 SCHESSER, J., ABBOTT, S.M., EASTON, R.L., STIX, M.S.: 'Design requirements for the current generation of undersea cable systems'. *AT&T Technical J.*, **74**, no. 1, Jan./Feb. 1995, pp. 16-32.

26 MORTENSON, R.L., JACKSON, B.S., SHAPIRO, S., SIROCKY, W.F.: 'Undersea optically amplified repeatered technology, products, and challenges'. *AT&T Technical J.*, **74**, no. 1, Jan./Feb. 1995, pp. 33-46.

27 STAFFORD, E.K., MARIANO, J., SANDERS, M.M.: 'Undersea non-repeatered technologies, challenges, and products'. *AT&T Technical J.*, **74**, no. 1, Jan./Feb. 1995, pp. 47-59.

28 KORDAHI, M.E., GLEASON, R.F., CHIEN, TA-MU: 'Installation and maintenance technology for undersea cable systems'. *AT&T Technical J.*, **74**, no. 1, Jan./Feb. 1995, pp. 60-74.

29 LISS, J.M., KUREK, K.A.: 'Network planning, operation, and maintenance practices for undersea communication systems'. *AT&T Technical J.*, **74**, no. 1, Jan./Feb. 1995, pp. 75-82.

30 LYNCH, R.L., MAYBACH, R.L., WALCH, P.F.: 'Design and deployment of optically amplified undersea systems'. *AT&T Technical J.*, **74**, no. 1, Jan./Feb. 1995, pp. 83-92.

31 KERFOOT, F.W., RUNGE, P.K.: 'Future directions for undersea communications'. *AT&T Technical J.*, **74**, no. 1, Jan./Feb. 1995, pp. 93-.

32 JEAL, A.J., LILLY, C.J.: *'Long-haul systems — technology trends'. Suboptic '93, Second International Conf. on Optical Fiber Submarine Telecommunication Systems*, Mar./Apr. 1993, Versailles, France.

33 HIRST, I.J., JEAL, A.J., BRANNAN, J.: 'Performance monitoring of long chains of optical amplifiers'. *Electronics Lett.*, **29**, no. 3, 4 Feb. 1993, pp. 255-6.

34 WAKEFIELD, K.S.: 'Submarine cables and cable ships'. *Electronics and Power*, Sep. 1985, pp. 652-4.

35 LEE, Sir G.: *Oliver Heaviside — the man*. The Heaviside Centenary Volume, IEE, 1950, pp. 10-7.

36 BARTY-KING, H.: *Girdle round the earth*, the story of Cable and Wireless'. Heinemann, London, 1979.

37 DAVIDZUIK, B.M., HOWARD, P.S.: *The impact of optical technology on the submarine system market*. IEE Colloquium Digest 1988/45.

38 TAYLOR, S.A., HOGGE, D.M.: Overload instability of submarine cable repeater systems'. *P.O.E.E.J.* **62**, no. 3, Oct. 1969, pp. 181-7.

39 *Submarine optical transmission systems*. IEE Colloquium Digest, 1988/45, 24 Mar. 1988.

40 WHITTINGTON, R.: 'UK-Belgium No. 5, Part 1, Marine aspects of the route selection'. *Br. Telecomm. Engineering*, **5**, Jul. 1986, pp. 133-7.

41 CHANNON, R.D.: 'UK-Belgium No. 5 Part 2, Technical and installation aspects'. *Br. Telecomm. Engineering*, **5**, Jul. 1986, pp. 138-47.

42 TEW, A.J.: 'Integrated circuits for submarine communication systems, their design and development'. *Br. Telecomm. Engineering*, **5**, Jul. 1986, pp. 84-93.

IEE Colloquium 1996/067 *Transoceanic cable communications TAT 12 and 13 heralds a new era'*. London, 25 March, 1996, will include:

43 MARLE, G., ANDREWS, I., GREEN, M.: 'TAT 12/13 cable network — the need for the ring.

44 speaker from AT&T, *Project overview*.

45 LEMAIRE, V., BALLARD, G., BOURRET, G.: *System design and integration issues for the first transatlantic 5 Gbit/s submarine cable*.

46 WRIGHT, J.: *'Numerical modelling of transoceanic cable systems'*.

47 LORD, A.: *Polarisation effects*.

48 BARNES, S.: *Transmission experiments carried on during installation of TAT 12*.

49 THORMINE, J.B., PIRIO, F., AUBIN, G.: *Future trends for high capacity optically amplified submarine systems*.

16
ASPECTS OF MEASUREMENT IN OFDLS

'... how to create a consensus on what should be measured, in what way, and to what standard.'

(*Financial Times* survey 'International Standards', 14 Oct. 1994)

16.1 INTRODUCTION

The preceding three chapters described many aspects of ofdls system design, including operational requirements, design philosophy, specifications, and several examples of both regenerative and optically amplified repeater systems. The hardware associated with these systems has been mentioned in detail to indicate the complexity of any operational transmission system. Earlier chapters had dealt with the lasers, photodetectors and their associated electronics in transmitters and receivers for use in repeaters and in terminal equipment. Sprinkled throughout these chapters there has been reference to measurements of one or more aspects of component or subsystem performance, especially in the chapters on optical receivers. The aim is to highlight some other aspects of a very large field, within the limitations of a single chapter.

Measurement techniques and equipment for use in optical fibre digital line systems have been under intensive development since the early days of fibre measurements when discrepancies of up to 2 dB/km were noted between loss measurements made by different laboratories on the same piece of fibre. Indeed, consistent measurements had to await a better understanding of the influence of launch conditions.

The major manufacturers of test and measuring equipment and instruments for use in ofdls are well known, and their evolving products cover all the necessary requirements up to the highest bit rates of interest (10 Gbit/s in the early 1990s). By the mid-1990s equipment will be available for use with SDH/SONET systems on a world-wide scale.

The body responsible for setting electrical standards worldwide is the International Electrotechnical Commission (IEC). The corresponding body in Europe is the Brussels-based CENELEC (European Committee for Electrotechnical Standards) whose policy is to adopt IEC standards as CENELEC standards as far as possible.

CENELEC is organised into technical committees which prepare European standards within the scope of the committee. For telecommunications, the bodies are the European Telecommunications Standards Institute (ETSI), TI in the US and the Telecommunications Technology Committee in Japan. In the UK the British Standards Institution transposes all CENELEC standards into BS EN standards, as well as transposing important ISO and IEC standards into British standards.

National bodies such as the National Bureau of Standards and the Electronic Industries Association in the United States cooperated to produce draft standards designed to provide repeatable, accurate and practical measurement techniques for the transmission and physical properties of fibres that are independent of the unit length of the fibre. Other national bodies, such as the British Standards Institution (BSI) and the VDE which issues the DIN standards in Germany, contribute to solutions of the problems.

In the European Union (formerly European Community), Directives are issued from time-to-time; for instance, the EC Second Stage Telecommunications Terminal Equipment Directive which came into force in November 1992, and which provided a significant step forward for the industry towards a single European market. Telecommunications terminal equipment (TTE) manufacturers can find detailed information on this Directive and on standards bodies, committees and telecoms standards in the *Telecommunications Terminal Equipment* handbook available from BSI.

International bodies such as the International Standards Organisation (ISO), to which more than 80 organisations belong, and the CCITT, whose work was outlined in Chapter 1, have formulated a range of standards for both the physical and transmission properties of optical fibre systems. ISO began to introduce some order into a chaotic situation by introducing the International Classification for Standards (ICS) in the early 1990s.

International comparisons have also taken place from time-to-time, e.g. the interlaboratory comparison of mode-field diameter measurements involving 14 European laboratories within the framework of Working Group 1 of the COST 217 Programme (optical measurement techniques for advanced optical fibre systems and devices) in the late 1980s. Participants were asked to meet or exceed the measurement requirements laid down in the relevant CCITT test procedure G.650.

OFDLS test equipment is used in manufacture and acceptance, on-site maintenance and in repair centres. By 1994 a great deal of specialist measuring equipment had been developed by various manufacturers and was commercially available for the characterisation of fibres, sources and detectors and for other components used in ofdls. Information on the equipment and appropriate test methods can be obtained from the manufacturers. For example, test equipment for use with SDH systems was becoming available in the early 1990s.

16.1.1 General philosophy

This section summarises the general philosophy adopted by British Telecom in the early days of optical systems, as given in [1] as an example, and the basic outline of methods and procedures required for both the optical fibre cables and the traffic path units, each divided into: factory work, installation and maintenance.

The aim was to reduce field testing as much as possible by factory-based assessments whenever practicable. Installation tests checked that the system arrived on site in working order and was assembled correctly. This approach had worked very well on digital radio relay systems and for satellite systems. Maintenance could be limited to checking supervisory signals, or to monitoring test points where such signals were not applicable. Preventive maintenance should be avoided where it required the traffic path to be interrupted. Many practices and procedures evolved for coaxial line systems could be modified to suit optical transmission, thus reducing the need for the retraining of staff.

The philosophy for dealing with faults on the system was to provide sufficient equipment to trace a fault to a particular card or unit, then to replace the faulty item to keep any down-time to a minimum. Detailed examination and repair could be carried out in fully equipped area repair centres. Let us look briefly at the three areas in turn.

16.1.2 Equipment test methods: factory, installation and maintenance

Factory production

Bearing in mind the desire to keep testing in the field to a minimum, and to restrict factory procedures to only those which were necessary, the following would typically be carried out at the factory stage:

(a) all set-up procedures and fitting of any 'adjust-on-test' or 'select-on-test' components

(b) any margin or tolerance measurements on individual cards or units should be performed as early as possible

(c) full system tests, insofar as they are practicable with the equipment available.

Performance tests using a traffic simulator can induce line errors, frame slips and any other desired fault conditions. Alternately, concatenated fibres on drums can provide a measured amount of dispersion, and the loss can be increased to the repeater section maximum value by means of fixed optical attenuators.

Installation

Much of the following is consistent with other types of line systems, and it aims primarily at proving the on-site assembly work. Typical operations would be as follows.

(a) Check power supplies and the operation of in-station alarms and alarm concentrator at the terminals and shelf alarms.

(b) Loop the terminal transmit side to the receive side of the traffic shelf either electrically or optically, then assess the error rate using a prbs data generator and detector.

(c) Note any reference levels and check any test points and agc levels if required.

(d) By changing the input data format, ensure that the supervisory subsystem is functioning, and that its indicators are visible where applicable.

(e) For systems employing dependent repeaters, ensure that the power feed system is working. Connect the terminal under test into the route, and loop the first dependent repeater (two-way) by a suitable optical attenuator. Check for power feed and supervisory connections being present, and measure the value of any monitor/test points on the repeater. Ensure that the error rate performance, measured at the terminal with a prbs data generator and error detector, of the looped route is satisfactory. If so, proceed down the route to include the next dependent repeater in the looped route, and check the items as before. By installing from the LTEs at each end until the full route is included, one can avoid possible degradation of the signal.

Note that if, in the system under test, the supervisory subsystem is designed so that all of the route must be equipped before it will work satisfactorily, a non-dedicated supervisory system must be available over the installation period, if the operations described above are to be followed. Examples of supervisory systems are to be found in Chapters 14 and 15.

Any jitter measurements, if carried out, would be over the complete looped route, on the basis that they would be acceptable if the specification were met. At this stage, assessment of the working system margin would not normally be done, assuming that the traffic cards and optical fibre cable performance had been measured to lie within specification, apart from the measurement of received power. A stability trial, lasting perhaps a week, would then be carried out to prove the system, bearing in mind that a new criterion may be needed for acceptance, because of the higher defined error rates than in metallic line systems.

Maintenance and repair

In accordance with the premise that the traffic path should be interrupted as infrequently as possible, no preventive maintenance was envisaged. One of the key results of a 1994 paper on reliability aspects of optical fibre systems and networks was that 'human interdiction can dramatically reduce reliability' [13] of Chapter 10. An overall reduction in network performance of some 50% was found to be a reasonable expectation and was 'supported by practical experience across a broad range of equipment, systems and network types, and was applicable to all forms of line plant, radio, electronic and photonic transmission, switching and computing hardware.'

Automatic interrogation and display of fault conditions were dealt with on 140 Mbit/s sytems by the comprehensive supervisory subsystem. At the lower rates of 34 and 8 Mbit/s, where UK practice was not to fit supervisory subsystems, further use was made of monitor and test points to gain information on the status of the system. Whenever a fault was indicated a simple procedure could be initiated.

16.2 MEASUREMENT STANDARDS

In this area, as in many others connected with optical systems, the aim is to refine the measurement techniques so that consistency and traceability to national standards are achieved. In the United States the National Bureau of Standards and the Electronic Industries Association have combined their efforts by first setting up draft standards, which were then refined into standards giving very minute measurement technique details in appropriate cases. Similar work has been carried out by other nations, and of course there are the interactions with international bodies, whose work and interrelations were described in Chapter 1.

There are accuracy considerations. One of the fundamental measurements in any communications system is of power. Received power determines the snr or ber of the system; either of these, together with the path loss, determine the transmitter power needed. In ofdls the power measurements needed to evaluate a system include source output power, fibre attenuation, splice and connector losses and received power. Many of these can be measurements of power relative to a suitable level, but for complete characterisation at least one must be an absolute measurement traceable to fundamental standards. It is important to realise that inaccuracy in measurement has financial and technical implications. For instance, if the cost of a source is proportional to the amount of power it can couple into a system fibre, then a 20% shortfall in specified power which is undetected because of measurement inaccuracy is equivalent to a hidden cost increase of 20%. At the receiver end, a received power which is 1 dB below the measured value can mean an actual ber of 10^{-8} rather than 10^{-10}.

Let us now turn to general considerations on power measurement in ofdls. Instruments for this purpose can be divided into two categories: thermal and quantum. Examples of the latter include pin and avalanche photodiodes which produce current proportional to the incident optical power. Their advantages include speed, sensitivity and ease of use. Disadvantages include the difficulty of achieving good absolute accuracy, since variations of between 10 and 50% can occur when these devices are compared with more accurate techniques. Another drawback

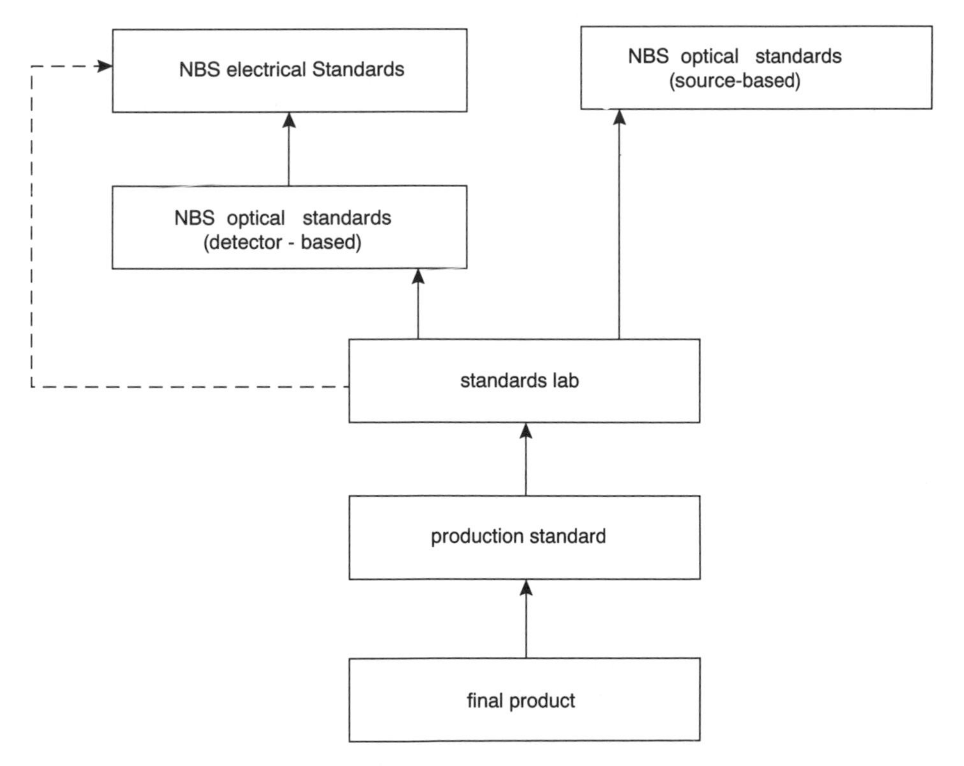

Figure 16.1 Traceability for an optical power meter (H.P. notes, Figure 5). Reproduced by permission of Hewlett-Packard

results from drift with time and temperature, for example 2–3% per annum and 1–2% per degree celsius. In order to use these devices for absolute measurements it is necessary to maintain their accuracy by careful and periodic recalibration against a more accurate and stable instrument based on thermal techniques.

Thermal devices convert incident optical radiation into heat, which in turn causes a change in a parameter such as temperature, resistance, etc., which can be measured accurately and made proportional to the incident power. Thermal techniques cover an enormous spectral range, from RF to ultra-violet. Although thermal devices also drift with temperature and time, they can, unlike quantum devices, be combined with a method known as electrical substitution to obtain very good and reliable accuracy, independent of changes in device characteristics. Of the three major classes of thermal detectors which can be used with the substitution method, thermistors are the most convenient for use in optical system measurements. This preference is due to their compatability with fibre optics, ease of use, ruggedness and low cost.

Ultimately, all measurement devices must be traceable back to fundamental standards, and for optical power meters the flow chart for the United States as given in one test equipment manufacturer's information is shown in Figure 16.1.

The product is calibrated against a production standard, which in turn is calibrated by a standards laboratory. There are three choices which this laboratory can make to obtain traceability back to the NBS, as indicated in the diagram.

16.3 MEASUREMENTS ON FIBRES, CABLES AND JOINTS

A number of examples of measurements are found in Chapter 5 and in several specific references, for instance 'Chromatic dispersion measurement over a 120 km dispersion-shifted single-mode fibre in the 1.5 μm region' ([10] of Chapter 5).

Accuracy and reproducibility are difficult to achieve in measurements of the properties of optical fibres because they depend on so many variables, notably launch conditions, measurements of absolute optical power levels, interference from ambient light and uncertainties when optical connectors are used. To minimise these difficulties, CCITT recommendations are based on reference test methods (RTMs) which are intended to yield accurate reproducible results, which can provide a common reference point. Unfortunately some RTMs are expensive and time-consuming in use, and may only be suitable for use in a controlled environment such as a test laboratory. In such cases alternative tests are employed for factory and field applications; such methods produce results which can be related to the corresponding RTM.

The situation in the late 1980s was covered in a special issue of the *Journal of Lightwave Technology* in August 1989, which discussed the main research topics in fibre measurements. The most relevant topics (cited in the call for papers) for this chapter were listed in two groups.

There were factors affecting transmission quality, namely: microbending, polarisation fluctuation under various environmental conditions, and ageing of optical fibres. In the issue itself, only the second topic was treated.

The call for papers also mentioned measurement techniques for the following properties of optical fibres: attenuation and backscatter, cut-off wavelength, mode field diameter, chromatic dispersion, refractive index profile, and polarisation. All of these were addressed in the issue, but are of less direct concern in this chapter.

The control and measurement of fibre dimensions is considered as part of the production process, as is much of the characterisation of the fibre, and will receive little attention here. Attenuation and dispersion, however, are of interest before and after installation of the cable, and of course must be within the relevant specification. Maintenance will cover the accurate location of any fault on the cable and the ability to provide on-site repair, if necessary.

Specifications on the properties of single-mode fibres can be divided into three groups, an example of which is given in Table 16.1. Definitions and test methods

Table 16.1 Single-mode fibre specifications (SM-02-R) (manufacturers data sheet/British Telecom Spec. CW1505E)

	SM2A SM2A-42S/30T-R	SM2A/P SM2A-37S/25T-R
Optical properties		
Attenuation at 1300 nm (dB/km)	≤ 0.42	≤ 0.37
-OR-		
Attenuation over range (dB/km) 1275–1325 nm	≤ 0.45	≤ 0.40
Attenuation at 1550 nm (dB/km)	≤ 0.30	≤ 0.25
-OR-		
Attenuation over range (dB/km) 1475–1575 nm	≤ 0.35	≤ 0.30
Mode-field diameter (μm)	9.75 ± 0.5	
Cut-off wavelength (nm)	1150–1250	
Dispersion (ps/nm.km)		
1285–1330 nm	≤ 3.8	
1270–1350 nm	≤ 6.0	
1550 nm	≤ 18.0	
Geometrical properties		
Reference surface diameter (μm)	125 ± 2	
Glass concentricity error (μm)	≤ 0.8 median ≤ 0.4	
Glass non-circularity (%)		
Core	< 6	
Reference surface	< 2	
Coating diameter (μm)	250 ± 15	
Coating concentricity error (μm)	< 15	
Coating, minimum thickness (μm)	40	
Coating non-circularity (%)	< 10	
Mechanical properties		
Bend test (60 mm diameter mandrel, 100 turns, loss at 1550 nm in dB and dB/m)	< 1.0, < 0.05	
Proof test (%)	0.75	
Standard lengths (km)	4.4, 6.6, 8.8, 11.0, 12.6, 13.2, 17.6, 25.2	

NOTES:
1. The attenuation values given for single wavelengths or wavelength ranges are alternative specifications.
2. Attenuation values can be selected down to 0.33 dB/km at 1300 nm and 0.18 dB/km at 1550 nm.
3. Fibre proof tested to a level of 1.0% is usually available from stock.
4. The median guarantee for glass concentricity error applies for orders in excess of 500 km.

for these properties were listed on the manufacturers data sheet and they constitute the tests carried out at the factory production stage.

16.3.1 Microbending

The results of an investigation of microbending sensitivity in fibres was reported in a 1994 paper ([19] of Chapter 4) which stated that serious difficulties arise from the practical determination and experimental simulation of the irregular curvature of a fibre in a cable and the characterisation of its sensitivity to microbendings. Two different test methods were used; these were the expandable drum test and the sieve test. Sensitivity was calculated as a function of the product of mode field diameter and cut-off wavelength. Investigation of microbending effects in buffer tubes and cabled ribbon stacks verified that these test methods provided realistic simulation. Further information on microbending can be found in Chapter 4, Section 4.6.

16.3.2 Polarisation fluctuations under various environmental conditions

Polarisation fluctuations can cause degradations in imdd systems wherever polarisation sensitive components are deployed; for example, couplers forming part of fibre amplifiers in repeatered systems. There are several references to this topic in Chapter 5. Polarisation fluctuations are of more concern and importance in coherent systems, but such systems are outside the scope of this book.

Measurements on inland network cables of the polarisation fluctuation observed at 1550 nm on samples of cable that Telecom Australia had installed in its inter-exchange network were reported in [2]. The rate of polarisation fluctuations was found to be slow, in the order of hours. The fluctuation in the polarisation angle was typically in the range of 2°–10° each day, with some changes up to about 25°. These results indicated that there were no significant limitations on implementing polarisation control in a coherent system receiver.

Real-time polarisation fluctuations measurements in an optical fibre submarine cable during laying and recovery in a 6000 m deep-sea trial were investigated, and the results were reported in [3]. State-of-polarisation (SOP) variation measurements in real time were achieved by the Stokes parameter analyser using an electro-optic $LiNbO_3$ device. It was found that the SOP variations during cable laying change randomly, whereas those during cable recovery change uniformly. The maximum power spectra of the polarisation fluctuations during cable installation were found experimentally to be less than about 50 Hz.

16.3.3 Ageing of optical fibres

Studies have been carried out on cables for inland and for submerged system applications. As an example of the former, strength and lifetime measurements were carried out on fibres ([9] of Chapter 4). Measured results of residual tensile stress at the surface were included in the theory of fibre strength and lifetime prediction. The second example is entitled 'Considerations for the lifetime of submarine optical cables' by authors from BTRL, UK ([9] of Chapter 4). Long-term stability of the attenuation of dispersion-shifted sm fibres exposed to hydrogen was discussed in ([6] of Chapter 5).

16.3.4 Attenuation and backscatter

The need for accurate measurements of loss in fibres and splices in production is paramount. For example, an accuracy of ±0.1 dB may be quoted in a laboratory environment, and will define the per-unit loss of a 5 km length to ±0.02 dB/km, but ±0.1 dB will be inadequate for short lengths of, say, 1 km or less, and for splices. The repeatability of loss measurements on the same piece of fibre seems to be good; for instance, a spread of only ±0.015 dB has been quoted.

A method of determining the backscatter and reflection response of an optical fibre and reflector to an arbitrary input was developed and described in [4]. Two specific cases were considered: a continuous wave input and a rectangular pulse input. From the rectangular pulse input response the equation was derived to compute the reflectance of a discrete component from an otdr measurement. Precautions were given to accurately perform reflectance measurements using an otdr. Two methods of determining the backscatter level of the fibre type under test were presented, and their importance in reflection measurements was demonstrated.

Advances in optical time-domain reflectometry such as enlargement of dynamic range, enhancement in resolution, reduction of noise intrinsic to single-mode fibres, and increased user-friendliness of the equipment were reviewed in the same special issue [5]. Future optical network diagnostics were also discussed.

Automatic fibre break location for duplex and diplex systems

Until the late 1980s fibre break location in systems had relied on the use of repeater and terminal-based supervision to identify the particular span at fault. This was followed by manual otdr testing to give a relatively precise location. However, the precision of the otdr technique was severely limited by the nature of a fibre break and its relative index matching. Reflections from breaks with the fibre in air and water can range from 14 to 60 dB, and to this must be added the round trip loss of the fibre. Thus for a mid-span break the total can exceed 40 dB, whereas then currently available otdr equipment only offered one-way dynamic ranges of 18–22 dB. The trend towards longer repeater spans and the possible future deployment of coherent systems will exacerbate the problems. Fibre break location is thus not always straightforward, and is generally both tedious and time-consuming [6]. A fibre break location scheme embedded within the transmission system would therefore offer operational advantages, and it was shown that this could be achieved when either duplex or diplex operation was employed. The concepts of an automatic fibre break location scheme were discussed and the results of an early laboratory trial on a standard BT 34 Mbit/s duplex system were presented. The mean values agreed very closely with the fibre break distances measured by otdr. The paper also described how such a scheme could be integrated into both existing duplex transmission systems and into future systems operating within a totally synchronous network.

OTDR on optically amplified systems

Linear optical amplifiers replace only a smaller part of the conventional optoelectronic regenerative repeater, the larger part of which normally consists of electronics to implement fault-location and supervisory functions. Simplified supervisory techniques applicable to optically amplified systems were thus of great interest for their potential to minimise the hardware required in repeaters. The use of otdr on long-span systems containing in-line optical amplifiers was investigated and described in a 1989 paper [7]. Experimental results were presented which demonstrated that established odtr techniques could be used on optically amplified systems, provided that there were no isolators between stages. The main requirements on the otdr equipment were first discussed, followed by a brief summary of those characteristics of an optical amplifier which affect the operation of otdr. Results of system measurements on links containing two or three cascaded amplifiers were presented. Moreover, the authors demonstrated that the technique could be used not only for fault location but also for detailed study of the individual characteristics of the optical amplifiers, as well as system impairments, arising from the interaction between adjacent gain stages. The performance of the equipment was summarised and the range of possible future improvements was identified, from which an estimate was made of system lengths up to 1000 km over which otdr from both terminals could be used for fault location and supervision.

Since the work was done in the late 1980s, semiconductor laser amplifiers were used in the repeaters of systems up to 300 km long, but the techniques were also applicable to fibre amplifiers. The technique was shown to work on systems operating under conditions where error-free data transmission would have been impossible, thus satisfying the requirement that a supervisory channel must still provide meaningful information when the signal channel has already deteriorated sufficiently to become unusable.

16.3.5 Cut-off wavelength

The cut-off wavelength of the LP_{11} mode determines the wavelength region of single-mode operation, and enters into the prediction of other fibre properties such as dispersion. Unfortunately it is not a quantity which can be accurately measured.

Measurement of equalisation wavelength for fibre characterisation

A different choice of wavelength for fibre characterisation was proposed in 1987 [8]. The equalisation wavelength λ_E of the LP_{01} and LP_{11} modes of an sm fibre operated in the dual wavelength regime occurs at a well-defined v-value, and can be measured precisely, and moreover factors such as detector linearity and bends in the fibre do not affect the measured value.

The fibre under test was operated in the dual-mode wavelength region by launching the light slightly off-axis so that both modes are excited. Provided that the group delay difference between the modes was shorter than the coherence time of the light, interference will be observed. The equalisation wavelength was defined as the wavelength at which the group velocities of the two modes are the same. The experimental arrangement and procedure were described and were relatively simple, so it was more acceptable as a factory-type test. A typical output is shown in Figure 16.2. The range of visible fringes was from 1050 to 1170 nm, and as can be seen from the curve obtained with a white light tungsten halogen source, the wavelength of interest can be located to within about 2 nm. The authors point out that a combination of light-emitting diodes covering the waveband of interest

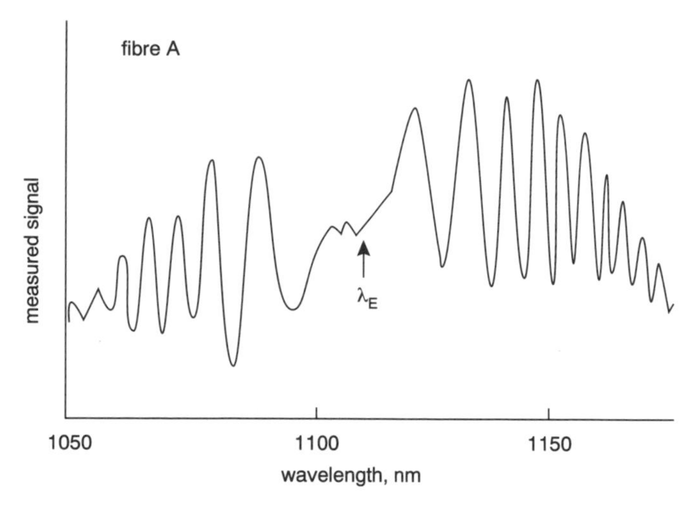

Figure 16.2 White light interferogram showing typical measured signal versus wavelength and visible fringes between 1050 and 1170 nm ([8], Figure 2). Reproduced by IEE

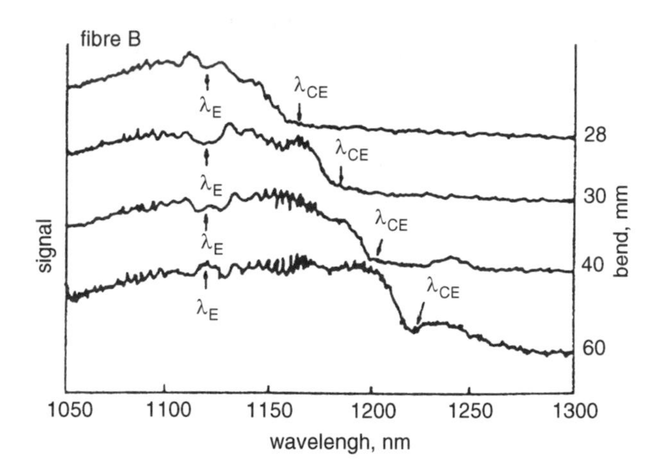

Figure 16.3 Signal versus wavelength showing dependence of λ_E and λ_{CE} on bend radius ([8], Figure 3). Reproduced by permission of IEE

would be preferred to white light, but that such sources are not always available for the band of interest. The most important perturbing effect results from bends, so the sensitivity of the measurements of the effective cut-off wavelength λ_{CE} and the equalisation wavelength to fibre bending was investigated. The results are shown in Figure 16.3.

Four samples of the results are illustrated, corresponding to four different drum or bend radii, and they clearly demonstrate the relative insensitivity of λ_E to bending, compared with the other wavelength. In addition, the ratio of the equalisation to the theoretical cut-off wavelengths was insensitive to profile variations for typical quasi-step index fibre of matched cladding design; therefore the authors propose that the equalisation wavelength be used in place of the cut-off wavelength in the characterisation of such fibres.

Cut-off wavelength from mode spot size

A simple, accurate and rapid method for measuring the mode spot size was described in 1983 by authors from the French firm Thomson-CSF [9]. Mode spot size measurement is a useful method to define the cut-off wavelength of sm fibres, the esi parameters and to predict the modal dispersion properties. The results of the mode spot size versus wavelength for a dispersion-shifted fibre were shown, and they lay on a smooth curve. Values of the esi parameters calculated from this data agreed with values evaluated by perturbative analysis from the index profile shape measured by the rnf technique. The value for the LP_{11} mode cut-off was also in good agreement with that obtained from the wavelength dependence of losses caused by bending the fibre on a circular arc. The measurements can be implemented rapidly using conventional spectral loss-measuring apparatus without additional handling of the fibre.

16.3.6 Mode field diameter (mfd)

This quantity determines the transmission coefficient of a splice and is related to other fibre parameters such as microbending loss and waveguide dispersion. Among the specified parameters of sm fibres, the measurement of mode-field diameter is therefore important. The results of a European interlaboratory comparison of measurements at 1300 and 1550 nm on six sm fibres of both conventional and dispersion-modified designs were reported at the end of 1991 in [10]. Fourteen laboratories participated and five different measurement techniques were used. Results showed clear systematic differences in the measured mfd between the methods. These differences could be explained in terms of the theory of the methods. The authors suggested that the most accurate values of mode-field diameter were obtained using the far-field scanning technique, which is thus the most suitable technique to use as a reference test method (RTM). Both laboratory and commercial measurement equipment were used.

An American interlaboratory comparison of far-field measurement methods to determine the mode-field diameter of sm fibres was conducted among members of the Electronics Industries Association and the results were presented in [11]. Measurements were made on dispersion-unshifted and dispersion-shifted fibres at 1300 and 1550 nm. Results were calculated using both Petermann and Gaussian definitions. The Petermann definition gave better agreement than the Gaussian in all cases. A systematic offset of 0.52 μm was observed between methods when applied to dispersion-shifted fibres. Such an offset may be caused by limited angular collection.

The role of the mode field diameter in the characterisation of various types of single-mode fibres was examined in [12]. The most relevant definitions of this parameter were reviewed and a comparative analysis of its measurement methods was performed. Emphasis was also given to the requirements posed by new fibre designs, such as polarisation-maintaining structures.

One of the techniques used in [10] was the the variable aperture in the far-field method to measure the sm fibre mode-field diameter. A modified form which provided accurate measurements of dispersion-shifted fibres was described in [13]. The system's numerical aperture (na) was increased by using a spherical mirror and a large aperture wheel with 23 apertures out to $0.556 \times$ na.

The Petermann II mfd definition was calculated with Simpson's rule for numerical integration. The new industrialised 'high NA VAMFF' system was accurate to within one percent, with a measurement standard deviation of less than 0.5 percent. Flexibility for measuring various fibre designs over a range of wavelengths was a key feature.

The relationship between the dynamic range and the measurement accuracy of the mode-field diameter in sm fibres was investigated theoretically [14]. This study clarified the fact that the required dynamic range in which the mode field can be measured within a given accuracy depended upon the index profiles and operating v-value.

Measurement of mfd by transverse offset and variable-aperture far-field methods

These two methods produced apparent differences in the value for the same fibre. Comparison with the results of calculations based on measured refractive index profiles showed that such discrepancies could be attributed, with confidence, to the deviation from Gaussian shape of the electromagnetic field in the fibre [15]. The author was from a company which was (1987) particularly interested because their principal customers favoured one method and their principal suppliers the other. Experiments using standard (matched-cladding) and dispersion-shifted fibres were performed at 1300 and 1550 nm, and simulations provide results for the former type. The apparatus for both techniques was described in detail. The results confirmed that the Petermann II definition of mfd was preferred.

16.3.7 Chromatic dispersion

This property is one of those treated in Subsection 5.4.1. A novel method for the measurement of chromatic dispersion in a single-mode optical fibre was proposed and demonstrated in [16]. This 1989 method adopted the technique of external modulation of the dfb laser output and direct optical frequency sweep range monitoring by coherent heterodyne detection. This method was also applicable to the future optical communication systems with narrow-band optical elements, such as the bandpass filters used with fibre amplifiers to limit the amplified spontaneous noise where measurement was impossible by then state-of-the-art methods.

16.3.8 Refractive index profile

1989 state-of-the-art refractive index profiling was described in [17]. Recent refracted near-field comparisons had highlighted calibration problems associated with the technique, with 10–15% discrepancies not uncommon. Results tend to be grouped according to profiler type and calibration method. Problems associated with the different calibration techniques, and some instrumental effects which can determine instrumental accuracy, were considered. It was shown how transmission effects could be observed and used to generate a correction curve for refracted near-field instruments. High-quality fibre slices needed for axial interferometry were used to obtain refractive index differences on suitable fibres to $\pm 1\%$, with the measurements traceable to national standards. Interferometry was used to calibrate a multiple-step fibre which was available in 14 m lengths with a certificate of calibration. The limitations of the technique were discussed and preliminary results were given of a study on the effects of residual stress and stress relief on measurements. Comparisons were made between measurements made by axial interferometry and refracted near-field on a range of fibres. It was shown that, with care, commercial profilers could produce reliable measurements of refractive index difference, profile shape and numerical aperture which agreed well with measurements by other techniques.

A comparison of three methods of measuring the refractive index profiles of optical fibres was presented in [18]. Refracted near-field, and axial interferometry measurements were compared on a step index fibre, and parabolic profile fibres which had large areas of uniform refractive index at the core. Only the maximum refractive index differences were compared so that the interferometry could be more accurately and easily performed. Differences between the measurements, possible difficulties with the calibration methods and uncertainties were discussed. An RNF calibration error, owing to changes in transmittivity of the instrument with changes in angle of incidence of the refracted rays, was measured. Measurements made using axial interferometry were considered to have an uncertainty not exceeding $\pm 1\%$, and should be suitable for calibrating step index calibration fibres. A modified refracted near-field design was proposed and was under development, which was capable of absolute calibration, thus overcoming some of the difficulties inherent in other instruments.

Simultaneous refracted and transmitted near field measurements have been made so that accurate differences could be easily observed. The effects of fibre length and radii of fibre loops are presented, and measurements showed the theoretical leaky-mode correction factors to be too large. With a parabolic profile, measurements of profile and core radius were in best overall agreement with RNF results when a 2 m length of fibre was used.

The measurement of the refractive index profile (rip) of an optical fibre preform by means of transverse illumination makes an important contribution to the manufacture of high-quality fibres. A useful introduction to preform profiling techniques (1982) was given in a review paper [19]. Two particular implementations of the transverse illumination (ti) method were in widespread use; both operated by measuring the angle of refraction of a ray of light, passing transversely through the preform, as a function of position, to obtain a deflection function which was then numerically transformed to give the refractive index profile. A model was developed to describe the spatial resolution function of a focused laser ti technique for the measurement of the rip. Results showed that the model predicted both the quantitative and qualitative behaviour of the resolution function correctly, and enabled the resolutions of different preform profiling techniques to be compared, as well as assisting in the study of anomalies in measurement [20].

The simple spatial resolution model had the same characteristics as those of the focused laser beam used in the measurement, and the model demonstrated excellent agreement with measured data for a range of preforms and beam diameters found in practice. It provided a sensitive and unambiguous means of assessing the resolution performance of the focused laser transverse illumination technique for refractive index profiling. An important result was the relative insensitivity of the resolution function to details of the interaction of the beam with the bulk of the preform, and this was expected to lead to a greater understanding of the problems related to high-quality refractive index profiling of optical fibre preforms.

16.3.9 Polarisation

In 1989 an overview of current measurement techniques on conventional and polarisation-maintaining single-mode fibres was presented in [21]. The various methods were discussed and classified with respect to the relevant polarisation parameters. Applicability ranges and resolution were pointed out for various types of fibres.

16.3.10 Determination of enhanced ESI parameters

By 1988 there were two trends in characterising single-mode fibres, namely:

(a) from refractive index profile (rip) measurements
(b) from mode-field diameter (mfd) measurements.

The equivalent step index (esi) method lies between these, and the two esi parameters used (the equivalent core radius and equivalent na) can be found from the rip or the mfd. With the esi parameters and the mfd known, most of the essential characteristics of sm fibres, such as bending loss, splice loss, microbending loss, waveguide dispersion, and therefore total dispersion, can be predicted. Unfortunately, there are several definitions and measurement techniques for both esi and mfd quantities. In 1988 a theory was proposed to unify these two trends [22]. Measurements of refractive index profile and mode-field diameter were made on mcvd fibres to support the theory.

In general, the rip of a real fibre differs from the ideal one, and the problem is then to determine the propagation characteristics of the real fibre with an arbitrary index profile. The equivalent step index method is an attempt to do this, but it has not been entirely satisfactory. However, one particular esi method can be enhanced readily, and this is called the enhanced esi, or (e)esi technique. This model is based on the moments of the refractive index profile, and it augments the two parameters given earlier by a third, called the enhancement factor.

Mode-field diameter (= twice spot size) measurement techniques are also not without difficulty and ambiguity, but that based on the moments of the near field (the Petermann II definition) appears to be gaining more widespread acceptance.

The cut-off wavelength is also essential in characterising sm fibres, but again there are difficulties, and care must be exercised in the measurements and their interpretation.

Profile measurements

Commercially available equipment was used to measure the rip of the preform by means of the spatial filtering technique, and that of the fibre by use of the refracted near-field technique. The resulting profiles for the sample fibre are illustrated in Figure 16.4. The (e)esi parameters were derived from a moment analysis of these profiles, and the problems caused by uncertainties in core radius and refractive index values were ameliorated by defining the profiles at the six points shown on the insets to the figures. The same core radii were used for corresponding points on the two inserts, and circular symmetry was assumed. The parameters derived from these measurements were averaged, as they differed slightly from one choice of profile to another, and it is the average results which are listed in Table 16.2.

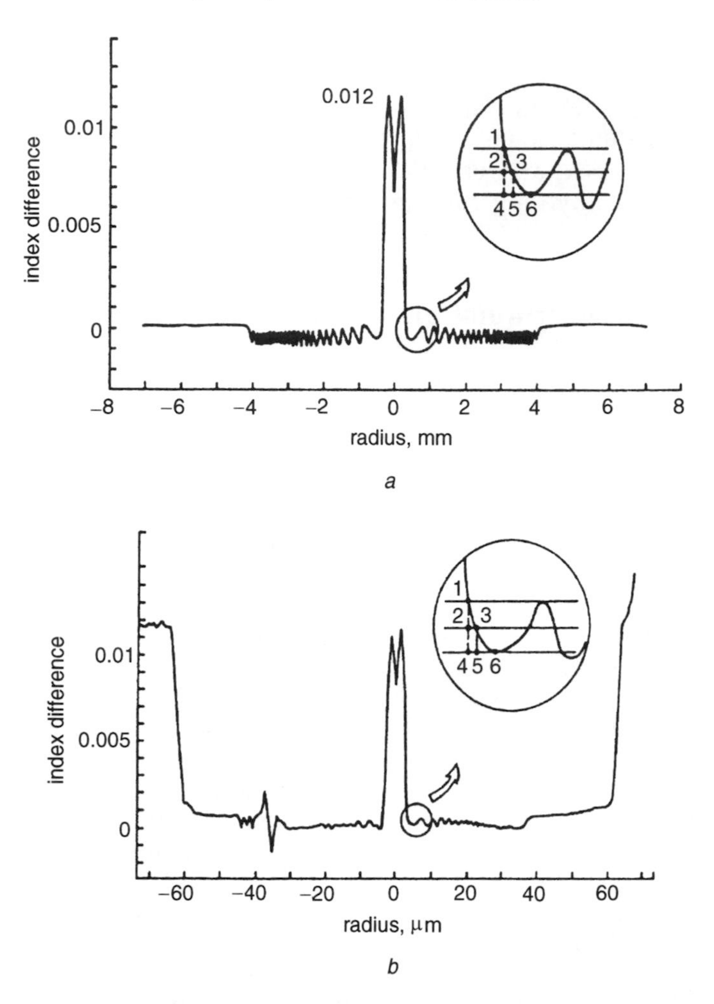

Figure 16.4 Index difference versus radius showing refractive index profile of: (*a*) preform, (*b*) fibre; insets show the six different core sizes and cladding levels used ([22], Figure 4). Reproduced by permission of IEE

Table 16.2 Average values of (e)esi parameters derived from measurements ([22], Table 1). Reproduced by permission of IEE

Parameter	Profile measurements	
	Fibre	Preform
λ_{CO} (nm)	1226 ± 25	1217 ± 11
a_E (μm)	2.86 ± 0.12	2.71 ± 0.1
na_E	0.164 ± 0.005	0.1722 ± 0.0007
$\|\Delta\Omega_4\|$	0.08 ± 0.02	0.03 ± 0.01

(Average theoretical cutoff is $\lambda_{CO} = 1222$ nm). The uncertainty quoted is the standard deviation

Mode-field diameter measurements

The near-field Petermann mfd was obtained from the inverse of the r.m.s. far-field width as measured using commercial equipment employing the variable aperture method. The cut-off wavelength was measured by two techniques:

(1) The Millar procedure from the variation of the mfd as a function of wavelength.

(2) The transmitted power technique given in CCITT Recommendation G.652.

The results obtained from measurements are shown in Table 16.3, with four rather than two different wavelengths being chosen to provide pairs of values from which slightly different results are obtained. These are then averaged as before. The averaged values of the three parameters derived from the above measurements at the two different cut-off wavelengths are given in Table 16.4. The reader can see that the agreement between the parameter values in Tables 16.3 and 16.4 is poor. The discrepancies are due to the use of measured cut-off wavelengths; if instead the average value (1222 nm) of the two obtained from the profile measurements is used, much better agreement is achieved, as can be seen from Table 16.5.

Table 16.3 Near-field Petermann mfd at four wavelengths ([22], Table 2). Reproduced by permission of IEE

λ (nm)	Petermann's MFD (μm)
1300	$6.47 \pm 1\%$
1350	$6.64 \pm 1\%$
1400	$6.85 \pm 1\%$
1450	$7.07 \pm 1\%$

Table 16.4 Average (e)esi parameters from mfd measurements ([22], Table 3). Reproduced by permission of IEE

Parameter	$\lambda_{CO1} = 1275 \pm 25$ (nm)	$\lambda_{CO2} = 1295 \pm 25$ (nm)
a_E (μm)	3.16 ± 0.12	3.22 ± 0.14
na_E	0.155 ± 0.006	0.154 ± 0.007
$\lvert\Delta\Omega_4\rvert$	0.13 ± 0.08	0.14 ± 0.09

λ_{CO1} was measured using the transmitted power technique. λ_{CO2} was measured using the variation of MFD as a function of wavelength

Table 16.5 Average (e)esi parameters based on average cut-off wavelength ([22], Table 5). Reproduced by permission of IEE

Parameter	(E)ESI	ESI
a_E (μm)	3.0 ± 0.07	2.84 ± 0.01
na_E	0.156 ± 0.003	0.1649 ± 0.0004
$\lvert\Delta\Omega_4\rvert$	0.09 ± 0.05	—

The authors give the expression for the theoretical cut-off wavelength as

$$\lambda_{CO} = \frac{2\pi a_E \times na_E}{2.405} \tag{16.1}$$

where the subscript E means effective.

16.3.11 Measurements of chromatic dispersion in single-mode fibres

A number of methods have emerged for this purpose, including:

(1) measurement using a laser source

(2) measurement of relative time-of-flight

(3) measurement of phase shift using led or laser sources.

A dispersion measurement technique for installed sm fibres using an eled and a laser operating at nominal wavelengths of 1300 and 1550 nm, respectively, with the phase-shift method was described in 1988 [23]. In the 1300 nm window, material dispersion is usually the dominant influence on pulse spreading, so the variations in group delay ($\tau(\lambda)$) can be described by a three-term Sellmeier equation and the dispersion can be found from:

$$\delta D(\lambda) = 1 - \frac{{\lambda_0}^4}{\lambda^4} \times \frac{S_0\lambda}{4} \tag{16.2}$$

with $\delta D = 0$ at $\lambda = \lambda_0$, and $S_0 = dD/d\lambda$ evaluated at λ_0. The fibre dispersion is thus specified through the quantities λ_0 and S_0.

Throughout the measurement of dispersion using a dispersive element such as a monochromator, an error arises from the finite size of the slit used to select narrow windows from the led spectrum. Likewise, another source of error lies in the centre wavelength not coinciding with the zero dispersion wavelength of the fibre under test. The paper provides graphs of the change in measured λ_0 for various monochromator slit resolutions (5, 10 or 20 nm) as λ_C of the led is changed in 5 nm steps from -30 to $+30$ nm about the nominal centre. The changes in slope of the dispersion characteristic S_0 at λ_0 as a function of λ_C, with the three slit widths as parameter was also illustrated. The phase shift is measured correctly at two wavelengths, the zero and the led centre values.

Temperature also has an influence on group delay times, through the temperature coefficient of the refractive index and the fibre length, being about 30 ps $km^{-1}K^{-1}$ in sm fibre. This effect was compensated for in the measuring apparatus used by the author to obtain values for the chromatic dispersion of two matched-cladding sm fibres in an installed cable. Each fibre was 28 km long, and five measurements were made on each fibre. A standard (σ) error of 0.03 nm in λ_0, which was in the vicinity of 1315 nm, resulted. Systematic error from the monochromator was estimated to be no more than 0.5 nm. The slope of the dispersion curve at the zero wavelength could be measured with a reproducibility of 0.2% at the σ level.

Direct measurement of chromatic dispersion

A technique for the direct measurement of chromatic dispersion in sm fibres was described in [24], and it used wavelength modulation and a double demodulation method to provide a differential fibre chromatic delay signal from which chromatic dispersion was obtained directly. The system was described in detail, and practical measurement results demonstrated the high accuracy of the technique and its versatility in use with all types of fibres. This followed from the fact that the group delay of real fibres was always a smooth function of the wavelength; therefore one obtains, to a very good approximation:

$$D_1(\lambda) = \frac{d\tau_1(\lambda)}{d\lambda} \simeq \frac{\delta\tau_1}{\delta\lambda} \qquad (16.3)$$

where, as usual, the subscript 1 denotes per-unit quantities, and δ denotes an increment. Since fibres of use in telecommunications must have a low material absorption loss, the material (chromatic) dispersion will always be a smooth continuous function, close to zero in value, in present and future fibres; hence the well-behaved group delay.

The technique used double demodulation of simultaneous intensity and wavelength modulation of light passed through the fibre under test to obtain the differential chromatic delay between adjacent wavelengths, giving the dispersion directly from equation (16.3). The method was applicable to all types of sm fibres, including dispersion-flattened designs, which are not described by the usual Sellmeier equation, and was capable of extremely high repeatability and stability. The authors claim that it was a strong candidate for use in fibre research, production quality control, cabling and installation.

Intermodal and intramodal dispersion of 1300 nm fibres at 850 nm

The application of sm fibres with zero dispersion wavelengths near 1300 nm in future broadband subscriber networks stimulated interest (late 1980s) in low-cost high-reliability component technology as an overlay on such systems for additional capacity (e.g. [25]). One option for such an overlay was the use of 850 nm GaA1As sources and Si detectors, together with fibre which was single-mode at 1300 nm but was overmoded at 850 nm. Evidently the traffic capacity would be limited by intermodal dispersion, and since the fibre was operating well away from its wavelength of zero dispersion, the group delay would be significant and wavelength-dependent for high bit rates and sources with wide spectra. The magnitudes of these effects were investigated by measurements on five commercial sm fibres and the results were reported in [26].

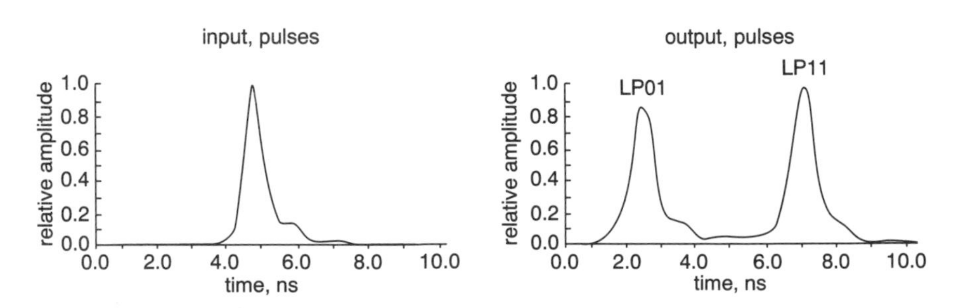

Figure 16.5 Typical waveforms of input pulse and output mode groups LP_{01} and LP_{11} ([26], Figure 2). Reproduced by permission of IEE

Table 16.6 Measured and predicted differential delays between LPO1 and LP11 modes for five different fibre designs ([26], Table 1). Reproduced by permission of IEE

Fibre no.	Fibre type	Differential mode delay		V-value
		Measured (ns/km)	Predicted (ns/km)	@ 0.85 μm
1	Rod in tube	−2.4	−2.9	4.1
2	Depressed cladding	−2.5	−2.2	4.5
3	Matched cladding, v. small dip	−1.5	−1.5	4.7
4	Matched cladding, large dip	−3.5	−4.2	4.8
5	Matched cladding, medium dip	−2.6	−2.1	3.9

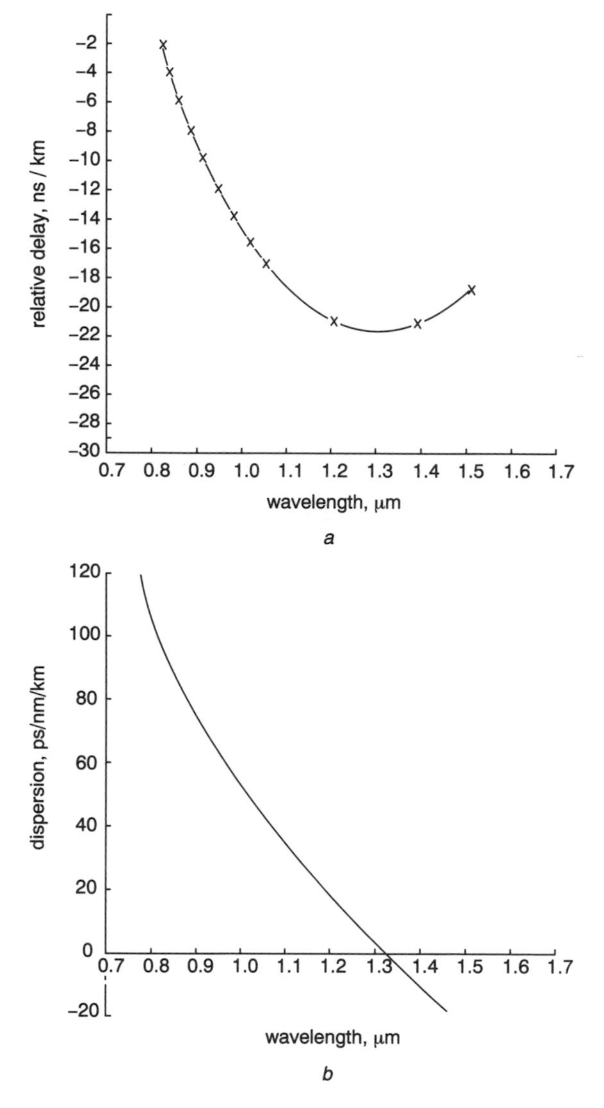

Figure 16.6 Typical results for intramodal dispersion in a sm 1300 nm fibre and the lowest order mode: (*a*) relative delay versus wavelength, (*b*) dispersion versus wavelength. Dispersion at 850 nm is 80 ps $nm^{-1}km^{-1}$ ([26], Figure 7). Reproduced by permission of IEE

Short pulses from a 850 nm laser were launched into the fibre under test in a suitable experimental arrangement, and the output pulse width was compared to that launched at the input to the fibre. Typical results are illustrated in Figure 16.5.

The mode groups were identified by traversing the far-field output of the fibre with the detector. At high angles the detected signal has a larger contribution from the higher-order modes, and the identification was verified by a different technique. All five fibres were measured by means of this technique, and in all five the higher-order modes were delayed relative to the lowest-order mode, which indicates an under-compensated profile. A summary of the measured results and values of delay computed from a finite-element analysis are listed in Table 16.6.

The intramodal dispersion of these fibres was determined by measuring the relative group delay as a function of wavelength over the range 750–1500 nm. The dispersion was found by fitting the points by a curve then plotting the slope versus wavelength. A typical result is shown in Figure 16.6, for one fibre and the lowest-order mode. At 850 nm the dispersion was 80 ps $nm^{-1}km^{-1}$ and, as can be seen from Figure 16.6*b*, the zero of dispersion lies at 1320 nm. The value of dispersion varied little from fibre to fibre, as one would expect, because it was almost entirely due to the material dispersion of the GeO_2 doped silica.

16.4 MEASUREMENTS ON SPLICES AND CONNECTORS

The use of otdr for the measurement of splice loss of installed optical fibre cables was long established by the end of the 1980s. However, because of fibre backscatter parameter variations these measurements had to be carried out from both ends of a fibre link and the splice measurements were averaged to obtain the absolute splice loss.

16.4.1 Single-ended otdr measurements of splice loss in local and junction systems

The possibility of field splice loss measurements by otdr from only one end of an optical fibre system was clearly of interest. [27] described a field and laboratory experiment to determine the feasibility of using a pulse reflection method for single-ended splice loss evaluation using an otdr. Results show that although small errors are obtained in the splice measurement (directly related to the level of reflection caused by the otdr) the method is reasonably accurate and practical for use in short-haul optical systems.

Figure 16.7 shows a singe-ended otdr trace for a simple system with a reflecting mechanism at one end.

The primary area A to B is a conventional trace for a link with a single fusion splice, and it shows a loss feature called the primary image. The area between B and C results from the superposition of two backscattered components and it shows two loss features. The first is due to the original otdr pulse being reflected at the mirrored end, giving a 'real image'. The second feature is caused by re-reflection of the mirror-reflected pulse from the otdr end, and it is named the secondary 'virtual image'. It can be shown that averaging the primary image loss and the secondary real image loss gives the absolute splice loss.

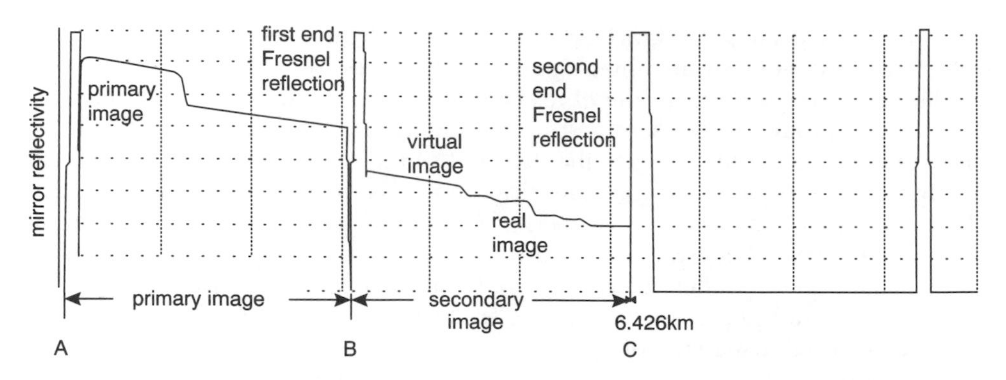

Figure 16.7 OTDR trace of practical two-fibre link with mirrored end ([27], Figure 3). Reproduced by permission of IEE

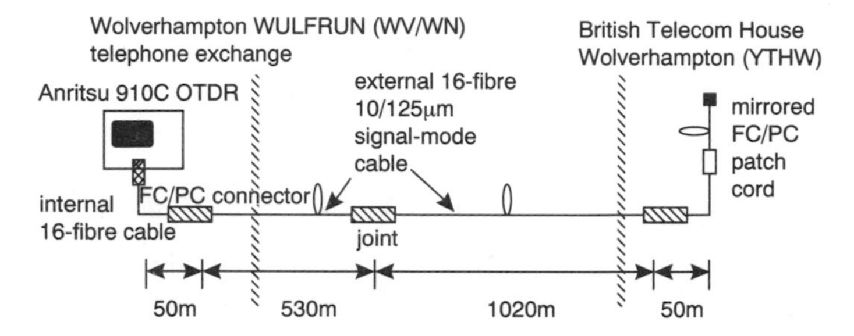

Figure 16.8 Single-ended otdr trial site ([27], Figure 1). Reproduced by permission of IEE

Table 16.7 Results of one- and two-way measurements on installed link ([27], Table 1). Reproduced by permission of IEE

Splice	Primary loss from Telecom House	Primary loss from exchange	Absolute loss, two-way method	Real image loss from Telecom House	Absolute loss, one-way method	Error in one-way method
	dB	dB	dB	dB	dB	dB
1	0.30	0.27	0.285	0.27	0.285	0
2	0.05	−0.01	0.02	−0.02	0.015	−0.005
3	0.02	−0.02	0	−0.02	0	0
4	0.31	0.26	0.285	0.22	0.265	−0.02
5	0.18	0.05	0.115	−0.02	0.08	−0.035
6	0.46	0.44	0.45	0.25	0.355	−0.095
7	0.08	0.14	0.11	0.14	0.11	0
8	0.21	0.18	0.195	0.13	0.17	−0.025
9	0.33	0.33	0.33	0.21	0.27	−0.06
10	0.02	0.04	0.03	0.04	0.03	0
11	0.36	0.35	0.355	0.28	0.32	−0.035

After giving the theory behind the method, the field experiment was described with the aid of Figure 16.8. Measurements were made at 1550 nm from both ends of the link to establish a two-way loss measurement. The results of conventional two-way measurements and of one-way pulse reflection measurements are listed in Table 16.7.

The authors suggested that a measurement method based on the principle should prove effective in reducing installation and maintenance costs of short optical links in the local and junction environment.

Since the widespread adoption of single-mode fibre technology in the mid-1980s, otdr is commonly used in fibre and cable production control as well as field installations. However, the technique cannot produce the splice loss directly, since the effects of statistical variations in scattering properties among fibres produces a false contribution to the splice loss. One remedy for this difficulty is to measure the link from both ends, and define the loss of a given splice as the average of the two measurements.

Splice losses measured by means of an otdr are well known to be affected by differences in the backscattering properties of the spliced fibres under test. Several other methods have been proposed to overcome this effect; one is to make a second measurement from the far end and average the two results to find the splice loss. A second method involves the prior measurement of the backscattering coefficient of each individual fibre, which compensates for the discontinuity in the received backscattered power due to the difference in the coefficients. A third technique is to use an ultra-sensitive odtr in conjunction with a reflector at the far end of the link under test. None of these techniques is attractive for use in the field, particularly if access to the far end of the fibre is needed.

Various methods had been tried by 1988 to obtain true values from one-way measurements, but none was very practical. As already mentioned, there are statistical variations in fibre properties which affect splice loss, and also a very large amount of data from splice loss measurements has been accumulated on the many sm systems installed. Thus it is useful to ask for the probability of finding that the actual loss of a given splice is less than some maximum value based on a simple one-way otdr reading. Based on data collected from about 500 bidirectionally measured splices in the field, the five parameters of a chosen two-dimensional probability distribution function with the actual splice loss and the one-way otdr discontinuity as variables were determined numerically from the data [28]. The probability of keeping the loss below a target value based on the one-way otdr reading was calculated and the results could be displayed graphically or in a table. Satisfactory agreement between predicted and experimental values was obtained for standard matched-cladding sm fibres measured at 1300 nm.

This information was expected to make it easier for operators to decide whether a splice measured by one-way otdr was acceptable or not. If unacceptable some considerations were given on enhancing the acceptance probability by repeated splicing, and various acceptance criteria were discussed. The method becomes more reliable as the number of splices in a link increase.

Splicing dissimilar fibres can lead to large values of scattering factor ratio and high values of splice loss. A table of typical values was provided for certain common types of fibre which might be of importance in some situations, e.g. splicing matched cladding fibre pigtails to dispersion-shifted fibres or when a repair section of different type of fibre is used; this is reproduced as Table 16.8.

Table 16.8 Ranges of typical values of splice loss and of backscattering factor ratio obtained from splicing dissimilar fibres ([28], Table IV. Reproduced by permission of IEEE, © 1989

Combination of fibres	1300 nm wavelength		1550 nm wavelength	
	loss	ratio	loss	ratio
dc/mc	0.05–0.50	0.25–1.50	—	—
ds/mc	0.4–0.6	4.1–4.8	0.2–0.4	3.0–3.4
ds/dc	0.3–0.5	3.2–3.9	≃0.2	≃2.3
	dB	dB	dB	dB

Key: dc depressed cladding, mc matched cladding, ds dispersion-shifted fibre.

This paper [28], by a member of staff of a cable company in Norway, was partially supported by the Norwegian Telecommunications Administration, and it dealt with the factors affecting fibre scattering, splice losses, a description of the probability model, the essential points concerning data collection and experimental procedures and results. Practical applications were then discussed, followed by some results obtained when splicing dissimilar fibres, and the finally main conclusions were stated.

Of the four basic techniques for measuring splice loss, one has already been mentioned (optical time-domain reflectometry); the remaining three are the cut-back method, the visual core alignment technique and the wet fusion method. The last two only estimate the loss at best and are therefore unsatisfactory.

A new technique for measuring absolute splice loss was described by authors from Bell Northern Research in [29], based on local launch and detect. When a fibre is bent, some light is radiated out of the fibre completely, and the amount is proportional to the power propagating in the core. If then a detector were to be placed symmetrically with respect to the bend, as shown in Figure 16.9, and if the fibre were uniform, the detected power would be independent of the direction of transmission. The detector can be calibrated, so that the power propagating in the core in one direction can be inferred from a reading of a known power level

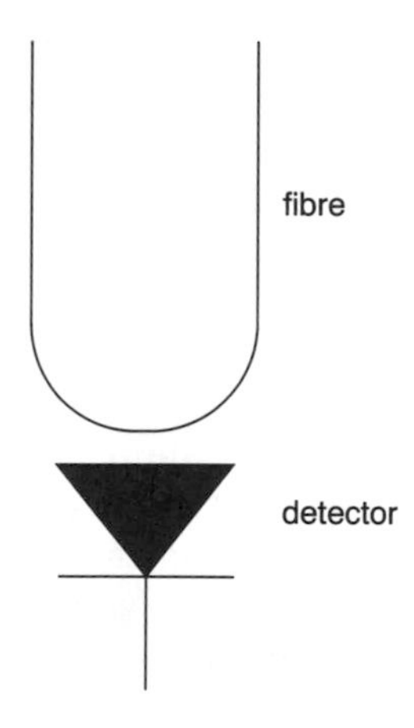

Figure 16.9 Fibre bent in a semicircle and symmetrically placed detector ([29], Figure 1). Reproduced by permission of IEEE, © 1987

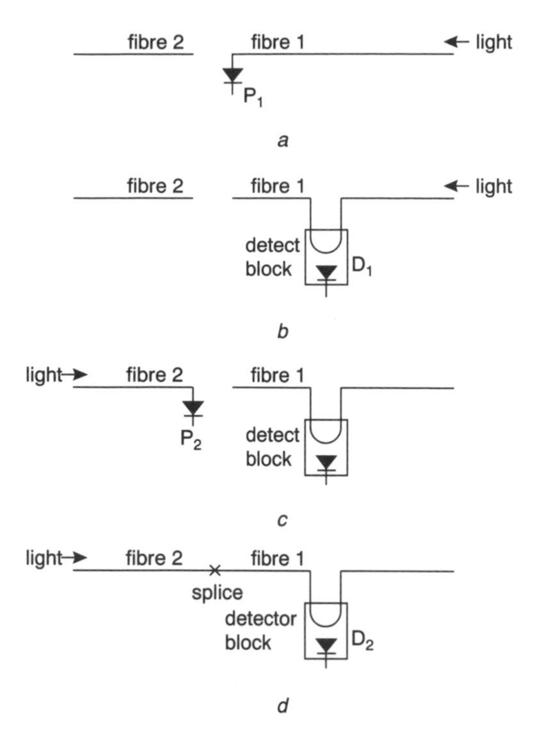

Figure 16.10 Diagrams illustrating technique for measuring absolute splice loss: (*a*) power level in fibre 1 before splicing; (*b*) detected power from bend after insertion of fibre 1 into semicircular groove; (*c*) power level in fibre 2 before splicing; (*d*) detected power from bend after splicing ([29], Figure 2). Reproduced by permission of IEEE, © 1987

of light propagating in the opposite direction. Based upon this principle, a suitable detector was constructed using a Ge photodiode with an active area diameter of 5 mm, and a bend radius of about 2 mm was employed. The arrangement for measuring splice loss is illustrated in Figure 16.10. The measurement procedure was then given, followed by the calibration procedure described in the paper, where it was shown that the factor P_C could be determined entirely from measurements on a pair of fibres involving the symmetrical detector. The authors found the value to be independent of position along the length of two sample fibres to within the accuracy of their measurements of ±0.05 dB, although the values themselves varied, as can be seen from Table 16.9.

Comparative measurements were made on fibre 1, which was cut and spliced several times. The loss for each splice was measured using the present method and the cut-back method, and the results obtained with P_C equal to 0.02 dB can be compared in Table 16.10. Six measurements of power were required: four for the present method and two for the cut-back method, and their accuracy of ±0.03 dB was the major contributor to the maximum deviation of 0.15 dB in the measured values. Fibre 2 was also measured and it gave results with a mean difference of 0.01 dB and standard deviation of 0.05 dB.

The authors concluded by noting the potential of the method as an invaluable tool for the measurement of splice loss, and suggested that the detector module could be used with local launch and detect schemes to permit low-loss splicing and measurement of the splice loss.

Table 16.9 Calibration factor P_C for two single-mode fibres ([29], Table 1). Reproduced by permission of IEEE, © 1987

Fibre 1	Fibre 2
0.01	0.07
0.07	0.03
−0.04	0.00
0.06	0.03
0.00	0.06
0.02	0.04
	0.03
	0.04
mean = 0.02	mean = 0.04
s. dev. = 0.04	s. dev. = 0.03

Table 16.10 Values of splice loss obtained by two methods ([29], Table 2). Reproduced by permission of IEEE, © 1987

Symmetric detector	Cutback	Difference
0.41	0.49	−0.08
0.30	0.45	−0.15
0.23	0.32	−0.09
0.16	0.02	0.14
0.15	0.03	0.12
0.05	0.11	−0.06
0.19	0.04	0.15
0.16	0.03	0.13
0.09	0.24	−0.15
		mean = 0.0011
		s. dev. = 0.13

16.4.2 Technique for reducing splicing loss between standard sm fibre and erbium-doped fibre

Erbium-doped fibre amplifiers (edfa) have been deployed as booster amplifiers to raise transmitted power levels, as repeaters and as preamplifiers for optical receivers in systems operating at 1550 nm. Extremely high-efficiency devices are achieved by means of small-core, high numerical aperture doped fibre, leading to smaller spot sizes than in standard fibre. The corresponding mismatch between the field distributions in these two types of fibre leads to a large splice loss, which in turn degrades the amplifier gain, noise figure and output power. A simple and practical splicing technique to reduce this mismatch was described in [30].

Two methods had been proposed earlier to overcome the problem but the results they achieved were much poorer than those obtained by the present technique.

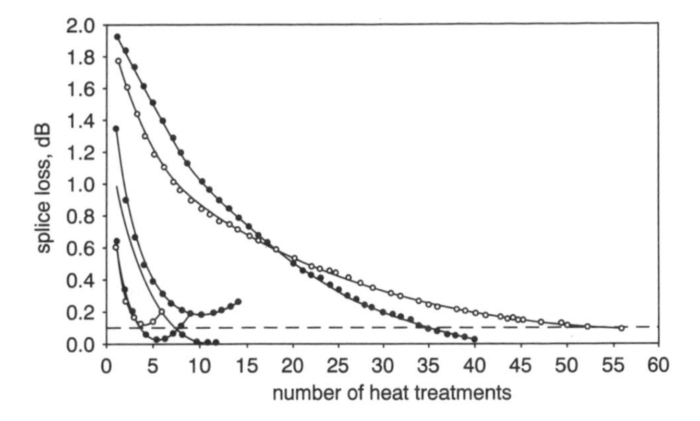

Figure 16.11 Measured splice loss versus number of heat treatments using programs 1, 2 and 3 ([26], Figure 2). Reproduced by permission of IEE

Table 16.11 Comparison of measured values of splice loss between standard sm fibre and edfa using the three methods. Reproduced by permission of IEE

		Er-doped fibre parameters		
Splicing technique	Splice loss	Core radius	*NA*	Reference
	dB	μm		
2 matching fibres	~ 0.48†	~1.19	~ 0.29	4
Thermal tapering	0.7–1.0*	1.34	~ 0.34	7
Fusion splice	0.3–0.7*	2.70	0.175	8
Arc fusion tapering	< 0.1*	1.85	0.24	This work

*Measured values
†Theoretical value

This achieved a reduction in splice loss from 1.8 to less than 0.1 dB. Feasibility studies demonstrated that a substantial increase in core size at a reduced fusion time by working at a higher temperature with a high na edfa and that this could be achieved with a commercial product. The splicing operation took place in four steps: withdrawal to form a gap; preheating; fusion; annealing, and by repeatedly applying the heat treatment the splice loss in both directions was reduced, as illustrated in Figure 16.11.

The splice losses between standard fibre and erbium-doped fibre obtained by the three methods are summarised in Table 16.11. An optimum splice is achieved when there is no further reduction in loss after an additional fusion process. By this method the diffusion of Ge can be controlled to increase the spot size of the edfa until it matches that of the standard sm fibre.

16.5 MEASUREMENTS ON LASERS

Lasers have been treated in Chapters 6–8. Many of the references in these chapters have titles which include the performance aspect. Such references generally include figures showing the experimental arrangements and give supporting text. In some specific instances, the title contains the word 'measurement', e.g. 'Measurement of very high speed photodetectors with picosecond InGaAsP film lasers' ([11] of Chapter 10). The frequency response of a certain kind of apd was described in a 1989 paper ([20] of Chapter 10).

Linewidth and linewidth enhancement factor are two of the important parameters of lasers in ofdls applications, and am and fm responses for analogue traffic. All four can be measured simultaneously by means of a commercially available computer-controlled spectrum analyser with tracking generator, optical input section and optical delay line arranged as described in [31] published in 1995. The determination of linewidth also provides information on the frequency noise density and α, the linewidth enhancement factor provides information on nonlinear gain. With an optical input power of 0 dBm, laser linewidth less than 50 MHz and laser modulation response without cut-off, am modulation index values exceeding 0.01%, and fm deviations greater than about 10 MHz can be detected. The measurement arrangement illustrated in Figure 16.12, is the well-known homodyne set-up for measuring linewidth and provides a simple, reasonably fast method for use in production. The laser dut can be modulated by the tracking generator. The measurement system requires that the delay line length be longer than about three times the coherence length of the laser, that the polarisation states of the source and delayed light be approximately matched, and the frequency band of interest be lower than the maximum frequency of the microwave spectrum analyser equipment.

The theory of the am and fm response measurements assumes a laser with only one dominant (time-averaged) mode, but for devices with several significant modes averaged response values are obtained. Alternatively, the responses for individual modes can be measured if each mode is selected by a suitable filter. The maximum

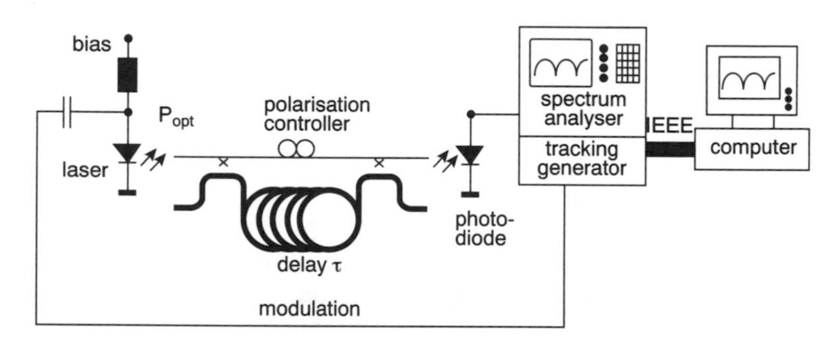

Figure 16.12 Experimental arrangement for the simultaneous measurement of linewidth, linewidth enhancement factor α, and fm and am response of a semiconductor laser ([31], Figure 1). Reproduced by permission of IEEE, © 1995

Table 16.12 Statistical data from measurements on a RCA C30902 ([31], Table 1). Reproduced by permission of IEEE, © 1995

no	laser type	from	Σ	Δf_L/ MHz	m (200 MHz) /1/mA	Δf (200 MHz) /MHz/mA	α	f_g/GHz
1	TTG/BH-type	Siemens	1	20-35	0.030-0.038	430	5.8-7.0	3.3-3.4
2	TTG/MCRW-type	Siemens	1	9.4-11.4	0.012-0.013	150	9.1-9.4	1.1-1.2
3	DFB/1528 nm	HHI	1	120	0.030	500	10.4	1.5
4	DFB/Modul	Fujitsu	2	13.4-14.1	0.030-0.037	120-210	5.1-5.8	1.3-1.6
5	DFB/FLD 150 F2 RH	Fujitsu	3	18.8-20.4	0.012-0.030	90-200	6.8-9.4	1.0-2.5
6	DFB/1550 nm	Siemens	4	4.8-12.8	0.018-0.030	100-150	5.6-8.4	1.7-2.1

frequency of measurement for multimode devices must be smaller than half the mode spacing to avoid overlap. Mode spacing is described in the chapters on lasers.

The results of some measurements on devices from different manufacturers are summarised in Table 16.12.

This subsection will mention several other papers not included elsewhere. The first two refer to frequency response measurements and they appeared in 1989 and 1990, respectively.

Novel measurement technique for α factor in dfb lasers

A simple and accurate method for measuring the linewidth enhancement factor α in dfb semiconductor lasers was proposed. This method, based on the principle of external optical injection locking, does not require knowledge of the absolute value of optical injection level [32].

High resolution direct measurements of laser spectral linewidth

These can be made by several techniques, including:

(a) delayed self-heterodyne
(b) heterodyne
(c) delayed self-homodyne
(d) fibre optic ring resonator.

All of these suffer from an number of disadvantages, and a new simple technique was developed, demonstrated and described in 1988 [33] which was based on mixing the carrier and direct-modulation sidebands in an all-fibre delayed self-homodyne scheme. The detected signal contained the laser spectral noise at the modulation frequency, thereby avoiding the need for optical modulators or large-index modulation in other frequency shifting techniques. The scheme used a passive all-fibre single-coupler interferometer. The principle can be understood from the experimental arrangement shown schematically in Figure 16.13. A single coupler was used to split the signal from the laser into the two arms and then recombine the two signals. Although the circuit has the appearance of a ring resonator, the fact that the transmission loss of the long arm is relatively large means that it acts like a

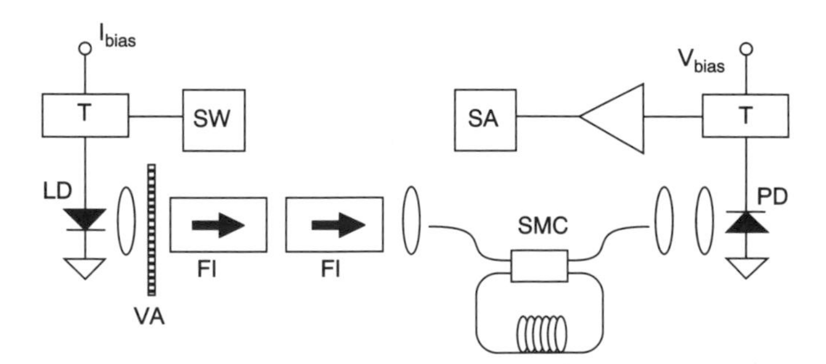

Figure 16.13 Experimental arrangement ([33], Figure 1). Reproduced by permission of IEE

single-pass interferometer. dc modulation of the laser results in both am (intensity modulation) and fm. The optical carrier field can be represented by:

$$E = E_0 \exp[\mathrm{j}2\pi ft + \phi(t)] \tag{16.4}$$

where f is the optical frequency and ϕ is the phase fluctuation. Under small-signal conditions when the modulation indices are very much less than one, and by neglecting intensity noise, the modulated field can be approximated by:

$$E = E_0 \left[1 + \left(\frac{m}{2} + \mathrm{j}\beta\right) \cos 2\pi f_{\mathrm{M}} t\right] \exp[\mathrm{j}2\pi ft + \phi(t)] \tag{16.5}$$

where f_{M} is the modulation frequency, and the terms in m and β represent the two am and two fm sidebands about f, respectively. The detected optical field E_{D} is the sum of the delayed and undelayed signals, and the authors showed that, under certain simplifying assumptions, the two-sided power spectral density of the detected photocurrent contained a quasi-Lorentzian component given by a fairly complicated expression. One can infer from this expression that the technique can also be used for pure am or pure fm in cases where the simple simultaneous am and fm cannot be applied.

The arrangement included an sm coupler, 800 m of sm fibre, a photodiode with a 3 dB bandwidth exceeding 1 GHz, Faraday isolators (F1,F2) and a variable attenuator to reduce optical feedback to negligible proportions. The ratio of the delayed to undelayed powers was about 0.3. A synthesised sweeper modulated the laser via a bias tee at 450 MHz and the resulting demodulated RF output from the pd is displayed on a spectrum analyser. Typical spectra observed with the delayed and undelayed arms of the interferometer open (coupled), with and without modulation are shown in curves a and b, respectively. The spikes at 450 MHz are the first-order am sidebands, and the broad peak in curve a centered on 450 MHz is caused by the laser phase noise term, which is the quasi-Lorentzian expression mentioned above. The fwhm of this peak gave the laser spectral linewidth. The broadening about this frequency in curve was due to up converted laser intensity noise, and is evidently

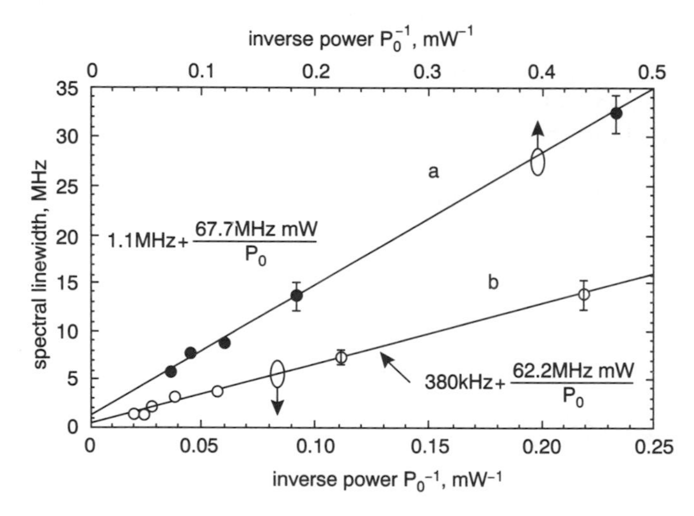

Figure 16.14 Spectral linewidth versus reciprocal of average single-facet output power showing linear dependence: (*a*) 300 μm cavity bulk laser; (*b*) 750 μm cavity quantum well laser ([33], Figure 3). Reproduced by permission of IEE

more than 25 dB below the phase noise. The peak at 900 MHz was larger than is used in the measurements, simply to make it obvious; in practice, the RF drive level to the laser was adjusted until this peak was at least 20 dB below the peak at 450 MHz, to ensure small-signal operation.

The results of the measurements of the spectral linewidth on two different types of laser are shown in Figure 16.14, for various single-facet output power levels, and as can be seen the curves follow the expected inverse dependence on this quantity. The linewidth of the quantum-well device was also measured by means of the delayed self-heterodyne technique, and the results were found to agree within the limits of experimental error.

16.6 MEASUREMENTS ON DETECTORS

Optical receivers were treated in Chapters 9–11. Many of the references in these chapters have titles which include the performance aspect. Such publications generally include figures showing the experimental arrangements and they give supporting text. In some specific instances, the title contains the word measurement, e.g. 'Measurement of very high speed photodetectors with picosecond InGaAsP film lasers' ([11] of Chapter 10). The frequency response of a certain kind of apd was described in a 1989 paper ([18] of Chapter 10).

This subsection will mention several other papers not included elsewhere. The first two refer to frequency response measurements and appeared in 1989 and 1990, respectively.

The frequency response of optical receivers was accurately calibrated by measuring a heterodyne signal generated by mixing two Nd:YAG ring lasers [34]. This heterodyne system offered more than 50 dB of dynamic range. Calibration of optical phase and amplitude modulators was achieved by down converting a sideband of the modulated optical carrier to a fixed intermediate frequency with another laser. This technique eliminated the need for a high-speed receiver.

In the second paper white optical noise (spontaneous–spontaneous beat noise) generated by amplified spontaneous emission from a semiconductor optical amplifier was used to measure the frequency response of very wide bandwidth photodetectors and optical receivers [35]. This novel technique could be used to characterise optoelectronic components of arbitrarily wide bandwidths.

In the early 1980s an automatic testing system for silicon apds was described by workers from the Italian organisation CSELT in [36] which could provide all the main characteristics with a single measuring procedure. Measurements of the following quantities were described:

(a) dark current
(b) responsivity
(c) noise current spectral density

and the determination of the excess noise factor and the effective constant k_{EFF}, all at 900 nm. All these quantities have been introduced in Chapters 9–11. The computer-controlled measuring system achieved a good balance between accuracy, simplicity and speed of operation. The system accuracy was about ±7% between 400 and 900 nm, being determined mainly by the absolute accuracy of the light emitting diode source, which was periodically checked by a radiometer. To check

Table 16.13 Repeatability test: statistical parameters referring to 20 independent measurements carried out on the same avalanche photodetector type RCA C30902. (Bias voltage 198V, corresponding to M $\simeq$ 150; input optical power: 100 nW). Reproduced by permission of CSELT

	Mean value	Standard dev.	
Dark current (nA)	20.8	1.2	(6%)
Total responsivity (A/W)	58.7	1.9	(3%)
Gain	149.5	7.7	(5%)
Noise current (pA/$\sqrt{\text{Hz}}$)	37.6	2	(5%)
Excess noise factor	5	0.075	(1.5%)
K_{eff} coefficient	0.020	0.001	(5%)
Primary responsivity	0.395	0.006	(1.5%)

the repeatability, one particular apd was remeasured 20 times, each time using the full standard procedure. The results of these measurements are listed in Table 16.13.

The measurements were carried out at room temperature, occupying several hours on a number of different days, and the estimated variation of temperature was about 4°C, which is probably the main cause of the dispersion in the results. Other considerations enter into the results reported in [36], which can be consulted for details.

16.7 JITTER MEASUREMENT TECHNIQUES AND REQUIREMENTS

Jitter origin, accumulation and control are vitally important in digital systems and networks, and they form an important part of relevant performance specifications. The principal properties of jitter were summarised and the philosophy behind mid-1980s international specifications were explained in an excellent paper published in mid-1987 [37]. A review of measurement requirements and techniques for testing the jitter performance were presented. Some aspects of jitter were discussed in Section 13.7 and in the context of several systems, e.g. in Chapter 15 the optically amplified OAL 32 system jitter performance conforms to CCITT Rec. G.958.

16.7.1 Jitter measurement methods

Measurement of timing jitter basically involves the detection and measurement of timing displacements of the leading and/or trailing edges of digital signals and of the zero-crossings of sinusoidal or rectangular clock pulses. The processing and display of these measured data forms an important part of any jitter measuring equipment.

Measurements may generally be made directly using time or phase measurement techniques, or indirectly using another parameter, e.g. frequency, followed by computation of the required quantity. In almost all applications the jitter to be measured will cover a considerable bandwidth and may contain components down to dc, thus necessitating extremely good low-frequency performance of the equipment. Some specifications require the measurement of jitter in particular frequency bands and suitable filters for this purpose are also an essential requirement. In all of these cases some sort of averaging measurement is needed and it is usual to display the r.m.s. value; however, in other cases where the effects of jitter depend on peak jitter amplitudes, it is usual to include an indication of peak-to-peak jitter.

Other general requirements for the measurement of timing jitter are:

(a) the resolution and accuracy needed for measurements to fractions of 1% of a unit interval (digit period)

(b) a dynamic range to accommodate the wide range of jitter amplitudes found within wideband jitter and associated with specification templates

(c) the capability to measuring jitter under live traffic conditions

(d) the acceptance of all standard transmission and interface codes.

Concerning (a), it is necessary to be able to measure the instantaneous jitter at the output of a single (regenerative) repeater where values of less than 0.01 UI are typical. At the other extreme, the output of a demux may contain low-frequency jitter of several UIs in amplitude, hence (b).

16.7.2 Measurement principles and problems

Under this heading the author of [37] mentioned that jitter can be detected and measured as either phase or frequency modulation, with the choice of technique determined by whichever representation was capable of giving the most accurate results in a given situation. Other factors were mentioned as possibly influencing the choice.

When phase modulation is appropriate, the most common technique is to 'demodulate' the signal and apply the jitter to a baseband calibrated phase demodulator. The requirements of this technique were given and a number of problems which can occur with phase demodulators were listed.

Direct measurement techniques were then discussed and the corresponding types of equipment mentioned, e.g. good-quality high-frequency oscilloscopes, calibrated phase demodulators, digital processing oscilloscopes and sampling oscilloscopes. Mention was made of the suitability or otherwise of the method for field or laboratory use. This was followed by a paragraph on direct measurement techniques.

The penultimate section dealt with jitter measurements for CCITT specifications under three subheadings: input jitter tolerance assessment; maximum permissible jitter; and jitter transfer characteristic.

16.8 STATUS OF IC DESIGN-FOR-TESTABILITY (END 1980s)

An overview of techniques for ic design-for-testability and highlights of the developments which by 1988 were beginning to have a significant impact were given in [38] which was an updated version of a keynote address given at the 1988 Custom Microelectronics Conference, Heathrow, England. The four main techniques then in use were outlined, ranging from completely ignoring testability during the design stage to the design of self-testing ics. A view of the future of ic design-for-testability was offered which emphasised the role of standards which were (and are) recognised as increasingly important and which were expected to affect the design of ics into the 1990s.

The ic design-for-testability approaches current in 1988 were all then in use somewhere in the integrated circuit industry and could be summarised thus:

(1) do nothing
(2) issue guidelines for ic design-for-testability
(3) scan design
(4) self-test.

Each of these approaches was described from a very practical and experienced viewpoint and the responses of organisations and their design staff and test engineers to them highlighted.

Looking to the future the author posed the questions 'what will be the major changes in design-for-test technology and the way that testability is achieved for integrated circuits?'. He offered the following answers.

The 'built-in' approach: testability will be increasingly be built into the design environment and into ASIC libraries in order to fully realise savings possible through techniques such as scan design and self-test.

Table 16.14 Main items of test equipment for ofdls

Description	Manuf./Accept.	On-site Main.	Rep. Centre
Pattern generator	✓	✓	✓
Error detector	✓	✓	✓
Oscilloscope, acces.	✓	✓	✓
Digital freq. meter	✓	✓	✓
Digital voltmeter	✓	✓	✓
Jitter generator	✓	—	✓
Jitter detector	✓	—	✓
Var. freq. generator	✓	—	✓
Stable osc. at line rate	✓	—	✓
Logic analyser	✓	—	✓
Route simulator (traffic)	✓	—	✓
Power feed trip test set	✓	—	✓
Supervisory test set	✓	—	✓
Optical transmitter	✓	—	✓
Optical receiver	✓	—	✓
Optical attenuator	✓	✓	✓
Opt. att. meas. set	✓	✓	✓
Opt. band. meas. set (pulse disp., swept freq.)	✓	—	✓
Optical backscatter (otdr)	✓	✓	✓
Opt. power meter, acces.	✓	✓	✓

Abbreviations:

acces. = accessories
var. = variable
osc. = oscillator
opt. = optical
att. = attenuator
meas. = measuring
band. = bandwidth

Standardised testabililty approach: standards were then currently being developed by the IEEE P1149 Testability Bus and P1149.1 Test Access Port and Boundary Scan Architecture Standard Committees which were to become a common feature of integrated circuits.

16.9 MEASURING EQUIPMENT REQUIRED FOR OFDLS

Both electrical and optical measurements are called for, so that appropriate test equipment is needed for both categories. A list of equipment compiled in the early 1980s is given in Table 16.14.

16.10 REFERENCES

1 FOX, S.: 'Measurement techniques for operational optical-fibre systems'. *IEE Electronics & Power*, Feb. 1983, pp. 159-162.

2 NICHOLSON, G., and TEMPLE, D.J.: 'Polarisation fluctuation measurements on installed single-mode optical fibre cables'. *IEEE/OSA J. Lightwave Technol.*, **7**, no. 8, Aug. 1989, pp. 1197-1200.

3 NAMIHIRA, Y., and WAKABAYASHI, H.: 'Real-time measurements of polarisation fluctuations in an optical fibre submarine cable in a deep-sea trial using electro-optic $LiNbO_3$ device'. *IEEE/OSA J. Lightwave Technol.*, **7**, no. 8, Aug. 1989, pp. 1201-1206.

4 KAPRON, F.P., ADAMS, B.P., THOMAS, E.A., and PETERS, J.W.: 'Fibre-optic reflection measurements using OCWR and OTDR techniques'. *IEEE/OSA J. Lightwave Technol.*, **7**, no. 8, Aug. 1989, pp. 1234-1241.

5 TATEDA, M., and HORIGUCHI, T.: 'Advances in optical time domain reflectometery'. *IEEE/OSA J. Lightwave Technol.*, **7**, no. 8, Aug. 1989, pp. 1217-1223.

6 ROSHER, P.A., FENNING, S.C., COCHRANE, P., and HUNWICKS, A.R.: 'An automatic optical fibre break location scheme for duplex and diplex transmission systems'. *Br. Telecom Technol. J.*, **6**, no. 1, Jan. 1988, pp. 54-59.

7 BLANK, L.C., and COX, J.D.: 'Optical time domain reflectometry on optical amplifier systems'. *J. Lightwave Technol.*, **7**, no. 10, Oct. 1989, pp. 1549-1555.

8 BOUCOUVALAS, A.C., and ROBERTSON, S.C.: 'Simple technique for measurement of equalisation wavelength in single-mode optical fibres'. *Electronics Lett.*, **23**, no. 5, 26 Feb. 1987, pp. 215-216.

9 POCHOLLE, J.P., and AUGE, J.: 'New simple method for measuring the mode spot size in monomode fibres'. *Electronics Lett.*, **19**, no. 6, 17 Mar. 1983, pp. 191-193.

10 COST 217 GROUP: 'COST 217 intercomparison and analysis of fibre mode field diameter measurements'. *IEE Proc. J*, **138**, no. 6, Dec. 1991, pp. 373-378.

11 DRAPELA, T.J., FRANZEN, D.L., CHERIN, A.H., and SMITH, R.J.: 'A comparison of far-field methods for determining mode field diameter of single-mode fibres using both Gaussian and Petermann definitions'. *IEEE/OSA J. Lightwave Technol.*, **7**, no. 8, Aug. 1989, pp. 1153-1157.

12 ARTIGLIA, M., COPPA, G., DI VITA, P., POTENZA, M, and SHARMA, A.: 'Mode field diameter measurements in single-mode fibres'. *IEEE/OSA J. Lightwave Technol.*, **7**, no. 8, Aug. 1989, pp. 1139-1152.

13 PARTON, J.R.: 'Improvements to the variable aperture method for measuring the mode-field diameter of dispersion-shifted fibres'. *IEEE/OSA J. Lightwave Technol.*, **7**, no. 8, Aug. 1989, pp. 1158-1161.

14 OHASHI, M., SHIBATA, N., and SATO, K.: 'Mode field diameter measurement conditions in single-mode fibres'. *Electronics Lett.*, **25**, no. 9, 13 Apr. 1989, pp. 493-495.

15 FOX, M.: 'Experience with mode-field radius measurements'. *IEE Proc.*, **135**, Pt. J., no. 3, June 1988, pp. 211-214.

16 RYU, S., HORIUCHI, Y., and MOCHIZUKI, K.: 'Novel chromatic dispersion measurement method over continuous gigahertz tuning range'. *IEEE/OSA J. Lightwave Technol.*, **7**, no. 8, Aug. 1989, pp. 1177-1180.

17 RAINE, K.W., BAINES, J.G.N., and PUTLAND, D.E.: 'Refractive index profiling — state-of-the-art'. *IEEE/OSA J. Lightwave Technol.*, **7**, no. 8, Aug. 1989, pp. 1162-1169.

18 RAINE, K.W., BAINES, J.G., and KING, R.J.: 'Comparison of refractive index measurements of optical fibres by three methods'. *IEE Proc.*, **135**, Pt. J, no. 3, June 1988, pp. 190-195.

19 STEWART, W.J.: 'Optical fiber and perform profiling technology'. *IEEE Quantum Electronics*, **QE-18**, no. 10, Oct. 1982, pp. 1451-1466.

20 SVENDSEN, D.A.: 'Resolution model for measurement of optical fibre performs by focused laser transverse illumination technique'. *IEE Proc.* **135**, Pt. J, no. 3, June 1988, pp. 196-201.

21 CALVANI, R., CAPONI, R., and CISTERNINO, F.: 'Polarisation measurements on single-mode fibres'. *IEEE/OSA J. Lightwave Technol.*, **7**, no. 8, Aug. 1989, pp. 1187-1196.

22 MARTINEZ, F., and HUSSEY, C.D.: '(E)ESI determination from mode-field diameter and refractive index profile measurements on single-mode fibres'. *IEE Proc.*, **135**, Pt. J, no. 3, June 1988, pp. 202-210.

23 BOOTHROYD, S.A.: 'Measurement of chromatic dispersion in installed single-mode fibre'. *IEE Proc.*, **135**, Pt. J, no. 3, June 1988, pp. 215-219.

24 BARLOW, A.J., JONES, R.S., and FORSYTH, K.W.: 'Technique for direct measurement of single-mode fiber chromatic dispersion'. *IEEE J. Lightwave Technol.*, **LT-5**, no. 9, Sept. 1987, pp. 1207-1216.

25 SO, V.C.Y., JIANG, J., CLEGG, D.D., VALIN, P., HUSZARIK, F.A., and VELLA, P.J.: 'Multiple wavelength bidirectional transmission for subscriber loop applications'. *Electronics Lett.*, **25**, no. 1, 5 Jan. 1989, pp. 16-18.

26 BYRON, K.C., and ASHWORTH, D.M.: 'Intermodal and intramodal dispersion measurements on 1.3 μm single-mode fibre for use at 0.85 μm'. *IEE Proc.*, **135**, Pt. J., no. 3, June 1988, pp. 220-222.

27 PEACOCK, J., SCARFE, J., REID, J., and MALLINSON, S.R.: 'Field measurement of fusion splice loss using pulse reflection method'. *Electronics Lett.*, **25**, no. 9, 27 Apr. 1989, pp. 602-603.

28 BJERKAN, L.: 'Optical fiber splice loss predictions from one-way otdr measurements based on a probability model'. *IEEE/OSA J. Lightwave Technol.*, **7**, no. 3, Mar. 1989, pp. 490-499.

29 SO, V.C.Y., HUGHES, R.P., LAMONT, J.B., and VELLA, P.J.: 'Splice loss measurement using local launch and detect'. *IEEE/OSA J. Lightwave Technol.*, **LT-5**, no. 12, Dec. 1987, pp. 1663-1666.

30 TAM, H.Y.: 'Simple fusion splicing technique for reducing splicing loss between standard singlemode fibres and erbium-doped fibre. *Electronics Lett.*, **27**, no. 17, 15 Aug. 1992, pp. 1597-1599.

31 KRÜGER, U, and KRÜGER, K.: 'Simultaneous measurement of the linewidth, linewidth enhancement factor α, and FM and AM response of a semiconductor laser'. *J. Lightwave Technol.*, **13**, no. 4, Apr. 1995, 592-597.

32 HUI, R., MECOZZI, A., DOTTAVI, A., and SPANO, P.: 'Novel measurement technique of α factor in dfb semiconductor lasers by injection locking'. *Electronics Lett.*, **26**, no. 14, 5 July 1990, pp. 997-998.

33 ESMAN, R.D., and GOLDBERG, L.: 'Simple measurement of laser diode spectral linewidth using modulation sidebands'. *Electronics Lett.* **24**, no. 22, 27 Oct. 1988, pp. 1393-1395.

34 TAN, T.S., JUNGERMAN, R.L., and ELLIOT, S.S.: 'Optical receiver and modulator frequency response measurements with a Nd:YAG ring laser heterodyne technique'. *IEEE Trans, Microwave Theory and Techniques*, **37**, no. 8, Aug. 1989, pp. 1217-1222.

35 EICHEN, E., SCHAFLER, J., RIDEOUT, W., and McCABE, J.: 'Wide-bandwidth receiver/photodetector frequency measurements using amplified spontaneous emission from a semiconductor laser amplifier'. *IEEE/OSA J. Lightwave Technol.*, **8**, no. 6, June 1990, pp. 912–916.
36 PULEO, M., and VEZZONI, E.: 'An automatic testing system for avalanche photodetectors'. *CSELT Rapporti tecnici*, **XI**, no. 3, June 1983, pp. 163–170.
37 COCK, C.C.: 'Assessment of timing jitter in digital telecommunications transmission systems'. *IEE Proc.*, **134**, Pt. F, no. 5, Aug. 1987, pp. 464–473.
38 MAUNDER, C.: 'Status of IC design-for-testability'. *Br Telecom Technol J.*, **7**, no. 1, Jan. 1989, pp.44–49.

AUTHOR INDEX

SUBJECT INDEX